Organic Conformational Analysis and Stereochemistry from Circular Dichroism Spectroscopy

Methods in Stereochemical Analysis

Other Books in the Series:

John G. Verkade and Louis D. Quin (editors)
Phosphorus-31 NMR Spectroscopy in Stereochemical Analysis: Organic Compounds and Metal Complexes

Gary E. Martin and Andrew S. Zektzer
Two-Dimensional NMR Methods for Establishing Molecular Connectivity: A Chemist's Guide to Experiment Selection, Performance, and Interpretation

David Neuhaus and Michael Williamson
The Nuclear Overhauser Effect in Structural and Conformational Analysis

Eiji Ōsawa and Osamu Yonemitsu
Carbocyclic Cage Compounds: Chemistry and Applications

Janet S. Splitter and Frantisek Turecek (editors)
Applications of Mass Spectrometry to Organic Stereochemistry

William R. Croasmun and Robert M. K. Carlson (editors)
Two-dimensional NMR Spectroscopy: Applications for Chemists and Biochemists. Second Edition

Jenny P. Glusker with Mitchell Lewis and Miriam Rossi
Crystal Structure Analysis for Chemists and Biologists

Kalevi Pihlaja and Erich Kleinpeter
Carbon-13 NMR Chemical Shifts in Structural and Stereochemical Analysis

Louis D. Quin and John G. Verkade (editors)
Phosphorus-31 NMR Spectral Properties in Compound Characterization and Structural Analysis

Eusebio Juaristi (editor)
Conformational Behavior of Six-Membered Rings: Analysis, Dynamics and Stereoelectronic Effects

Motohiro Nishio, Minoni Hirota, and Yoji Umezawa
The CH-Π Interaction Evidence: Nature and Consequences

Organic Conformational Analysis and Stereochemistry from Circular Dichroism Spectroscopy

David A. Lightner

R.C. Fuson Professor of Chemistry
Department of Chemistry
University of Nevada
Reno, Nevada 89557-0020

Jerome E. Gurst

Professor of Chemistry
Department of Chemistry
University of West Florida
Pensacola, Florida 32514

WILEY-VCH

A John Wiley & Sons, Inc., Publication

New York • Chichester • Weinheim • Brisbane • Singapore • Toronto

This book is printed on acid-free paper. ♾

Published simultaneously in Canada.

For ordering and customer service, call 1-800-CALL-WILEY.

Library of Congress Cataloging-in-Publication Data:

Lightner, David A.
Organic conformational analysis and stereochemstry from circular dicroism spectroscopy. / David A. Lightner, Jerome E. Gurst.
p. cm.—(Methods in stereochemical analysis)
Includes index.
ISBN 0-471-35405-8 (alk. paper)
1. Circular dichroism. 2. Organic compounds—Analysis.
I. Gurst, Jerome E. II. Title. III. Series.
QD473.L44 2000
547'.30858—dc21 99-38864
CIP

Printed in the United States of America

10 9 8 7 6 5 4 3 2 1

The book is dedicated to those who taught us organic stereochemistry, optical rotatory dispersion, and circular dichroism spectroscopy: Professor Carl Djerassi at Stanford University and the late Professor Albert Moscowitz of the University of Minnesota.

Contents

Preface

The objective of this book is to show how circular dichroism (CD) measurements can lead to stereochemical information. It is written for the organic chemist who has a familiarity with elementary stereochemical principles. Special emphasis is placed on the use of CD to obtain conformational information—an area not heretofore treated adequately in books on CD and stereochemistry. Because most work on conformational analysis by CD has involved variable temperature measurements, that topic will be the major unifying theme. Wherever instructive, molecular mechanics calculations and the use of other spectroscopic information, especially NMR, will be worked into discussions of conformational analysis by CD. That overlap, expressed through specific examples, should provide valuable links between CD and other, perhaps more widely known, spectroscopic methods used in organic stereochemistry.

Measurements of optical activity are among the earliest spectroscopic determinations of organic compounds and easily predate most of those now considered to be routine in organic chemistry: nuclear magnetic resonance (NMR), mass spectrometry, infrared, even ultraviolet-visible. Long before the era of recording spectrophotometers, chemists measured optical rotations at selected wavelengths of light, typically at 589 nm (sodium D-line), but also at other accessible (Hg lines) wavelengths. In the 1930s optical rotatory dispersion (ORD) and CD spectra were plotted from point-wise determinations through various electronic transitions in order to observe anomalous dispersion (or Cotton effects). But it was not until the late 1950s that ORD measurements over a wide spectral region could be obtained more rapidly, and in the early 1960s that reliable commercial ORD and CD instruments became available. This introduction led to an explosion of ORD and CD measurements and to the formulation of various chiroptical rules, *e.g.*, the octant rule, to correlate such data with the absolute configuration or conformation of a molecule. It is this latter aspect of organic stereochemistry that is the focus of this book: organic conformational analysis.

In the early 1960s, Carl Djerassi and Keith Wellman at Stanford University showed how one might routinely measure CD spectra over a wide range of temperatures, from −178° to +100°C. Thus, CD spectra were often found to change in intensity and/or sign as the sample temperature was lowered and the population of conformational isomers changed toward selection of the most stable isomer. From these seminal studies it became clear that analysis of conformation could be accomplished by variable temperature CD spectroscopy, even where there were only small energy differences separating such isomers. Since most of the published data have dealt with optically active ketones and the carbonyl chromophores, they are the major focus of this book. Thus, we include from the original literature many CD spectra run at varying temperatures, typically from room temperature down to liquid nitrogen temperature. For this collection of original spectra alone, the book provides a valuable resource.

So far as we are aware, there has been no systematic summary presentation of CD spectroscopy used for organic conformational analyses, nor have the advantages of measurements by variable temperature CD been widely recognized. Rather, much of organic conformational analysis is due to NMR spectroscopy and its measurement at variable temperature. We therefore integrate our CD discussion with considerable NMR data, as well as information from X-ray and electron diffraction and other spectroscopic measurements from which stereochemistry may be interpreted.

In recent years computer-assisted molecular mechanics calculations have become very reliable predictors of conformational structure and energies, and used widely in organic chemistry. Inexpensive programs for desk-top computers have made such calculations routine and often indispensable. In this book we blend the results of molecular mechanics calculations with those from variable temperature CD and NMR in our attempt to make CD spectroscopy more understandable and attractive.

Understanding the stereochemistry of cyclohexane and its derivatives is a logical starting point for any discussion of organic conformational analysis. Thus, our discussion begins with the structure of cyclohexane and with the seminal conclusion, based on electron diffraction measurements, that cyclohexane resides in a chair conformation (Chapter 2). The important temperature-dependent NMR measurements of cyclohexane and substituted cyclohexanes are discussed; then, conformational analyses of these systems by molecular mechanics calculations is introduced. From cyclohexanes, conformational analysis using the same techniques proceeds to cyclohexanone and its derivatives. For example, the nearly invariant CD curves of 2(S)-methylcyclohexanone at +25°C and −192°C may be taken as an indication that the ketone is essentially conformationally homogeneous at both temperatures.

Chiroptical measurements of ORD and CD are introduced in Chapter 3 and used in connection with one of the most successful chiroptical sector rules, the octant rule, which is treated in depth in Chapter 4. Subsequent chapters treat conformational analysis of simple cyclic ketones: cyclohexanones (Chapter 5) and other monocyclic ketones (Chapter 6). Conformational analysis of polycyclics begins with the decalins and decalones (Chapter 7), hydrindanones and other bicyclo[m.n.o]alkanones—all drawing from available CD data. From two-ring systems, conformational analysis is extended to polycyclics (Chapter 8). Some of the most extensive studies of variable temperature CD spectroscopy come from the relatively more recent research on isotopic substitution, where it is shown that this spectroscopic tool is an extraordinarily sensitive probe of conformation and its perturbation by, *e.g.*, deuterium (Chapter 9). Conjugated and homoconjugated systems are treated in Chapters 10 and 11. Serious concerns about the potential for interchromophoric interaction (even over long distances) is an important central theme, and it is noted that one of the earliest detections of such interaction was by optical rotations and their enormous enhancements. Chapters 12 and 13 leave discussions of ketone stereochemistry to focus on the stereochemistry of the carbon-carbon double bond, treating chiropticity of simple olefins, dienes and their conformational analysis, the benzene chromophore, biphenyls and helicenes. Our final chapter (14) treats stereochemistry from exciton coupling of two (or more) chromophores—an area of considerable current interest and exploration. It is shown that absolute configuration can readily be obtained and that chromophores may interact over incredibly long distances (50 Å).

No attempt will be made here to render an exhaustive account of CD spectroscopy. In treating organic conformational analysis by CD spectroscopy, it is beyond the scope of this book to provide a comprehensive treatment of all chromophores and all systems. The inclusion or exclusion of particular topics is more a reflection of the authors' tastes and inclinations than it is of the importance of the subject matter to the field of organic chemistry. The presence or absence of a particular reference is less indicative of its inherent worth than it is of the authors' narrowness or ignorance.

This manuscript could not have been prepared and completed without the invaluable assistance of Ms. Nancy Olson, who typed the entire document and created almost all of the non-spectral graphics—a Herculean task for any one person. We sincerely thank Dr. Stefan E. Boiadjiev for reading and constructively commenting on the manuscript, the many graduate students at UNR who used and improved the chapter material in their stereochemistry course, and Mr. Michael T. Huggins for measuring the NMR spectra and creating several figures. We are especially appreciative of the many graduate students and postdoctoral fellows who, in the Lightner lab, contributed some of the important research discussed in this book; Professor Aage E. Hansen and the late Professor Thomas D. Bouman for decades of fruitful collaboration and gemütlichkeit; and Professor Carl Djerassi and the late Professor Albert Moscowitz for their training, inspiration and friendship.

David A. Lightner
Reno, Nevada

Jerome E. Gurst
Pensacola, Florida

Organic Conformational Analysis and Stereochemistry from Circular Dichroism Spectroscopy

CHAPTER

1

Introduction

Stereochemistry and conformational analysis are so much a part of modern organic chemistry that it seems incredible, for a discipline over 150 years old, that conformational analysis is but a relatively new component—having been recognized with the Nobel prize only in 1969 and 1975. Since its beginnings, organic chemistry has been deeply concerned with determining the structure of molecules. It was recognized at a very early date that to understand how a substance behaves, both in a chemical sense and in a physical sense, one had to unravel and understand its molecular structure. Thus, it was necessary to know the *chemical composition* (i.e., which elements are incorporated into a particular substance and the relative amounts of those elements) and the *connectivity* of the atoms in the molecule.

By the 1870s, it had become apparent that this structural information, although sufficient to define the *constitutional structure* of a molecule, was often insufficient to describe the molecule's chemical behavior. It became important to think of molecular structure in three dimensions, thus giving birth to stereochemistry. At first, this meant understanding what is now known as the *configurational structure* of a molecule. Most chemists are quite familiar with terms such as *cis*- and *trans*-configuration, *relative* configuration, and *absolute* configuration. More importantly, they know that configurational isomerization at a carbon atom of a molecule is usually a high-energy process, typically requiring one or more chemical reactions. Thus, bond breaking and bond making are integral to any discussion of the isomerization of molecular configurations. Current stereochemical convention,[1,2] which we will use throughout, defines the term "configuration" to mean the three-dimensional arrangement of the atoms in a molecule that distinguishes stereoisomers, except those stereoisomers that are interconvertible through rotation about single bonds.

More recently, it was recognized that to fully define molecular stereochemistry, one must also take into account changes in the spatial arrangements of atoms that occur by low-energy processes involving rotations about single bonds, namely, changes in conformation. *Conformational structure* is thus concerned with rotations about single bonds that alter the positions of atoms in three dimensions to yield distinguishable stereoisomers called conformations. *Conformational analysis* is a relatively new field of study within organic chemistry, having been recognized with the Nobel prize in 1969 (to D. H. R. Barton and O. Hassel) and again in 1975 (J. W. Cornforth and V. Prelog). For example, the stereochemistry of ethane may be discussed in terms of two recognizably distinct conformations—staggered and eclipsed—in which the relative spatial arrangements of the atoms differ. The staggered and eclipsed conformations are interconverted by rotation about the C–C (single) bond with an energy difference of around 3 kcal/mole. The higher energy eclipsed conformation has more torsional strain than the lower energy staggered conformation. In most instances, many conformations may be used to describe the stereochemistry of a compound. Indeed, it could be said that an infinite number of conformations are available for ethane, since every increment of rotation over a 60° range (however minute) about the C–C bond leads to a new stereoisomer. While the number of possible conformations may be infinite, only those lying at potential-energy minima are defined as *conformers*. Thus, the staggered conformation of ethane is a conformer, while the eclipsed conformation is not.

Therefore, it can be said that a compound generally exists as a mixture of conformations at equilibrium, with the mole fraction of each conformation determined by its energy (ΔG) relative to the energy of the other possible conformations.

$$\Delta G_T^\circ = \Delta H^\circ - T\Delta S^\circ = -RT \ln K_{eq}, \qquad (1\text{–}1)$$

Equation 1-1 relates ΔG to K_{eq}. For a hypothetical two-component equilibrium, $A \rightleftarrows B$, $K_{eq} = X_B (1 - X_B)^{-1}$, where X_B is the mole fraction of component B and X_A ($= 1 - X_B$) is the mole fraction of component A. Thus, equation 1-1 can be written: $\Delta G_T^\circ = -RT \ln[X_B(1 - X_B)^{-1}]$, where R is the gas constant and T is the temperature in °K. The relationship between ΔG° and X_B (where B is more stable than A) at various different values of T is plotted in Figure 1-1.

The equation can be used to determine the mole fractions of the various components of a mixture: From Figure 1–1, one can determine that at 25°C, around 99% of an ethane sample should be in the staggered conformation at any point in time, but at

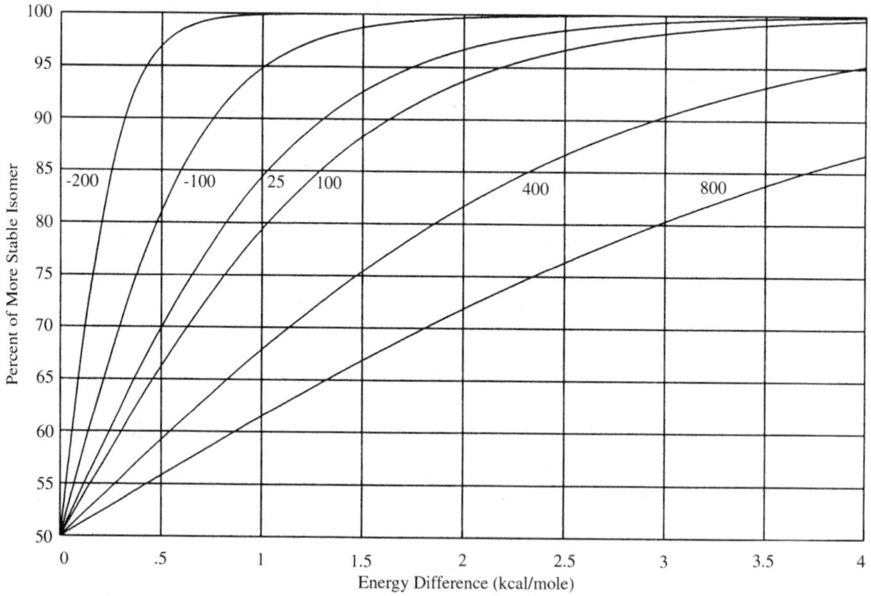

Figure 1–1. Composition of a binary mixture at equilibrium as a function of energy difference (ΔG°) at the temperatures (°C) indicated on the curves.

800°C, some 80% is in that conformation and 20% is in the eclipsed conformation, based on an energy difference of 3 kcal/mole.

Much of what we know and understand of organic stereochemistry is based fundamentally on the dimensionality and direction of bonding at carbon atoms hybridized typically as sp, sp^2, and sp^3. From an understanding of σ molecular orbitals, one can recognize that rotations about single bonds have relatively low energy because the highest electron density of the bond lies between the two nuclei of the bond and remains undisturbed by the rotation. This contrasts with rotations about a pi (π) bond (as a component of a double bond), wherein the molecular orbitals are located above and below the plane containing the two carbon atoms of the double bond. For example, the configurational change that takes 2(***E***)-butene to 2(***Z***)-butene requires the breaking and remaking of a chemical π-bond—a high-energy process.

Once we define the chemical composition of a molecule (which will always involve a qualitative question—which elements?—and a quantitative question—how many?) and the connectivity of the atoms, the question of the molecule's configuration arises. If the three-dimensional structure is to be taken into account, the configuration must be defined. If the structure is chiral, the absolute configuration should be defined. Conformational considerations come to the fore whenever single-bond rotations lead to new stereoisomers. There are few organic molecules in which one would not have to consider their conformation for a complete description of their stereochemistry. Although the configuration of ethylene does not change much with temperature, the conformational population of ethane does. The configuration of a rigid molecule typically does not change on dissolution in a solvent (unless a chemical reaction occurs), but again, the molecule's conformational population might easily change. The configuration of a molecule is often recognized as a static property, whereas its conformational mix is typically a dynamic property. Given the small energies often associated with changes in conformation, the equilibrium structure may be discussed as a mixture of conformations. And the equilibrium is related to the energy differences between stereoisomers, either configurational or conformational, with the familiar equation 1-1 allowing us to generate Figure 1-1.

Most chemists are well aware of the multitude of experimental techniques available for studying the structure of organic molecules. Over time, earlier techniques have been displaced by newer procedures. Some experiments have been discredited. Other than X-ray crystallography, no one experiment can tell us everything (i.e., the composition, the configuration, and the conformation). But even crystallography suffers: We must have a crystal suitable for study, and we learn the conformation of the molecule only in that crystal.

In this book, we will concentrate on the interplay of three methods used in determining stereochemistry. We will learn how information determined by one method may (or may not) corroborate stereochemical conclusions drawn from another or from other data. Two of the methods are spectroscopic:

(*i*) *Chiroptical spectroscopy*: Circular dichroism (CD) or optical rotatory dispersion (ORD) can afford significant insight into the three-dimensional structure of molecules and their absolute configuration.

(*ii*) *Nuclear magnetic resonance spectroscopy*: Both ^{1}H-NMR and ^{13}C-NMR provide a wealth of information on the structure of organic molecules.

A third method is computational: Molecular modeling or molecular mechanics has become a technique that is available even to the (almost) computer illiterate! Each of these three techniques is essentially empirically based. Numerous compounds whose structures were laboriously determined by what are now called classical methods serve as the models from which correlation tables and other empirically based "rules" have been constructed.

Much of what we know of conformational analysis comes from investigations of carbocyclic ring systems. And given the historical importance of natural products in organic chemistry, the most investigated ring systems were based on cyclohexane, whose stereochemistry is now well understood and widely appreciated. It is with cyclohexane that we begin our discussions of stereochemistry and conformational analysis.

References

1. See Moss, G. P., "Basic Terminology of Stereochemistry (IUPAC Recommendations) 1996," *Pure and Applied Chem.* **68** (1996) 2193–2222.
2. Eliel, E. L., and Wilen, S. H., "Stereochemistry of Organic Compounds," John Wiley & Sons, Inc., New York, 1994.

CHAPTER

2

Cyclohexane and Cyclohexanone

2.1. Cyclohexane

The *constitutional* structure of cyclohexane (C_6H_{12}; Figure 2-1A) became evident nearly 100 years ago following its synthesis from 1,6-dibromohexane (by the action of sodium),[1] from cyclohexyl iodide (by the action of zinc),[2] and by the reduction of benzene.[3] At the time, cyclohexane was called hexamethylene and hexahydrobenzene. Its shape was taken to be hexagonal and planar, a belief apparently influenced by its early synthesis from benzene, which is hexagonal and planar. However, evidence for the structure, let alone the shape of cyclohexane, was indirect and uncertain—a situation that remained largely undisturbed until the advent of spectroscopy. As they do today, molecular models then often guided thinking on organic stereochemistry. The models apparently served to provide a rationale for the planar structure of cyclohexane, which could be constructed from the Kekulé models (employing spheres with metal rods directed to the corners of a tetrahedron), modified, and used by Adolf Baeyer.[4,5] Of course, such models could just as easily have provided a rationale for chair cyclohexane—and planar cyclohexane can readily be made from most molecular

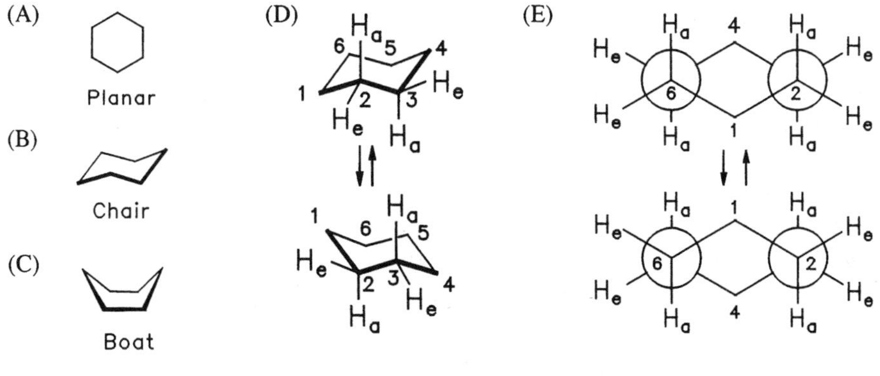

Figure 2-1. (A) Constitutional structure of cyclohexane. A carbon with two hydrogens attached lies at each vertex of the hexagon. (B) Chair conformation. (C) Boat conformation. (D) Interconverting chair conformations of cyclohexane showing the interconversion of axial (H_a) and equatorial (H_e) hydrogens at C(2) and C(3). (E) Corresponding chair conformations drawn in Newman projections.

models currently used—usually with some angle distortion. From our current perspective, it may seem odd that the planar structure of cyclohexane was so widely accepted for decades, considering that the valency of saturated carbon was already recognized as tetrahedral[4] at the time of its first rational synthesis.[1,2] The belief persisted, nonetheless, possibly due to Baeyer's prestige, persuasiveness, or force of argument—because it figured so prominently in Baeyer ring strain theory,[5] an early contribution to physical organic chemistry that correctly predicted qualitative differences in stability (heats of combustion) of small rings due to angle strain (the deviation from the tetrahedral valence angle, 109° 28′). [In the cyclic series $(CH_2)_n$, cyclohexane was the smallest ring to show a negative valence deviation, $d = 1/2$ (109° 28′ minus the internal bond angle of the planar ring) = −5° 16′, implying a negative ring strain.][6,7] Or possibly the belief endured because it sufficed to explain most of the organic chemistry of cyclohexane for many years.

The acceptance of the planar cyclohexane conformation, which we now know is highly unstable, was not uniform. As early as 1890, using models prepared from tetrahedral carbons, Sachse[8] predicted *two strain-free multiplanar shapes: a chair (Figure 2-1B) and a boat (Figure 2-1C).*[4] This insight was not generally acknowledged, however, until some 60 years later, possibly because Sachse suggested that such conformational isomers might be isolable.[7] Subsequently, Mohr[9] applied Sachse's thinking to propose two isomeric (*cis* and *trans*) decalins in which the component cyclohexane rings were nonplanar. These insights notwithstanding, organic chemists found it convenient to think of cyclohexane either as a planar molecule or as a net average planar molecule of two nonplanar forms in equilibrium. Evidence for nonplanar forms was indirect and nonspectroscopic.

Emerging, but limited, spectroscopic evidence on the three-dimensional structure of cyclohexane was controversial. X-ray crystallographic studies on β-benzenehexachloride in the 1920s indicated a boat conformation for the cyclohexane carbons,[10] and Raman studies of monosubstituted cyclohexanes in 1936 supported a chair conformation.[11] A subsequent Raman spectral analysis of liquid cyclohexane that favored the planar conformation[12] was reinterpreted in terms of the chair conformation.[13] However, the early spectroscopic evidence that seemed to favor the chair as the dominant conformation, as opposed to the planar conformation (or the boat), was slow to be recognized by organic chemists—possibly because Raman spectroscopy was not widely appreciated. This situation remained largely unchanged into the 1940s, probably due to preoccupation with World War II, despite Hassel's 1943 seminal publication[14] that drew from his gas-phase electron diffraction spectroscopic analyses of substituted cyclohexanes and Rasmussen's infrared analysis of liquid cyclohexane[15] published in the same year. Rasmussen's normal-coordinate analysis favored the chair as the dominant conformation in the liquid state. Hassel concluded that (i) the chair cyclohexane conformation was favored in gaseous chlorocyclohexane, with the chlorine in an equatorial configuration, and, importantly, (ii) a chair-to-chair conformational inversion (Figures 2-1D and 2-1E) would convert an equatorial chlorine substituent into an axial one—thus defining configurational stereochemistry on a conformationally mobile ring. Hassel illustrated these concepts with line drawings of the chair conformation, while showing equatorial and axial sites, which he called ε and κ, respectively. (But in 1918, Mohr[9] had also represented cyclohexane in strain-free conformational drawings.)

Hassel's subsequent electron diffraction spectroscopic analysis of gaseous cyclohexane confirmed the earlier findings with the liquid[13,15] and defined the chair as the most stable conformer.[16,17] This conclusion was reached independently by Pitzer et al.[18] at Berkeley in 1947 on the basis of thermochemical analyses of alkyl cyclohexanes. Pitzer also illustrated the stereochemical differentiation of axial and equatorial sites (which he called polar and equatorial, respectively) on chair cyclohexane and indicated that the chair-to-chair interconversion passes through a higher energy boat conformation.[19] Thus, he recognized the implications of configurational stereodifferentiation and interconversion on chair cyclohexane, and, using data for the internal rotation barrier of ethane,[20] he determined that the boat conformation lies approximately 5.6 kcal/mole higher in energy than the chair.

The implications of the chair cyclohexane geometry for organic chemistry were recognized and expressed by Barton in 1950,[21] marking the birth of organic conformational analysis and leading to its subsequent wide acceptance and application. In the late 1940s Barton was apparently one of the few organic chemists to recognize the importance of Hassel's[14,16,17] and Pitzer's[18] work in understanding the physical properties, and in predicting the conformational and configurational stability and chemical reactivity, of organic molecules, especially in the chemistry of natural products. By generalizing the work of Hassel and Pitzer to other cyclohexane systems, Barton[22] was able to predict the conformational and configurational stability, as well as the stereochemistry and relative rates of reaction, of decalins, terpenes, and steroids.

Direct evidence for the three-dimensional structure of cyclohexane comes almost exclusively from spectroscopic methods. Normal-coordinate analysis, coupled with the liquid-phase Raman[13] and infrared spectra,[15] was consistent with the D_{3d} symmetry of the chair conformation (Figure 2-2A), rather than D_{6h} (of the planar).[6] Gas-phase electron diffraction studies gave tetrahedral bond angles at carbon that favored the chair as the lowest energy conformation and that rendered the planar shape untenable. Subsequent improvements in collecting and reducing electron diffraction data gave more refined values for bond angles and bond lengths (Table 2-1).[23–26] Finally, a low-temperature X-ray study of cyclohexane crystals confirmed the chair conformation in the solid.[27]

In the earlier spectroscopic studies,[13–17] the presence of some boat conformation (Figure 2-2B) could not be excluded. The boat has a different shape and symmetry (C_{2v}) than the chair (D_{3d}). Unlike the chair conformation, in which all bonds are staggered (thus yielding an energy minimum in nonbonded steric repulsions), the boat has considerable bond eclipsing, in addition to destabilizing nonbonded steric repulsions (Figure 2-2B). Pitzer[18] put this on a quantitative basis. Using rotation barriers from ethane, he predicted that the boat would be much less stable (~5.6 kcal/mole) than the chair. This early attempt at conformational analysis by what we now call molecular mechanics was followed shortly by one of Barton,[28] whose semiempirical calculations[29,30] gave the boat some 1.3–6.9 kcal/mole more energy than the chair. In

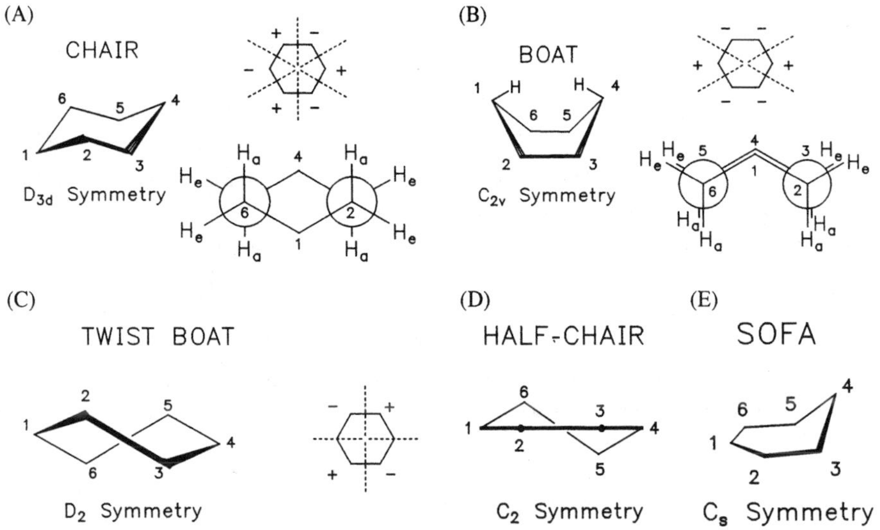

Figure 2-2. Conformations of cyclohexane and their symmetry. In (A), (B), and (C), the internal C–C–C angles are strainless. In the chair conformation of (A), all C–C and C–H bonds are staggered (*gauche*), and torsional and angle strain[6,7] are minimized. The chair conformation is well defined and more "rigid" than the others shown. The boat conformation of (B) suffers destabilizing 1,4-transannular bowsprit-flagpole H–H interactions, in addition to eclipsing of the C(1)–C(2) bond with the C(3)–C(4) and of the C(1)–C(6) bond with the C(4)–C(5), and eclipsing between the pairs of C–H bonds at C(2) with C(3), and C(5) with C(6). The boat conformation easily pseudorotates from one boat to another via a twist-boat intermediate (C). The twist boat is lower in energy than the boat due to partial relief of 1,4-transannular nonbonded steric interactions and eclipsed butane interactions. Pseudorotation occurs by concerted alteration of the C–C–C–C torsion angles ψ (e.g., in (B), ψ (C_1–C_2–C_3–C_4) is 0°). The signs + and – in (A), (B), and (C) signify areas above (+) and below (–) the average plane passing through the molecule.

agreement with Pitzer, Oosterhoff[31] shortly thereafter calculated that the boat would be 5.5 kcal/mole less stable than the chair, but Turner[32] calculated that it would lie some 7.2–10.6 kcal/mole above the chair. The early spectroscopic studies and attempts at conformational analysis of cyclohexane were evaluated by Dauben and Pitzer,[33] who concluded that 6 kcal/mole would be the minimum energy difference between the boat and the more stable chair. Cyclohexane conformation was also evaluated by Orloff,[34] who calculated that planar cyclohexane would lie 31 kcal/mole above the chair.

Oosterhoff's conformational analysis[31] was particularly insightful, for he recognized that one difficulty in evaluating the conformational energy of the boat was its ease of deformation by rotation about the carbon–carbon bonds (pseudorotation) into a twist-boat conformation (Figure 2-2C), in which partial relief of the destabilizing interactions in the boat is achieved. Pseudorotation converts one boat into another through the twist-boat conformation. Hendrickson, at UCLA and Brandeis, computed

Table 2-1. Cyclohexane Bond Angles and Lengths from Electron Diffraction (ED) Spectroscopy and X-ray Crystallography

Favored Conformation	Method (Year)	Bond Angle (°)	Bond Length (Å)		Torsion Angle (°)
		C–C–C	C–C	C–H	C–C–C–C
	ED (1947)[a]	111	1.54	1.10	56
	ED (1963)[b]	111.5 ± 1.5	1.53 ± 0.01	1.09 ± 0.02	55 ± 4
	ED (1963)[c]	111.55 ± 0.15	1.528 ± 0.005	1.104 ± 0.005	54.5 ± 0.4
	ED (1971)[d]	111.05 ± 0.12	1.528 ± 0.003	1.119 ± 0.003	55.9 ± 0.35
	ED (1974)[e]	111.4 ± 0.4	1.536 ± 0.002	1.121 ± 0.004	54.9 ± 0.4
Chair	ED (1976)[f]	111.34 ± 0.24	1.5335 ± 0.002	1.111 ± 0.004	55.1 ± 0.4
	X ray (1973)[g]	111.34	1.526	—	55.1
Ideal chair		109° 28′	—	—	60°

[a] Refs. 16 and 17.
[b] Alekseev, N. V., Kitaigorodskii, A. I., *Zh. Struct. Chem.* **1963**, 163–166.
[c] Ref. 23.
[d] Ref. 24, ∠H–C–H = 110.1° ± 2.5°.
[e] Ref. 25, ∠H–C–H = 107.5° ± 1.5°.
[f] Ref. 26, ∠H–C–H = 105.3° ± 2.3°.
[g] Ref. 27.

that the twist boat was 1.6 kcal/mole more stable than the boat, which in his calculations lies some 5.33 kcal/ mole above the chair.[35a] His computations also showed that the conformational change from chair directly to twist boat passes through a high-energy conformation (transition state) resembling a half-chair (Figure 2-2D) lying some 12.7 kcal/mole above the chair. In the half-chair conformation, four contiguous ring atoms become coplanar, introducing both angle and torsional strain. An even higher energy conformation (Figure 2-2E) lying some 14.1 kcal/mole above the chair was computed for the sofa transition state, wherein the chair goes directly to the boat. (The sofa conformation has five contiguous coplanar ring atoms.) A planar conformation, with all six ring atoms co-planar, would have an even higher energy due to a multitude of eclipsing (torsional) interactions and angle strain at every internal angle. Subsequently, Hendrickson reevaluated the transition state energies downward, coming up with the half-chair at 11.0 kcal/mole and the sofa at 11.3 kcal/mole.[35b] The various cyclohexane conformations and their potential energies (Figure 2-3) may be computed reliably and readily nowadays using molecular mechanics desktop computer programs such as PCMODEL,[36] which employs a modification of Allinger's MM2 force field.[37]

An experimental measurement of the difference in conformational energy of the chair versus the boat was achieved at UCLA by Squillacote, Sheridan, Chapman, and Anet.[38] Assuming an energy difference of 5–6 kcal/mole, the amount of twist boat near room temperature would be only about 0.1%. (See Figure 1-1.) However, at high temperatures, the amount of twist boat was expected to increase. With a force-field-computed $\Delta H^\circ = 5.7$ kcal/mole and $\Delta S^\circ = 4$ e.u., the cyclohexane conformational population was calculated (assuming a two-component equilibrium) to contain approximately 30% of the twist boat at 800°C (Figure 1-1). By heating cyclohexane vapor (0.2 mm Hg pressure) to 800°C (~10 msec contact time) and then depositing it on a CsI plate at 10°K, the presence of the twist-boat conformer could be detected by infrared spectroscopy (Figure 2–4). Upon warming of the sample, the disappearance of the twist-boat infrared bands could be followed over time at three different temperatures (72.5°, 73.0°, and 74.0°K), and a $\Delta G^{\ddagger} = 5.27 \pm 0.05$ kcal/mole was calculated. By using certain limiting assumptions (e.g., $\Delta G^{\ddagger} \sim \Delta H^{\ddagger}$ and $\Delta H^{\ddagger} =$ 10.7–10.8 kcal/mole),[39] the enthalpy difference between chair and twist boat was calculated to be 5.5 kcal/mole—incredibly, the same value computed by Oosterhoff in 1951.[31] More recent molecular mechanics force field[37] computations using PCMODEL[36] (Figure 2-3) are in excellent agreement with the experimental value. The experimentally determined enthalpy difference (5.5 kcal/mole) is also close to the values determined experimentally much earlier by the combustion of locked-ring systems (5.5 kcal/mole)[40] and by the interconversion of *cis* and *trans*-1,3-di-*tert*-butyl-cyclohexane (5.9 kcal/mole).[41]

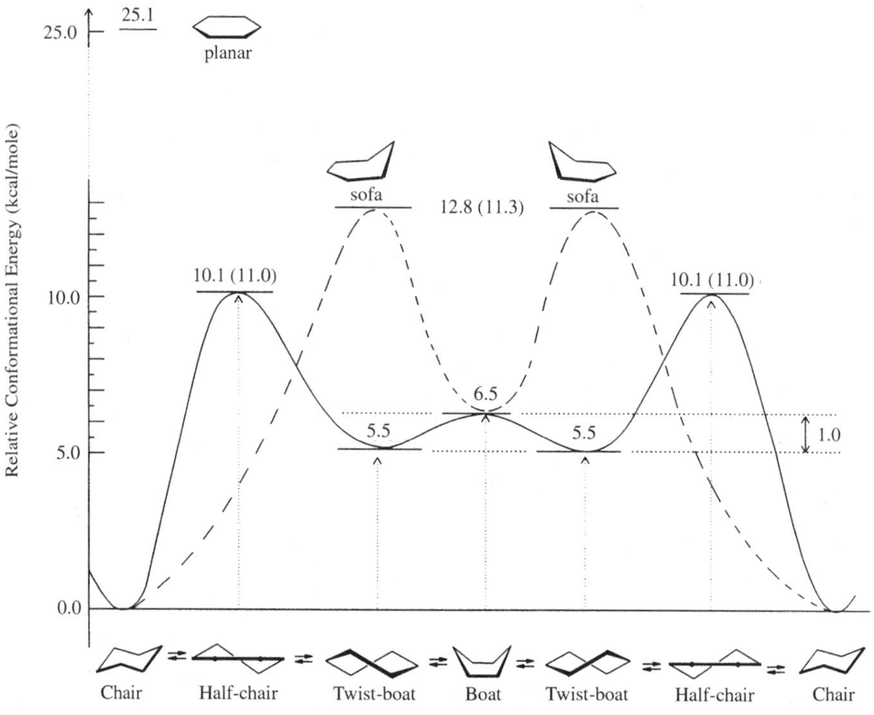

Figure 2-3. Potential-energy diagram for equilibrating cyclohexane conformations. The interrelationships are relative to the chair conformation set at 0.0 kcal/mole, and the twist-boat ⇄ boat pseudorotation barrier is 1.0 kcal/mole. The energies are from PCMODEL.[36] (Those in parentheses are from Hendrickson's force field.)[35b]

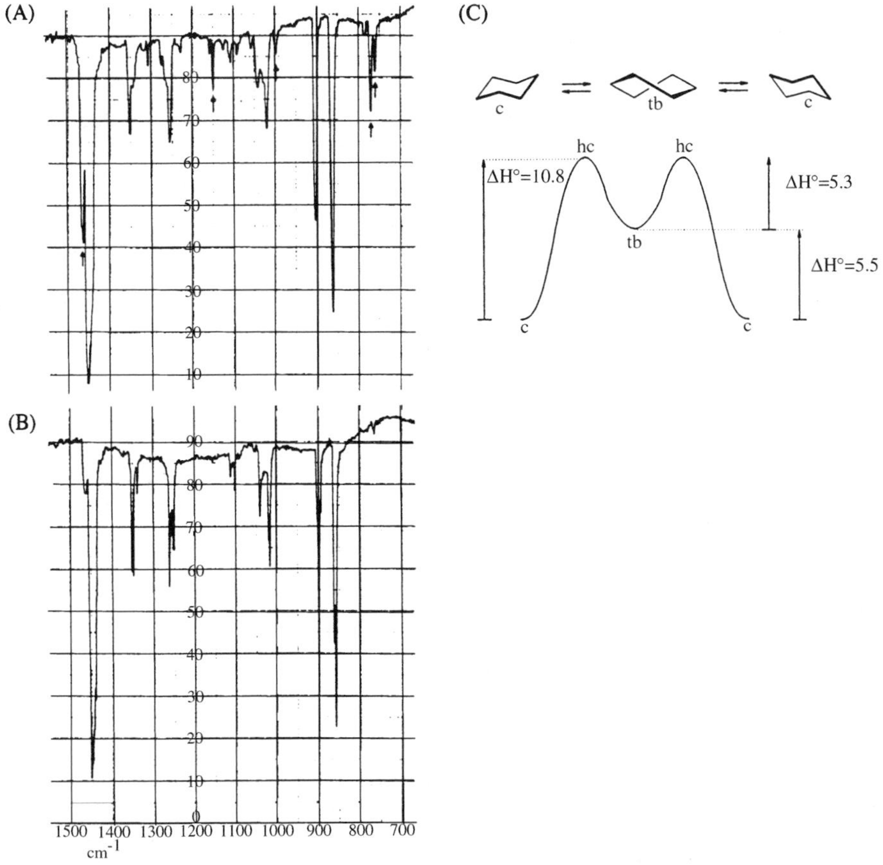

Figure 2-4. (A) Infrared spectrum between 700 and 1500 cm^{-1} at 10°K of cyclohexane deposited from the vapor at 800°C. Small arrows point to bands near 1150 and 770 cm^{-1} assigned to the twist-boat conformation. (B) The same sample as in (A) after warming to 74°K for several hours before the spectrum at 10°K was remeasured. (C) Potential-energy diagram for the chair-(c) to-chair (c) interconversion, which passes over a half-chair (hc) transition state and through a twist-boat (tb) intermediate. [(A) and (B) are reprinted with permission from ref. 38. Copyright© 1975 American Chemical Society.]

From the perspective of organic stereochemistry, the most important dynamic process in cyclohexane is the chair ⇄ chair conformational inversion—a process that interconverts axial and equatorial substituent sites. And from the viewpoint of the organic chemist, the most important and far-reaching early spectroscopic study of the stable conformation of cyclohexane in solution made use of nuclear magnetic resonance (NMR) spectroscopy. In their seminal investigations at Berkeley in the late 1950s and early 1960s, Jensen, Noyce, Sederholm, and Berlin[42] measured the proton NMR spectrum (^{1}H-NMR) of cyclohexane at 60 MHz in carbon disulfide solvent from

(A)

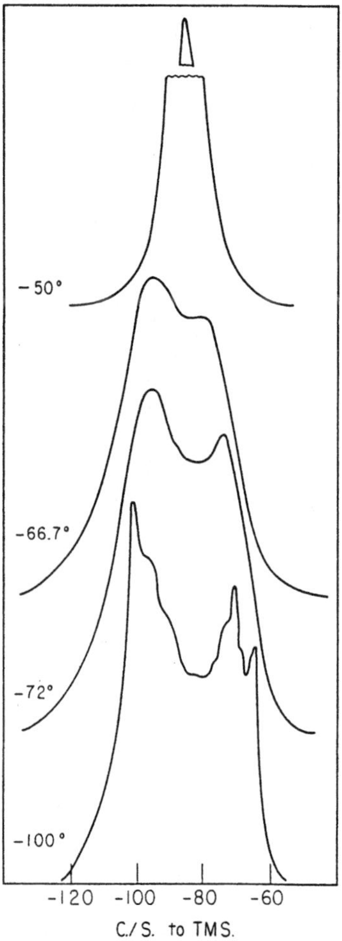

(B)

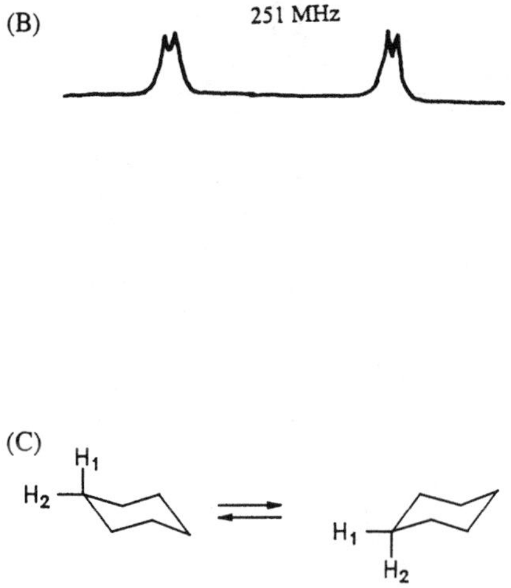

(D)

Axial (δ_a) and Equatorial (δ_e) Proton Resonances for Cyclohexane

Temp. °C	δ_a	δ_e	$(\delta_a+\delta_e)/2$	$2(\delta_a$-$\delta_e)$
−55	(Time avg. signal)		1.42	—
−64.8	1.34	1.51	1.42	0.17
−66.6	1.32	1.53	1.42	0.21
−68.0	1.27	1.58	1.42	0.31
−82	1.21	1.63	1.42	0.42
−105	1.20	1.65	1.42	0.45

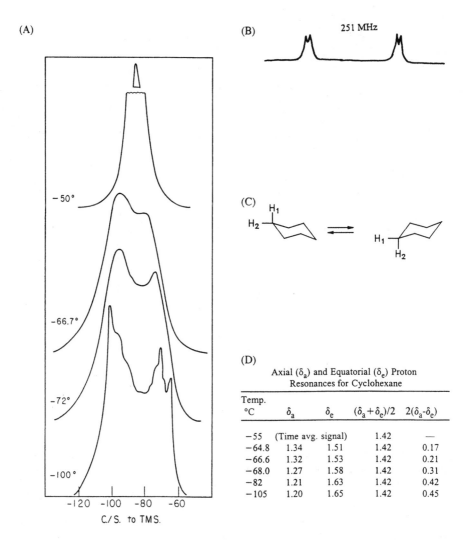

Figure 2-5. (A) Changes in cyclohexane ^{1}H-NMR spectrum between –50° and –100 °C. Spectra were run on 3 M solutions in CS_2 solvent at 60 MHz and reported in cycles/sec relative to tetramethylsilane (TMS), 0.04 mL/mL of solution. Here, 60 cps = 1 ppm (δ). (B) 251 MHz ^{1}H-NMR spectrum at –100°C in CS_2 solvent by Anet and Basus, reported in 1975 (redrawn from ref. 39). (C) Cyclohexane chair ⇄ chair conformer interconversion illustrating the change from axial (H_1, left) to equatorial (H_1, right) configuration associated with ring inversion. (D) Variation of the NMR resonances for axial (δ_a) and equatorial (δ_e) hydrogens with decreasing temperature (from ref. 42). [(A) is reprinted with permission from ref. 42. Copyright© 1962 American Chemical Society.]

room temperature down to −100°C. (See Figure 2-5.) At temperatures above −50°C, only one signal can be detected for all 12 hydrogens on cyclohexane, but as the temperature is lowered, at −66.7°C the resonance begins to split, and at −100°C two signals of approximately equal intensity are clearly visible (Figure 2-5A). In this experiment, which was extraordinary in its day, the resonances for all six axial hydrogens could be distinguished from the six equatorial hydrogens at a sufficiently low temperature, whereas at higher temperatures the signals coalesced. The observation is consistent only with a chair cyclohexane conformation, which has six axial and six equatorial hydrogens (Figure 2-1). When the coalesced proton resonance is observed, site exchange during the chair $\rightleftarrows$ chair interconversion equilibrium is fast on the time scale of the NMR measurement. At low temperatures, where the signals separate, the site exchange equilibrium is slow on the NMR time scale. The rate constant for interconversion at −66.7 ±0.5°C (the temperature of half-peak separation) was determined to be 52.5 ±5 sec^{-1}; the activation barrier ($\Delta G^{\ddagger}$) was determined to be 10.1 ±0.1 kcal/mole at the same temperature. The enthalpy of activation, $\Delta H^{\ddagger}$, was determined to be 11.5 ±2 kcal/mole by the NMR line-broadening method—or 11.1 kcal/mole from the experimentally determined $\Delta G^{\ddagger}$ and a $\Delta S^{\ddagger}$ value estimated by symmetry considerations to be 4.9 e.u. The enthalpy barrier was previously estimated to be 9–10 kcal/mole[43] and 14 kcal/mole.[18]

2.2. Cyclohexane and Dynamic NMR

The success of the NMR studies of Jensen et al.[42] in detecting the slow exchange of hydrogens between the axial and equatorial sites on the chair cyclohexane led to the widespread use of NMR as a spectroscopic method of major importance in organic conformational analysis. The dynamic NMR method,[44] with its characteristics illustrated in Figure 2-6, has been used successfully and repeatedly to assess dynamic processes associated with the site exchange of nuclei in the conformational analysis of organic molecules. Following the studies of Jensen et al.,[42] Anet and Bourn[45] at UCLA studied the conformational dynamics of cyclohexane-d_{11} ($C_6D_{11}H$), a simple substituted cyclohexane with only one hydrogen. The dynamics of the chair-to-chair interconversion in this compound, in which the single proton can be observed at sufficiently low temperatures either at the axial or the equatorial site, was studied by coalescence and line-shape analysis. Refined values for the interconversion were thus determined: $\Delta G^{\ddagger}$ = 10.25 kcal/mole (at −50° to −60°C), $\Delta H^{\ddagger}$ = 10.71 kcal/mole, and $\Delta S^{\ddagger}$ = 2.2 e.u. More recently, the same $\Delta G^{\ddagger}$ value was determined in CD_2Cl_2 at 600 MHz, but a larger $\Delta G^{\ddagger}$ (10.55 ± 0.05 kcal/mole) was found for cyclohexane-d_{11} encapsulated in a "jelly donut" host dimer.[46]

In 1981, Aydin and Günther[47] reported work on the chair-to-chair interconversion of cyclohexane-d_1 by ^{1}H-decoupled ^{13}C-NMR. The carbon bearing the lone deuterium is shielded relative to the other five carbons (Figure 2-7) and, at 25°C, appears as a three-line pattern due to deuterium coupling. Upon lowering the temperature, one finds that the three-line carbon signal broadens and then separates below −80°C into two

NMR Line Shape	Process	Life-Time(t) (1/k), sec	Characteristic Temperature for $\Delta G^{\ddagger}$(kcal/mole)		
			9.6	14.3	19.1
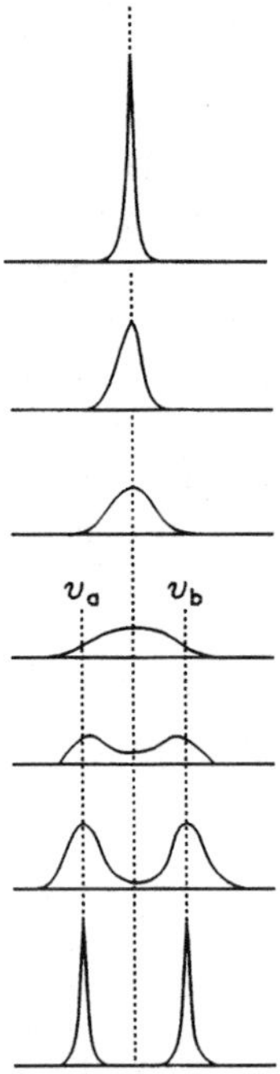	fast exchange $t \ll \frac{1}{2\pi(\nu_b - \nu_a)}$	0.001	−50°	50°	160°
	coalescence $t = \frac{\sqrt{2}}{2\pi(\nu_b - \nu_a)}$	0.01	−70°	20°	120°
		0.1	−80°	0°	90°
	slow exchange $t \gg \frac{1}{2\pi(\nu_b - \nu_a)}$				

Figure 2-6. (Left) Expected NMR line shapes for two-nuclei exchange between two equally populated sites and (right) lifetimes and temperatures associated with several different activation barriers ($\Delta G^{\ddagger}$). The rate constant (k_c) at the coalescence temperature (T_c) is given by $k_c = \pi\Delta\nu/\sqrt{2}$, where k_c is $k/2$, since k is the total rate constant for the forward and reverse directions. The activation barrier at the coalescence temperature is given by $\Delta G_c^{\ddagger} = 2.303\ RT_c\ [10.32 + \log(T_c/k_c)]$; see ref. 44.

well-resolved sets of triplets at –93°C. This means that the ^{13}C chemical shift of a carbon bearing an axial deuterium is different from that with an equatorial deuterium. The assignment of the more shielded carbon (18.89 ppm versus 19.49 ppm) to that bearing an axial deuterium was made on the basis of the former's smaller C–D coupling constant.[47] The two sets of triplets coalesced over a narrow (4°) temperature range, and $\Delta G^{\ddagger}$ was determined to be 10.20 kcal/mole at –75°C, a value in excellent agreement with the best value (10.24 kcal/mole) from the literature cited by Anet.[39] From their studies, Aydin and Günther[47] were also able to determine that the chair conformer with an axial deuterium predominates slightly in the equilibrium. That is, an axial deuterium is more stable than an equatorial or by about 50 cal/mole.

This finding was counter to expectations and opposite to that predicted by molecular mechanics calculations.[48] Nevertheless, although the energy difference between the

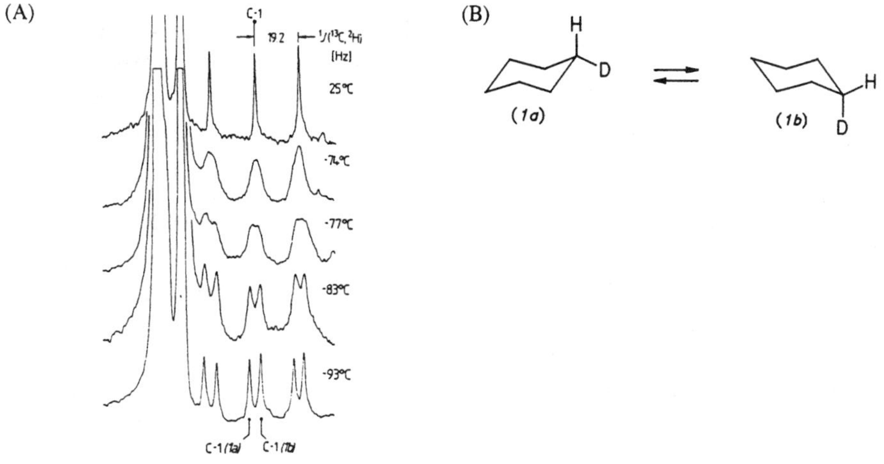

Figure 2-7. (A) Variable-temperature ^{13}C-NMR spectrum of cyclohexane-d_1 at 100.61 MHz with ^{1}H broadband decoupling. (B) Chair-to-chair interconversion of cyclohexane-d_1 that converts deuterium in an equatorial configuration (left) to an axial one (right). The conformer on the right is slightly more stable than that on the left. [(A) is reprinted with permission from ref. 47. Copyright© 1981 Wiley-VCH Verlag GmbH.]

axial and equatorial deuterium conformations was determined to be an order of magnitude too large by Anet and Kopelevich[49] on the basis of their dynamic ^{1}H-NMR studies of 1,2,2,3,3,4,5,5,6,6-decadeuteriocyclohexane, the qualitative result remained: Deuterium prefers the axial position. Anet and Kopelevich determined an energy difference of 6.3 ±1.5 cal/mole and supported this value with their own force field calculations, which were in excellent agreement (6.7 cal/mole). These conclusions also were supported qualitatively by Hartree–Fock calculations at the 3-21G level.[50]

Although solvent effects on the height of the cyclohexane ring inversion barrier appear to be small,[39,51] pressure effects are not, with the ring inversion rates being faster at higher pressure than at lower pressures.[51] In keeping with these observations, the activation enthalpy and entropy were found to be larger ($\Delta H^\ddagger$ = 12.1 ±0.5 kcal/mole and $\Delta S^\ddagger$ = 5.7 ±0.5 e.u.)[52] for cyclohexane ring inversion in the gas phase than in solution ($\Delta H^\ddagger$ = 10.71 ±0.04 kcal/mole and $\Delta S^\ddagger$ = 2.2 ±0.2 e.u.).[39,51] Hence, interestingly, the cyclohexane chair-to-chair ring inversion has been determined to be slightly slower in the gas phase ($\Delta G^\ddagger$ = 10.4 ±0.2 kcal/mole) than in CS_2–CD_2Cl_2 solution ($\Delta G^\ddagger$ = 10.25 kcal/ mole) at 25°C.

The three-dimensional structure of cyclohexane thus seems clear: The chair is the most stable conformer, and a chair-to-chair conformational inversion passes through the next most stable conformation, the twist boat, which lies some 5.5 kcal/mole higher. Hydrogens and other substituents may occupy either of two distinguishable positions: axial or equatorial (Figure 2-1). These results are important, because cyclohexane is the prototypic six-membered ring, and it is found in a wide variety of natural and synthetic organic compounds. But what of conformational changes in the cyclohexane ring caused by nonbonded interactions between attached groups larger than hydrogen, as is seen in substituted cyclohexanes? And what conformational changes might follow the replacement of a ring $-CH_2-$ group with a ketone carbonyl, as in cyclohexanone?

2.3. Substituted Cyclohexanes and Dynamic NMR

Consider the effects of substitution on cyclohexane. With the chair as the most stable ring conformation, there are two possible isomers—one with the substituent at an axial site, the other with the substituent at an equatorial site. These two isomers may interconvert through a chair-to-chair ring inversion process (illustrated in Figure 2-8 for methylcyclohexane). Typically, the equatorial isomer is the more stable. The axial chair conformation is destabilized relative to the equatorial by nonbonded steric repulsions between the axial substituent and the $-CH_2-$ groups at C(3) and C(5). Such steric repulsions belong to a general class called *gauche* butane interactions[6,7] (Figure 2-8C). In addition to *gauche* butane interactions within the cyclohexane ring, an axial substituent, as in methylcyclohexane, imposes two more *gauche* butane interactions on the structure. Since one such interaction is worth 0.8 kcal/mole in destabilization energy, as determined from measurements of heat capacity[33,53] and electron diffraction,[54] the axial isomer of methylcyclohexane is expected to be approximately 1.6 kcal/mole less stable than the equatorial. This approximation contains a number of

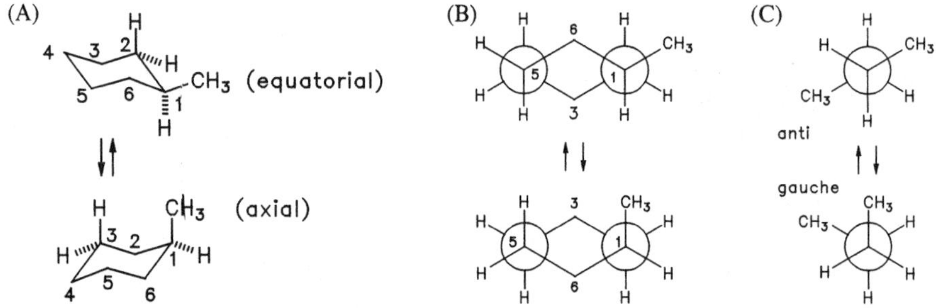

Figure 2-8. (A) and (B) Chair-to-chair conformational inversion in methylcyclohexane, showing the change in methyl configuration from equatorial (upper) to axial (lower). The axial isomer is destabilized relative to the equatorial by approximately 1.7 kcal/mole, with destabilization coming from *gauche* butane-type nonbonded steric repulsions between the CH_3 group and CH_2 groups at C(3) and C(5) and from 1,3-diaxial interactions between the CH_3 and axial hydrogens also at C(3) and C(5). Those elements of conformational destabilization are absent in the equatorial isomer. (C) Newman projections for butane, seen looking down the 2,3-carbon–carbon bond, showing the change from the *anti* conformation (upper) to the *gauche* conformation (lower). The *gauche* conformation is less stable than the *anti* by about 0.8 kcal/mole. The *anti* conformation is found in the equatorial methylcyclohexane, the *gauche* in the axial.

assumptions, including the assumption that the chair cyclohexane conformation is not deformed as a result of substitution.

The energy difference between axial and equatorial methylcyclohexane could not be measured directly by ^{1}H-NMR. In 1971, Garbisch et al. at Minnesota could not observe the axial isomer at low temperatures, but nevertheless estimated that $\Delta G° =$ 1.7 kcal/mole on the basis of a temperature dependence of the hydrogen chemical shifts.[55] Later, Anet et al. determined the difference in conformational free energy by ^{13}C-NMR.[56] At –110°C, weak carbon signals for (i) the CH_2 carbons at C(3) and C(5) in the axial isomer and (ii) the axial CH_3 could be seen lying approximately 6 ppm upfield from the corresponding $-CH_2-$ and CH_3 signals in the equatorial isomer (Figure 2-9). (The shielding is due to the "γ-gauche" steric effect in ^{13}C-NMR.)[57] From this study, a value of approximately 1.6 kcal/mole was determined for $\Delta G°$.

Shortly thereafter, Anet and Squillacote[58] heated a sample of methylcyclohexane to 500°C, deposited it onto a surface cooled to –175°C, and observed its ^{1}H-NMR at –160°C in 1:1 $CHClF_2$–CCl_2F_2 (Figure 2-9). These researchers observed an axial methyl doublet at 1.00 ppm (downfield from the equatorial methyl doublet) and determined that the amount of axial isomer present could range from 10–25%. A more accurate percentage could not be determined due to overlapping of the methyl signals with the ring methylene signals. Because earlier ultrasonic relaxation measurements on pure liquid methylcyclohexane[59] had given $\Delta H° = +2.9 \pm 0.5$ kcal/mole, $\Delta S° = +3.2$ ± 0.6 e.u., $\Delta H^{\ddagger} = +10.3 \pm 0.6$ kcal/mole, and $\Delta S^{\ddagger} = 2.0 \pm 2.0$ e.u. for the methylcyclohexane conformational equilibrium depicted in Figure 2-9A, the mole fraction of axial isomer is calculated to be about 0.43 at 500°C. The NMR results of Anet and Squillacote are in modest agreement with the photoacoustic results.

The most accurate values for the energy difference between axial and equatorial isomers of methylcyclohexane come from low-temperature ^{13}C-NMR studies. In the late 1970s, Booth and Everett studied methylcyclohexane with 91% ^{13}C enrichment in the methyl group.[60a] Like Anet,[56] although with considerably greater ease, they observed the proton-decoupled upfield methyl carbon resonance of the axial isomer at 172°K, determining a 1:165.2 or 1:163.2 ratio of axial-to-equatorial methyl signals at 17.43 δ and 23.47 δ, respectively. From these ratios, they calculated that $\Delta G° = +1.74$ $\pm$ 0.06 kcal/mole at 172°K for the equilibrium shown in Figure 2-9A, assuming the same spin-lattice relaxation times for the axial and equatorial methyls. More refined values, presented in Table 2-2, were calculated subsequently[60b] and include equilibrium thermodynamic data for the axial and equatorial isomers of ethyl and isopropylcyclohexane. In 1999, Wiberg et al. remeasured these values by ^{13}C-NMR (Table 2-2) and supported them by ab initio calculations.[61] The entropic term for the equilibrium in isopropylcyclohexane was determined previously to be 2.2 e.u., and the activation barrier for conformational inversion, $\Delta G^{\ddagger}$, was found to be 4.1 ±0.1 kcal/mole.[62]

With an energy difference between axial and equatorial methylcyclohexane of approximately 1.7 kcal/mole, the mole fraction of the equatorial conformation at 22°C is about 0.95 (Figure 1-1), and with a predominantly equatorial conformation, the axial and equatorial ring hydrogens should be distinguishable. They can be detected by H,C-COSY NMR, as shown in Figure 2-10, where H–C coupling is detected between

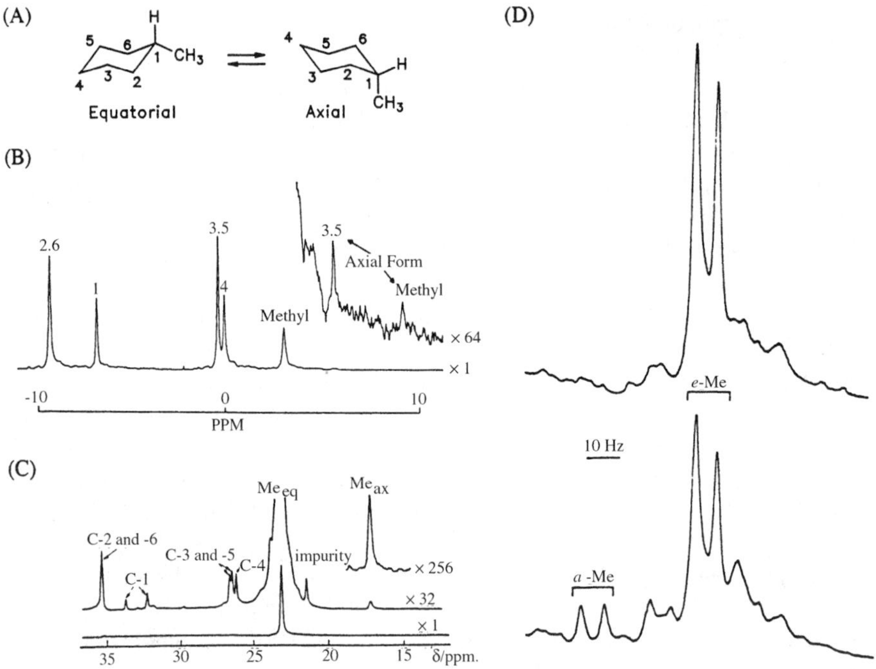

Figure 2-9. (A) Equatorial ⇄ axial methylcyclohexane equilibrium as seen by (B) 63.1-MHz proton-decoupled ^{13}C-NMR spectroscopy at −110°C, (C) ^{13}C-NMR of $^{13}CH_3$-enriched material at −101°C, and (D) 251-MHz ^{1}H-NMR high field region at −160°C (bottom). [(B) and (D) are reprinted with permission from refs. 56 and 58. Copyright© 1971, 1975 American Chemical Society. (C) is reprinted from ref. 60a with permission from the Royal Society of Chemistry.]

the C(2) and C(6) carbons at 35.3 ppm in the ^{13}C-NMR spectrum and the attached equatorial hydrogens (buried) near 1.7 ppm and axial hydrogens near 0.9 ppm in the ^{1}H-NMR spectrum.

Does substitution of the cyclohexane alter the ring conformation? Large deformations are typically encountered when two or more large groups are substituted so as to create special destabilization, as in *trans*-1,3-di-*tert*-butylcyclohexane.[41,60b] In order to adopt the chair conformation, one *tert*-butyl must be axial—at a cost computed by PCMODEL[36] to be approximately 0.8 kcal/mole relative to a twist-boat conformation in which nonbonded steric interactions are at a minimum. The twist-boat cyclohexane ring is thus thought to be favored over the chair in this case.

More subtle ring deformations, such as regional flattening or puckering, are thought to occur with the substitution of chair cyclohexane. However, such changes are difficult to measure directly. Analyzing vicinal coupling constants by ^{1}H-NMR spectroscopy[44] is one way to determine ring H–C–C–H torsion angles and, therefore (indirectly), the internal C–C–C–C torsion angles. The magnitude of a vicinal (or 3-bond) coupling constant (^{3}J) depends on the torsion angle ϕ (H–C–C–H) separating the two hydro-

Table 2-2. Thermodynamic Parameters (ΔG, ΔH in kcal/mole; ΔS in e.u.) from ^{13}C-NMR Spectroscopy for the Axial ⇄ Equatorial Inversion on Chair Cyclohexane[56,60,61]

Methylcyclohexane			Isopropylcyclohexane		
$-\Delta G°$	$-\Delta H°$	$\Delta S°$	$-\Delta G°$	$-\Delta H°$	$\Delta S°$
1.74[a]	1.75 ± 0.05[a,c]	−0.03 ± 0.25[a]	2.61[a]	1.35 ± 0.49[a,c]	4.19 ± 3.04[a]
1.80 ± 0.02[b]	1.76 ± 0.10[b]	0.2 ± 0.02[b]	1.96 ± 0.02[b]	1.40 ± 0.15	3.5 ± 0.9
Ethyl in 4-Methylethylcyclohexane			**Isopropyl in 4-Methylisopropylcyclohexane**		
$-\Delta G°$	$-\Delta H°$	$\Delta S°$	$-\Delta G°$	$-\Delta H°$	$\Delta S°$
1.79[a]	1.60 ± 0.06[a,c]	0.64 ± 0.35[a]	2.21[a]	1.52 ± 0.06[a]	2.31 ± 0.38[a]
1.75 ± 0.02[b]	1.54 ± 0.12[b]	1.3 ± 0.8[b]			

[a] In $CFCl_3$-$CDCl_3$, refs. 56 and 60, ΔG°_{300}.

[b] In $CBrF_3$-CD_2Cl_2, ref. 61, ΔG°_{157}, ethylcyclohexane.

[c] Computed by PCMODEL[36] to be 1.78 kcal/mole (methyl), 1.72 kcal/mole (isopropyl), and 1.81 kcal/mole (ethyl).

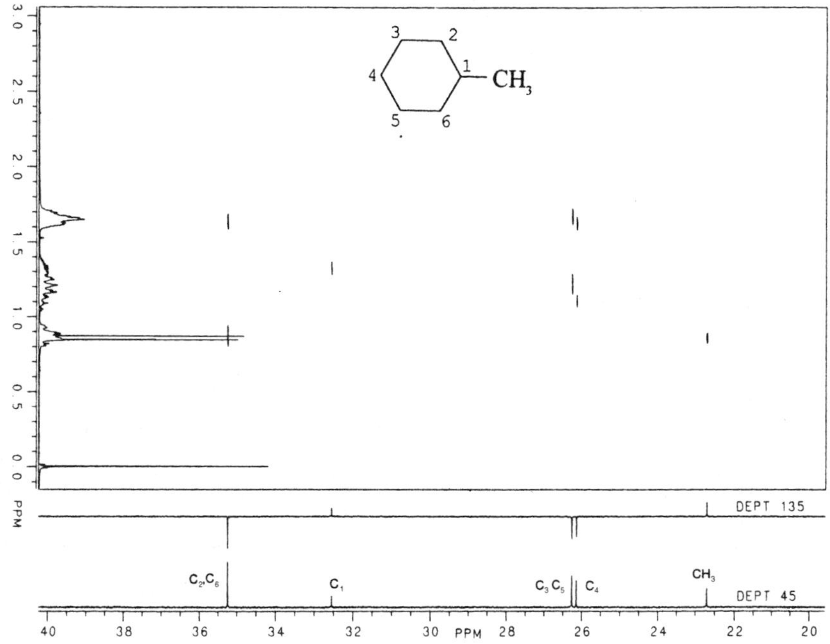

Figure 2-10. H,C-COSY NMR spectrum of methylcyclohexane at 22°C. The DEPT ^{13}C-NMR on the horizontal axis distinguishes CH_2 carbons (down) from CH and CH_3 carbons (up). The ^{1}H-NMR spectrum is on the vertical axis. H–C coupling can be detected between C(2) and C(6) and both an equatorial H (near 1.7 ppm) and an axial H (near 0.9 ppm). Similar couplings can be seen for C(3), C(5), and C(4).

gens. Using valence bond calculations, Karplus defined the theoretical relationship between ϕ and ^3J (Hz). This relationship is diagrammed in Figure 2-11. Since the publication of the original reports of the Karplus equations,[63] a half dozen or more Karplus-type equations have been suggested and have found extensive use in analyzing conformation.

Conformational distortion in cyclohexanes may be detected by NMR and quantified by a method proposed by Lambert at Northwestern.[69] For a rapid chair-to-chair conformational inversion, the *cis* and *trans* vicinal (3-bond) coupling constants (^3J) in an X–CH_2–CH_2–Y fragment are seen as averaged values, J_{cis} and J_{trans} (Figure 2-12). The ratio $J^\circ_{trans}/J^\circ_{cis}$ (where J° is the coupling constant with no dependence on the electronegativity of X and Y) is defined as R, and the relationship between R and the internal dihedral angle ψ (X–C–C–Y) is given by the relationship $\cos\psi = [3/(2 + 4R)]^{1/2}$. From 1,1,4,4-tetradeuteriocyclohexane, Lambert[70] determined that $J_{trans} = 8.07$ and $J_{cis} = 3.73$, giving $R = 2.16$ and $\psi = 57.9°$. In an independent experiment, a Dutch group[71] determined that $J_{aa} = 13.12$ Hz, $J_{ee} = 2.96$ Hz, and $J_{ae} = 3.65$ Hz at −100°C. From these data, one calculates $R = 2.20$ and $\psi = 58.2°$. Using PCMODEL,[36] from the atomic coordinates of the energy-minimum chair, we find that the computed ψ is 56.4° and the computed[67] coupling constants are $J_{aa} = 13.06$ Hz, $J_{ee} = 2.84$ Hz, and $J_{ae} = 3.5$ Hz. (With these computed coupling constants, R and ψ are calculated to be 2.27 and 58.6°, respectively.) The variously determined ψ values are in good agreement, but differ slightly from the experimental values of Table 2-1. From Figure 2-12C, it can be seen that ring distortion caused by flattening closes ψ while opening the torsion angle $\phi(H_e–C–C–H_e)$ and closing $\phi(H_a–C–C–H_a)$. The associated change

(A)

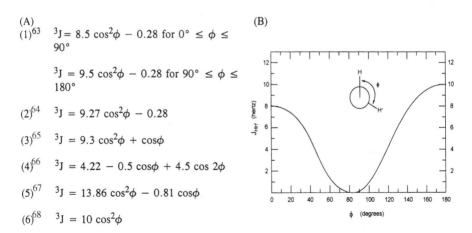

(1)[63] $^3J = 8.5\cos^2\phi - 0.28$ for $0° \le \phi \le 90°$

$^3J = 9.5\cos^2\phi - 0.28$ for $90° \le \phi \le 180°$

(2)[64] $^3J = 9.27\cos^2\phi - 0.28$

(3)[65] $^3J = 9.3\cos^2\phi + \cos\phi$

(4)[66] $^3J = 4.22 - 0.5\cos\phi + 4.5\cos 2\phi$

(5)[67] $^3J = 13.86\cos^2\phi - 0.81\cos\phi$

(6)[68] $^3J = 10\cos^2\phi$

(B)

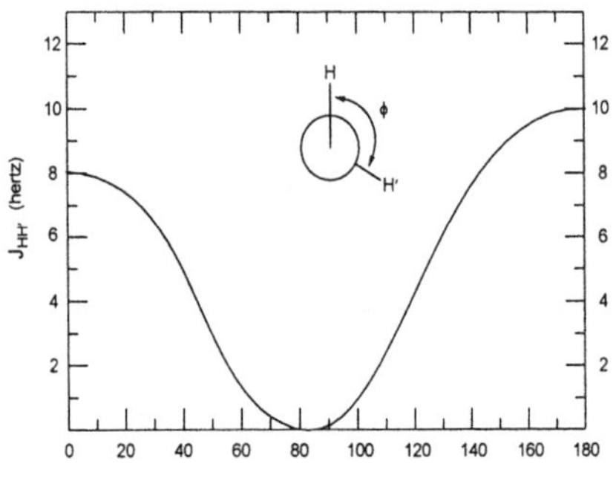

Figure 2-11. (A) Karplus-type equations and their (superscripted) literature references. (B) Karplus diagram for the theoretical dependence of H–H coupling constants (^3J) on the torsion angle ϕ (H–C–C–H) separating the two hydrogens. The Karplus equations used are $^3J = 8.5\cos^2\phi - 0.28$ (Hz) for $0° \le \phi \le 90°$ and $^3J = 9.5\cos^2\phi - 0.28$ (Hz) for $90° \le \phi \le 180°$. [(B) is redrawn from Lambert, J. B., Shurvell, H. F., Lightner, D. A., and Cooks, R. G., *Organic Structural Spectroscopy*, Prentice-Hall, Inc., Upper Saddle River, NJ, 1998.]

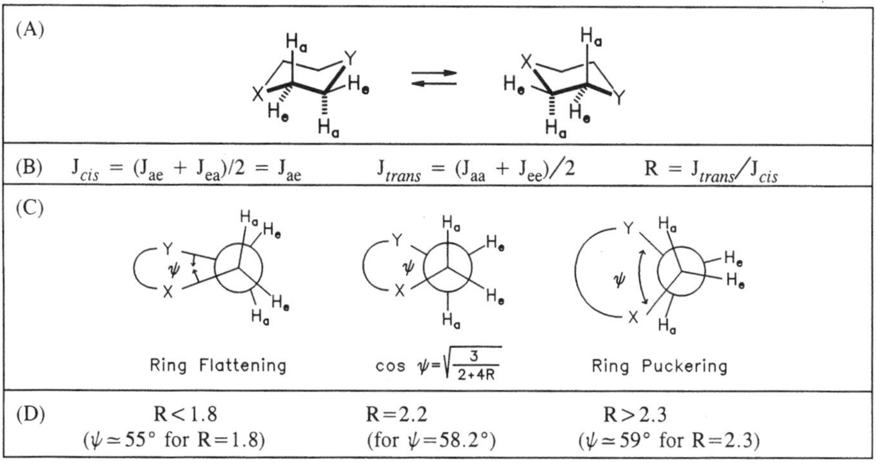

Figure 2-12. (A) Cyclohexane system chair-to-chair interconversion showing the site exchange of axial and equatorial hydrogens in an X–CH_2–CH_2–Y fragment. (B) The average vicinal coupling constants *J* for *cis* and *trans* hydrogens, and their ratio defined as *R*.[69] (C) Changes in internal torsion angle ψ(X–C–C–Y) and external torsion angle ϕ(H–C–C–H) with ring flattening and puckering, and the definition of the relationship between ψ and *R*.[69] (D) Definitions of ring flattening and ring puckering in terms of *R*.

in coupling constants is in the direction that makes *R* smaller. On the other hand, ring distortion caused by puckering opens ψ while closing the torsion angles ϕ(H_e–C–C–H_e) and ϕ(H_a–C–C–H_a). Here, the coupling constants change to make *R* larger. For very large conformational distortions, such as that leading to a twist-boat conformation, the coupling constants give $R < 1.5$ (where ψ ~ 52°). For example, 1,4-cyclohexanedione (where X and Y are C=O of Figure 2-12A) "prefers" the twist-boat conformation.[69] From its coupling constants, $J_{trans} = 8.05$ Hz and $J_{cis} = 6.23$ Hz, *R* and ψ are calculated to be 1.29 and 50°, respectively. PCMODEL[36] computes ψ = 55.5°. The relationship also fits well for six-membered ring heterocycles where X = Y = NH ($R = 2.15$, ψ = 57.9°), X = Y = O (R ~2.20, ψ = 58.2°), and X = Y = S ($R = 3.90$, ψ = 65.6°).[69] In comparison, PCMODEL gives ψ = 56.6°, 58.3°, and 70.7°, respectively.

From an analysis of vicinal coupling constants in methylcyclohexane[71] using the Karplus equation modified by Garbisch[72] for cyclohexane ($^3J = 12.95 \cos^2 \phi - 0.02 \cos \phi$), the ϕ (H–C–C–H) torsion angles are found to differ from those in cyclohexane itself by less than 1°, suggesting that there is no ring conformational distortion induced by an equatorial methyl group. By way of contrast, in *tert*-butylcyclohexane, from an analysis of vicinal coupling constants, ϕ(H_{1a}–C_1–C_2–H_{2e}) is found to be enlarged by approximately 3°, and ϕ(H_3–C_3–C_4–H_4) is found to be closed by 1°, both figures compared with cyclohexane. These results indicate that near the equatorial *tert*-butyl group the ring is slightly puckered relative to methylcyclohexane (or cyclohexane) and that slight ring flattening occurs across the ring, near C(4). Apparently, equatorial alkyl substituents—even the bulky *tert*-butyl—have at most a very minor influence on ring conformation. Whether an axial methyl substituent would leave an undistorted cyclohexane ring has not been evaluated experimentally, but can be estimated from molecular mechanics force field calculations.

2.4. Conformation from Molecular Mechanics

Following the pioneering studies by Pitzer, who showed that the barrier to rotation in ethane is about 3 kcal/mole,[73,74] predictions of the energetics associated with conformational changes emerged in the 1940s and 1950s, leading to what are now called molecular mechanics force field calculations.[37] Such calculations take into account the total energy of the molecule from bond angle deformation (Baeyer strain), bond stretching and compression, torsional energy (Pitzer strain), nonbonded van der Waals attraction and repulsion, electrostatics (including hydrogen bonding), and solvation. Readily available programs written for desktop computers have made molecular mechanics analyses easy and convenient. Those programs, including the PCMODEL versions 4.0–7.0[36] used in this work, are typically based on the most widely used force fields: MM2[37] and MM3[75] from Allinger at Wayne State and the University of Georgia.[76] The computed (PCMODEL)[36] and experimentally derived bond and angle parameters of cyclohexane are compared in Table 2-3. Given the limiting approximations or assumptions built into each method, the agreement is excellent.

Computed ring torsion angles for axial and equatorial methyl- and *tert*-butylcyclohexane are given in Table 2-4. The ring C–C–C–C torsion angles in equatorial methyl

Table 2-3. Comparison of Computed and Experimentally Derived Torsion Angles (ψ and ϕ), Bond Angles ($\angle$) and Bond Lengths (d) for Chair Cyclohexane

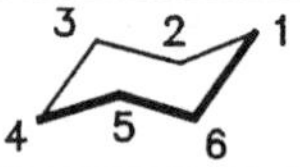	Method				
	PCMODEL[a]	MM3[b]	Electron Diffraction[c]	X-Ray[d]	^{1}H-NMR[e]
ψ (C_1–C_2–C_3–C_4)	56.4°	55.3°	55.1 ± 0.4°	55.1°	58.2°
ϕ (H_e–C–C–H_e)	60.1°				60.8°
ϕ (H_a–C–C–H_e)	57.2°				57.2°
ϕ (H_a–C–C–H_a)	174.6°				180.0°
$\angle$ (C–C–C)	110.9°	111.3°	111.34° ± 0.24°	111.34°	
$\angle$ (H–C–H)	107.1°	106.7°	105.3° ± 2.3 Å		
d (C–C)	1.536 Å	1.536 Å	1.5335 ± 0.002 Å	1.526 Å	
d (C–H)	1.116 Å	1.1145 Å	1.1099 ± 0.004 Å		

[a] Values from PCMODEL (ref. 36) and MM2 (ref. 37).

[b] Values from ref. 76.

[c] Values from ref. 26.

[d] Values from ref. 27.

[e] H–C–C–H torsion angles from coupling constants J_{aa} = 13.12 Hz, J_{ae} = 3.65, and J_{ee} = 2.96 (ref. 71) and a Karplus equation ($J = 12.95 \cos^2\phi - 0.02 \cos\phi$) modified by Garbisch for cyclohexane (ref. 72). C–C–C–C torsion angle from Fig. 2-12.

Table 2-4. Computed Chair Cyclohexane Ring Torsion Angles, ψ (C–C–C–C) for Methyl and *tert*-Butylcyclohexanes[a]

R=	Ring Torsion Angles (ψ, °)[a]					
	C_6–C_1–C_2–C_3	C_2–C_1–C_6–C_5	C_1–C_2–C_3–C_4	C_1–C_6–C_5–C_4	C_2–C_3–C_4–C_5	C_3–C_4–C_5–C_6
Equatorial –CH_3	56.1	–56.1	–56.6	56.6	56.4	–56.4
Axial –CH_3	53.4	–53.4	–55.4	55.4	56.1	–56.1
Equatorial –$C(CH_3)_3$	55.7	–55.6	–57.4	57.4	56.6	–56.6
Axial –$C(CH_3)_3$	51.7	–48.7	–57.5	51.9	57.2	–54.7

[a] PCMODEL, ref. 36. The ring torsion angles (ψ) in cyclohexane are computed to be 56.4°.

chair cyclohexane are essentially the same as that (56.4°) computed for cyclohexane. In equatorial *tert*-butylcyclohexane, the ring is very slightly flattened near C(1). Although equatorial methyl and *tert*-butyl groups leave the chair cyclohexane essentially unperturbed, when the groups are axial, larger deformations are found. Flattening is seen with the axial isomers. In methylcyclohexane, the ring is approximately 3° flatter near C(1) bearing an axial methyl than C(1) bearing an equatorial. In *tert*-butylcyclohexane, the ring is approximately 4–8° flatter near C(1) bearing an axial *tert*-butyl. Flattening is transmitted across the entire ring, although it is much less pronounced near C(4). An axial *tert*-butyl also induces a nonsymmetric ring deformation, leaving the chair in a somewhat twisted conformation.

Although axial *tert*-butyl groups are seldom found on a chair cyclohexane, in *trans*-1,3-di-*tert*-butylcyclohexane one *tert*-butyl group is required to be in the axial configuration when cyclohexane adopts the chair conformation. However, an alternative ring conformation, a slightly twisted boat, has been computed to be more stable than the chair by approximately 0.4 kcal/mole.[77] The small energy difference suggests that (i) a substantial mole fraction of the axial *tert*-butyl chair conformation remains at room temperature and (ii) the available ring conformations of *trans*-1,3-di-*tert*-butylcyclohexane are all high-energy conformations. Since a twist-boat cyclohexane is computed (PCMODEL)[36] to lie about 5.3 kcal/mole above the chair, an axial *tert*-butyl substituent would be expected to contribute approximately 6 kcal/mole more in destabilization energy to the chair conformation than to the twist boat, assuming that the equatorial *tert*-butyl group has a comparable effect on each conformation. In an earlier experiment in which the *trans* and more stable *cis*-1,3-di-*tert*-butylcyclohexane isomers were equilibrated, Allinger[78] determined an energy difference of $\Delta H° =$ 4.6 kcal/mole. This value would correspond roughly to the energy difference between a chair (*cis*) and twist boat (*trans*).

2.5. Cyclohexanone

What changes in cyclohexane ring conformation might accompany the replacement of a single sp^3-hybridized cyclohexane ring carbon by an sp^2 carbon? In cyclohexanone, for example, the ketone carbonyl appears to have little effect on the overall shape of the ring. Thus, cyclohexanone conformations are analogous to those of cyclohexane and consist of the chair, (flexible) boat, and twist boat (Figure 2-13). Although there is only one chair cyclohexanone conformation, there are two twist-boats and two boats. As with cyclohexane, the chair is the minimum-energy conformation. The twist boats lie higher in energy and the boats even higher. Significantly, the energy difference between chair and twist-boat *cyclohexanone* is substantially less (by about 50%) than the energy difference between chair and twist-boat *cyclohexane*, suggesting that flexible conformations might be more common in cyclohexanones than in cyclohexanes.

Chair cyclohexanone can be expected to be flattened somewhat in the vicinity of the carbonyl group, due to the change from sp^3 to sp^2 hybridization of a ring carbon and to an associated change in the C–C–C bond angle. As predicted by the change in hybridization, electron diffraction and molecular mechanics calculations (Table 2-5) show that the ring is flattened near the carbonyl group (by ~6°) relative to cyclohexane (Table 2-3), and puckered slightly (by ~1°) across the ring from the carbonyl, near C(4). These changes can be expected to increase the strain in chair cyclohexanone and presumably account for at least part of the smaller energy difference between the chair and twist-boat cyclohexanone conformations, compared with those of cyclohexane. Other elements of destabilization might come from an eclipsing interaction between the carbonyl oxygen and the α-equatorial hydrogens at C(2) and C(6).[79] The elements of destabilization characteristic of the higher energy conformations of cyclohexanone have not yet been identified.

Using dynamic NMR, Anet et al. determined the activation energy for the chair-to-chair conformational inversion of 3,3,4,5,5-pentadeuteriocyclohexanone (Figure 2-14A).[80] The deuterium-decoupled 251-MHz ^{1}H-NMR spectrum (Figure 2-14B), measured from room temperature to –170°C, consisted of two lines: one signal from the four α-hydrogens and one signal from the lone hydrogen at C(4). The singlet signal for the C(4) hydrogen broadens significantly below –180°C and splits into two lines below –184°C, as shown in Figure 2-14B. At –183°C, the NMR study gives a rate constant of 130 sec^{-1} for ring inversion. Thus, $\Delta G^{\ddagger}$ is calculated to be 4.1 kcal/mole for the chair-to-chair ring inversion, a process in which the twist boat is presumably an intermediate. The value is in good agreement with that (3.9 kcal/mole) calculated earlier by molecular mechanics,[79] and both are substantially lower than the experimen-

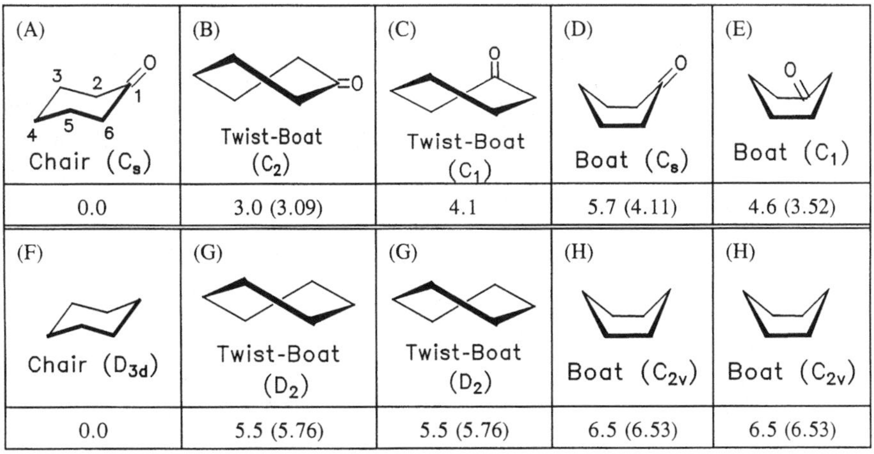

(A)
3
2
O
1
4
5
6
Chair (C_s)
0.0
(B)
O
Twist-Boat
(C_2)
3.0 (3.09)
(C)
O
Twist-Boat
(C_1)
4.1
(D)
O
Boat (C_s)
5.7 (4.11)
(E)
O
Boat (C_1)
4.6 (3.52)
(F)
Chair (D_{3d})
0.0
(G)
Twist-Boat
(D_2)
5.5 (5.76)
(G)
Twist-Boat
(D_2)
5.5 (5.76)
(H)
Boat (C_{2v})
6.5 (6.53)
(H)
Boat (C_{2v})
6.5 (6.53)

Figure 2-13. (Upper) Cyclohexanone conformations and their symmetry. The computed relative energies (kcal/mole) below each conformation are referenced to the chair as that conformation with the lowest energy. (Lower) Corresponding cyclohexane conformations and their symmetry. The computed (PCMODEL)[36] relative energies (kcal/mole) shown below each conformation are referenced to the chair as that conformation with the lowest energy. Values in parentheses are from MM3 (refs. 75, 76).

Table 2-5. Comparison of Torsion Angles and Bond Angles in Chair Cyclohexanone Determined by Experiment and by Molecular Mechanics Calculations

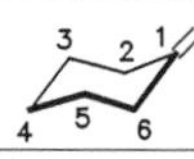	Method			
	Electron Diffraction[a]	Microwave[b]	PCMODEL[c]	MM3[d]
ψ (C_6–C_1–C_2–C_3)	51.7°		49.8°	48.2°
ψ (C_1–C_2–C_3–C_4)	−53.0°		−52.8°	51.5°
ψ (C_2–C_3–C_4–C_5)	56.3°		57.7°	56.8°
ϕ (H_{2e}–C_2–C_3–H_{3e})			62.9°	
ϕ (H_{2a}–C_2–C_3–H_{3e})			−55.6°	
ϕ (H_{3e}–C_3–C_4–H_{4e})			58.9°	
ϕ (C_{3a}–C_3–C_4–C_{4e})			58.5°	
$\angle$ (C_2–C_1–C_6)	115.3° ± 0.3°	116.20°	115.5°	116.08°
$\angle$ (C_1–C_2–C_3)	111.5° ± 0.1°	110.40°	111.3°	111.59°
$\angle$ (C_2–C_3–C_4)	110.8° ± 0.2°	114.60°	111.1°	111.66°
$\angle$ (C_3–C_4–C_5)	110.8° ± 0.2°	110.70°	110.8°	111.06°
d (C_1–C_2)	1.503 ± 0.002 Å	1.516 Å	1.536 Å	1.5199 Å
d (C_2–C_3)	1.542 ± 0.002 Å	1.535 Å	1.534 Å	1.5362 Å
d (C_3–C_4)	1.545 Å	—	1.535 Å	1.5345 Å
d (C=O)	1.229 ± 0.003 Å	1.222 Å	1.211 Å	1.2108 Å

[a] Electron diffraction : Dillen, J., and Geise, H. J.,*J. Molec. Struct.* **69** (1980), 137–144. (See ref. 75 for other references.)
[b] Microwave spectroscopy: Ohnishi, Y., and Kozima, K.,*K. Bull. Chem. Soc. Jpn.* **41** (1968), 1323–1325.)
[c] PCMODEL, ref. 36.
[d] MM3, ref. 75.

tally determined barrier for cyclohexane chair interconversion ($\Delta G^{\ddagger}$ = 10.2 kcal/mole).[45] They are also lower than the value for the barrier in other cyclohexene analogs with one sp^2 ring carbon: methylenecyclohexane ($\Delta G^{\ddagger}$ = 8.4 ±0.1 kcal/mole,[81] calculated $\Delta G^{\ddagger}$ = 8.1 kcal/mole)[82] and cyclohexanone oxime methyl ether ($\Delta G^{\ddagger}$ = 5.6 ±0.5 kcal/mole).[81,83]

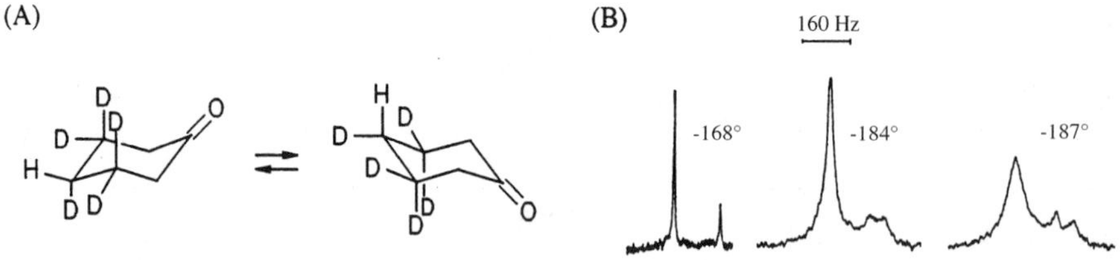

Figure 2-14. (A) Chair-to-chair interconversion in 3,3,4,5,5-pentadeuteriocyclohexanone. (B) Deuterium-decoupled 251-MHz ^{1}H-NMR spectrum measured in $CHClF_2$–$CHCl_2F$ (5:1) at −168°, −184°C, and −187°C. The smaller upfield signal due to the hydrogen at C(4) splits into two signals for axial and equatorial hydrogens. The observed difference in chemical shifts (0.24 ppm) is only one-half of that (0.47) seen in cyclohexane. [(B) is reprinted with permission from ref. 80. Copyright© 1973 American Chemical Society.]

2.6. Substituted Cyclohexanones

Due to its greater ease of deformation and more accessible twist-boat conformations, chair cyclohexanone should be more easily deformed by substitution than cyclohexane is. For example, although a substituent at an α-axial position on chair cyclohexanone lies in essentially the same steric environment as it does in chair cyclohexane, an α-equatorial group lies nearly eclipsed with the carbonyl group. This α-substituent-carbonyl nonbonded eclipsing interaction can easily be alleviated by flattening the cyclohexanone ring near the carbonyl group. However, as indicated by molecular mechanics calculations (Table 2-6), a comparison of ring torsion angles (ψ) of 2-methylcyclohexanone with those of cyclohexanone itself shows that the introduction of an equatorial methyl group at C(2) leaves the cyclohexanone ring in an essentially unperturbed chair (as does the introduction of equatorial methyl groups at C(3) or C(4)). Thus, it would appear that eclipsing an α-equatorial methyl group with a carbonyl group is not particularly destabilizing, because the cyclohexanone ring does not deform to alleviate such eclipsing. Indeed, eclipsing may even provide stabilization. For example, microwave[84] and NMR[85,86] studies have shown that eclipsed conformations are the minimum-energy conformations in propionaldehyde, with the lowest energy having the methyl and carbonyl groups eclipsed (Figure 2-15A). In addition, electron diffraction studies have shown that the most stable conformation of 3-pentanone has the C(1) and C(5) methyl groups in a *syn*-periplanar orientation to (or nearly eclipsed with) the carbonyl (Figure 2-15C).[87]

Table 2-6. Computed[a] Ring Torsion Angles and Heats of Formation for Chair Cyclohexanone and its Monoalkylated Analogs

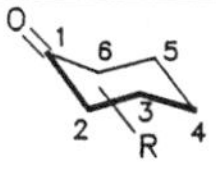

R=	Computed[a] Ring Torsion Angles (°) $\psi(C_3{-}C_2{-}C_1{-}C_6)$ $[\psi(C_2{-}C_1{-}C_6{-}C_5)]$	$\psi(C_1{-}C_2{-}C_3{-}C_4)$ $[\psi(C_1{-}C_6{-}C_5{-}C_4)]$	$\psi(C_2{-}C_3{-}C_4{-}C_5)$ $[\psi(C_3{-}C_4{-}C_5{-}C_6)]$	Heat of Formation (kcal/mole)
H	49.8 [–49.8]	–52.8 [52.8]	57.7 [–57.7]	–53.72
2(e)–CH_3	50.8 [–51.4]	–52.9 [53.4]	57.6 [–51.4]	–61.08
3(e)–CH_3	50.0 [–49.8]	–52.5 [52.8]	57.4 [–58.0]	–61.24
4(e)–CH_3	49.7 [–49.7]	–53.0 [53.0]	57.5 [–57.5]	–61.21
2(a)–CH_3	45.9 [–48.1]	–49.5 [52.9]	56.6 [–57.7]	–58.79
3(a)–CH_3	52.2 [–51.1]	–52.6 [51.3]	55.5 [–55.2]	–59.85
4(a)–CH_3	48.2 [–48.2]	–51.4 [51.4]	55.3 [–55.3]	–59.53
2(e)–CH_2CH_3	51.2 [–52.9]	–52.3 [54.2]	57.1 [–57.3]	–65.77
2(a)–CH_2CH_3	48.1 [–49.9]	–50.7 [53.0]	56.6 [–57.0]	–64.19
2(e)–$CH(CH_3)_2$	55.1 [–56.7]	–53.6 [54.9]	56.6 [–56.2]	–71.32
2(a)–$CH(CH_3)_2$	51.9 [–51.9]	–54.1 [52.0]	57.9 [–55.7]	–70.22
2(e)–$C(CH_3)_3$	54.7 [–56.1]	–54.4 [54.6]	57.5 [–56.3]	–77.13
2(a)–$C(CH_3)_3$	39.3 [–46.9]	–42.1 [55.1]	53.6 [–59.0]	–74.83

[a] PCMODEL (ref. 36); ψ values are ±0.5°.

(A) (B) (C)

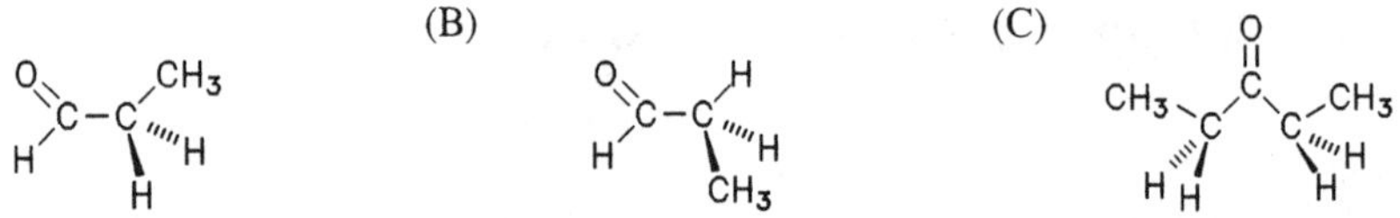

Figure 2-15. (A) Eclipsed conformation of propionaldehyde with CH_3 eclipsing C=O. (B) Eclipsed conformation of propionaldehyde with H eclipsing C=O. The conformations in (A) and (B) are minimum-energy conformations, but that in (A) is lower in energy by 0.8–1.0 kcal/mole over that in (B). (C) Most stable conformation of 3-pentanone, with CH_3 groups eclipsing the C=O.

Analyzing α-alkyl-cyclohexanone conformation by ^{1}H-NMR is difficult due to overlapping signals, even at 300 MHz.[88] However, Servis et al.,[89a] using lanthanide shift (LIS) reagents, carried out a study of 2-methyl, 2-ethyl, 2-isopropyl, and 2-*tert*-butylcyclohexanones at 100 MHz. From this work, conducted at 31.9°C, 2-methylcyclohexanone was predicted to contain some 81–98% of the chair conformer with the methyl in an equatorial position. The work also indicated 79% equatorial ethyl chair cyclohexanone. Subsequent LIS NMR measurements in $CDCl_3$ by Abraham et al.[90] showed 95% of the equatorial conformer in 2-methylcyclohexanone. More recent ^{1}H-NMR studies by Lambert et al.[91] indicate a value between 0 and 50%, but the inherent experimental error is stated to be large.

Some 40 years ago, however, eclipsing an α-equatorial alkyl group with the adjacent ketone carbonyl of 2-alkylcyclohexanone was thought to introduce a severe nonbonded steric repulsion of about 1 kcal/mole. As it turned out, this value was arrived at on the basis of misleading data. (The eclipsing interaction energy term was called the "2-alkyl ketone effect.")[92,93] Thus, the introduction of an alkyl group at the 2-equatorial position of cyclohexanone would raise the energy of the molecule by 1 kcal/mole more than the introduction of an equatorial methyl group into cyclohexane, where no serious eclipsing interactions are present. Assuming that the introduction of a 2-axial alkyl group leads to nearly identical steric interactions on cyclohexanone, as is experienced with an axial alkyl on cyclohexane, the energy difference between axial and equatorial 2-alkylcyclohexanones would be reduced by 1 kcal/mole (again, relative to the energy difference between axial and equatorial alkyl cyclohexane). If the energy difference between axial and equatorial methyl cyclohexane is $\Delta H°= 1.75$ kcal/mole (Table 2-2) and $\Delta G^{\circ}_{300} = 1.74$ kcal/mole,[58,60] then the energy difference between axial and equatorial 2-methylcyclohexanone is predicted to be $\Delta H° \sim \Delta G^{\circ}_{300} \sim 0.75$ kcal/mole on the basis of the "2-alkyl ketone effect." To put this result in the context of isomer populations, it means 95% equatorial methylcyclohexane versus only 78% equatorial methylcyclohexanone. (See Figure 1-1.)

The "2-alkyl ketone effect" was subsequently investigated in several laboratories[94,95] by equilibration studies. Several of these investigations cast doubts on whether the "2-alkyl ketone effect" is important. For example, Allinger and Blatter[95] equilibrated 2-alkyl-4-*tert*-butylcyclohexanones (Table 2-7) and found that essentially no "2-alkyl ketone effect" could be found for an α-methyl group. Subsequently, Cotterill

Table 2-7. Epimeric Equilibration of *cis* and *trans* 2-Alkyl-4-*tert*- butylcyclohexanones and Associated Energies for the Equilibrium at 25°C

(A)

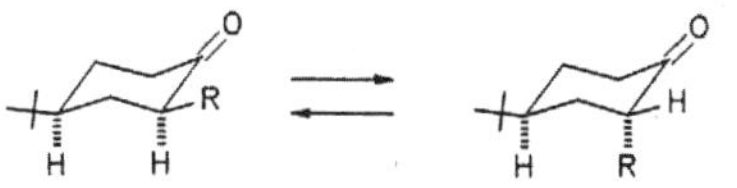

(*cis*) (*trans*)

(B)

(*cis*) (*trans*)

R	System	Author	$\Delta H°$ (kcal/mole)	$\Delta S°$ (e.u.)	$\Delta G°$ (kcal/mole)
CH_3	A	Allinger[a]	−1.57 ± 0.21	−0.1 ± 0.6	−1.60
	A	Robinson[b]	−2.16 ± 0.11	−0.8 ± 0.2	−1.92
	B	Robinson[b]	−2.18 ± 0.11	−1.8 ± 0.2	−2.05[d]
	B	Rickborn[c]	−1.95 ± 0.26	−1.8 ± 0.8	−1.82 ± 0.3[d]
CH_2CH_3	A	Allinger[a]	−1.06	±0	−1.06
	B	Robinson[b]	−1.26	+0.8	−1.91
	B	Rickborn[c]	−1.28 ± 0.16	−1.6 ± 0.5	−1.21 ± 0.03[d]
$CH(CH_3)_2$	A	Allinger[a]	−0.44 ± 0.12	+0.5 ± 0.3	−0.29
	B	Rickborn[c]	+0.03 ± 0.07	+0.6 ± 0.2	−0.56 ± 0.03[d]
$C(CH_3)_3$	A	Allinger[a]	−2.39 ± 0.38	−2.5 ± 1.0	−3.14
	B	Rickborn[c]	−2.02 ± 0.36	−3.1 ± 1.0	−1.52 ± 0.03[d]

[a] Ref. 95.
[b] Ref. 94.
[c] Rickborn, B., *J. Am. Chem. Soc.* **84** (1962), 2414–2417. [d] $\Delta G = \Delta G(25°) - RT \ln 2(0.41\text{ kcal})$, to correct for optical isomerism of the *trans* ketone.

and Robinson[94] reexamined the base-catalyzed α-methyl epimerization in 2-methyl-4-*tert*-butylcyclohexanone, finding that $\Delta H_{e\to a} = 2.16 \pm 0.11$ kcal/mole and $\Delta S_{e\to a} = +0.8 \pm 0.2$ eu. In confirmation of these values, they also studied the α-methyl epimerization in 2,6-dimethylcyclohexanone, determining that $\Delta H = 2.18 \pm 0.1$ kcal/mole and $\Delta S = 1.8 \pm 0.2$ eu; and in 2,4-dimethylcyclohexanone, they found that $\Delta H = 2.02 \pm 0.10$ kcal/mole and $\Delta S = 1.6 \pm 0.2$ eu. Significantly, their average value for ΔH (~2.1 kcal/mole) is larger than the 1.75 kcal/mole observed in methylcyclohexane (Table 2-2). Cotterill and Robinson concluded in favor of an *inverse* "2-alkyl ketone effect" (−0.4 kcal/mole rather than +1 kcal/mole), with an α-methyl "preferring" to be eclipsed with the adjacent carbonyl by occupying the equatorial position.

In contrast to the large $\Delta H°$ measured for the equilibration of axial and equatorial 2-methyl-4-*tert*-butylcyclohexanones (Table 2-7), Allinger and Blatter[95] found smaller $\Delta H°$ values for the α-ethyl and α-isopropyl analogs. Unlike $\Delta H°$ for the α-methyl analog, where ΔH is larger than the $\Delta H°$ difference between axial and equatorial methylcyclohexane (1.75 kcal/mole, Table 2-2), $\Delta H°$ for the α-ethyl ketone is significantly smaller than that separating axial and equatorial ethylcyclohexane ($\Delta G° \sim 1.75$ kcal/mole),[96] suggesting a positive "2-alkyl ketone effect." With isopropyl, the difference between the ketone $\Delta G°$ and $\Delta G°$ determined for isopropylcyclohexane (2.15 kcal/mole;[96] cf. 2.61 kcal/mole, Table 2-2) is far larger, suggesting an unusually large "2-alkyl ketone effect" (1.8–2.3 kcal/mole) for isopropyl. Clearly the "2-alkyl ketone effect" still needs further clarification. In the studies to date, there are at least a couple of complications in the experimental design: (i) $\Delta G°$ determined by NMR comes from a different solvent from that for which ΔG is determined by equilibration, and (ii) the NMR studies were carried out on cyclohexanes with one alkyl group, whereas the equilibration data come from cyclohexanones with two alkyl groups.

In the case of an α-*tert*-butyl, in 2,4-di-*tert*-butylcyclohexanone (Table 2-7), the larger value of ΔS suggests more disorder than the simple change in configuration at C(2), implying that nonchair conformations have intruded—a potentially general problem in the studies. The underlying assumptions in the study are (i) that the 4-*tert*-butyl group serves as a conformational anchor for the chair conformation and (ii) that the chair cyclohexanone ring geometry remains unperturbed when α-axial or α-equatorial alkyl substituents are placed on the ring. However, since *trans*-1,3-di-*tert*-butylcyclohexane is thought to adopt a twist-boat conformation,[41] a similar large deformation of the chair geometry should probably be expected in *trans*-2,4-di-*tert*-butylcyclohexanone—and might even be expected with other α-axial alkyls, (e.g., isopropyl).

Ring distortion in *cis* and *trans*-2-alkyl-4-*tert*-butylcyclohexanone conformations may be assessed by molecular mechanics calculations (Table 2-8). Significant distortion of the chair appears only in the *trans*-di-*tert*-butyl isomer with an α-axial-*tert*-butyl group, compared with other α-alkyls, and contrasts with an essentially normal chair in the α-equatorial (*cis*) isomer. The origin of the conformational distortion lies with the α-axial *tert*-butyl group and does not depend on the presence of the 4-*tert*-

Table 2-8. Computed[a] Ring Torsion Angles and Heats of Formation for Minimum-Energy Conformations of *cis* and *trans*-2-Alkyl-4-*tert*-butylcyclohexanones

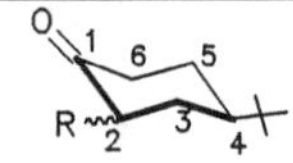 R=	Computed[a] Ring Torsion Angles (°)			
	$\psi(C_3–C_2–C_1–C_6)$ [$\psi(C_2–C_1–C_6–C_5)$]	$\psi(C_1–C_2–C_3–C_4)$ [$\psi(C_1–C_6–C_5–C_4)$]	$\psi(C_2–C_3–C_4–C_5)$ [$\psi(C_3–C_4–C_5–C_6)$]	Heat of Formation (kcal/mole)
H	52.4 [–52.2]	–55.1 [54.4]	56.7 [–56.1]	–77.69
2(e)–CH_3	51.9 [–52.0]	–54.8 [53.9]	57.1 [–56.0]	–85.10
2(a)–CH_3	46.2 [–48.3]	–50.8 [53.8]	56.3 [–57.1]	–82.81
2(e)–CH_2CH_3	50.4 [–52.8]	–52.4 [55.9]	56.4 [–57.5]	–89.82[b]
	54.8 [–54.7]	–56.2 [54.3]	57.1 [–55.2]	–89.76[c]
2(a)–CH_2CH_3	48.3 [–48.4]	–54.0 [52.5]	58.5 [–56.6]	–87.80[b]
	47.4 [–49.7]	–51.2 [54.6]	56.3 [–57.4]	–88.22[c]
2(e)–$CH(CH_3)_2$	56.7 [–58.3]	–54.8 [55.9]	55.0 [–54.5]	–94.93
2(a)–$CH(CH_3)_2$	51.5 [–50.9]	–56.1 [52.3]	58.8 [–55.5]	–94.50
2(e)–$C(CH_3)_3$	54.8 [–56.1]	–55.7 [55.4]	57.3 [–55.7]	–101.26
2(a)–$C(CH_3)_3$	39.8 [–61.3]	–50.3 [50.6]	60.1 [–58.5]	–98.26

[a] PCMODEL, (ref. 36); ψ values are ±0.5°.

[b] CH_3 rotated away from C=O.

[c] CH_3 rotated toward C=O.

butyl, because axial 2-*tert*-butylcyclohexanone shows a similar large distortion. Although chair cyclohexanone is more easily distorted than chair cyclohexane, most of the alkyl groups are only marginally effective in exerting conformational distortion.

The influence of a 3-alkyl substituent on cyclohexanone was also considered some 40 years ago.[92,93] It was noted that on one face of chair cyclohexanone, there are only two axial positions, while there are three on the opposite face. In contrast, cyclohexane has three axial positions on each face. It was expected that this difference would lead to a decreased destabilization interaction for an axial substituent at C(3) or C(5) of cyclohexanone, since there is only one 1,3-diaxial alkyl–hydrogen interaction versus two in cyclohexane. It was thought that the energy difference between axial and equatorial conformations should therefore be reduced by one-half, namely, from a value of about 1.70 kcal/mole[96] in methylcyclohexane to 0.85 kcal/mole in 3-methylcyclohexanone. However, Allinger and Freiberg[97] estimated the "3-alkyl ketone effect" to be only 0.3–0.4 kcal/mole, on the basis of the equilibration of *cis* and *trans*-3,5-dimethylcyclohexanone. As computed by PCMODEL,[36] the energy difference between *cis* (with two equatorial methyls) and *trans* (with one axial and one equatorial methyl) is 1.4 kcal/mole. Since the energy difference between axial and equatorial methylcyclohexane is ΔH ~1.75 kcal/mole (Table 2-2), the "3-alkyl ketone effect" is computed to be 0.35 kcal/mole—in good agreement with Allinger and Freiberg's result. As before, whether the chair conformation predominates in the more easily deformed cyclohexanone ring (compared with the cyclohexane ring) is a matter of speculation. Molecular mechanics computations[36] (Table 2-6) show that there is greater twisting in 3-axial methylcyclohexanone, compared with 3-equatorial, and the computed conformational energy difference (ΔH = 1.39 kcal/mole) between 3-axial and 3-equatorial methyl cyclohexanone is a significant percentage of the energy difference (ΔH = 3.0 kcal/mole, Figure 2-13) between the chair and lowest energy twist-boat conformation of cyclohexanone.

2.7. Substituted Cyclohexanones, Optical Activity, and Circular Dichroism

Unlike chair cyclohexanone, 2-alkyl and 3-alkyl chair cyclohexanones are chiral molecules and thus are capable of exhibiting optical activity. Optical activity is typically detected by a nonzero rotation α of the plane of polarized light. Although this method can detect optical activity per se, which might even be due to an optically active contaminant in the sample, no stereochemical information can be deduced a priori. A related spectroscopic measurement, achieved through circular dichroism spectroscopy, can, however, often be used to elicit structural information from optically active molecules. For example, 2-methylcyclohexanone has two enantiomeric mirror-image chair conformations (Figure 2-16A) that are expected to be the most stable conformations. 2-Methylcyclohexanone can, of course, adopt higher energy twist-boat or boat conformations to some extent. However, as predicted by molecular mechanics

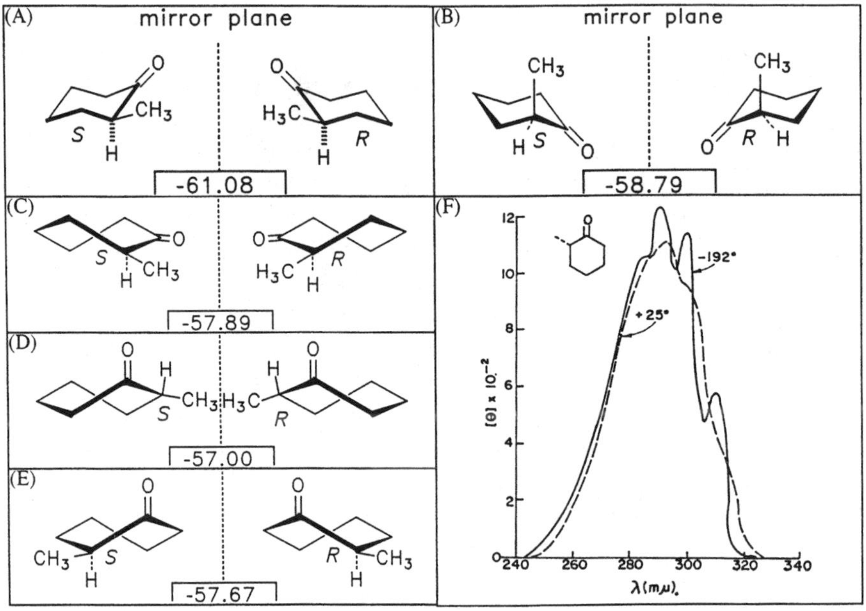

Figure 2-16. Mirror-image enantiomeric conformations of 2(*S*)-methylcyclohexanone (left) and 2(*R*)-methylcyclohexanone (right) in the chair cyclohexanone conformation of the equatorial isomer (A) and axial isomer (B). Three different twist-boat conformations are represented in (C), (D), and (E). The boxed numbers in (A)–(E) are the computed (PCMODEL, ref. 36) heats of formation in kcal/mole. (F) Circular dichroism spectra of 2(*S*)-methylcyclohexanone at 25°C (– – – –) and at –192°C (————) in ether–isopentane–ethanol (5:5:2) solvent. [(F) is reprinted with permission from ref. 98. Copyright© 1965 American Chemical Society.]

calculations,[36] the chair is more stable than the twist boat by 3–4 kcal/mole (Figure 2-16).

Although some spectroscopic techniques can distinguish between chair and twist-boat conformations, very few can distinguish between mirror-image conformations. Circular dichroism (CD) spectroscopy is one such technique. The CD spectrum of 2(*S*)-methylcyclohexanone in the region of the ketone $n \rightarrow \pi^*$ electronic transition shown in Figure 2-16F is reminiscent of the compound's ultraviolet absorption spectrum. However, since a CD spectrum is a difference spectrum, the difference in absorbance (ΔA) or difference in absorbance coefficient ($\Delta\varepsilon$) plotted on the vertical axis may be either positively signed (as shown in Figure 2-16F) or negatively signed (as it is for the enantiomeric 2(*R*)-methylcyclohexanone). The sign of the $n \rightarrow \pi^*$ CD curve may be correlated with the absolute configuration of the ketone if the conformation is known. Or the conformation may be deduced for a ketone of known absolute configuration. In the case of 2(*S*)-methylcyclohexanone, it has been shown that CD is insensitive to temperature, with CD spectra measured at +25°C, –5°C, –29°C, –41°C, –74°C, and –192°C, all being essentially identical after correcting for contraction of the solvent.[98] These data suggest that (i) 2-methylcyclohexanone is in its most stable conformation at 25°C and (ii) high-energy conformations have inherently different CD spectra and make no significant contribution to the isomer population. The possibility that high-energy conformations make a contribution to the CD spectrum identical to that of the lowest energy conformation is improbable because the contributions to the $n \rightarrow \pi^*$ CD come from the orientation of the extrachromophoric parts of the molecule relative to the ketone carbonyl chromophore. As indicated earlier, from studies of cyclohexanone conformation, one would predict that the chair would be the lowest energy conformation. And from what we know of the differing influence on conformational stability of axial and equatorial methyl groups, one would predict that the chair cyclohexanone with an α-equatorial methyl would be the most stable conformation. (Figure 2-16A).

For reasons that will become clear in Chapters 4 and 5, one would predict an intense CD from the axial chair conformation (Figure 2-16B) and the twist-boat conformations (Figures 2-16C and D), but only a weak CD from the chair conformation of Figure 2-16A and the twist boat of Figure 2-16E. Given the higher energy of the twist-boat conformations relative to the chair, the chair is the more probable conformation. And given the greater stability of an α-equatorial methyl over an α-axial methyl, chair cyclohexanone with an α-equatorial methyl is predicted to be the most stable structure. The weak $n \rightarrow \pi^*$ CD spectrum (Figure 2-16F) confirms this analysis, and the fact that the CD sign is positive correlates best with the 2(*S*) enantiomer. On the other hand, if the absolute configuration were known to be 2(*S*), then one would predict a predominance of the chair conformation with an equatorial methyl. Small populations of twist boats or even the chair conformation with an α-axial methyl group cannot be rigorously excluded, and neither can deformation (e.g., flattening of the chair), unfortunately, without recourse to CD spectra of conformationally well-defined (rigid) ketones.

CD spectroscopy thus offers a powerful tool for stereochemical and conformational analysis. For ketone conformation analysis, the rule governing the correlation between the $n \rightarrow \pi^*$ CD transition and the absolute stereochemistry and conformation of the molecule is called the *octant rule*, which will be discussed in Chapter 4. In Chapter 3, we discuss the origin of optical activity and circular dichroism.

References

1. Haworth, E., and Perkin, W. H., Jr., *J. Chem. Soc.* **65** (1894), 591–602.
2. Von Baeyer, J. F. W. A., *Liebig's Ann. Chem.* **278** (1894), 88–116.
3. Baeyer (*Liebig's Ann. Chem.* **155** (1870), 266–281) refers to Berthelot, *Bull. Soc. Chim. Paris* **VII** (1867), 53, **IX** (1868), for the reduction of benzene with HI to give what was then called hexahydrobenzene and later discovered to be methylcyclopentane. See Wreden, F., *Liebig's Ann. Chem.* **187** (1877), 153–177.
4. For an interesting history of stereochemistry and the shape of cyclohexane, see Ramsay, O. B., *Stereochemistry*, Heyden & Sons, Ltd., Philadelphia, 1981.
5. Von Baeyer, J. F. W. A., *Chem. Ber.* **18** 1885, 2269–2281.
6. Mislow, K., *Introduction to Stereochemistry*, W. A. Benjamin, 1965. This reference work also contains a useful introduction to symmetry designations in organic stereochemistry.

7. Eliel, E. L., and Wilen, S. H., *Stereochemistry of Organic Compounds*, John Wiley & Sons, Inc., New York, 1994.
8. Sachse, H., *Chem. Ber.* **23** (1890), 1363–1370.
9. Mohr, E., *J. Prakt. Chem.* [2] **98** (1918), 315–353.
10. Hendricks, S. B., and Bilicke, C., *J. Am. Chem. Soc.* **48** (1926), 3007–3015.
11. Kohlrausche, K. W. F., and Stockmair, W., *Z. Physik. Chem.* **B31** (1936), 382–401.
12. Langseth, A., and Bok, B., *J. Chem. Phys.* **8** (1940), 403–409.
13. Kohlrausche, K. W. F., and Wittek, H., *Z. Physik. Chem.* **B48** (1941), 177–187.
14. Hassel, O., *Tidssk. Kjemi. Bergv. Metall.* **3** (1943), 32–34. For an English translation, see Hedberg, K., *Topics in Stereochem.* **6** (1971), 11–17; or ref. 4.
15. Rasmussen, R., *J. Chem. Phys.* **11** (1943), 249–252.
16. Hassel, O., and Viervoll, H., *Acta Chem. Scand.* **1** (1947), 149–168.
17. For a summary, see Hassel, O., and Ottar, B., *Acta Chem. Scand.* **1** (1947), 929–942.
18. Beckett, C. W., Pitzer, K. S., and Spitzer, R., *J. Am. Chem. Soc.* **69** (1947), 2488–2495.
19. Pitzer, K. S., *Science* **101** (1945), 672.
20. Pitzer, K. S., *Chem. Rev.* **27** (1940), 39–57.
21. Barton, D. H. R., *Experientia*, **6** (1950), 316–320.
22. See Barton's autobiography: Barton, D. H. R., *Some Recollections of Gap Jumping*, American Chemical Society, Washington, DC, 1991.
23. Davis, M., and Hassel, O., *Acta Chem. Scand.* **17** (1963), 1181.
24. Geise, H. J., Buys, H. R., and Mijlhoff, F. C., *J. Molec. Struct.* **9** (1971), 447–454.
25. Bastiansen, O., Fernholt, L., Seip, H. M., Kambara, H., and Kuchitsu, K., *J. Molec. Struct.* **18** (1974), 163–168.
26. Eubank, J. D., Kirsch, G., and Schäfer, L., *J. Molec. Struct.* **31** (1976), 39–45.
27. Kahn, R., Fourme, R., André, D., and Renaud, M., *Acta Crystallogr.* **B29** (1973), 131–138.
28. Barton, D. H. R., *J. Chem. Soc.* (1948), 340–342.
29. Westheimer, F. H., and Mayer, J. E., *J. Chem. Phys.* **14** (1946), 733–738.
30. Dostrovsky, I., Hughes, E. D., and Ingold, C. K., *J. Chem. Soc.* (1946), 173–194.
31. Hazebroek, P., and Oosterhoff, L. T., *Disc. Faraday Soc.* **10** (1951), 87–93.
32. Turner, R. B., *J. Am. Chem. Soc.* **74** (1952), 2118–2119.
33. Dauben, W. G., and Pitzer, K. S., in *Steric Effects in Organic Chemistry* (Newman, M. S., ed.), J. Wiley, New York, 1956.
34. Orloff, H. D., *Chem. Rev.* **54** (1954), 347–447.
35. (a) Hendrickson, J. B., *J. Am. Chem. Soc.* **83** (1961), 4537–4547. (b) Hendrickson, J. B., *J. Am. Chem. Soc.* **89** (1967), 7047–7061.
36. PCMODEL version 4.0 to 7.0, Serena Software, Inc., Bloomington, IN 47402-3076. Many molecular mechanics programs are commercially available. PCMODEL uses the MMX force field, a variant of Allinger's MM2 force field. We chose PCMODEL for its reliability in reproducing cyclohexane and cyclohexanone conformations and energy differences. The software is easy to install and use on desktop computers and is inexpensive.
37. Burkert, U., and Allinger, N. L., *Molecular Mechanics* (ACS Monograph 177), American Chemical Society, Washington, DC, 1982.
38. Squillacote, M., Sheridan, R. S., Chapman, O. L., and Anet, F. A. L., *J. Am. Chem. Soc.* **97** (1975), 3244–3246.
39. Anet, F. A. L., and Anet, R., in *Dynamic Nuclear Magnetic Resonance Spectroscopy* (Jackman, L. M., and Cotton, F. A., eds.), Academic Press, New York, 1975, Chap. 14.
40. Johnson, W. S., Bauer, V. J., Margrave, J. L., Frisch, M. A., Dreger, L. H., and Hubbard, W. N., *J. Am. Chem. Soc.* **83** (1961), 606–614.
41. Allinger, N. L., and Freiberg, L. A., *J. Am. Chem. Soc.* **82** (1960), 2393–2394.
42. Jensen, F. R., Noyce, D. S., Sederholm, C. H., and Berlin, A. J., *J. Am. Chem. Soc.* **84** (1962), 386–389.
43. Shoppee, C. W., *J. Chem. Soc.* 1946, 1138–1147.

44. For a useful discussion of measuring dynamic processes by NMR, see Sandström, J., *Dynamic NMR Spectroscopy*, Academic Press, New York, 1982.
45. Anet, F. A. L., and Bourn, A. J., *J. Am. Chem. Soc.* **89** (1967), 760–768.
46. O'Leary, B. M., Grotzfeld, R. M., and Rebek, J., Jr., *J. Am. Chem. Soc.* **119** (1997), 11701–11702.
47. Aydin, R., and Günther, H., *Angew. Chem. Int. Ed. Engl.* **20** (1981), 985–986.
48. Ellison, S. L. R., Fellows, M. S., Robinson, M. J. T., and Widgery, M. J., *J. Chem. Soc. Chem. Commun.* 1984, 1069–1070.
49. Anet, F. A. L., and Kopelevich, M., *J. Am. Chem. Soc.* **108** (1986), 1355–1356.
50. Williams, I. H., *J. Chem. Soc. Chem. Commun.* 1986, 627–628.
51. Hasha, D. L., Eguchi, T., and Jonas, J., *J. Am. Chem. Soc.* **104** (1982), 2290–2296.
52. Ross, B. D., and True, N. S., *J. Am. Chem. Soc.* **105** (1983), 4871–4875.
53. Pitzer, K. S., *J. Chem. Phys.* **8** (1940), 711–720.
54. Bartell, L. S., and Kohl, D. A., *J. Chem. Phys.* **39** (1963), 3097–3105.
55. Garbisch, E. W., Jr., Hawkins, B. L., and Mackay, K. D., in *Conformational Analysis* (Chiurdoglu, C., ed.), Academic Press, New York, 1971.
56. Anet, F. A. L., Bradley, C. H., and Buchanan, G. H., *J. Am. Chem. Soc.* **93** (1971), 258–259.
57. Stothers, J. B., *Carbon-13 NMR Spectroscopy*, Academic Press, Inc., New York, 1972.
58. Anet, F. A. L., and Squillacote, M., *J. Am. Chem. Soc.* **97** (1975), 3243–3244.
59. Piercy, J. E., and Subramanyan, S. V., *J. Chem. Phys.* **42** (1965), 4011–4017.
60. (a) Booth, H., and Everett, J. R., *J. Chem. Soc. Chem. Commun.* 1976, 278–279. (b) Booth, H., and Everett, J. R., *J. Chem. Soc. Perkin Trans. 2* (1980), 255–259.
61. Wiberg, K. B., Hammer, J. D., Castejon, H., Bailey, W. F., DeLeon, E. L., and Jarret, R. M., *J. Org. Chem.* **64** (1999), 2085–2095.
62. Squillacote, M. E., *J. Chem. Soc. Chem. Commun.* **105** (1983), 4871–4875.
63. Karplus, M., *J. Chem. Phys.* **30** (1959), 11–15.
64. Abraham, R. J., Cooper, M. A., Salmon, J. R., and Whittaker, D., *Org. Magn. Reson.* **4** (1972), 489–507.
65. Baretta, A. J., Jefford, C. W., and Waegell, B., *Bull. Soc. Chim. France* (1970), 3899–3908, 3985–3993.
66. Karplus, M., *J. Am. Chem. Soc.* **85** (1963), 2870–2871.
67. Haasnot, C. A. G., DeLeuw, F. A. A. M., and Altona, C. *Tetrahedron*, **36** (1980), 2783–2792.
68. Williamson, K. L., and Johnson, W. S., *J. Am. Chem. Soc.* **83** (1961), 4623–4627.
69. Lambert, J. B., *Acc. Chem. Res.*, **4** (1971), 87–94.
70. Lambert, J. B., *J. Am. Chem. Soc.* **91** (1969), 3567–3571.
71. Remijnse, J. D., von Bekkum, H., and Wepster, B. M., *Recl. Trav. Chim. Pays-Bas* **89** (1970), 658–666; **90** (1971), 779–790.
72. Garbisch, E. W., Jr., *J. Am. Chem. Soc.* **90** (1968), 6543–6544.
73. Kemp, J. D., and Pitzer, K. S., *J. Chem. Phys.* **4** (1936), 749.
74. The best value for the rotation barrier in ethane is 2.875 ± 0.125 kcal/mole; see Pitzer, K. S., *Disc. Faraday Soc.* **10** (1951), 66–73.
75. Allinger, N. L., Chen, K., Rahman, M., and Pathiaseril, A., *J. Am. Chem. Soc.* **113** (1991), 4505–4517.
76. Allinger, N. L., Yuh, Y. H., and Lii, J.-H., *J. Am. Chem. Soc.* **111** (1989), 8551–8566.
77. Allinger, N. L., Miller, M. A., Van-Catledge, F. A., and Hirsch, J. A., *J. Am. Chem. Soc.* **89** (1967), 4345–4357.
78. Allinger, N. L., *J. Am. Chem. Soc.* **81** (1959), 5727–5733.
79. Allinger, N. L., Tribble, M., and Miller, M. A., *Tetrahedron* **28** (1972), 1173–1190.
80. Anet, F. A. L., Chmurny, G. N., and Krane, J., *J. Am. Chem. Soc.* **95** (1973), 4423–4424.
81. Jensen, F. R., and Beck, B. H., *J. Am. Chem. Soc.* **90** (1968), 1066–1067.
82. Anet, F. A. L., and Yavari, I., *Tetrahedron* **34** (1978), 2879–2886.
83. Gerig, J. T., and Rimerman, R. A., *J. Am. Chem. Soc.* **92** (1970), 1219–1224.
84. Butcher, S. S., and Wilson, E. B., Jr., *J. Chem. Phys.* **40** (1964), 1671–1678.
85. Abraham, R. J., and Pople, J. A., *Mol. Phys.* **3** (1960), 609–611.

86. Karabatsos, G. J., and Hsi, N., *J. Am. Chem. Soc.* **87** (1965), 2864–2870.
87. Romers, C., and Creutzberg, J. E. G., *Recl. Trav. Chim. Pays-Bas* **75** (1956), 331–345.
88. Gurst, J. E., *J. Chem. Educ.* **69** (1992), 774–775.
89. (a) Servis, K. L., Bowler, D. J., and Ishii, C., *J. Am. Chem. Soc.* **97** (1975), 73–80, 80–88.
90. Abraham, R. J., Chadwick, D. J., Griffiths, L., and Sancassan, F., *J. Am. Chem. Soc.* **102** (1980), 5128–5130. (b) We have rerun the ^{1}H-NMR spectra of 2-*tert*-butylcyclohexanone at 300, 500 and even 800 MHz, but the data do not adequately lend themselves to further interpretation of the conformational equilibrium discussed.
91. Basso, E. A., Kaiser, C., Rittner, R., and Lambert, J. B., *J. Org. Chem.* **58** (1993), 7865–7869.
92. Klyne, W., *Experientia* **12** (1956), 119–124.
93. Robins, P. A., and Walker, J., *J. Chem. Soc.* (1955), 1789–1790.
94. For leading references, see Cotterill, W. D., and Robinson, M. J. T., *Tetrahedron* **20** (1964), 765–775, 777–790.
95. Allinger, N. L., and Blatter, H. M., *J. Am. Chem. Soc.* **83** (1961), 994–995.
96. Hirsch, J. A., *Topics in Stereochemistry*, Vol. 1, John Wiley and Sons, New York, 1967, pp. 199–222.
97. Allinger, N. L., and Freiberg, L. A., *J. Am. Chem. Soc.* **84** (1962), 2201–2203.
98. Wellman, K. M., Briggs, W. S., and Djerassi, C., *J. Am. Chem. Soc.* **87** (1965), 73–81.

CHAPTER

3

Optical Activity

3.1. Absorption of Light and Its Relationship to Optical Activity

All organic molecules absorb light in the ultraviolet (UV)–visible region of the electromagnetic spectrum, which stretches from the far UV (below 200 nm) to the near infrared (800–1000 nm). In UV–visible spectroscopy, one records the absorption of light in the spectral region from about 200 to approximately 800 nm by measuring both the energy and the probability associated with exciting a molecule from its ground electronic state to an electronically excited state. A UV–visible absorption spectrum (Figure 3-1A) thus typically plots the absorption of light energy (usually displayed as the wavelength λ in organic spectroscopy) on the horizontal axis against the intensity of absorbance (A) or the molar absorptivity (ε) on the vertical axis. A and ε are related by Beers law,

$$A = \varepsilon Cl, \tag{3-1}$$

where ε is sometimes called the molar extinction coefficient, C is the concentration in mol liter^{-1}, and l is the thickness of the sample (or path length of the cell) in cm. A is defined as

$$A = \log(I_0/I), \tag{3-2}$$

where I_0 is the incident intensity of a light beam impinging upon a medium containing C mol liter^{-1} of absorbing material and I is the exit intensity of the beam after it has traveled 1 cm of that medium. Other expressions relate I and I_0 to the molar absorptivity ε, viz.,

$$I = I_0\, 10^{-\varepsilon Cl}, \tag{3-3}$$

to the absorption coefficient k, namely,

$$I = I_0\, e^{-kl}, \tag{3-4}$$

and to the absorption index κ, to wit,

$$I = I_0\, e^{-(4\pi\kappa/\lambda)l}. \tag{3-5}$$

Consequently,

$$k = 4\pi\kappa/\lambda = 2.303\varepsilon C. \tag{3-6}$$

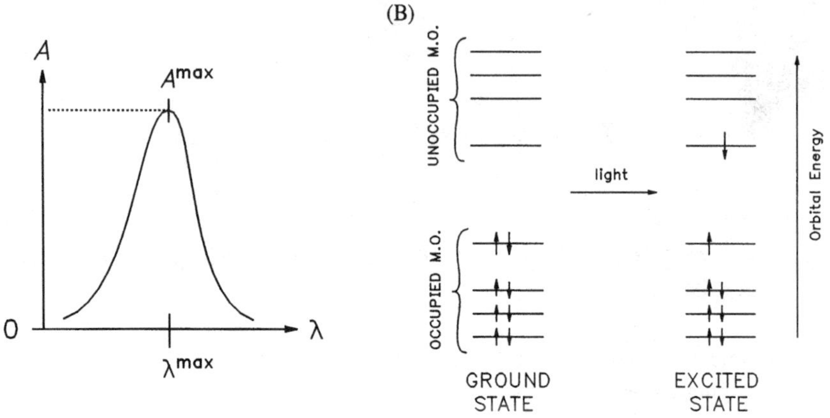

Figure 3-1. (A) An idealized representation of the UV–visible absorption spectrum corresponding to the electronic transition shown on the right. In organic chemistry, excitation energy is typically plotted as wavelength (λ) in nanometers (nm) on the horizontal axis, which represents the energy of the excitation process, since λ is related to energy (E) by Planck's law, $E = hc/\lambda$, where h is Planck's constant and c is the speed of light. The absorption of light is plotted on the vertical axis, typically as absorbance (A), which is a measure of the probability of the electronic transition actually occurring when light of the right energy strikes the molecule. For a solution of the sample, the absorption could just as well have been plotted in path length and concentration-normalized units (ε), since ε (molar absorptivity) is related directly to A by Beer's law, $A = \varepsilon \cdot C \cdot l$, where l is the sample (cuvette) path length in cm and C is the molar concentration of the sample. The wavelength corresponding to the energy of maximum light absorption (A^{max}) is called λ^{max}. (B) Idealized representation of an electronic excitation in a molecule by the absorption of UV–visible light. Occupied and unoccupied molecular orbitals (M.O.'s) and idealized electronic configurations are shown for the electronic ground and excited states. Here, the molecule is elevated into an electronically excited state by the promotion of an electron from the highest occupied molecular orbital to the lowest unoccupied molecular orbital. The electrons and their relative spin orientations are represented by small arrows. The ground and excited states shown are singlet states, in which all electron spins are paired.

A, ε, k, and κ are measures of the probability of converting a molecule from its ground to its excited state—a process that involves the promotion of an electron—usually a valence electron—from an occupied molecular orbital to an unoccupied molecular orbital (Figure 3-1B). Ordinarily, the absorption phenomenon can be said to originate from within a particular component or group in a molecule, called a chromophore, from which many different electronic excitations may arise.

Most organic chemists are familiar with electronic absorption spectroscopy (as UV–visible spectroscopy). When a molecule is elevated from its electronic ground state to an electronically excited state, the movement of an electron from an occupied molecular orbital to a higher energy, unoccupied orbital (Figure 3-1B) creates a momentary dipole moment called the *electric transition dipole moment* ($\boldsymbol{\mu}_e$). Since an organic molecule typically has numerous occupied and unoccupied molecular orbitals, many different electronic excitations are possible, each with a corresponding electric transition dipole moment. The electronic transitions of greatest importance in organic chemistry are typically the low-energy excitations (e.g., those involving the promotion of an electron from the highest occupied bonding or nonbonding orbital to the lowest (or a lower) unoccupied molecular orbital). UV–visible spectroscopy offers a way to understand the nature of the ground and electronic states and the excitation process connecting them. And from a unique type of UV–visible *difference spectroscopy*, called *circular dichroism* (CD), which is sensitive only to chiral molecules, much can be learned about the structure of a molecule in addition to understanding electronic structure.

Circular dichroism and UV–visible spectroscopy have their origins in the same photophysical process: raising an electronic ground state to an electronically excited state by movement of an electron. Consequently, their spectra closely resemble each other (Figure 3-2). However, because CD involves measuring a difference in absorption, CD curves can take on either a positive or an inverted negative bell shape (Figures 3-2B and C). In contrast, UV–visible spectra are, by convention, always positive (Figure 3-2A). In UV–visible spectroscopy, the light that is used is ordinary (or isotropic) light. In CD spectroscopy, the light is anisotropic and circularly polarized, which is achieved by filtering ordinary light through a polarizer to give linearly (or plane) polarized light. The linearly polarized light is then separated into its left and right circularly polarized components by passing it through a stress plate modulator. When the difference in absorption of left and right circularly polarized light is nonzero ($\Delta A = A_L - A_R \neq 0$), circular dichroism (meaning the unequal absorption of left and right circularly polarized light; $\Delta A = \Delta\varepsilon Cl \neq 0$) can be measured and recorded in the region of the UV–visible absorption curve for the electronic transition. Natural circular dichroism is seen only with chiral molecules and recorded only with optically active

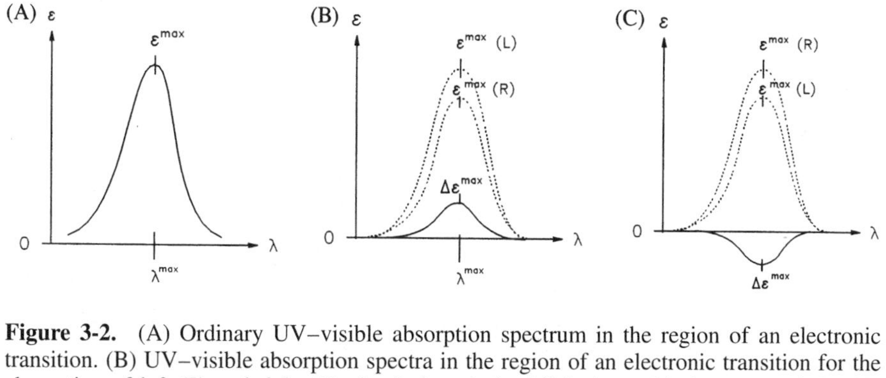

Figure 3-2. (A) Ordinary UV–visible absorption spectrum in the region of an electronic transition. (B) UV–visible absorption spectra in the region of an electronic transition for the absorption of left (*L*) and right (*R*) circularly polarized light by a chiral compound. In this example, left circularly polarized light is absorbed to a greater extent than right circularly polarized light ($\varepsilon_L > \varepsilon_R$), so the difference curve plotted as $\Delta\varepsilon$ versus λ ($\Delta\varepsilon$ is defined as $\varepsilon_L - \varepsilon_R$) is positive ($\Delta\varepsilon > 0$) and is called a positive circular dichroism curve. (C) When right circularly polarized light is absorbed to a greater extent than left circularly polarized light ($\varepsilon_R > \varepsilon_L$), a plot of $\Delta\varepsilon$ versus λ gives a negative circular dichroism curve ($\Delta\varepsilon < 0$).

samples. In the next section we discuss the phenomenon of optical activity as it relates to the measurement of CD and to molecular electronic structure.

3.2. The Origin of Optical Activity

Natural optical activity has its origins in microscopic phenomena, such as molecular structure and chromophores, and is observed through macroscopic phenomena, such as chiroptical spectroscopic measurements.[1,2,3,4] The latter include circular dichroism (CD), the familiar optical rotation (α), and optical rotatory dispersion (ORD), all of which are interrelated difference measurements involving linearly polarized and circularly polarized light. To understand the basis for these chiroptical measurements, recall that light is electromagnetic radiation, which can be considered to behave as two wave motions: electric and magnetic. Electric and magnetic waves (or fields), propagated in time, are generated by oscillating electric and magnetic vectors (dipoles) that lie perpendicular to one another. As illustrated in Figure 3-3 for electromagnetic fields propagated along the z-axis, the electric wave oscillates in the xz-plane and the magnetic field oscillates in the yz-plane. Although the energies associated with the electric and magnetic waves are equal, most optical measurements (UV–visible, CD, etc.) are concerned only with the electric field, as we shall see in this chapter. (By way of calibration, it has been estimated that a 100-watt lightbulb emitting a beam with a cross section of one square meter generates an electric field of about 300 volts/meter and a magnetic field of approximately 10^{-6} tesla. The earth's magnetic field is about 5×10^{-5} tesla.)[5]

Ordinary electromagnetic radiation is *unpolarized* (or isotropic), meaning that the electric field (vectors) associated with the light oscillates in all directions perpendicular to the direction of propagation (Figure 3-4A). Thus, ordinary light consists of different wavelengths vibrating in many different planes. However, if the light is filtered so as to remove all oscillations or waves other than in one direction (Figure 3-4B), such as

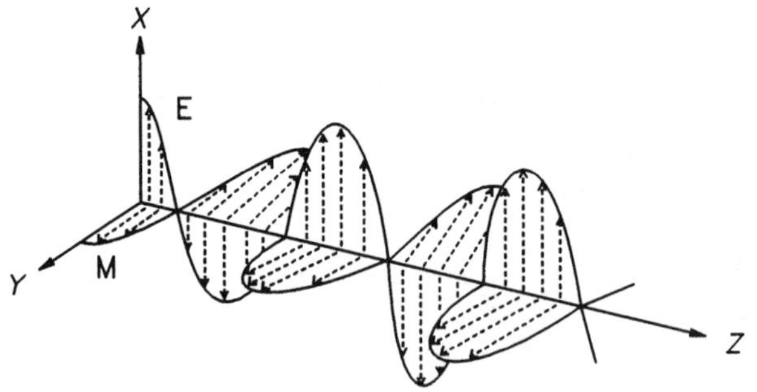

Figure 3-3. Electromagnetic radiation is made up of two wave motions perpendicular to each other, as shown. An electric (E) wave lying in the xz-plane and a magnetic (M) wave lying in the yz-plane are propagated along the z-direction.

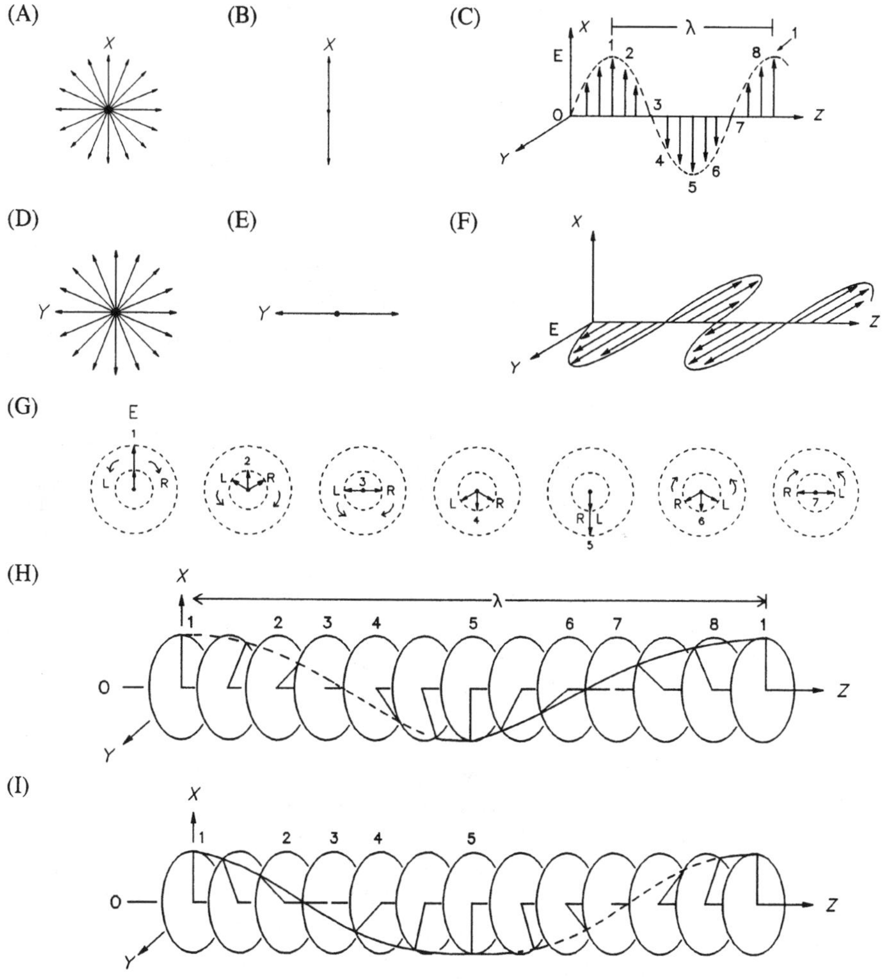

Figure 3-4. Schematic representation of the electric field vectors (A) and (D) in an unpolarized light beam moving (along the z-axis) toward the observer, (B) in plane (or linearly) polarized light (the vertical xz-plane is selected) moving in the direction of the observer, (C) viewed from the side, and (E) and (F) in linearly polarized light (with the horizontal yz-plane selected). Arrows denote the instantaneous direction and magnitude of the electric field vector as a monochromatic linearly polarized light wave progresses along the z-axis. (G) Linearly polarized light (traveling in the xz-plane) decomposed into left (**L**) and right (**R**) circularly polarized component vectors. The vector **L** moves in a counterclockwise direction, the vector **R** in a clockwise direction. The linearly polarized vector (**E**) at any point (e.g., 1–7), in the wave is the vector sum of its two circularly polarized component vectors. (H) Left circularly polarized light whose electric field vector travels along a left-handed helical path from 0 to z. (I) Right circularly polarized light traveling along a right-handed helical path.

that lying in the *xz*-plane of Figure 3-4C, the light is anisotropic and *linearly polarized* (sometimes less rigorously referred to as *plane polarized* light; sunglasses with Polaroid™ lenses filter light in this way).

In the example shown in Figures 3-4B and C, light is linearly and vertically polarized (in the *xz*-plane). A perpendicular (horizontal) plane of polarization (the *yz*-plane) could just as easily have been selected (Figures 3-4E and F). Two linearly polarized light waves whose planes of polarization are perpendicular, as in Figure 3-4C (vertical) and Figure 3-4F (horizontal) are said to be in *orthogonal* polarization states. Radiation in any arbitrary state may be projected onto these two orthogonal states (or, for that matter, onto any two orthogonal polarization states).

Another pair of orthogonal polarization states comprises left (L) and right (R) circularly polarized light that comes from the decomposition of linearly polarized light, as shown in Figure 3-4G. In this figure, the plane of polarization of the linearly polarized light is vertical, and the linearly polarized vectors (e.g., **E** of Figure 3-4C) are the resultant vectors from vector addition of the left and right circular polarization vectors. In circularly polarized light, the electric field vector, which is always perpendicular to the direction of propagation of the radiation (along the *z*-axis), revolves around the propagation axis, either clockwise (*R*) or counterclockwise (*L*) once every wavelength λ—as in Figure 3-4G. That figure, of course, represents the vectors as viewed from *z* to 0 along the 0*z* axis for the light wave. The resultant vectors (**E**) of linearly polarized light travel in a plane (the *xz*-plane), but the circularly polarized vectors travel along helical paths. The left circularly polarized light vector traces out a left-handed helix (Figure 3-4H), whereas the right circularly polarized light vector traces out a right-handed helix (Figure 3-4I).

Circularly polarized light can be produced from unpolarized (isotropic) light by first passing the beam through a polarizing filter (a linear polarizer, from which only one plane of linearly polarized or anisotropic light emerges) and then through a birefringent plate. (e.g., a very thin quartz plate cut parallel to its optical axis), which has two axes—one fast and one slow. In Figure 3-5, linearly polarized light is passed through a birefringent plate. The two in-phase circularly polarized components (Figure 3-4G) pass through with different velocities and emerge with a phase difference δ proportional to the thickness *d* and the birefringence of the plate. The applicable equation is $\delta = 2\pi d(n_{\text{slow}} - n_{\text{fast}})/\lambda$. A quarter-wave plate introduces a phase difference $\delta = \pi/2$ and is oriented so that the fast and slow axes are at ±45° with respect to the plane of incident light (e.g., the *xz*-plane of Figure 3-4C). Linearly polarized light is then projected equally onto the fast and slow axes of the quarter-wave plate to *phase shift* the two components either ahead or back by a quarter of a wavelength (±λ/4), thus yielding either left or right circularly polarized light. If the quarter-wave plate modulator is rotated through 90°, the circular polarization state of the light beam can be modulated between left and right circular polarization states, transmitting alternately left and right circularly polarized light toward the sample. The same separation or modulation is achieved in most CD instruments nowadays by the use of a photoelastic modulator, in which stress on a quartz crystal at a modulation frequency in the kilohertz

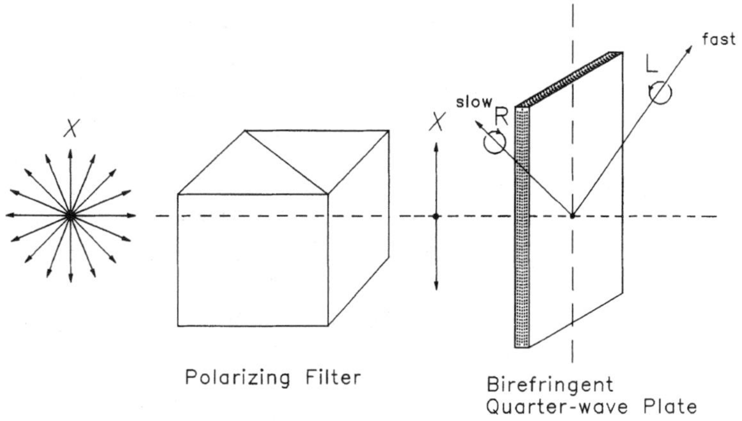

Figure 3-5. Isotropic light passes through a polarizing filter, and linearly polarized light is emitted and passed through a birefringent quarter-wave plate, with fast and slow axes oriented at ±45° with respect to the plane of incident light. Linearly polarized light projected equally onto the fast and slow axes yields either left (*L*) or right (*R*) circularly polarized light.

range causes retardation of the two orthogonal linearly polarized components that are to be varied sinusoidally.

As a microscopic phenomenon, optical activity relates to molecular stereochemistry. For UV–visible-range chiroptical spectroscopy, the electrons in a molecule are most important, and from this perspective, one might view a chiral molecule as a static dissymmetric field of electrons. Given that each valence electron behaves as a small harmonic oscillator, there are certain natural frequencies (ν_0) associated with energy-level spacings ($\Delta E = h\nu_0$) in any molecule (Figure 3-1B). When a beam of light passes through a molecule, the electromagnetic field (frequency $\nu = c/\lambda$, where c is the speed of light and λ is the wavelength of the light; see Figure 3-3) subjects the electrons to forced oscillations. If the damping coefficient is weak and ν is far removed from ν_0 (or far from the region of a UV–visible absorption band), the motion of the electrons is slight and remains in phase with the electric field. But as the frequency of the light (ν) approaches one of the natural frequencies (ν_0) of the molecule, the interaction between the light wave and the molecule becomes more pronounced, and the displacement of any electron can be large (e.g., from an occupied molecular orbital to an unoccupied molecular orbital—as in the formation of an excited state measured by the absorption of light (Figure 3-1A).

The electric field of the light beam thus alters the electronic description of the molecule by disturbing electrons from their equilibrium position and creating a momentary charge redistribution, called an induced electric dipole ($\boldsymbol{\mu}_e$). (The ease of inducing the dipole is characterized by the dielectric constant, which is related to the index of refraction (n) by Maxwell's equation[4]; thus, n is related to the amplitude of motion of electrons in a molecule.) Because the light waves are periodic in time, so are the electron distributions, and most of the energy is periodically returned to the wave. When the interaction between the light and the electrons is weak (as is characteristic of regions remote from the UV–visible absorption bands), the molecule has an extremely low probability of being in an excited state. And since the electron spends little time in an excited state, the chance for nonradiative dissipation of energy is small, and there is no notable energy change because the electron gives back in one-half period the energy the field gave it during the preceding one-half period. This give-and-take process produces a change in the velocity of the light through the sample compared with the velocity of light in a vacuum (c). That is, $n \neq 1$, and a plot of n versus the wavelength (λ) of light gives a typical *dispersion* curve, shown in Figure 3-6.

In an optically active sample, left and right circularly polarized (anisotropic) light beams are transmitted with different velocities. That is, the refractive indices for left (n_L) and right (n_R) circularly polarized light are affected to different extents in an optically active molecule because it has a dissymmetric field of electrons, and $\Delta n = n_L - n_R \neq 0$. The sample is said to exhibit *circular birefringence*, and one may observe a rotation of the original plane of polarization of linearly polarized light, or *optical rotatory dispersion* (ORD), for a spectrum recording the variation in the angle of rotation versus the wavelength of light. On the other hand, in an achiral molecule (an isotropic sample), the electron field is symmetric, and left and right circularly polarized light are transmitted equally. In this case, $n_L = n_R$ over all wavelengths of light, and, as we shall learn, the rotation of the plane of polarization of linearly polarized light is zero over all wavelengths.

When the light energy approaches that corresponding to a UV–visible absorption band (Figure 3-6, where the natural frequency is ν_0), the probability becomes high that an electron will be promoted to an unoccupied orbital and a particular excited state will become dominant in the electronic description of the molecule. That is, the probability is high that an electron will be raised to an energy level long enough to lose its excess energy in dissipative processes (emission, chemical reaction, etc.), with a corresponding reduction in the intensity of the light beam as it emerges after passing through the sample. Again, with an optically active molecule, which possesses a dissymmetric field of electrons, the coefficients of absorption for left and right

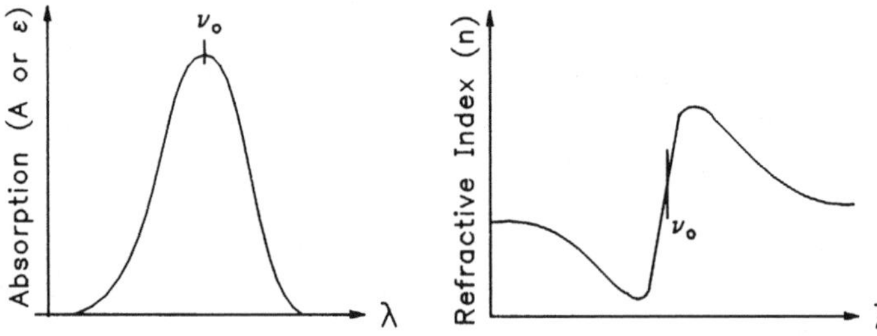

Figure 3-6. Comparison of the interaction of ordinary light with matter, as seen in wavelength-dependent absorption (left) and in the dispersion, or refractive index (right), curve. ν_0 is the natural frequency of the excitation associated with the energy-level spacing shown in Figure 3-1B.

circularly polarized light, κ_L and κ_R (Eq. 3-5), respectively, are not equal. Hence, $\Delta\varepsilon = \varepsilon_L - \varepsilon_R \neq 0$, and circular dichroism is observed (Figures 3-2B and C).

As is characteristic of phenomena arising from the interaction of isotropic light with matter, where there is both a dispersive aspect and an absorptive aspect (Figure 3-6), so can the differential behavior of polarized (anisotropic) light with an optically active (anisotropic) sample be seen by a differential absorption (CD) and differential refraction (circular birefringence, as measured by ORD) of left and right circularly polarized light. Phenomenologically, this means a difference in the complex indices of refraction, N_L and N_R, which are related to what is called the complex rotatory power (Φ) by

$$\Phi = (\pi/\lambda)(N_L - N_R) = (\pi/\lambda)[(n_L - n_R) - i(\kappa_L - \kappa_R)]. \tag{3-7}$$

As noted earlier, n_L and n_R are the indices of refraction for, respectively, left and right circularly polarized light, $i = \sqrt{-1}$, and κ_L and κ_R are the respective absorption indices for left and right circularly polarized light. Thus, Φ contains both refractive index and absorption coefficient terms and can be expressed simply as

$$\Phi = \phi - i\theta, \tag{3-8}$$

where $\phi = (\pi/\lambda)(n_L - n_R)$ is the angle of rotation (α, of the plane of polarized light) per unit length and $\theta = (\pi/\lambda)(\kappa_L - \kappa_R)$ is the ellipticity per unit length. The angle ϕ is explained in Section 3.3 and θ in Section 3.4.

3.3. Angle of Rotation (ϕ, α) and ORD[1]

The rotation of plane polarized light (optical rotation) can typically be detected at almost every wavelength for an optically active compound, even at wavelengths far removed from a UV–visible absorption band (e.g., at the sodium D-line (589 nm) for a colorless substance). Since an anisotropic medium does not transmit left and right circularly polarized light with equal velocity, the initial phase relationship between these two beams is not maintained when the light emerges from the sample, so the resultant plane of polarization is rotated by an angle α from its original orientation (Figure 3-7). That is, for an optically active substance, the indices of refraction for left (n_L) and right (n_R) circularly polarized light are not equal (circular birefringence). The angle of rotation per unit length of plane (linearly) polarized light is given by

$$\phi = (\pi/\lambda)(n_L - n_R), \tag{3-9}$$

where λ is the wavelength of the light used and the units of ϕ are radians per centimeter. Converting ϕ to the more familiar units of degrees per decimeter (α) is achieved by multiplying Eq. (3-9) by $1800/\pi$:

$$\alpha = (1800/\pi)\phi = 1800(n_L - n_R)/\lambda(\text{cm}). \tag{3-10}$$

For an observed rotation $\alpha = 1°$ at 300 nm in a 1-dm cell, $\Delta n = n_L - n_R = 1.67 \times 10^{-8}$. Typical values of the observed rotation α are 0.001° to 1° (measured in a 1-dm cell);

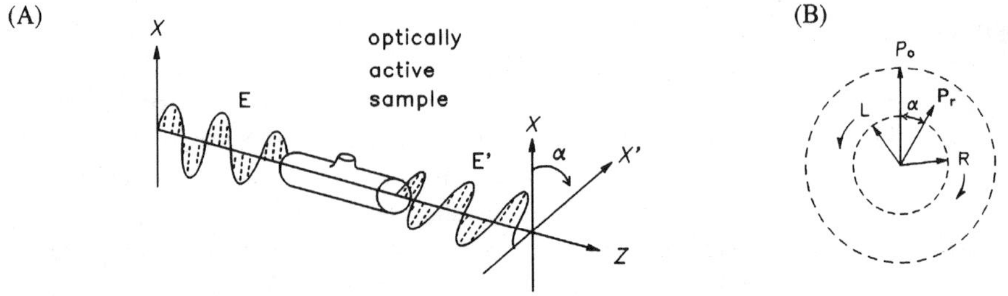

Figure 3-7. (A) Rotation of the plane (P_0) of linearly polarized light from the *xz*-plane (see Figure 3-4C) to the *x′z*-plane when the light is passed through an optically active sample. (B) The resultant plane $\mathbf{P}_r$ is rotated from P_0 by the angle of rotation, α, when left (*L*) and right (*R*) circularly polarized light beams are transmitted with different velocities through an optically active medium. Here, **L**, the electric vector for left circularly polarized light, moves slower than **R**, the electric vector for right circularly polarized light; thus, $n_L > n_R$. The resultant plane of linearly polarized light ($\mathbf{P}_r$) is thus rotated by +α° from the original plane of polarization (P_0).

thus, the differences in refractive indices ($n_L - n_R$) in ORD or α measurements are extremely small—on the order of 10^{-6} to 10^{-9} percent.

When reporting rotations of optically active solutions, one typically normalizes α to the sample concentration (*c′* in g/cc) and path length (*l* in dm) and cites the specific rotation as a function of temperature (*T* in °C) and wavelength (λ in nm):

$$[\alpha]_\lambda^T = \alpha/(c'l') = \frac{\text{observed rotation (°)}}{\text{conc(g/cc)} \times \text{path length(dm)}}. \tag{3-11}$$

In ORD, α is usually expressed as molar rotation [ϕ], which is the specific rotation multiplied by the molecular weight (*M*) and divided by 100:

$$[\phi]_\lambda^T = \frac{[\alpha]_\lambda^T (M)}{100}. \tag{3-12}$$

When one plots the behavior of α, [α], or [ϕ] versus λ far away from a region of *absorption* of light (e.g., away from electronic transitions), one finds a rather simple, smooth plain dispersion curve (Figure 3-8A), either positive or negative, since α can be either positive or negative. However, in the region of light absorption, the curve typically ascends or descends rapidly and then turns over, becoming anomalous. This anomalous dispersion effect is called the *Cotton effect*, in this case an ORD Cotton effect. When the peak appears at a longer wavelength than the trough, as shown in curve 1, the ORD curve is said to have a positive Cotton effect. The mirror-image curve with the trough at the longer wavelength has a negative Cotton effect.

The ORD curve of an optically active ketone might look like that of Figure 3-8B. In one region, from about 400 nm on past the sodium D-line at 589 nm, the ORD curve is called a plain dispersion curve. But in the region of the lowest energy ($n \rightarrow \pi^*$) carbonyl electronic transition near 300 nm, the curve rapidly ascends to a peak and then descends to a trough. The peak-to-trough vertical distance, measured in [ϕ] units,

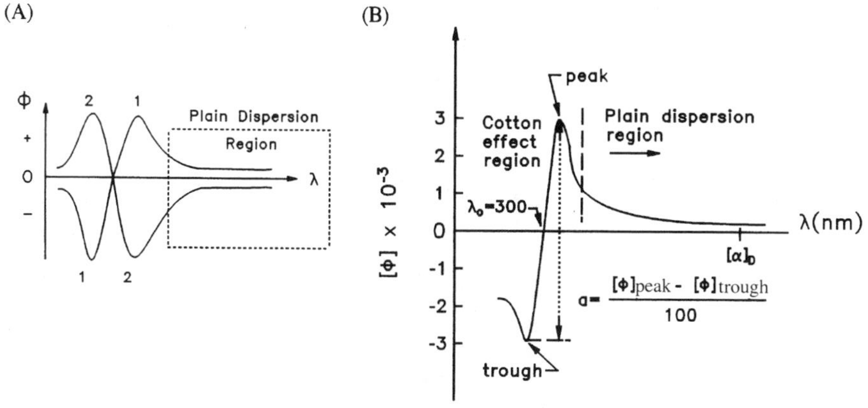

Figure 3-8. (A) Oppositely signed mirror-image optical rotatory dispersion (ORD) curves. At wavelengths far from the region of electronic transitions, the ORD tails are said to be *plain dispersion curves.* Tails 1 and 2 are positive and negative plain dispersion curves, respectively. In the region of an electronic transition, an "anomalous" behavior (the Cotton effect) is found in the ORD curves, where in curve 1 exhibits a positive Cotton effect and curve 2 exhibits a negative Cotton effect. (B) ORD curve of a typical optically active saturated ketone. An "anomalous" dispersion, or Cotton, effect occurs in the region of the ketone carbonyl $n \rightarrow \pi^*$ electronic transition at 300 nm. The example shown exhibits a positive Cotton effect. The crossover point from positive to negative rotational values, $\lambda_0 = 300$ nm, corresponds roughly to the absorption maximum of the ketone. The intensity of the ORD Cotton effect is called the amplitude (a), which is defined as shown and can be converted to CD $\Delta\varepsilon$ units by the equation $a \sim 40.27 \times \Delta\varepsilon$.

is called the amplitude (a) of the ORD Cotton effect. The ORD amplitude is used as a quantitative measure of the ORD transition.

3.4. Angle of Ellipticity (θ) and Circular Dichroism (Δε)

As explained earlier, in the region of an absorption band of an optically active substance, left and right circularly polarized light beams are absorbed to an unequal degree ($A_L \neq A_R$, hence circular dichroism), in addition to being transmitted to an unequal extent.[1,2,4] When this occurs, (e.g., when the linearly polarized light passes through a UV–visible absorption band of an optically active compound), the incident linearly or plane polarized light emerges elliptically polarized, with the resultant electric vector $\mathbf{P}_r$ tracing out an ellipse (Figure 3-9) rather than remaining in a plane (Figure 3-7), and the tilt of the major axis of the ellipse (a, on x') represents the angle of rotation (α or ϕ). The change from plane polarized light to elliptically polarized light is measured by the eccentricity of the ellipse, ψ, defined as the angle whose tangent is the ratio of the minor axis (b) to the major axis (a) of the ellipse:

$$\psi \text{ (radians)} = \tan^{-1} (b/a), \text{ or } \tan \psi = b/a. \qquad (3\text{-}13)$$

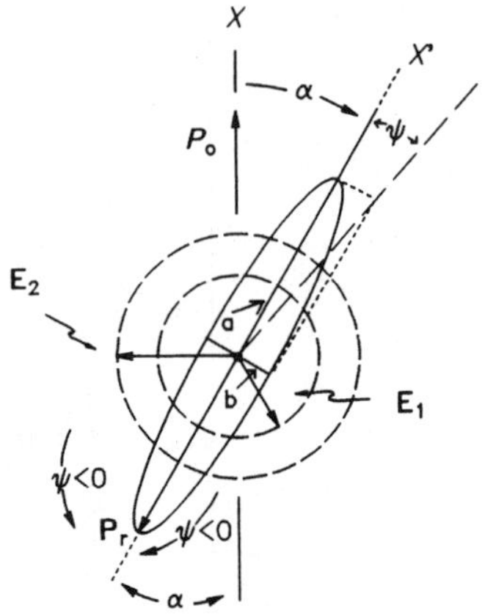

Figure 3-9. Elliptical polarization of light due to unequal absorption and unequal transmission (velocity) of left (**L**) and right (**R**) circularly polarized light passing through an optically active medium. Ellipticity is measured by the plane angle ψ, which is defined as the tangent of the ratio of the minor axis (b) to the major axis (a) of the ellipse, viz., $\tan \psi = b/a = (\mathbf{R} - \mathbf{L})/(\mathbf{R} + \mathbf{L})$. If $\mathbf{E}_1 = \mathbf{L}$ and $\mathbf{E}_2 = \mathbf{R}$, ψ is positive, and the resultant vector $\mathbf{P}_r$ describes a clockwise direction when viewed against the direction of propagation of the light (z of Figure 3-4) or the plane of polarization of light (P_0) before entering the medium. If $\mathbf{E}_1 = \mathbf{R}$ and $\mathbf{E}_2 = \mathbf{L}$, ψ is negative, and $\mathbf{P}_r$ describes a counterclockwise rotation. The angle of rotation, α, of the plane polarization from P_0 to $\mathbf{P}_r$ is given by the displacement of the major axis of the ellipse.

The ellipticity, or elliptical angle, is usually expressed in degrees:

$$\theta \text{ (degrees)} = 360\ \psi/2\pi. \tag{3-14}$$

However, ψ and θ can also be expressed in terms of the unequal absorption of left and right circularly polarized light (ΔA), because the eccentricity of the ellipse is determined by the electric field vectors for left (**L**) and right (**R**) circularly polarized light. In Figure 3-9, if $\mathbf{E}_1 = \mathbf{L}$ and $\mathbf{E}_2 = \mathbf{R}$, then $\psi > 0$ and ΔA is negative; if $\mathbf{E}_1 = \mathbf{R}$ and $\mathbf{E}_2 = \mathbf{L}$, then $\psi < 0$ and ΔA is positive. That is,

$$\tan \psi = (\mathbf{R} - \mathbf{L})/(\mathbf{R} + \mathbf{L}). \tag{3-15}$$

Expressing Eq. (3-15) in terms (Eq. 3-16) of absorption coefficients (k_L, k_R), absorption indices (κ_L, κ_R), absorbance (A_L, A_R) and molar absorptivities (ε_L, ε_R) for left and right circularly polarized light[3,5] gives

$$\begin{aligned}\tan \psi &= \tanh\,[(k_L - k_R)l/4] = \tanh\,[(\kappa_L - \kappa_R)\pi l/\lambda] \\ &= \tanh\,[(A_L - A_R)\,2.303/4] = \tanh\,[(\varepsilon_L - \varepsilon_R)2.303\ lC/4].\end{aligned} \tag{3-16}$$

Just as $\Delta n = n_L - n_R$ is small in magnitude compared with the mean index of refraction $n = (n_L + n_R)/2$, $\Delta A = A_L - A_R$ is only a small fraction (rarely more than a few hundredths) of the mean absorbance $A = (A_L + A_R)/2$. Similarly, $\Delta\varepsilon = \varepsilon_L - \varepsilon_R$ is typically on the order of 10^{-2} to 10^{-4} the magnitude of $\varepsilon = (\varepsilon_L + \varepsilon_R)/2$, and $k_L - k_R$ is rarely greater than 10^{-6} or 10^{-7}. Thus, in practice, the ellipse is always extremely elongate, and Δk, $\Delta\kappa$, ΔA, and $\Delta\varepsilon$ per unit length may be approximated by

$$\psi = (k_L - k_R)/4 = (\kappa_L - \kappa_R)\pi/\lambda$$
$$= (A_L - A_R)2.303/4l = (\varepsilon_L - \varepsilon_R)2.303C/4, \quad (3\text{-}17)$$

which, in terms of κ, has the same form as Eq. (3-9), [Cf. also Eq. (3-7).] Converting ψ from radians to degrees (θ) gives

$$\theta = (1800/4\pi)(k_L - k_R) = (1800/\lambda\ (\text{cm}))(\kappa_L - \kappa_R)$$
$$= (1800/\pi)(2.303/4l)(A_L - A_R) = (1800/\pi)(2.303C/4)(\varepsilon_L - \varepsilon_R). \quad (3\text{-}18)$$

The form of Eq. (3-18) in κ is again similar to that of Eq. (3-10) in n. In CD work, the molar ellipticity $[\theta]$ and $\Delta\varepsilon$ are commonly used,[6] where $[\theta]$, with units of degrees $\text{cm}^2\ \text{dmol}^{-1}$, has a form similar to that of $[\phi]$ in Eq. (3-12). That is,

$$[\theta] = (\theta/c'l')(M/100) = \frac{\text{observed ellipticity}(^\circ) \times \text{molec. wt.}}{\text{conc}(\text{g/cc}) \times \text{pathlength}(\text{dm}) \times 100}. \quad (3\text{-}19)$$

In terms of $\Delta\varepsilon$,

$$[\theta] = \frac{1800(2.303)CM\Delta\varepsilon}{4\pi c'l'100} = 2.303(4500/\pi)\Delta\varepsilon \sim 3300\ \Delta\varepsilon \sim \frac{3300\Delta A}{Cl}. \quad (3\text{-}20)$$

Thus, Eqs. (3-17) through (3-20) relate the angle of ellipticity—ψ or θ or $[\theta]$—of Figure 3-9 to $\Delta\varepsilon$ and the difference in absorbance, ΔA.

The relationship between positive and negative CD and ORD Cotton effects is displayed in Figure 3-10. A CD spectrum plots the difference in the absorbance

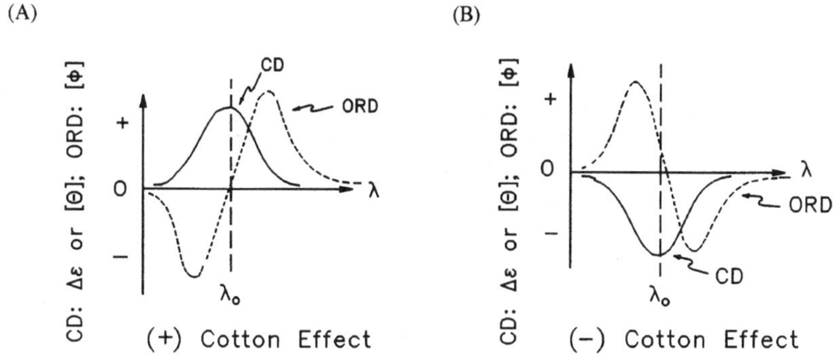

Figure 3-10. (A) Positive and (B) negative circular dichroism (———) and corresponding optical rotatory dispersion (- - -) curves in the region of a UV–visible absorption band. The crossover point (λ_0) of the ORD curves corresponds closely to λ at $\Delta\varepsilon^{\text{max}}$.

($\Delta A = A_L - A_R$) or absorption coefficient difference ($\Delta\varepsilon = \varepsilon_L - \varepsilon_R$) versus the wavelength (λ) of incident light. Both A_L and A_R (or ε_L and ε_R) may be measured directly, but ΔA is typically very small (0.1 to 0.01). Consequently, it is sometimes easier experimentally to measure the change in the angle of ellipticity (θ) over the spectrum (wavelength range) and convert it to ΔA or $\Delta\varepsilon$ units ($\theta \sim 33\ \Delta A$ for a 1-cm path length). For example, the measured value of θ at 300 nm for a sample in a 1-cm cuvette might be on the order of 1° for $A \sim 0.03$. The sign and magnitude of the CD $\Delta\varepsilon$ (or θ) can often be correlated with the absolute configuration or conformation of chiral molecules.

As with A_L and A_R in CD, n_L and n_R in ORD are seldom measured directly, because Δn is typically very small ($\sim 10^{-8}$ to 10^{-11}). It is easier to measure the angle of rotation (α) of the emergent plane of polarization relative to the plane of polarization of the incident light, for α values are often on the order of 1°. An ORD spectrum (Figures 3-8 and 3-10) plots the variation in the rotation of the plane of polarization (of plane polarized light) as a function of wavelength. The signed order of an ORD curve [e.g., a peak of long wavelength and a trough of short wavelength] can often be correlated with the stereochemistry of a chiral molecule. However, CD curves have definite advantages over ORD curves for quantitative measurements. For optically active samples, although ΔA and θ (and hence CD) can be detected only in the region of an absorption band, $n_L - n_R$, or the change in optical rotation (α), is everywhere finite and thus can be detected throughout the electromagnetic spectrum. Hence, CD spectroscopy offers the distinct advantage of few overlapping signals, whereas in ORD all of the transitions overlap at every wavelength. Consequently, CD has become the method of choice for detecting optically active transitions in the UV–visible spectrum.

3.5. Quantitative Relationship Between CD and ORD

Since absorptive and dispersive phenomena (Figure 3-6) have their origins in the same charge displacements (Section 3-2), the real and imaginary parts of Eq. (3-7), and hence ORD and CD, can be related quantitatively. Kronig and Kramers showed that, from a knowledge of dispersion as a function of wavelength over the entire spectral range ($\lambda = 0$ to $\lambda = \infty$), the corresponding absorption curve (Figure 3-6) may be predicted, and vice versa.[1,3,4,7] The result, known as the Kronig–Kramers theorem, may be expressed quantitatively as the Kronig–Kramers reciprocal relationships for ORD and CD, thus relating the curves of Figure 3-10.

$$[\theta_k(\lambda_0)] = -2(\pi\lambda_0)^{-1} \int_0^\infty [\phi_k(\lambda)] \frac{\lambda^2}{(\lambda_0^2 - \lambda^2)}\, d\lambda \qquad (3\text{-}21)$$

or

$$\Delta\varepsilon(\lambda_0) = -1.9303 \times 10^{-4}(\lambda_0)^{-1} \int_0^\infty [\phi_k(\lambda)] \frac{\lambda^2}{(\lambda_0^2 - \lambda^2)}\, d\lambda$$

and

$$[\phi_k(\lambda_0)] = 2\pi^{-1} \int_0^\infty [\theta_k(\lambda)] \frac{\lambda}{(\lambda_0^2 - \lambda^2)} d\lambda \tag{3-22}$$

or

$$[\phi_k(\lambda_0)] = 2099.6 \int_0^\infty \Delta\varepsilon(\lambda) \frac{\lambda}{(\lambda_0^2 - \lambda^2)} d\lambda,$$

Where $[\theta_k(\lambda_0)]$ and $[\phi_k(\lambda_0)]$ are the familiar molar ellipticity and molar rotation at a specified wavelength (λ_0), respectively, for the kth electronic transition. Using Eqs. (3-21) and (3-22), one may convert an ORD curve into a CD curve and vice versa (Figure 3-10). For the 300-nm electronic transition of ketones, one derives the following relationship between the amplitude (a) of an ORD curve (Figure 3-8B) and $\Delta\varepsilon$ or $[\theta]$ of a CD curve:

$$a \sim 40.28(\Delta\varepsilon^{max}) \sim 0.0122([\theta]^{max}). \tag{3-23}$$

Here, a is the difference, divided by 100, between $[\phi]$ at the peak and at the trough of the ORD curve.

3.6. The Magnitude of Optical Activity, Cotton Effects, and Rotatory Strength

CD curves are typically plotted in $\Delta\varepsilon$ (or ΔA, θ, or $[\theta]$) units ,as shown in Figures 3-2 and 3-10.[1–4] In terms of Cotton effect magnitudes (Figure 3-10), the intensity of an ORD curve, measured by its amplitude a (Figure 3-8), is related to the intensity of the CD magnitude ($\Delta\varepsilon$) by the expression $a \sim 40.28\ \Delta\varepsilon^{max}$. However, the intensity of a CD transition is usually only approximated by $\Delta\varepsilon^{max}$, just as the intensity of a UV–visible transition is only approximated by ε^{max}. It can be measured more exactly by the rotational (or rotatory) strength, the wavelength-weighted area under a CD curve,[1–4,7] where

$$\boldsymbol{R} = 2.297 \times 10^{-39} \int \frac{\Delta\varepsilon}{\lambda} d\lambda, \tag{3-24}$$

For approximately Gaussian-shaped curves,

$$\boldsymbol{R} \sim 2.445 \times 10^{-39} (\Delta\varepsilon^{max})(\lambda_{1/2})(\lambda^{max}), \tag{3-25}$$

where $\lambda_{1/2}$ is the half-width of the band. [For a numerical integration of a CD curve, see the appendix to the chapter.]

The expression for $\boldsymbol{R}$ as a measure of the probability or intensity of a CD transition has its parallel in the probability or intensity of an ordinary electronic transition, measured by the dipole strength, which $\boldsymbol{D}$ is the wavelength-weighted area under a UV–visible absorption curve for the electronic transition studied. The dipole strength is given by

$$\boldsymbol{D} = \frac{300hc}{8\pi^3 N}\int \frac{\varepsilon}{\lambda}\, d\lambda = 9.184 \times 10^{-39} \int \frac{\varepsilon}{\lambda}\, d\lambda, \tag{3-26}$$

where h is Planck's constant, $\boldsymbol{c}$ is the speed of light in vacuum, N is Avogadro's number, ε is the molar absorption coefficient (L cm^1 mol^{-1}) for the electronic transition, and λ is the wavelength of light. The dipole strength usually varies from 10^{-34} to 10^{-38} erg · cm^3.

The intensity of the given UV–visible transition may also be expressed as the oscillator strength, (f), a dimensionless quantity typically with a value lying between zero and unity and given by

$$f = \frac{2.303mc^2}{N\pi e^2}\int \varepsilon\, d\nu = 4.315 \times 10^{-9} \int \varepsilon\, d\nu, \tag{3-27}$$

where ν is the frequency of the light in cm^{-1}, m and e are, respectively, the mass and charge of the electron, and ε and c are as described earlier. Solving for ε and approximately by using differences, we obtain

$$\varepsilon \sim 0.464 \times 10^9 f/\Delta\nu \tag{3-28}$$

3.7. Optical Activity of Chromophores

For purposes of detecting optical activity, we focus on the light-absorbing unit (the chromophore) in the molecule.[1,3,4,8] Stated in terms of molecular structure, for optical activity to occur—meaning a nonvanishing rotatory strength ($\boldsymbol{R}$)—the molecule cannot possess a rotation reflection symmetry axis (i.e., it cannot have a reflection plane or a center of symmetry). From the perspective of the molecule undergoing excitation by the light wave, both the rotatory strength and the dipole strength ($\boldsymbol{D}$) associated with the promotion of an electron from a ground-state molecular orbital to an unoccupied molecular orbital can be expressed in terms of the electric dipole transition moment ($\boldsymbol{\mu}_e$): $\boldsymbol{D} = \boldsymbol{\mu}_e \cdot \boldsymbol{\mu}_e$ and $\boldsymbol{R} = \boldsymbol{\mu}_e \cdot \boldsymbol{\mu}_m$. An important distinction between $\boldsymbol{D}$ and $\boldsymbol{R}$ is drawn here because only the *electric* dipole transition moment ($\boldsymbol{\mu}_e$) appears in the dipole strength expression for the absorption of ordinary light, but both the *electric* moment and the *magnetic* moment ($\boldsymbol{\mu}_m$) appear in the rotatory strength expression. Thus, in promoting a molecule from its ground electronic state to an excited state, $\boldsymbol{D}$ depends only on $\boldsymbol{\mu}_e$, but $\boldsymbol{R}$ depends on both $\boldsymbol{\mu}_e$ and $\boldsymbol{\mu}_m$. Of course, for an *achiral* molecule, $\boldsymbol{R}$ must be zero even when the dipole strength is nonzero; and even a chiral molecule may have $\boldsymbol{R} = 0$ if either $\boldsymbol{\mu}_e$ or $\boldsymbol{\mu}_m$ is forbidden by spectroscopic selection rules.[9] Typically, both $\boldsymbol{\mu}_e$ and $\boldsymbol{\mu}_m$ are nonzero in molecules that exhibit *strong* optical activity ($\boldsymbol{R} > \sim 10^{-40}$ cgs units), whereas molecules that exhibit *weak* optical activity

($\boldsymbol{R} \leq 10^{-40}$ cgs units) have either $\boldsymbol{\mu}_e = 0$ or $\boldsymbol{\mu}_m = 0$ (or both = 0). However, even when optical activity is predicted to be zero, CD may be detected due to a breakdown in the spectroscopic selection rules (e.g., by vibronic coupling or intensity borrowing when $\boldsymbol{\mu}_e = 0$). A specific case in point is the $n \rightarrow \pi^*$ transition of an optically active saturated alkyl ketone, for which $\boldsymbol{\mu}_e = 0$ and $\boldsymbol{\mu}_m = 1$ Bohr-magneton. Nonetheless, a weak UV transition can usually be measured, with $\varepsilon^{max} \sim 10$–$100$ L cm mol^{-1}, (i.e., $\boldsymbol{D} \neq 0$). And a CD curve can also usually be recorded, with $\Delta\varepsilon^{max} \sim 1$ L cm mol^{-1} (i.e., $\boldsymbol{R} \neq 0$). The phenomenon of optical activity, together with its associated measurable parameters in CD and ORD spectroscopy, is thus directly related to the chromophore(s) in a molecule.

3.8. Limiting Classification of Chromophores as Inherently Symmetric or Inherently Dissymmetric

Moffitt and Moscowitz[10] classified optically active chromophores in terms of two limiting types: (i) inherently dissymmetric and (ii) inherently symmetric, but dissymmetrically perturbed.[7] This classification has proven to be of practical value for understanding the orders of magnitude encountered in optical activity data and, most importantly, for interpreting CD (or ORD) data in a stereochemically informative way. The *inherently dissymmetric* chromophore is a chromophore such that the geometry of the chromophoric grouping lacks a rotation–reflection (S_n) axis and both $\boldsymbol{\mu}_e$ and $\boldsymbol{\mu}_m \neq 0$. Thus, even in isolation, its electronic transitions will manifest optical activity. For transitions that are both electric dipole and magnetic dipole allowed, if $\boldsymbol{\mu}_e$ is taken to be on the order of 1 Debye ($\sim 10^{-18}$ cgs units) and $\boldsymbol{\mu}_m$ on the order of 1 Bohr-magneton, then R ($= \boldsymbol{\mu}_e \cdot \boldsymbol{\mu}_m$) will be on the order of 10^{-38} cgs units. The archetypical example is hexahelicene (Figure 3-11), which exhibits multiple, intense electronic transitions between 220 and 380 nm and very large CD Cotton effects near 330 and 245 nm, with $\Delta\varepsilon^{max} \sim +200$ and -220 L cm mol^{-1}, respectively, and associated rotatory strengths on the order of 10^{-38} cgs.[11]

Even in a dissymmetric molecular environment, the contributions to the signed magnitude of the large R values of inherently dissymmetric chromophores are usually determined primarily by the chirality of the chromophore itself. Hence, qualitative stereochemical questions, the answers to which derive from the chiral nature of the chromophore geometry, can be posed and answered without recourse to considerations of dissymmetric vicinal action. When extrachromophoric contributions weigh in heavily, however, the dissymmetry of the inherently dissymmetric chromophore is not necessarily the controlling factor, and a detailed examination of all major contributions is required—as turned out to be necessary for the twisted 1,3-diene chromophore.

In the *inherently symmetric* chromophore, the symmetry of the *isolated* chromophore is sufficiently high to preclude optical activity (e.g., $\boldsymbol{\mu}_e$ or $\boldsymbol{\mu}_m = 0$, and $R = 0$). In order for optical activity ($R \neq 0$) to be observed, the chromophore must be perturbed by dissymmetric vicinal action, in which case the signed magnitudes of the associated rotatory strengths provide information on the stereochemical nature of the extrachro-

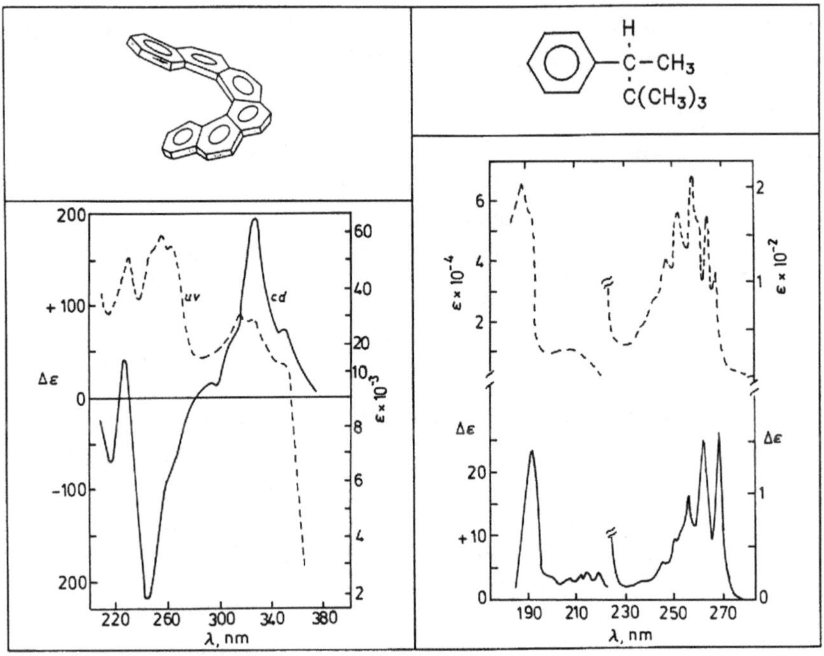

Figure 3-11. (Left) Absolute configuration (top) and circular dichroism (———) and UV–visible (- - - -) spectra of (*R*)-(+)-hexahelicene in CH_3OH solvent. (Right) Absolute configuration (top) and circular dichroism (———) and UV–visible (- - - -) spectra of (*S*)-(+)-2-phenyl-3,3-dimethylbutane in methylcyclohexane–isopentane (1:3, vol:vol) and *n*-heptane solvents, respectively. [Redrawn and used from Lambert, J. B., Shurvell, H. F., Lightner, D. A., and Cooks, R. G., *Organic Structural Spectroscopy*, Prentice Hall, Inc., Upper Saddle River, NJ, 1998.]

mophoric molecular environment and its disposition relative to the symmetry planes of the chromophore (It is as if the chromophore were a window for observing its surrounding molecular structure.) The aromatic chromophore found in the optically active alkyl benzene of Figure 3-11 is a case in point. The monoalkylated benzene chromophore (for the isolated benzene or toluene chromophores, $\boldsymbol{\mu}_e \neq 0$ $\boldsymbol{\mu}_m = 0$) is symmetric in local symmetry, but is dissymmetrically perturbed by the chiral center of the alkyl chain. Weak to strong transitions ($\varepsilon^{max} = 200$ and 70,000) in the UV–visible spectrum lying between 190 and 270 nm give corresponding weak to moderate CD transitions with $\Delta\varepsilon^{max} \sim +1.5$ and $\Delta\varepsilon^{max} \sim +24$, whose signs correlate with the absolute stereochemistry of the *extrachromophoric* part (the chiral center) of the molecule.[12]

The most successful correlation of Cotton effect sign with extrachromophoric dissymmetry may be found in the octant rule for the $n \rightarrow \pi^*$ transition of saturated alkyl ketones.[4,8] In the inherently symmetric, isolated carbonyl chromophore of saturated alkyl ketones, $\boldsymbol{\mu}_e \neq 0$ and $\boldsymbol{\mu}_m = 1$ Bohr-magneton for the $n \rightarrow \pi^*$ transition, which involves a circular movement of charge; and $\boldsymbol{\mu}_e \neq 0$ and $\boldsymbol{\mu}_m = 0$ for the $\pi \rightarrow \pi^*$ transition, which involves a vertical movement of charge (Figure 3-12). Taking into account just the n and π orbitals and electrons, we would describe the ground state as $(n)^2(\pi)^2$, where there are two electrons in the n orbital and two electrons in the π orbital. The orbitals transform as $(b_2)^2(b_1)^2$ $(= n^2\pi^2)$, and thus, the ground state transforms as A_1. That is, the product $b_2b_2b_1b_1$ has the same symmetry as A_1. The $n \rightarrow \pi^*$ excited state would be described as $(n)(\pi^*)(\pi)^2$ (or $b_2b_1^*b_1^2$). The symmetry of the $n \rightarrow \pi^*$ excited state therefore transforms as A_2. Hence, we see that the excitation from the $(n^2\pi^2)$ ground state into the $n \rightarrow \pi^*$ excited state $[(n)(\pi^*)(\pi^2)]$ involves a change of symmetry, from A_1 to A_2, that transforms as A_2 and is electric dipole forbidden. The transformation $A_1 \rightarrow A_2$ (which takes the ground state (A_1) ketone to its $n \rightarrow \pi^*$ excited state (A_2) by promoting an electron from an n orbital in the ground state to an unoccupied π^* orbital) is symmetry forbidden because the electric dipole moment operator is a symmetric operator. That is, the transformation that converts the A_1 ground state to the A_2 excited state requires an operator that inverts the symmetry. But the electric dipole operator, $\boldsymbol{\mu}_e$, is a symmetric operator, and thus, the $n \rightarrow \pi^*$ transition is an electric dipole forbidden process—forbidden by symmetry; consequently, $\langle A_1 \mid \boldsymbol{\mu}_e \mid A_2 \rangle = 0$. In contrast, the magnetic dipole moment operator, $\boldsymbol{\mu}_m$, is an antisymmetric operator; so the $n \rightarrow \pi^*$ transition, which transforms the A_1-symmetry ground state to the A_2-symmetry $n \rightarrow \pi^*$ excited state, is a magnetic dipole-allowed transition; thus, $\langle A_1 \mid \boldsymbol{\mu}_m \mid A_2 \rangle \neq 0$. The $n \rightarrow \pi^*$ magnetic dipole transition transforms as A_1, with the induced magnetic moment, $\boldsymbol{\mu}_z$, lying along the C=O bond, or Z-axis. With no $\boldsymbol{\mu}_z$ component coming from the electric dipole, $\boldsymbol{\mu}_e \cdot \boldsymbol{\mu}_e = 0$ and $\boldsymbol{\mu}_e \cdot \boldsymbol{\mu}_m = 0$; that is, both the dipole strength and rotatory strength are zero. However, a very weak UV–visible absorption is typically observed

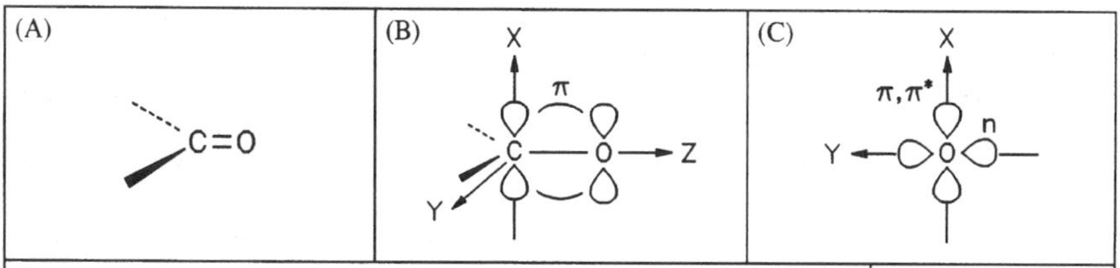

(D)

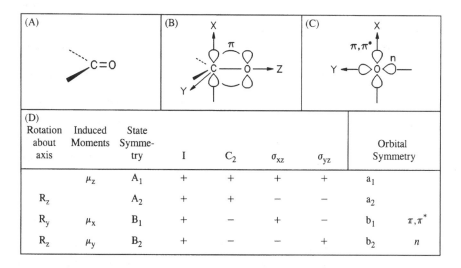

Rotation about axis	Induced Moments	State Symmetry	I	C_2	σ_{xz}	σ_{yz}	Orbital Symmetry	
	μ_z	A_1	+	+	+	+	a_1	
R_z		A_2	+	+	−	−	a_2	
R_y	μ_x	B_1	+	−	+	−	b_1	π,π^*
R_z	μ_y	B_2	+	−	−	+	b_2	n

Figure 3-12. (A) Carbonyl group, C_{2v} symmetry. (B) Coordinates for carbonyl chromophore, showing π–p orbitals on carbon and oxygen. (C) n and π or π^* orbitals on oxygen, as viewed down the Z-axis. (D) Symmetry character table for the C_{2v} symmetry carbonyl chromophore (ref. 9).

due to vibrational coupling from extrachromophoric parts, so optically active ketones typically show an $n \rightarrow \pi^*$ Cotton effect ($R \neq 0$).

In the case of the ketone $\pi \rightarrow \pi^*$ transition, the A_1-symmetry ground state $[(n)^2(\pi)^2$ or $(b_2)^2 (b_1)^2)]$ is promoted to an A_1-symmetry excited state $[(n)^2 (\pi) (\pi^*)$ or $(b_2)^2 (b_1)(b_1^*)]$. This excitation is symmetry allowed for the symmetric electric dipole operator $\boldsymbol{\mu}_e$, ($<A_1 | \boldsymbol{\mu}_e | A_2> \neq 0$), but symmetry forbidden for the antisymmetric magnetic dipole operator $\boldsymbol{\mu}_m$ ($<A_1 | \boldsymbol{\mu}_m | A_1> = 0$). The induced electric moment for the $\pi \rightarrow \pi^*$ (vertical) transition lies along the Z-axis, and a $\pi \rightarrow \pi^*$ excitation can be expected to show a strong UV–visible absorption. But in the CD, the rotatory strength would be zero, unless the transition "borrowed" magnetic dipole intensity (lying along the Z-axis) from extrachromophoric parts. Thus, although a ketone may have a strong $\pi \rightarrow \pi^*$ UV absorption, the $\pi \rightarrow \pi^*$ CD transition may be weak.

References

1. Moscowitz, A., in *Optical Rotatory Dispersion* (Djerassi, C., ed.), McGraw-Hill, New York 1960, Chap. 12.
2. For a glossary of optical-activity terminology, see the provisional IUPAC rules on chiro-optical spectroscopy in *Chiroptical Techniques, Nomenclature, Symbols, Units* (Snatzke, G.), 1987. This document is being rewritten by the IUPAC Commission I.5 (Chairman, Prof. John Bertie, Univ. Alberta, Canada) for the IUPAC Interdivision Committee on Nomenclature and Symbols.
3. Moscowitz, A., *Adv. Chem. Phys.* **4** (1962), 67–112.
4. Charney, E., *The Molecular Basis of Optical Activity*, Wiley-Interscience, New York, 1979.
5. Campbell, I. D., and Dwek, R. A., *Biological Spectroscopy*, Benjamin/Cummings Publ. Co., Menlo Park, CA, 1984.
6. In some CD instruments (e.g., Jasco) the $\Delta\varepsilon$ vertical axis is called "mol. CD."
7. Deutsche, C. W., Lightner, D. A., Woody, R. W., and Moscowitz, A., *Ann. Rev. of Phys. Chem.* **20** (1969), 407–448.
8. Moffitt, W., Woodward, R. B., Moscowitz, A., Klyne, W., and Djerassi, C., *J. Am. Chem. Soc.* **83** (1961), 4013–4018.
9. Jaffé, H. H., and Orchin, M., *Theory and Applications of Ultraviolet Spectroscopy*, J. Wiley, New York, 1962.
10. Moffitt, W., and Moscowitz, A., *J. Chem. Phys.* **30** (1959), 648–660.
11. Newman, M. S., Darlak, R. S., and Tsai, L., *J. Am. Chem. Soc.* **89** (1967), 6191–6193.
12. Salvadori, P., Lardicci, L., Menicagli, R., and Bertucci, C., *J. Am. Chem. Soc.* **94** (1972), 8598–8600.

Appendix: Experimental Rotatory Strengths

The observed rotational strength R_0^T associated with a particular absorption band at temperature T is conveniently gauged by the wavelength-weighted area under the corresponding circular dichroism (CD) curve (Figure 3-13) according to the equation

$$R_0^T \sim 0.696 \times 10^{-42} \times 3300 \int \frac{\Delta\varepsilon}{\lambda}\, d\lambda \text{ (cgs units)}, \tag{3-29}$$

which is a form of Eq. (3-24). When the CD curve is available in digital form, the integration is fairly straightforward, and one of us (JEG) has written programs to calculate rotatory strengths from such data using a desktop computer.

When the CD curve is not available in digitized form, which is characteristic of published CD curves, the curve may be digitized using several available computer programs, such as Un-Scan-It or Un-Plot-It (Silk Scientific, Orem, UT). Alternatively, a number of "paper" methods are available for determining experimental rotatory strengths. The simplest method assumes that the CD curve has a Gaussian shape. In that case,

$$R_0^T \sim 0.696 \times 10^{-42} \sqrt{\pi}\, 3300(\Delta\varepsilon^{\max})\,(\Delta_\lambda/\lambda^{\max}), \tag{3-30}$$

where Δ_λ is the half-width (nm) at $1/e$ times the height ($0.368\Delta\varepsilon^{\max}$), and $\lambda^{\max}$ is the wavelength in nm at $\Delta\varepsilon^{\max}$ (Figure 3-13).

However, the CD curve does not always have a Gaussian shape, and in such cases graphical integration methods are used. One method shows how the area of the CD curve may be approximated by a series of rectangles, each of whose area is $\Delta\varepsilon_n \times \Delta\lambda_n$ (Figure 3-14). Since the rotatory strength is the wavelength-weighted area, each of the rectangles must be divided by the appropriate average wavelength λ_n to give the desired rotatory strength increment. The observed rotatory strength is then obtained by summing the increments:

$$R_0^T \sim 0.696 \times 10^{-42} \times 3300 \sum^{n} \frac{\Delta\varepsilon_n \times \Delta\lambda_n}{\lambda_n}. \tag{3-31}$$

A third graphical integration method uses a planimeter calibrated in square centimeters and the plotted CD curve. As before, the CD spectrum is decomposed into suitable increments (usually, $\Delta\lambda = 10$ nm). The area of each increment is determined with the planimeter. Since $\Delta\varepsilon$ and the wavelength are typically measured in terms of millimeters (1 mm = 1 nm) from the machine trace, the areas obtained using the planimeter (square centimeters) must be converted to square millimeters by multiply-

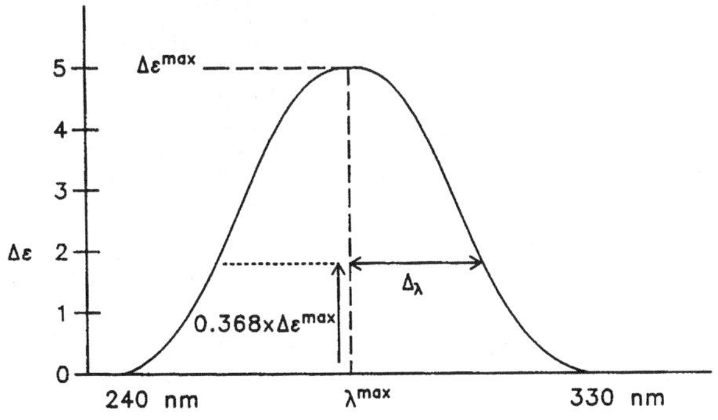

Figure 3-13. Arbitrary circular dichroism curve with a positive Cotton effect.

ing by 100. Thus, the area of each increment is taken with the planimeter, multiplied by 10, and divided by the appropriate average wavelength. The rotatory strength is then given by the equation

$$R_0^T \sim 0.695 \times 10^{-42} \times 3300\ \Delta\varepsilon \times V_{25}^T \times \sum^{n} \frac{(\mathrm{cm}^2)_n \times 100}{\lambda_n}. \tag{3-32}$$

where V_{25}^T corrects for the change in density of the solution due to solvent contraction upon cooling. Table 3-1 provides some useful density corrections for the two commonly used low-temperature, glass-forming solvents: ether-isopentane-ethanol (EPA, 5:5:2 by volume) and isopentane-methylcyclohexane (P5M1, 5:1 by volume).

Calculation of the rotatory strength from the $n \rightarrow \pi^*$ CD Cotton effect of epicamphor is illustrated in the following. In Figure 3-15, the CD spectrum of 2.98 mg of epicamphor in 5 mL of isooctane measured at 21°C in a 1-cm path length at a sensitivity (Sens) of 5×10^{-4} ΔA/cm is plotted at ΔA versus λ (nm). Since $\Delta\varepsilon = \Delta A / lc$ (where $\Delta A = A_L - A_R$, l is the cell path length in cm, and c is the concentration in moles/liter), and since $\Delta A = n \times$ Sens (where n is the number of cm on the CD spectrum and Sens is the scale, for example, on the epicamphor spectrum; for ΔA at 307 nm, $n \sim 13.5$ cm), Eq. (3-29) can be written as

$$R_0^T = 22.9_{68} \times 10^{-40} \times \frac{\text{Sens}}{l \cdot c} \int_{\lambda} \frac{n}{\lambda}\, d\lambda\ (\text{cgs}), \tag{3-33}$$

because Sens, l, and c are constant in the CD band measurement, but n varies with λ.

To estimate the integral part of Eq. (3-33), the CD spectrum is summed graphically as follows (for large bands one may sum over 10-nm units (blocks), for smaller bands over smaller units, for example, 5-nm or 2-nm blocks; the technique is illustrated for the epicamphor CD spectrum):

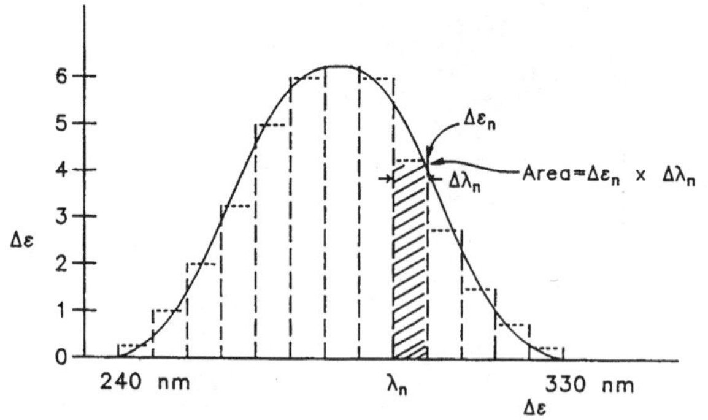

Figure 3-14. Example of graphical integration of a circular dichroism curve by dividing it into a series of rectangles.

Table 3-1. Density Correction Factors for EPA and P5M1:[a]

$$V_{25}^{T} = \frac{\text{Vol. at } T°\text{C}}{\text{Vol. at } 25°\text{C}} = \frac{\text{Density at } 25°\text{C}}{\text{Density at } T°\text{C}}$$

	Density Correction at Temperature (°C)					
Solvent	25°C	–5°C	–29°C	–41°C	–74°C	–192°C
EPA	1.000	0.956	0.925	0.911	0.874	0.798
P5M1	1.000	0.957	0.926	0.911	0.874	0.793

[a]Taken from the data of Passerini, R., Ross, I. G., Ramsey, W., and Foster, R., (*J. Sci. Instrum.* **30**, 274 (**1953**), and adjusted to a 25°C base.

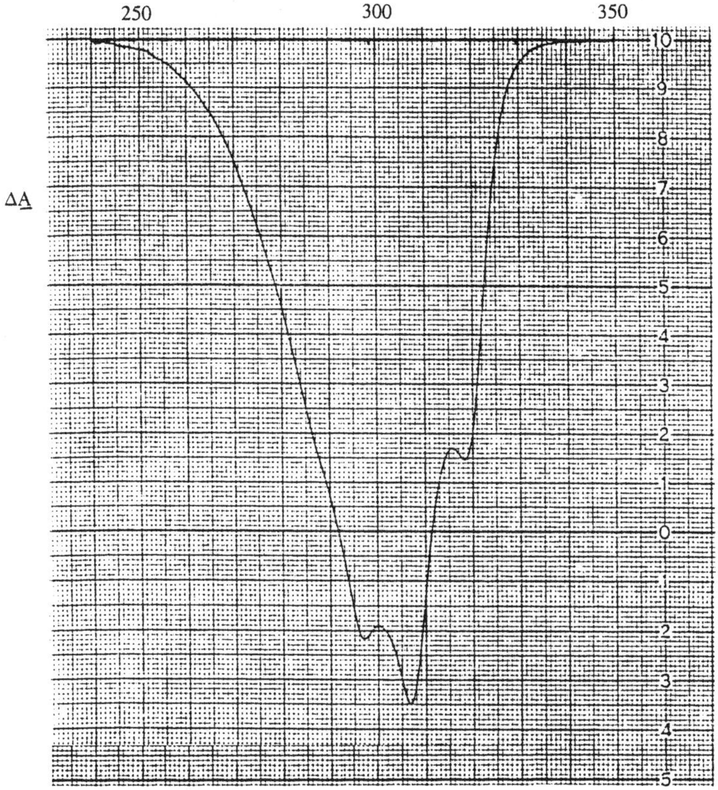

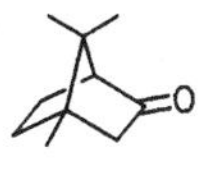

epicamphor

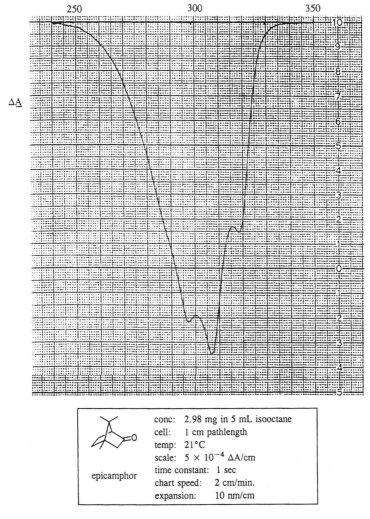

conc: 2.98 mg in 5 mL isooctane
cell: 1 cm pathlength
temp: 21°C
scale: 5×10^{-4} ΔA/cm
time constant: 1 sec
chart speed: 2 cm/min.
expansion: 10 nm/cm

Figure 3-15. $n \rightarrow \pi^*$ Circular dichroism of epicamphor in isooctane.

(i) Count the number of square millimeters (small squares) in each 1-cm^2 unit, reading across the baseline from $\lambda = 243$ to $\lambda = 338$ nm. This operation gives the first row of figures (in cm^2) in Table 3-2.

(ii) Count as in (i) for each successive 1-cm row, descending for negative Cotton effects and ascending for positive Cotton effects.

(iii) Sum each column.

(iv) Determine λ_{av} for each column (e.g., since the third column summed to λ between 270 and 280, the average is 275 nm).

(v) Divide 10 times each sum by λ_{av}. The use of 10 times the sum is to correct the dimensions of the block to "cm" by "mm"; cm corresponds to n, mm to λ.

(vi) Sum the data from (v) to obtain the graphically determined value of the integral in Eq. (3-33); in this case, $\Sigma = -1.65_{04}$.

From the data given, the molecular weight of epicamphor (152), and Eq. (3-33), it follows that

$$R_0^T = 22.9_{68} \times 10^{-40} \times \frac{5 \times 10^{-4}}{(1)\left(\dfrac{2.98}{152} \times \dfrac{1}{5}\right)} \int_\lambda \frac{n}{\lambda}\, d\lambda \text{ (cgs).} \tag{3-34}$$

Therefore,

$$R_0^T = 2.93 \times 10^{-40}\ (\Sigma = -1.65_{04}) = -4.83 \times 10^{-40} \text{ (cgs).} \tag{3-35}$$

The value of R_0^T can be checked independently by using the Gaussian approximation of Equation (3-30). Since the height (depth) of the epicamphor band is 13.5 cm, the half-width is determined at 13.5/e cm, or $0.368 \times 13.5 = 4.9_7$ cm. The width (nm) at $1/e$ times $\Delta\varepsilon^{max}$ is 43.6 mm, which corresponds to 43.6 nm. The band half-width is therefore 21.8 nm, and we have

$$R_0^T = 0.696 \times 10^{-42} \sqrt{\pi}\ 3300 \times \frac{-13.5 \times 5 \times 10^{-4}}{(1)\left(\dfrac{2.98}{152} \times \dfrac{1}{5}\right)} \times \frac{21.8}{307} = -4.9_8 \times 10^{-40}, \tag{3-36}$$

in cgs units. This value is suitably close to that calculated by integration (Eq. 3-35).

Table 3-2. Graphical Determination of the Integral in Equation (3-34). Determination of the number of square centimeters in each block of the epicamphor $n \rightarrow \pi^*$ CD spectrum.

ROW No. (each row is 1 cm deep)	COLUMN NUMBER (each column is 1 cm wide)									
	1	2	3	4	5	6	7	8	9	10
1	0.09	0.50	0.99	1.00	1.00	1.00	1.00	1.00	0.94	0.10
2			0.52	1.00	1.00	1.00	1.00	1.00	0.63	
3			0.05	0.95	1.00	1.00	1.00	1.00	0.50	
4				0.60	1.00	1.00	1.00	1.00	0.36	
5				0.28	1.00	1.00	1.00	1.00	0.28	
6				0.02	0.97	1.00	1.00	1.00	0.20	
7					0.73	1.00	1.00	1.00	0.12	
8					0.50	1.00	1.00	1.00	0.03	
9					0.20	1.00	1.00	0.60		
10					0.01	0.96	1.00	0.20		
11						0.70	1.00	0.06		
12						0.50	0.97			
13						0.03	0.46			
14							0.09			
Sum cm^2	0.09	0.50	1.56	3.85	7.41	11.19	12.52	8.86	3.06	0.10
λ_{av}	247	255	265	275	285	295	305	315	325	334
$10 \times cm^2/\lambda_{av}$	0.0036	0.0196	0.0589	0.14	0.26	0.3793	0.4105	0.2813	0.0942	0.0030

CHAPTER 4

The Octant Rule

The octant rule was formulated by Djerassi et al.[1,2,3] 40 years ago to determine the absolute configuration of a saturated ketone when its conformation is known—or, conversely, to determine its conformation when the absolute configuration is known. The octant rule was the first sector rule in organic stereochemistry, and it is probably the most important, most widely known, and most successful of the many different chiroptical sector rules formulated for various chromophores or systems of chromophores.[4,5,6] Such sector rules focus on chromophores in a molecule and relate the CD or ORD Cotton effects accompanying the chromophores' electronic transitions to the chirality of the extrachromophoric environment, the chirality of the chromophore, or both. In some chiral molecules, the chromophore is achiral in local symmetry, as in the case of the ketone carbonyl of (2*S*)-methylcyclohexanone or the phenyl group of (*S*)-2-phenyl-3,3-dimethylbutane (Figure 4-1). In other chiral molecules, the chromophore itself is chiral, as in the classic case of hexahelicene, in which the molecule is a chromophore, or in the case of a nonplanar diene, such as (5*S*)-methyl-1,3-cyclohexadiene, which has both a chiral chromophore and chiral extrachromophoric parts. The distinction between two limiting types of chromophores—inherently chiral (dissymmetric) and inherently achiral (symmetric)—is discussed in Chapter 3 and was proposed first by Moffitt and Moscowitz[7] and later by Djerassi, Mislow, Moscowitz, and coworkers[8] as a useful classification to account for the origin of the Cotton effects. Thus, CD from an inherently dissymmetric chromophore has its origins within the (dissymmetry of the) chromophore itself, whereas an inherently symmetric chromophore cannot exhibit CD without dissymmetric vicinal perturbation.[9]

These distinctions are often reflected in the magnitude ($\Delta\varepsilon$) of the CD Cotton effects. Those from inherently dissymmetric chromophores are typically larger by one or two orders of magnitude than those from inherently symmetric, but dissymmetrically perturbed, chromophores. For example, the carbonyl chromophore, which is one of the most studied by CD spectroscopy, has $\Delta\varepsilon$ values for the $n \rightarrow \pi^*$ transition

(A) (B) (C) (D)

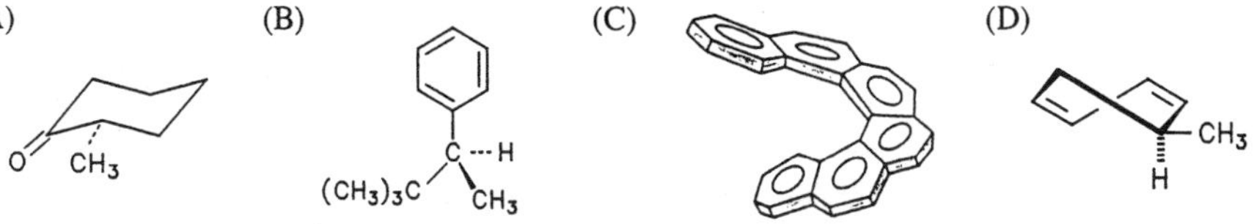

Figure 4-1. Stereochemical drawings of (A) 2(*S*)-methylcyclohexanone, (B) (*S*)-2-phenyl-3,3-dimethylbutane, (C) ***P***-hexahelicene, and (D) (5*S*)-methylcyclohexa-1,3-diene.

typically on the order of unity. In (2*S*)-methylcyclohexanone, for instance, Δε ~ +0.3 near λ^{max} ~ 290 nm, which may be contrasted with Δε ~ +200 for the long-wavelength ($\pi \rightarrow \pi^*$) transition near λ^{max} ~ 330 nm of ***P***-hexahelicene (Figure 3-11). Because optical activity of the carbonyl chromophore originates from chiral perturbers within the molecule, but lying distant from the chromophore, the latter can serve as a very sensitive chiroptical probe, viz., a window for detecting extrachromophoric stereochemistry, as described in the octant rule. However, as with any sector rule, the applicability of the octant rule depends on the extent to which the chromophore in question retains its identity in the molecule as a whole—that is, the extent to which the remainder of the molecule can be considered as a perturbation in determining the chiroptical properties of the chromophoric electronic transition.

4.1. Optical Activity of the Ketone *n* → π* Electronic Transition

The relevant UV–visible absorption for the octant rule is the $n \rightarrow \pi^*$ transition near 290 nm. An $n \rightarrow \pi^*$ transition involves the rotation of charge—as shown in Figure 4-2A, the movement of an electron from an $n(p_y)$ oxygen nonbonding orbital to a π^* antibonding orbital composed of a linear combination of oxygen and carbon p_x orbitals (Figures 4-2B, C). Although such a circular movement of charge leads to a large induced magnetic dipole moment (μ_m = 1 Bohr-magneton) along the C=O bond (*Z*-axis), it does not induce an electric dipole moment ($\mu_e = 0$) (translation of charge) in the same direction. That is, as indicated in Section 3.7, on the basis of local symmetry, the $n \rightarrow \pi^*$ transition is an electric dipole-forbidden ($\mu_e = 0$), magnetic dipole-allowed ($\mu_m \neq 0$), transition. Of course, weak UV absorbance (ε ~ 10–100) is typically observed for ketones near 300 nm, but the electric dipole intensity is

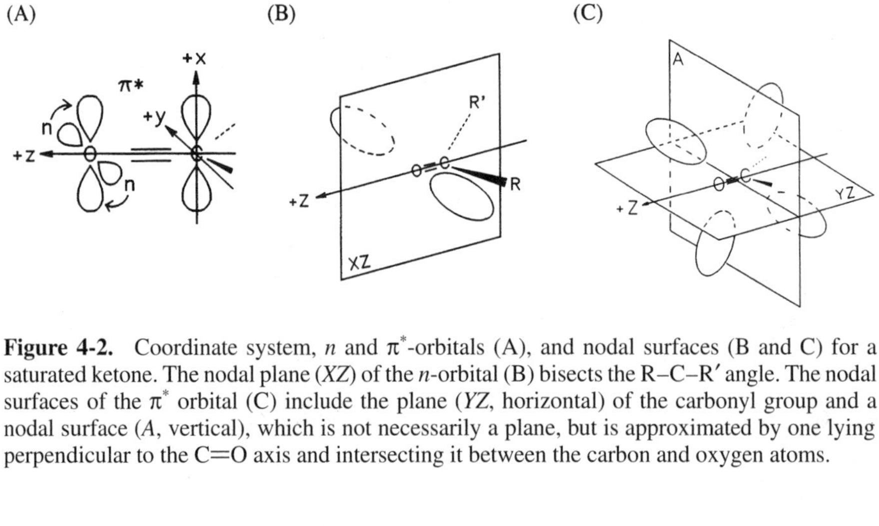

Figure 4-2. Coordinate system, *n* and π^*-orbitals (A), and nodal surfaces (B and C) for a saturated ketone. The nodal plane (*XZ*) of the *n*-orbital (B) bisects the R–C–R′ angle. The nodal surfaces of the π^* orbital (C) include the plane (*YZ*, horizontal) of the carbonyl group and a nodal surface (*A*, vertical), which is not necessarily a plane, but is approximated by one lying perpendicular to the C=O axis and intersecting it between the carbon and oxygen atoms.

"borrowed" by vibronic coupling from electric dipole-allowed transitions of higher energy—for example, $\pi \rightarrow \pi^*$ or $\sigma \rightarrow \sigma^*$. (The $\pi \rightarrow \pi^*$ transition is polarized along the *Z*-axis of Figure 4-2, but is magnetic dipole forbidden.) Such intensity borrowing provides a mechanism for observing weak $n \rightarrow \pi^*$ absorption in the UV spectrum, but does not lead to optical activity of the chromophore, which still possesses a plane of symmetry. However, when the carbonyl group is located in a chiral environment, as in 2(*S*)-methylcyclohexanone (Figures 2-14 and 4-1), dissymmetric perturbations of the $n \rightarrow \pi^*$ transition lead to a nonzero rotatory strength ***R*** and, hence, a nonzero CD (Figure 2-16F).

4.2. Formulation of the Octant Rule from Carbonyl $n \rightarrow \pi^*$ Nodal Planes

Since its formulation in the late 1950s, the octant rule[1,2,3,10,11] for the $n \rightarrow \pi^*$ transition of saturated alkyl ketones has become one of the most important chirality rules for extracting stereochemical information from optically active ketones. According to the octant rule, all space surrounding the carbonyl chromophore is divided into eight regions (octants), and the octant occupied by a particular perturber (e.g., the methyl group of 2(*S*)-methylcyclohexanone) determines the sign of its contribution to the intensity (rotatory strength) of the $n \rightarrow \pi^*$ transition. The octants under discussion are derived, in part, from the local symmetry (C_{2v}) of the carbonyl group and a consideration of the relevant orbitals of the $n \rightarrow \pi^*$ transition. (See Section 3-7.) Two well-defined carbonyl symmetry planes (*XZ* and *YZ*; see Figures 4-2 and 4-3)—the *n* and π^* nodal planes—divide all space about the C=O group into quadrants, generating a quadrant rule. A third, poorly defined non-symmetry-derived surface further divides quadrant space into octants—hence the octant rule. Purely for convenience rather than on any theoretical basis, Moffitt et al.[2] approximated this third surface to be a plane bisecting the C=O bond. Indeed, they specifically cautioned that this surface was very probably *not* a plane. As the octant rule came under closer scrutiny, the shape of the third nodal surface became a major area of investigation.

The reflection of a perturber across either of the carbonyl symmetry planes (*XZ* and *YZ* in Figure 4-3) leads to a mirror-image molecular fragment and hence one with oppositely signed contributions to the rotatory strength. However, since the third nodal surface (*A*) does not follow from symmetry, "reflection" across it does not correspond to a mirror-image situation, and hence, the weight given to a perturber in a front octant is not the same as for a like position in a back octant. The signs of the octant contributions for atoms such as C, H, Cl, Br, and I are shown in Figure 4-3. Atoms lying in symmetry planes offer no contribution, and atoms symmetrically disposed across the carbonyl symmetry planes will exert no net effect on the CD, due to cancellation. Usually, the *sign* made by an octant perturber to the observed CD of the $n \rightarrow \pi^*$ transition varies as the inverse *sign* of the product $x \cdot y \cdot z$ of its atomic coordinates. In the right-handed coordinate system chosen for Figures 4-2 and 4-3, a perturber in an upper left back octant (coordinates $+x$, $+y$, $-z$) would make a $-(x \cdot y \cdot$

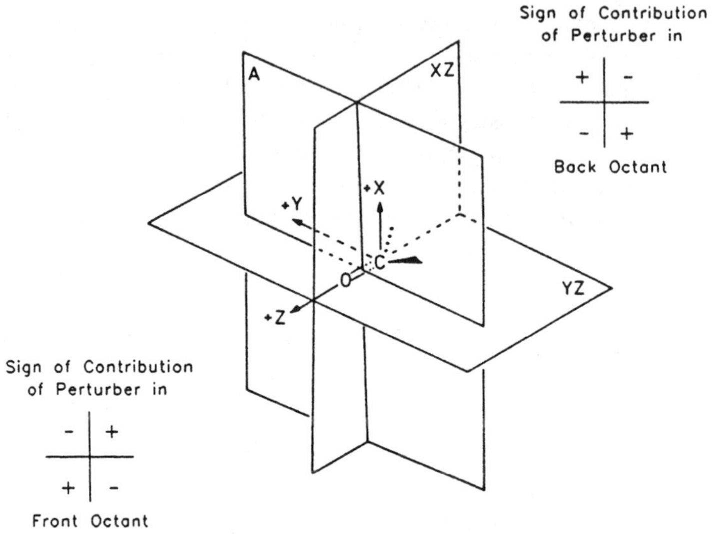

Figure 4-3. Octant rule diagram for the ketone carbonyl $n \rightarrow \pi^*$ transition. Local symmetry-derived, orthogonal octant planes *XZ* and *YZ* divide all space into quadrants, and a non-symmetry-derived third nodal surface (*A*) is approximated by an orthogonal plane bisecting the C=O bond. "Front" octants are those nearer an observer along the +*Z* axis, while "back" octants lie towards –*Z*. [Reprinted with permission from ref. 12. Copyright© 1986 American Chemical Society.]

$-z) = +(x \cdot y \cdot z)$, or positive, contribution. The magnitudes of the contributions vary with the nature of the perturber, while falling off rapidly with increasing distance from the carbonyl chromophore or closeness to the nodal surfaces. Contributions are assumed to be additive. This geometrical rule, so simple and straightforward to apply, has served well in establishing absolute configurations and elucidating conformations in a large number of ketones.[1,2,3,5,10,11]

4.3. How to Use the Octant Rule to Determine Absolute Stereochemistry and Conformation

In order to use the octant rule to determine the absolute configuration of a ketone, the conformation of the ketone must be known. Consider, for example, the cyclohexanones discussed in Chapter 2. We assumed a predominance of the chair conformation, and now we position chair cyclohexanone as shown in Figure 4-4 so that the local symmetry planes of its carbonyl chromophore are coincident with those shown on the octant diagram of Figure 4-3. In the chair conformation of cyclohexanone, which is symmetric and achiral, carbons 1, 2, 4, and 6 lie on symmetry planes *XZ* and *YZ* and thus make no contribution to the CD. Carbons 3 and 5 (and their hydrogens) lie equally disposed across the *XZ* symmetry plane in the upper left positive (+) and upper right

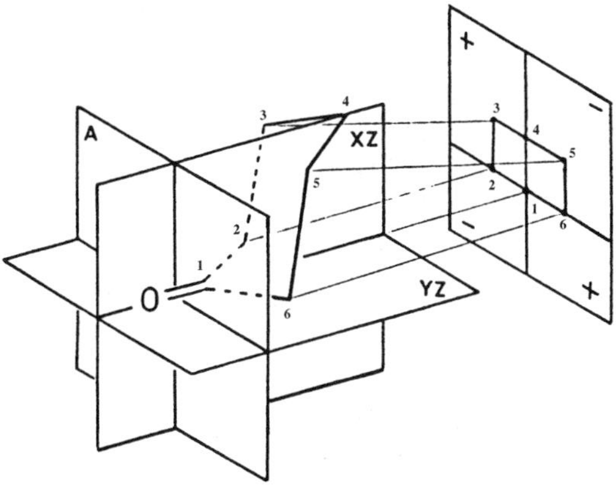

Figure 4-4. Octants and octant projection diagram applied to cyclohexanone. The C=O group is aligned along the intersection of the *XZ*- and *YZ*-planes and is viewed from O to C. The cyclohexanone carbons are projected into the back octant projection diagram, with signs (+, –) of the four back octants shown.

negative (–) back octants and thus make no net contribution to the CD. The net octant contributions are thus zero, and the CD of cyclohexanone is therefore predicted to be zero, which, of course, it must be for such a symmetric molecule. The symmetry of cyclohexanone may be broken by introducing a substituent at carbon 2 (or 6) or 3 (or 5), leaving the molecule capable of exhibiting optical activity and CD. This property was shown previously for 2(*S*)-methylcyclohexanone (Section 2.7) and may also be seen for 3(*R*)-methylcyclohexanone. In both ketones, the more stable chair conformation has an equatorial methyl. In 2(*S*)-methylcyclohexanone (Figure 4-5), the methyl group lies slightly in an upper left (+) back octant (Figure 4-5B); hence, the net contribution to the CD is predicted to be *positive*, since all other carbon atoms lie in a carbonyl symmetry plane or are cancelling. Strictly speaking, one weighs the contribution of the equatorial methyl carbon and its three hydrogens lying in the upper left (+) back octant vs. an equatorial hydrogen lying in an upper right (–) back octant. More weight is given to the larger group, and thus, a net positive $n \rightarrow \pi^*$ CD Cotton effect is predicted. The experimentally determined CD is in fact *weakly positive*. If the axial methyl isomer were to predominate in the equilibrium, the CD would be expected to exhibit a *strongly positive* Cotton effect, which is not observed. And the fact that the magnitude remains essentially invariant down to –192°C suggests that the axial conformer makes little or no contribution to the net population of isomers.

In 3(*R*)-methylcyclohexanone, the lone methyl perturber also lies in a (+) back octant when it is equatorial on chair cyclohexanone (Figure 4-6). Consequently, the net contribution to the CD $n \rightarrow \pi^*$ Cotton effect is predicted to be positive, which is in fact observed experimentally (Figure 4-6). The greater magnitude of the 3(*R*)-methyl isomer compared with 2(*S*)-methylcyclohexanone is related to the fact that, although the 2(*S*) equatorial methyl lies closer to the carbonyl group (and should thus have a stronger influence on the Cotton effect), it also lies very close to an octant nodal plane.

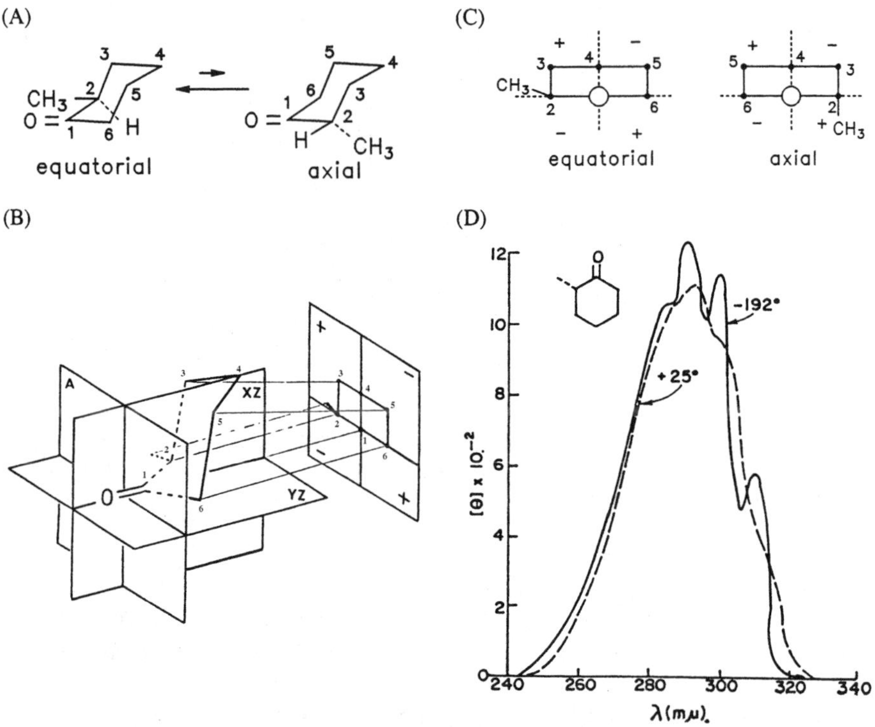

Figure 4-5. (A) Interconverting chair conformations of 2(*S*)-methylcyclohexanone. The equatorial methyl isomer is thought to be more stable. (B) Octant rule applied to equatorial 2(*S*)-methylcyclohexanone. (C) Octant projection diagrams for the equatorial and axial conformers. (D) Circular dichroism spectra of 2(*S*)-methylcyclohexanone in ether–isopentane–ethanol (EPA, 5:5:2 by vol.) at 25°C and −192°C. [(D) is reprinted with permission from Wellman, K. M., Briggs, W. S., and Djerassi, C., *J. Am. Chem. Soc.* **87** (1965), 73–81. Copyright© 1965 American Chemical Society.]

If it lay in a nodal plane, it would be predicted to make zero contribution to the Cotton effect. (This concept will be taken up at a later point in the chapter.) Unlike 2(*S*)-methylcyclohexanone, whose CD intensities were essentially invariant between +25° and −192°C, the CD of 3(*R*)-methylcyclohexanone becomes more positive with decreasing temperature. Conformational inversion to afford the axial methyl isomer appears to be an important consideration. With an experimental and computed conformational energy difference of approximately 1.4 kcal/mole (Section 2.6), one might expect about 9% of the axial conformer to be present at room temperature (Figure 1-1)—a conformer predicted (Figure 4-6C) to have a (−) $n \rightarrow \pi^*$ Cotton effect. So, upon lowering the temperature to −192°, the amount of axial conformer can be expected to be vanishingly small, thus leading to an increased positive CD.

One may assess the validity of the octant contributions of an α- or β-methyl perturber in cyclohexanone by comparing its CD with the CD from structurally more

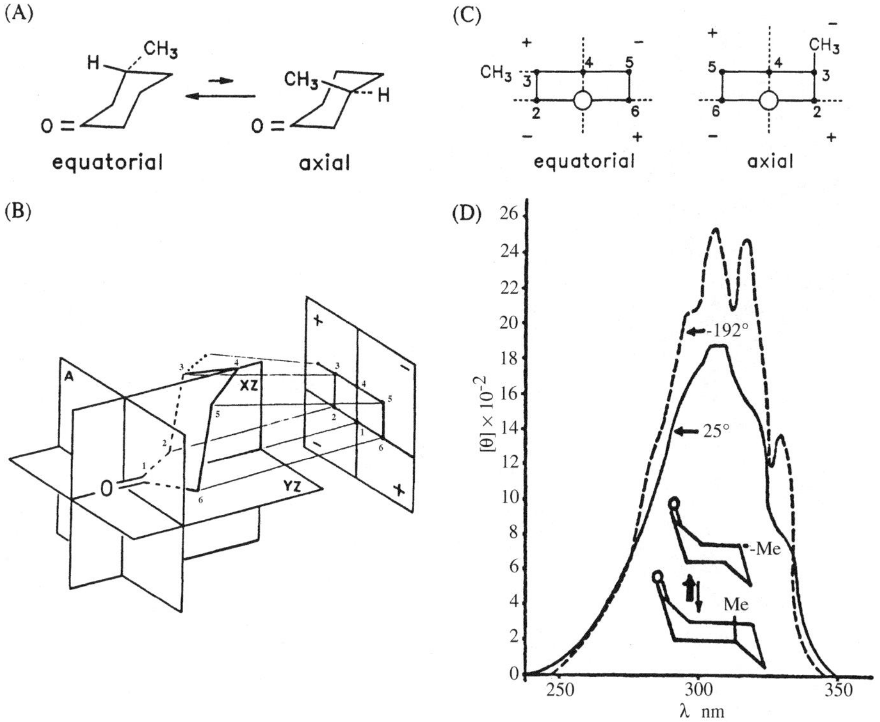

Figure 4-6. (A) Interconverting chair conformations of 3(*R*)-methylcyclohexanone. The equatorial methyl isomer is thought to be the more stable. (B) Octant rule applied to equatorial 3(*R*)-methylcyclohexanone. (C) Octant projection diagrams for the equatorial and axial conformers. (D) Circular dichroism spectra of 3(*R*)-methylcyclohexanone in ether–isopentane–ethanol (EPA, 5:5:2 by vol.) at 25°C and −192°C. [(D) is reprinted with permission from Wellman, K. M., Bunnenberg, E., and Djerassi, C., *J. Am. Chem. Soc.* **85** (1963), 1870–1872. Copyright© 1965 American Chemical Society.]

"rigid" or conformationally well-defined cyclohexanones. For example, the observed $n \rightarrow \pi^*$ CD sign and magnitude ($\Delta\varepsilon_{284}^{max}$ +0.57) of 3(*R*)-methylcyclohexanone is quite similar to that observed for (1*S*,3*R*)-4(e)-methyladamantan-2-one ($\Delta\varepsilon_{284}^{max}$ +0.67) in ethyl ether–isopentane–ethanol (5:5:2, v/v/v) solvent at room temperature.[12] And the $\Delta\varepsilon$ = +0.3–0.4 for the $n \rightarrow \pi^*$ CD Cotton effect of 2(S)-methylcyclohexanone (Figure 4-5) compares favorably with the $|\Delta\varepsilon|$ = 0.4 observed for the $n \rightarrow \pi^*$ CD of 2(*R*)-methyl-4(*R*)-*tert*-butylcyclohexanone (Table 4-1).[13] In adamantanone, the cyclohexanone moiety is locked into a chair conformation, and thus, the methyl perturber is frozen in either the equatorial position (shown in Fig. 4-7) or the axial position. In 2-methyl-4-*tert*-butylcyclohexanone, it is the *tert*-butyl group that locks the conformation in one chair, as discussed in Section 2.6. Interestingly, in the adamantanone system, with its well-defined stereochemistry, replacing the methyl perturber with larger alkyl substituents (e.g., ethyl, isopropyl, *tert*-butyl, or neohexyl) causes very little change in the

Table 4-1. Chair Conformation (Upper), Approximate Δε (Middle), and Octant Projection (Lower) Diagrams for α and β-Methylcyclohexanones. The Bottom Row Displays the Corresponding Structures Used to Lock in the Cyclohexanone Chair Conformation and Methyl Configurations Shown.

α-Equatorial CH_3	α-Axial CH_3	β-Equatorial CH_3	β-Axial CH_3
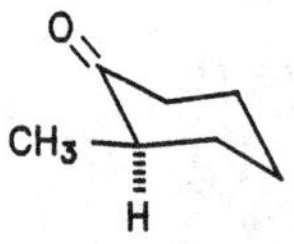	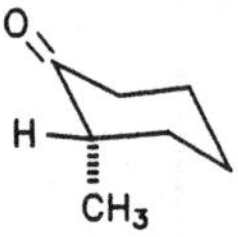	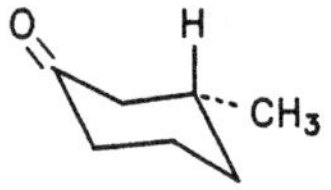	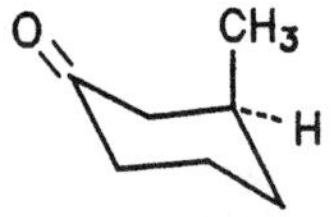
Δε ~ −0.3	Δε ~ +1.4[a]	Δε ~ +0.6[b]	Δε ~ −0.1[c]
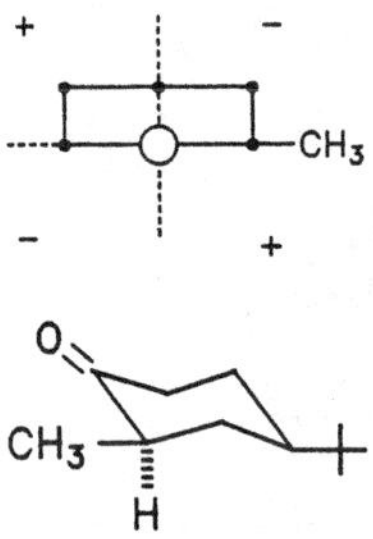	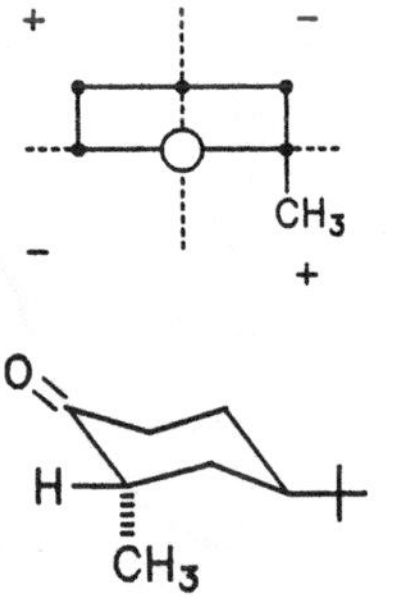	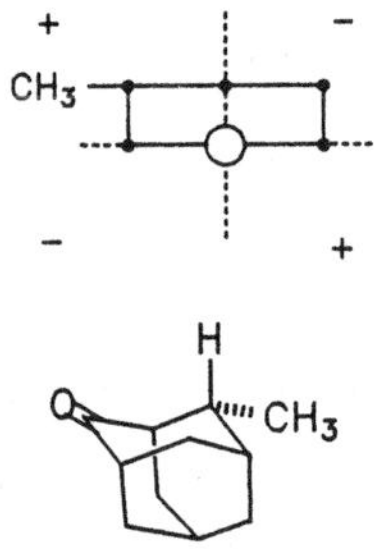	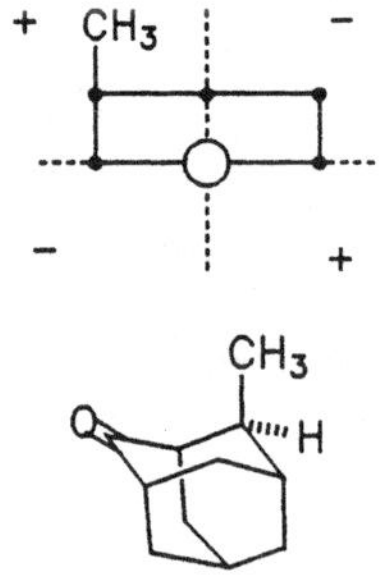

[a] On 4-*tert*-butylcyclohexanone, ref. 13. [b] Refs. 10,38. [c] On adamantanone, ref. 12.

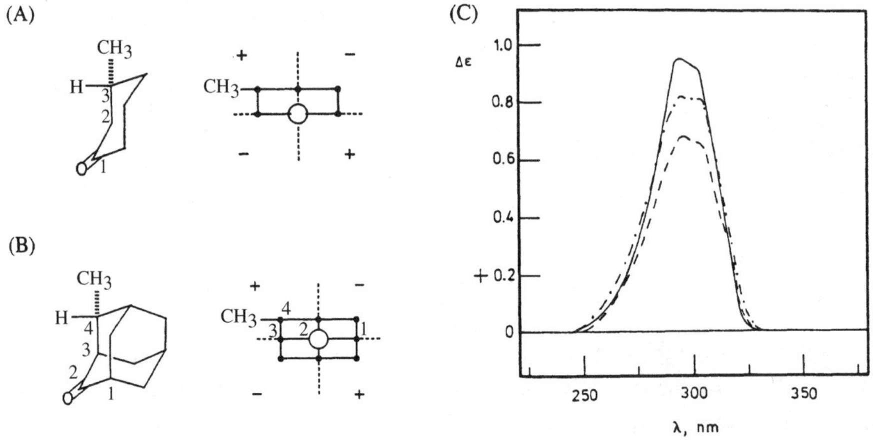

Figure 4-7. (A) 3(*R*)-Methylcyclohexanone and its octant projection diagram. (B) 1(*S*),3(*R*)-4(e)-Methyladamantan-2-one and its octant projection diagram. (C) Circular dichroism spectra of 10^{-3} *M* 1(*S*),3(*R*)-4(e)-substituted adamantan-2-ones in ether–isopentane–ethanol (EPA, 5:5:2, by vol.) run at 25°C and corrected to 100% e.e.The equatorial substituents are methyl (- - - -), $\Delta\varepsilon^{max}_{295} = +0.67$; ethyl (— · — · —), $\Delta\varepsilon^{max}_{295} = +0.81$; isopropyl (approx. — · — · —), $\Delta\varepsilon^{max}_{295} = +0.80$; *tert*-butyl (approx. — · — · —), $\Delta\varepsilon^{max}_{295} = +0.771$; and neohexyl (—), $\Delta\varepsilon^{max}_{295} = +0.92$. [(C) is reprinted with permission from ref. 12. Copyright© 1986 American Chemical Society.]

adamantanone $n \rightarrow \pi^*$ CD. So the major contribution appears to be made by the first carbon in the alkyl perturber chain, and the magnitude of an octant contribution falls off rapidly with increasing distance from the carbonyl chromophore.

In general, the magnitude of the contribution to the Cotton effect made by an octant perturber falls off with increasing distance from the carbonyl chromophore, but contributions also fall off with increasing proximity to an octant surface (Figure 4-3). These concepts are illustrated by comparing the $n \rightarrow \pi^*$ Cotton effects from α-equatorial and α-axial methylcyclohexanone and from α-axial methyl- and β-equatorial methylcyclohexanone (Table 4-1). In the first set, the α-equatorial methyl lies closer to the carbonyl chromophore, but also lies closer to an octant plane and thus makes a weaker contribution to the $n \rightarrow \pi^*$ Cotton effect. When the octant perturber lies close to an octant symmetry plane, even if it also lies close to the carbonyl chromophore (as in α-equatorial methylcyclohexanone), the magnitude of the octant contribution is significantly diminished. In the second set, the methyl perturber lies far from octant nodal planes, but the α-axial methyl lies closer to the carbonyl chromophore and thus makes the larger octant contribution.

A β-axial methyl does not lie near an octant symmetry plane, nor does it lie at a considerable distance from the carbonyl chromophore; so it is expected to be an ordinary octant perturber. The β-axial methylcyclohexanone shown in Table 4-1 was predicted to give a moderately strong (+) $n \rightarrow \pi^*$ Cotton effect. However, quite surprisingly, and contrary to expectations, it was discovered[14,15] that a β-*axial* methyl perturber in the rigid adamantanone analog did not obey the octant rule, but gave only a very weak (–) octant contribution—weaker even than α-equatorial. Since a β-axial methyl does not lie near a C=O nodal plane (Figure 4-6), its oppositely signed ("antioctant") contribution was a source of much concern and led to a reevaluation of the octant rule and the location of the third nodal surface (*A* of Figure 4-4).[16]

4.4. The Question of Front Octants

The experimental evidence on which the octant rule rests comes largely and convincingly from numerous examples in which the dissymmetric elements (the groups perturbing the carbonyl chromophore in a nonsymmetric way—e.g., the equatorial methyl group of 3(*R*)-methylcyclohexanone) are invariably located behind the carbon of the carbonyl group (*back octants*), as viewed from oxygen toward carbon.[1–6,10–12,16] There are few examples in which dissymmetric perturbers are located in front of the carbonyl carbon or oxygen, and almost all the known rare compounds having such perturbers in front octants also have other perturbers nonsymmetrically located in back octants. The question of the existence of, or need for, front octants had not escaped attention altogether, however, for it was quite obvious at an early stage in the development of the octant rule that some atoms would occasionally lie in front octants (e.g., in 1-oxo, 7-oxo-, and 11-oxosteroids), although the effects of atoms lying in back octants always appeared to dominate the sign of the Cotton effect.[1,2] Djerassi and Klyne discussed examples of contributors entering front octants,[17] but these examples also had back-octant as well as front-octant contributions and therefore did not clearly test the existence of front octants. To overcome this difficulty, Kirk, Klyne, and Mose[18,19] prepared structurally related D-homo and D-nor 7-ketosteroids. By subtracting the back-octant Cotton effect of the D-norsteroid from the Cotton effect of the D-homosteroid (determined by contributions from the same back-octant perturbers and from new front-octant perturbers), they concluded in favor of an octant rule rather than a quadrant rule. At the same time, CD spectra of potentially cleaner examples—*cis*- and *trans*-6-methylspiro[4.4]nonan-1-ones—supported the notion of front octants, but the analysis was complicated by ring conformational changes.[20] Without an unambiguous proof of the existence of front octants, an "octant" rule remained unproven, and a quadrant rule was analyzed and proposed as an alternative.[21,22,23] With a quadrant rule, the third nodal surface (*A* of Figure 4-3) is deleted, and the quadrants are signed the same as back octants.

In the decade after the octant rule was postulated,[1,2,3] there was considerable debate over whether "an octant" rule was even necessary—whether a quadrant rule was sufficient. Schellman[21,22] had elegantly shown how group theory could be used to generate the sign-determining regions of inherently symmetric chromophores. Quadrants are the minimum number of spatial subdivisions based on the intersecting symmetry planes of the isolated C=O chromophore of ketones. The symmetry planes must also coincide with the symmetry-derived nodal surfaces of the electronic wave functions. However, the possibility that additional nodal surfaces exist that are not symmetry derived was explored by Bouman and Moscowitz.[24] In support of the octant rule, Bouman and Moscowitz showed, in a theoretical treatment using a limited basis set, that quadrant contributions are suppressed by assuming delocalized *n*-orbitals and that the octant set gives larger contributions than the quadrant set. Thus, if only the $n \rightarrow \pi^*$ and $\pi \rightarrow \pi^*$ states are mixed by incomplete coulombic screening, quadrant behavior is obtained for a localized *n*-orbital; but if the *n*-orbital is delocalized, octant behavior is predicted, but with the wrong signs. On the other hand, Bouman and Moscowitz found that mixing of the $n \rightarrow \pi^*$ state with *D* states generated from 3*d* orbitals leads to octant behavior with the correct sign for hydrogen and carbon perturbers. In either case, the contribution to the CD intensity from the *D* states is in most instances about an order of magnitude greater than that obtained from mixing $n \rightarrow \pi^*$ states with $\pi \rightarrow \pi^*$ states. Consequently, any quadrant contributions that come into play would be suppressed by octant contributions.

In 1974, the existence of front octants was shown unequivocally by the synthesis and CD of two spiroketones: *syn*-(1′*R*)-spiro[cyclobutan-2-one-1,2′-(4′(a)-methyladamantane)] (**1**) and its *anti* isomer (**2**) (Figure 4-8), prepared from (–)-(1*R*,3*S*)-4(*R*)(a)-methyladamantan-2-one (Table 4-1).[25] In **1**, the methyl group lies in *front* of the carbonyl oxygen; in **2**, it lies well behind. With a time-averaged planar cyclobutanone conformation, the plane of the cyclobutanone ring bisects the spirofused adamantane skeleton. That is, in the absence of the methyl group, the plane of the cyclobutanone ring would lie on a plane of symmetry (*YZ* of Figure 4-2) passing through the molecule. Consequently, the methyl group, which does not lie on a symmetry or nodal plane, represents the lone dissymmetric perturber of the ketone carbonyl chromophore. In a quadrant rule, the methyl perturbers

(A)

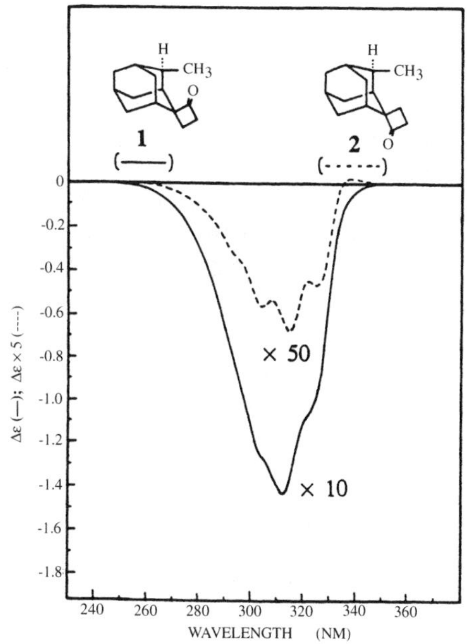

(B)

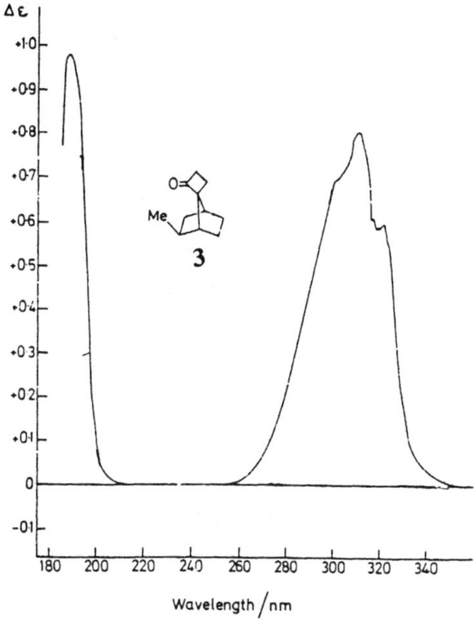

Figure 4-8. Circular dichroism spectra of *syn*-(1′*R*)-spiro[cyclobutan-2-one-1,2′-(4′(a)-methyladamantane)] (**1**) (—) and *anti*-(1′*R*)-spiro[cyclobutan-2-one-1,2′-(4′(a)-methyladamantane (**2**) (- - -) in isopentane at 20° (A) and *syn*-(1′*R*)-spiro[cyclobutan-2-one-1,7′-(2′-*exo*-methylnorbornane)] (**3**) in isopentane at 20° (B). Corrections are made to 100% optical purity. [(B) Reprinted with permission from ref. 26. Copyright© 1974 The Royal Society of Chemistry. (A) Reprinted with permission from ref. 25. Copyright© 1974 American Chemical Society.]

of **1** and **2** would lie in oppositely signed quadrants—upper left or lower right (+) for **1** and upper right or lower left (–) for **2**—since quadrants take on the same signs as back octants (Figure 4-3). However, the observed $n \rightarrow \pi^*$ CD Cotton effects of **1** and **2** have the same (–) sign. The (–) CD sign of **2** is in agreement with both a quadrant rule and the octant rule, because the methyl perturber lies in a (–) back octant. In contrast, the (–) CD sign of **1** is in agreement only with an octant rule and *not with* a quadrant rule, since the methyl perturber of **1** lies in an upper left or lower right (–) *front* octant. Additional support for front octants comes from the (+) $n \rightarrow \pi^*$ CD[26] of a spiroketone(**3**) (Figure 4-8) with a different carbon framework. Again, the lone dissymmetric methyl perturber lies in front of the carbonyl oxygen in an upper right or lower left (+) front octant. Ketones **1** and **3** thus provided the previously missing unequivocal experimental proof for the existence of front octants. Even in analyses that consider puckering of the cyclobutanone ring and attendant changes in octant locations of the perturbers,[27] the $n \rightarrow \pi^*$ Cotton effect is dominated by a methyl perturber in a *front octant*.

4.5. The Location and Shape of the Third Nodal Surface: "Antioctant" Effects

When formulating the octant rule, Moffitt et al.[2] recognized several potential problems in applying it. These include, among other things, the (then) uncertain assignment of hydrogen atom contributions and the effects of unforeseen distortions from the "idealized" geometries employed in applying the rule. But perhaps the most widely discussed reservation that was originally expressed concerned the existence and shape of the "third surface." (A quadrant rule is the minimum sector rule for the carbonyl chromophore,[21,22] with the quadrants being defined by the intersecting local symmetry planes. Further subdivision into octants depends on the nature and nodal properties of the wave functions associated with the $n \rightarrow \pi^*$ transition.) Although the existence of front octants had been established by 1974,[25,26] the shape of the third nodal surface remained ill defined. Moffitt et al.[2] approximated it as a plane (*A* in Figure 4-3) bisecting the C=O bond purely for convenience rather than on any theoretical ground, while specifically cautioning that it was very probably *not* a plane.

The failure to locate the third nodal surface led to situations in which ordinarily well-behaved octant perturbers did not behave normally in certain octant locations, giving rise to what were called "antioctant" effects. Shortly after, the octant rule was postulated,[1] in 1966, Pao and Santry[28] used a Gaussian orbital calculation to derive the octant rule for various methyl-substituted chair cyclohexanones. Their results agreed with the predictions of Moscowitz's original theoretical derivation of the octant rule[29] for *all* methyl configurations except 3-axial,[30] as did later calculations using an extended Hückel treatment.[31] Intimations that the third nodal surface might be curved came first from the calculations of Pao and Santry, who suggested that it envelops the unsubstituted molecule and that substituents may penetrate it. At nearly the same time, Snatzke and coworkers[14,15,32,33] published the first experimental verification that the 3-axial position of chair cyclohexanone did not follow the octant rule. In particular, 1(*R*),3(*S*)-4(*R*)(a)-methyladamantan-2-one (mirror image of that shown in Table 4-1) gave a weak, *positive* CD Cotton effect in ethanol or dioxane solvent.[14] The observed Cotton effect sign was thus opposite to that predicted by the octant rule: a *negative* Cotton effect for the methyl perturber in a lower left or upper right back octant. The significance of this surprising observation was clouded somewhat by the fact that a weak negative Cotton effect was observed for the same ketone in isooctane. However, adamantanones with other β-axial perturbers (e.g., Cl, Br, I, and N_3)[14,15] and SCN, ONO_2, OAc, and OCO_2CH_3[32] exhibited "antioctant" CD Cotton effects that did not change sign. Other apparent "antioctant" effects were claimed[34,35,36] shortly thereafter. Perhaps the most revealing of these came out of the work of Coulombeau and Rassat,[34] who analyzed CD and ORD data for a number of ketones and made the interesting proposal that the third nodal surface was convex, curving sharply *away* from the carbonyl oxygen. Thus, alleged "antioctant" behavior could be explained by the substituent actually lying *in front of* the surface, as newly defined. Tocanne[37] drew a similar conclusion from his work on cyclopropylketones.

Despite these few interesting observations and calculations of "antioctant" effects, there was no unambiguous experimental demonstration from carbon and hydrogen as static dissymmetric perturbers, except for the work of Snatzke with β-axial methyladamantanone[14,15] in which a peculiar solvent effect had been noted (*vide supra*). This changed in 1974, when it was shown that the stereochemically rigid and well-defined 1(*R*)-*endo*-2-methyl-7-norbornanone (**4**) and 1(*R*)-*exo*-2-methyl-7-norbornanone (**5**) gave (+) and (–) $n \rightarrow \pi^*$ CD Cotton effects, respectively, in isopentane (Figure 4-9).[38] Like adamantanone, 7-norbornanone is achiral, and thus, the methyl perturbers in **4** and **5** are the lone dissymmetric octant perturbers. In both compounds, they lie *behind* the third nodal surface (*A* in Figure 4-4) and in a (+) upper left or lower right octant. The interpretation given for the "antioctant" effect in **5** was that it should be interpreted rather as a *front-octant* effect, with the methyl perturber of **5** lying in a (–) front octant. Consistent with earlier work,[28,34] it was further suggested that the apparent "antioctant" effects of other alkyl and possibly other perturbers[14,15,34–36,39] were more likely due to the perturbers lying in front octants, all of which required a nonplanar third nodal surface.[38]

In 1976, following a detailed theoretical analysis and CNDO/S calculations of the known "antioctant" compounds and a long series of decalones, Bouman and Lightner[16] revised the octant rule to define the third, non-symmetry-derived, nodal surface as concave from the perspective of the viewers looking down the C=O bond, from O to C (Figure 4-10). The third nodal surface bends outward in the +*Z* direction. At first glance, this would appear to contradict the empirical results. However, surface *B* cuts just behind the 3-axial position, thus locating the erstwhile "antioctant" β-axial methyl group of adamantanone and 2-*exo*-methyl group of 7-norbornanone in front octants. In support of this notion, it was found that when the methyl perturber is replaced by larger alkyl groups that project even farther into

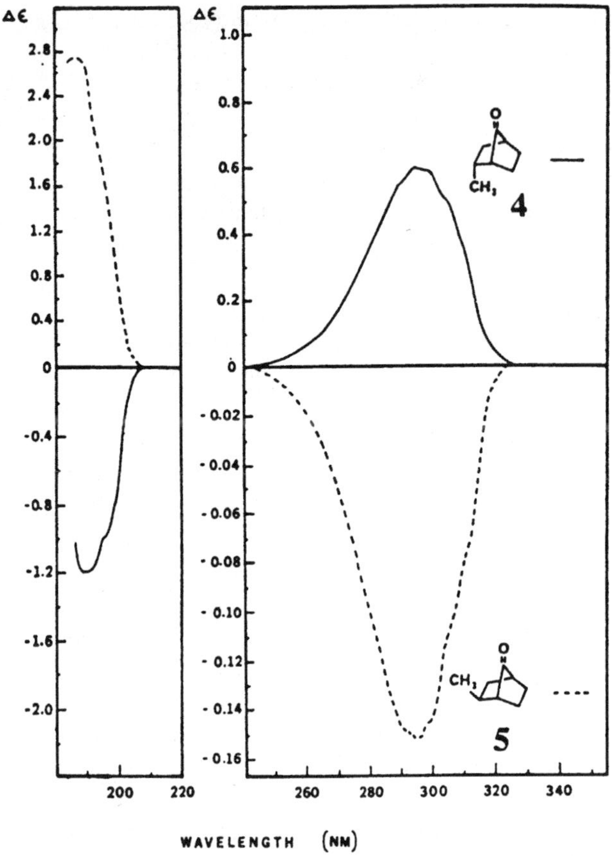

Figure 4-9. (Right) $n \rightarrow \pi^*$ and (Left) short-wavelength circular dichroism spectra of (1*R*)-*endo*-2-methylbicyclo[2.2.1]heptan-7-one (**4**) (——) and (1*R*)-*exo*-2-methylbicyclo[2.2.1]heptan-7-one (**5**) (– – –) in isopentane at 20°C. Corrections are made to 100% optical purity. [Reprinted with permission from ref. 38. Copyright© 1974 American Chemical Society.]

front octants, as in 1(*S*),3(*R*)-4(a)-ethyl or *tert*-butyladamantan-2-one (see Table 4-1 for the absolute configuration), the relatively weak Cotton effect due to a front-octant methyl is magnified (Figures 4-11 and 4-12). Since the adamantanone perturbers lie in an upper left (or lower right) front octant, they will be expected to make a (–) contribution to the $n \rightarrow \pi^*$ CD, as observed. However, a methyl perturber, as noted previously,[14,15] is usually a negative perturber (Figure 4-11), but occasionally is a positive perturber (Figure 4-12).[12,14,15] In contrast, the $n \rightarrow \pi^*$ Cotton effects are strongly *negative* for methyl and other alkyl perturbers at low temperatures, as may be seen in Table 4-2. The pronounced temperature and solvent effects in the CD of β-axial-methyladamantanone, whose methyl perturber lies close to the third nodal surface, have been ascribed to the importance of restricted rotation and

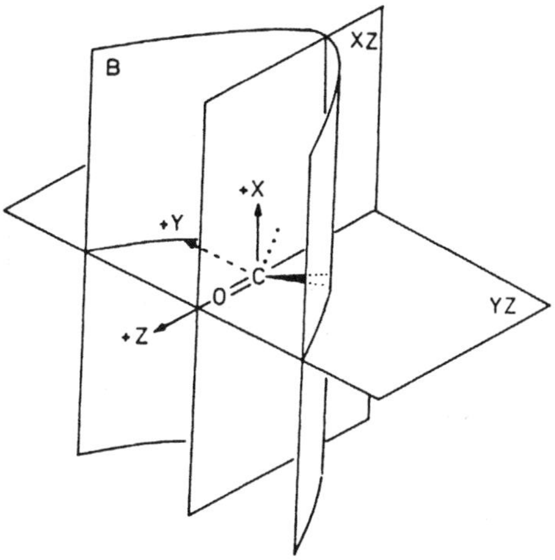

Figure 4-10. Revised octant rule (ref. 16) with symmetry-determined octant planes *XZ* and *YZ*. (See Figure 4-3.) A third nodal surface (*B*), defined theoretically as concave (viewed from O toward C) cuts behind C and intersects the *YZ* plane at approximately $Y = Z + 2$. [Reprinted with permission from ref. 12. Copyright© 1986 American Chemical Society.]

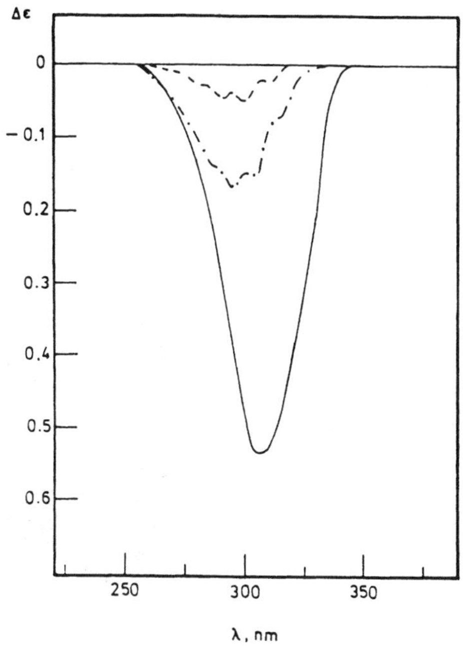

Figure 4-11. Circular dichroism spectra of 10^{-3} *M* 1(*S*),3(*R*)-4(a)-substituted adamantan-2-ones in ether–isopentane–ethanol (EPA; 5:5:2, by vol.) run at 25°C and corrected to 100% e.e. The axial substituents are methyl (- - -), $\Delta\varepsilon_{306}^{max} = -0.046$; ethyl (approx. - · · -), $\Delta\varepsilon_{296}^{max} = -0.15$; isopropyl (- · · - · -), $\Delta\varepsilon_{297}^{max} = +0.17$; *tert*-butyl (—), $\Delta\varepsilon_{296}^{max} = -0.54$; and neohexyl (- · · - - · -), $\Delta\varepsilon_{296}^{max} = -0.17$. [Reprinted with permission from ref. 12. Copyright© 1986 American Chemical Society.]

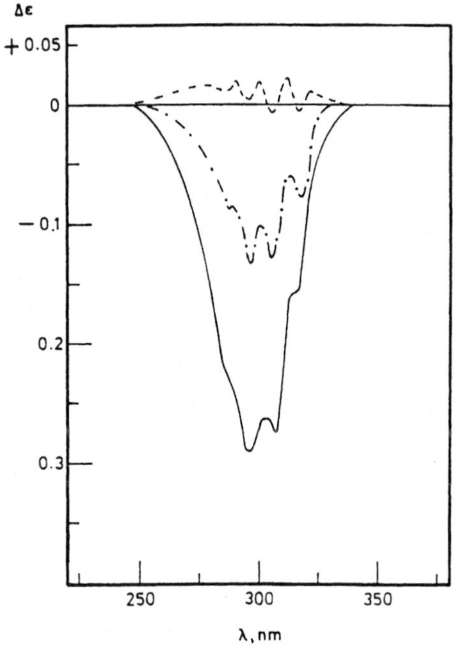

Figure 4-12. Circular dichroism spectra of 10^{-3} *M* 1(*S*),3(*R*)-4(a)-substituted adamantan-2-ones in methylcyclohexane–isopentane (MI; 4:1, by vol.) run at 25°C and corrected to 100% e.e. The axial substituents are methyl (- - -), $\Delta\varepsilon_{313}^{max} = +0.025$; ethyl (approx. - · · -), $\Delta\varepsilon_{301}^{max} = -0.10$; isopropyl (- · · -), $\Delta\varepsilon_{297}^{max} = +0.15$; *tert*-butyl (—), $\Delta\varepsilon_{296}^{max} = -0.29$; and neohexyl (approx. - · · -), $\Delta\varepsilon_{297}^{max} = -0.10$. [Reprinted with permission from ref. 12. Copyright© 1986 American Chemical Society.]

Table 4-2. Reduced Rotatory Strengths[a] of (1*S*,3*R*)-4(a)-Alkyladamantan-2-ones

β-Axial Substituent	Solvent	$[R]^{25}$	$[R]^{0}$	$[R]^{-100}$	$[R]^{-175}$
CH_3	EPA[b]	−0.100	−0.120	−0.492	−0.916
	MI[c]	+0.0784	−0.095	−0.436	−0.758
CH_2CH_3	EPA	−0.362	−0.375	−0.652	−0.880
	MI	−0.200	−0.252	−0.404	−0.950
$CH(CH_3)_2$	EPA	−0.453	−0.462	−0.637	−0.969
	MI	−0.320	−0.354	−0.590	−0.901
$C(CH_3)_3$	EPA	−1.670	−1.709	−1.752	−1.703
	MI	−0.978	−0.980	−0.905	−0.907
$CH_2CH_2C(CH_3)_3$	EPA	−0.580	−0.571	−0.768	−1.072
	MI	−0.300	−0.302	−0.598	−1.248

[a] Data are from ref. 12. [*R*] is rotatory strength (cgs units), × 1.08 × 10^{40}. Values are corrected to 100% e.e. and for contraction of the solvent. The superscript numbers are temperatures,±2, in °C.

[b] EPA = diethyl ether–isopentane–ethanol, 5:5:2 (v/v/v).

[c] MI = methylcyclohexane–isopentane, 4:1 (v/v).

solventperturbationoftheC=Ochromophoreorvibroniceffects.[12] Larger perturbers do not show such a pronounced sensitivity to temperature, but the trend is typically toward more intense Cotton effects at −175°C than at 25°C. In fact, the intensities at −175°C are essentially the same for all of the alkyl substituents measured. This surprising result suggests that the substituent carbon attached to the adamantanone framework in effect acts as the major-octant, front-octant contributor—except for the *tert*-butyl in EPA.[40]

4.6. Perturbers Lying in Octant Symmetry Planes: Qualitative Completeness and the Octant Rule

Throughout a long period of evaluation and investigation, the octant rule has consistently stated that atoms with counterparts symmetrically placed across the carbonyl symmetry planes and *atoms lying in symmetry planes* will exert no effect on the CD. Although not rigorously correct, this rule has nonetheless been practically useful. In addition to the undefined shape of the third nodal surface (approximated as a plane), among the problems recognized quite early by Moffitt et al.[2] in applying the octant rule was the effect of unforeseen distortions from the "idealized" geometries typically employed.

Problems with distortion from the "idealized" geometry have been less well recognized or investigated than those related to the location of the third nodal surface. In particular, the octant symmetry planes (*XZ* and *YZ* in Figures 4-4 and 4-10) are derived from the local symmetry (C_{2v}) of the C=O chromophore and are therefore only approximations when the molecular symmetry does not coincide with the local symmetry of the chromophore. In chair cyclohexanone, only the *XZ* octant plane coincides with both a C=O local symmetry plane and the molecular symmetry plane. In contrast, the *YZ* octant plane is no longer a molecular symmetry plane; thus, the C=O local symmetry begins to break down, and the surface *YZ* is only *approximated* by a plane. Similarly, for 3(e)-methylcyclohexanone, which has no planes of symmetry (and therefore no molecular symmetry coincident with the local symmetry of the C=O group), the *XZ* octant surface is, strictly speaking, only approximately a plane. Distortion from the planarity of the C=O local symmetry-derived octant surfaces (*XZ* and *YZ*; see Figures 4-4 and 4-10) attends all chiral molecules, but the octant rule works because most perturbers lie far from the octant "planes." When a perturber lies close to an octant surface, "anomalous" behavior can be anticipated.[41] A good example is (1*S*,5*S*)-dimethyladamantan-2-one (**6**) (Figure 4-13), which has its methyl perturbers lying in octant symmetry planes. This compound is optically active, yet no (zero) Cotton effect is predicted by the octant rule for the $n \rightarrow \pi^*$ transition.[42,43] Nonetheless, the data clearly reveal a weak monosignate (–) Cotton effect in two very different solvents. The CD spectrum of racemic **6** serves as the baseline; hence, any observed CD must be uniquely due to optically active **6**. The Cotton effect remains negative, with an essentially invariant rotatory strength down to −150°C in isopentane–methylcyclohexane and in EPA glasses.

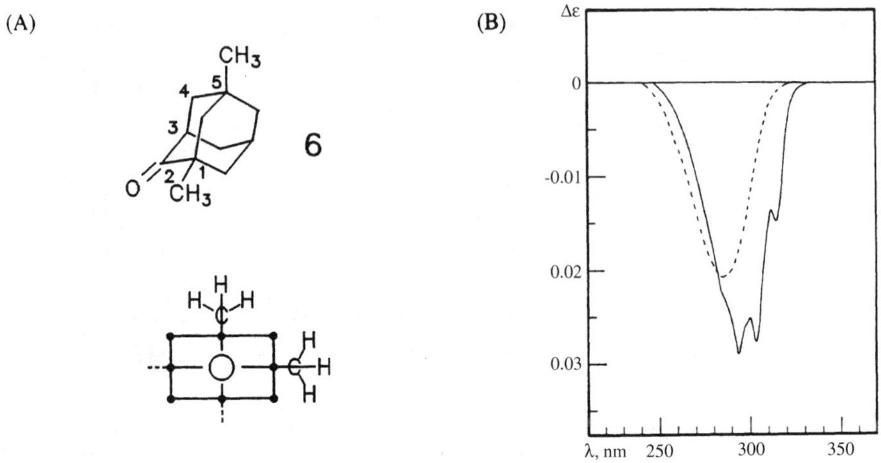

Figure 4-13. (A) 1(S),5(S)-Dimethyladamantan-2-one (**6**) and its octant projection diagram. (B) Circular dichroism spectra of **6** in cyclopentane (—) and 2,2,2-trifluoroethanol (- - -) at 18°C, with the $\Delta\varepsilon = 0$ baseline being the CD of racemic ketone. Sample concentrations are 0.03–0.04 M (1-cm cell path length) for the region 220–370 nm. CDs corrected to 100% e.e. [(B) is reprinted from ref. 42, Copyright© 1987, with permission from Elsevier Science.]

This apparent discrepancy with the predictions of the octant rule was anticipated 10 years previously in a theoretical paper of Yeh and Richardson,[44] who provided an analysis and insight into how **6** and related adamantanones (e.g., 4(e),8(e),9(e)-trimethyladamantan-2-one) might be reconciled with the octant rule. They pointed out that the one-electron perturbation model of $n \rightarrow \pi^*$ optical activity in chiral ketones, which provides the simplest and most direct rationalization for the octant rule, may be insufficient to account for optical activity when carried out just to first order. In first-order perturbation, only (*additive*) pairwise interactions between the C=O group and dissymmetric perturbers are considered, and in this sense, the octant rule lacks qualitative completeness. When the one-electron model is carried to higher order perturbation, *multiplicative* terms contribute to the $n \rightarrow \pi^*$ rotatory strength. Thus, ketones such as **6**, which, in first-order one-electron perturbation theory, are predicted by the octant rule to have zero optical activity, are in fact predicted to be optically active in second-order perturbation, which accounts for three-way interactions among the two CH_3 perturbers and the C=O chromophore. Qualitatively, this means that each CH_3 group of **6** destroys a plane of symmetry in the C_{2v} C=O chromophore and, hence, an octant symmetry plane. Consequently, except for molecules with C_{2v} symmetry, such as adamantanone, the octant or quadrant symmetry planes are only approximate planes. The deviation from planarity will depend on the location and nature of the perturbation. While such deviations are not important for most qualitative applications of the octant rule, they can in fact be detected and analyzed in molecules like **6**.

4.7. Octant Rule Theory and Analysis

Although there have been many theoretical analyses of ketone $n \rightarrow \pi^*$ circular dichroism,[1,4,6,16,21–24,28,31] the most comprehensive recent analysis is that of Bouman and Hansen,[12] using the random-phase approximation (RPA).[45,46,47] The intensity (rotatory strengths) of the CD transitions are computed using the RPA, which includes, in part, electron correlation effects on excitation properties and is designed to give a linear response to properties such as excitation energies and transition moments. The results of the RPA computations are analyzed in a localized orbital scheme that relates contributions to the $n \rightarrow \pi^*$ CD to molecular structure features. Such an approach allowed, for the first time, a detailed analysis of the mechanisms responsible for the CD intensity of ketones important to our understanding of the octant rule.[12,46,48]

The $n \rightarrow \pi^*$ CD Cotton effect may be defined in terms of the sign and magnitude of its rotatory strength. The rotatory strength R_{0q} for a transition from the ground state $|0\rangle$ to an excited state $|q\rangle$ is given by the dipole velocity expression (in atomic units)[49]

$$R_{0q}^{\nabla} = (2c\Delta E_{0q})^{-1} \langle 0|\hat{\nabla}|q\rangle \cdot \langle 0|\boldsymbol{r} \times \hat{\nabla}|q\rangle, \tag{4-1}$$

where ΔE_{0q} is the excitation energy and $\langle 0|\hat{\nabla}|q\rangle$ and $\langle 0|\mathbf{r}\times \hat{\nabla}|q\rangle$ are, respectively, the electric and magnetic dipole transition moment vectors. Other expressions are available, but lend themselves less readily to the analysis that follows.

The transition moments and excitation energies in Eq. (4-1) are computed in the RPA,[47,50] which is discussed extensively elsewhere.[45,46,49] Here, only the relations needed to define the notation are presented. In a basis of occupied (ϕ_a, ϕ_b, . . .) and virtual Hartree–Fock molecular orbitals (ϕ_m, ϕ_n, . . .), the RPA prescription for a transition moment of a Hermitian one-electron operator (such as **r**) may be written

$$\langle 0|\mathbf{r}|\, q\rangle = 2^{1/2} \sum_a \sum_m \langle \phi_a|\mathbf{r}|\, \phi_m\rangle\, S_{am,q}, \tag{4-2}$$

while the relevant anti-Hermitian one-electron operators yield

$$\langle 0|\hat{\nabla}|q\rangle = 2^{1/2} \sum_a \sum_m \langle \phi_a|\hat{\nabla}|\phi_m\rangle\, T_{am,q} \tag{4-3}$$

and

$$\langle 0|\mathbf{r}\times \hat{\nabla}|q\rangle = 2^{1/2} \sum_a \sum_m \langle \phi_a|\mathbf{r}\times \hat{\nabla}|\phi_m\rangle\, T_{am,q}. \tag{4-4}$$

The coefficients $S_{am,q}$ and $T_{am,q}$, together with the excitation energies ΔE_{0q}, are determined from the coupled sets of linear RPA equations, which are easily solved by standard methods.[49] The (generally small) differences between the S and T coefficients turn out to be crucial for an adequate description of oscillator and rotatory strengths.

Assuming now that the occupied orbitals ϕ_a, ϕ_b, ··· have been localized, an excitation-characteristic "bond" (bond or lone pair) electric dipole transition moment from Eq. (4-3) may be defined by summing over all the virtual orbitals:

$$\nabla_{a,q} = \sqrt{2} \sum_m \langle \phi_a | \hat{\nabla} | \phi_m \rangle \, T_{am,q} \tag{4-5}$$

A similar definition for the magnetic dipole transition moment from Eq. (4-4) is

$$\mathrm{I}_{a,q} = \sqrt{2} \sum_m \langle \phi_a | \mathbf{r} \times \hat{\nabla} | \phi_m \rangle \, T_{am,q}. \tag{4-6}$$

Expanding $\mathbf{I}_{a,q}$ relative to p_a, the centroid position of orbital ϕ_a— that is,

$$\mathrm{I}_{a,q} = \sqrt{2} \sum_m \langle \phi_a | \mathbf{r} - \boldsymbol{\rho}_a) \times \hat{\nabla} | \phi_m \rangle \, T_{am,q} + \boldsymbol{\rho}_a \times \nabla_{a,q} \tag{4-7}$$

$$\equiv \mathrm{I}'_{a,q} + \boldsymbol{\rho}_a \times \nabla_{a,q}, \tag{4-8}$$

we obtain $\mathbf{I}'_{a,q}$ as an inherent bond magnetic dipole transition moment, while $\boldsymbol{\rho}_a \times \nabla_{a,q}$ is a "moment of momentum" term. Substituting Eqs. (4-2) through (4-7) into Eq. (4-1) yields

$$R_{0q} = \sum_a \sum_b R^q ab = \sum_a \sum_b \{R^q(\mu_a, m_b) + R^q(\mu_b, m_a) + R^q(\mu_a, \mu_b)\}, \tag{4-9}$$

where

$$R^q(\mu_a, m_b) = (4c\Delta E_{0q})^{-1} \nabla_{a,q} \cdot \mathbf{I}_{b,q}, \tag{4-10}$$

$$R^q(\mu_b, m_a) = (4c\Delta E_{0q})^{-1} \nabla_{b,q} \cdot \mathbf{I}_{a,q}, \tag{4-11}$$

and

$$R^q(\mu_a, \mu_b) = (4c\Delta E_{0q})^{-1} (\boldsymbol{\rho}_a - \boldsymbol{\rho}_b) \cdot (\nabla_{a,q} \times \nabla_{b,q}). \tag{4-12}$$

A diagonal term ($a = b$) represents the "one-electron," or static, perturbation mechanism, while the off-diagonal terms $R(\mu_a, m_b)$ and $R(\mu_b, m_a)$ in Eqs. (4-10) and (4-11) represent two physically distinct electric dipole–magnetic dipole (μ–m) terms.[21,22] Equation (4-12) corresponds to the μ–μ, or polarizability, term of Kirkwood.[51] Each term is independent of the origin and so can be assigned a physical meaning, in contrast to earlier decompositions in which electric and magnetic dipole contributions could not be separated.[52,53] The analysis applies as long as the transition moments can be written in the form of Eq. (4-3), regardless of how the coefficients $T_{am,q}$ are computed, and no approximations are introduced beyond those entering into that computation. In the present formalism, all references to individual virtual orbitals have disappeared, so that the entire intensity can be discussed in terms of structural features (i.e., "bonds" or "lone pairs" ϕ_a, ϕ_b).

A static perturbation, or "one-electron," mechanism[24] would be reflected in terms $R(\mu_a, m_a)$ involving the orbitals associated with an isolated carbonyl chromophore—that is, the two oxygen lone pairs, the C=O double-bond orbitals, and (perhaps) the remaining two σ bonds from the carbonyl carbon. The μ–*m* mechanism,[21,22] on the other hand, is expected to manifest itself through large $R(\mu_a, m_b)$ terms, where ϕ_b is a lone-pair orbital on oxygen. $R(\mu_a, \mu_b)$ terms would couple pairs of local electric dipole bond transition moments in a chiral manner.

While this analysis will yield a complete mechanistic picture of the couplings responsible for the overall optical activity, the picture may well be too detailed to be useful, especially if a large fraction of the $n_{occ}(n_{occ} + 1)/2$ bond pair contributions are similar in magnitude. A somewhat coarser grained picture can be obtained by condensing all contributions involving a particular orbital ϕ_a into one term,

$$R_{0q} = \sum_a [R^q_{aa} + 1/2 \sum_b (R^q_{ab} + R^q_{ba})] \equiv \sum_a R^q_{a,\mathrm{eff}} \tag{4-13}$$

where *q* is again a reminder that this type of term is specific to a particular excitation. When displayed on a structural diagram, these effective, gross bond contributions exhibit the net effect of a given bond on the total rotatory intensity, *in the presence of the rest of the molecule.*

Computed $n \rightarrow \pi^*$ excitation energies and rotatory strengths for β-equatorial and β-axial methyladamantanone are shown in Table 4-3. The experimental results are reproduced reasonably well by the RPA method using idealized molecular geometries. The computed results may be decomposed into the five major coupling terms of the form of Eq. (4-9). These terms are listed in descending order of importance for β-equatorial and β-axial methyl adamantanone (Table 4-4) together with the corresponding mechanistic type and the values of the terms for each quadrant in the

Table 4-3. Comparison of Computed and Experimental $n \rightarrow \pi^*$ Excitation Energies (λ_{max}) and Reduced Rotatory Strengths. The Octant Projection Diagram Is Shown (inset) with Numbering for Quadrants Determined by C_{2v} Symmetry Planes

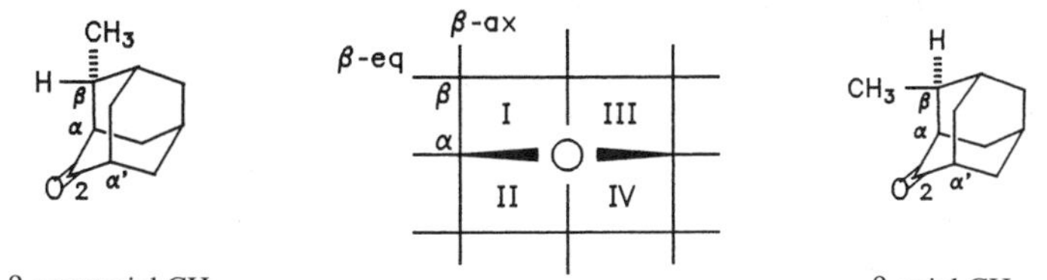

β-equatorial CH_3	β-axial CH_3
Computed: λ_{max}289 nm, $[R^{\nabla}] = +1.24$	Computed: λ_{max}289 nm, $[R^{\nabla}] = -0.38$
Exptl.: λ_{max}295 nm, $[R]^{-175°C} = +1.96$	Exptl.: λ_{max}306 nm, $[R]^{-175°C} = -0.84$

Table 4-4. Major Coupling Types Computed for Generating $n \rightarrow \pi^*$ Optical Activity (C_2 Is the Carbonyl Carbon, and O (π_y) is the In-plane π-type Oxygen Lone-pair Orbital) and Values of Computed Coupling Terms ([R] units) in each Quadrant for β-Equatorial and β-Axial Methyl Adamantanones in the Idealized Geometry.[a]

Adaman-tanone	Type	Bond a	Bond b	Mechanism	Quadrant[b] I	II	III	IV
β-eq.CH_3	1	$C_\alpha–C_\beta$	O (π_y)	$\mu_a–m_b$[c]	5.65	–5.45	–5.50	5.42
	2	$C_\beta–H_{\beta\text{-eq.}}$	O (π_y)	$\mu_a–m_b$	3.28	–3.02	–2.96	2.97
	3	$C_\alpha–C_\beta$	$C_2–C_{\alpha'}$	~60% $\mu_a–m_b$ + ~40% $\mu_a–\mu_b$	3.19	–3.06	–3.11	3.06
	4	$C_\beta–H_{\beta\text{-eq.}}$	$C_2–C_{\alpha'}$	~55% $\mu_a–m_b$ + ~45% $\mu_a–\mu_b$	1.97	–1.83	1.79	1.79
	5	$C_\alpha–C_\beta$	$C_2–C_\alpha$	$\mu_a–m_b$	0.77	–0.73	–0.75	0.72
β-ax.CH_3	1	$C_\alpha–C_\beta$	O (π_y)	$\mu_a–m_b$[2]	5.56	–5.39	–5.44	5.40
	2	$C_\beta–H_{\beta\text{-eq.}}$	O (π_y)	$\mu_a–m_b$	2.64	–3.01	–2.94	3.00
	3	$C_\alpha–C_\beta$	$C_2–C_{\alpha'}$	~60% $\mu_a–m_b$ + ~40% $\mu_a–\mu_b$	–3.11	–3.02	–3.03	3.03
	4	$C_\beta–H_{\beta\text{-eq.}}$	$C_2–C_{\alpha'}$	~55% $\mu_a–m_b$ + ~45% $\mu_a–\mu_b$	1.61	–1.83	–1.76	1.81
	5	$C_\alpha–C_\beta$	$C_2–C_\alpha$	$\mu_a–m_b$	0.75	–0.71	–0.72	0.72

[a] From ref. 12. [b]See Table 4-3. [c]Greater than 90% a single type.

molecule. Electric dipole couplings with the magnetic dipole moment of the oxygen "π_y" lone-pair type of component are numerically the largest.

Overall, in terms of the mechanistic contributions in Eq. (4-9), about 79% of the total rotatory intensity comes from off-diagonal μ–m terms [Eqs. (4-10) and (4-11)], while 15% comes from μ–μ couplings [Eq. (4-12)]. The diagonal μ–m, or "one-

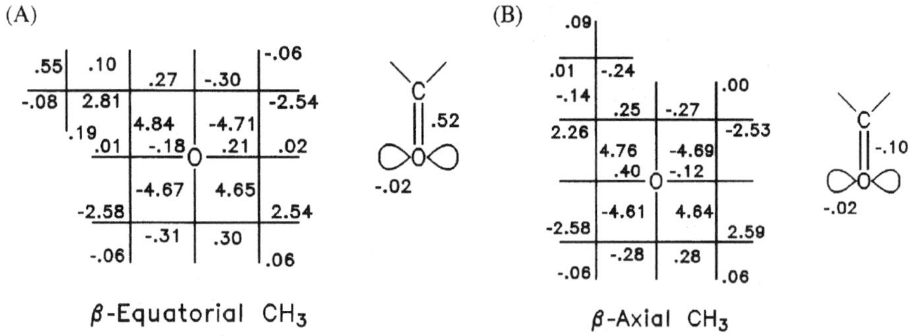

Figure 4-14. Computed effective bond contributions to [R^∇] [Eq. (4-13)] displayed on octant projections, with carbonyl group contributions shown at right for (A) 4(e)-methyladamantan-2-one and (B) 4(a)-methyladamantan-2-one of Table 4-3. Data from ref. 12.

electron" terms [Eq. (4-10) and (4-11)] contribute very little—just 6%—although their signs are the same as those of the overall rotatory strength in the methyladamantanones.

Figure 4-14 shows a condensation of the foregoing fine-grained analysis according to Eq. (4-13), where all contributions involving a given orbital ϕ_a have been collapsed to a single term. Notice that all the larger bond contributions are consignate with the octant rule prescription for the respective octants. Since the effective bond contributions sum to the net rotatory strength of the molecule, these numbers can be compared, at least approximately, with the increment $\delta\Delta\varepsilon$ values derived by Kirk.[10,39] Note, however, that both the R_a^{eff} terms [Eq. (4-13)] and the $\delta\Delta\varepsilon$ values are only approximately independent of their surroundings, and a given term (e.g., C_α–C_β) may be perturbed significantly by other structural features that vary in a class of molecules.

4.8. The Octant Rule and Octant Surfaces

Table 4-4 and Figure 4-14 show not only which couplings or bond contributions are large, but also that there are a number of nontrivial symmetry-breaking (chiral) contributions from pairs of bonds that are related by the erstwhile symmetry of the skeleton. In fact, these symmetry-breaking skeleton contributions are at least as important as direct contributions from the perturber, in marked contrast to the assumptions of the original octant rule.

For the surfaces, we note that the reflection of, say, Figure 4-14B in the vertical symmetry plane of the skeleton would give the mirror image of the numerical value of all bond contributions and change all signs at the same time. This, therefore, represents a quadrant rule effect, albeit one that includes skeleton contributions. The experimental Cotton effects are consistent with the picture (Figure 4-10) of a concave third nodal surface that cuts behind the carbonyl carbon and extends outward toward oxygen,[16] but, as discussed previously, the relation of the various coupling terms to this picture is not readily apparent. Some couplings involving the β-substituent do change sign between axial and equatorial forms, but other changes within the skeleton are also significant. It turns out that these changes in sign are *not* due to changed couplings on the axial methyl, but rather to small changes throughout the skeleton.

Figure 4-15 shows a plot of the transition density of the $n \rightarrow \pi^*$ excitation of the β-axial methyladamantanone of Table 4-3, in a plane one atomic unit above the plane of the C=O group. The dissymmetry of the transition density is not apparent on this scale, a fact that is also reflected in the computed oscillator strength of less than 10^{-4}. Since the transition density determines the electric dipole transition moment, an overall quadrant rule distribution is apparent from the density. The nodes of the n and π^* orbitals do not yield a single third nodal surface, but rather introduce only narrow regions of changes in sign. The third surface thus remains as an *a posteriori* construction derived from the overall rotatory strengths, and no simple rationale for its existence is apparent at this time.

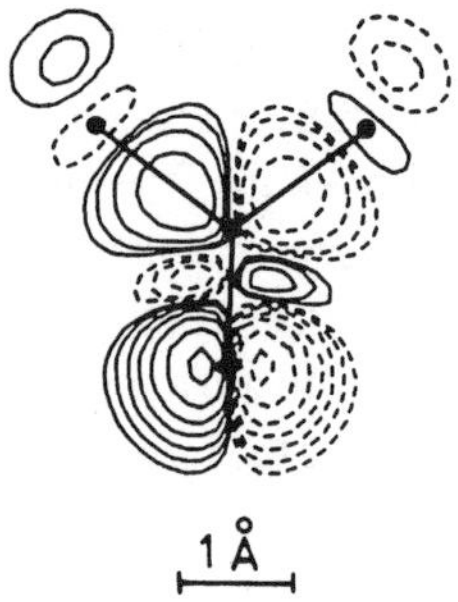

Figure 4-15. Transition density[51] for $n \rightarrow \pi^*$ excitation of the β-axial methyladamantanone of Table 4-3, 1.0 atomic unit (0.53 Å) above ($x > 0$) the *YZ*-plane. Successive contours differ by a factor of two. The carbonyl group skeleton is superimposed for clarity. Solid and dashed contours indicate regions of (arbitrary) positive and negative phase, respectively. [Reprinted with permission from ref. 12. Copyright© 1986 American Chemical Society.]

4.9. Mechanism of Optical Activity and CD from Octant Consignate and Dissignate *W*-Coupling Paths

Kirk and Klyne[54] defined as octant *consignate* those groups or locations wherein the octant rule is obeyed and octant *dissignate* wherein it is not obeyed ("antioctant"). They also proposed a somewhat different model for observed CD signs and magnitudes, based on an extensive analysis of decalones and their extended analogs.[8,39] Kirk and Klyne adopted the view of Hudec and coworkers[55,56] (cf. also Howell)[53] that interactions within the hydrocarbon chains outside the chromophore, rather than direct perturbational action on the carbonyl group itself, dominate the contributions to the induced rotatory power. Further, they assert that the "through-bond" interactions are quite sensitive to chain conformations, reaching appreciable values only when a planar *zigzag* (*W*-shaped) path can be traced along the bonds from the carbonyl group to the dissymmetrically placed substituent. Within this framework, Kirk and Klyne were able to integrate the data for *cis*- and *trans*-decalones, as well as Snatzke's[14,15] "antioctant" compounds, into a single empirical scheme. The shape of a third surface in this analysis, however, is more difficult to assess.[10]

Perturbers lying on a primary zigzag are said to make octant consignate contributions to the ketone $n \rightarrow \pi^*$ CD.[10,39] Thus, they obey the octant rule. Perturbers not lying on a primary zigzag make octant dissignate contributions; they give rise to "antioctant" or front-octant effects. The definitions can be seen in Figure 4-16, in which the primary zigzag paths (*Z* or *W* paths) are illustrated. This analysis places considerable importance on bond couplings in the framework and a lesser importance on octant "perturbers" and therefore offers a complementary perspective. An instructive, comprehensive analysis of polycyclic ketones is provided by Kirk.[10]

Analysis shows that dynamic μ–*m* coupling contributions dominate the resulting ketone $n \rightarrow \pi^*$ optical activity and that the most important of these involve couplings between the localized nonbonding oxygen orbital and a zigzag pattern of bonds

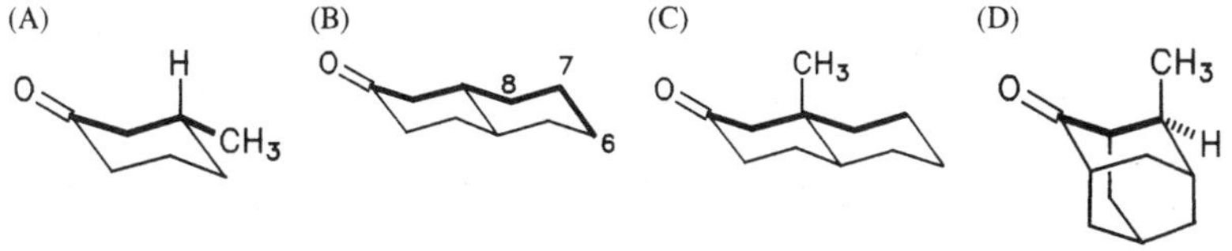

Figure 4-16. Examples of zigzag carbon-coupling paths (heavy lines) leading to octant consignate perturbers (CH_3 of (A) and C(6), C(7), and C(8) of (B). Octant dissignate perturbers (the angular CH_3 of (C)) and axial CH_3 of (D) do not lie on a primary zigzag path.

extending away from the carbonyl group.[52] Static perturbation (one-electron) contributions play a very small role, while the overall polarizability contributions (μ–μ couplings) are small, but not quite negligible. In fact, individual μ–μ terms can be large, but occur in nearly canceling sets, showing that the chiral perturbations of the skeleton contributions are much larger for the μ–m terms than for the μ–μ terms. From the analysis described here, the current picture of the mechanism of ketone optical activity is primarily one of μ–m couplings controlled by *W*-shaped paths extending from the carbonyl group, but with net bond contributions that obey the octant rule for perturbing groups. Unfortunately, this detailed analysis does not seem to yield a picture of a third nodal surface in any simple way.

4.10. Ketone Carbonyl Short-Wavelength Transitions

The ketone carbonyl $n \rightarrow \pi^*$ transition is an excellent probe of the extrachromophoric environment. There are no close-lying transitions that might mix with the $n \rightarrow \pi^*$ and thereby complicate correlations of $n \rightarrow \pi^*$ CD or ORD data with molecular structure. The octant rule follows largely and simply from the symmetry of the n and π^* orbitals, but this simplicity would be lost for mixed transitions, and extracting structural information would become much more difficult, in principle. Such is the case with higher energy ketone carbonyl transitions for which strong CD Cotton effects can be detected near 190 nm, a region rich in electronic transitions, where mixing is likely to occur.

In 1972, Kirk, Klyne, and Mose[57] reported CD Cotton effects at 185–195 nm for 27 steroid ketones (Table 4-5) in *n*-hexane solution and proposed a simple octant-like rule (Figure 4-17) to rationalize their observations. In their octant-like rule for the ketone 190-nm transition, α-axial or β-axial methyl or methylene substituents on a cyclohexanone ring make large octant-sense contributions to the Cotton effect, whereas α- or β-equatorial groups make relatively small contributions. Front-octant effects were not considered.

Subsequently, Kirk[58] refined his earlier data and expanded the scope of study of the approximately 190-nm transition to include 152 ketones (mainly steroids), with $\Delta\varepsilon$ values determined in *n*-hexane (Table 4-6). A detailed octant-like rule for six-membered ring ketones was postulated (Figure 4-18). For various reasons,[58] the approxi-

Table 4-5. Circular Dichroism Cotton Effects in the 185–195-nm Region of Selected Steroid Ketones[a]

Steroid Ketone	Δε	Steroid Ketone	Δε
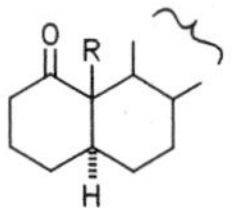 R=CH_3 R=H	+3.7 ~0	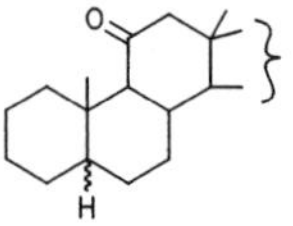 5α or 5β	+4.5 to +5.6
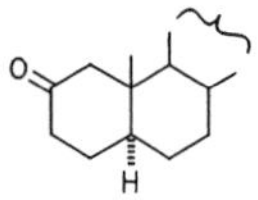	+3.7 to +5.2	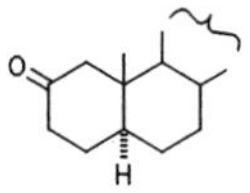	−8.7
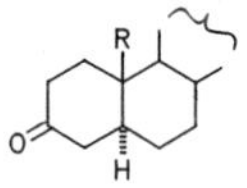 R=CH_3 R=H	~0 to +0.5 ~0 to +0.5	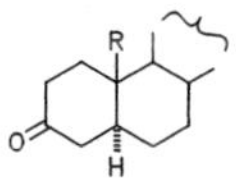 R=CH_3 5β-CH_3 R=H	+0.9 −4.7 ~0
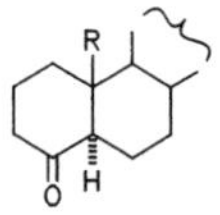	−4.8 to −5.6	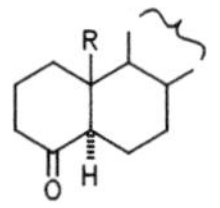	+15.9
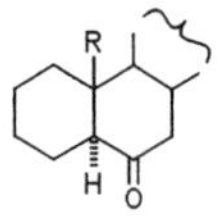 R=CH_3 R=H	+4.4 to +5.5 −3.7	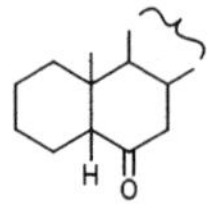	+5.3
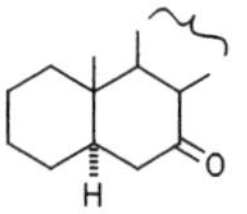	0 to −2.6	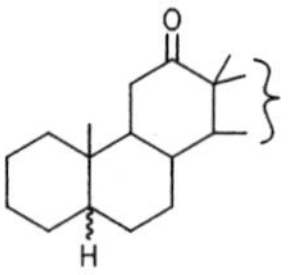 5α or 5β	+4.3 to +10.5

[a] Ref. 57.

(A) (B)

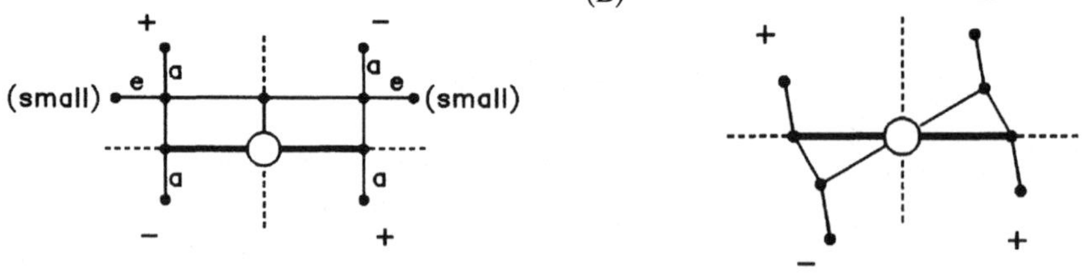

Figure 4-17. Octant-type projection diagrams for the 190-nm transition of alkyl ketones. Depicted are methyl and methylene substituents in back octants of (A) cyclohexanones, showing axial and equatorial groups, and (B) cyclopentanones, showing pseudoaxial groups. In (B), the skewed ring makes a positive contribution to the Cotton effect. Diagrams as in ref. 57.

Table 4-6. Solvent Dependence of the Circular Dichroism Cotton Effect Data ($\Delta\varepsilon$) for the approximately 190-nm Transition of Selected Ketones.[a]

	$\Delta\varepsilon$ (λ^{max}, nm)			$\Delta\varepsilon$ (λ^{max}, nm)	
Ketone	*n*-Hexane	CF_3CH_2OH	Ketone	*n*-Hexane	CF_3CH_2OH
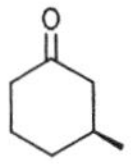	+1.0(185)	−0.5(188)	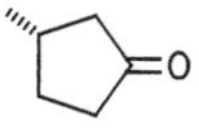	−5.7(190)	−1.2(185)
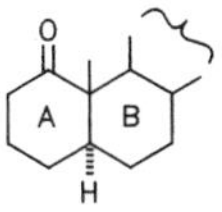	+6.8(189)	+6.0(198)	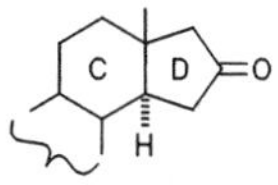	+3.5(193)	+3.6(187)
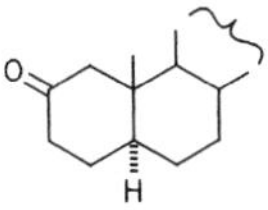	+4.9(188)	−5.0(194)	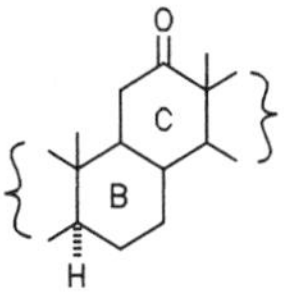	+10.5(188)	+7.6(196)
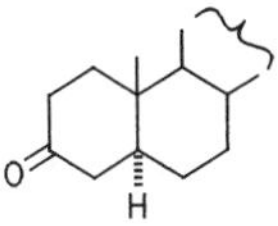	−0.5(192)	−2.4(198)	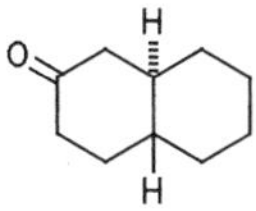	−0.5(190)	+1.0(190)
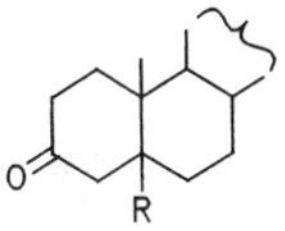 R=H R=CH_3	+0.9(190) −3.3(188)	+0.8(190) +1.0(198)	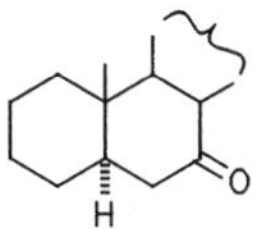	−0.8(185)	0.0
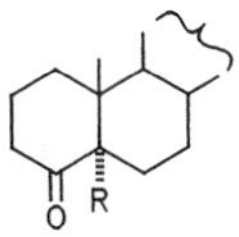 R=H R=CH_3	−5.6(191) 0.0	+3.0(201) +0.8(201)	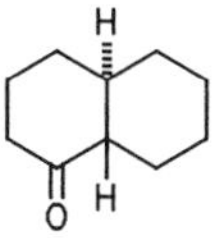	+3.6(188)	−1.9(189)
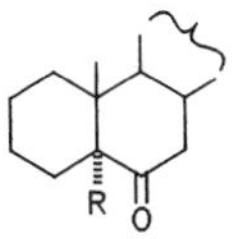 R=H R=CH_3	+5.2(192) 0.0	+3.0(200) +5.5(190)	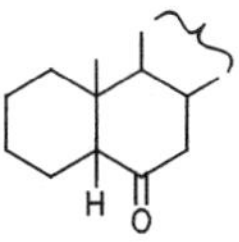	+9.0(188)	+4.0(190)

[a] Selected from refs. 58 and 59.

mately 190-nm transition was then thought to be an $n \rightarrow \sigma^*$ or $n \rightarrow 3s$ Rydberg transition (involving the promotion of a carbonyl n electron into a molecular Rydberg orbital extending over the entire molecule) or a mixture of the two. As in the earlier study,[57] octant consignate contributions of α-axial or β-axial substituents, or any α-axial or β-axial bonds that were part of a second ring, generally dominated the observed Cotton effect. The various contributions were combined into a single

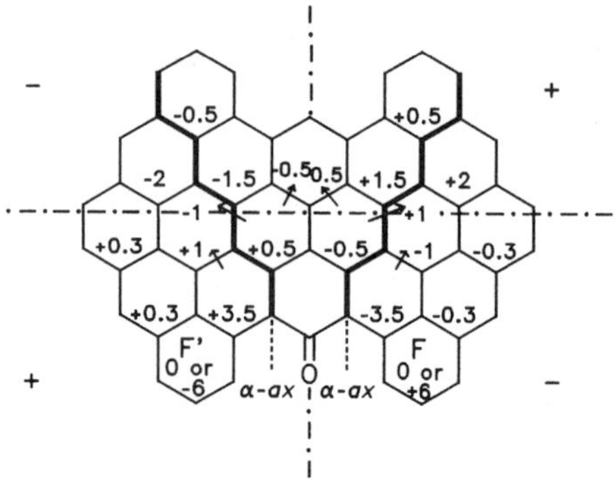

Figure 4-18. 190-nm CD contributions of cyclohexane rings in an "all-*trans*" array of polycyclic systems representing extended *trans*-decalones. (Arrows indicate the manner of connection to adjoining rings where different values apply.) The line (— · — · —) indicates apparent positions of sector boundaries: Rings *F* and *F'* (in front octants) make no contributions when the carbonyl group is in a terminal ring, as in 5α-cholestan-1-one, but add 6 when the carbonyl group lies in a middle ring, as in 5α-cholestan-11-one. Data from ref. 57.

octant-like sector rule (Figure 4-18) that divided space behind the carbonyl group into eight sectors instead of the usual four octants. Figure 4-18 shows the projection of the upper four of these sectors onto the horizontal (*YZ*; see Figures 4-2 and 4-3) symmetry plane of the carbonyl group. The four corresponding lower sectors have signs that were reversed. In Kirk's sector rule, a fourth boundary surface (in addition to the three octant surfaces) cuts through the second row of rings, dividing each back octant into a near region of consignate contributions and a far region of dissignate contributions, as shown in the figure. As with the $n \rightarrow \pi^*$ transition, any rings not lying on one side or the other of a primary zigzag (heavy lines) make the smallest contributions. The significance of the fourth boundary surface was not clear, and Kirk cautioned that it might be simply a chance coincidence of the presence of two overlapping transitions near 190 nm.

This cautionary note proved to be perceptive, as Kirk subsequently reported[59] that CD spectra run in 2,2,2-trifluoroethanol (TFE) in the 185–195-nm region often had different Cotton effect signs. Trifluoroethanol (TFE) is a strongly hydrogen-bonding solvent, as can be seen, for example, in 5α-cholestan-3-one, where, in the ^{13}C-NMR, $\delta_{C=O} = 206.1$ in cyclohexane-d_{12}, but shifts to 214.0 with 5% TFE.[59] Hydrogen bonding causes a blue shift of the ketone $n \rightarrow \pi^*$ transition steroid CD from about 296 nm in hexane to approximately 283 nm in TFE,[59] due partly to stabilization of the *n*-electrons through hydrogen bonding and partly to an inability of the solvent shell to relax around the excited molecule in the short lifetime of the electronic transition. Other transitions involving *n*-electrons (e.g., $n \rightarrow \sigma^*$ and $n \rightarrow 3s$ Rydberg) would be expected to undergo blue shifts. For instance, the 185-nm electronic transition of acetone found in hexane solvent (196 nm in the gas phase) is shifted to 171 nm in TFE.[60] Consequently, in TFE, any $n \rightarrow \sigma^*$ or $n \rightarrow 3s$ Rydberg transitions, either overlapped or admixed with the approximately 190-nm CD transition seen in hexane would be expected to be blue-shifted out of the (then) accessible CD wavelength range to fall at $\lambda < 185$ nm.

Interestingly, in TFE, the approximately 190-nm band is typically reduced in intensity, often with an inverted sign and a slight bathochromic shift, compared with that in *n*-hexane (Table 4-6). Although Kirk[59] could not assign the transition corresponding to the observed CD near 190 nm in TFE (he suggested only that it might belong to a $\sigma \rightarrow \pi^*$ type),[61] he did postulate two new sector rules, one for the CD in TFE and (by difference) one for the erstwhile $n \rightarrow \sigma^*$ or $n \rightarrow 3s$ Rydberg component of the 190-nm CD in hexane (Figure 4-19).

Subsequent to these studies, Gedanken et al.[62] measured the vacuum UV CD of camphor using synchrotron radiation. They assigned positive Cotton effect peaks at 198 and 202 nm to an $n \rightarrow 3s$ Rydberg transition, a second, more intense, positive Cotton effect near 187 nm to an $n \rightarrow 3p$ Rydberg, and a third, even more intense, *negative* Cotton effect near 173 nm to an $n_- \rightarrow \sigma^*$ transition (where n_- is the second lone pair of the carbonyl oxygen). Gedanken et al. concluded that the reason that Kirk's approximately 190-nm Cotton effect may have opposite signs in hexane and TFE is that in the former it is dominated by $n \rightarrow 3s$, whereas in TFE the $n_- \rightarrow \pi^*$ is revealed and the $n \rightarrow 3s$ is buried. RPA calculations[61] indicate that the second excited valence

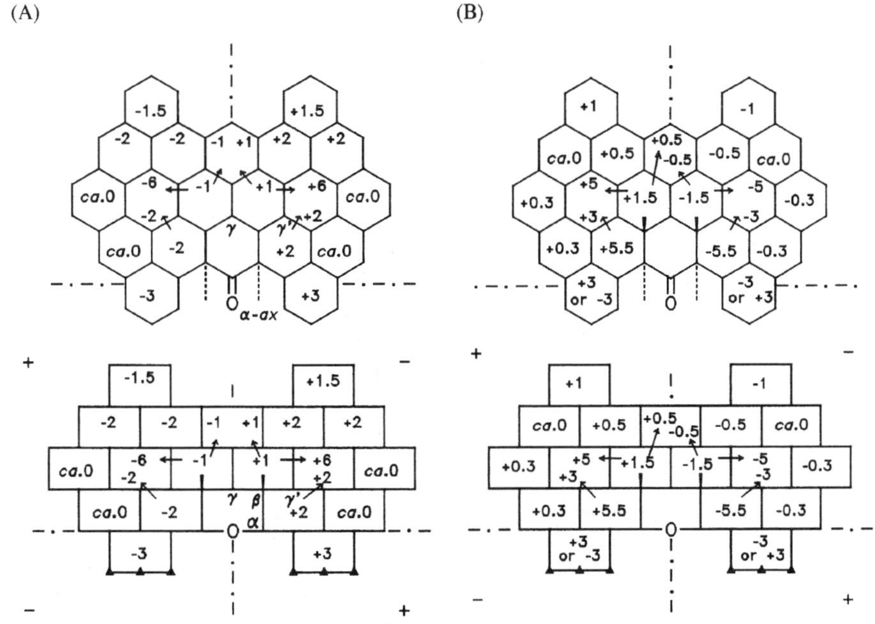

Figure 4-19. (A) Sector rule for the approximately 190-nm CD transition of ketones measured in 2,2,2-trifluoroethanol (TFE) solvent. Back octant CD contributions of cyclohexane rings with classical octant projection. (See Figure 4-18.) ▲ indicates a carbon atom in a front octant. (B) Sector rule for ring contribution differences in *n*-hexane and TFE ($\Delta\varepsilon_{\text{hexane}} - \Delta\varepsilon_{\text{TFE}}$). Increments for front octant rings: First value is applicable when the oxo group is in a terminal ring; second value is applicable when the oxo group is in a middle ring. Corresponding values when the front octant ring contains five members are 0 and 5 units. Data from ref. 59.

state in saturated ketones has a large magnetic moment and a weak electric moment (weaker than $\pi \rightarrow \pi^*$, but stronger than $n \rightarrow \pi^*$) and may be assigned a mixed $\sigma \rightarrow \pi^*$ transition (where σ is composed of n_-, σ_{CO}, and $\sigma_{CC\alpha}$).

More recently, the vacuum UV and CD of camphor, norcamphor, and fenchone were (re)measured and recalculated.[63] The calculations indicated the existence of the well-established $n \rightarrow \pi^*$ transition (near 280 nm) that is isolated from a progression of higher energy absorptions, all except one ($\sigma \rightarrow \pi^*$) attributed to Rydberg transitions: $n \rightarrow 3s$, $n \rightarrow 3p$, $\sigma \rightarrow \pi'^*$, and $n \rightarrow 3d$ excitations (in order of increasing excitation energy).

The sector rules for the approximately 190-nm transition(s) of saturated alkyl ketones lack the simplicity embodied in the octant rule for the $n \rightarrow \pi^*$ transition at approximately 290 nm; thus there is greater difficulty in extracting stereochemical information from them. For example, the $n \rightarrow \pi^*$ CD transition of (1*R*)-*endo*-2-methylbicyclo[2.2.1]heptan-7-one in isopentane is positive, as predicted by the octant rule; its approximately 190-nm CD transition (Figure 4-9) in isopentane is strongly negative,[38] in accordance with Kirk's sector rule:[58] a weakly dissignate contribution for a β-equatorial-like group. In the epimeric (1*R*)-*exo*-2-methyl-bicyclo[2.2.1]heptan-7-one, the methyl perturber lies in a front octant, and in isopentane the $n \rightarrow \pi^*$ CD Cotton effect is weak and negative (Figure 4-9), as predicted by the octant rule (for front octants). As with the *endo*-methyl epimer, the approximately 190-nm CD Cotton effect is oppositely signed and now large and positive ($\Delta\varepsilon \sim +2.8$), in accordance with Kirk's sector rule:[58] a consignate contribution from a β-axial-like methyl group. Unfortunately, short-wavelength CD data for this compound in TFE are not available. As was shown earlier in the chapter for the $n \rightarrow \pi^*$ transition, a systematic experimental–theoretical study of the approximately 190-nm ketone carbonyl CD transitions of model compounds such as 7-norbornanones, adamantanones, and α-methyl ketones should lead to a clearer indication of the assignment of the relevant transitions and a more refined sector rule. At present, however, for CD spectral–structural correlations in ketones, the octant rule for the $n \rightarrow \pi^*$ transition is adequate and easy to apply.

References

1. The octant rule is a generalization of the axial haloketone rule and was presented, along with many examples, in C. Djerassi, *Optical Rotatory Dispersion and Its Applications to Organic Chemistry*, McGraw Hill, New York, 1960, Chapter 13.
2. Moffitt, W., Woodward, W. B., Moscowitz, A., Klyne, W., and Djerassi, C., *J. Am. Chem. Soc.* **83** (1961), 4013–4018.
3. Moffitt, W., and Moscowitz, A., *Abstr. Pap.-Am. Chem. Soc.* **133d** (1958), Abstr. No. 1.
4. Snatzke, G., *Angew. Chem. Int. Ed. Engl.* **18** (1979), 363–377.
5. LeGrand, M., and Rougier, M. J., "Application of Optical Activity to Stereochemical Determinations," in *Stereochemistry. Fundamentals and Methods; Vol 2: Determination of Configurations by Dipole Moments, CD or ORD* (Kagan, H. B., ed.), Georg Thieme, Stuttgart, 1977.
6. Charney, E., *The Molecular Basis of Optical Activity: Optical Rotatory Dispersion and Circular Dichroism*, John Wiley & Sons, Inc., New York, 1979.
7. Moffitt, W., and Moscowitz, A., *J. Chem. Phys.* **30** (1959), 648–660.

8. For leading references, see Deutsche, C. W., Lightner, D. A., Woody, R. W., and Moscowitz, A., *Ann. Rev. Phys. Chem.* **20** (1969), 407–448.
9. The extent to which this condition of separability into chromophoric and nonchromophoric moieties is satisfied depends not only on the nature of the chromophore itself, but also on the nature and disposition of the extrachromophoric parts of the molecule. The problem, therefore, is to be able to recognize a priori when, for all practical purposes, the separability condition is met.
10. Kirk, D. N., *Tetrahedron* **42** (1986), 777–818.
11. For a review of the literature from 1977 to mid-1995, see Boiadjiev, S., and Lightner, D. A., "Chiroptical Properties of Compounds Containing C=O Groups," in *Supplement A3: The Chemistry of Double-Bonded Functional Groups* (Patai, S., ed.), John Wiley & Sons, Ltd., London, 1997, Chap. 5. For earlier work, see Crabbé, P., *ORD and CD in Chemistry and Biochemistry*, Academic Press, New York, 1972.
12. Lightner, D. A., Bouman, T. D., Wijekoon, W. M. D., and Hansen, Aa. E., *J. Am. Chem. Soc.* **108** (1986), 4484–4497.
13. Konopelski, J. P., Sundararaman, P., Barth, G., and Djerassi, C., *J. Am. Chem. Soc.* **102** (1980), 2737–2745.
14. Snatzke, G., and Eckhardt, G., *Tetrahedron* **24** (1968), 4543–4556.
15. Snatzke, G., Ehrig, B., and Klein, H., *Tetrahedron* **25** (1969), 5601–5609, and references therein.
16. Bouman, T. D., and Lightner, D. A., *J. Am. Chem. Soc.* **98** (1976), 3145–3154.
17. Djerassi, C., and Klyne, W., *J. Chem. Soc.* (1963), 2390–2402.
18. Kirk, D. N., Klyne, W., and Mose, W. P., *Tetrahedron Lett.* (1972), 1315–1318.
19. Klyne, W., and Kirk, D. N., in *Fundamental Aspects and Recent Developments in Optical Rotatory Dispersion and Circular Dichroism* (Salvadori, P., and Ciardelli, F., eds.), Heyden & Sons, New York, 1973, Chap. 3, pp. 101–102.
20. Lightner, D. A., and Christiansen, G. D., *Tetrahedron Lett.* (1972), 883–886.
21. Schellman, J. A., *J. Chem. Phys.* **44** (1966), 55–63.
22. Schellman, J. A., *Acc. Chem. Res.* **1** (1968), 144–151.
23. Wagnière, G., *J. Am. Chem. Soc.* **88** (1966), 3937–3940.
24. Bouman, T. D., and Moscowitz, A., *J. Chem. Phys.* **48** (1968), 3115–3120.
25. Lightner, D. A., and Chang, T. C., *J. Am. Chem. Soc.* **96** (1974), 3015–3016.
26. Jackman, D. E., and Lightner, D. A., *J.C.S. Chem. Commun.* (1974), 344–345.
27. Lightner, D. A., Chang, T. C., Hefelfinger, D. T., Jackman, D. E., Wijekoon, W. M. D., and Givens, J. W., III, *J. Am. Chem. Soc.* **107** (1985), 7499–7508.
28. Pao, Y. H., and Santry, D. P., *J. Am. Chem. Soc.* **88** (1966), 4157–4163.
29. Moscowitz, A., *Adv. Chem. Phys.* **4** (1962), 67–112; see also ref. 24.
30. The 3-axial methyl configuration of chair cyclohexanone was not included in the original theoretical derivation of the octant rule (Moscowitz, A., personal communication); see also ref. 29.
31. Gould, R. R., and Hoffmann, R., *J. Am. Chem. Soc.* **92** (1970), 1813–1818.
32. Snatzke, G., and Eckhardt, G., *Tetrahedron* **26** (1970), 1143–1155.
33. Snatzke, G., and Marquarding, D., *Chem. Ber.* **100** (1967), 1710–1724.
34. Coulombeau, C., and Rassat, A., *Bull. Soc. Chim. Fr.* **71** (1971), 516–526.
35. McDonald, R. N., and Steppel, R. N., *J. Am. Chem. Soc.* **92** (1970), 5664–5670.
36. Becket, A. H., Khokhar, A. Q., Powell, G. P., and Hudec, J., *J. Chem. Soc. Chem. Commun.* (1971), 326–327.
37. Tocanne, J. F., *Tetrahedron* **28** (1972), 389–416.
38. Lightner, D. A., and Jackman, D. E., *J. Am. Chem. Soc.* **96** (1974), 1938–1939, and references therein.
39. Kirk, D. N., and Klyne, W., *J. Chem. Soc. Perkin Trans.* **1** (1974), 1076–1103.
40. Wijekoon, W. M. D., and Lightner, D. A., *J. Am. Chem. Soc.* **107** (1985), 2815–2817.
41. Lightner, D. A., Crist, B. V., Kalyanam, N., May, L. M., and Jackman, D. E., *J. Org. Chem.* **50** (1985), 3867–3878.
42. Lightner, D. A., and Toan, V. V., *Tetrahedron* **43** (1987), 4905–4916.
43. Lightner, D. A., and Toan, V. V., *J. Chem. Soc. Chem. Commun.* (1987), 210–211.
44. Yeh, C-Y., and Richardson, F. S., *Theor. Chim. Acta* **43** (1977), 253–260.

45. Hansen, Aa. E., and Bouman, T. D., *Adv. Chem. Phys.* **44** (1980), 545–644.
46. Hansen, Aa. E., and Bouman, T. D., *J. Am. Chem. Soc.* **107** (1985), 4828–4839.
47. For a recent review, see Oddershede, J., Jørgensen, P., and Yeager, D. L., *Comput. Phys. Repts.* **2** (1984), 33–92.
48. Lightner, D. A., Bouman, T. D., Crist, B. V., Rodgers, S. L., Knobeloch, M. A., and Jones, A. M., *J. Am. Chem. Soc.* **109** (1987), 6248–6259.
49. Bouman, T. D., Hansen, Aa. E., Voigt, B., and Rettrup, S., *Int. J. Quant. Chem.* **23** (1983), 595; *QCPE Bull.* **3** (1983), 64 (Program 459).
50. McCurdy, C. W., Rescigno, T. N., Yeager, D. L., and McKoy, V., in *Methods in Electronic Structure Theory* (Schaefer, H. F., ed.), Plenum Press, New York, 1977, ch. 9.
51. Kirkwood, J. G., *J. Chem. Phys.* **5** (1937), 497; **7** (1939), 139.
52. Lightner, D. A., Bouman, T. D., Wijekoon, W. M. D., and Hansen, Aa. E., *J. Am. Chem. Soc.* **106** (1984), 934–944.
53. Howell, J. M., *J. Chem. Phys.* **53** (1970), 4152–4160.
54. Klyne, W., and Kirk, D. N., *Tetrahedron Lett.* (1973), 1483–1486.
55. Hudec, J., *J. Chem. Soc. Chem. Commun.* (1970), 829–831; Hughes, M. T., and Hudec, J., *ibid.*, 831–832.
56. Ernstbrunner, E., and Hudec, J., *J. Am. Chem. Soc.* **96** (1974), 7106–7108.
57. Kirk, D. N., Klyne, W., and Mose, W. P., *J. Chem. Soc. Chem. Commun.* (1972), 35–36.
58. Kirk, D. N., *J. Chem. Soc. Perkin Trans.* **1** (1980), 787–803.
59. Kirk, D. N., *J. Chem. Soc. Perkin Trans.* **1** (1980), 1810–1819.
60. Meister, T. G., Zelinka, G. Ya., Golovina, L. V., and Fesik, V. V., *Opt. Spektrosk.* **37** (1974), 903–911.
61. Bouman, T. D., Hansen, Aa. E., and Voigt, B., *J. Am. Chem. Soc.* **101** (1979), 550–558.
62. Gedanken, A., Lagier, H. D., Schiller, J., Klein, A., and Hôrmes, J., *J. Am. Chem. Soc.* **108** (1986), 5342–5344.
63. Pulm, F., Schramm, J., Hormes, J., Grimme, S., and Peyerimhoff, S. D., *Chem. Phys.* **224** (1997), 143–155.

CHAPTER 5

Conformational Analysis of Substituted Cyclohexanones

5.1. 2-Methylcyclohexanone and 2-Methyl-4-*tert*-butylcyclohexanone

Previously, we noted that the magnitude of the positive $n \rightarrow \pi^*$ Cotton effect of 2(*S*)-methylcyclohexanone (Figure 5-1) remained essentially invariant from +25°C to −192°C. (See Figure 2-16F.)[1] That is, the rotatory strength remained constant (R = $+1.01 \times 10^{-40}$ cgs) over the indicated temperature range, and $\Delta\varepsilon$ remained constant at approximately +0.30. These data indicate that at room temperature the ketone is predominantly in its minimum-energy conformation, which, according to the octant rule and molecular mechanics calculations, is a chair with an equatorial methyl. This conclusion also seems reasonable on intuitive grounds, because the energy difference between axial and equatorial 2-methylcyclohexanone was found to be large and positive ($\Delta G^\circ \sim +2.2$ kcal/mole),[2] as calculated from the *cis* → *trans* epimerization of 2,5-dimethylcyclohexanone and 2-methyl-5-*tert*-butylcyclohexanone, processes which convert an α-equatorial methyl to an α-axial methyl. The invariant CD magnitude over a wide range of temperature supports the conclusion that neither the higher energy α-axial chair conformation nor higher energy twist-boat conformations make a contribution to the equilibrium of Figure 5-1A. Although a positive Cotton effect is predicted by the octant rule for both the equatorial and the axial isomers, the small

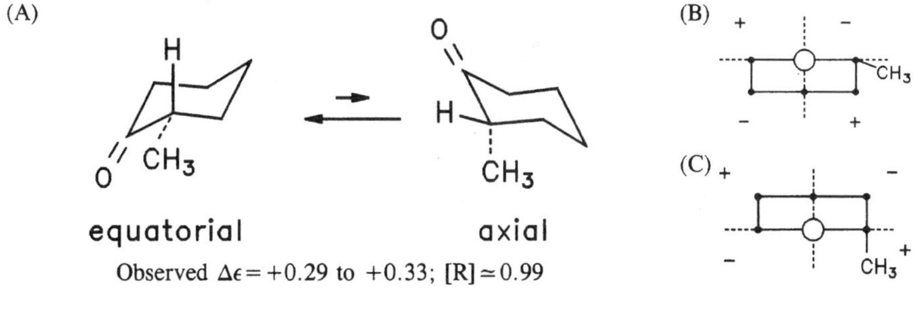

Figure 5-1. (A) Chair-to-chair conformational equilibrium in 2(*S*)-methylcyclohexanone, with observed $\Delta\varepsilon$ and reduced rotatory strength [*R*] values (ref. 1). (B) Octant projection diagram for the equatorial isomer. (C) Octant projection diagram for the axial isomer.

magnitude ($\Delta\varepsilon \sim +0.3$) found experimentally is more consistent with the methyl perturber being located close to an octant symmetry plane, as is found in the α-equatorial isomer (Figure 5-1B), rather than in an α-axial isomer (Figure 5-1C). Since molecular mechanics calculations[3] predict an essentially undistorted chair cyclohexanone (Table 2-6) when the α-methyl substituent is equatorial, one is tempted to assign $|\Delta\varepsilon| = 0.3$ to the magnitude of the octant contribution of an α-equatorial methyl perturber. Other attempts to quantify the octant contribution of an α-equatorial methyl using different model systems reach similar values, but it is also clear that small changes in the $CH_3–C_\alpha–C = O$ torsion angle can affect the value of Δε.[4] Of course, desymmetrization of the cyclohexanone ring will cause the ring atoms to make octant contributions, and these are difficult to assess quantitatively.

A particularly interesting "conformationally fixed" model for 2-methylcyclohexanone is *cis*-2(*R*)-methyl-4(*R*)-*tert*-butylcyclohexanone (Table 5-1), in which the *tert*-butyl group acts as a conformational anchor to inhibit chair-to-chair interconversion. Djerassi et al.[5] at Stanford measured the $n \rightarrow \pi^*$ CD Cotton effect of the compound at room temperature and at –196°C in two different low-temperature glass-forming solvents (EPA, 5:5:2 ether–isopentane–ethanol, by volume and MI, 1:4 methylcyclohexane-isopentane, by volume) (Figure 5-2). Very little change in the rotatory strength is noted over the temperature range, indicating that at room temperature the ketone is almost exclusively in its most stable conformation: a chair cyclohexanone with equatorial methyl and equatorial *tert*-butyl groups, as shown in Table 5-1. Although the reduced rotatory strengths and Δε values compare favorably in magnitude with those of 2(*S*)-methylcyclohexanone ([*R*] ~ 0.99, Δε ~ +0.29–0.33), they are still about 50% higher (for what is assumed to be the same enantiomeric excess). These differences are puzzling, because the two ketones appear to adopt nearly identical ring

Table 5-1. Reduced Rotatory Strengths [*R*][a] and Δε Values for the CD Spectra of *cis*-(+)-2(*R*)-Methyl-4(*R*)-*tert*-butylcyclohexanone at Room Temperature and at –196°C in EPA and MI[b]

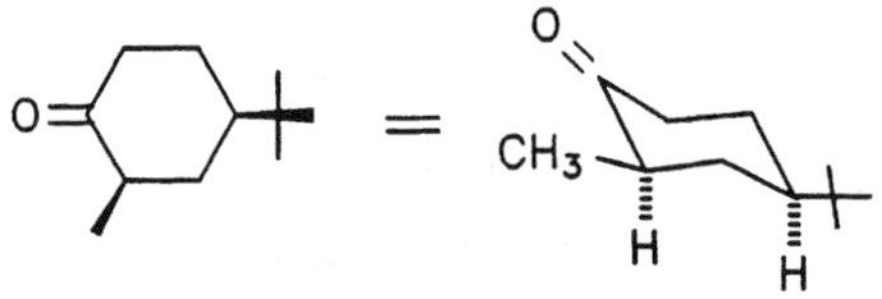

Solvent (temperature)	[*R*][a]	Δε
EPA (room temp)	–1.42	–0.47
EPA (–196°C)	–1.34	
MI (room temp)	–1.52	–0.46
MI (–196°C)	–1.57	

[a] Reduced rotatory strength = $1.078 \times 10^{40} \times$ rotational strength (cgs).

[b] Data from ref. 5.

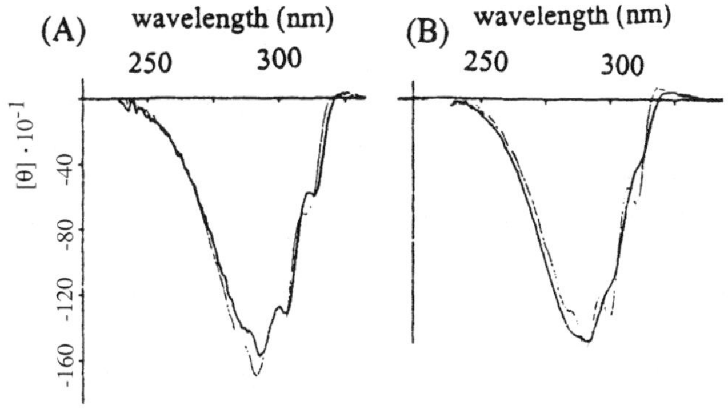

Figure 5-2. Circular dichroism spectra of *cis*-(+)-2(*R*)-methyl-4(*R*)-*tert*-butylcyclohexanone in (A) EPA (ether–isopentane–ethanol, 5:5:2, by vol.) and (B) MI (methylcyclohexane–isopentane, 1:4, by vol.) at room temperature (dark line) and −196°C (faint line). [Reprinted with permission from ref. 5. Copyright© 1980 American Chemical Society.]

conformations, as computed by molecular mechanics calculations (Table 5-2). However, the slight differences in ring conformation near the C=O group appear to cause a small change in orientation of the equatorial methyl. (Cf. the (CH_3)–C_2–C_1=O torsion angles listed in the table). Such a change in octant location of the methyl perturber might be the cause of the differing CD magnitudes.

The Stanford group[5] considered a different explanation. Their molecular mechanics calculations predicted considerable ring distortion of a nonsymmetric nature about the

Table 5-2. Computed[a] Ring Torsion Angles, CH_3–C_α–C=O Torsion Angles, and Heats of Formation for Equatorial 2-Methylcyclohexanone and *cis*-2-Methyl-4-*tert*-butylcyclohexanone

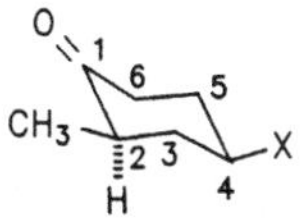 		Computed Torsion Angles (°)			
		C_3–C_2–C_1–C_6	C_1–C_2–C_3–C_4	C_2–C_3–C_4–C_5	
		C_2–C_1–C_6–C_5	C_1–C_6–C_5–C_4	C_3–C_4–C_5–C_6	
X =	(CH_3)–C_2–C_1=O	Dissymmetry[b]	Dissymmetry[b]	Dissymmetry[b]	Heat of Formation (kcal/mole)
H	-5.55	50.8	−52.9	57.6	−61.08
		−51.4	53.4	−57.4	
		[0.6]	[0.5]	[0.2]	
tert-Butyl	-4.44	51.9	−54.8	57.1	−85.10
		−52.0	53.9	−56.0	
		[0.1]	[0.9]	[1.1]	

[a] By PCMODEL, ref. 3; values are approximately ±0.5.

[b] Ring dissymmetry [value] as determined by the difference in absolute values of corresponding ring torsion angles.

carbonyl group, thereby desymmetrizing the stable chair conformation. Such deformations would be expected to produce octant contributions both from some ring atoms and from the *tert*-butyl group, which are ordinarily counted as noncontributors. However, distortion of the chair geometry seems to be overestimated. More recent MM2 molecular mechanics calculations[4] (Table 5-2) indicate only slight ring distortion in equatorial 2-methylcyclohexanone and in *cis*-2-methyl-4-*tert*-butylcyclohexanone. Whether the slightly greater ring distortion near C(4) in the latter is the cause of its more intense $n \rightarrow \pi^*$ Cotton effect is unclear, because it is currently difficult to assess the octant contributions made by the ring atoms in a nonsymmetric chair cyclohexanone and by the attached *tert*-butyl group. So far as can be determined, although the octant location of the equatorial methyl perturber is expected to be the same in both molecules, the CH_3–C_α–C=O torsion angles are not identical. The current assessment is that slight nonsymmetric ring distortions and small differences in octant locations of the α-equatorial methyl perturber are the bases for the differences in [*R*] and Δε. Collectively, they indicate the exquisite sensitivity of CD to small conformational changes. Since the CD of the *cis*-ketone is insensitive to changes in solvent (Table 5-1), it would appear that asymmetric solvation[6,7] is an insufficient rationale to explain the difference in rotatory strength.

As noted earlier, 2-methylcyclohexanone adopts a chair conformation with an equatorial methyl orientation. The corresponding axial isomer, which is thought to lie some 2.2 kcal/mole above the equatorial,[2] apparently makes little or no contribution to the conformational equilibrium of Figure 5-1. It is predicted to exhibit the same-sign, but a larger octant contribution. However, since the axial isomer cannot be isolated as such, a *tert*-butyl anchoring group was attached at C(4), as in 2(*S*)-methyl-4(*R*)-*tert*-butylcyclohexanone (Table 5-3), in order to lock the conformation into the unfavorable axial-methyl chair. The observed large positive $n \rightarrow \pi^*$ Cotton effect (Figure 5-3 and Table 5-3) follows the predictions of the octant rule and may be contrasted with the much smaller negative Cotton effect of the epimeric *cis* analog (Figure 5-2 and Table 5-1). Unlike those of the *cis* isomer, however, the $n \rightarrow \pi^*$ Cotton effect intensities of the *trans* isomer at room temperature drop some 15–17% from +25°C to –196°C, suggesting the presence of higher energy conformations at room temperature. Those could be twist-boat conformations with inherently more positive octant contributions than are found in the axial chair isomer.

In their molecular mechanics analysis of *trans*-2-methyl-4-*tert*-butylcyclohexanone, the Stanford group[5] determined that a severely twisted chair conformation would be the most stable. More recent calculations[4] also find a nonsymmetrically distorted chair, but one not nearly as severely distorted (Table 5-4). Interestingly, the parent axial 2-methylcyclohexanone appears to adopt a slightly more, but probably not significantly more, distorted chair than the *trans*-2-methyl-4-*tert*-butylcyclohexanone—a behavior not seen with the *cis*-ketone and its parent equatorial 2-methylcyclohexanone (Table 5-2). The cyclohexanone ring of the α-axial methyl isomers is, however, clearly flatter near the C=O group than is that of the α-equatorial, and the ring dissymmetry is larger. Apparently, the *tert*-butyl "conformational anchor" does not facilitate nonsymmetric distortion of chair cyclohexanone, although it does arrest the chair ⇄ chair conformational inversion. These results are initially surprising, as

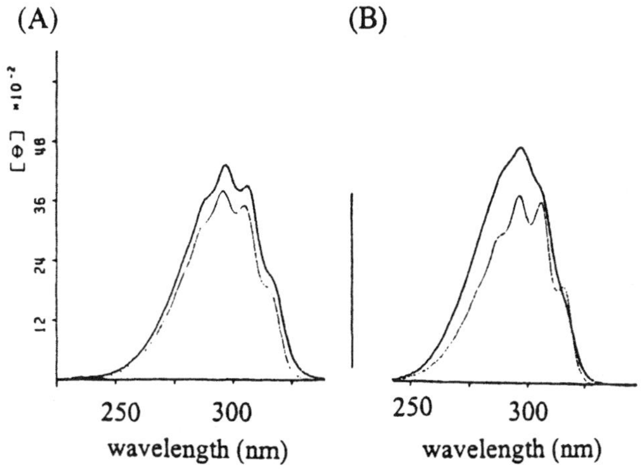

Figure 5-3. Circular dichroism spectra of *trans*-(+)-2(*S*)-methyl-4(*R*)-*tert*-butylcyclohexanone in (A) EPA (ether–isopentane–ethanol, 5:5:2, by vol.) and (B) MI (methylcyclohexan–isopentane, 1:4, by vol.) at room temperature (dark line) and –196°C (faint line). [Reprinted with permission from ref. 5. Copyright© 1980 American Chemical Society.]

previous criticism of the *tert*-butyl anchoring group revolved around its inducing a ring-flattening distortion of the chair geometry. As judged from the data of Tables 5-2 and 5-4, however, such criticism may be unfounded.

Interestingly, the computed differences in enthalpy between the axial and equatorial isomers of 2-methylcyclohexanone ($\Delta\Delta H$ = 2.3 kcal/mole) and of *cis* and *trans*-2-methyl-4-*tert*-butylcyclohexanone ($\Delta\Delta H$ = 2.3 kcal/mole) are identical—and essen-

Table 5-3. Reduced Rotatory Strengths $[R]^a$ and $\Delta\varepsilon$ Values for the CD Spectra of *trans*-(+)-2(*S*)-Methyl-4(*R*)-*tert*-butylcyclohexanone at Room Temperature and at –196°C in EPA and MI[b]

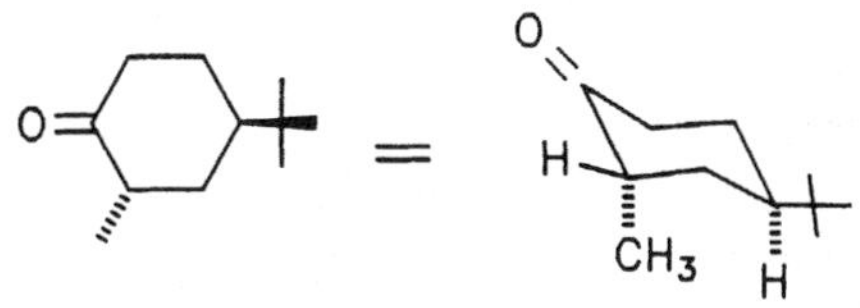

Solvent (temperature)	$[R]^a$	$\Delta\varepsilon$
EPA (room temp)	+4.71	+1.2
EPA (–196°C)	+3.53	
MI (room temp)	+4.41	+1.2
MI (–196°C)	+3.72	

[a] Reduced rotatory strength = $1.078 \times 10^{40} \times$ rotational strength (cgs).

[b] Data from ref. 5.

Table 5-4. Computed[a] Ring Torsion Angles, CH_3–C_α–C=O Torsion Angles, and Heats of Formation for Axial 2-Methylcyclohexanone and *trans*-2-Methyl-4-*tert*-butylcyclohexanone

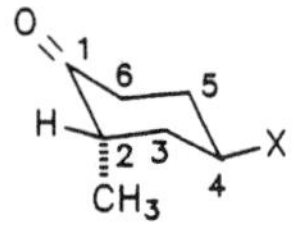

		Computed Torsion Angles (°)			
		C_3–C_2–C_1–C_6	C_1–C_2–C_3–C_4	C_2–C_3–C_4–C_5	
		C_2–C_1–C_6–C_5	C_1–C_6–C_5–C_4	C_3–C_4–C_5–C_6	Heat of Formation
X =	(CH_3)–C_2–C_1=O	Dissymmetry[b]	Dissymmetry[b]	Dissymmetry[b]	(kcal/mole)
H	98.5	45.9	–49.5	56.6	–58.79
		–48.1	52.9	–57.8	
		[2.2]	[3.4]	[1.2]	
tert-Butyl	98.6	46.2	–50.8	57.3	–82.81
		–48.3	53.8	–57.1	
		[2.1]	[3.0]	[0.8]	

[a] Computed by PCMODEL, ref. 3; values are approximately±0.5.

[b] Ring dissymmetry [value] as determined by the difference in absolute values of corresponding ring torsion angles.

tially the same as those determined experimentally (ΔΔH ~ 2.16–2.18 kcal/mole) for 2-methylcyclohexanone.[2] Assuming a negligible ΔS, one would calculate the presence of approximately 2% of the axial isomer in 2-methylcyclohexanone at room temperature, or about 2% of *trans*-2-methyl-4-*tert*-butylcyclohexanone in an epimerically equilibrated sample of the *cis* isomer. An experimental determination of the percent of *trans* isomer follows from the observation that the reduced rotatory strength of the *cis* isomer in methanol drops from –1.39 to –0.80 upon base-catalyzed equilibration.[5] Given a reduced rotatory strength of +5.30 for the *trans* isomer in methanol and +0.30 for the *cis* isomer in methanol,[5] one calculates (*i*) 8.8% of *trans* at equilibrium and (*ii*) $\Delta G° = +1.38$ kcal/mole for the equatorial ⇄ axial equilibrium. This value of $\Delta G°$ is not far from that determined by Allinger ($\Delta G° = 1.54 \pm 0.21$; see Table 2-7), but is distant from that ($\Delta H° = 2.16$–2.18 kcal/mole) determined by Cotterill and Robinson.[2] The calculated 8.8% of *trans* isomer appears to be too high. Using this value to approximate the percent axial isomer of 2-methylcyclohexanone that is present at room temperature, one might expect to observe a diminution in Δε and rotatory strength upon lowering the temperature from room temperature to –196°C, but this is not observed. So the 2% solution appears the more likely in the absence of special solvent effects. (See Section 2.6.)

5.2. Other Conformationally Immobilized α-Methylcyclohexanones

On a grander molecular scale, 2α-methyl-5α-cholestan-3-one represents a useful model for equatorial 2-methylcyclohexanone. If one assumes that the introduction of a 2-equatorial methyl onto the 5α-cholestan-3-one framework induces no changes in the octant positions of the remaining atoms in the cholestanone unit—a potentially risky assumption in light of the preceding discussion of the conformation and CD of 2(*S*)-methylcyclohexanone and its 4-*tert*-butyl analog—then subtraction of Δε for the $n \rightarrow \pi^*$ Cotton effect of 5α-cholestan-3-one from 2α-methyl-5α-cholestan-3-one should yield the octant contribution of an α-equatorial methyl on cyclohexanone. The results of such a CD study[8] on axial and equatorial methyls are shown in Table 5-5. Considering the approximations made, the agreement with the CD data from 2-methylcyclohexanone and *cis*-2-methyl-4-*tert*-butylcyclohexanone is rather good. In the equatorial methyl examples, the values derived from 2α-methyl-5α-cholestan-3-one are quite close to that of 2(*R*)-methylcyclohexanone (Δε = –0.30) and to its 4-*tert*-butyl analog (Δε = –0.47). The agreement is not as good with the 3β-methyl-5α-cholestan-2-one, for reasons that are unclear, as one would not expect any special conformational distortion in 5α-cholestan-2-one upon introducing an equatorial 3β-methyl. On the other hand, ring distortion might be expected in the axial 2β-methyl-5α-cholestan-3-one, due to a severe 1,3-diaxial methyl-methyl nonbonded steric repulsion with the C(19) angular methyl. And so here, the poorer agreement with the "expected" octant contribution value of an α-axial methyl is understandable. Such a nonbonded repulsion

Table 5-5. Octant Contributions of α-Methyl Perturbers Obtained by Subtracting Δε of the Parent Steroid Ketone from Methyl Ketones and a Comparison with the Values Derived from *cis* and *trans*-2-Methyl-4-*tert*-butylcyclohexanones

Steroid Ketone	$n \rightarrow \pi^*$ CD Cotton Effect in EPA ΔΔε (25°C)	$n \rightarrow \pi^*$ CD Cotton Effect in EPA ΔΔε (–175°C)	Δε (room temperature) 2-Methyl-4-*tert*-butylcyclohexanones
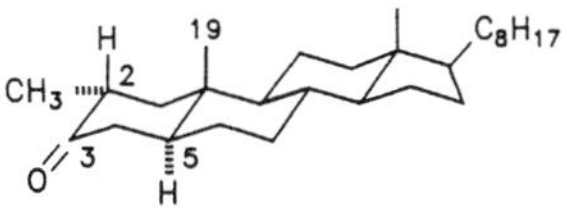	–0.32	–0.40	–0.47
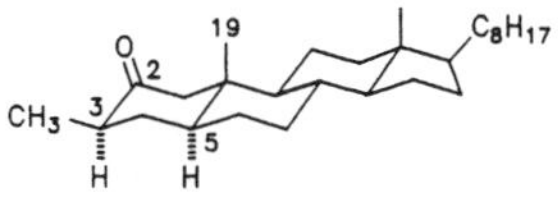	–0.12	–0.092	
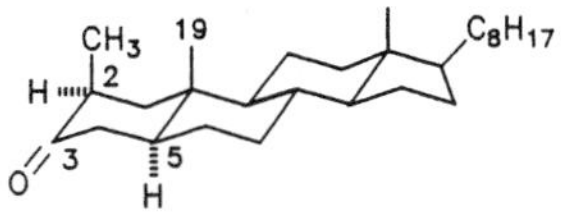	+0.60	+0.73	+1.2
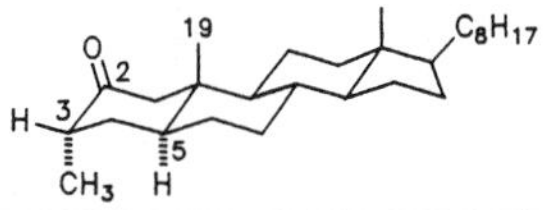	+0.86	+1.1	

does not exist in axial 3α-methyl-5α-cholestan-2-one, so in that case the agreement is better.

Other studies of the octant contributions of axial and equatorial methyl perturbers have made use of symmetric cyclohexanone templates that are held conformationally immobile by bridging groups, as in bicyclo[3.2.1]octan-3-one and bicyclo[3.1.1]heptan-3-one.[4] Although such bridging dictates a large measure of conformational rigidity, the erstwhile cyclohexanone chair conformation is symmetrically distorted somewhat (Table 5-6) in the former by an "anti-reflex" effect[9,10,11] and is very considerably distorted in the latter to afford something close to the sofa conformation of cyclohexanone (Figure 2-2). Thus, the one- to two-carbon bridge flattens the cyclohexanone ring torsion angles near the C=O group and pinches those across the cyclohexanone ring, while leaving the midsection less strongly distorted. Such symmetric distortions of the cyclohexanone ring bring about reorientations of the traditional axial and equatorial configurations. In α-methyl ketones, the methyl groups are thereby moved into new and different octant locations, while the ring atoms still make no octant contribution.

The latter phenomenon may be seen in the following: The methyl group of 2-methylcyclohexanone moves upward from an axial into an equatorial orientation during the chair-to-chair conformational ring inversion (Figure 5-4), with the CH_3–C_α–C=O torsion angle undergoing large changes—from about 100° to approximately 5°. According to the octant rule, an α-equatorial methyl will make a significantly smaller contribution than an α-axial one to the $n \to \pi^*$ Cotton effect, because it lies close to an octant symmetry plane. This prediction is in fact observed experimentally.[4,5] In the conformational inversion described in Figure 5-4, the methyl perturber remains in the same octant, and thus, no changes in sign are expected or observed.

As the cyclohexanone ring inverts, the α-methyl group rotates upward from approximately 100° to about 5° through intermediate angles (e.g., 60° in the sofa

Table 5-6. Comparison of Cyclohexanone Ring Torsion Angles and the Anti-reflex Action Influence of Bridging

Ketone	Ring Torsion Angles (°)		
	C_3–C_2–C_1–C_6	C_1–C_2–C_3–C_4	C_2–C_3–C_4–C_5
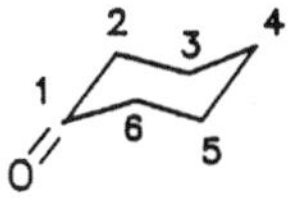	51.5	–53.3	57.5
	39.4	–57.3	72.3
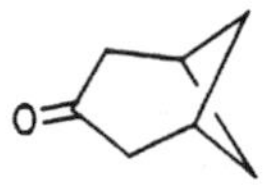	0.18	–47.4	82.7

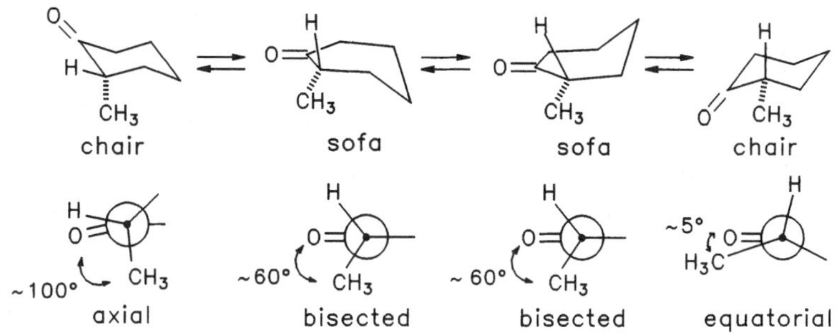

Figure 5-4. (Upper) Changes in the CH_3–C_α–C=O torsion angle accompanying the change from axial to equatorial methyl in the chair ⇄ chair conformational inversion of 2-methylcyclohexanone shown passing through a high-energy sofa conformation. (Lower) Torsion angle values are indicated on the Newman projection diagrams corresponding to each conformation above.

conformation). The methyl orientation corresponding to selected torsion angles lying between 100° and 5° can be stabilized by bridging the cyclohexanones, as in bicyclo[3.2.1]octan-3-one and in bicyclo[3.1.1]heptan-3-one (Table 5-6). Thus, in the former system, the axial (*exo*) methyl group is slightly distorted from the axial orientation—from 98° to 91° as measured by the CH_3–C_α–C=O torsion angle (Table 5-7)—and the equatorial (*endo*) methyl torsion angle is somewhat more distorted, to 14° from about 5°. In contrast, in the bicyclo[3.1.1]heptan-3-one system, as found in pinanones, the CH_3–C_α–C=O torsion angles lie very far from the axial and equatorial orientations. In the pinanones, the octant rule is not obeyed.

Although one is tempted to predict a gradual change in the magnitude of the octant contribution as the CH_3–C_α–C=O torsion angle changes from approximately 98° in the axial configuration to around 5° in the equatorial one, this is not borne out by experiment. As the methyl perturber moves upward from the axial toward the equatorial orientation, the magnitude of its octant contribution decreases, but not in the expected way. At the intermediate CH_3–C_α–C=O torsion angles represented in the pinanones, the $n \rightarrow \pi^*$ Cotton effect sign inverts—from positive to negative for the absolute configuration shown in Table 5-7 and Figure 5-4. (The pinanones are constrained to adopt a sofa conformation and thus have their α-methyl groups in a bisected orientation.) As the torsion angle continues to decrease, approaching the equatorial, the sign inverts again, becoming positive, and the magnitude increases to that of the equatorial methyl. The sign inversion, from positive to negative, is not predicted by the octant rule, indicating that as the conformation changes from axial to equatorial, the Cotton effect intensity mechanism and the relative importance of "perturbers" and framework bond couplings change.[4] A theoretical analysis[4] shows a delicate balance of different framework and methyl contributions in the bisected configurations of the pinanones, compared with the "classical" axial and equatorial configurations on chair cyclohexanone, in which the dominant contributions come from the methyl perturbers. As a consequence, the octant rule (which is based on

Table 5-7. Correlation of Circular Dichroism Cotton Effects and Methyl NMR Chemical Shifts with CH_3–C_2–C_1=O Torsion Angle in 2-Methylcyclohexanone and Its Analogs[a]

Ketone	Torsion Angle CH_3–C_α–C=O[b]	$n \rightarrow \pi^*$ Cotton Effect $\Delta\varepsilon$[c]	CH_3 Chemical Shifts[d] ^{13}C	CH_3 Chemical Shifts[d] ^{1}H
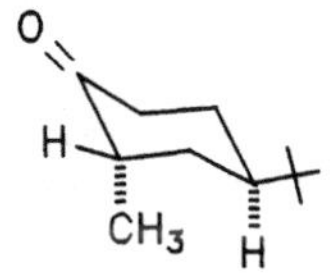	98°	+1.5	16.8	1.15
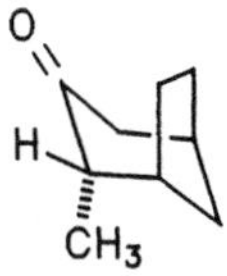	91°	+0.82	17.5	1.16
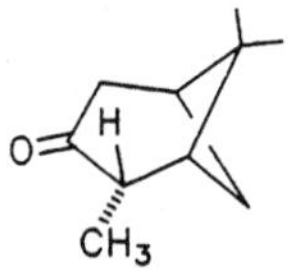	43°	–0.19	15.1	1.21
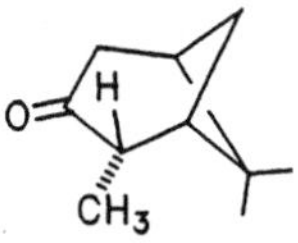	35°	–0.42	16.7	1.24
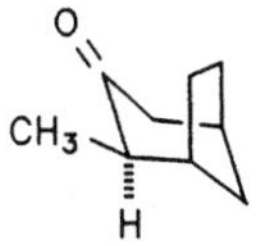	14°	+0.11	12.7	1.01
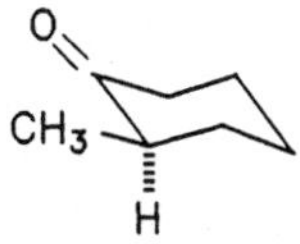	5.5°	+0.30	14.3	1.01
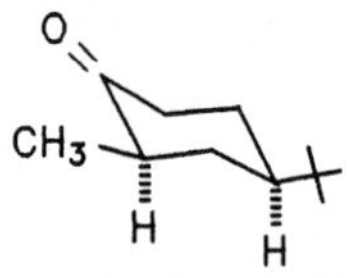	4.4°	+0.42	14.5	1.03

[a] Data from ref. 4.

[b] From PCMODEL calculation, ref. 3; values are approximately±0.50.

[c] In L · $mole^{-1}$ · cm^{-1}.

[d] In δ (ppm) downfield from $(CH_3)_4Si$.

contributions from "dissymmetric perturbers") becomes inadequate to represent ketones, such as pinanones, for which structural features of the "symmetric" framework take on unusual importance.

Table 5-7 also reveals the relative ^{1}H-NMR and ^{13}C-NMR shieldings of the α-methyls. Consistent with other work, an equatorial methyl is relatively more shielded than an axial. This is a useful qualitative distinction, but as with the $n \rightarrow \pi^*$ Cotton effect, there is no gradual upfield shift in going from the axial to the equatorial orientation. The bisected methyl configuration of the pinanones again leads to anomalies: The ^{1}H-NMR δ values are larger than those of either axial or equatorial methyls.

5.3. 2-Isopropyl and 2-*tert*-Butylcyclohexanone

Similar to that seen with 2(*S*)-methylcyclohexanone, which is thought to exist mainly in the equatorial chair conformation, a positive $n \rightarrow \pi^*$ Cotton effect (Figures 5-5A and B) is observed for 2-isopropyl and 2-*tert*-butylcyclohexanone of the same absolute (*R*) configuration series.[12] And again, since the isopropyl and *tert*-butyl groups, whether equatorial or axial, lie in a positive back octant, the absolute configuration is

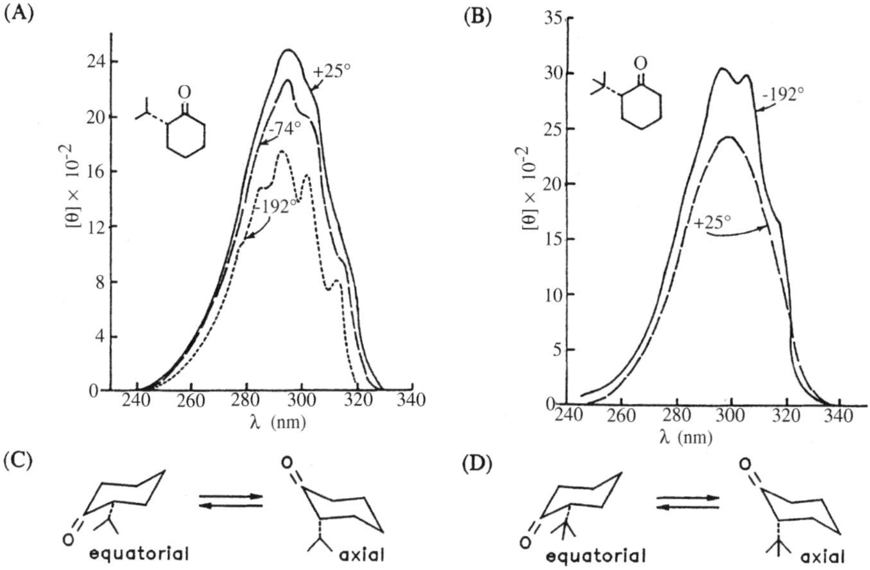

Figure 5-5. Variable temperature circular dichroism spectra of (A) 2(*R*)-isopropylcyclohexanone and (B) 2(*R*)-*tert*-butylcyclohexanone in EPA (ether–isopentane–ethanol, 5:5:2, by vol.) at the temperature noted on the curves. Chair ⇄ chair conformational equilibria are shown for (C) 2(*R*)-isopropylcyclohexanone and (D) 2(*R*)-*tert*-butylcyclohexanone. [(A) and (B) are reprinted with permission from ref. 12. Copyright© 1965 American Chemical Society.]

Table 5-8. Rotatory Strengths of α-Alkylcyclohexanones in EPA Solvent.[a,b]

Rotatory Strength (cgs) at Temperatures (°C)	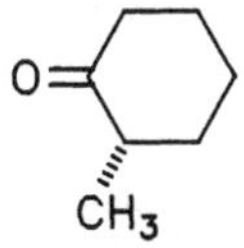	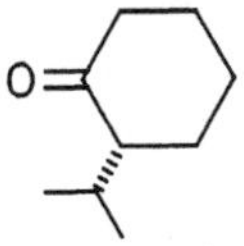	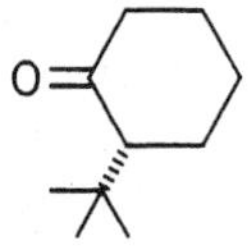
$R^{25°} \times 10^{40}$	+1.05[c,d]	+2.32[e]	+2.16[e,f]
$R^{-192°} \times 10^{40}$	+1.10[c,d]	+1.50[e]	+2.75[e,f]

[a] Ether–isopentane–ethanol, 5:5:2, by vol.
[b] Computed following digitization of published (ref. 12) CD curves.
[c] Value from ref. 12 is $+1.01 \times 10^{-40}$.
[d] Computed from CD curves of Figure 4-5D.
[e] Computed from the CD curves of Figure 5-5.
[f] Values from ref. 12 are $+2.34 \times 10^{-40}$ at 25°C and $+3.15 \times 10^{-40}$ at −192°C.

confirmed. Here, too, the most stable conformation is predicted and computed[3] to be a chair with an equatorial alkyl (Figures 5-5C and D). Unlike the α-methylcyclohexanone, however, the α-isopropyl and α-*tert*-butyl analogs exhibit more intense and temperature-dependent Cotton effects. (Cf. Figure 2-16F with Figure 5-5 and Table 5-8.) The increase in intensity is not entirely surprising, as there are more octant contributors in the isopropyl and *tert*-butyl groups than in a methyl group. Since the dissymmetry of a cyclohexanone ring is computed to be small in the case of the equatorial conformers (Table 5-9) and only slightly larger than that computed for 2-methylcyclohexanone (Table 5-2), it would appear unlikely that ring distortion

Table 5-9. Computed[a] Ring Torsion Angles, R–C_α–C=O Torsion Angles, and Heats of Formation for Equatorial 2(*R*)-Isopropylcyclohexanone and 2(*R*)-*tert*-butylcyclohexanone.

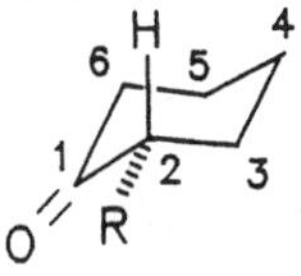 R =	R–C_2–C_1=O	Computed Torsion Angles (°) C_3–C_2–C_1–C_6 C_2–C_1–C_6–C_5 Dissymmetry[b]	 C_1–C_2–C_3–C_4 C_1–C_6–C_5–C_4 Dissymmetry[b]	 C_2–C_3–C_4–C_5 C_3–C_4–C_5–C_6 Dissymmetry[b]	Heat of Formation (kcal/mole)
Isopropyl	2.5	−55.1 56.7 [1.6]	53.6 −54.9 [1.3]	−56.6 56.2 [0.4]	−71.32
tert-Butyl	5.3	−54.7 56.1 [1.4]	54.4 −54.6 [0.2]	−57.5 56.3 [1.2]	−77.13

[a] Computed using PCMODEL, ref. 3; values are approximately±0.5.
[b] Ring dissymmetry [value] as determined by the difference in absolute values of corresponding ring torsion angles.

contributes very much to the larger $n \rightarrow \pi^*$ Cotton effects observed for 2(*R*)-isopropyl and 2(*R*)-*tert*-butylcyclohexanone. In addition to a possible chair–chair conformational equilibrium, other, probably more relevant, conformational changes are made possible through rotations of the isopropyl and *tert*-butyl groups. Such rotations bring the β-methyl perturbers into different octant locations, and one might thus expect to see a temperature-dependent CD. However, a precise description of the energetically most favored rotameric conformers may not be possible on the basis of the CD data alone. Interestingly, molecular mechanics calculations predict that the three staggered rotational isomers differ very little in energy, but that the barrier for interconversion of rotational isomers is surprisingly high: 15–20 kcal/mole for *tert*-butyl and 5–15 kcal/mole for isopropyl.[3]

The greater temperature dependence of the $n \rightarrow \pi^*$ Cotton effects of 2(*R*)-isopropyl and 2(*R*)-*tert*-butylcyclohexanone (Figure 5-5) relative to 2(*S*)-methylcyclohexanone (Figure 2-16F) is at first surprising if one assumes that the larger groups are better at stabilizing the chair equatorial cyclohexanone conformation. Two explanations for the temperature dependence come to mind: (i) distortion of the chair cyclohexanone ring, raising its energy toward that of traditionally higher energy conformations (e.g., the twist boat), and (ii) rotational isomerization about the C–C bond extending from the α-carbon to the isopropyl or *tert*-butyl groups. In 2-isopropylcyclohexanone, the computed[3] energy difference between equatorial and axial chair conformations is only 1.1 kcal/mole, much less than that in 2-methylcyclohexanone (~ 2.3 kcal/mole). Using the experimental data of Table 2-7, one predicts 39% of the axial conformation at 25°C and only 8% at –192°C. This shift could account for the reduction in magnitude of the Cotton effect (Figure 5-5A) in going from 25°C to –192°C, as an axial isopropyl group is predicted by the octant rule to make a larger positive contribution to the Cotton effect than does an equatorial group. Although chair 2-isopropylcyclohexanone is predicted to be more puckered near the C=O group than cyclohexanone or 2-methylcyclohexanone (Table 2-6), relevant ring torsion angles (ψ = 55.1° to 56.7°) are comparable to those of cyclohexane (ψ = 56.4°), and the remainder of the ring is essentially unperturbed. This distortion, however, is apparently sufficient to narrow the energy difference between equatorial and axial 2-isopropylcyclohexanones. On the other hand, the introduction of an equatorial isopropyl at C(2) on chair cyclohexanone causes very little ring dissymmetry (Table 5-9).

Wellman, Briggs, and Djerassi[12] considered the influence of rotational isomerization about the α-isopropyl group to be a major contributor to the temperature-dependence of the Cotton effects. Support for this explanation is found in the following: Similar to 2-methylcyclohexanone, which exhibits no temperature dependence of its $n \rightarrow \pi^*$ rotatory strength (Figure 2-16F, Table 5-8), 2α-methyl-5α-cholestan-3-one exhibits very little temperature dependence on its rotatory strength[8,13] (Table 5-10). In contrast, a large temperature dependence is seen in 2α-isopropyl-5α-cholestan-3-one and in 2α-isopropyl-19-nor-5α-androstan-3-one. Ring conformational changes are discounted because they are not seen in the 2α-methyl analog and because the same temperature dependence is seen in both the cholestanone and 19-nor-androstanone, where the latter has a more flexible ring *A*. On the other hand, rotation about the C_2–$CH(CH_3)_2$ bond

Table 5-10. Rotatory Strengths of 2-Alkyl-5α-Cholestan-3-ones and 19-nor-5α-Androstan-3-ones Measured in EPA[a] at Varying Temperatures

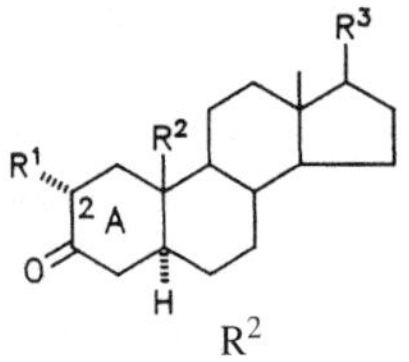

			Rotatory Strengths ($\times 10^{40}$)[b] at			% Change Between +25°C and –192°C
R^1	R^2	R^3	25°C	–74°C	–192°C	
2α-CH_3	CH_3	C_8H_{17}	+2.23	+2.28[c]	+2.46[d]	+10%
2α-$CH(CH_3)_2$	CH_3	C_8H_{17}	+1.53	+1.60	+2.13	+39%
2α-$CH(CH_3)_2$	H	H	+1.90	+1.99	+2.47	+30%
2β-$CH(CH_3)_2$	H	H	+10.3	—	+9.95	–3%

[a] Ether–isopentane–ethanol, 5:5:2 (by vol.). Data from Refs. 8, 12, and 13.

[b] c.g.s. units.

[c] Measured at –100°C.

[d] Measured at –175°C.

places the methyl perturbers in disparate octant locations, with the result that different Cotton effects are predicted for the three distinct staggered rotational isomers. By changing the rotamer population with decreasing temperature, the magnitude of the Cotton effect is expected to change. In marked contrast, 2β-isopropyl-19-nor-androstane-3-one, with an axial isopropyl, exhibits essentially no temperature dependence on the $n \rightarrow \pi^*$ Cotton effects. In that case, it is thought that C_2–$CH(CH_3)_2$ rotamers are restricted to one rotameric conformation.

Although 2(*R*)-*tert*-butylcyclohexanone also exhibits a temperature dependence on its $n \rightarrow \pi^*$ Cotton effects, unlike 2(*R*)-isopropylcyclohexanone, the magnitude increases upon lowering the temperature (Figure 5-5B). Also unlike 2-isopropylcyclohexanone, the computed energy difference between axial and equatorial conformations is large (2.3 kcal/mole) and close to that (ΔH = 2.39 ± 0.38 kcal/mole, Table 2-7) determined experimentally. Unexpectedly, it is nearly the same as that of 2-methylcyclohexanone. Using the experimentally determined $\Delta H°$ and $\Delta S°$ (Table 2-7), one calculates < 1% axial 2-*tert*-butylcyclohexanone at 25°C and << 1% at –192°C. Unless the contribution of an α-axial *tert*-butyl group is enormous ($\Delta\varepsilon \sim -100$), the presence of any α-axial chair conformation is unlikely. More likely is either (i) contributions from C_2–$C(CH_3)_3$ rotational isomers[12] or (ii) a contribution from a twist-boat conformation and octant contributions from ring atoms. In support of (ii), one computes[3] (Table 5-11) large changes in ring torsion angles (ψ) and a large dissymmetry of the cyclohexanone ring in the "axial" *tert*-butyl conformation. And in contrast, a surprisingly very small ring dissymmetry is found in axial chair α-isopropylcyclohexanone. The much more pronounced ring distortion influenced by an α-axial *tert*-butyl is probably due to the fact that one methyl of the *tert*-butyl must always be placed over the ring and be sterically repulsed by the ring CH_2s at 4 and 6.

Table 5-11. Computed[a] Ring Torsion Angles, R–C_α–C=O Torsion Angles, and Heats of Formation for Axial 2(*R*)-Isopropylcyclohexanone and 2(*R*)-*tert*-Butylcyclohexanone.

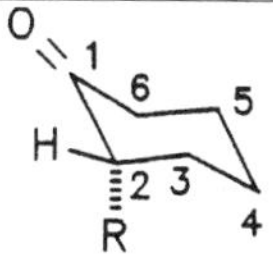

		Computed Torsion Angles (°)			
		C_3–C_2–C_1–C_6	C_1–C_2–C_3–C_4	C_2–C_3–C_4–C_5	
		C_2–C_1–C_6–C_5	C_1–C_6–C_5–C_4	C_3–C_4–C_5–C_6	Heat of Formation
R =	R–C_2–C_1=O	Dissymmetry[b]	Dissymmetry[b]	Dissymmetry[b]	(kcal/mole)
Isopropyl	103.2	51.9	–54.1	57.9	–70.22
		–51.9	52.0	–55.7	
		[0.0]	[2.1]	[2.2]	
tert-Butyl	83.4	39.3	–42.1	53.6	–74.83
		–46.9	55.1	–59.0	
		[7.6]	[13.0]	[5.4]	

[a] Computed using PCMODEL, ref. 3; values are approximately±0.5.

[b] Ring dissymmetry [value] as determined by the difference in absolute values of corresponding ring torsion angles.

5.4. 3-Methylcyclohexanone

Previously, we discussed the chair–chair conformational equilibrium of 3(*R*)-methylcyclohexanone in connection with the octant rule. In view of the observed temperature dependence of that compound's $n \rightarrow \pi^*$ CD (Figure 5-6),[13] one is led to think in terms of a predominance of the equatorial isomer, with the presence of other isomers contributing less positively (or even absolutely negatively) to the CD at room temperature. One such "other isomer" is the axial isomer. In early studies, it was thought that a 3-axial methyl would be an intense back-octant perturber. However, subsequent work indicates that the 3-axial methyl lies very near the third nodal surface of the octant rule or actually protrudes into a front octant and is thus a very weak octant perturber at room temperature.[14] Consequently, the presence of a moderate concentration of the axial isomer would not be expected to have a profound effect on the net CD, contrary to what was believed earlier.

Analysis of the variable-temperature $n \rightarrow \pi^*$ Cotton effects in terms of a two-component equilibrium indicated the presence of 90% of the major isomer at room temperature in ether–isopentane–ethanol (EPA; 5:5:2, by vol.).[15] The reduced rotatory strengths of the major and minor isomers (major, [*R*] = +2.8; minor, [*R*] ~ –0.2) are close to those of the corresponding β-methyladamantanones (equatorial, [*R*] = +2.1; axial, [*R*] = –0.1). The major isomer is almost certainly the equatorial chair; the minor isomer is probably the axial chair. However, the equilibrium data obtained from variable-temperature CD ($\Delta G°$ = 0.5 kcal/mole) is very different from that ($\Delta G°$ = 1.36 kcal/mole) estimated from the equilibration of *cis* and *trans*-3,5-dimethylcyclohex-

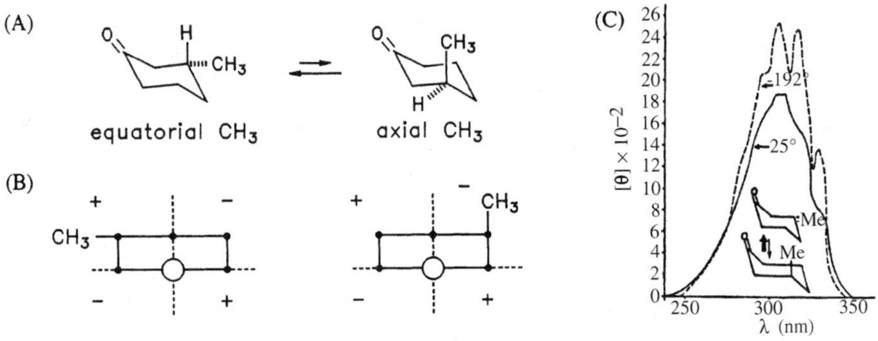

Figure 5-6. (A) Equatorial ⇄ axial configurational inversion during the chair ⇄ chair conformational equilibrium of 3(*R*)-methylcyclohexanone. (B) Octant diagrams for the equatorial (left) and axial (right) isomers in (A). (C) Variable temperature circular dichroism spectra of 3(*R*)-methylcyclohexanone in EPA solvent (ether–isopentane–ethanol, 5:5:2, by vol.); $[R]^{25°}$ = +1.78 and $[R]^{-192°}$ = +2.34. [(C) is reprinted with permission from ref. 13. Copyright© 1963 American Chemical Society.]

anone over Pd(C).[16] In contrast, 3(*R*)-*tert*-butylcyclohexanone shows essentially no change in its CD between +25°C ([*R*] = +2.18) and −192°C ([*R*] = +2.34),[13] suggesting that the molecule remains exclusively in the equatorial chair at room temperature. It is interesting to note that at −192°C, a temperature at which the most stable equatorial chair conformation is present almost exclusively, the [*R*] values for 3(*R*)-*tert*-butylcyclohexanone (+2.34) and 3(*R*)-methylcyclohexanone (+2.32) are basically identical.

Table 5-12. Computed[a] Ring Torsion Angles and Heats of Formation for 3-Methylcyclohexanone Chair Isomers

Ketone	Computed Torsion Angles (ψ,°)			Heat of Formation (kcal/mole)
	C_3–C_2–C_1–C_6 C_2–C_1–C_6–C_5 Dissymmetry[b]	C_1–C_2–C_3–C_4 C_1–C_6–C_5–C_4 Dissymmetry[b]	C_2–C_3–C_4–C_5 C_3–C_4–C_5–C_6 Dissymmetry[b]	
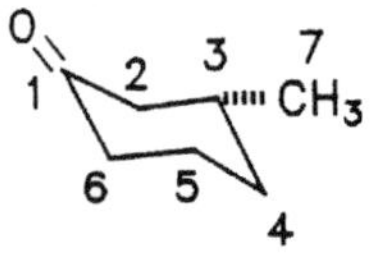	50.0 −49.8 [0.2]	−52.5 52.8 [0.3]	57.4 −58.0 [0.6]	−61.24
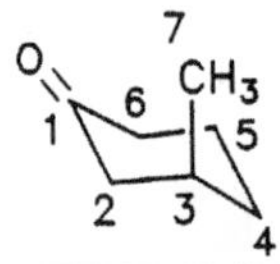	52.2 −51.2 [1.0]	−52.6 51.3 [1.3]	55.5 −55.2 [0.3]	−59.85

[a] Computed using PCMODEL, ref. 3.
[b] Ring dissymmetry [value] as determined by the difference in absolute values of corresponding ring torsion angles.

One would conclude that the octant contribution from equatorial 3-methyl and 3-*tert*-butyl groups are nearly the same. This is surprising, for it would appear that the methyls of the *tert*-butyl group make very little contribution to the Cotton effect. More likely, they do make contributions, but there may also be cancellation from front-octant contributions.

Molecular mechanics calculations predict an energy difference of 1.4 kcal/mole between the axial and equatorial isomers (Table 5-12) and much less ring distortion in the 3-axial isomer than in the 2-axial (Table 5-2). The introduction of a 3-equatorial methyl induces a very minor dissymmetry in the cyclohexanone ring. The introduction of a 3-axial methyl induces only a slightly larger ring dissymmetry, but it tends to cause a small amount of puckering near the C=O and slight flattening near C(4). The computed energy difference ($\Delta H°$) for the equilibrium is comparable to that ($\Delta G°$) determined experimentally ($\Delta G° = 1.36$ kcal/mole)[16] by catalyzed equilibration, but rather different from that ($\Delta G° = 0.5$ kcal/mole)[15] determined by analyzing variable-temperature CD data. The reason for the large discrepancy in the latter is unclear.

5.5. *trans*-2-Bromo- and 2-Chloro-5-methylcyclohexanone

One of the earliest examples of conformational analysis by ORD and CD comes from studies on the two α-haloketones *trans*-2(*R*)-bromo-5(*R*)-methylcyclohexanone and *trans*-2(*R*)-chloro-5(*R*)-methylcyclohexanone. Two different chair cyclohexanone conformational isomers are considered to be in equilibrium, one with diequatorial, and the other with diaxial, substituents (Figure 5-7). By the usual stereochemical considerations, the diequatorial one is predicted to be the more stable. And according to the octant rule, the diequatorial isomer is predicted to exhibit a moderate positive Cotton effect, whereas the diaxial one is predicted to exhibit a strong negative Cotton effect (Figure 5-7B). These predictions are borne out in the ORD spectra of the chloroketone, which shows a positive Cotton effect in methanol. Here, the diequatorial conformation predominates. In contrast, the bromoketone shows a weaker negative ORD Cotton effect. There is also a powerful solvent effect on the ORD spectra: Both ketones exhibit negative ORD Cotton effects in isooctane solvent (Figure 5-7).[17,18] Since the diaxial conformation is the source of negative octant contributions, the change from methanol to isooctane appears to force a change in the position of equilibrium toward an increased concentration of diaxial conformer. In contrast, more polar methanol solvent aligns with its dipoles around the polar bonds of the solute, effectively shielding the local bond dipoles in the diequatorial conformer from repulsion. The solvent effect is rationalized as follows: The hydrocarbon solvent is predicted to support the *anti* alignment of the C–X and C=O bond dipoles of the diaxial isomer in order to minimize the net dipole moment of the molecule. Thus, the equilibrium shifts from predominantly diequatorial in methanol to (more) diaxial in hydrocarbon solvent. This transformation, however, does not necessarily mean a complete shift from all diequatorial to all diaxial, as the two isomers are not predicted to have Cotton effects of equal magnitude. Support for these conclusions comes from analyses of the dependence of

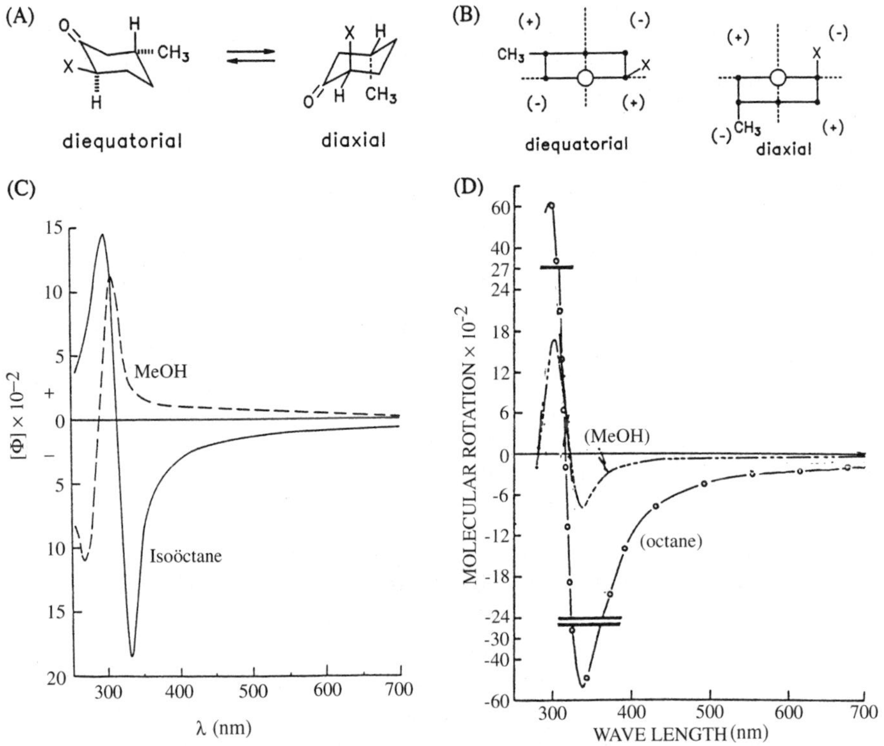

Figure 5-7. (A) Interconverting chair conformations of *trans*-2(*R*)-bromo-5(*R*)-methylcyclohexanone (X = Br) and *trans*-2(*R*)-chloro-5(*R*)-methylcyclohexanone (X = Cl). (B) Their octant diagrams. ORD curves of (C) *trans*-2(*R*)-chloro-5(*R*)-methylcyclohexanone and (D) *trans*-2(*R*)-bromo-5(*R*)-methylcyclohexanone. [(C) is reprinted with permission from Crabbé, P., *Optical Rotatory Dispersion and Circular Dichroism in Organic Chemistry*, Holden-Day, Inc. Boca Raton, FL, 1965. Copyright© Holden-Day, Inc. (D) is reproduced in modified form from ref. 17. Copyright© 1960 American Chemical Society.]

the solvent on the UV and IR spectra, which suggest an exclusively diequatorial population in methanol, but 20–30% diaxial in isooctane.[19] In contrast, for the bromoketone, dipole moment, UV and IR measurements indicate 39–45% diaxial in heptane and 11% diaxial in methanol.[20]

In assessing the octant contributions from the diequatorial isomer, one assumes that (i) there are no contributions from the ideal chair cyclohexanone ring atoms and (ii) the predicted positive contribution from the equatorial methyl at C(5) overcomes an expected weaker negative contribution from the equatorial halogen at C(2) (Figure 5-7B). Since the octant contribution from the β-equatorial methyl in 3(*R*)-methylcyclohexanone and in the corresponding methyl adamantanone has a value $\Delta\varepsilon \sim +0.6$ (Table 4-1), in order to get a net positive Cotton effect for the diequatorial isomer, an α-equatorial chlorine or bromine would be expected to contribute $\Delta\varepsilon > -0.6$. However, the exact magnitude of the octant contribution from an α-equatorial halogen has not yet been established. In the case of the chloroketone, it can be estimated to be $\Delta\varepsilon \sim -0.15$ from the $\Delta\varepsilon$ value ($\sim +0.45$)[21] of *trans*-2(*R*)-chloro-5(*R*)-methylcyclohexanone in methanol solvent, wherein the diequatorial chair conformation is thought to predominate (97–100%).[22] However, it is more difficult to estimate the octant contribution of an α-equatorial bromine, because the bromoketone conformation is apparently less homogeneous in methanol than the chloroketone is, with only 83–89% diequatorial isomer present.[20] Optically active α-chloro- and α-bromocyclohexanones with a chair conformation anchored by a *tert*-butyl group at C(4) or with an ethylene bridge connecting C(3) and C(5) (as in bicyclo[3.2.1]octan-3-one) are not known or have not been submitted to CD analysis.

In assessing the magnitude of the octant contributions from the diaxial isomer, one again assumes that the ring atoms make no contribution—that is, the cyclohexanone adopts an undistorted chair. This approximation may not appear as reasonable in the diaxial as in the diequatorial isomer, because a β-axial methyl group has several destabilizing *gauche*-butane interactions, as does an α-axial halogen, in addition to dipole–dipole interactions between the carbon–halogen and C=O groups. According to molecular mechanics calculations,[3] the cyclohexanone ring of the diequatorial isomers is barely altered from that of chair cyclohexanone and only slightly desymmetrized (Table 5-13). In contrast, in the diaxial isomers, the chair is flatter and somewhat desymmetrized, especially in the bromoketone. Yet the energy difference between diequatorial and diaxial isomers is quite small: 0.64 kcal/mole in the chloroketone and 0.17 kcal/mole in the bromoketone. Although octant contributions are difficult to assess from ring atoms of a desymmetrized cyclohexanone ring, in the diaxial isomer (Figure 5-7) a strong negative Cotton effect is predicted, with its origins mainly from the α-axial halogen octant perturber, which lies in a negative back octant. The β-axial methyl is thought to lie barely in a positive front octant, where it makes only a very weak contribution. As with an equatorial chlorine or bromine perturber, however, it is difficult to know quantitatively the octant contribution of an α-axial halogen, because no direct measurements have been made on stereochemically rigid α-halo-cyclohexanones.

An indirect determination of the rotatory strength of diaxial *trans*-2(*R*)-chloro-5(*R*)-methylcyclohexanone was made by Moscowitz, Wellman, and Djerassi,[22] who analyzed the diequatorial ⇄ diaxial configurational/conformational inversion represented in Figure 5-8A for the chloroketone by variable-temperature CD (Figure 5-8B). A negative $n \rightarrow \pi^*$ CD Cotton effect is clearly observed in isooctane and in CCl_4, whereas a bisignate Cotton effect is seen in ether–isopentane–ethyl alcohol (EPA; 5:5:2 by vol.) at room temperature. In EPA, the negative component of the CD curve is due to the presence of the diaxial isomer, the positive component to the diequatorial, and the summing of negative and positive Cotton effects gives the observed bisignate CD of Figure 5-8B. As the temperature is lowered to −192°C, the equilibrium shifts exclusively toward the more stable diequatorial isomer, and only a positive Cotton effect is observed. The following values were determined by analyzing the variable-temperature CD in EPA: $\Delta G° = 1.72$ kcal/ mole for the equilibrium, with $R = +2.34 \times 10^{-40}$ cgs and $\Delta\varepsilon = +0.76$ for the diequatorial isomer; and $R = -34.7 \times 10^{-40}$ cgs and $\Delta\varepsilon = -9.7$ for the diaxial isomer. These data, together with the values $R = +1.95 \times 10^{-40}$

Table 5-13. Computed[a] Ring Torsion Angles, X–C_α–C=O Torsion Angle, Dipole Moments, and Heats of Formation for 2(*R*)-Bromo and 2(*R*)-Chloro-5(*R*)-methylcyclohexanone Chair Isomers Compared with Cyclohexanone

		Computed Torsion Angles (°)				
		C_3–C_2–C_1–C_6	C_1–C_2–C_3–C_4	C_2–C_3–C_4–C_5		
		C_2–C_1–C_6–C_5	C_1–C_6–C_5–C_4	C_3–C_4–C_5–C_6	Dipole Moment	Heat of Formation
Ketone	X–C_α–C=O	Dissymmetry[b]	Dissymmetry[b]	Dissymmetry[b]	(Debye)	(kcal/mole)
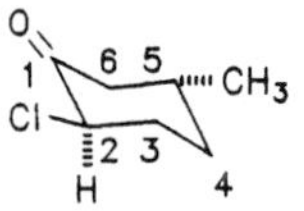	5.44	50.8 –51.5 [0.7]	–53.0 53.8 [0.8]	57.6 –57.4 [0.2]	4.19	–70.97
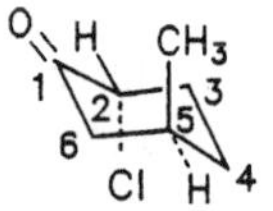	–110	–52.0 54.0 [2.0]	50.5 –52.9 [2.4]	–53.8 54.3 [0.5]	2.58	–70.33
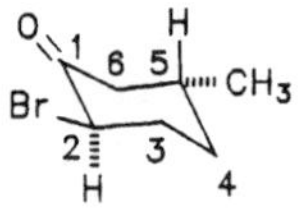	5.81	50.4 –51.2 [0.8]	–52.8 53.7 [0.9]	57.6 –57.6 [0.0]	4.07	–60.33
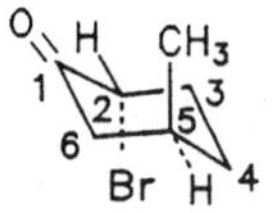	–103	–47.9 50.8 [2.9]	48.4 –52.6 [4.2]	–54.0 55.4 [1.4]	2.73	–60.16
		51.5 –51.5 [0.0]	–53.3 53.3 [0.0]	57.5 –57.5 [0.0]	2.8	–53.7

[a] Computed using PCMODEL, ref. 3.

[b] Ring dissymmetry [value] determined by the difference in absolute values of corresponding ring torsion angles.

cgs and $\Delta\varepsilon = +0.63$ for a 3-equatorial methyl perturber,[14] suggest a weak *positive* octant contribution from the α-equatorial chlorine of Figures 5-7 and 5-8, by difference: $R \sim +0.4 \times 10^{-40}$ cgs and $\Delta\varepsilon \sim +0.13$. Yet, according to the octant rule and to earlier estimates, α-equatorial chlorine should make a weak *negative* octant contribution ($\Delta\varepsilon \sim -0.15$). By way of comparison, an α-equatorial methyl at the same octant location would contribute $\Delta\varepsilon \sim -0.3$.[1] In contrast, using the preceding data for the diaxial isomer in EPA and taking into account the octant contribution from a β-axial methyl perturber ($R = +0.84 \times 10^{-40}$ and $\Delta\varepsilon = +0.28$),[14] one finds that an α-axial chlorine would exhibit a strong negative octant contribution: $R = -35.5 \times 10^{-40}$ cgs and $\Delta\varepsilon = -10$.[23] In comparison, the octant contribution of an α-axial methyl ($\Delta\varepsilon \sim -1.2$) is nearly an order of magnitude smaller.

(A)

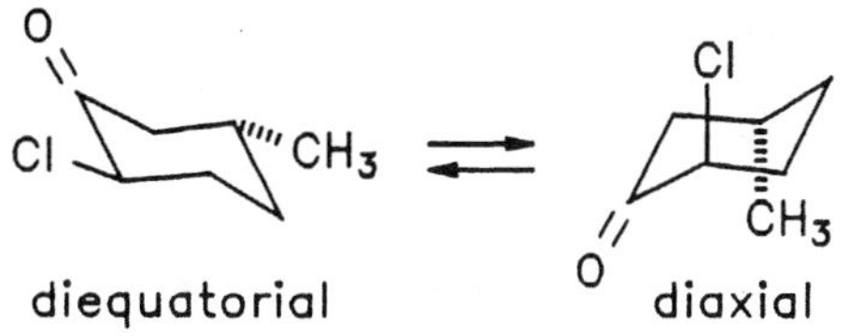

(B)

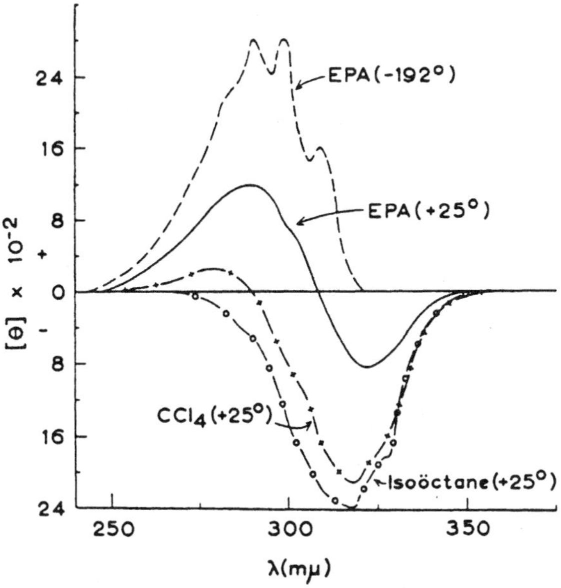

Figure 5-8. (A) Diequatorial chair ⇄ diaxial chair conformational inversion equilibrium of *trans*-2(*R*)-chloro-5(*R*)-methylcyclohexanone. (B) Variable-temperature circular dichroism spectra of *trans*-2(*R*)-chloro-5(*R*)-methylcyclohexanone in ether–isopentane–ethanol (EPA; 5:5:2, by vol.), compared with CD spectra at 25°C in CCl_4 and isooctane solvents. [(B) is reprinted with permission from Crabbé, P., *Optical Rotatory Dispersion and Circular Dichroism in Organic Chemistry*, Holden-Day, Inc., Boca Raton, FL, 1965. Copyright© Holden-Day, Inc.]

No CD measurements, variable temperature or otherwise, have been carried out on 2(*R*)-bromo-5(*R*)-methylcyclohexanone. As nearly as one can judge from the ORD[17] and other spectral analyses,[20] more of the diaxial isomer is present at equilibrium in hydrocarbon and in methanol solvents, compared with the corresponding chloroketone. This is surprising, because the computed dipole moments of corresponding diaxial isomers (Table 5-3) are very similar, as they are for the corresponding diequatorial isomers. Yet, in methanol, the negative ORD Cotton effect (Figure 5-7D) is dominated by the presence of an estimated 17% of the diaxial isomer. Further studies of this ketone, as well as of the corresponding 4-*tert*-butyl isomers, seem warranted.

5.6. Conformer Populations and ΔG from Variable-Temperature CD

In order to extract quantitative information from CD spectra, investigators make use of the quantitative measure of a CD transition, e.g., the rotatory strength (Chapter 3), or the wavelength-normalized area under a CD curve. Rotatory strengths are determined most conveniently with a computer program. Alternatively, as described in the appendix to Chapter 3, graphical (paper) methods may be used for calculating the rotatory strength. Several examples are presented next.

For an equilibrium between *two* species, $\boldsymbol{a} \rightleftarrows \boldsymbol{b}$, both of which absorb in the same spectral region, the observed rotatory strength will be the population-weighted average of the rotatory strength, from each species as expressed in the equation

$$R_0^T = X_a R_a + X_b R_b, \tag{5-1}$$

where X_a and X_b are the mole fractions of ***a*** and ***b***, and R_a and R_b are the rotatory strengths of the pure species ***a*** and ***b***. If ***b*** is the less stable, then, assuming a Boltzman distribution of the two conformers, it can readily be shown[22] that the observed rotatory strength is given by the relation

$$R_0^T = (R_a - R_b)\frac{1}{1 + \exp(-\Delta G^\circ/RT)} + R_b, \tag{5-2}$$

where ΔG° is the free energy for the equilibrium process, R is the gas constant, and T is the temperature (°K).

At first glance, solving this equation appears formidable, since there are three unknowns: R_a, R_b, andΔG°. On the other hand, note that Eq. (5-2) represents a straight line ($y = mx + b$), where $(R_a - R_b)$ is the slope and R_b is the intercept, when R_0^T is plotted against $1/[1 + \exp(-\Delta G^\circ/RT)]$.

Thus, if the rotatory strengths of at least three (or better yet, four or five) temperatures are known, one need merely plot R_0^T vs. $1/[1 + \exp(-\Delta G^\circ/RT)]$, using various *arbitrarily chosen* values of ΔG° until a straight line is obtained. The value of ΔG° that *uniquely* satisfies Eq. (5-2) (i.e., the value of ΔG° that gives a straight line) will be the difference in free energy between ***a*** and ***b***.

A specific example is the analysis of temperature-dependent rotatory strengths of (+)-*trans*-2(*R*)-chloro-5(*R*)-methylcyclohexanone[22] in EPA (Figure 5-8). In one run, the rotatory strengths obtained were $R_0^{25°C} = 0.445 \times 10^{-40}$, $R_0^{-29°C} = 1.310 \times 10^{-40}$, and $R_0^{-74°C} = 1.872 \times 10^{-40}$, cgs. By referring to a table of values $1/[1 + \exp(-\Delta G^\circ/RT)]$ for various values of ΔG° (e.g., Table 5-14 contains a few such values), it is a simple matter to plot R_0^T vs. $1/[1+ + \exp(-\Delta G^\circ/RT)]$ using arbitrary values of ΔG°. (In this

Table 5-14. Examples of Calculated Values of $1/[1 + \exp(\Delta G°/RT)]$ at Selected Temperatures (°K)

$\Delta G°$ (kcal)	298°K	244°K	199°K
1.630	0.94005	0.96647	0.98404
1.680	0.94463	0.96966	0.98591
1.730	0.94888	0.97255	0.98757

case, values of $\Delta G°$ = 1.630, 1.680, and 1.730 kcal were chosen.) From the plots of Figure 5-9, it can be seen that only $\Delta G°$ = 1.68 kcal yields a straight line. Other values of $\Delta G°$ will give lines with some degree of curvature, depending on the deviation from the correct value.

Since the slope of the straight line is $R_a - R_b$, Eq. (5-2) can be rearranged to give

$$R_b = R_0^T - \text{slope}\,[1 + \exp(-\Delta G°/RT)] \tag{5-3}$$

and R_a = slope + R_b. Substituting the appropriate values of R_0^T, $\Delta G°$, and T gives $R_b = -32.2 \times 10^{-40}$ cgs and, therefore, $R_a = 2.36 \times 10^{-40}$ cgs. R_a and R_b represent the rotatory strengths of the diequatorial (***a***) and diaxial (***b***) conformers, respectively, *even though neither species can be isolated as such.* Actually, the assignment of R_a and R_b

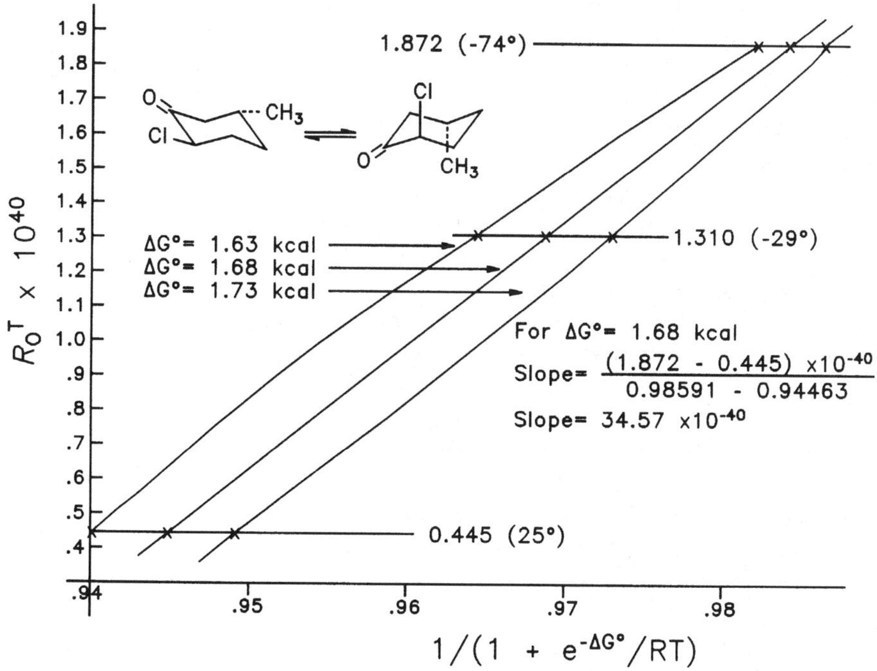

Figure 5-9. Plots of Rotatory Strength vs. $(1 + e^{-\Delta G°}/RT)^{-1}$ for given values of $\Delta G°$.

to conformers ***a*** and ***b*** is clearly not a direct result of the calculation, since the derivation merely assumes that ***b*** is the less stable species. However, since the absolute configurations of ***a*** and ***b*** are as shown in Figure 5-8, both the sign and the magnitude of R_a and R_b require the assignments to be made as indicated (by the octant rule).

5.7. Dimethyl- and Trimethylcyclohexanones

In a study designed to reveal the octant contributions of methyl perturbers on cyclohexanones, Djerassi et al.[24] pursued an analysis of cyclohexanone conformations, with a focus on probable twist conformations. Since that study assumed an additivity of contributions from octant perturbers, it was essential to use exact ORD or CD $n \rightarrow \pi^*$ Cotton effect magnitudes, corrected to 100% enantiomeric excess. One of the earliest estimates[25] that the octant contribution of an α-equatorial methyl was zero was made by subtracting the ORD Cotton effect amplitude *a* of 3(*R*)-methylcyclohexanone ($a =$ +25) from that of *trans*-2(*R*),5(*R*)-dimethylcyclohexanone ($a = +25$) (Table 5-15). The method assumes that the β-equatorial methyl group in each ketone makes the same octant contribution and that each ketone adopts the (identical) chair conformation. The results of this study gave rise to the belief that an α-equatorial methyl group would make no contribution, but this is now known to be incorrect.[1] In fact, the ORD ($a =$ -12) or CD ($\Delta\varepsilon \sim -0.30$) $n \rightarrow \pi^*$ Cotton effect of 2(*R*)-methylcyclohexanone is nonzero. And since it undergoes essentially no change down to −192°C (Figure 2-16F), one can assume that the experimentally determined value does not originate from a small population of higher energy conformations. The data suggest that either (i) the ORD *a* value of *trans*-2(*R*),5(*R*)-dimethylcyclohexanone, which is > 99% diequatorial, is too high or (ii) the compound's cyclohexanone ring does not adopt the same chair conformation as either of the corresponding monomethyl cyclohexanones, which are known to adopt an undistorted equatorial chair conformation almost exclusively. The latter explanation seems only remotely possible, because in *trans*-2,5-dimethylcyclohexanone both methyls are equatorial and molecular mechanics calculations[3] show that the ring conformation is very similar to that of chair cyclohexanone (Table 5-16). Clearly, some clarification is needed. The ORD of *trans*-2(*R*),5(*R*)-dimethylcyclohexanone should be rerun, and its CD spectrum should be measured, as, apparently, it has never been recorded.

The octant contribution ($a = +56$) of an α-axial methyl group was also determined at an early date[25] from the ORD Cotton effects of 2,2,5(*R*)-trimethylcyclohexanone (a $= +81$) (Table 5-15) and 2(*R*),5(*R*)-dimethylcyclohexanone ($a = +25$) by subtraction. The trimethyl ketone has both an α-equatorial methyl and a β-equatorial methyl, the same as in the dimethyl ketone, and it was assumed that the two ketones adopt the same chair cyclohexanone conformation. The derived value for an α-axial methyl octant contributor ($a = +56$) was, however, found to be too low[24] compared to the octant contribution of the α-axial methyl in 2(*S*)-methyl-4(*R*)-*tert*-butylcyclohexanone ($a =$ $+71$ or $\Delta\varepsilon = +1.60$; see Table 5-15.) The lack of agreement suggested the presence at equilibrium of a small percent of chair conformation with two axial methyl groups in

Table 5-15. Ketone $n \rightarrow \pi^*$ Cotton Effect Intensities as Determined by ORD Amplitudes a in Methanol and CD $\Delta\varepsilon$ Values in EPA[a] at Room Temperature[b]

Ketone	ORD Amplitude a[c]	Calculated $\Delta\varepsilon$[d]	Experimental $\Delta\varepsilon$[a]	Chair ⇄ Chair Conformational Equilibrium	$\Delta\Delta H_f$[e]
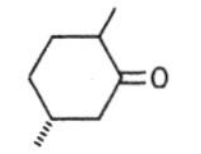	+25	+0.62	N/A	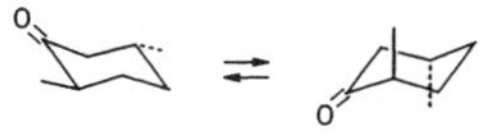	3.70
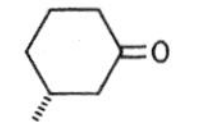	+25	+0.62	+0.60		2.39
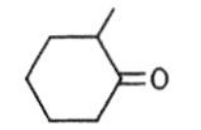	−12	−0.30	−0.30	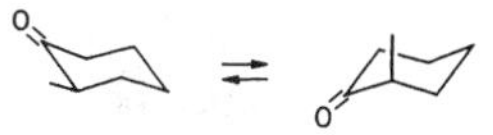	2.29
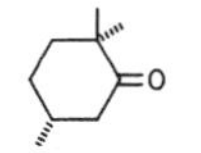	+81	+2.0	N/A		1.38
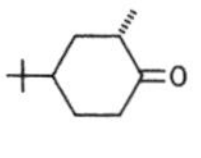	+71	$+1.7_6$	+1.6	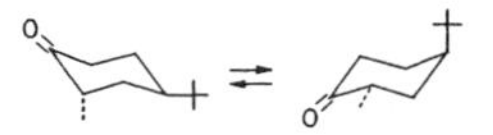	2.56
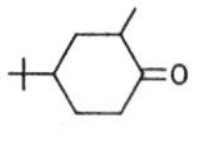	−17.5	−0.43	−0.42	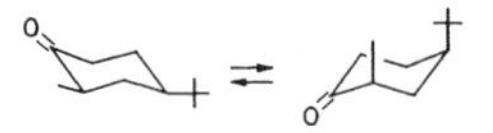	10.34
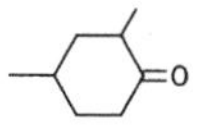	−16.5	−0.41	N/A	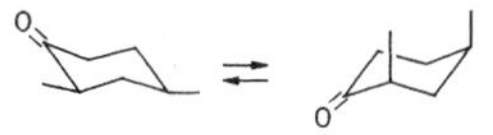	5.51
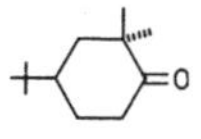	+64	+1.6	+1.35 +1.59[d]	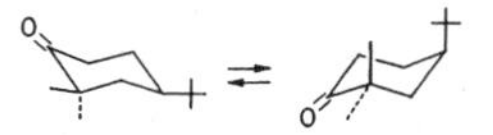	9.01
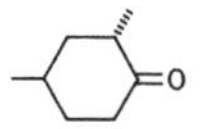	+35	+0.87	N/A		0.56
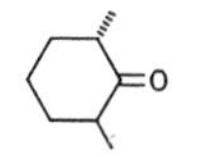	+75	$+1.8_6$	N/A		0.00
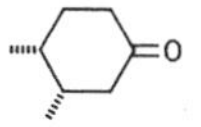	+16	+0.38	N/A	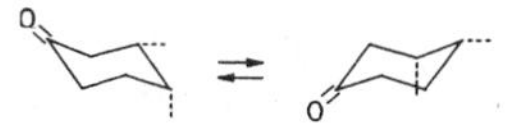	−0.22
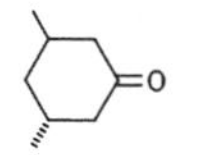	+37 +23[a]	+0.92	+0.58	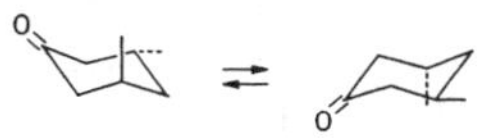	0.00
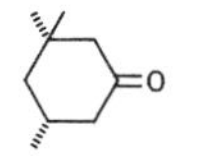	+1[a]	+0.02	+0.03	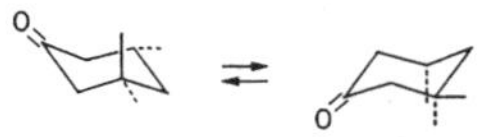	3.70

[a] In ether–isopentane–ethanol (EPA; 5:5:2, by vol.);

[b] Experimentally determined ORD values are taken from refs. 24–29; experimentally determined CD values are from refs. 4, 5, 14, 28, and 29;

[c] Calculated from $a = 40.28\ \Delta\varepsilon$, as recommended in ref. 5;

[d] In CH_3OH;

[e] Differences in the computed (PCMODEL, ref. 3) heats of formation ($\Delta\Delta H_f$) are reported in kcal/mole; $+\Delta\Delta H_f$ means that the isomer on the right is less stable.

the trimethylcyclohexanone.[24] This may at first seem unlikely; however, given an energy difference between the two chair conformations of the trimethylcyclohexanone isomers computed by molecular mechanics[3] to be only 1.38 kcal/mole, the mole fraction of the diaxial conformation is calculated to be approximately 0.1 at room temperature—an amount sufficient to reduce the expected value of *a* by approximately 20%, from about +71 to about +57.

Alternatively, the anomalously low derived value of *a* for an α-axial methyl octant perturber appears to support the notion that value of *a* for *trans*-2(*R*),5(*R*)-dimethylcyclohexanone found in the literature[26] (25; see Table 5-15) is too low. In fact, it may be estimated to be *a* = +13, assuming additivity and the correct *a* values for the octant contributions of a β-equatorial methyl (*a* = +25, from 3(*R*)-methylcyclohexanone) and an α-equatorial methyl (*a* = −12, from 2(*R*)-methylcyclohexanone). An alternative estimate (*a* = +8) may be made, assuming that an α-equatorial methyl perturber contributes *a* = −17 (the average value from 2(*R*),4(*R*)-dimethylcyclohexanone (*a* = −16.5) and 2(*S*)-methyl-4(*R*)-*tert*-butylcyclohexanone (*a* = −17.5). Using the first estimate (*a* = +13) for 2(*R*),5(*R*)-dimethylcyclohexanone and the experimental value (*a* = +81) for 2,2,5(*R*)-trimethylcyclohexanone, one arrives at a value *a* = +68 for the octant contribution of an α-axial methyl. Using the second estimate (*a* = +8), one finds that *a* = +73 for an α-axial methyl group. Both predicted values are very close to that measured for 2(*S*)-methyl-4-*tert*-butylcyclohexanone (*a* = +71, Table 5-15).

2,2,5(*R*)-Trimethylcyclohexanone is probably a flawed model, however, and should be reinvestigated using low-temperature CD measurements. If molecular mechanics calculations[3] (Table 5-16) accurately predict both the large ring distortion (from chair cyclohexanone) and dissymmetry, the result is an anomalously large CH_3–C_α–C=O torsion angle (~ −15°) for an α-equatorial methyl. [Cf. ~ −5° in 2-methylcyclohexanone (Table 5-2).] One might therefore anticipate difficulty in using the trimethylcyclohexanone model, as the magnitude and sign of the α-methyl perturber has been shown to be highly dependent on the CH_3–C_α–C=O torsion angle[4] (Table 5-7). A potentially better model, 2,2-dimethyl-4(*R*)-*tert*-butylcyclohexanone,[5] was not available during the early studies, but its *a* value (+64) is also not very close to what one would calculate (*a* = +53.5) by summing the contributions from an α-axial methyl (*a* = +71) and an α-equatorial methyl (*a* = −17.5) on 4-*tert*-butylcyclohexanone. The lack of agreement is apparently due to ring distortion (Table 5-16), which significantly alters the location of the equatorial methyl, in addition to bringing in octant contributions from the ring atoms. The *a* = +64 value for 2,2,5(*R*)-trimethylcyclohexanone is, however, close to that (*a* = +67) suggested[24] for an α-axial methyl perturber and is arrived at by subtracting the *a* value of *trans*-9(*S*)-1-decalone from that of *trans*-9(*S*)-methyl-1-decalone.[24]

Despite the apparent reconciliation of $n \rightarrow \pi^*$ Cotton effects, additional complications arise when one uses the experimental ORD Cotton effect contribution of α-equatorial (| *a* | = 12 to 17) and α-axial (| *a* | = 71) methyls for predicting the ORD Cotton effect of 2(*S*),6(*S*)-dimethylcyclohexanone (Table 5-15), which has one α-axial and one α-equatorial methyl. Since the observed value (*a* = +75) is different from the predicted value (*a* = +83 to 88), one might suspect ring distortion, especially since the

Table 5-16. Computed[a] Ring Torsion Angles, CH_3–C_α–C=O Torsion Angles, and Heats of Formation for Methylated Cyclohexanones

		Computed Torsion Angles (°C)			
Ketone	CH_3–C_α–C=O	C_3–C_2–C_1–C_6 C_2–C_1–C_6–C_5 Dissymmetry[b]	C_1–C_2–C_3–C_4 C_1–C_6–C_5–C_4 Dissymmetry[b]	C_2–C_3–C_4–C_5 C_3–C_4–C_5–C_6 Dissymmetry[b]	Heat of Formation (kcal/mole)
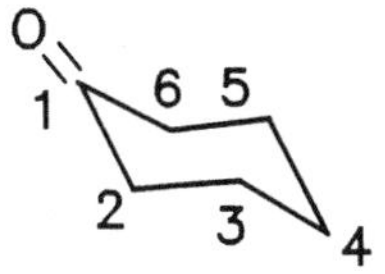		51.5 –51.5 [0.0]	–53.3 53.3 [0.0]	57.5 –57.5 [0.0]	–53.7
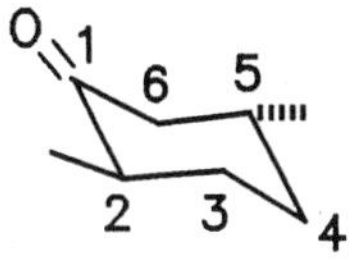	–5.43	50.9 –51.8 [0.9]	–52.9 53.3 [0.4]	57.8 –57.2 [0.6]	–68.61
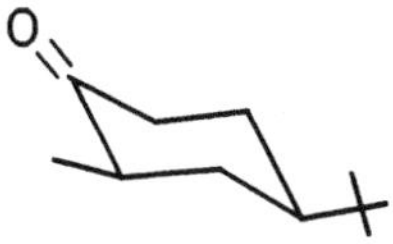	–4.44	51.9 –52.0 [0.1]	–54.8 53.9 [0.9]	57.1 –56.0 [1.1]	–85.10
	98.6	46.2 –48.3 [2.1]	–50.8 53.8 [3.0]	56.3 –57.1 [0.8]	–82.82
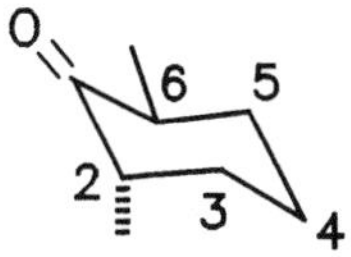	–7.61 (eq) –99.4 (ax)	46.9 –48.7 [1.8]	–49.7 53.2 [3.5]	56.4 –58.1 [1.7]	–66.09
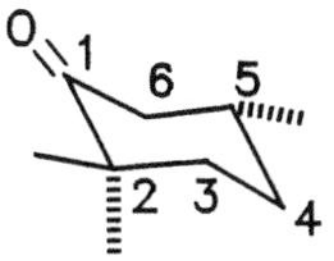	–15.2 (eq) 103.2 (ax)	46.4 –49.9 [3.5]	–49.4 53.2 [3.8]	57.0 –57.1 [0.1]	–73.37
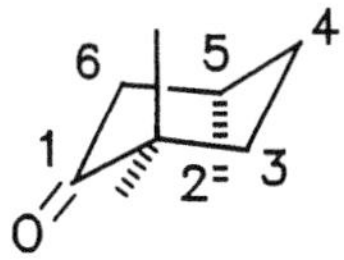	14.6 (eq) –103.7 (ax)	–46.6 51.3 [4.7]	47.3 –53.3 [6.0]	–54.5 55.8 [1.3]	–71.99
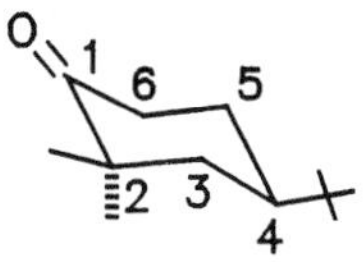	–15.0 (eq) 103.0 (ax)	46.4 –50.3 [3.9]	–49.9 55.0 [5.1]	55.9 –56.9 [1.0]	–89.84

(Continued)

Table 5-16. (Continued)

Ketone	$CH_3{-}C_\alpha{-}C=O$	Computed Torsion Angles (°C) $C_3{-}C_2{-}C_1{-}C_6$ $C_2{-}C_1{-}C_6{-}C_5$ Dissymmetry[b]	$C_1{-}C_2{-}C_3{-}C_4$ $C_1{-}C_6{-}C_5{-}C_4$ Dissymmetry[b]	$C_2{-}C_3{-}C_4{-}C_5$ $C_3{-}C_4{-}C_5{-}C_6$ Dissymmetry[b]	Heat of Formation (kcal/mole)
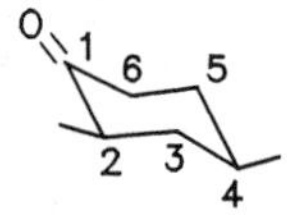	−5.71	50.6 −51.4 [0.8]	−53.0 53.7 [0.7]	57.4 −57.3 [0.1]	−68.59
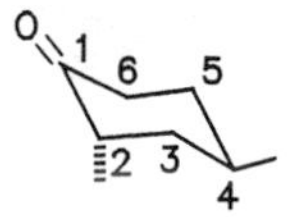	97.52	45.0 −47.5 [2.1]	−49.3 53.2 [3.9]	56.6 −57.9 [1.3]	−66.29
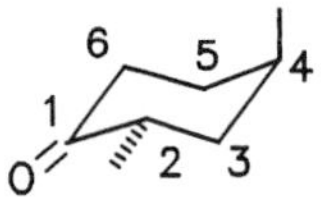	−6.72	−49.7 50.5 [0.66]	51.6 −52.3 [0.63]	−54.8 54.6 [0.17]	−66.85
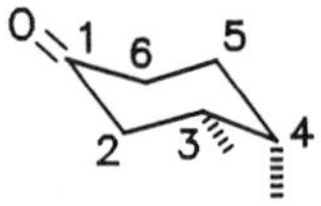	—	47.8 −47.9 [0.1]	−51.5 52.2 [0.7]	56.1 −56.8 [0.7]	−66.02
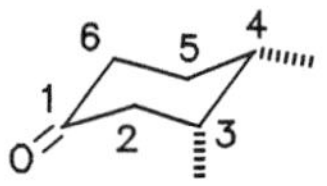	—	−53.1 50.7 [2.4]	54.3 −51.4 [2.9]	−56.5 55.2 [0.7]	−66.24
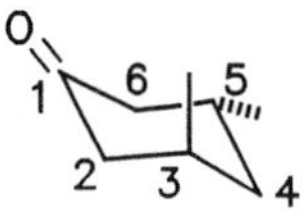	—	52.5 −51.7 [0.8]	−52.6 51.2 [1.4]	52.6 −55.0 [2.4]	−67.35
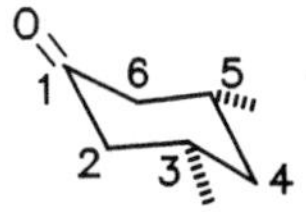	—	50.5 −50.5 [0.0]	−52.7 52.7 [0.0]	57.6 −57.6 [0.0]	−68.78
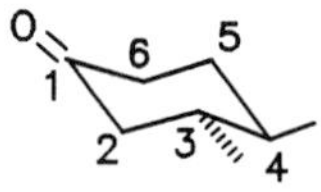	—	50.5 −50.2 [0.3]	−52.8 53.5 [0.7]	56.7 −57.8 [1.1]	−67.47
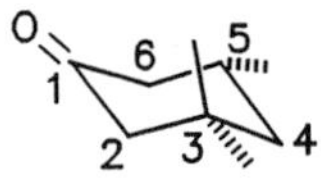	—	53.0 −51.8 [1.2]	−52.4 50.9 [1.5]	55.1 −54.7 [0.4]	−75.63
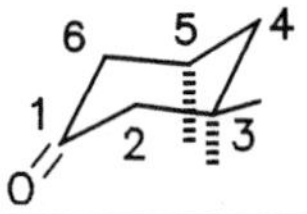	—	−56.9 56.2 [0.7]	49.5 −49.2 [0.3]	−47.6 47.8 [0.2]	−71.93

[a] Computed using PCMODEL, ref. 3.

[b] Ring dissymmetry [value] as determined by the difference in absolute values of corresponding ring torsion angles.

chair ⇄ chair conformational inversion (Table 5-15) leaves the molecule unchanged. (There is no net change in the number and location of axial and equatorial methyls.) Molecular mechanics calculations[3] in fact show ring distortion and nonsymmetric flattening near the C=O group in the midsection, both resulting in a larger-than-normal α-equatorial CH_3–C_α–C=O torsion angle (–7.6°). (Cf. Table 5-7.) It might also be argued that desymmetrization of the ring leads to octant contributions from cyclohexanone ring atoms, but the ring twisting is probably not as great as that previously suggested.[24] Low-temperature CD spectroscopy applied to 2(*S*),6(S)-dimethylcyclohexanone might be useful in distinguishing high-energy conformations and in yielding a reliable Δε for the chair.

A more interesting conformational problem arises with *trans*-2(*S*),4(*R*)-dimethylcyclohexanone. Here, the chair ⇄ chair conformational inversion (Table 5-15) interconverts two distinct isomers, one with an α-axial methyl and the other with an α-equatorial methyl. Assuming that there is no cyclohexanone ring distortion and no contribution from the C(4) axial or equatorial methyl, one can determine the position of the conformational equilibrium from the observed $n \rightarrow \pi^*$ ORD Cotton effect (a = +35) and the predicted values for the two different chair conformers: a = +17 for the conformer with an α-equatorial methyl and a = +71 for the α-axial methyl. (The value a = +17 is the average of the a values for the *enantiomers* of 2(*R*)-methyl-4-*tert*-butylcyclohexanone and 2(*R*),4(*R*)-dimethylcyclohexanone, respectively, from Table 5-15.) Thus, $35 = 17(X_e) + 71\,(1 - X_e)$, where the mole fraction X_e of an α-equatorial methyl conformer is approximately 0.67 and the mole fraction $1 - X_e$ of α-axial isomer is around 0.33. According to these mole fractions, the energy difference is predicted to be $\Delta G°$ = 0.41 kcal/mole, which is moderately close to the ΔH = 0.56 predicted by molecular mechanics calculations.[3] These data indicate moderate ring distortion, similar to, but larger than, that seen with 2(*S*)-methyl-4(*R*)-*tert*-butylcyclohexanone (Table 5-16). Here, too, variable-temperature CD might be expected to clarify the equilibrium by eliminating contributions from high-energy conformations at the lowest temperatures used in the CD measurements.

Given a secure value for the octant contribution of a β-equatorial methyl group (| a | = 25, | Δε | = 0.6), efforts were made to determine the octant contribution of a β-axial methyl by using cyclohexanone models. In one study, the chair ⇄ chair cyclohexanone ring inversion (Table 5-15) was investigated in *cis*-3(*R*),4(*S*)-dimethylcyclohexanone, which has two different chair cyclohexanones, one with an axial β-methyl (and equatorial C(4) methyl) and the other with an equatorial β-methyl (and axial C(4) methyl). These two conformations are thought to be in dynamic equilibrium, with a net Cotton effect magnitude a = +16[27] (Table 5-15). The authors[27] determined the equilibrium to be 70:30 at room temperature, with the β-axial isomer predominating. This seems qualitatively correct, as an enthalpy difference of –0.22 kcal/mole is computed by molecular mechanics[3] (Table 5-16) for the equilibrium of Table 5-15. Assuming that the methyl at C(4) makes no contribution to the Cotton effect (i.e., that the ring remains in a chair without dissymmetry that would cause the C(4) methyl to leave the octant nodal plane), one would predict that only the β-methyls would make octant contributions in these isomers. Since one also would predict an ORD Cotton effect $a = +25$ for the β-equatorial isomer, the calculated value of a for the conformer with a β-axial methyl is +12. This result follows from the expression $16 = 0.3(25) + 0.7a$, where the observed value of a is +16 for the dimethylcyclohexanone at (70:30) equilibrium (Table 5-15). The calculated value of a is positive, which is opposite to that predicted by the octant rule for back octants, but is in agreement with studies of β-axial methyl adamantanones and other rigid ketones with well-defined β-axial stereochemistry, where the methyl projects into a front octant. However, the magnitude is unusually large, suggesting ring distortion. Moderate ring distortion is, in fact, predicted by molecular mechanics calculations[3] for the β-axial isomer, but not for the β-equatorial one. The relevant equilibrium shown in Table 5-15 could easily be reinvestigated by variable-temperature CD, from which both an equilibrium constant and the $\Delta\varepsilon$ (or a) values of the two isomers can be derived.[12,22]

A different attempt to determine the octant contribution of a β-axial methyl group was made in a study of 3(*R*),5(*R*)-dimethylcyclohexanone and 3,3,5(*R*)-trimethylcyclohexanone[28] (Table 5-15). In the chair ⇄ chair conformational equilibrium of the former, there is no net change in the number of axial and equatorial methyl groups. Assuming a chair cyclohexanone, the compound has one β-equatorial methyl and one β-axial methyl. With an experimental $a = +37$ in methanol, one calculates that $a = +12$ for the β-axial methyl, given that $a = +25$ for the β-equatorial methyl. The magnitude of the calculated value for a β-axial methyl is the same as that derived (see earlier) from 3,4-dimethylcyclohexanone, and again, the sign is opposite to that predicted by the octant rule for back octants. Here, however, possible ring distortion is less pronounced (Table 5-16).

In the chair ⇄ chair conformational equilibrium defined in Table 5-15 for 3,3,5(*R*)-trimethylcyclohexanone, the isomer with two β-axial methyl groups is predicted to be much less stable than that with one β-axial methyl. If the latter adopts an undistorted chair, the octant contributions of two β-equatorial methyls are expected to cancel due to internal compensation, leaving a β-axial methyl as the lone dissymmetric perturber. The observed[28] $a = +1$ from the ORD Cotton effect and $\Delta\varepsilon = +0.03$ for the CD Cotton effect can thus be assigned as the contribution from the β-axial methyl. This is opposite to that predicted by the octant rule for back octants, but very much in accord with the prediction for a β-axial methyl in a front octant. And the magnitude (including the trend to larger values at lower temperatures) is very much what one would expect, based on studies of β-methyladamantanone[14] (Table 4-2).

5.8. Bisignate $n \rightarrow \pi^*$ Cotton Effects: Menthone and Isomenthone

In one of the most detailed analyses of conformations of an optically active cyclohexanone, Wellman et al.[29] examined *trans*-2(*S*)-isopropyl-5(*R*)-methylcyclohexanone ((−)-menthone) and *cis*-2(*R*)-isopropyl-5(*R*)-methylcyclohexanone ((+)-isomenthone) by CD in a wide variety of solvents (Figure 5-10). The positive $n \rightarrow \pi^*$ Cotton effects of (+) isomenthone appeared relatively insensitive to solvent, over a wide range

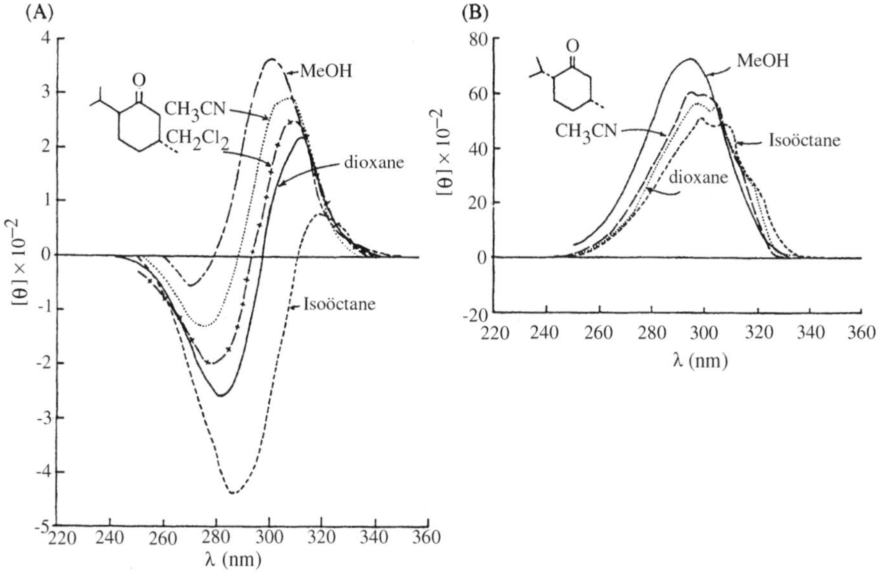

Figure 5-10. Influence of solvent on the $n \rightarrow \pi^*$ Cotton effects in the CD spectra of (A) (–)-menthone and (B) (+)-isomenthone at 25°C. [Reprinted with permission from ref. 29. Copyright © 1965 American Chemical Society.]

of polarity, from methanol to isooctane, at 25°C. In contrast, quite surprisingly, strongly solvent-dependent bisignate $n \rightarrow \pi^*$ CD Cotton effects were found for (–)-menthone, with a largely positive Cotton effect in methanol and a largely negative Cotton effect in isooctane. Previously, solvent-dependent bisignate $n \rightarrow \pi^*$ CD Cotton effects had been noted in *trans*-2(*R*)-chloro-5(*R*)-methylcyclohexanone. (See Section 5.5.) These observations were explained on the basis of a chair ⇄ chair conformational equilibrium and a dependence of the solvent on the orientation of the C=O and C–Cl dipoles in the diaxial and diequatorial conformers. However, no such sets of dipoles are as obvious in (–)-menthone. And if they were important, one is left to wonder why (+)-isomenthone does not also exhibit a strong solvent-dependent CD. Bisignate Cotton effects in the CD spectra of (–)-menthone may have their origin in a chair ⇄ chair conformational equilibrium, but just how the solvent acts to affect the equilibrium is not immediately clear.

Bisignate Cotton effects had also been observed previously for the $n \rightarrow \pi^*$ transition of *structurally rigid* ketones such as isofenchone.[7] In those studies, two CD Cotton effects of opposite sign were observed, and their intensities were found to be solvent-dependent. This unusual behavior was interpreted in terms of a solvational equilibrium consisting of at least two species with different degrees of solvation and oppositely signed $n \rightarrow \pi^*$ Cotton effects.[6,7,29] So in the case of (–)-menthone, which is not restricted to one conformation, as are isofenchone, camphor, etc., the origin of the

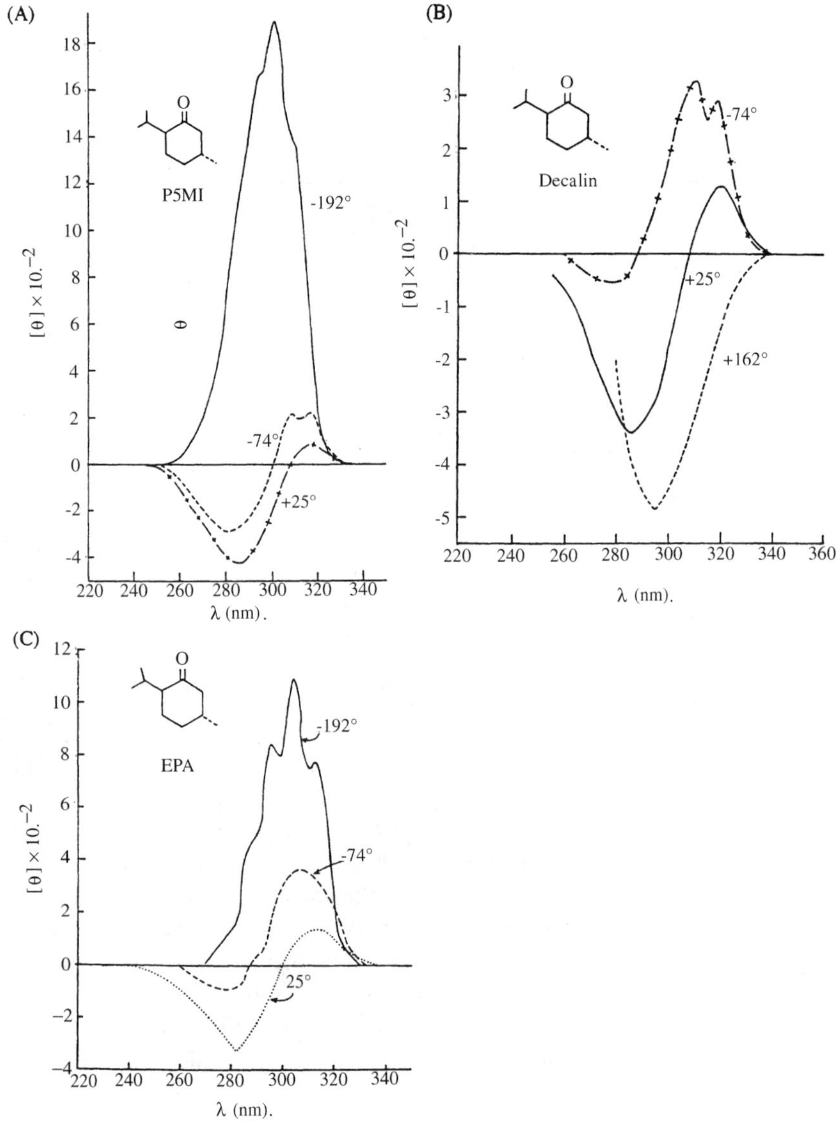

Figure 5-11. Temperature-dependent CD curves for the $n \rightarrow \pi^*$ transition of (−)-menthone in (A) P5M1 methylcyclohexane–isopentane (1:5, by vol.), (B) decalin, and (C) ether–isopentane–ethanol (EPA), 5:5:2, by vol.). Temperatures (°C) are indicated on the curves. [Reproduced with permission from ref. 29. Copyright © 1965 American Chemical Society.]

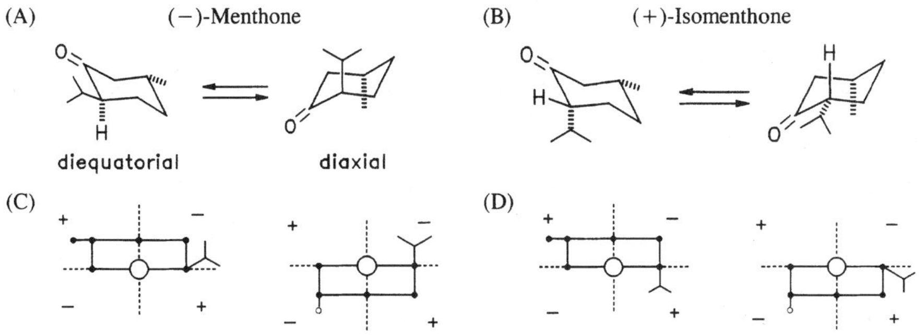

Figure 5-12. Chair ⇄ chair conformational equilibria in (A) (−)-menthone and (B) (+)-isomenthone. (C) Octant projection diagrams for the diequatorial (left) and diaxial (right) conformations of (−)-menthone. (D) Octant projection diagrams for the conformations of (+)-isomenthone.

bisignate Cotton effects might come from solvational equilibria. One consequence of the solvational equilibrium hypothesis is an increase in the population of complexed (or more highly solvated) species upon lowering the temperature of the solution. In the CD spectrum, this means that the short-wavelength band should increase at the expense of the longer wavelength band, as is observed in isofenchone and even norcamphor.[6] However, as may be seen in the temperature-dependent CD spectra of (−)-menthone, it is the *longer wavelength* component that grows upon lowering the temperature (Figure 5-11). Consequently, it appears more likely that the origin of the bisignate Cotton effect comes from the presence of at least two conformations that have oppositely-signed Cotton effects.

For both (−)-menthone and (+)-isomenthone, the chair ⇄ chair cyclohexanone ring inversion equilibrium (Figure 5-12) is especially relevant, as it produces conformational isomers with oppositely signed Cotton effects predicted by the octant rule. In (−)-menthone, the diequatorial conformer is expected to be more stable than the diaxial one and is predicted by the octant rule to exhibit a positive $n \rightarrow \pi^*$ Cotton effect. The diaxial conformer is predicted to give a negative Cotton effect, but the relative intensities of the two Cotton effects can be predicted only roughly, with the magnitude of the diaxial effect expected to be much greater than that of the diequatorial one. Wellman et al.[29] assumed that the β-methyl and α-isopropyl octant contributions were additive in order to estimate ORD amplitudes and obtained $a \sim +12$ for the diequatorial effect and $a \sim -131$ for the diaxial effect. However, those are only crude approximations, and the diaxial value is probably overestimated because the contribution of an axial β-methyl was only assumed ($a \geq -33$) and is now known to be much weaker ($a \sim +1$). Consequently, for the diaxial isomer, $a \sim +97$. Using the predicted values of a, Wellman et al.[29] calculated the mole fraction of diaxial isomer as 0.028 from the observed[30] amplitude ($a = +8$) for (−)-menthone. With the revised a value for the diaxial isomer, its mole fraction becomes 0.037—still very small. Nonetheless, even after correction, the qualitative difference between the expected Cotton effect magnitudes from diequatorial and diaxial conformers are substantial, suggesting that only a minor component of the diaxial isomer in the equilibrium (Figure 5-12) can exert a considerable contribution to the net Cotton effect. This notion is illustrated in the fact that if one approximates $\Delta G^\circ \sim \Delta H^\circ$ for the conformational equilibrium (Figure 5-12A), then the $\Delta H \sim 2.55$ kcal/mole approximated by molecular mechanics computations[3] translates into a 1.3% population of diaxial isomer at 25°C, a 10^{-7}% population at −192°C, and a 5% population at +162°C. Other workers[23] have analyzed the conformations further, focusing on the varying octant contributions of the equatorial α-isopropyl group.

Although this overall qualitative picture suffices to explain the origin of the two oppositely signed CD Cotton effects of Figure 5-10A, it does not explain the origin of the solvent effect on the equilibrium of Figure 5-12A, viz., why there is apparently relatively more diequatorial conformer in a polar solvent such as methanol, compared with the hydrocarbon isooctane. Nor does it explain why the resultant CD curve should have a bisignate shape. An explanation for the bisignate shape is of fundamental importance and has been explained as follows:[29] When two oppositely signed CD curves overlap, there may be complete cancellation (and the disappearance) of one curve, or there may be only partial cancellation of each. The former occurs when λ^{max} is the same for the two curves, the latter when λ^{max} differs—even very slightly (e.g., 1 nm). This effect can be shown by assuming that CD curves can be approximated by Gaussian curves of the form $A_i \exp[-(\lambda - \lambda_i)^2/\Delta^2]$, where A_i is the amplitude of the ith curve, λ_i is the value of λ at the center of the curve (λ^{max}), and Δ is the half-width at $1/e$ times the height. The superposition $F(\lambda)$ of two oppositely signed Gaussian CD curves gives a resultant curve with a new λ^{max}. The new curve $F(\lambda)$ takes on a bisignate shape for any set of A_i, λ_i, and Δ_i, and the new (observed) λ^{max} can be obtained by inspection. The equation of the curve is

$$F(\lambda) = A_1 e^{-(\lambda-\lambda_1)^2/\Delta_1^2} - A_2 e^{-(\lambda-\lambda_2)^2/\Delta_2^2} \tag{5-4}$$

Now, consider the case where $A_1 = A_2$ and $\Delta_1 = \Delta_2 = \Delta$. If one sets $(\delta F/\delta\lambda) = 0$, it can be shown[29] that, for $(\lambda_2 - \lambda_1)$ small compared with Δ,

$$\delta\lambda^{max} \sim \Delta\sqrt{2}, \tag{5-5}$$

where $\delta\lambda^{max} = \lambda_2^{max} - \lambda_1^{max}$, the difference in wavelength between the extrema (λ^{max}) of the summation curve from Eq. (5-4). Interestingly, to a first approximation within the constraints indicated, $\delta\lambda^{max}$ is independent of $(\lambda_2 - \lambda_1)$. Since $\Delta \sim 20$ nm for many $n \rightarrow \pi^*$ Cotton effect curves,

$$\delta\lambda^{max} \sim 30 \text{ nm}. \tag{5-6}$$

This interesting prediction has been confirmed graphically[29] and holds true for $A_1 = A_2$, $A_1 = 2A_2$, and even $A_1 = 5A_2$ over a range of $(\lambda_2 - \lambda_1)$ from 1–20 nm. This important and striking result confirms that *separation between* λ^{max} *of the* $n \rightarrow \pi^*$ *bisignate Cotton effects is typically about 30 nm.* Examples of overlapping, oppositely signed CD curves (approximated by Gaussian curves) can be seen in Figure 5-13.

Unlike (–)-menthone, (+)-isomenthone shows the more normal monosignate CD Cotton effects, which are relatively insensitive to changes in solvent (Figure 5-10B). Even at very low temperatures, the positive $n \rightarrow \pi^*$ Cotton effect seen at 25°C for (+)-isomenthone remains positive and grows very little, suggesting that the predominant isomer present at –192°C is present at roughly the same percent at room

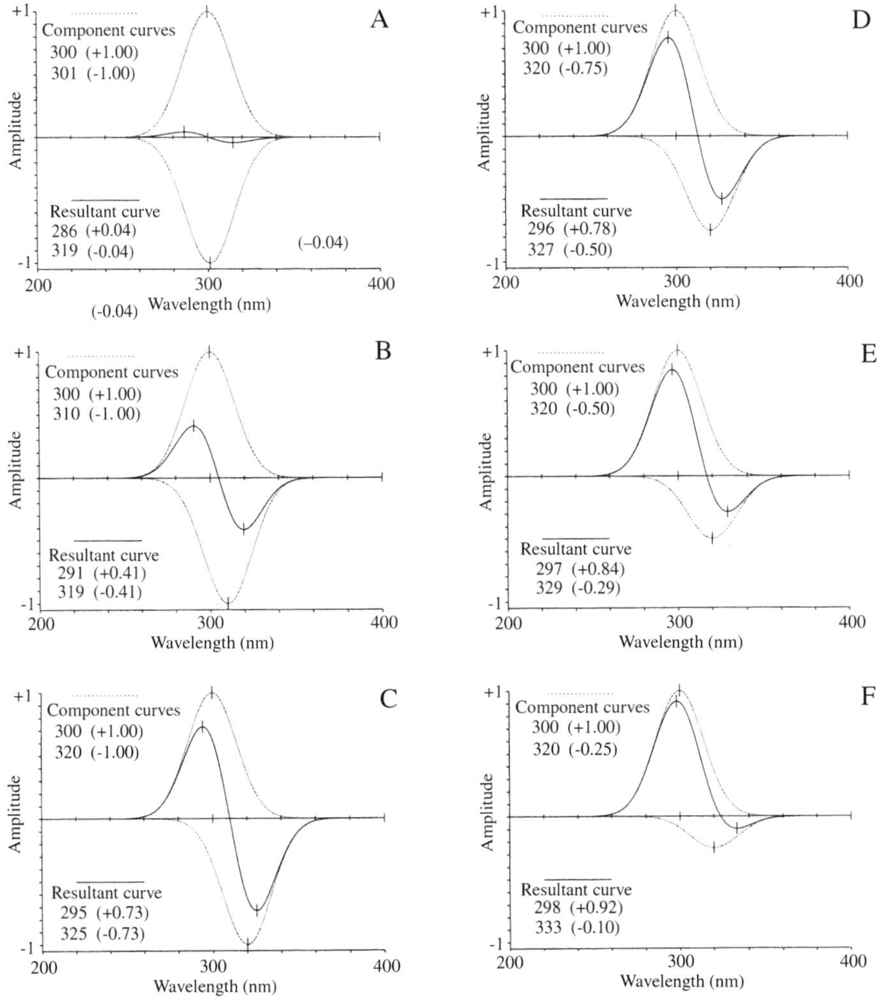

Figure 5-13. Addition of two oppositely signed Gaussian curves. Each diagram has a band centered at 300 nm with an intensity of +1.0. The half-bandwidth at $1/e$ times the height for all curves is set at 20 nm. In **A**–**C**, the intensity of the second curve is –1.0, and the bands are centered at 301 nm (**A**), 310 nm (**B**), and 320 nm (**C**). In **D**, **E**, and **F**, the curves are set at 300 and 320 nm with amplitudes of –0.75, –0.50, and –0.25 for the smaller curve, respectively. The resultant bisignate curve would be the observed curve with the peak positions and amplitudes indicated.

temperature (Figure 5-14A). With regard to chair cyclohexanone isomers (Figure 5-12B), that with an α-axial isopropyl is predicted by the octant rule to give a strongly positive Cotton effect, while that with an α-equatorial isopropyl is predicted to give a weakly positive Cotton effect. One reaches the surprising qualitative conclusion, therefore, that the conformation with the axial isopropyl is the more stable, and perhaps the predominant, isomer in the equilibrium.

The equilibrium constant may be estimated for the equilibrium of Figure 5-12B, assuming that octant contributions from β-methyl and α-isopropyl groups are additive (Table 5-17). From the observed $\Delta\varepsilon \sim +1.7$ of (+)-isomenthone, one computes a mole fraction of axial isopropyl isomer, $X_a \sim 0.33$, from the equation $+1.7 = 4.15(X_a) + 0.48(1 - X_a)$. These data suggest that the α-axial isopropyl isomer is the less prevalent isomer at room temperature and even at −192°C ($X_a \sim 0.46$), contrary to one element of the qualitative conclusion reached in this section. The data suggest that $K_{eq} \sim 1$—that the energies of the two isomers are comparable. The equilibrium constant at 25°C is calculated to be $K_{eq} \sim 0.49$, and $K_{eq} = 0.85$ at −192°C, with $\Delta G^{25^\circ} \sim 0.4$ kcal/mole and $\Delta G^{-192^\circ} \sim 0.03$ kcal/mole.

In support of small equilibrium values of ΔG°, a very small value of ΔH (~ −0.17 kcal/mole) is calculated for the equilibrium from molecular mechanics computations,[3] which predict that the α-axial isopropyl is slightly more stable than the α-equatorial one—opposite to the results from CD analysis. However, molecular mechanics calculations also indicate considerable ring distortion (Table 5-18) relative to cyclohexanone, especially with puckering near the carbonyl group in the α-equatorial isopropyl isomer. The nonsymmetric distortions render the additivity approach tenuous for

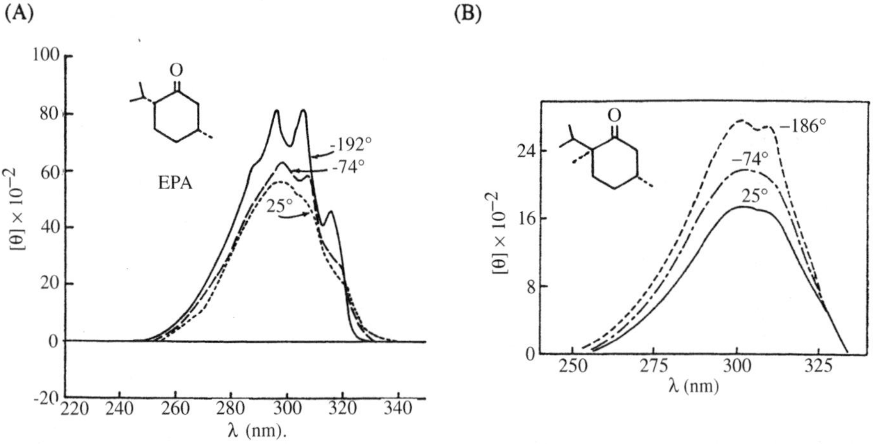

Figure 5-14. Temperature-dependent CD curves for the $n \to \pi^*$ transition of (A) (+)-isomenthone in EPA (ether–isopentane–ethanol, 5:5:2, by vol.) and (B) 2(*S*)-isopropyl-2,5(*R*)-dimethylcyclohexanone in methylcyclohexane–isopentane (1:3, by vol.). [(A) is reprinted with permission from ref. 29. Copyright© 1965 American Chemical Society. (B) is reprinted from ref. 31 with permission from The Chemical Society of Japan. Copyright© 1973.]

Table 5-17. Additivity of Methyl and Isopropyl Octant Perturbers for Predicting Cotton Effect Δε Values of (+)-Isomenthone and (–)-Menthone[a]

β-Methyl Octant Perturber		α-Isopropyl Octant Perturber	
equatorial	axial	equatorial	axial
Δε +0.63 (*R*)[b]	Δε –0.05 (*R*)[c]	Δε +0.53 (*R*)	Δε +3.52 (*R*)
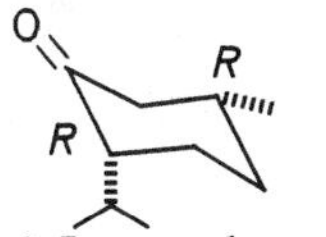 (+)-Isomenthone	Predicted net Δε +4.15	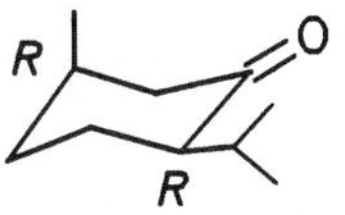 (+)-Isomenthone	Predicted net Δε +0.48
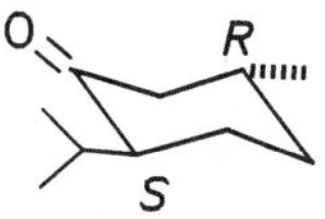 (–)-Menthone	Predicted net Δε +0.10	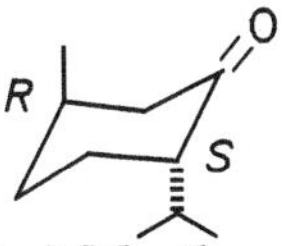 (–)-Menthone	Predicted net Δε –3.57

[a] Additivity values for EPA. Data from ref. 29.
[b] In equatorial β-methyladamantanone, Δε = +0.67 (Figure 4-7).
[c] Data from axial β-methyladamantanone (Figure 4-11).

estimating the net Δε values for the isomer (Table 5-17). The ring dissymmetry predicted by molecular mechanics for the α-equatorial isomer of (+)-isomenthone is larger than that predicted for α-isopropylcyclohexanone (Table 5-9), due apparently to the influence of the β-axial methyl in the former. In the α-axial equatorial isomer, surprisingly little ring distortion is calculated near the C=O group, relative to cyclohexanone, but significant ring dissymmetry is predicted across the ring, near C(4). Similar distortion is found for the α-axial isopropyl isomer of (–)-menthone (Table 5-18), but dissymmetry of the α-equatorial isopropyl isomer is not as great as that found for (+)-isomenthone. As with (+)-isomenthone, ring dissymmetrization may explain why the predicted Δε value (Table 5-17) for the more stable α-equatorial isopropyl isomer of (–)-menthone is somewhat different from that (Δε ~ +0.33) determined from the $n \rightarrow \pi^*$ CD curve at –192°C (Figure 5-10A).

The surprising bisignate $n \rightarrow \pi^*$ CD Cotton effects of (–)-menthone return to more normal behavior when an α-methyl group is added (Figure 5-14B).[31] Thus, 2(*S*)-isopropyl-2,5(*R*)-dimethylcyclohexanone shows monosignate positive Cotton effects at +25°C, –74°C, and –190°C, with Δε ~ +0.42, +0.63, and +1.13, respectively, in EPA.[31] In addition, no unusual solvent effects are detected; the CD curves are all monosignate and positive in methanol (Δε ~ +0.28), dioxane (Δε ~ +0.44), carbon tetrachloride (Δε ~ +0.60), and isooctane (Δε ~ +0.46) at 25°C. Like the chair ⇄ chair conformational equilibrium of (–)-menthone (Figure 5-12A), here the equilibrium is between diequatorial and diaxial isomers (Figure 5-15); and as in (–)-menthone, it might be assumed that the former is more stable than the latter. The diequatorial isomer has an

Table 5-18. Computed[a] Ring Torsion Angles, $(CH_3)_2CH–C_\alpha–C{=}O$ Torsion Angles, and Heats of Formation for (–)-Menthone and (+)-Isomenthone

Ketone	R–Cα–C=O	Computed Torsion Angles (°)			Heat of Formation (kcal/mole)
		$C_3–C_2–C_1–C_6$ $C_2–C_1–C_6–C_5$ Dissymmetry[b]	$C_1–C_2–C_3–C_4$ $C_1–C_6–C_5–C_4$ Dissymmetry[b]	$C_2–C_3–C_4–C_5$ $C_3–C_4–C_5–C_6$ Dissymmetry[b]	
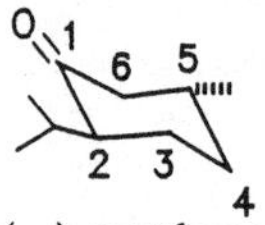 (−)-menthone	–2.64	55.3 –57.3 [2.0]	–53.6 54.9 [1.3]	56.8 –56.1 [0.7]	–78.89
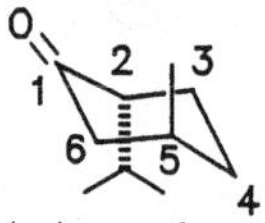 (−)-menthone	–104.5	–53.0 53.7 [0.7]	52.8 –51.3 [1.5]	–55.9 53.6 [2.3]	–76.34
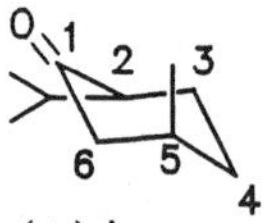 (+)-isomen-thone	+3.10	–55.6 58.4 [2.8]	52.0 –54.6 [2.6]	–54.6 54.4 [0.2]	–77.64
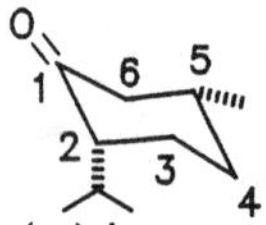 (+)-isomen-thone	+103.5	52.2 –52.1 [0.1]	–54.4 51.5 [2.9]	58.2 –55.3 [2.9]	–77.81
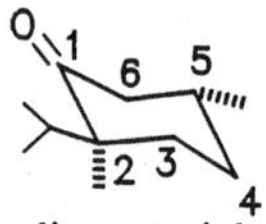 diequatorial	–8.69 (eq) +112.6 (ax)	50.8 –56.0 [5.2]	–49.6 54.7 [5.1]	55.2 –55.4 [0.1]	–81.04
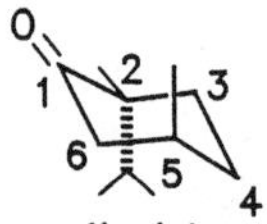 diaxial	+14.8 (eq) –106.7 (ax)	–50.2 54.7 [4.5]	49.5 –53.2 [3.7]	–55.0 54.2 [0.8]	–81.10
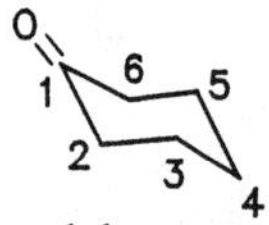 cyclohexanone		51.5 –51.5 [0.0]	–53.3 53.3 [0.0]	57.5 –57.5 [0.0]	–53.7

[a] Computed using PCMODEL, ref. 3.

[b] Ring dissymmetry [value] as determined by the difference in absolute values of corresponding ring torsion angles.

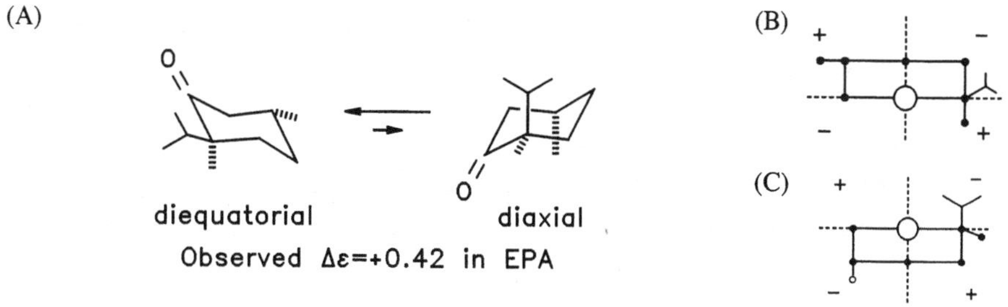

Figure 5-15. (A) Chair ⇄ chair conformational inversion of 2(*S*)-isopropyl-2,5(*R*)-dimethylcyclohexanone. (B) Octant projection diagram for the diequatorial isomer. (C) Octant projection diagram for the diaxial isomer.

α-axial methyl group, which is expected to contribute significantly to the $n \rightarrow \pi^*$ Cotton effect, and the diaxial isomer has an α-equatorial methyl, which is not expected to make a large contribution thereto. The observed Cotton effect magnitude at −186°C (Figure 5-14B) exceeds that of (−)-menthone (Figure 5-11A) by approximately +0.25 Δε units, which one might estimate to be due to the contribution of an α-axial methyl, assuming that the cyclohexanone rings in both isomers have the same geometry and that the presence of an α-methyl does not perturb the population of α-equatorial isopropyl rotamers. However, the Δε ~ +0.25 value is far smaller than the Δε ~ +1.2 value found in 2-axial-methyl-4-*tert*-butylcyclohexanone (Table 5-3), suggesting an incomplete displacement of the chair ⇄ chair equilibrium at −186°C to −192°C, or conformational distortion.

In fact, molecular mechanics calculations (Table 5-18) show considerable ring dissymmetry in and around the carbonyl chromophore of both the diequatorial and diaxial conformation. None of the chair conformations of (−)-menthone, (+)-isomenthone, α-methylcyclohexanone, 3-methylcyclohexanone, or 2-isopropylcyclohexanone show such a large ring dissymmetry. This results in α-equatorial isopropyl and methyl groups with R–C_α–C=O torsion angles that are much larger than normal and ring atoms that no longer lie in, or even close to, octant symmetry planes.

Equally surprising, the two chair conformations (Figure 5-15) are computed[3] to have the same heats of formation. In the absence of large entropy effects, the equilibrium would be insensitive to temperature, and one would predict the same $n \rightarrow \pi^*$ Cotton effect intensities at high and at low temperatures. Although this is not observed (Figure 5-14B), the CD curves are also not very sensitive to lowering the temperature. In principle, analysis of the variable-temperature CD, as in Section 5.6, should afford both the mole fraction of each component (and hence $\Delta G°$) of the assumed two-component equilibrium (Figure 5-15) and the Δε values of each component. However, this analysis has not been carried out, so one is left at present with (i) the insights provided by molecular mechanics computations (Table 5-18) that predict distorted chair conformations of essentially the same energy and (ii) a deceptively simple low-temperature CD behavior (Figure 5-14B) that indicates a predominance at room temperature of a conformation with a positive Cotton effect, more positive than that of the minor isomer.

5.9. Intramolecular Hydrogen Bonding and 2-Oxo-1-*p*-menthanol

2-Oxo-1-*p*-menthanol (2-hydroxy-2(*S*)-methyl-5(*R*)-isopropylcyclohexanone) presents yet another example of a bisignate $n \rightarrow \pi^*$ CD Cotton effect, which, like that of (−)-menthone, is dependent on the solvent (Figure 5-16A).[12] The origin of the bisignate Cotton effects is apparently not associated with asymmetric solvation,[6,7] as the short-wavelength component decreases with increasing polarity of the solvent.[12] Rather, the conformational equilibrium expressed in Figure 5-16B is the probable cause. The chair conformation with equatorial alkyl groups is predicted by the octant rule to give a very strongly positive Cotton effect; the diaxial isomer is predicted to exhibit a strongly negative Cotton effect (Figure 5-16C). Hence, one is tempted to attribute the positive and negative components of the observed CD to the superposition of the CDs of these two conformations. Although the diaxial conformer is predicted to be less stable than the diequatorial one, intramolecular hydrogen bonding probably acts to stabilize this conformation—especially in solvents that do not interfere with hydrogen bonding, such as isooctane and the low-temperature glass-forming mixture isopentane–methylcyclohexane (5:1, by vol.).

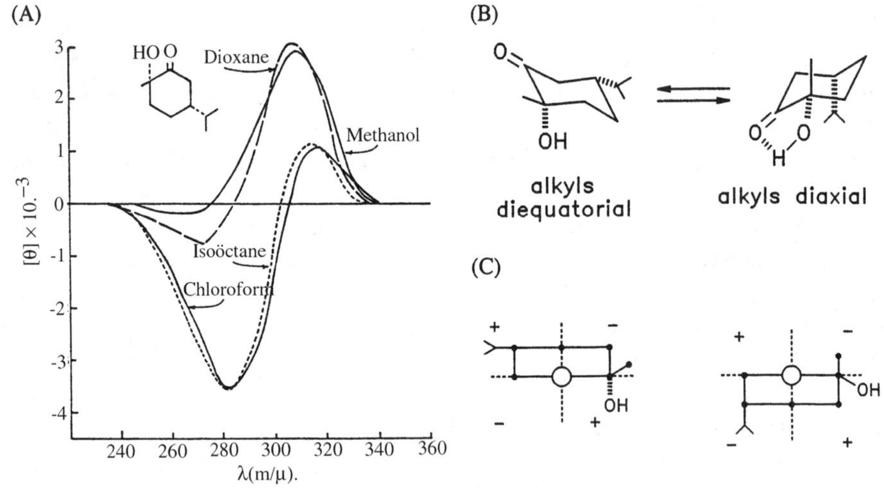

Figure 5-16. (A) Solvent-dependent CD curves of 2-oxo-1-*p*-menthanol at 25°C. (B) Chair ⇄ chair conformational equilibrium. In the diaxial conformation, the equatorial hydroxyl lies in a position suitable for intramolecular hydrogen bonding. (C) Octant projection diagrams for the diequatorial (left) and diaxial (right) conformations of (B). [(A) is reprinted with permission from ref. 12. Copyright© 1965 American Chemical Society.]

At low temperatures, even in a hydrocarbon solvent, the bisignate CD shifts to positive, monosignate (Figure 5-17) CD, indicating that the conformation contributing to the negative CD is of higher energy than that associated with the positive one.[12] This qualitative conclusion about energy differences is supported by molecular mechanics calculations,[3] which predict an enthalpy difference of approximately 0.4 kcal/mole, with the diequatorial conformer being of lower energy than the most stable diaxial conformer. The latter, which is intramolecularly hydrogen bonded, is more stable than its non-hydrogen-bonded counterpart by about 3 kcal/mole. Interestingly, in both the diequatorial and the diaxial conformation, considerable ring distortion is noted (Table

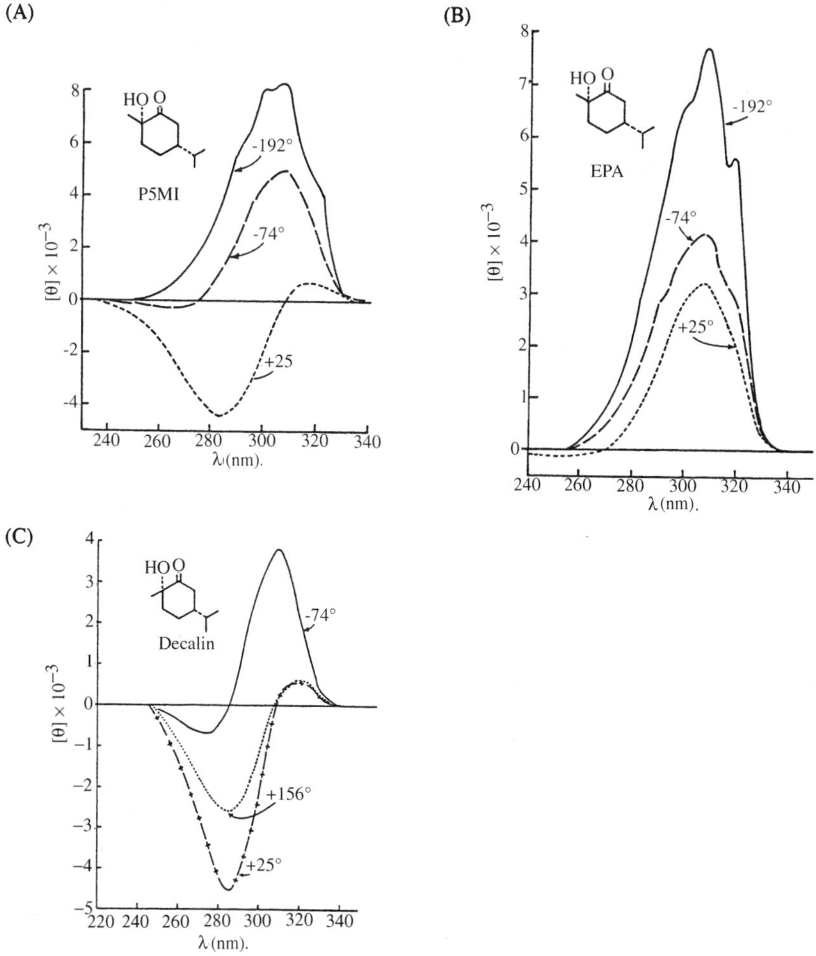

Figure 5-17. Variable temperature CD spectra for the $n \rightarrow \pi^*$ transition of 2-oxo-1-*p*-menthanol in (A) P5MI (isopentane–methylcyclohexane, 5:1, by vol.), (B) EPA (ether–isopentane–ethanol, 5:5:2, by vol.), and (C) decalin. Temperatures (°C) are indicated on the CD curves. [Reprinted with permission from ref. 12. Copyright© 1965 American Chemical Society.]

Table 5-19. Computed[a] Ring Torsion Angles, ψ(C–C–C–C), $CH_3–C_\alpha–C = O$, and $HO–C_\alpha–C=O$ Torsion Angles, and Heats of Formation of 2-Oxo-1-*p*-menthanol

		Computed Torsion Angles (°)			
		$C_3–C_2–C_1–C_6$	$C_1–C_2–C_3–C_4$	$C_2–C_3–C_4–C_5$	
Conformation	$R–C_\alpha–C=O$	$C_2–C_1–C_6–C_5$ Dissymmetry[b]	$C_1–C_6–C_5–C_4$ Dissymmetry[b]	$C_3–C_4–C_5–C_6$ Dissymmetry[b]	Heat of Formation (kcal/mole)
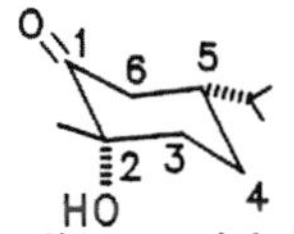 diequatorial	–0.65 (eq) +122.0 (ax)	57.6 –56.3 [1.3]	–56.8 49.8 [7.0]	57.6 –51.7 [4.9]	–120.14
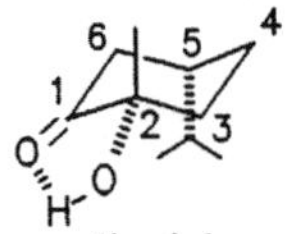 diaxial	+14.7 (eq) –102.5 (ax)	–46.2 55.0 [3.8]	45.4 –58.9 [13.5]	–55.5 60.3 [4.8]	–120.52
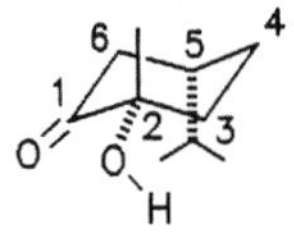	+15.8 (eq) –101.4 (ax)	–45.4 53.6 [8.2]	45.1 –57.8 [12.7]	–55.7 60.3 [4.6]	–117.45
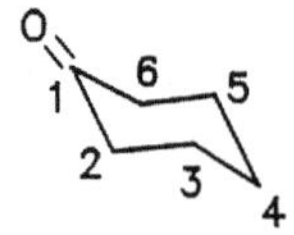		51.5 –51.5 [0.0]	–53.3 53.3 [0.0]	57.5 –57.5 [0.0]	–53.7

[a] Computed using PCMODEL, ref. 3.

[b] Ring dissymmetry [value] as determined by the difference in absolute values of corresponding ring torsion angles.

5-19), particularly in the midsection of the molecule. The ring dissymmetry is greatest in the intramolecularly hydrogen-bonded diaxial conformation; here, the molecule comes close to adopting a twist-boat conformation.

References

1. Beard, C., Djerassi, C., Elliott, T., and Tao, R. C. C., *J. Am. Chem. Soc.* **84** (1962), 874–875. The value Δε = +0.30 in EPA for the $n \rightarrow \pi^*$ Cotton effect of 2(*S*)-methylcyclohexanone was confirmed independently in Cheer, C. J., and Djerassi, C., *Tetrahedron Lett.* (1976), 3877–3878, for 2(*R*)-methylcyclohexanone; Δε = –0.30 in CH_3OH.
2. Cotterill, W. D., and Robinson, M. J. T., *Tetrahedron* **20** (1964), 765–775.
3. Using PCMODEL versions 4.0 to 7.0, Serena Software, Inc., Bloomington, IN 47402-3076. PCMODEL employs the MMX force field.

4. Lightner, D. A., Bouman, T. D., Crist, B. V., Rodgers, S. L., Knobeloch, M. A., and Jones, A. M., *J. Am. Chem. Soc.* **109** (1987), 6248–6259.
5. Konopelski, J. P., Sundararaman, P., Barth, G., and Djerassi, C., *J. Am. Chem. Soc.* **102** (1980), 2737–2745.
6. Moscowitz, A., Wellman, K. M., and Djerassi, C., *Proc. Natl. Acad. Sci. (U.S.)* **50** (1963), 799–804.
7. For a comprehensive set of examples of solvent-dependent bisignate CD Cotton effects in rigid bicyclo[2.2.1]heptanone derivatives, see Coulombeau, C., and Rassat, A., *Bull. Soc. Chim. France* (1966), 3752–3762.
8. Lightner, D. A., and Eng, F. P. C., *Steroids* **35** (1980), 189–207.
9. Crist, B. V., Rodgers, S. L., and Lightner, D. A., *J. Am. Chem. Soc.* **104** (1982), 6040–6045, and references therein.
10. Fournier, J., *F. Mol. Struct.* **27** (1975), 177–183.
11. Reise, J., Piccini-Leopardi, C., Zahra, J. P., Waegell, B., and Fournier, J., *J. Org. Magn. Res.* **9** (1977), 512–517.
12. Wellman, K. M., Briggs, W. S., and Djerassi, C., *J. Am. Chem. Soc.* **87** (1965), 73–81.
13. Wellman, W. S., Bunnenberg, E., and Djerassi, C., *J. Am. Chem. Soc.* **85** (1963), 1870–1872.
14. Lightner, D. A., Bouman, T. D., Wijekoon, W. M. D., and Hansen, Aa. E., *J. Am. Chem. Soc.* **108** (1986), 4484–4497.
15. Lightner, D. A., and Crist, B. V., *Applied Spectrosc.* **33** (1979), 307–310.
16. Allinger, N. L., and Freiberg, L. A., *J. Am. Chem. Soc.* **84** (1962), 2201–2203.
17. Djerassi, C., Geller, L. E., and Eisenbraun, E. J., *J. Org. Chem.* **25** (1960), 1–6.
18. Djerassi, C., *Proc. Chem. Soc.* (1964), 314–330.
19. Allinger, J., Allinger, N. L., Geller, L. E., and Djerassi, C., *J. Org. Chem.* **26** (1961), 3521–3523.
20. Allinger, N. L., Allinger, J., Geller, L. E., and Djerassi, C., *J. Org. Chem.* **25** (1960), 6–12.
21. Calculated from the empirical relationship $[R] = 3.32\ \Delta\varepsilon$ (Table I, footnote *c*, of ref. 6), where $R^{25°} = +1.40 \times 10^{-40}$ c.g.s. (or $[R] = +1.51$) from ref. 22.
22. Moscowitz, A., Wellman, K. M., and Djerassi, C., *J. Am. Chem. Soc.* **85** (1963), 3515–3516.
23. See also Wilson, S. R., and Au, W., *J. Org. Chem.* **54** (1989), 6047–6055.
24. Beard, C., Djerassi, C., Sicher, J., Šipoš, F., and Tichý, M., *Tetrahedron* **19** (1963), 919–928.
25. Moffitt, W., Woodward, R. B., Moscowitz, A., Klyne, W., and Djerassi, C., *J. Am. Chem. Soc.* **83** (1961), 4013–4018.
26. Djerassi, C., Mitscher, L. A., and Mitscher, B. J., *J. Am. Chem. Soc.* **81** (1959), 947–955.
27. Milhavet, J.-C., Sablayrolles, C., Girard, J.-P., and Chapat, J.-P., *J. Chem. Res. (M)* (1980), 1901–1913; (*S*) (1980), 134–135.
28. Allinger, N. L., and Riew, C. K., *J. Org. Chem.* **40** (1975), 1316–1321.
29. Wellman, K. M., Laur, P. H. A., Briggs, W. S., Moscowitz, A., and Djerassi, C., *J. Am. Chem. Soc.* **87** (1965), 66–72.
30. Ohloff, G., Osiecki, J., and Djerassi, C., *Chem. Ber.* **95** (1962), 1400–1408.
31. Watanabe, S., *Bull. Chem. Soc. Jpn.* **46** (1973), 1546–1549.

CHAPTER

6

Other Cycloalkanones

Until 1881, it was generally believed that carbocyclic rings larger or smaller than cyclohexane could not exist. There were theoretical grounds for this opinion related to violating the 109° 28′ tetrahedral carbon bond angles, and small rings had not been found in nature. Cyclopropane and cyclobutane were thought to be much too strained to exist, at least until the first cyclobutane, cyclobutane-1,3-dicarboxylic acid, formed by heating β-chloropropionic acid with dry sodium ethoxide, was reported in 1881 by Markownikoff and Krestownikoff[1] at the University of Moscow. Very shortly thereafter, in 1882, Freund[2] reported the first synthesis of cyclopropane by reacting 1,3-dibromopropane with sodium. Syntheses of other carbocycles followed rapidly.

6.1. Cyclopropane and Cyclopropanone

Cyclopropane boils at a low temperature (–30°C) and has been used as an anesthetic. The carbocyclic structure of cyclopropane (D_{3h} symmetry) is planar and nonflexible, of course, with C–C–C bond angles of 60°, as required by geometric considerations and suggesting considerable angle strain (relative to the normal tetrahedral angle of 109° 28′).[3] Electron diffraction studies[4,5] of cyclopropane indicate that the compound has shorter C–C bonds (1.5127 ± 0.0020 Å),[5] shorter C–H bonds (1.0840 ± 0.0020 Å),[5] and larger H–C–H bond angles (114.5 ± 0.9°) than cyclohexane. (See Table 2-1: r_{CC} = 1.5335 ± 0.002 Å, r_{CH} = 1.111 ± 0.004 Å, and ∠H–C–H = 105.3° ± 2.3°.) The MM3 force field reproduces the cyclopropane parameters well, with r_{CC}= 1.512 Å, r_{CH}= 1.087, and H–C–H = 115.8°, as does the MM2 force field.[6]

The first unequivocal synthesis of the very reactive cyclopropanone[7] was reported in 1966 by Turro and Hammond[7a] at Columbia University and Schaafsma et al.[7b] at the University of Amsterdam, some 80 years after the first synthesis of its parent cyclopropane.[2] Microwave studies[8] gave $r_{C=O}$ = 1.18 Å, $r_{C_1C_2}$ = 1.49 Å, $r_{C_2C_3}$ = 1.58 Å, r_{CH} = 1.085 Å, ∠H–C–H = 117° 35′, and $\angle C_2C_1C_3$ = 64° for this planar, nonflexible ketone. Cyclopropanone spectral data[9] indicate a strongly shifted infrared $\nu_{C=O}$ (1813 cm^{-1}) and UV $n \rightarrow \pi^*$ absorption (ε_{310}^{max} ~23, λ_{330}^{sh}). The only cyclopropanone to have had its CD (or ORD) measured, *trans*-2,3-di-*tert*-butylcyclopropanone[10] (structure at right) exhibits a large Cotton effect ($\Delta\varepsilon_{354}^{max}$+ 2.92) in isooctane (corrected to 100% from 9% e.e.) and a strongly bathochromically shifted UV $n \rightarrow \pi^*$ transition (ε_{354}^{max} = 33). Since the Cotton effect is positive, one may deduce the 2(*R*),3(*R*) configuration of the compound by using the octant rule (Chapter 4).

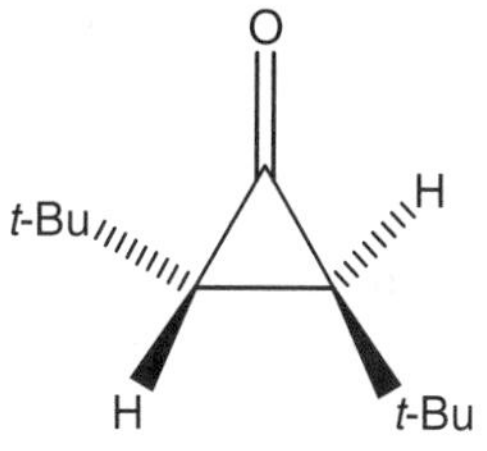

6.2. Cyclobutane and Cyclobutanones

Cyclobutane[11,12,13] is the simplest conformationally flexible carbocyclic ring, but compared with cyclohexane, its conformation has been studied only relatively recently. In one of the earliest spectroscopic studies of cyclobutane, carried out by Raman and infrared spectroscopy, Wilson[14] concluded in 1943 that the molecule was planar. Subsequently, gas-phase electron diffraction analyses of Dunitz and Schomaker[15] in 1952 at Cal Tech indicated a longer-than-normal C–C bond length ($1.56_8 \pm 0.02$ Å), with the ring either permanently bent or planar with large-amplitude out-of-plane ring-puckering motions. At nearly the same time at the University of California at Berkeley, Pitzer et al.[16] reinvestigated the infrared and Raman spectra of cyclobutane, which supported a puckered form with a sufficiently low barrier to ring inversion such that an appreciable fraction of the molecules would lie in a planar form. At about the same time, Carter and Templeton's[17] X-ray diffraction of cyclobutane crystals at −100°C yielded no molecular parameters, due to rotational disorder (either static or dynamic). Not quite a decade later, Almenningen, Bastiansen, and Skancke[18] confirmed by electron diffraction that the C–C bond lengths in cyclobutane were 0.010–0.015 Å longer than in linear alkanes and that the ring is static nonplanar. However, the ring-puckering angle (β) could not be determined with any reasonable accuracy.[18] If cyclobutane were planar, all four CH_2 groups would lie in an unstable eclipsed ethane orientation with all ring torsion angles[19] ψ(C–C–C–C) being 0°. Partial relief of this nonbonded torsional strain is achieved by modest distortion of the torsion angles (ψ = 20–25°), resulting in puckering (Figure 6-1). However, the cost of puckering is slightly increased internal angle strain, from 90° in the planar conformation to approximately 88° in the puckered. On balance, however, the ring is puckered, and the planar conformation thus represents the conformation at the barrier to ring inversion in cyclobutane, as shown in the figure.

The ring-puckering angle (β) and the energy associated with ring inversion were first determined by Raman and IR spectroscopy. In 1968, Ueda and Shimanouchi found cyclobutane to be puckered,[20] with a dihedral angle β = 33° 22′ ± 30′, and an inversion barrier of 448.1 ± 18 cm^{-1} (1.28 kcal/mole). NMR investigations of cyclobutane in a (nematic) liquid crystal by Meiboom and Snyder[21] favored a puckered conformation with β = 23–27° and a conformational lifetime of less than 10^{-6} sec. Later reinvestigations led to an improved value for the barrier: 515.8 ± 4.4 cm^{-1} (~1.47 kcal/mole).[13,22] More recently, an IR analysis of the puckering potential by Kuchitsu et al.[23] confirmed the value of the ring inversion barrier, 510 ± 2 cm^{-1} (1.46 kcal/ mole), but gave a smaller puckering angle β = 28.8 ± 1.1°. In 1987, by electron diffraction and FTIR analysis, Kuchitsu et al.[24] reported a zero-point average dihedral angle β = 27.9 ± 1.6° and gave refined bond angles and distances (Table 6-1).

PCMODEL molecular mechanics calculations[25] using the MMX force field, a variation of Allinger's MM2 force field, also predict that the most stable cyclobutane conformation is puckered, with β ~ 29° and an inversion barrier of about 0.9 kcal/mole. Although PCMODEL underestimates the experimentally derived energy difference

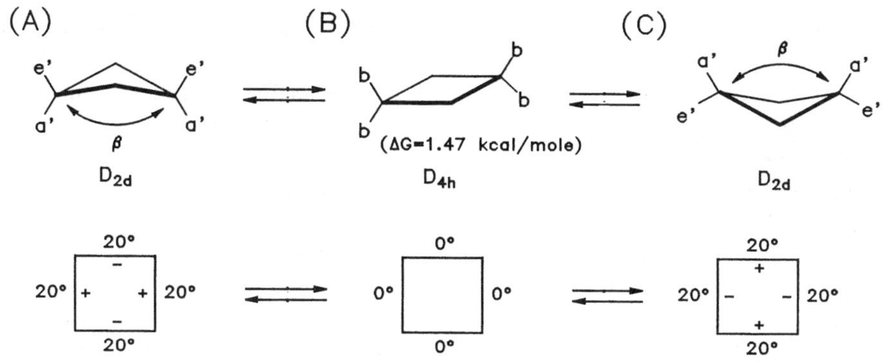

Figure 6-1. Bent or puckered conformations [(A) and (C)] interconverting through the higher energy planar conformation (B). In (A) and (C), the interplanar puckering angle (β) is 152.1 ± 1.3° (ref. 21), and in (B) it is 180°. Puckered conformers have D_{2d} symmetry; the planar conformer has D_{4h} and lies 1.47 kcal/mole above the puckered one (ref. 22). The exocyclic bonds in the planar conformation have a bisectional (*b*) orientation, whereas those in the puckered conformation are pseudoaxial (*a*′) or pseudoequatorial (*e*′). Ring inversion interconverts the *a*′ and *e*′ configurations. Shown below each conformation are the Bucourt diagrams (ref. 19), which note the sign and magnitude of the internal ψ(C–C–C–C) torsion angles of the conformation. The cited torsion angles come from molecular mechanics calculations (Table 6-1).

Table 6-1. Comparison of Computed and Experimentally Derived Inversion Barrier (IB, kcal/mole), Pucker Angle (β), Torsion Angles (ψ and ϕ), Bond Angles (∠), and Bond Lengths (*d*) of Puckered Cyclobutane with PCMODEL Data for Chair Cyclohexane

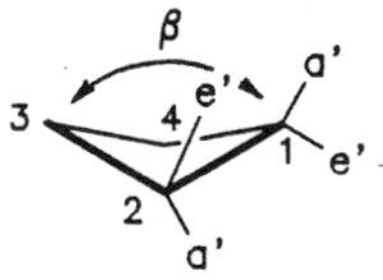	Method				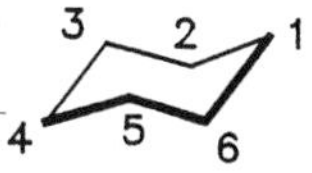
	PCMODEL[a]	MM3[b]	ED[c]	ED + FTIR[d]	
IB	0.92	1.37			10.1
β	151°	147.7°		152.1 ± 1.3°	129°
ψ (C_1–C_2–C_3–C_4)	20.1°	22.5°			56.4°
ϕ (H_e–C–C–H_e)	107.2°				60.1°
ϕ (H_a–C–C–H_e)	23.1°				57.2°
ϕ (H_a–C–C–H_a)	153.3°				174.6°
∠ (C–C–C)	88.2°				110.9°
∠ (H–C–H)	112.7°			106.4 ± 1.3°	107.1°
d (C–C)	1.548 Å	1.557 Å	1.438 (8) Å	1.554 (1) Å	1.536 Å
d (C–H)	1.115 Å	1.112 Å	1.092 (10) Å	1.109 (3) Å	1.116 Å

[a] Values from PCMODEL (ref. 25).

[b] Values from ref. 26.

[c] Electron diffraction (ED) values from ref. 18.

[d] Electron diffraction (ED)–Fourier transform infrared (FTIR) values from ref. 24.

between puckered and planar conformations, Allinger's more recent MM3 force field gives a better estimate (1.37 kcal/mole), but predicts a slightly more puckered shape (β ~ 32°).[26] The various parameters determined by molecular mechanics calculations are compared with the experimentally determined values in Table 6-1.

One of the consequences of ring puckering is the existence of two types of exocyclic bonds (Figure 6-1): pseudoaxial (a') and pseudoequatorial (e'). Although different in orientation from the classical axial and equatorial bond orientations in cyclohexane (see Table 6-1), a' and e' are closer in orientation to the bisectional bonds (*b*) of the planar conformation (Figure 6-1B). Gas-phase electron diffraction[12] and microwave studies[27] indicate that, for halogen substituents, the pseudoequatorial position is energetically favored.[11] Other substituents also favor the pseudoequatorial site,[11] although not necessarily as strongly as in cyclohexane. For example, in equilibration studies at 100°C, ethyl *cis*-3-*tert*-butylcyclobutanecarboxylate (with pseudoequatorial groups) was shown to be more stable than *trans* by $\Delta G° \sim -0.58 \pm 0.02$ kcal/mole ($\Delta H° = -0.8 \pm 0.2$ kcal/mole and $\Delta S° = -0.7 \pm 0.5$ cal/deg/mole).[28] In comparison, methyl *trans*-4-*tert*-butylcyclohexanecarboxylate (with equatorial groups) is more stable than *cis* by $\Delta G° = -1.26$ kcal/mole ($\Delta H° = -1.09 \pm 0.03$ kcal/mole and $\Delta S° = +0.4 \pm 0.1$ cal/mole/deg) at 100°C.[29] Thus, comparatively, an equatorial group is much more stable than an axial group in cyclohexane than in cyclobutane. Although the enthalpic terms are comparable, the entropic terms are reversed, suggesting that *trans* cyclobutane is more flexible than the *cis* form. The data indicate that one should be cautious in using intuition derived from cyclohexane for cyclobutane conformational analysis, because in the latter the difference in conformational energies is much smaller.

Puckered and planar cyclobutane conformations lie in a broad potential well (Figure 6-2). Although molecular mechanics calculations[25] predict a puckered conformation at the global minimum with internal torsion angles ψ(C–C–C–C) ~ 20° and puckering angle β ~ 28°, the energy of the planar conformation (or interconversion barrier) is probably no higher than that of a more puckered conformation with ψ ~ 30°. However, as ring puckering increases, the energy rises steeply.

The overall picture is altered only somewhat by the change in hybridization of one ring carbon from sp^3 to sp^2, as in cyclobutanone.[11–13] In cyclobutanone, the walls of the potential well rise more steeply, but most significantly, as determined by infrared[30] and microwave[31] spectroscopy, the planar conformation lies only about 5 cm^{-1} (~0.01 kcal/ mole)[30] or 7.6 cm^{-1} (~0.02 kcal/mole)[31] above the puckered confirmation—well below the zero-point energy barrier of 16.8 cm^{-1} (0.048 kcal/mole).[13] More recently, the barrier was redetermined (1.2 ± 1.5 cm^{-1}, or approximately 0 kcal/mole) from electron diffraction and spectroscopic data.[32] PCMODEL[25] and MM3[33] molecular mechanics calculations support a planar minimum-energy conformation lying in a broad potential well[34] (Figure 6-2). The various parameters derived for cyclobutanone from molecular mechanics are compared with experimentally derived values in Table 6-2.

Cyclobutanone may thus be viewed as a planar molecule, but one easily deformed into a puckered conformation (Figure 6-2). As in cyclobutane (Figure 6-1B), the planar conformation has bisectional exocyclic bonds, but in the puckered conformation they

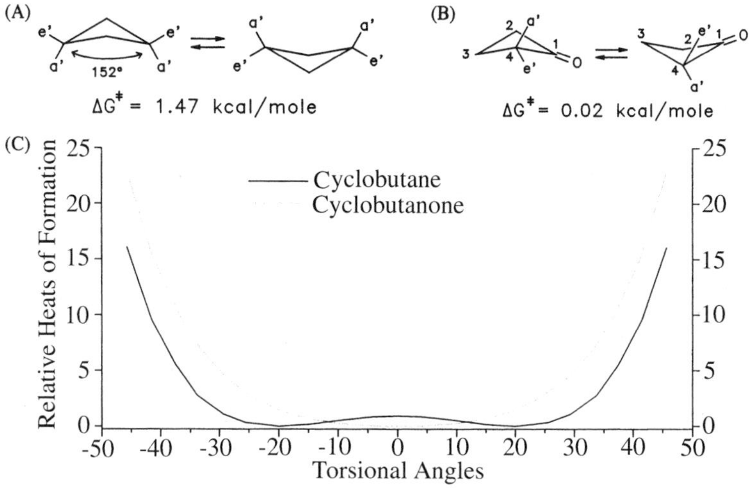

Figure 6-2. Puckered-ring conformations of cyclobutane (A) and cyclobutanone (B) and activation barriers (ΔG‡) for ring inversion. In the cyclobutanone, there is little to distinguish the puckered conformation from the planar. (C) Molecular mechanics (ref. 25) computed changes in cyclobutane (———) and cyclobutanone (- - - -) relative conformational energy ($\Delta\Delta H_f$, kcal/mole) with ring puckering, as determined by ring torsion angle ψ(1–2–3–4). When $\psi = 0°$, the ring is planar.

deform slightly to the pseudoaxial (a') or pseudoequatorial (e') orientations (Figure 6-2B). In substituted cyclobutanones, nonbonded steric interactions associated with the substituents influence the conformation of the ring and typically cause the puckered conformation to be more stable than the planar. However, conformational analysis based on the relative stability of pseudoequatorial and pseudoaxial groups is not nearly as clear in cyclobutanones as it is in cyclohexanones—as will be seen in the following discussion from an examination of substituted cyclobutanones by circular dichroism spectroscopy and molecular mechanics calculations.

The simplest optically active alkylcyclobutanones, 2(*R*) and 2(*S*)-methylcyclobutanone, were prepared in 1981 from 1,3-dibromobutane by reaction with (+)-neomenthylsulfonylmethyl isocyanide.[35] The 2(*R*) isomer exhibited a negative $n \rightarrow \pi^*$ Cotton effect (Figure 6-3A), with $\Delta\varepsilon_{306}^{max} = -0.31$ (100% e.e.) in isooctane, and a UV $\varepsilon_{293}^{max} = 22$. The negative Cotton effect of the 2(*R*)-enantiomer is consistent with predictions of the octant rule for the planar or for either puckered conformation. And the weak magnitude is consistent with that seen in pinanones (Table 5-7), in which the α-methyl group lies nearly in a bisected orientation. Although the conformational equilibrium (Figure 6-3B) between the two puckered cyclobutanone conformations is predicted by molecular mechanics calculations[25] to lie mainly toward the pseudoequatorial methyl isomer, both conformers are expected to exhibit comparably modest Cotton effects. The equilibrium position may be estimated by molecular mechanics calculations,[25] which predict that the pseudoequatorial isomer is more stable than the pseudoaxial one by about 0.4 kcal/mole (Table 6-3). If one assumes that, for the equilibrium (Figure

Table 6-2. Comparison of Computed and Experimentally Derived Inversion Barrier (IB, kcal/mole), Pucker Angle (β), Torsion Angles (ψ and ϕ), Bond Angles (∠), and Bond Lengths (*d*) of Puckered Cyclobutanone with PCMODEL Data for Chair Cyclohexanone

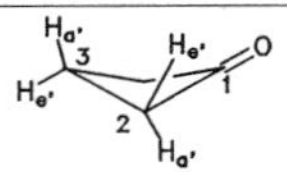	Method					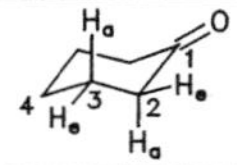
	PCMODEL[a]	MM3[b]	Electron Diffraction[c]	Microwave[d]	ED + MW[e]	
IB	0.00	0.0		0.02	~0.00	3.9
β	0.0°	0.0°			169.6 ± 2.7°	135°
ψ (C_1–C_2–C_3–C_4)	0.82°					52.72°
ϕ (H_e–C_2–C_3–H_a)	4.64°					54.29°
ϕ (H_e–C_2–C_3–H_e)	132.66°					62.93°
ϕ (H_a–C_2–C_3–H_e)	2.69°					55.58°
ϕ (H_a–C_2–C_3–H_a)	130.71°					172.80°
ϕ (H_e–C–C=O)	64.1°					8.92°
ϕ (H_a–C–C=O)	–65.8°					–107.92°
∠ (C_α–C_1–$C_{\alpha'}$)	94.73°	95.53°	92.2 (5)°	93.1 (3)°	92.8 (3)°	115.5°
∠ (C_1–C_2–C_3)	86.56°	88.17°	89.1 (6)°	88.0 (3)°	88.3 (2)°	111.3°
∠ (C_2–C_3–C_4)	92.13°	90.13°	89.4 (8)°	90.9 (3)°	90.3 (4)°	111.1°
d (C=O)	1.205 Å	1.2041 Å	1.202 (2) Å	1.202 (2) Å	1.202 (4) Å	1.211 Å
d (C_1–C_2)	1.520 Å	1.5290 Å	1.533 (5) Å	1.527 (3) Å	1.534 (3) Å	1.536 Å
d (C_2–C_3)	1.553 Å	1.5375 Å	1.569 (5) Å	1.556 (1) Å	1.567 (5) Å	1.534 Å
d (C–H) av	1.115 Å	1.1119 Å	1.101 (5) Å	—	1.100 (4) Å	

[a] Values from PCMODEL, ref. 25.
[b] Values from ref. 33.
[c] Values from ref. 32.
[d] Values from ref. 31.
[e] Values from ref. 32.

6-3B), the computed $\Delta H^\circ \sim \Delta G^\circ$, then 66% of the pseudoequatorial isomer is predicted at 25°C. The conformational equilibrium of this simple ketone has not yet been analyzed by low-temperature CD, in which one would predict (*vide infra*) that the Cotton effect would become less negative as the temperature is lowered, assuming a smaller octant contribution of a pseudoequatorial methyl based on its proximity to an octant plane. Low-temperature CD measurements, however, have been carried out on more highly substituted cyclobutanones.

In 2,2,3(*S*)-trimethylcyclobutanone, the C(3)-methyl lies in an octant symmetry plane and thus should make no contribution to the CD. If the ring were planar, the α-methyl contributions would be equal and opposite (hence canceling), and the net CD Cotton effect would be approximately zero. However, the observed positive $n \rightarrow \pi^*$ Cotton effect ($\Delta\varepsilon$ + 0.24 in cyclohexane)[36] suggests a predominance of the puckered ring conformer with pseudoequatorial methyl at C(3) (Table 6-3). In this conformation, the pseudoaxial α-methyl lies in a (+) octant, whereas the pseudoequatorial lies in a (–) octant. The predicted net Cotton effect is weakly (+). In contrast, the conformer with a pseudoaxial C(3) methyl is predicted to have a small negative Cotton effect. Quantitative analysis of the equilibrium between these two puckered conformers might have been determined by variable low-temperature CD using the analysis illustrated in Section 5.6, but that was not attempted. The available low-temperature CD data

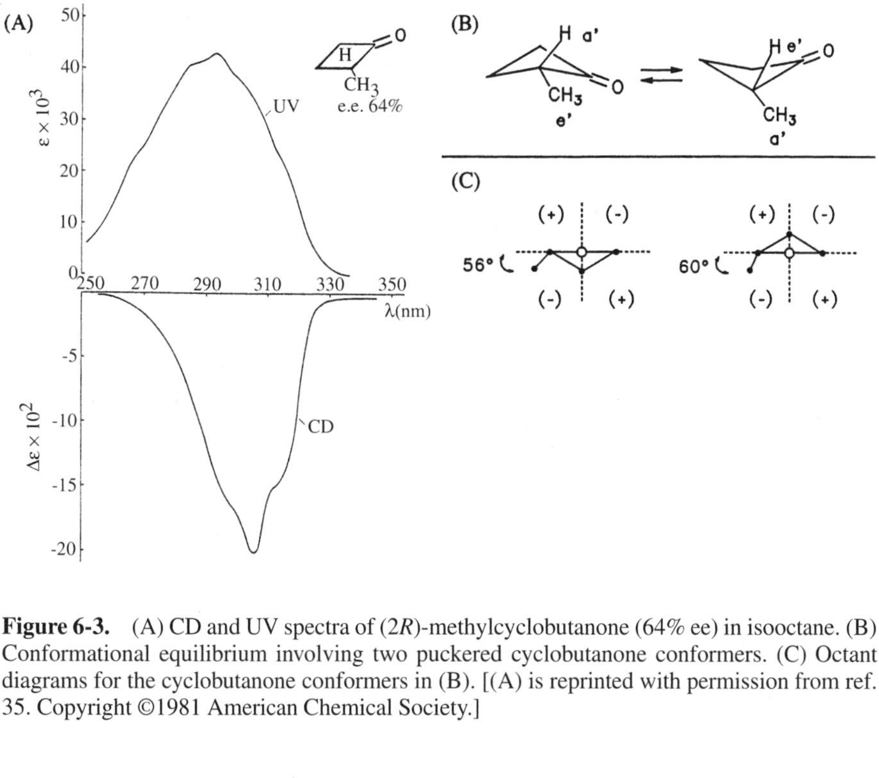

Figure 6-3. (A) CD and UV spectra of (2*R*)-methylcyclobutanone (64% ee) in isooctane. (B) Conformational equilibrium involving two puckered cyclobutanone conformers. (C) Octant diagrams for the cyclobutanone conformers in (B). [(A) is reprinted with permission from ref. 35. Copyright ©1981 American Chemical Society.]

Table 6-3. Computed Torsion Angles and $\Delta\Delta H_f$ Values for the Cyclobutanone Ring-Puckering Equilibrium and Experimental CD Intensities for the $n \to \pi^*$ Transition

Cyclobutanone	CD[a] $\Delta\varepsilon$ (λ, nm)	X–C_α–C=O Torsion Angle[b] (°)	Conformational Equilibrium and (C_1–C_2–C_3–C_4) Torsion Angle (°)	Computed $\Delta\Delta H_f^c$
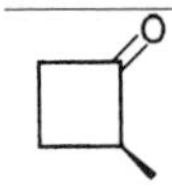	−0.31[d](306)	55.9 (CH_3) 68.5 (CH_3)	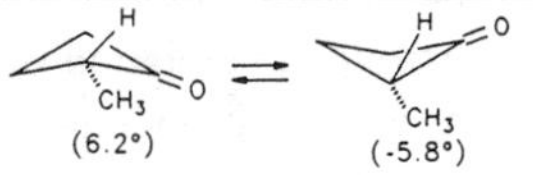	+0.39 kcal/mole
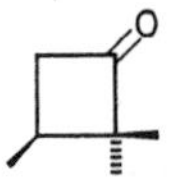	+0.24[e](300) +0.12[f](290)	67.0 (*cis* CH_3) 60.1 (*trans* CH_3) 56.4 (*cis* CH_3 70.6 (*trans* CH_3)	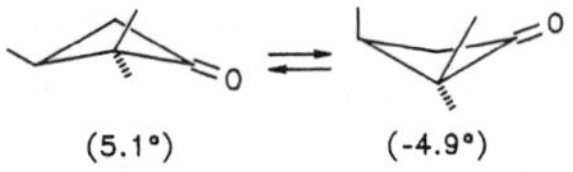	+0.18 kcal/mole
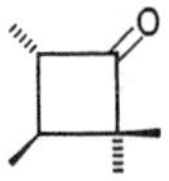	+0.036[e](304)	54.2 (CH_3) 56.8 (*cis* CH_3) 70.4 (*trans* CH_3) 69.4 (CH_3) 72.3 (*cis* CH_3) 54.6 (*trans* CH_3)	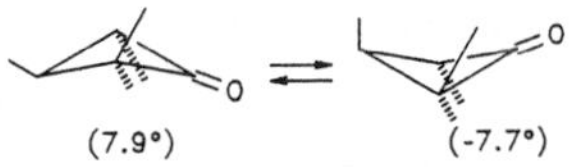	+0.96 kcal/mole
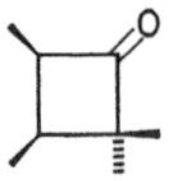	(−)weak(300)[d]	−67.2 (CH_3) 69.1 (*cis* CH_3) −58.0 (*trans* CH_3) −56.8 (CH_3) 52.7 (*cis* CH_3) −74.21 (*trans* CH_3)	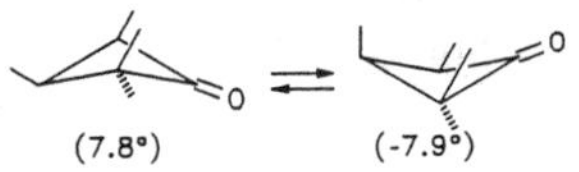	−0.16 kcal/mole
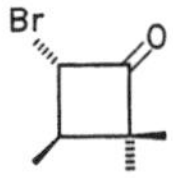	+1.97[e](318) +2.67[f](314)	53.9 (Br) 56.4 (*cis* CH_3) 70.8 (*trans* CH_3) 72.6 (Br) 73.3 (*cis* CH_3) 53.6 (*trans* CH_3)	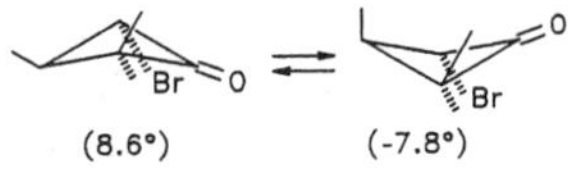	+0.19 kcal/mole
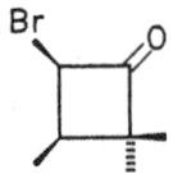	−1.22[e](322) −0.30[f](320–330)	70.6 (Br) −70.6 (*cis* CH_3) 56.8 (*trans* CH_3) 52.4 (Br) 52.1 (*cis* CH_3) 74.6 (*trans* CH_3)	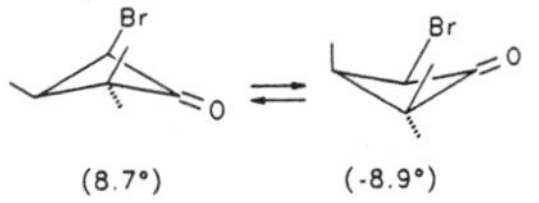	+1.14 kcal/mole
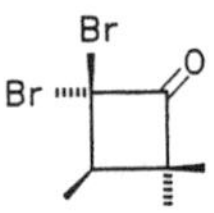	−0.64[e](335) −0.58[f](332)	72.1 (Br) 55.2 (Br) −72.1 (CH_3) 55.3 (CH_3) 52.9 (Br) −73.8 (Br) 73.8 (CH_3) 52.6 (CH_3)	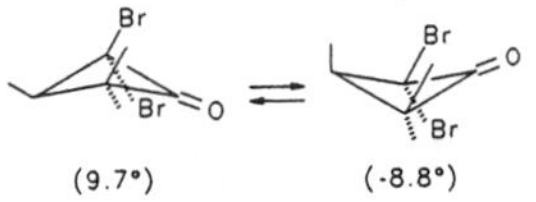	+1.14 kcal/mole

[a] Data from refs. 36 and 37.

[b] Upper data set corresponds to conformer on the left in the column headed "Conformational Eqilibrium"; lower data set corresponds to conformer on the right.

[c] With respect to the conformers shown in column 4, $\Delta\Delta H_f = \Delta H_f$ (right) – ΔH_f (left).

[d] Isooctane.

[e] Cyclohexane.

[f] Methanol.

(Figure 6-4A),[37] which show an increasingly positive $\Delta\varepsilon$ with decreasing temperature, are consistent with an equilibrium such that the isomer with a C(3) pseudoequatorial methyl is more stable than that with a pseudoaxial isomer. This conclusion is supported by molecular mechanics calculations,[25] which indicate (Table 6-3) a 0.18 kcal/mole preference for the puckered conformer with the C(3) pseudoequatorial methyl.

Further methyl substitution alters the CD in a qualitatively predictable way. For example, 2,2,3(*S*),4(*S*)-tetramethylcyclobutanone (Table 6-3) shows only a very weak positive $n \rightarrow \pi^*$ Cotton effect. The octant rule predicts a positive Cotton effect. If the ring adopts a puckered conformation, one would predict that the conformer with three pseudoequatorial methyls would be more stable than that with three pseudoaxial methyls, and this prediction is indeed supported by molecular mechanics calculations[25] (Table 6-3). Thus, it would appear that a fourth pseudoaxial α-methyl controls the sign and magnitude of the Cotton effect. Although the octant rule predicts a positive sign for this conformer, $\Delta\varepsilon$ is surprisingly weak, considering that a pseudoaxial α-methyl group with CH_3–C_α–C=O torsion angle approximately equal to 70° is the major contributor to the Cotton effect. It is difficult to find other α-methyl ketones with comparable torsion angles. The lack of qualitative agreement is puzzling, as one would not expect this ketone to exhibit a smaller $|\Delta\varepsilon|$ than that of 2-methylcyclobutanone.

Interestingly, the epimeric 2,2,3(*S*),4(*R*)-tetramethylcyclobutanone gives a negative $n \rightarrow \pi^*$ Cotton effect that is too weak to be measured exactly.[36] In this isomer, the conformational equilibrium between puckered rings involves conformers with two pseudoaxial and two pseudoequatorial methyls (Table 6-3). Molecular mechanics calculations[25] predict that the isomer with two pseudoaxial α-methyls is only slightly less stable than that with two pseudoequatorial α-methyls. The octant rule predicts a negative Cotton effect for the former and a positive Cotton effect for the latter, so it follows that a weak negative Cotton effect might be observed.

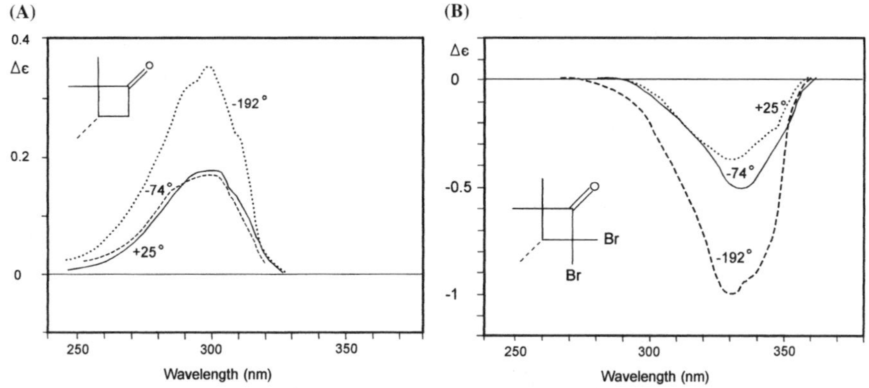

Figure 6-4. Variable-temperature CD spectra of cyclobutanone derivatives in ether–isopentane–ethanol (EPA; 5:5:2 by vol.). [Redrawn from ref. 37].

The conformational ambiguities attending cyclobutanones with α-methyl derivatives are also a concern with those carrying α-bromo groups. A large positive Cotton effect is observed for 2(*S*)-bromo-3(*R*),4,4-trimethylcyclobutanone, and a large negative Cotton effect is observed for the epimeric 2(*R*)-bromo analog (Table 6-3). The $\Delta\varepsilon$ values are much larger than those so far discussed. In the 2(*S*)-bromoketone, molecular mechanics[25] calculations predict a slight favoring of the puckered conformation with a pseudoequatorial bromine, but either conformer is predicted to give a positive Cotton effect, with the bromine controlling the sign and magnitude. In the case of the 2(*R*)-bromoketone, the favored conformer is decidedly that with a pseudoaxial bromine, but either conformer places the bromine in a negative octant, and the observed $\Delta\varepsilon$ is strongly negative. These data indicate the difficulty of using CD to extract conformational information involving interconverting puckered cyclobutanones; however, the assignment of the bromine configuration can readily be made by using the octant rule and assuming either puckered conformation—or even a planar conformation. Only the Br–C_α–C=O torsion angle varies with conformation; yet, with the range of torsion angles seen (54–73°), the bromine is sufficiently axial-like to elicit the typical UV λ_{max} shifts. In the epimeric α-bromoketones, the band is centered (λ^{max}) near 320 nm, whereas in the alkylcyclobutanones, it lies near 300–305 nm. The 15–20-nm bathochromic shift in the bromoketones is characteristic of that produced by an α-axial bromine on cyclohexanone,[38,39] whereas in the case of cyclohexanones, an α-equatorial bromine substituent has little influence on λ^{max}.

Although no attempts at conformational analysis of the mono-bromocyclobutanones by variable-temperature CD have been reported, low-temperature CD spectra of 2,2-dibromo-3(*R*),4,4-trimethylcyclobutanone (Figure 6-4B) has been published.[37] Molecular mechanics calculations[25] predict a large energy difference between the two puckered conformations (Table 6-3), with the 3-pseudoequatorial methyl conformation being decidedly more stable. Octant projection diagrams predict equal and oppositely signed Cotton effects for the two puckered conformers, with the methyl at C(3) making no contribution. In a 50:50 mixture of conformers, one would expect a vanishingly small Cotton effect. Yet the Cotton effect is negative at room temperature and becomes increasingly negative as the temperature is lowered. If one assumes that a pseudoaxial bromine makes a larger negative octant contribution than a pseudoequatorial one, then the CD data suggest that the puckered conformer with an equatorial methyl at C(3) is favored over that with the axial methyl. Unfortunately, the CD data have not yet been used to extract a conformational energy difference, which would be useful, given the paucity of experimentally determined conformational energy differences in cyclobutanones.

6.3. Cyclopentane

After cyclohexane, the five-membered carbocyclic ring is undoubtedly the most studied and most common. In contrast to conformational analysis of cyclohexanes and cyclobutanes, that of cyclopentane is more complex and is dominated by large-amplitude ring motion called *pseudorotation.*[12,13,40,41] In one of the earliest studies of the structure of cyclopentane, Wierl[42] determined the ring to be planar. This conformation did not seem unreasonable, as a planar five-membered ring with internal angles of 108° close to tetrahedral is nearly free of angle strain. (See Baeyer ring strain, Section 2.1.) Even well into the 1950s, the assumption of a planar ring was satisfactory in many studies involving cyclopentane stereochemistry.[40] However, in the early 1940s, it was recognized that a planar cyclopentane would have no fewer than five eclipsed nonbonded butane steric interactions (and a torsional strain of approximately 14 kcal/mole). It was then becoming clear to a few investigators that the planar structure could not be reconciled with calorimetric studies of cyclopentane.[43,44,45] In particular, Aston et al.[43,44] found that the predicted entropy (61.08 cal/deg/mole) of planar cyclopentane based on spectroscopic data and the symmetry number of the compound ($\sigma = 10$ for C_{5h} symmetry), was incompatible with the experimental value found (65.27 ± 0.15 cal/deg/mole). Rather, the experimental value was much closer to the entropy (65.65 cal/deg/mole) calculated for either of two nonplanar (puckered) conformations (Figure 6-5): the half-chair (C_2, $\sigma = 2$) and the envelope (C_s, $\sigma = 1$). The half-chair, with its C_2 rotation axis, has two enantiomers; the envelope, with its plane of symmetry, has only one isomer.

The concept of a puckered ring as the most stable conformation of cyclopentane was reinforced by thermodynamic studies carried out by Pitzer,[45] who estimated that ring puckering would result in an overall lowering of strain by about 3.6 kcal/mole.[41] At around the same time, Hassel and Viervoll[46] showed by gas-phase electron diffraction spectroscopy that cyclopentane might deviate from a symmetric pentagonal structure, but concluded that the deviation would be very limited. Subsequent electron diffraction studies by Bastiansen et al.[18] concluded that the cyclopentane ring deviates from planarity, but the degree of puckering was not determined. A more recent electron diffraction study by Adams, Geise, and Bartell[47] provided refined molecular parame-

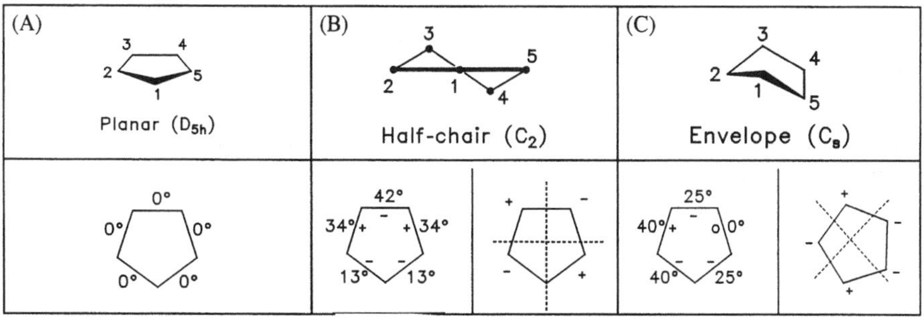

Figure 6-5. Structures of cyclopentane conformations. The half-chair and envelope are nearly isoenergetic and are some 5 kcal/mole more stable than the planar. The representations below each structure designate (left) the magnitudes and signs of the corresponding ring ψ(C–C–C–C) torsion angles and (right) the relative ring atom positions above (+) and below (–) the average plane passing through the ring.

ters for a puckered cyclopentane structure, with a puckering displacement calculated to be approximately 0.44 Å.

The half-chair and envelope puckered conformations easily interconvert by pseudorotation. Pitzer et al.[41] introduced that notion to account for the additional degree of freedom (i.e., in addition to puckering) required to explain the heat of formation and the different entropies derived from thermodynamic and spectroscopic measurements of cyclopentane.[16,39,40,41,48] In pseudorotation, puckering is transmitted around the ring through internal and concerted C–C bond rotations. Thus, according to Pitzer and Donath,[48] puckering rotates around the ring in a practically free pseudorotation, while the amplitude of puckering oscillates about a stable equilibrium value of about 0.48 Å. This motion interconverts the half-chair and envelope conformers, as is illustrated in Figure 6-6 using the ring torsion angle notation (+, –, o) of Bucourt.[19]

Unlike cyclobutane ring inversion, cyclopentane conformers interconvert without passing through the planar form, which is neither a transition state nor an intermediate in pseudorotation. However, it was not until 1968 that direct spectroscopic (mid-infrared) observation of pseudorotation in cyclopentane was made by Durig and Wertz.[49] Although microwave spectroscopic studies could not be performed on cyclopentane, since it has no permanent dipole, midrange IR and Raman studies indicated that cyclopentane undergoes nearly free pseudorotation.[50] In 1972, Carreira et al.[51] showed by Raman spectroscopy that the barrier to planarity in cyclopentane is $1824 \pm 50\ cm^{-1}$ (~5.2 kcal/ mole)—in good agreement with the 4.8 kcal/mole estimated by Pitzer and Donath[48] on the basis of thermodynamic data. The barrier to planarity is thus significantly higher than the barrier to pseudorotation. Anet[52] has suggested that when the barrier to pseudorotation is very low, or in the limit when pseudorotation is free, as in cyclopentane, there is actually only one stable conformation, and pseudorotation is simply a molecular vibration.

Molecular mechanics calculations using a diversity of force fields with different potential functions or purely structural data, semiempirical calculations, and ab initio calculations all conclude that puckered cyclopentane is more stable than the planar form (by approximately 5 kcal/ mole) and that the C_2 and C_s puckered conformers are essentially isoenergetic with a pseudorotation barrier of less than 0.1 kcal/mole.[40,53] More recent ab initio and MM2 calculations[54] indicate that the C_2 half-chair is slightly more stable (3.7–3.8 cal/mole) than the C_s envelope, which might be viewed as a saddle point on the conformational energy hypersurface.[55] Computed data for the C_2 half-chair and C_s envelope conformations from PCMODEL[25] calculations (Table 6-4) compare favorably with those of the most recent molecular mechanics calculations using MM3.[56] And both data sets compare favorably with corresponding data from electron diffraction spectroscopy.[46,47]

Cyclopentane is clearly flatter than chair cyclohexane (Table 6-4). And, unlike the dynamic NMR analysis of cyclohexane, that of cyclopentane does not lend itself to the detection of conformational interconversion. The only ^{1}H-NMR signal in cyclopentane (near 1.51 δ) remains a singlet down to very low temperatures,[57] as is expected from the very low barrier (~kT) to pseudorotation. That is, the internal energy of the cyclopentane molecule changes by less than RT (~600 cal/mole) at room temperature.

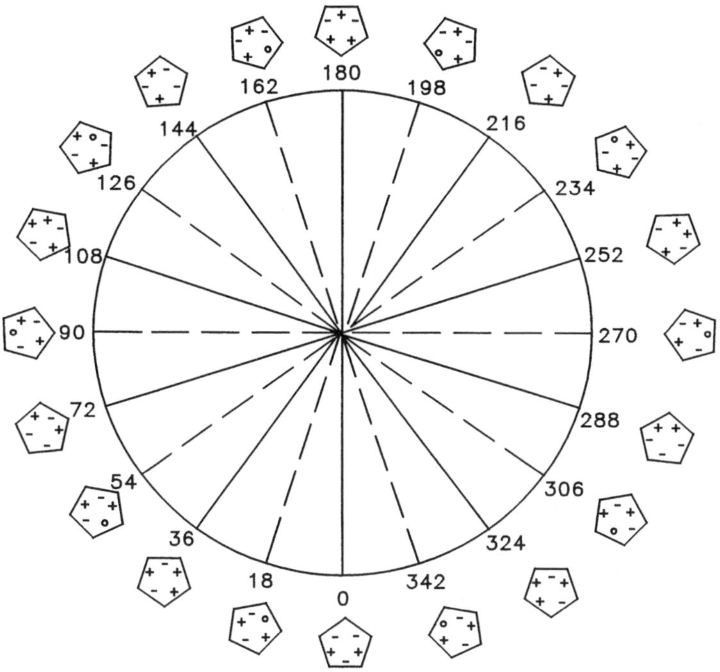

Figure 6-6. Full pseudorotation roulette wheel for cyclopentane. (See Figure 6-5.) Each point on the circle represents a specific value of the pseudorotation phase angle in 18° increments. There are 10 C_2 half-chair forms (at the solid diagonals) and 10 C_s envelope forms (at the dashed diagonals). Mirror-image conformations lie across the circle, connected by diagonal solid lines or diagonal dashed lines. Each 18° pseudorotation alternates the conformation between C_2 and C_s.

Unlike the situation with cyclohexane, no definite global energy minimum (as in chair cyclohexane) or local minima (as in twist-boat cyclohexane) or maxima (as in half-chair cyclohexane) are detected (Figure 2-3). However, as with cyclohexane, with cyclopentane the analysis of NMR vicinal coupling constants can provide information on conformations. The vicinal coupling constants derived from 1,1,2,2,3,3-hexadeuteriocyclopentane, viz., $^3J_{trans}$ = 6.30 Hz and $^3J_{cis}$ = 7.90 Hz,[58] can be used to calculate the ring torsion angle according to the procedure developed by Lambert and described in Section 2.3. The calculated Lambert R value is 0.80, which indicates considerable ring flattening (e.g., relative, to cyclohexane, for which R = 2.2) and a C–C–C–C torsion angle ψ ~ 40°. This value overestimates the average cyclopentane ring torsion angle computed by molecular mechanics (ψ ~ 26°) and that (ψ ~ 27°) determined by electron diffraction (Table 6-4). The failure is probably due to a breakdown of the threefold symmetry of projection angles in cyclopentane i.e., to a deviation of the H–C–H projection angle from the 120° value used in the Lambert calculation of ψ.

Table 6-4. Comparison of Computed and Experimentally Derived Ring Torsion Angles (ψ), Average Bond Angles (∠), and Average Bond Lengths (d) for Cyclopentane and Cyclohexane

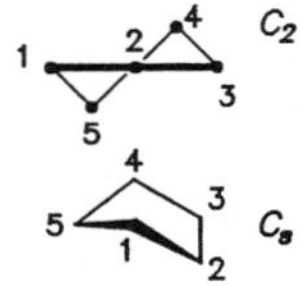	METHOD						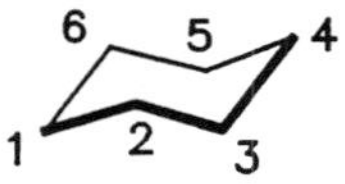	METHOD		
	PCMODEL[a]		MM3[b]		Electron Diffraction[c]					
	C_2	C_s	C_2	C_s	C_2	C_s		PCMODEL[a]	MM3[d]	Electron Diffraction[e]
$\psi\ (C_1–C_2–C_3–C_4)$	13.2°	0.0°	12.8°	0.0°	13.2°	0.0°	$\psi\ (C_1–C_2–C_3–C_4)$	56.4°	55.3°	55.1° ± 0.4°
$\psi\ (C_2–C_3–C_4–C_5)$	–34.5°	–25.1°	–33.5°	–24.2°	–34.3°	–25.0°	$\psi\ (C_2–C_3–C_4–C_5)$	–56.4°	–55.3°	–55.1° ± 0.4°
$\psi\ (C_3–C_4–C_5–C_1)$	42.5°	40.5°	41.4°	39.2°	42.3°	40.3°	$\psi\ (C_3–C_4–C_5–C_6)$	56.4°	55.3°	55.1° ± 0.4°
ψ (average)	27.5°	26.2°	26.8°	25.4°	27.5°	26.1°	ψ (average)	56.4°	55.3°	55.1° ± 0.4°
$\angle\ (C–C–C)_{av}$	104.3°	104.4°	104.7°	104.7°	104.5°	104.5°	∠ (C–C–C)	110.9°	111.3°	111.34° ± 0.24°
$\angle\ (H–C–H)_{av}$	108.1°	108.1°					∠ (H–C–H)	107.1°	106.7°	105.3° ± 2.3°
$d\ (C–C)_{av}$	1.537 Å	1.535 Å	1.545 Å	1.549 Å	$1.546 \pm 0.001_2$ Å		d (C–C)	1.536 Å	1.536 Å	1.5335 ± 0.002 Å
d (C–H)	1.116 Å	1.116 Å			$1.113_5 \pm 0.001_5$ Å		d (C–H)	1.116 Å	1.1145 Å	1.1099 ± 0.004 Å

[a] Values computed by PCMODEL (ref. 25).

[b] Values from ref. 56.

[c] Values from ref. 47.

[d] Values from ref. 26.

[e] See Table 2-3.

Nevertheless, NMR clearly indicates that the cyclopentane ring is less puckered than cyclohexane, but is not flat ($R = 0.25$ for $\psi = 0°$).

In the half-chair and envelope conformations of cyclopentane, as shown in Figure 6-7, the exocyclic bonds may be characterized as axial (*a*), equatorial (*e*), pseudoaxial (*a*′), pseudoequatorial, (*e*′) or bisectional (*b*). The characterization of a particular substituent thus changes during pseudorotation, and a substituted cyclopentane can easily adopt its most stable conformation(s), which may even differ from the half-chair or envelope. In substituted cyclopentanes, pseudorotation becomes more hindered, and the relative energies of the half-chair and envelope conformations may differ. But it is not necessarily easy to determine the most stable conformation. For example, calculations[48,53] showed the most stable conformation of methylcyclopentane would be an envelope with the methyl group occupying an equatorial position at the tip of the envelope—as in Figure 6-7B, with a methyl at position 1*e*. Other computations favored the half-chair with a bisectional methyl conformation (Figure 6-7A, CH_3 at 1*b*), but when the size of the substituent increased to that of isopropyl, the envelope conformation (Figure 6-7B) was favored, with the isopropyl group at position 1*e*.[40] Clearly, the analysis of substituted cyclopentanes is not as straightforward as that of substituted cyclohexanes.[40] The energy difference between axial and equatorial conformations of cyclopentanes is smaller than in cyclohexanes, thus complicating conformational analysis achieved by simple inspection. For example, the conformational analysis of 1,3-dimethylcyclopentane is more complex than that of its cyclohexane counterpart. Unlike 1,3-dimethylcyclohexane, *cis*-1,3-dimethylcyclopentane is more stable than the *trans* form by $\Delta H° = 0.53$ kcal/mole.[59] (In comparison, *cis*-1,3-dimethylcyclohexane is more stable than *trans* by $\Delta H° = 1.96$ kcal/mole.)[60] Although it is clear (Chapter 2) that the diequatorial chair conformer of 1,3-dimethylcyclohexane is the more stable conformation, it is less clear in the case of 1,3-dimethylcyclopentane. The data provide undisputed evidence against a planar cyclopentane conformation, but they were also thought to provide a convincing case for a *cis*-1,3-dimethylcyclopentane envelope conformation with equatorial methyls[59,60] (located, for example, at 2*e* and 5*e* in Figure 6-7B). However, that conclusion has been challenged by molecular mechanics calculations,[53] which correctly predict that the *cis* isomer is more stable than the *trans* (by $\Delta H° \sim 0.3$ kcal/mole) and also indicate that the half-chair conformation is preferred in the *cis* isomer. Two isoenergetic half-chairs are found for the *cis*, one with methyls at 3*e* and 5*e*′ (Figure 6-7A), the other with an equatorial methyl at 3*e* and a bisectional

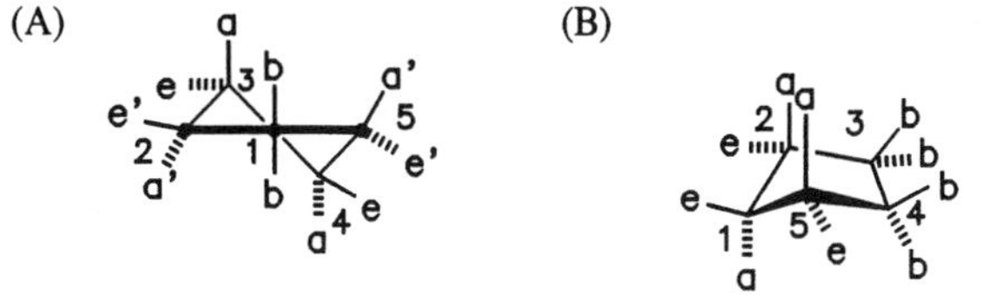

Figure 6-7. (A) Half-chair and (B) envelope conformations of cyclopentane, showing axial (*a*), equatorial (*e*), pseudoaxial (*a*′), pseudoequatorial (*e*′), and bisected (*b*) exocyclic positions.

methyl at *cis*-1*b* (Figure 6-7A). For the *trans*, two isoenergetic half-chair conformers are also found, one with methyls at 2*e*′ and 5*e*′, the other with methyls at 3*e* and *trans*-1*b*. Conformational analysis by inspection or by the use of molecular models is far less simple than in cyclohexanes, and the conformational energy differences are small; yet a considerable body of work has been published on the analysis of substituted cyclopentane conformation and reactivity.[40]

6.4. Cyclopentanones

Pitzer and Donath[48] predicted that if the hybridization of one ring atom were changed from sp^3 to sp^2, as in cyclopentanone,[12,13,40] (1) pseudorotation would be restricted (the barrier in cyclopentanone is 2.4 kcal/mole vs. 0.1 kcal/mole in cyclopentane), and (2) the C_2-symmetry half-chair twist conformation (Figure 6-8) would be adopted. These predictions were supported by experimental data from microwave,[61] electron diffraction,[62] Raman,[63] and far-infrared[64] spectroscopy. Thus, cyclopentanone adopts either of two mirror-image half-chair conformers that interconvert through the planar conformation over a barrier of 750 cm^{-1} (~2.15 kcal/mole) (Figure 6-8).[13] Although the half-chair lies at a global energy minimum, unlike the situation with cyclopentane, there are no minima corresponding to the envelope conformations. This makes conformational analysis of cyclopentanones implicitly simpler than that of cyclopentanes.

Molecular mechanics calculations using the MM2 force field[53] predict the envelope conformation to be 3.22 kcal/mole above the half-chair,[65] due to increased van der Waals and torsional energies associated with eclipsing in the C_2–C_3–C_4–C_5 ring segment. Similar findings result from PCMODEL[25] calculations, and the various structural parameters are compared in Table 6-5. Data from both sets of calculations—

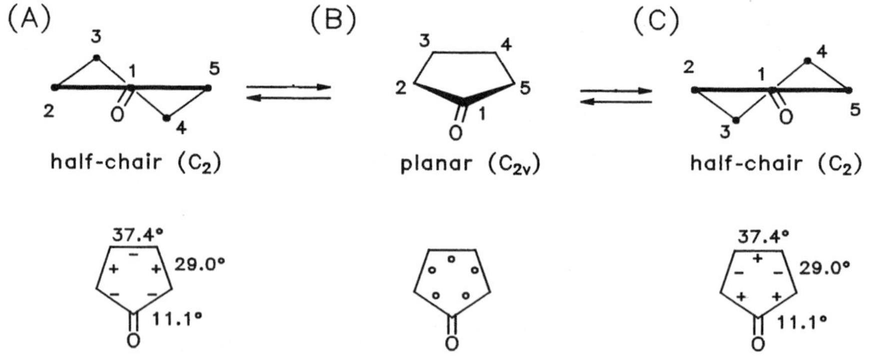

Figure 6-8. (A) and (C) Minimum-energy, enantiomeric, C_2-symmetry half-chair conformations of cyclopentanone interconvert through the planar conformer (B) with C_{2v}-symmetry over a barrier of 750 cm^{-1} (2.15 kcal/mole) (ref. 13). The ring torsion angles are shown on the structures in the bottom half of the figure for the conformational drawings shown in the top half.

PCMODEL and the more recent MM3[33]—are in good agreement with electron diffraction[62] and microwave data[61] and combined spectral data[66] for a stable C_2-symmetry half-chair conformation. These conclusions are supported by vicinal coupling-constant analyses of cyclopentanone in solution that point to a C_2 conformation with a $\psi(C_2–C_3–C_4–C_5)$ ring torsion angle of 34.5°[58]—compared with 37.4° from electron diffraction[62] analysis and 38.2° from PCMODEL calculations (Table 6-5).

Substitution does not appear to alter the conformational "preference" of cyclopentanone for the C_2-symmetry half-chair conformation. For example, on the basis of analyses of vibrational[67] and NMR spectroscopy,[68] *trans*-3,4-dimethylcyclopentanone is thought to "prefer" the C_2 half-chair conformation with equatorial methyls. PCMODEL molecular mechanics calculations[25] reach the same conclusion and predict an energy difference of approximately 1.9 kcal/mole between the more stable diequatorial and diaxial conformers and very little ring distortion in either relative to the parent cyclopentanone (Table 6-6).

Some of the earliest spectroscopic investigations of alkyl-substituted cyclopentanone conformation were carried out using optical rotatory dispersion (ORD).[69] The data were analyzed by the octant rule (Chapter 4) for the ketone carbonyl $n \rightarrow \pi^*$ transition and by force field calculations.[70] Although the conformational analysis is relatively straightforward, based upon the energetically favored C_2-symmetry half-chair or deformations thereof, the application of the octant rule is somewhat more complicated than in the case of chair cyclohexanone or cyclobutanone, both of which have C_s-symmetry and are thus optically inactive. In contrast, C_2-symmetry half-chair cyclopentanone adopts either of two enantiomeric conformations, and thus, a cyclopentanone molecule is chiral (Figure 6-9). Application of the octant rule to the C_2-half-chair enantiomeric conformations of cyclopentanone places ring atoms 1, 2, and 5 on an octant symmetry plane, but carbon atoms 3 and 4 both lie either in positive (+) or in negative (–) back octants, depending on the enantiomeric conformation. A further complication arises from the fact that the hydrogens at the α-carbons, C(2) and C(5), are not canceling, as they are for the α-hydrogens in chair cyclohexanone. In the octant rule, contributions from hydrogens are normally neglected, but Kirk[71] has suggested that octant-dissignate contributions from α-hydrogens dominate the $n \rightarrow \pi^*$ Cotton effect in cyclopentanones and in twisted or twist-boat cyclohexanones. We will return to this notion shortly.

Only a relatively modest number of optically active cyclopentanones have been studied by ORD or CD. The majority are shown in Table 6-7, along with ORD-based amplitudes (*a*) and CD-derived $\Delta\varepsilon^{max}$ values. Ouannes and Jacques[69] calculated the ORD *a* values for several of these alkylcyclopentanones on the basis of conformational analysis and a population mix of 20 conformers, including various C_2 and C_s cyclopentanone conformations. Although the calculated magnitudes are only a fair match to the experimental ones, the Cotton effect signs are correctly predicted in the analysis.

Molecular mechanics calculations[25] offer a way to analyze the ORD and CD data further. Since cyclopentanone is known to favor the C_2 half-chair conformation, a good starting point in the analysis of alkylcyclopentanone conformations is to consider the influence of alkyl groups on the half-chair $\rightleftarrows$ half-chair equilibrium (Figure 6-9). For

Table 6-5. Comparison of Computed and Experimentally Derived Torsion Angles (ψ and ϕ), Bond Angles (∠), and Bond Lengths (*d*) of Half-Chair Cyclopentanone with PCMODEL Computed Values for Chair Cyclohexanone

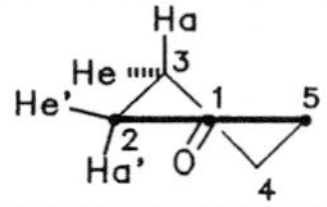	Method					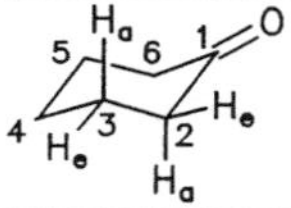
	PCMODEL[a]	MM3[b]	ED[c]	MW[d]	ED + MW[e]	
ψ (C_3–C_2–C_1–C_5)	−11.7°		11.1 (5)°	12.4 (2)°	12.4 (2)°	−49.8°
ψ (C_1–C_2–C_3–C_4)	30.4°		29.0°	32.5 (5)°	32.5 (5)°	52.7°
ψ (C_2–C_3–C_4–C_5)	−38.2°		−37.4°	−40.5 (6)°	−40.5 (6)°	−57.7°
ϕ (H_e–C–C=O)	46.6°					8.92°
ϕ (H_a–C–C=O)	−73.9°					172.80°
ϕ (H_e–C_2–C_3–H_e)	−34.7°					−62.93°
ϕ (H_e–C_2–C_3–H_a)	86.0°					54.92°
ϕ (H_a–C_2–C_3–H_e)	37.3°					55.58°
ϕ (H_a–C_2–C_3–H_a)	158.0°					172.80°
∠ (C_2–C_1–C_5)	109.8°	108.82°	112.4 (3)°	110.5 (7)°	108.6 (2)°	115.5°
∠ (C_1–C_2–C_3)	104.0°	104.84°	102.2 (3)°	104.5 (5)°	104.2 (1)°	111.3°
∠ (C_2–C_3–C_4)	104.1°	104.21°	105.1°	103.0 (5)°	103.4 (2)°	111.1°
d (C=O)	1.209 Å	1.2086 Å	1.226 (4) Å	1.215 (5) Å	1.213 (4) Å	1.211 Å
d (C_1–C_2)	1.518 Å	1.5259 Å	1.519 Å	1.504 (10) Å	1.531 (3) Å	1.536 Å
d (C_2–C_3)	1.536 Å	1.5374 Å	1.532 Å	1.557 (7) Å	1.542 (3) Å	1.534 Å
d (C_3–C_4)	1.536 Å	1.5415 Å	1.540 Å	1.557 (7) Å	1.542 (3) Å	1.535 Å
d $(C–H)_{av}$	1.116 Å	1.1159 Å	1.122 (6) Å	—	1.104 (4) Å	

[a] Values from PCMODEL (ref. 25); see Table 2-5 for cyclohexanone values.
[b] Values from ref. 33.
[c] Electron Diffraction (ED) values from ref. 62.
[d] Microwave (MW) values from ref. 61.
[e] Values from ref. 66.

Table 6-6. Comparison of Ring Torsion Angles and Heats of Formation in Cyclopentanone and *trans*-3,4-Dimethylcyclopentanone[a]

Ketone	Ring Torsion Angle (°)			H–C–C=O Torsion Angle (°)		Computed ΔH_f (kcal/mole)[a]
	C_3–C_2–C_1–C_5 C_2–C_1–C_5–C_4	C_1–C_2–C_3–C_4 C_1–C_5–C_4–C_3	C_2–C_3–C_4–C_5	$H_{e'}$–C_2–C=O $H_{e'}$–C_5–C=O	$H_{a'}$–C_2–C=O $H_{a'}$–C_5–C=O	
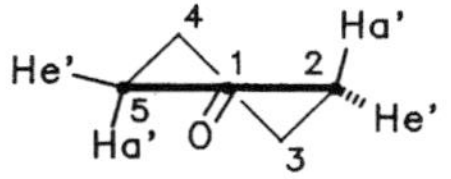	−11.86 −11.48	30.54 30.31	−38.19	46.44 46.84	−74.07 −73.66	−46.01
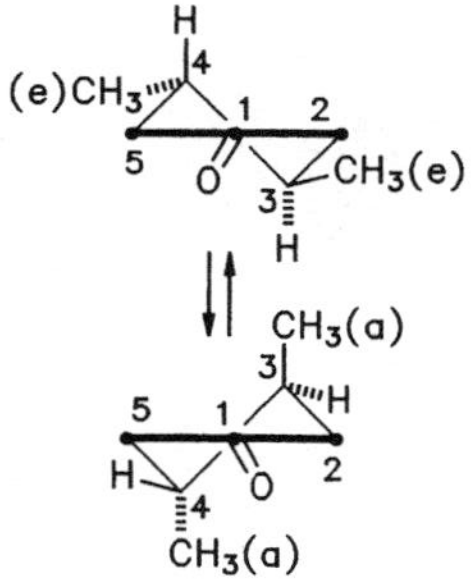	−12.00 −11.54	30.76 30.48	−38.28	46.26 46.70	−73.89 −73.41	−60.92
	11.32 10.19	−28.27 −27.59	34.92	−46.18 −47.26	73.18 72.04	−59.02

[a] Calculated using PCMODEL, ref. 25.

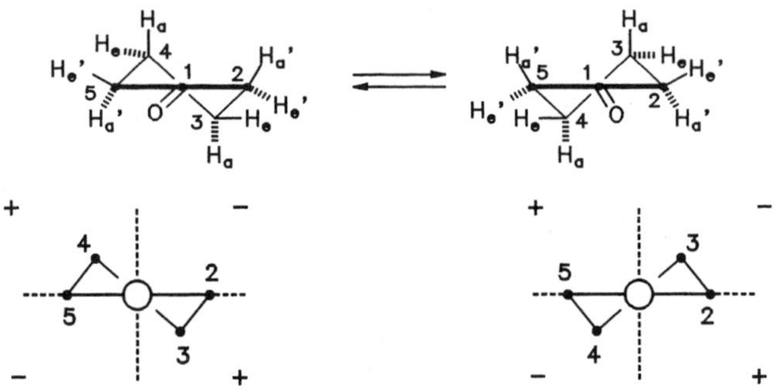

Figure 6-9. (Upper) Equilibrating conformational enantiomers of C_2-symmetry cyclopentanone and (lower) their corresponding octant diagrams.

example, does an α-methyl group prefer the pseudoaxial or pseudoequatorial position? And does a β-methyl group prefer an equatorial position, as it does in cyclohexane? Intuitively, one would predict that the cyclopentanone half-chair conformer with a pseudoequatorial α-methyl or an equatorial β-methyl would be more stable than that with corresponding pseudoaxial or axial methyls. In fact, these predictions are supported by molecular mechanics calculations[25] (Table 6-8), which indicate that 3-methylcyclopentanone "prefers" a half-chair cyclopentanone conformation with an equatorial methyl by about 1 kcal/mole over the half-chair with an axial methyl. There is little evidence for ring flattening toward the envelope conformation. Similarly, in 2-methylcyclopentanone, the half-chair conformation with a pseudoequatorial methyl is more stable (by approximately 1 kcal/mole) than the half-chair with a pseudoaxial methyl. There is some evidence for slight ring flattening in the pseudoequatorial methyl conformer [Cf. $\psi(C_3–C_2–C_1–C_5)$ and $\psi(C_2–C_1–C_5–C_4)$.] In a half-chair, they are normally both about 12°, but in this half-chair, one is much smaller than 12° and the other is much larger. In an envelope conformation, one would be approximately 0° and the other would be about 25°. So there is a tendency toward ring flattening in the pseudoequatorial 2-methylcyclopentanone half-chair conformation. Such ring distortion appears to alleviate a nonbonded steric interaction between the methyl group and the ring methylene at C_4.

Interestingly, in 3-methylcyclopentanone, the axial conformer shows only slight distortion from the half-chair. And even when the 3-axial methyl is replaced by an isopropenyl group or a *tert*-butyl group, only a slightly distorted half-chair is adopted. In the 3-axial isopropenyl conformer, the rotational isomer with the $=CH_2$ over the ring is predicted to be more stable by about 1 kcal/mole than that with the CH_3 over the ring, but both are more stable than the rotamer with neither over the ring.

Serious ring deformation of the half-chair cyclopentanone ring conformation occurs when substitution includes an α-pseudoaxial alkyl and a *cis*-alkyl β-axial alkyl—as in 2(*R*)-methyl-4(*S*)-isopropylcyclopentanone. Here, considerable ring flattening alleviates a 1,3-diaxial type of interaction, while leaving the ring conformation much more envelope-like. However, this conformation is approximately 4 kcal/mole

Table 6-7. Experimental[a] and Calculated[b] ORD Amplitudes (a) and CD $\Delta\varepsilon^{max}$ Values[c] for the Carbonyl $n \rightarrow \pi^*$ Transition of Alkylcyclopentanones

Substituted Cyclopentanone	Experimental ORD (a)	Calculated ORD (a)	CD $\Delta\varepsilon^{max}$	Substituted Cyclopentanone	Experimental ORD (a)	Calculated ORD (a)	CD $\Delta\varepsilon^{max}$
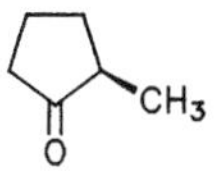	−46[d]	−23	−1.45[e]	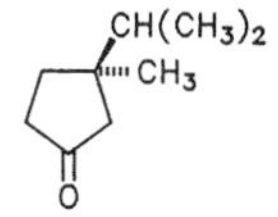	+97	—	+2.4
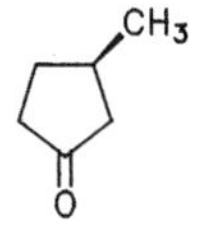	+86	+22	+2.1	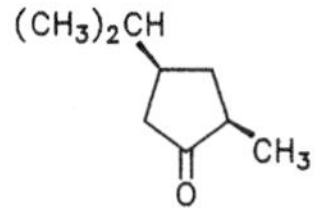	−80	−141	−2.0
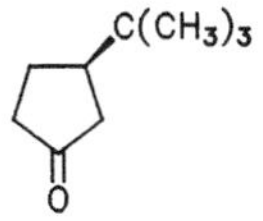	+130	—	+3.2	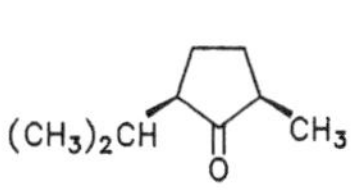	+17	—	+0.42
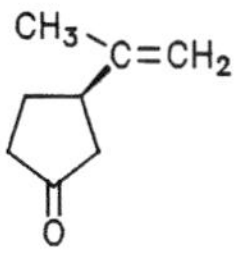	+91	—	+2.3	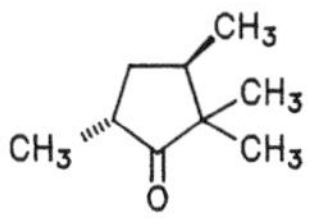	+20[d]	+17	+0.5f
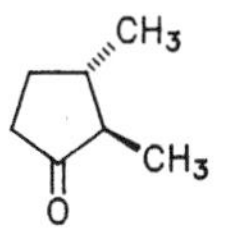	−69	−49	−1.7	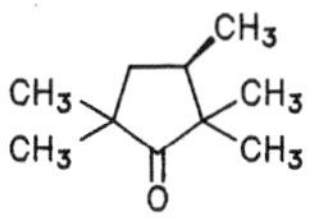	+48[d]	—	+1.2[f]

[a] In methanol at room temperature (ref. 69).
[b] Calculated (ref. 69) on the basis of a model with methyl substituents only.
[c] Calculated from experimental ORD a values using $\Delta\varepsilon^{max} = a/40.28$ (Chapter 3).
[d] Calculated from Collet, A., Jacques, J., Chion, B., and Lajzerowicz, J., *Tetrahedron* **31** (1975), 2243–2246.
[e] From CD measured by Partridge, J. J., Chadhar, N. K., and Uskokovic, M. R., *J. Am. Chem. Soc.* **95** (1973), 532–540.
[f] From CD measured by the authors of footnote *d*.

higher in energy than the half-chair with a pseudoequatorial α-methyl and an equatorial isopropyl and is thus unlikely to contribute to the equilibrium and the $n \rightarrow \pi^*$ Cotton effect. A similar situation obtains with pentamethylcyclopentanone.

The cis-α,α′-disubstituted cyclopentanone 2(*S*)-isopropyl-5(*S*)-methylcyclopentanone (Table 6-8) offers the most complex conformational analysis, for the energy differences between envelope and half-chair conformations are small. Two different half-chair conformers are possible with C_2-symmetry cyclopentanone rings, one with a pseudoaxial methyl and the other with a pseudoaxial isopropyl. The C_2-symmetry cyclopentanone rings of each adopt enantiomeric ring conformations. In each case (Table 6-9), the α-isopropyl group can adopt three distinct fully staggered rotameric conformations. And in each case, the choice of staggered isopropyl rotamer determines

Table 6-8. Torsion Angles and Heats of Formation Computed by Molecular Mechanics Calculations[a] for Stable Alkylcyclopentanone Conformations

Substituted Cyclopentanone	Conformational Equilibrium	Computed Torsion Angles (°)						ΔH_f (kcal/mole)
		C_3–C_2–C_1–C_5 C_2–C_1–C_5–C_4	C_1–C_2–C_3–C_4 C_1–C_5–C_4–C_3	C2–C_3–C_4–C_5	$H_{e'}$–C_2–C=O $H_{e'}$–C_5–C=O	$H_{a'}$–C_2–C=O $H_{a'}$–C_5–C=O	C_3–C_2–C=O C_4–C_5–C=O	
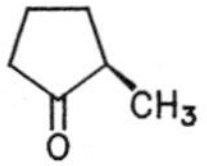	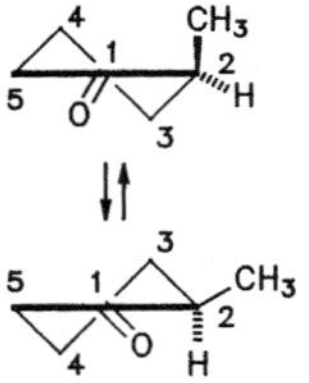	–5.69	25.93		56.71	–63.59[b]	174.57	
		–16.65	32.39	–36.70	41.55	–79.11	163.05	–52.25
		14.69	–32.53		–40.70[b]	80.22	–164.73	
		8.91	–29.08	38.69	–49.72	70.66	–171.70	–53.40
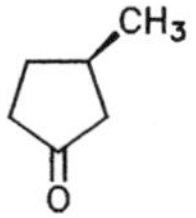	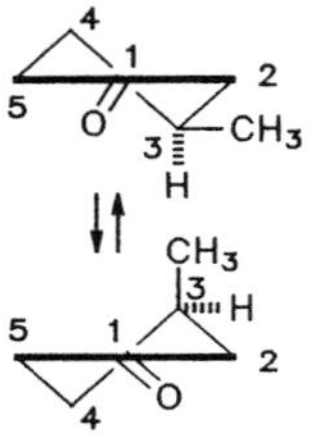	–12.04	30.90		46.34	–73.98	167.86	
		–11.61	30.72	–38.64	46.70	–73.82	168.48	–53.67
		13.08	–30.05		–44.35	75.37	–166.94	
		9.29	–28.13	36.46	–49.14	71.03	–170.68	–52.53

(Continued)

Table 6-8. *(Continued)*

Substituted Cyclopentanone	Conformational Equilibrium	Computed Torsion Angles (°) $C_3–C_2–C_1–C_5$ $C_2–C_1–C_5–C_4$	$C_1–C_2–C_3–C_4$ $C_1–C_5–C_4–C_3$	C2–$C_3–C_4–C_5$	$H_{e'}–C_2–C=O$ $H_{e'}–C_5–C=O$	$H_{a'}–C_2–C=O$ $H_{a'}–C_5–C=O$	$C_3–C_2–C=O$ $C_4–C_5–C=O$	ΔH_f (kcal/mole)
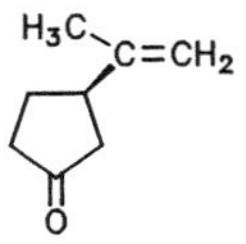	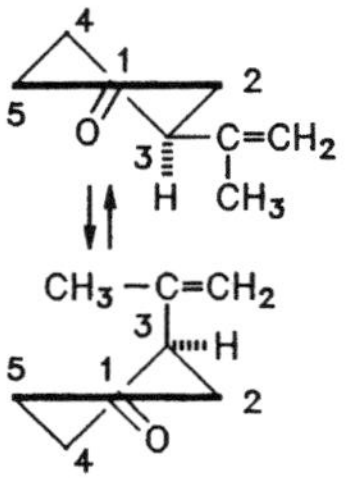	−12.26	32.12		45.76	−74.52	167.75	
		−12.63	32.67	−40.55	45.81	−74.85	167.37	−36.84
		17.95	−32.36		−38.00	80.53	−162.69	
		4.21	−24.96	35.84	−54.51	64.85	−175.55	−34.67[c]
		14.70	−32.49		−40.70	77.99	−165.20	
		9.31	−29.84	38.93	−49.29	71.14	−170.79	−35.59[d]
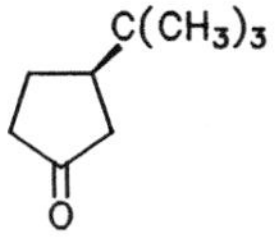	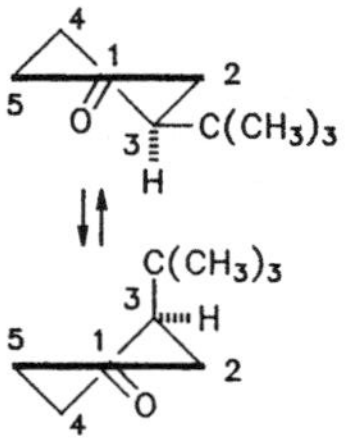	−11.88	30.67		46.41	−72.86	168.62	
		−11.77	30.90	−38.58	46.50	−73.92	167.80	−71.65
		11.93	−25.82		−42.99	74.10	−169.18	
		7.31	−24.02	31.05	−51.36	68.01	−171.78	−68.72

(Continued)

Table 6-8. *(Continued)*

Substituted Cyclopentanone	Conformational Equilibrium	Computed Torsion Angles (°)						ΔH_f (kcal/mole)
		$C_3–C_2–C_1–C_5$ $C_2–C_1–C_5–C_4$	$C_1–C_2–C_3–C_4$ $C_1–C_5–C_4–C_3$	$C2–C_3–C_4–C_5$	$H_{e'}–C_2–C{=}O$ $H_{e'}–C_5–C{=}O$	$H_{a'}–C_2–C{=}O$ $H_{a'}–C_5–C{=}O$	$C_3–C_2–C{=}O$ $C_4–C_5–C{=}O$	
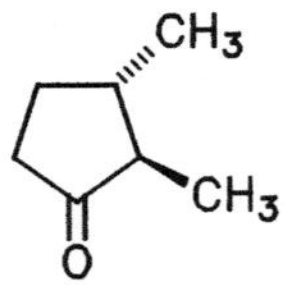	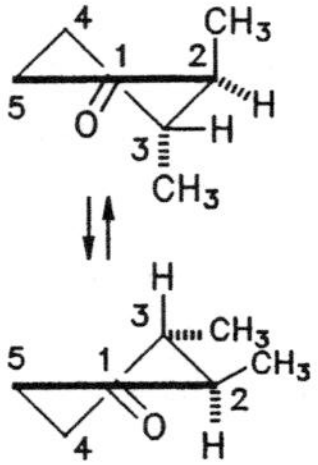	–6.12 –15.24	25.01 30.73	 –35.04	55.58 43.10	–63.87[b] –77.30	174.51 164.15	 –58.75
		13.62 9.93	–31.72 –29.66	 38.56	–41.78[b] –48.53	78.64 71.89	–166.12 –170.37	 –60.74
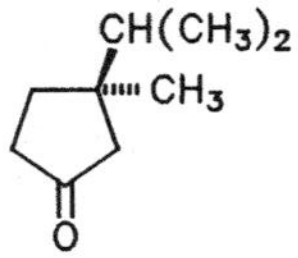	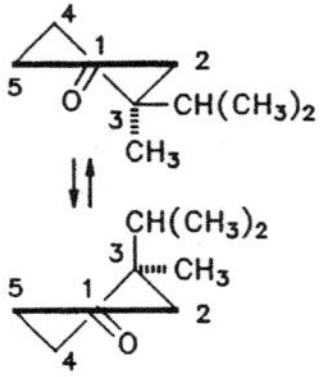	–13.77 –8.35	29.72 27.49	 –35.64	43.41 50.18	–75.15 –69.82	166.31 171.58	 –70.78
		16.18 9.32	–34.29 –31.61	 41.10	–40.00 –49.49	77.99 71.21	–163.89 –170.61	 –69.02

(Continued)

Table 6-8. *(Continued)*

Substituted Cyclopentanone	Conformational Equilibrium	Computed Torsion Angles (°)						ΔH_f (kcal/mole)
		$C_3–C_2–C_1–C_5$ $C_2–C_1–C_5–C_4$	$C_1–C_2–C_3–C_4$ $C_1–C_5–C_4–C_3$	C2–$C_3–C_4–C_5$	$H_{e'}–C_2–C{=}O$ $H_{e'}–C_5–C{=}O$	$H_{a'}–C_2–C{=}O$ $H_{a'}–C_5–C{=}O$	$C_3–C_2–C{=}O$ $C_4–C_5–C{=}O$	
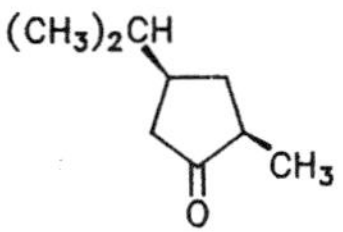	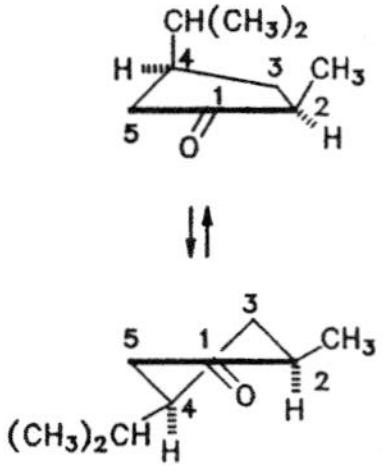	2.93	20.46		−53.53[b]	66.16	−177.01	
		−24.93	36.00	−35.37	−87.58	31.12	155.01	−68.81
isopropyl C-*H* over ring		14.78	−32.68		−40.80[b]	80.11	−164.94	
		8.77	−28.70	38.40	−49.55	70.24	−171.51	−73.18
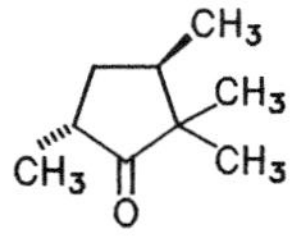	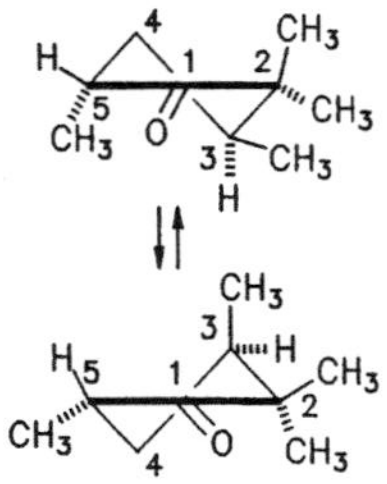	−17.06	32.00		43.43[b]	−76.63[b]	162.83	
		−4.43	24.46	−35.94	57.86	−62.34[b]	175.57	−72.51
		24.02	−32.29		−32.79[b]	86.61[b]	−156.17	
		−6.07	−14.77	29.91	−69.13	50.15[b]	174.12	−70.87

(Continued)

Table 6-8. *(Continued)*

Substituted Cyclopentanone	Conformational Equilibrium	Computed Torsion Angles (°) $C_3–C_2–C_1–C_5$ $C_2–C_1–C_5–C_4$	$C_1–C_2–C_3–C_4$ $C_1–C_5–C_4–C_3$	$C2–C_3–C_4–C_5$	$H_{e'}–C_2–C=O$ $H_{e'}–C_5–C=O$	$H_{a'}–C_2–C=O$ $H_{a'}–C_5–C=O$	$C_3–C_2–C=O$ $C_4–C_5–C=O$	ΔH_f (kcal/mole)
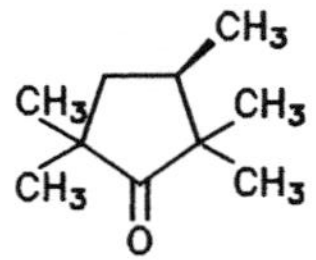	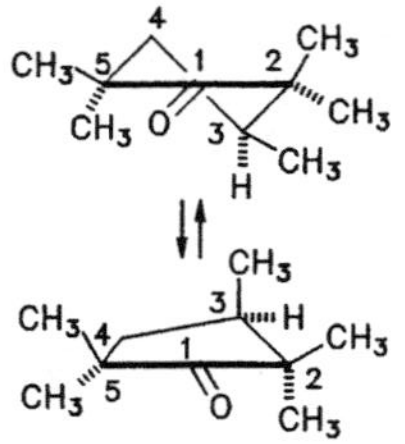	–11.38 –10.56	28.93 28.62	–36.82	49.14[b] 49.42[b]	–70.68[b] –71.81[b]	162.24 179.92	–80.13
		18.19 0.36	–28.45 –18.07	29.74	–38.38[b] –60.23[b]	80.50[b] 59.74[b]	–162.24 –179.92	–77.85
		12.76 10.74	–31.46 –30.06	39.00	–49.49 –45.71[b]	70.80[e] 75.05	–166.57 –169.92	–70.76
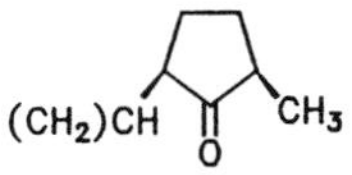	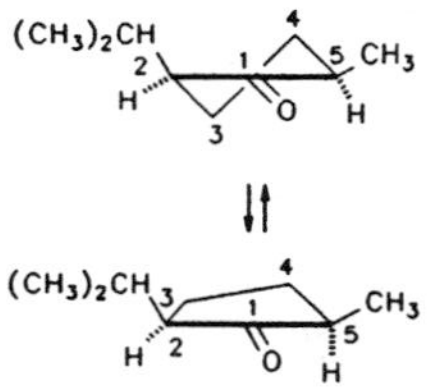	–0.34 20.70	–20.39 –32.70	33.53	–65.26 –35.71[b]	54.71[e] 85.35	–179.94 –159.69	–71.31
		–2.23 22.58	–19.26 –33.86	33.58	–67.00 –33.54[b]	51.06[e] 87.80	177.58 –157.23	–70.76

(Continued)

Table 6-8. *(Continued)*

Substituted Cyclopentanone	Conformational Equilibrium	Computed Torsion Angles (°) C_3–C_2–C_1–C_5 / C_2–C_1–C_5–C_4	C_1–C_2–C_3–C_4 / C_1–C_5–C_4–C_3	C2–C_3–C_4–C_5	$H_{e'}$–C_2–C=O / $H_{e'}$–C_5–C=O	$H_{a'}$–C_2–C=O / $H_{a'}$–C_5–C=O	C_3–C_2–C=O / C_4–C_5–C=O	ΔH_f (kcal/mole)
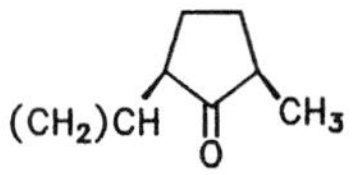	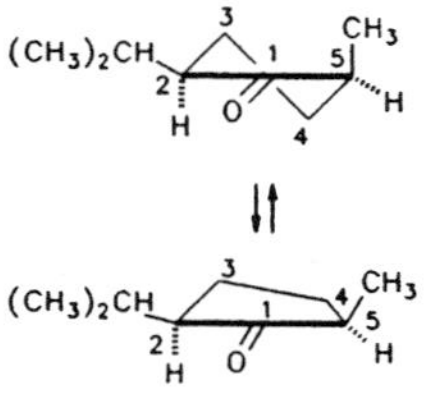	−16.62	32.68		37.21[e]	−84.27	163.38	
		−5.91	26.31	−37.18	56.08	−64.19[b]	174.08	−70.50
		14.50	31.25		37.54[e]	−81.68	165.24	
		−7.93	27.40	−36.97	54.25	−66.13[b]	172.32	−70.73
		21.83	35.97		32.34[e]	−89.26	157.63	
		−0.71	23.30	−37.67	61.95	−58.33[b]	179.82	−71.15

[a] PCMODEL (ref. 25).
[b] CH_3–C–C=O.
[c] CH_3 over ring.
[d] CH_2 over ring.
[e] $(CH_3)_2$CH–C–C=O.

not only the energy of the molecule, but also the conformation of the ring. Thus, six distinct conformations all lie within about 0.8 kcal/mole of each other, with the octant rule predicting that the two lowest energy conformations exhibit roughly opposite CD Cotton effects. Evidence that alkyl substitution forces a change from the half-chair toward the envelope conformation in other cyclopentanones was detected by molecular mechanics calculations on 2(*R*)-methyl-4(*S*)-isopropylcyclopentanone and on pentamethylcyclopentanone (Table 6-8).

In most of the examples in Tables 6-7 and 6-8, simple inspection is sufficient to determine which conformation is more stable and which make the largest contributions to the net CD Cotton effect. The $n \rightarrow \pi^*$ Cotton effect signs may thus be predicted by the octant rule (Table 6-10) and are in good agreement with the experimentally determined signs. In an attempt to place structural–chiroptical properties on a quantitative basis, Kirk[71] predicted the magnitude of each twisted conformer, based on the notion that the most important octant contributions come from octant dissignate α-hydrogens (or α-alkyl groups), rather than from the C(3)/C(4) carbons and their substituents. This concept led to the formulation of two equations that were used to predict the magnitudes of the octant contributions coming from the left and right sides (relative to the carbonyl) of the cyclopentanone:

$$\delta\Delta\varepsilon = -1.9 \sin^2 \omega - 5.4 \sin 2\omega \qquad (6\text{-}1)$$

for positive values of ω, which is the C–C–C=O ring torsion angle, and

$$\delta\Delta\varepsilon = +1.9 \sin^2 \omega - 5.4 \sin 2\omega \qquad (6\text{-}2)$$

Table 6-9. Effect of Staggered Isopropyl Rotamers on the Heats of Formation and Ring Conformation in 2(*R*)-Isopropyl-5(*R*)-methylcyclopentanone

Pseudoaxial isopropyl

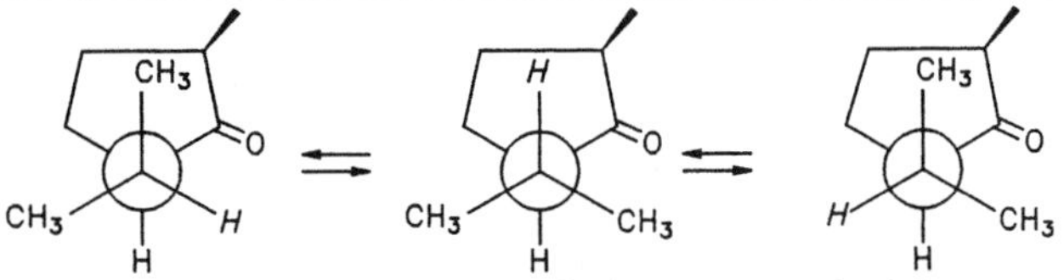

H orientation	(–)-synclinal	(+)-synclinal	antiperiplanar
ΔH_f (molecule)	–71.15	–70.50	–70.73 kcal/mole
Ring conformation	envelope	half-chair	envelope

Pseudoequatorial isopropyl

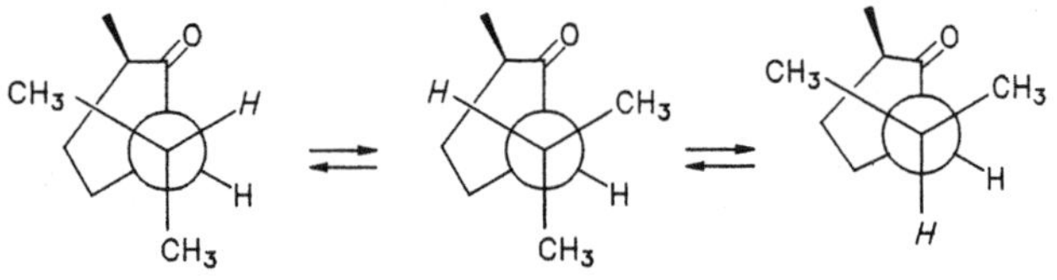

H orientation	(–)-synclinal	(+)-synclinal	(+)-anticlinal
ΔH_f (molecule)	–71.31	–70.76	–70.76 kcal/mole
Ring conformation	deformed half-chair	envelope	deformed half-chair

Table 6-10. Comparison of Experimental[a] and Predicted[b] $\Delta\varepsilon_{max}$ Values for the Carbonyl $n \rightarrow \pi^*$ Transition of Alkyl Cyclopentanones, with Conformational Equilibria and Computed[c] Equilibrium Energies ($\Delta H_{eq} = \Delta\Delta H_f$) Shown

Substituted Cyclopentanone	CD[a] $\Delta\varepsilon^{max}$	Predicted[b] CD $\Delta\varepsilon^{max}$	Conformational Equilibrium and Predicted[b] $n \rightarrow \pi^*$ Cotton Effect $\Delta\varepsilon^{max}$	ΔH_{eq} (kcal/mole)[c]	% Most Stable Isomer[d] Predicted by Octant Rule
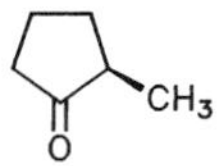	−1.45[e]	−3.16	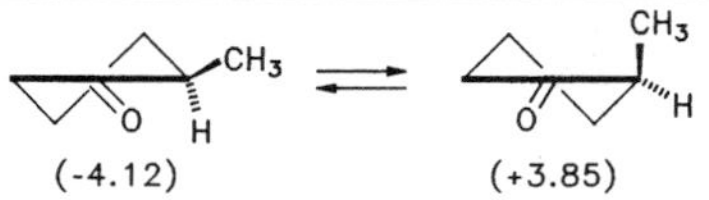	1.15	88
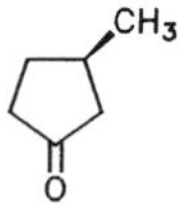	+2.1	+3.11	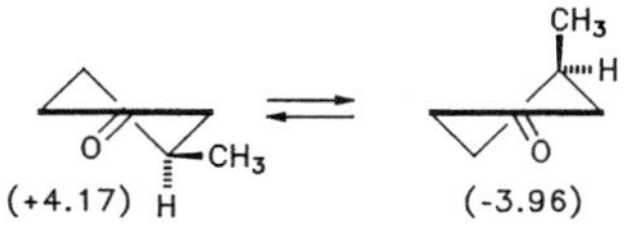	1.14	87
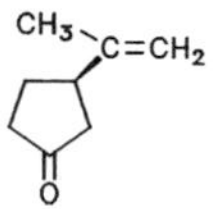	+2.3	+3.40	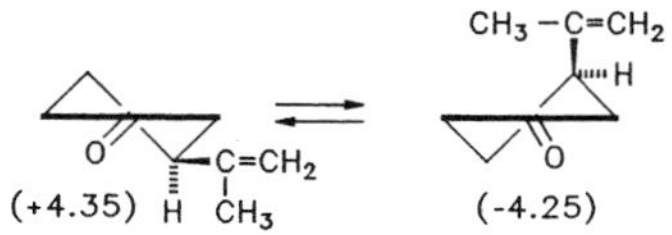	1.25	89
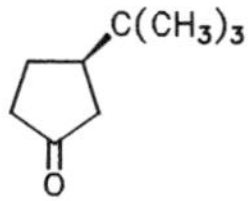	+3.2	+4.08	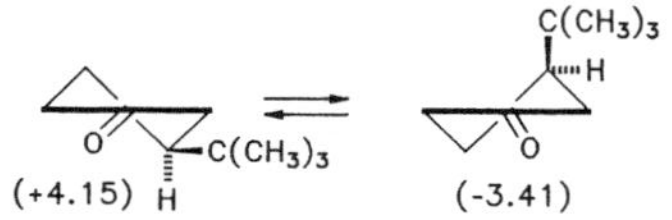	2.93	99

(Continued)

Table 6-10. *(Continued)*

Substituted Cyclopentanone	CD[a] $\Delta\varepsilon^{max}$	Predicted[b] CD $\Delta\varepsilon^{max}$	Conformational Equilibrium and Predicted[b] $n \rightarrow \pi^*$ Cotton Effect $\Delta\varepsilon^{max}$	ΔH_{eq} (kcal/mole)[c]	% Most Stable Isomer[d] Predicted by Octant Rule
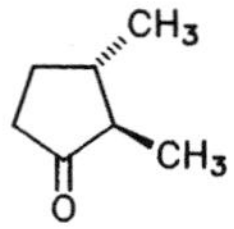	−1.7	−4.67	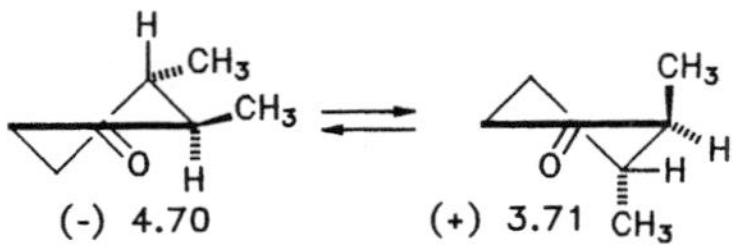	1.99	97
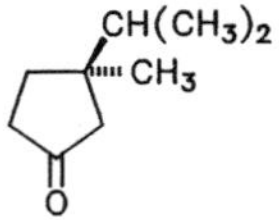	+2.4	+3.48	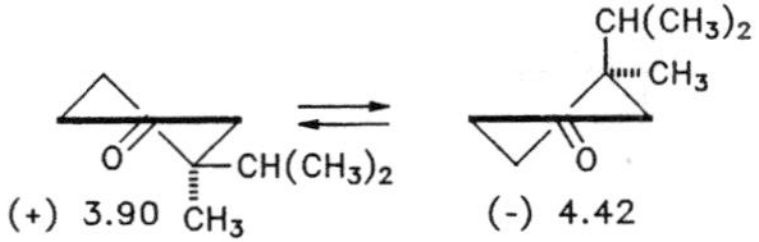	1.76	95
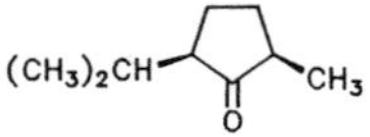	0.42	+0.62	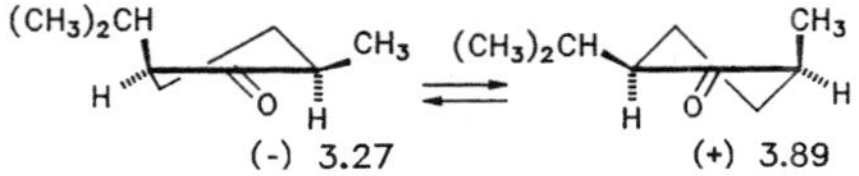	0.16	~50

(Continued)

Table 6-10. *(Continued)*

Substituted Cyclopentanone	CD[a] $\Delta\varepsilon^{max}$	Predicted[b] CD $\Delta\varepsilon^{max}$	Conformational Equilibrium and Predicted[b] $n \rightarrow \pi^*$ Cotton Effect $\Delta\varepsilon^{max}$	ΔH_{eq} (kcal/mole)[c]	% Most Stable Isomer[d] Predicted by Octant Rule
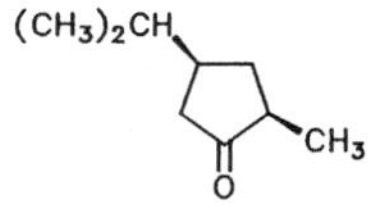	−2.0	−4.12	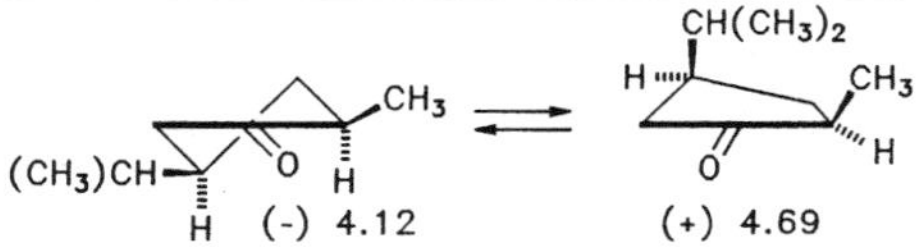	4.37	100
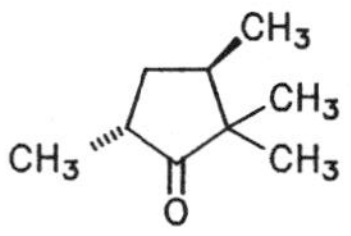	+0.5[f]	+3.19	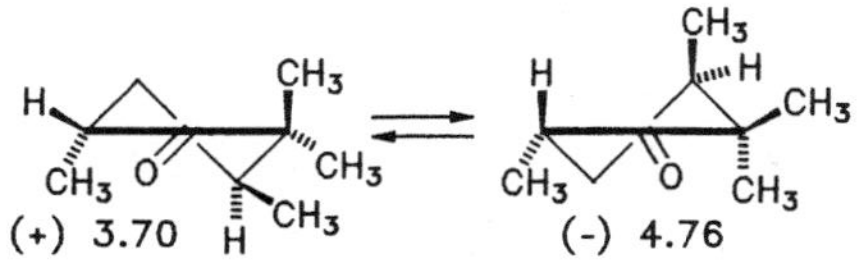	1.64	94
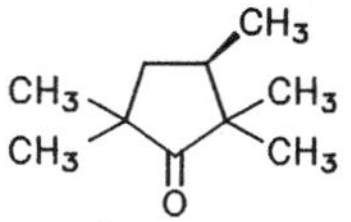	+1.2[f]	+3.07	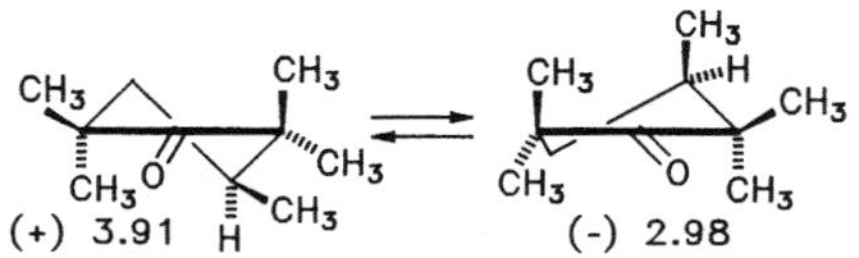	2.28	98

[a] Calculated from ORD *a* values (ref. 69), using $\Delta\varepsilon \sim a/40.28$.

[b] Using Kirk's equations (ref. 71) and the C_3–C_2–C=O and C_4–C_5–C=O torsion angles (Table 6-8) from PCMODEL (ref. 25).

[c] PCMODEL (ref. 25).

[d] Assuming that $\Delta H_{eq} = \Delta G_{eq}$.

[e] From Partridge, J. J., Chadhar, N. K., and Uskokovic, M. R., *J. Am. Chem. Soc.* **95** (1973), 532–540

[f] From Collet, A., Jacques, J., Chion, B., and Lajzerowicz, J., *Tetrahedron* **31** (1975), 2243–2246.

for negative values of ω, which now is the O=C–C_α–C_β ring torsion angle. The calculated Δε values for each conformer are shown in Table 6-10, along with the net population-weighted predicted value. The predicted values all carry the observed sign, but the magnitudes are higher than those observed. Whether that is due to inaccuracies in the way the conformer population was estimated or to failings of Kirk's formulas is unclear. However, it may be noted that if Kirk's values are correct, then the conformer populations are much closer to 50:50 than those estimated in the table.

One way to analyze the conformational equilibrium is to measure the CD spectra over a range of temperatures, as was carried out by Djerassi et al.[72,73] for several of the cyclopentanones of Table 6-10. Possibly the most recent variable low-temperature CD measurements of the $n \rightarrow \pi^*$ transition in a cyclopentanone was reported in 1981 for 2(*R*)-methylcyclopentanone (Figure 6-10). The octant rule predicts a strong negative Cotton effect for the pseudoequatorial conformer and a strong positive Cotton effect for the pseudoaxial one. As the temperature is lowered, the equilibrium shown in Figure 6-10 should shift toward the more stable conformer, a shift predicted by molecular

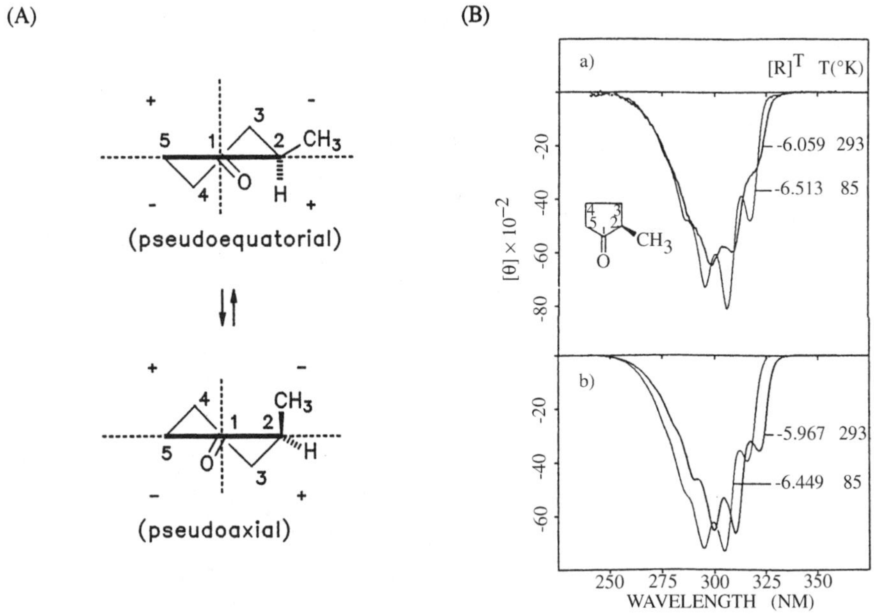

Figure 6-10. (A) Conformational equilibrium in 2(*R*)-methylcyclopentanone, showing enantiomeric C_2 conformers with (upper) pseudoequatorial and (lower) pseudoaxial isomers. The octant signs are projected onto the conformers. A negative $n \rightarrow \pi^*$ Cotton effect is predicted for the pseudoequatorial conformer, due to octant contributions from C_3, C_4, and CH_3. A positive Cotton effect is predicted for the pseudoaxial conformer, due to contributions from the same groups. (B) Circular dichroism spectra of 2(*R*)-methylcyclopentanone in (a) ether–isopentane–ethanol (EPA; 5:5:2, by vol.) and in (b) isopentane–methylcyclohexane (4:1, by vol.) at room temperature and 85°K. The reduced rotatory strengths $[R]^T$ are indicated on the spectra. [(B) is reprinted with permission from ref.73. Copyright © 1981 American Chemical Society.]

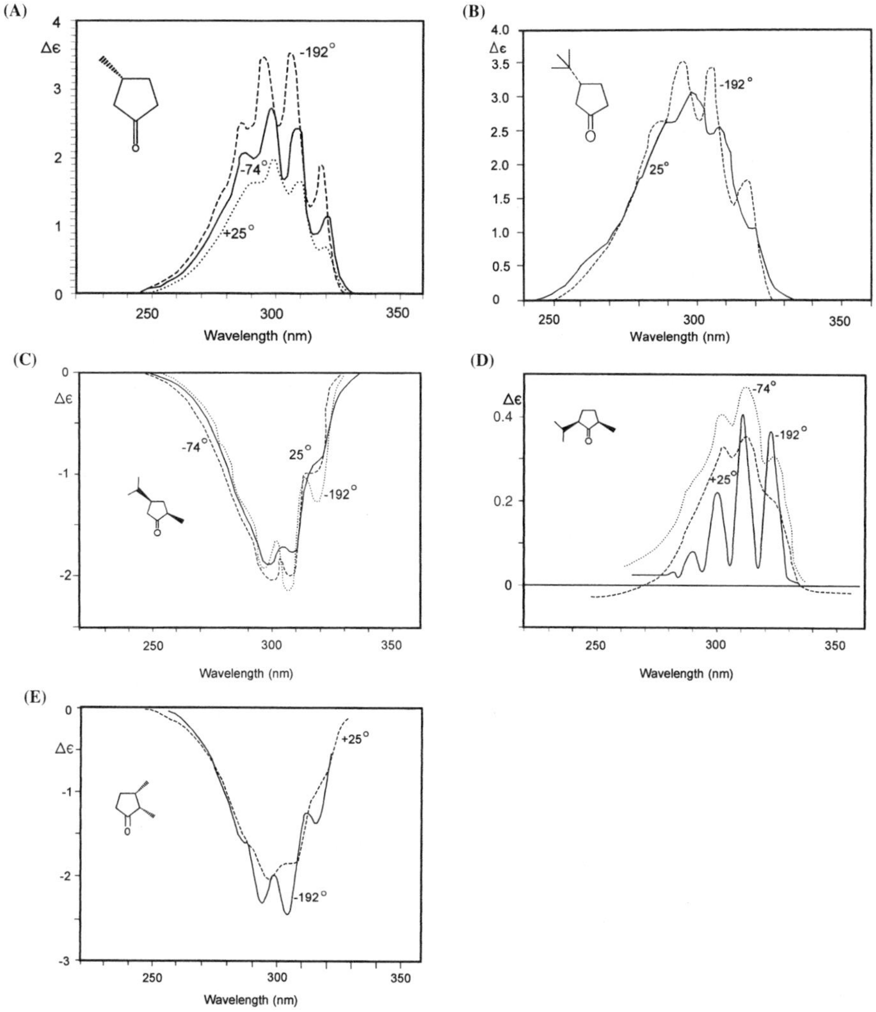

Figure 6-11. Room-temperature and low-temperature CD spectra of alkylcyclopentanones measured in EPA solvent. Temperatures are indicated on the CD curves. EPA is ether–isopentane–ethanol (EPA; 5:5:2, by vol.). [Redrawn from ref. 72.]

mechanics calculations[25] (Table 6-8) to be pseudoequatorial. In fact, the strongly negative Cotton effects observed at room temperature increase only slightly (~5–10%). The data are consistent with a predominance of the pseudoequatorial conformation at room temperature, as predicted. Earlier variable low-temperature CD spectra are summarized in Figure 6-11.

Evidence for the *tert*-butyl group as a conformational anchor in cyclopentanone comes from a comparison of $n \rightarrow \pi^*$ CD Cotton effects and associated rotatory strengths of 3(*R*)-*tert*-butylcyclopentanone at 25°C and –192°C (Figure 6-11B and Table 6-11). The very small (7%) increase in rotatory strength at –192°C indicates that even at 25°C, the compound is almost entirely in its most stable conformation, computed to be the half-chair with an equatorial *tert*-butyl (Table 6-10). The data suggest an inadequacy in Kirk's method[71] of approximating the magnitude of the Cotton effect. The CD data of Table 6-11 also indicate that, although 3(*R*)-methylcyclopentanone is mainly in its most stable (half-chair) conformation at room temperature, approximately 20% of the molecules adopt the conformation with an axial 3-methyl in an enantiomeric half-chair ring. Molecular mechanics calculations (Table 6-10) slightly overestimate the population at 88% of the most stable conformation; Kirk's calculations[71] considerably underestimate it at 53 ± 5%.

Low-temperature CD data (Table 6-11) for the *trans*-2,3-dimethyl and 2-methyl-4-isopropylcyclopentanones indicate essentially no population change, corresponding to an equilibrium lying nearly completely on the side of the most stable (half-chair) conformation (Table 6-10). Again, these findings are well supported by molecular mechanics[25] calculations, which suggest that these cyclopentanones lie more than 97% in a specific half-chair conformation, with alkyl groups occupying pseudoequatorial/equatorial sites. In contrast, the more complex *cis*-2-isopropyl-5-methylcyclopentanone showed an increased positive rotatory strength with a lowering of the temperature, suggesting a population shift toward the half-chair conformer, computed by molecular mechanics[25] (Table 6-10) to be the most stable conformer (albeit only slightly more stable than is predicted by the octant rule) and computed by Kirk[71] to exhibit a strong negative $n \rightarrow \pi^*$ Cotton effect.

Surprisingly, except for a study of 2(*R*)-methylcyclopentanone (Figure 6-10) by Djerassi et al.,[73] no other variable-temperature CD studies on monocyclic cyclopentanones have been reported since 1966.[72] But from the limited data available, it seems clear that the conformational analysis of cyclopentanones is relatively straightforward. Half-chair conformations dominate the equilibrium, and the deformation of the half-chair toward the envelope conformation does not become important, except to alleviate

Table 6-11. Rotatory Strengths[a] of Alkylcyclopentanones in EPA Solvent[b] Measured at 25°C and –192°C

Ketone	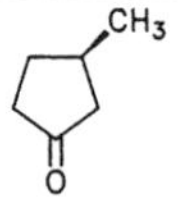	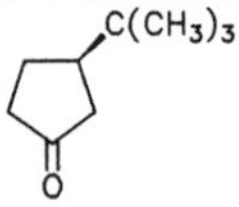	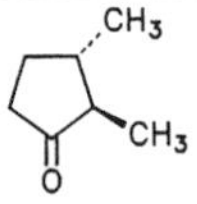	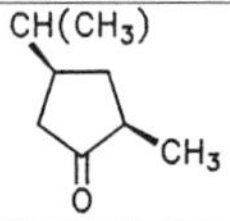	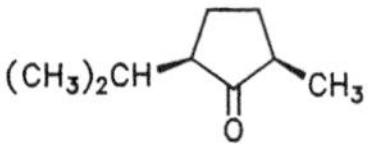
$R^{25°C}$	+6.02	+8.85	–5.92	–5.22	+0.849
$R^{-192°C}$	+9.88	+9.49	–6.04	–5.27	+1.25[c]

[a] $R \times 10^{-40}$ gs from ref. 72.

[b] Ether–isopentane–ethanol, 5:5:2 (by vol.).

[c] Value at –74°C. The reported value at –192° (+0.506) may reflect cracking of the EPA glass, leading to poor transmission of light.

nonbonded steric interactions across the ring between axial-like alkyl groups at the α-carbons and ring carbons or substituents at ring carbons 3 and 4.

Few attempts have been made to analyze or predict the chiroptical properties of substituted cyclopentanones. An ab initio method[74] using a minimal STO-3G basis set, including C=O *d*-orbitals, predicted (*i*) a large positive rotatory strength for the $n \rightarrow \pi^*$ transition of 3(*R*)-methylcyclopentanone in the half-chair with a 3-equatorial methyl and (*ii*) a large negative rotatory strength for the axial methyl conformer. An energy difference between the conformers was calculated to be 1.34 kcal/mole—close to the 1.14 kcal/ mole (Table 6-10) predicted by molecular mechanics calculations.

6.5. Cycloheptane, Cycloheptanones and Larger Rings

The structure analysis[52,53] of cycloheptanes and higher cycloalkanes by spectroscopic methods is complicated by the large number of parameters that must be considered. The rings are often quite flexible, so large numbers of different conformations coexist, and they often have little symmetry. Consequently, the earliest successful attempts at conformational analysis of cycloheptanes and higher cycloalkanes were carried out by means of molecular mechanics calculations.[53] In 1961, Hendrickson[75] was the first to perform a comprehensive analysis, following slightly earlier, more primitive analyses that considered only the torsional energy of the cycloheptane ring or only the interaction potential between hydrogen atoms. Hendrickson distinguished two general types of conformations in cycloheptane: the chair and the boat (Figure 6-12).

Like the boat in cyclohexane, both the chair and boat in cycloheptane are quite flexible and easily pseudorotate into C_2-symmetry twist-chair and twist-boat conformations, respectively (Figure 6-13). The latter were computed[26,53,54,55,76,77] to lie at the global minima on the corresponding chair and boat conformational energy hypersurfaces, while the C_s chair and C_s boat lie higher in energy and not at local minima. Rather, they represent the transition state for the respective twist-chair ⇄ twist-chair and twist-boat ⇄ twist-boat interconversions. As in cyclopentane, the family of C_2 twist-chair conformers interconverts by pseudorotation through the C_s chair, with a barrier now thought to be approximately 1 kcal/mole.[53,55] Similarly, a family of C_2 twist-boat conformers interconverts by pseudorotation through the C_s boat, with a barrier estimated variously as 0.0–0.8 kcal/mole.[53,55] The interconversion barrier between the chair and boat families is computed to be about 8 kcal/mole,[53] a value close to that (ΔH = 10.8 kcal/mole) of the interconversion barrier between chair and twist-boat cyclohexane (Chapter 2).

Hendrickson subsequently revised the computed energies,[75–77] and other force fields have been applied to the conformational analysis of cycloheptane,[26,53–55,65,78,79] giving a variety of energies (Table 6-12). All investigators agree that the twist-chair conformation is more stable than the chair, and most agree that the twist boat is more stable than the boat. These conclusions seem reasonable, since twisting the chair and boat conformations alleviates eclipsing interactions near carbons 3, 4, 5, and 6 (Figure 6-12). All researchers also agree that the twist-chair conformation lies at the global

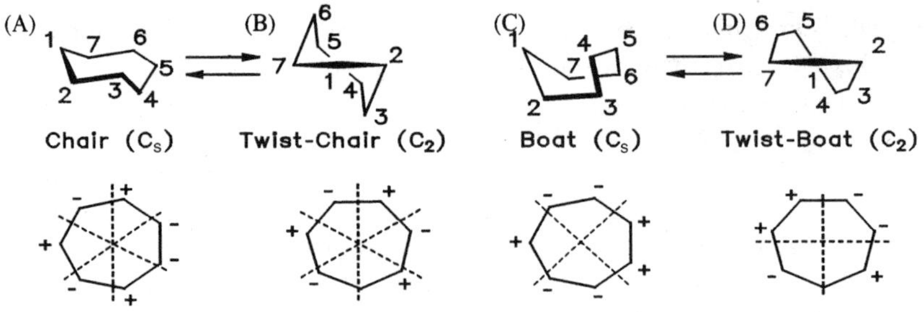

Figure 6-12. [Upper] (A)–(D): Cycloheptane conformations and their symmetry designation. [Lower] Positions of ring atoms above (+) and below (–) the average plane passing through the corresponding conformation shown above.

energy minimum; however, there has been some disagreement over the relative energies of the other conformers. The transition state between the twist-chair and boat conformers was computed to be 8.1[76] to 8.2[79] kcal/ mole and was predicted to have a geometry closer to the twist boat than the boat (Figure 6-13).

Experimental evidence on cycloheptane conformation is sparse. Analysis by gas-phase electron diffraction spectroscopy indicates a mixture of twist-chair and chair conformers,[80] with uncertainties arising due to the small differences in conformational energy. The interconversion of conformers is so rapid that only a single proton peak is

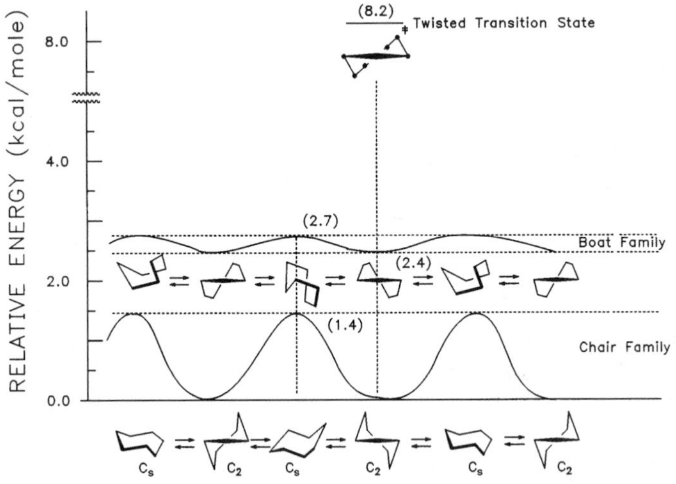

Figure 6-13. Potential-energy diagram for equilibrating cycloheptane conformations. The interrelationships are relative to the twist-chair conformation set at 0.0 kcal/mole. The twist-chair ⇄ chair and twist-boat ⇄ boat pseudorotation barriers are 1.4 and 0.3 kcal/mole, respectively. The energies are from PCMODEL (ref. 25); those in parentheses are from a different force field (ref. 76).

Table 6-12. Comparison of Relative Energies (kcal/mole) of Cycloheptane Conformers Computed by Different Molecular Mechanics Force Fields[a]

Molecular Mechanics Force Field		Twist Chair 	Chair 	Twist Boat 	Boat 
Hendrickson	(1961)[b]	0.0	2.16	2.49	3.02
Lifson	(1966)[c]	0.0	0.67	2.64	2.40
Hendrickson	(1967)[d]	0.0	1.40	2.40	2.70
Strauss	(1975)[e]	0.0	1.08	0.60	1.39
Allinger	(1977)[a]	0.0	1.01	3.15	3.15
Allinger	(1989)[f]	0.0	—	3.04	—
PCMODEL	(1994)[g]	0.0	1.01	3.15	3.15

[a] Ref. 53.
[b] Ref. 75.
[c] Ref. 78.
[d] Ref. 76.
[e] Ref. 79.
[f] Ref. 26.
[g] Ref. 25.

observed in the NMR spectrum, and no dynamic NMR has been observed.[52] Attempts to distinguish conformations by NMR have focused largely on substituted cycloheptanes. Pseudorotation interconverts bisected (*b*), axial (*a*), and equatorial (*e*) sites. For example, in the interconversion of twist-chair conformations, the C_2 axis of symmetry is moved sequentially by one carbon through pseudorotation (Figure 6-13). This process creates 7 identical twist-chair conformations of one type of handedness and 7 of the enantiomeric handedness, for a total of 14. Lying between these 14 twist-chair conformers are the same number of chair conformers (Figure 6-13).

Roberts et al.[81] at Cal Tech used the *gem*-dimethyl group to slow the pseudorotation and found that low-temperature ^{19}F-NMR spectra of *gem*-difluorocycloheptanes were different from room-temperature spectra only if *gem*-dimethyl[57] or *vic* bromines were also present. Subsequently, the Roberts group examined ^{13}C-NMR spectra of 1,1-dimethylcycloheptane and concluded that the *gem*-dimethyl group "preferred" the bisected orientation (Figure 6-14), as predicted by molecular mechanics calculations.[82] However, the ^{13}C-NMR spectra were not measured at low temperatures.

Unlike the situation with cyclopentane, changing one sp^3-hybridized ring carbon to sp^2-hybridized in cycloheptanone does not greatly simplify the conformational analysis. For, as in cycloheptane, there are many cycloheptanones of similar energy. Most conformational analyses of cycloheptanone have been carried out by molecular mechanics calculations.[33,53] Cycloheptanone conformers belong either to the chair pseudorotational family or to the higher energy (2.0–4.0 kcal/mole)[53] boat pseudoro-

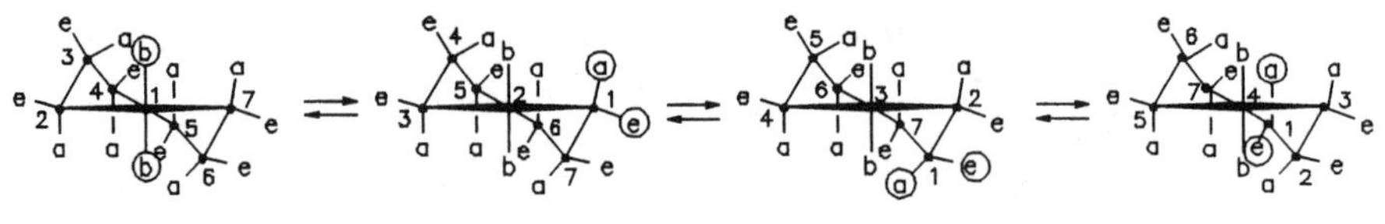

$\Delta H_f = -42.36$	$\Delta H_f = -40.19$	$\Delta H_f = -41.09$	$\Delta H_f = -40.96$
($1b, 1b = CH_3$)	($1a, 1e = CH_3$)	($1a, 1e = CH_3$)	($1a, 1e = CH_3$)

Figure 6-14. Interconversion of bisected (*b*), axial (*a*), and equatorial (*e*) sites on twist-chair cycloheptane through a pseudorotation that moves the C_2 axis of symmetry by one carbon. The ΔH_f values (kcal/mole) were computed by PCMODEL (ref. 25) for *gem*-dimethylcycloheptane as the *gem*-dimethyl group moves from C_1 to C_4 by pseudorotation.

tational family, with the twist chair thought to be lowest energy. There are four different twist-chair cycloheptanones, reflecting the differing locations of the carbonyl group on the ring (Figure 6-15). In 1972, using the MM2 force field Allinger et al.[65] determined the relative energies of these conformations, with the result that the asymmetric twist chair with the carbonyl at C_2 was found to be slightly more stable than the C_2-symmetry conformer with the carbonyl at C_1. Boat cycloheptanones are 2–4 kcal/mole higher in energy than the twist chair.

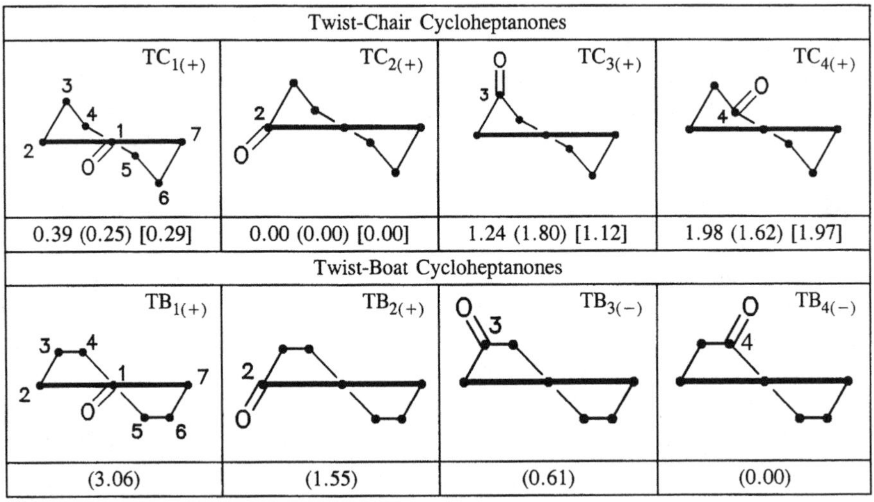
Twist-Chair Cycloheptanones
TC1(+)
TC2(+)
TC3(+)
TC4(+)
0.39 (0.25) [0.29]
0.00 (0.00) [0.00]
1.24 (1.80) [1.12]
1.98 (1.62) [1.97]
Twist-Boat Cycloheptanones
TB1(+)
TB2(+)
TB3(−)
TB4(−)
(3.06)
(1.55)
(0.61)
(0.00)

Figure 6-15. Cycloheptanone conformations and their relative heats of formation, computed by PCMODEL molecular mechanics calculations (ref. 25). The values in parentheses are from MM2 calculations (ref. 65); those in square brackets are from MM3 calculations (ref. 33). The $TC_{n(+)}$ designations in the upper right-hand corners are for the twist-chair (TC) conformations where *n* denotes the carbon at the rotation axis and (+) indicates the sign of the octants in which the contributing ring carbons lie.

More recently, using the MM3 force field, Allinger found qualitatively similar results (Figure 6-15).[33] In particular, the twist-chair cycloheptanone with the carbonyl at C_2 was found to be 0.29 kcal/mole more stable than that with the carbonyl at C_1. But in the MM3 force field, the conformer with the carbonyl at C_3 is somewhat less destabilized than in the MM2 force field, whereas the conformer with the carbonyl at C_4 is somewhat more destabilized. Gas–phase electron diffraction measurements at 98°C[83] support the general conclusions on cycloheptanone conformation, but the data are too complicated to point unambiguously to the twist-chair conformations.

As with cycloheptane, the low pseudorotation barriers and small energy differences between conformers (Figure 6-15) render spectroscopic analyses rather complicated. The only available studies of cycloheptanone conformational analysis by NMR are those of Roberts et al.[82] and St. Jacques et al.[84] The latter group examined the spectral changes of isomeric *gem*-dimethylcycloheptanones by ^{1}H-NMR at low temperatures and correlated the data with conformational energy calculations. The authors concluded that the twist-chair conformation with the carbonyl group at C_2 is the most stable.

Some of the earliest chiroptical investigations of cycloheptanones were carried out in 1959 and 1963 by Djerassi et al.,[85,86] using ORD. An optically active (+)-3(*R*)-methylcycloheptanone and (–)-4(*R*)-methylcycloheptanone, both prepared by the ring expansion of 3(*R*)-methylcyclohexanone with diazomethane, were observed to have nearly mirror-image ORD curves (Figure 6-16), which was thought to be unusual.

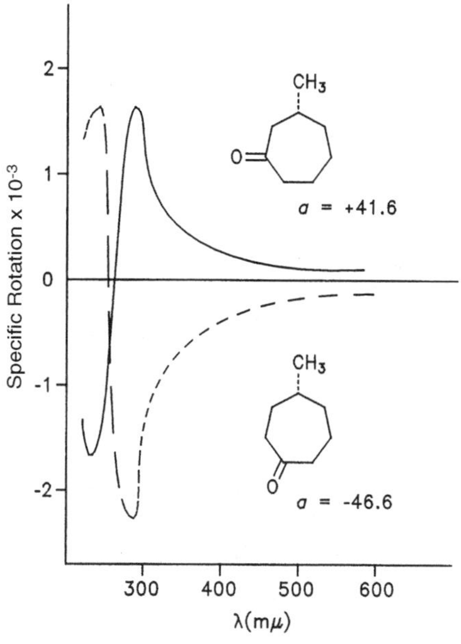

Figure 6-16. Optical rotatory dispersion curves of 3(*R*)-methylcycloheptanone and 4(*R*)-methylcycloheptanone in methanol at room temperature. Data from ref. 85.

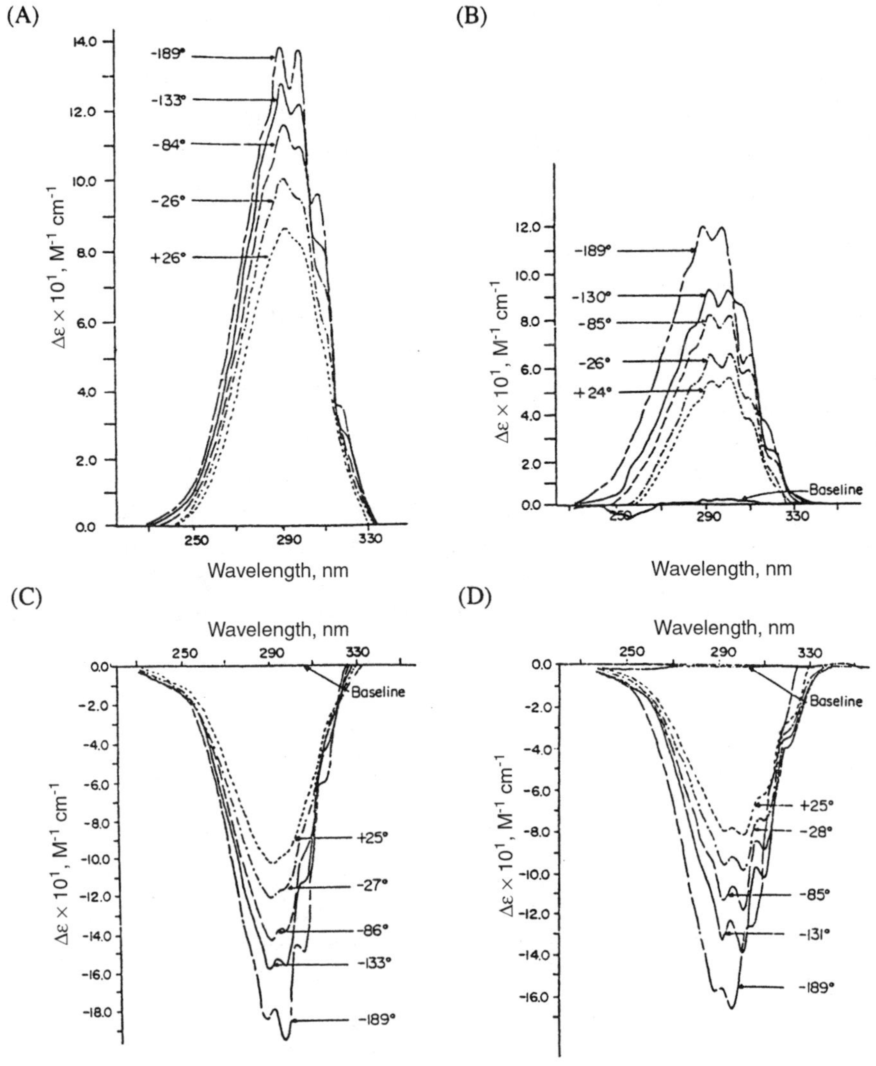

Figure 6-17. Variable low-temperature CD spectra of (+)-3(*R*)-methylcycloheptanone in (A) EPA and (B) P5M1 solvents and of (–)-4(*R*)-methylcycloheptanone in (C) EPA and (D) P5M1 solvents. Temperatures (°C) are indicated on the curves. EPA = ether–isopentane–ethanol (5:5:2, by vol.) and P5M1 = isopentane–methylcyclohexane (5:1, by vol.). The ketone concentrations were (A) 3.5×10^{-2} M, (B) 4.9×10^{-3} M, (C) 1.8×10^{-2} M, and (D) 9.7×10^{-3} M. [Reprinted from ref. 89, copyright © (1979), with permission of Elsevier Science.]

However, no attempt was made to incorporate the data into cycloheptanone conformational analysis until 1967, when Jones et al.[87] used molecular models to examine methylcycloheptanone conformations and evaluate the octant contributions from each. The relative importance of each conformation in the equilibrium was calibrated by Hendrickson's molecular mechanics calculations[75,76] on cycloheptane. The authors concluded that (1) only three conformers ($TC_{1(+)}$, $TC_{4(+)}$, and $TC_{7(-)}$; see Figure 6-15) would be expected to give rise to the large positive Cotton effect that was observed (Figure 6-16) for 3(*R*)-methylcycloheptanone and (2) only two conformers ($TC_{1(-)}$ and $TC_{7(+)}$), and possibly a third ($TC_{4(-)}$), would give rise to the large negative Cotton effect observed for 4(*R*)-methylcycloheptanone.

Subsequently, Kirk[88] calculated the expected CD $\Delta\varepsilon$ values for the $n \rightarrow \pi^*$ transitions of 3(*R*) and 4(*R*)-methylcycloheptanone. The twist-chair conformations were taken to be most stable; and on the basis of the molecular mechanics calculations of Allinger et al.[65] on cycloheptanone, two isoenergetic lowest energy twist-chair conformers (TC_2 and TC_7) were considered, along with the nearby TC_1 conformer. Kirk[88] calculated that, to reproduce the observed Cotton effects (Figure 6-16), the population of 3(*R*)-methylcycloheptanone would have to consist of 70 ± 5% of $TC_{1(+)}$ and 30 ± 5% of TC_2 and TC_7. If correct, these calculations would imply that the TC_1 isomer is approximately 0.9 kcal/mole ($\Delta G°$) more stable than TC_2 or TC_7 in 3-methylcycloheptanone (vs. TC_2 or TC_7 being computed to be 0.25–0.29 kcal/mole (ΔH) more stable than TC_1 in the parent cycloheptanone). For 4(*R*)-methylcycloheptanone, Kirk calculated a conformer population consisting of a 65:35 ratio of $TC_{1(-)}$ to ($TC_2 + TC_7$), with an estimated $\Delta G°$ of about 0.75 kcal/mole.

Kirk's analyses[88] were followed by the variable low-temperature CD spectroscopy of 3(*R*)- and 4(*R*)-methylcycloheptanone.[89] For both ketones, the magnitude of the $n \rightarrow \pi^*$ Cotton effect increased by 32–110% (Figure 6-17), reflecting an increasing population of the lowest energy conformer, which is the major sign-determining contributor. The data were interpreted in terms of a simplified conformational analysis that focused on twist-chair cycloheptanone conformations (Figure 6-15) as being of the lowest energy. Further simplification was achieved by excluding the higher energy twist chairs (TC_3, TC_4, TC_5, TC_6) as being relatively unimportant. The analysis thus concentrated on the TC_1, TC_2, and TC_7 conformations.[88,89] These conformations also appear as lowest energy conformations in molecular mechanics calculations.[25] Thus, as summarized in Table 6-13, conformations $TC_{2(-)}$, $TC_{7(-)}$, $TC_{1(+)}$, and $TC_{7(+)}$ are the predicted dominant contributors to the conformational isomer population of (+)-3(*R*)-methylcycloheptanone, whereas $TC_{1(-)}$, $TC_{2(+)}$, $TC_{7(+)}$, and $TC_{2(-)}$ are predicted to dominate in (–)-4(*R*)-methylcycloheptanone. The octant projection diagrams for these conformations are shown in Table 6-14. It is difficult to predict Cotton effect magnitudes for each conformation, but it seems clear that the TC_1 conformations should make the largest inherent contribution by virtue of greater contributions from ring atoms C_3 and C_4. In contrast, contributions from ring atoms in the TC_2 and TC_7 conformations are approximately canceling. Thus, the observed $n \rightarrow \pi^*$ CD Cotton effect signs (Figure 6-17)—(+) for 3(*R*)-methylcycloheptanone and (–) for 4(*R*)-

Table 6-13. Conformational Energy Differences ($\Delta\Delta H_f$, kcal/mole) Computed for the Twist Chair (TC)[a] of 3(*R*)-Methylcycloheptanone and 4(*R*)-Methylcycloheptanone by Molecular Mechanics Calculations[b]

	TC_1	TC_7	TC_6	TC_5	TC_4	TC_3	TC_2
				3(*R*)-Methylcycloheptanone			
(−)	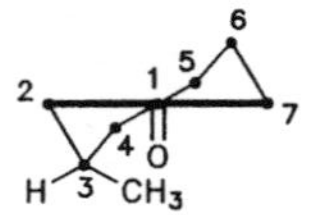	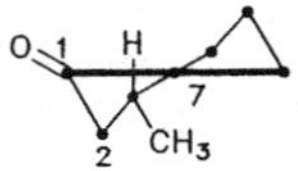	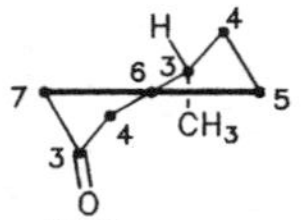	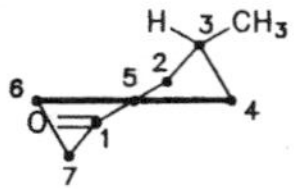	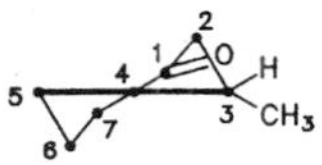	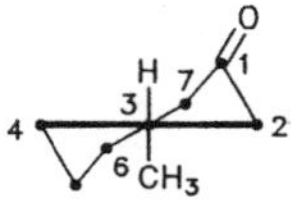	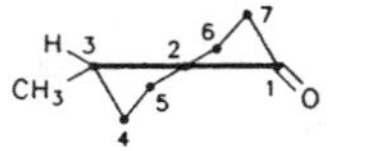
	1.03	0.00[c]	2.96	2.22	3.43	1.66	0.19
(+)	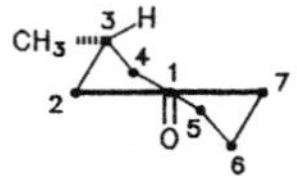	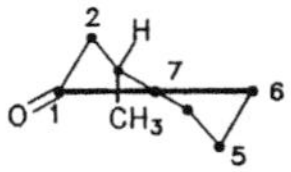	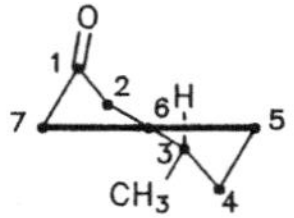	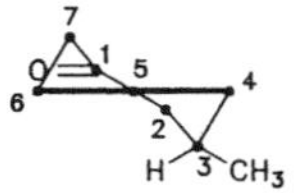	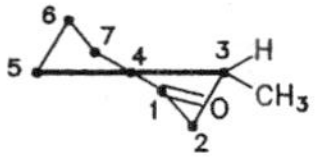	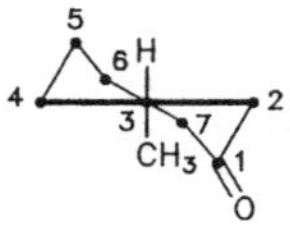	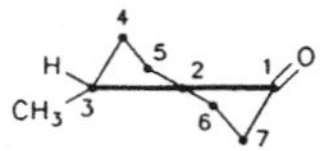
	0.72	1.01	1.20	2.22	2.11	1.51	2.78
				4(*R*)-Methylcycloheptanone			
(−)	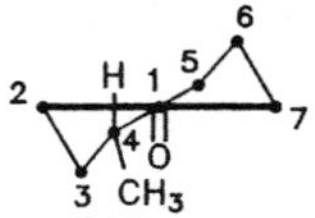	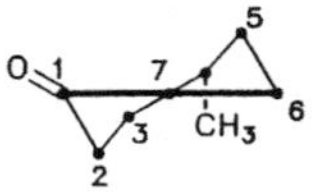	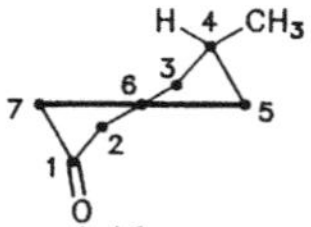	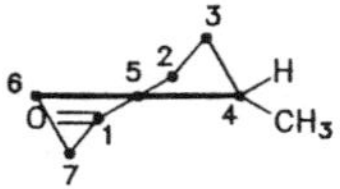	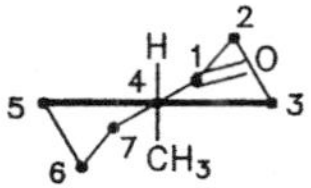	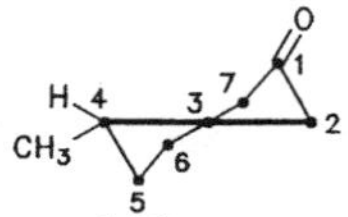	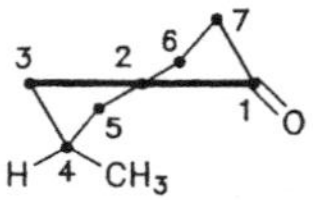
	0.41	1.99	1.46	4.51	2.40	1.38	1.20
(+)	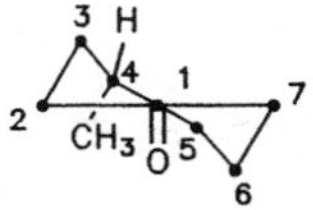	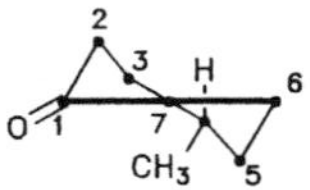	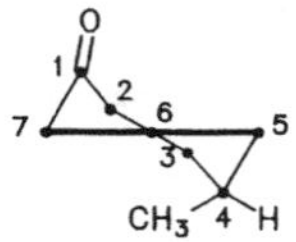	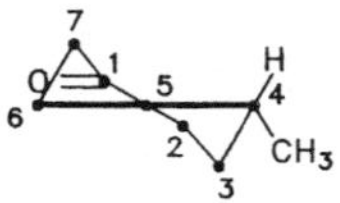	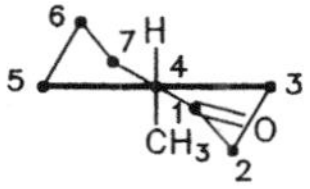	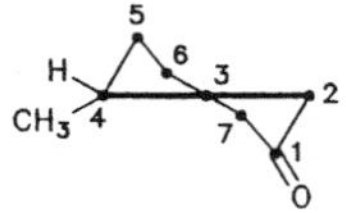	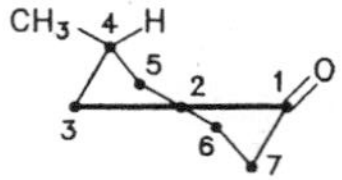
	2.39	0.00[d]	2.91	2.13	2.38	2.59	0.22

[a] The subscript in TC_n refers to the carbonyl location group relative to the C_2-symmetry axis. An adjacent sign in parentheses, (+) or (−), following the subscript refers to the location of the octant contributing ring atoms, as per Figure 6-15.

[b] PCMODEL, ref. 25.

[c] 3(*R*)-Methylcycloheptanone adopts the $TC_{7(-)}$ conformation.

[d] 4(*R*)-Methylcycloheptanone adopts the $TC_{7(+)}$ conformation.

methylcycloheptanone—correlate well with those predicted by the dominant TC_1 conformations.

At low temperatures, the observed Cotton effect intensities increase (Figure 6-17), suggesting that (1) negatively contributing conformations of 3(*R*)-methylcycloheptanone (e.g., $TC_{2(-)}$) are of higher energy than positively contributing conformations and (2) positively contributing conformations of 4(*R*)-methylcycloheptanone (e.g., $TC_{2(+)}$) are of higher energy than the negatively contributing ones. Although these relationships are predicted by molecular mechanics calculations[25] (Table 6-14), the same calculations also predict that the TC_1 conformations, which apparently dominate the Cotton effect, lie higher in energy than the corresponding TC_2 isomers that are predicted to give oppositely signed Cotton effects. Thus, if the model is valid, one might predict that the CD Cotton effect magnitudes of 3(*R*)- and 4(*R*)-methylcycloheptanone would decrease in magnitude, but not change sign. Contrary to this

Table 6-14. Low-energy[a] Twist-chair (TC) Conformations[b] of 3(*R*)-Methyl and 4(*R*)-Methylcycloheptanone and Their Octant Projection Diagrams and Predicted $n \rightarrow \pi^*$ CD Cotton Effect Signs

3(*R*)-Methylcycloheptanone			4(*R*)-Methylcycloheptanone		
Conformation	Octant Diagram	Cotton Effect Sign	Conformation	Octant Diagram	Cotton Effect Sign
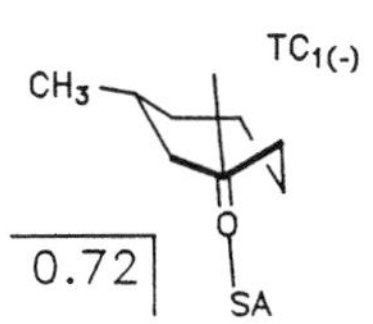 $TC_{1(-)}$ 0.72	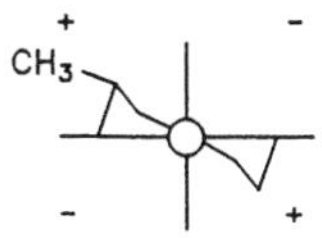	+++	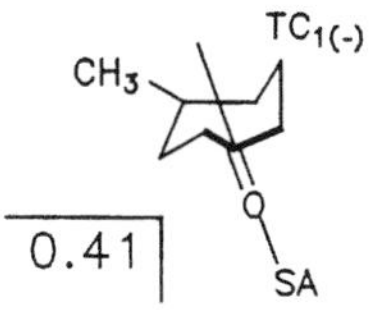 $TC_{1(-)}$ 0.41	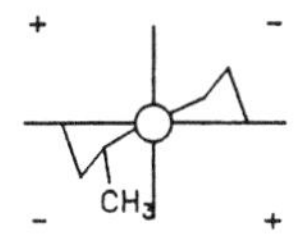	–
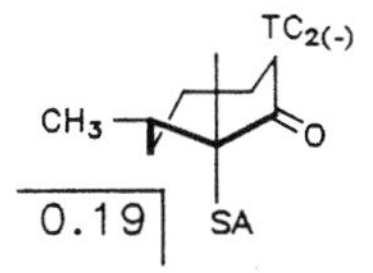 $TC_{2(-)}$ 0.19	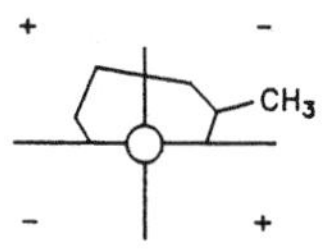	–	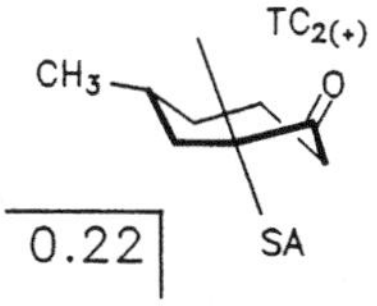 $TC_{2(+)}$ 0.22	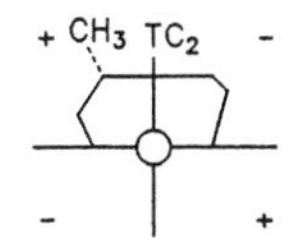	+
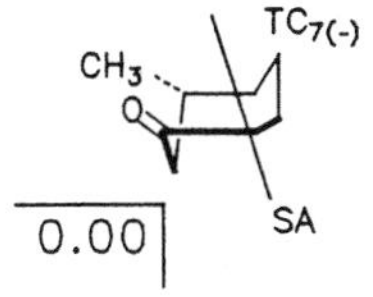 $TC_{7(-)}$ 0.00	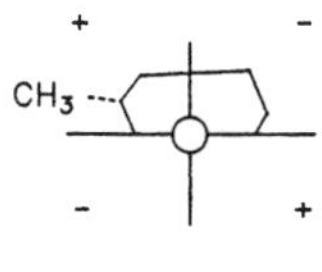	++	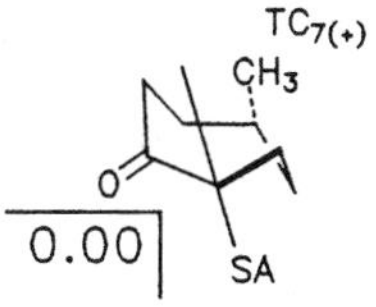 $TC_{7(+)}$ 0.00	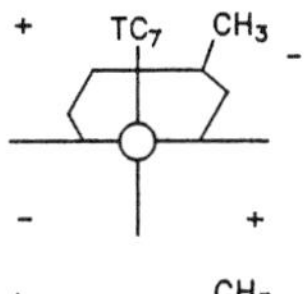	–
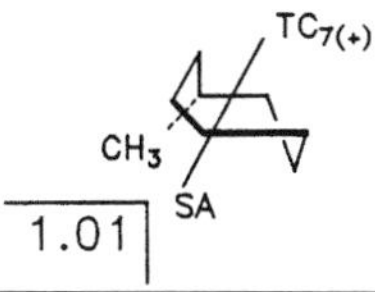 $TC_{7(+)}$ 1.01	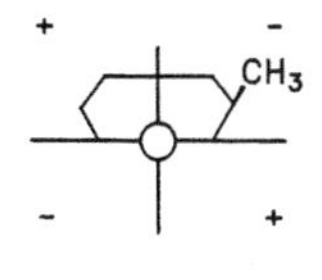	±	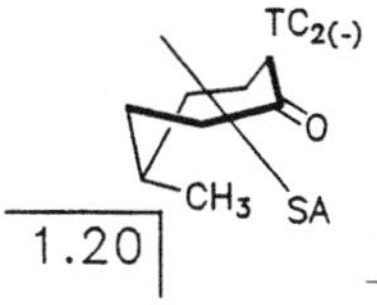 $TC_{2(-)}$ 1.20	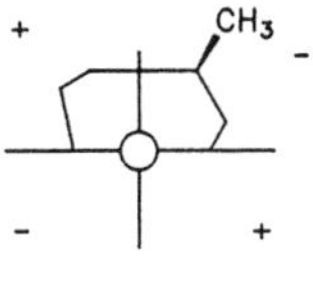	±

[a] Relative energies ($\Delta\Delta H_f$, kcal/mole) for each isomer computed by PCMODEL molecular mechanics program (ref. 25).

[b] SA = symmetry axis.

prediction, the Cotton effect magnitudes are observed to increase. Alternatively, the observed data are more consistent with TC_1 being lower in energy ($\Delta G°$) than TC_2 or TC_7, which is contrary to the enthalpy differences predicted by molecular mechanics calculations (Table 6-14).

Kirk calculated a different set of Cotton effects for the TC_1, TC_2, and TC_7 conformations on the basis of major contributions coming from C_α–H bonds and lesser contributions from C_α–C_β bonds and methyl groups.[88] For (+)-3(*R*)-methylcycloheptanone, the calculated $\Delta\varepsilon$ values are $TC_{1(-)}$ (+7), $TC_{2(-)}$ (−10), and $TC_{7(-)}$ (−9). For (+)-4(*R*)-methylcycloheptanone, the calculated $\Delta\varepsilon$ values are $TC_{1(-)}$ (−7), $TC_{2(+)}$ (+10), and $TC_{7(+)}$ (+10). From the observed $\Delta\varepsilon$ values at room temperature in methanol ($\Delta\varepsilon$ = +1.05 and $\Delta\varepsilon$ = −1.2 for 3(*R*)- and 4(*R*)-methylcycloheptanone, respectively), Kirk[88] calculated a 65:35 mixture of TC_1:(TC_2 + TC_7). However, this approach does not resolve the inconsistency noted in the previous paragraph. Kirk's data predict that TC_1 lies lower in energy than TC_2 or TC_7. It is not entirely clear how to reconcile the experimental CD data with the computed relative stability of the conformations of 3(*R*)- and 4(*R*)-methylcycloheptanone.

Although the conformations of cycloalkanes larger than cycloheptane and cycloheptanone have been analyzed by molecular mechanics calculations[26,33,53,55] and by NMR,[52] almost no conformational studies of cyclooctanones or larger cycloalkanones by ORD or CD spectroscopy have been reported. The lone instance is the study of an optically active methylcyclononanone, thought to be either 3(*R*) or 4(*R*), for which Djerassi and Krakower reported a partial ORD curve in 1959.[85] Given that the energy differences among cyclooctanones are larger than in cycloheptanones,[52,53] this would seem to be a potentially fruitful area for variable-temperature CD investigations.

References

1. Markownikoff, W., and Krestownikoff, A., *J. Leibigs Ann. Chem.* **208** (1881), 333–349.
2. Freund, A., *Monatsh.* **3** (1882), 626–635.
3. Eliel, L. L., and Wilen, S. H., *Stereochemistry of Organic Compounds,* J. Wiley & Sons, Inc., New York, 1994.
4. Bastiansen, O., Fritsch, F. N., and Hedberg, K., *Acta Crystallogr.* **17** (1964), 538–543.
5. Yamamoto, S., Nakata, M., Fukuyama, T., and Kuchitsu, K., *J. Phys. Chem.* **89** (1985), 3298–3302.
6. Aped, P., and Allinger, N. L., *J. Am. Chem. Soc.* **114** (1992), 1–16.
7. (a) Turro, N. J., and Hammond, W. B., *J. Am. Chem. Soc.* **88** (1966), 3672–3673. (b) Schaafsma, S. E., Steinberg, H., and DeBoer, Th. J., *Recl. Trav. Chim. Pays-Bas* **85** (1966), 1170–1172. (c) For a review of cyclopropanone, see Turro, N. J., *Acc. Chem. Res.* **2** (1969), 25–32.
8. Pochan, J. M., Baldwin, J. E., and Flygare, W. H., *J. Am. Chem. Soc.* **90** (1968), 1072–1073.
9. Turro, N. J., and Hammond, W. B., *Tetrahedron* **24** (1968), 6017–6028.
10. Pazos, J. F., Pacifici, J. G., Pierson, G. O., Sclove, D. B., and Greene, F. D., *J. Org. Chem.* **39** (1974), 1990–1995.
11. Moriarty, R. M., "Stereochemistry of Cyclobutane and Heterocyclic Analogs," in *Topics in Stereochemistry* (Eliel, E. L., and Allinger, N. L., eds.) **8** (1974), 271–421.
12. Bastiansen, O., Kveseth, K., and Møllendal, H., "Large Amplitude Motion in Molecules," in *Topics in Current Chemistry* (*Fortschritte der Chemische Forschung*) **11**, Springer-Verlag, Berlin, 1979, 53–185.

13. Malloy, T. B., Bauman, L. E., and Carreira, L. A., "Conformational Barriers in Small Ring Molecules," in *Topics in Stereochem.* (Allinger, N. L., and Eliel, E. L., eds.), **11** (1979), 53–185.
14. Wilson, T. P., *J. Chem. Phys.* **11** (1943), 369–378.
15. Dunitz, J. D., and Schomaker, V., *J. Chem. Phys.* **20** (1952), 1703–1707.
16. Rathjens, R. G., Jr., Freeman, N. K., Gwinn, W. D., and Pitzer, K. S., *J. Am. Chem. Soc.* **75** (1953), 5634–5642.
17. Carter, G. F., and Templeton, D. H., *Acta Crystallogr.* **6** (1953), 805.
18. Almenningen, A., Bastiansen, O., and Skancke, P. N., *Acta Chem. Scand.* **15** (1961), 711–712. A value for the ring puckering angle ($\beta \sim 35°$) is reported in Skancke, P. N., thesis, Norwegian Technical University (Trondheim), cited in ref. 12.
19. Bucourt, R., "The Torsion Angle Concept in Conformational Analysis," in *Topics in Stereochemistry*, Vol. 8 (Eliel, E. L., and Allinger, N. L., eds.), Wiley Interscience, New York, 1974, 159–224.
20. Ueda, T., and Shimanouchi, T., *J. Chem. Phys.* **49** (1968), 470–471.
21. Meiboom, S., and Snyder, L. C., *J. Chem. Phys.* **52** (1970), 3857–3863.
22. Malloy, T. B., Jr., and Lafferty, W. J., *J. Mol. Spectrosc.* **54** (1975), 20–38.
23. Egawa, T., Yamamoto, S., Ueda, T., and Kuchitsu, K., *J. Mol. Spectrosc.* **126** (1987), 231–239.
24. Egawa, T., Fukuyama, T., Yamamoto, S., Takabayashi, F., Kambara, H., Ueda, T., and Kuchitsu, K., *J. Chem. Phys.* **86** (1987), 6018–6026.
25. PCMODEL versions 4.0–7.0, Serena Software, Inc., Bloomington, IN 47402-3076. PCMODEL uses the MMX force field, a variation of Allinger's MM2 force field.
26. Allinger, N. L., Yuh, Y. H., and Lii, J.-H., *J. Am. Chem. Soc.* **111** (1989), 8551–8566.
27. Kim, H., and Gwinn, W. D., *J. Chem. Phys.* **44** (1966), 865–873.
28. Lampman, G., Hager, G. D., and Couchman, G. L., *J. Org. Chem.* **35** (1970), 2398–2402.
29. Ouellette, R. J., and Booth, G. D., *J. Org. Chem.* **31** (1966), 587–588.
30. Borgers, T. R., and Strauss, H. L., *J. Chem. Phys.* **45** (1966), 947–955.
31. Sharpen, L. H., and Laurie, V. W., *J. Chem. Phys.* **49** (1968), 221–228.
32. Tamagawa, K., and Hilderbrandt, R. L., *J. Phys. Chem.* **87** (1983), 5508–5516.
33. Allinger, N. L., Chen, K., Rahman, M., and Pathiaseril, A., *J. Am. Chem. Soc.* **113** (1991), 4505–4517.
34. Lightner, D. A., Chang, T. C., Hefelfinger, D. T., Jackman, D. E., Wijekoon, W. M. D., and Givens, J. W., III., *J. Am. Chem. Soc.* **107** (1985), 7499–7508.
35. Van Leusen, D., Rouwette, P. H. F. M., and Van Leusen, A. M., *J. Org. Chem.* **46** (1981), 5159–5163.
36. Conia, J-M., and Goré, J., *Bull. Soc. Chim. France* (1964), 1968–1975.
37. Goré, J., Djerassi, C., and Conia, J-M., *Bull. Soc. Chim. France* (1967), 950–955.
38. Cookson, R. C., *J. Chem. Soc.* (1954), 282–286.
39. Allinger, N. L., Allinger, J., Geller, L. E., and Djerassi, C., *J. Org. Chem.* **25** (1960), 6–12.
40. For an excellent treatment of conformational analysis in five-membered rings, see Fuchs, B., "Conformations of Five-Membered Rings," in *Topics in Stereochemistry*, Vol. 10 (Eliel, E. L., and Allinger, N. L., eds.), 1978, 1–94.
41. Kilpatrick, J. E., Pitzer, K. S., and Spitzer, R., *J. Am. Chem. Soc.* **69** (1947), 2483–2488.
42. (a) Wierl, R., *Ann. Physik.* **8** (1931), 521–564. (b) Wierl, R., *Ann. Physik.* **13** (1932), 453–482.
43. Aston, J. G., Schumann, S. C., Fink, H. L., and Doty, P. M., *J. Am. Chem. Soc.* **63** (1941), 2029–2030.
44. Aston, J. G., Fink, H. L., and Schumann, S. C., *J. Am. Chem. Soc.* **65** (1943), 341–346.
45. Pitzer, K. S., *Science* **101** (1945), 672.
46. Hassel, O., and Viervoll, *Acta Chem. Scand.* **1** (1947), 149–168.
47. Adams, W. J., Geise, H. J., and Bartell, L. S., *J. Am. Chem. Soc.* **92** (1970), 5013–5019.
48. Pitzer, K. S., and Donath, W. E., *J. Am. Chem. Soc.* **81** (1959), 3213–3218.
49. Durig, J. R., and Wertz, D. W., *J. Chem. Phys.* **49** (1968), 2118–2121.
50. Legon, A. C., *Chem. Rev.* **80** (1980), 231–262.
51. Carreira, L. A., Jiang, G. J., Person, W. B., and Willis, N. J., Jr., *J. Chem. Phys.* **56** (1972), 1440–1443.
52. Anet, F. A. L., and Anet, R., "Conformational Processes in Rings," in *Dynamic Nuclear Magnetic Resonance Spectroscopy* (Jackman, L. M., and Cotton, F. A., eds.), Academic Press, New York, 1975, Chapter 14.

53. Burkert, U., and Allinger, N. L., *Molecular Mechanics,* ACS Monograph 177, American Chemical Society, Washington, DC, 1982.
54. Ferguson, D. M., Gould, I. R., Clauser, W. A., Schroeder, S., and Kollman, P. A., *J. Comp. Chem.* **13** (1992), 525–532.
55. Kolossvary, I., and Guida, W. C., *J. Am. Chem. Soc.* **115** (1993), 2107–2119.
56. Allinger, N. L., Geise, H. J., Pyckhout, W., Paquette, L. A., and Galluci, J. C., *J. Am. Chem. Soc.* **111** (1989), 1106–1114.
57. Glazer, E. S., Knorr, R., Ganter, G., and Roberts, J. D., *J. Am. Chem. Soc.* **94** (1972), 6026–6032.
58. Lambert, J. B., Papay, J. J., Khan, S. A., Kappauf, K. A., and Magyar, E. S., *J. Am. Chem. Soc.* **96** (1974), 6112–6118.
59. Eliel, E. L., Allinger, N. L., Angyal, S. J., and Morrison, G. A., *Conformational Analysis,* Wiley Interscience, New York, 1965, data cited on p. 202.
60. Haresnape, J. N., *Chem. & Ind. (London)*, (1953), 1091–1092.
61. Kim, H., and Gwinn, W. D., *J. Chem. Phys.* **51** (1969), 1815–1819.
62. Geise, H. J., and Mijlhoff, F. C., *Recl. Trav. Chim. Pays Bas* **90** (1971), 577–583.
63. Durig, J. R., Coulter, G. L., and Wertz, D. W., *J. Mol. Spectrosc.* **27** (1968), 285–295.
64. Ikeda, T., and Lord, R. C., *J. Chem. Phys.* **56** (1972), 4450–4466.
65. Allinger, N. L., Tribble, M. T., and Miller, M. A., *Tetrahedron* **28** (1972), 1173–1190.
66. Tamagawa, K., Hilderbrandt, R. L., and Shen, Q., *J. Am. Chem. Soc.* **109** (1987), 1380–1383.
67. Bardet, L., Granger, R., Sablayrolles, C., and Girard, J. P., *J. Mol. Struct.* **13** (1972), 59–77.
68. Sablayrolles, C., Granger, R., Girard, J. P., Bodot, H., Aycard, J. P., and Bardet, L., *Org. Magn. Reson.* **6** (1974), 161–164.
69. Ouannes, C., and Jacques, J., *Bull. Soc. Chem. France* (1965), 3611–3623.
70. Ouannes, C., and Jacques, J., *Bull. Soc. Chim. France* (1965), 3601–3610.
71. Kirk, D. N., *J. Chem. Soc. Perkin Trans. 1* (1976), 2171–2177.
72. Djerassi, C., Records, R., Ouannes, C., and Jacques, J., *Bull. Soc. Chim. France* (1966), 2378–2381.
73. Sundararaman, P., Barth, G., and Djerassi, C., *J. Am. Chem. Soc.* **103** (1981), 5004–5007.
74. Flament, J. P., and Gervais, H. P., *Tetrahedron* **36** (1980), 1949–1952.
75. Hendrickson, J. B., *J. Am. Chem. Soc.* **83** (1961), 4537–4547.
76. Hendrickson, J. B., *J. Am. Chem. Soc.* **89** (1967), 7036–7043.
77. Hendrickson, J. B., *J. Am. Chem. Soc.* **89** (1967), 7047–7061.
78. Bixon, M., and Lifson, S., *Tetrahedron* **23** (1967), 769–784.
79. Bocian, D. F., Pickett, H. M., Rounds, T. C., and Strauss, H. L., *J. Am. Chem. Soc.* **97** (1975), 687–695.
80. Dillen, J., Geise, H., *J. Chem. Phys.* **70** (1979), 425–428.
81. Roberts, J. D., *Chem. Brit.* **2** (1966), 529–535.
82. Christl, M., and Roberts, J. D., *J. Org. Chem.* **37** (1972), 3443–3452.
83. Dillen, J., and Geise, H., *J. Mol. Struct.* **72** (1981), 247–255.
84. St. Jacques, M., Vaziri, C., Frenette, D. A., Goursot, A., and Fliszár, S., *J. Am. Chem. Soc.* **88** (1976), 5759–5765.
85. Djerassi, C., and Krakower, G. W., *J. Am. Chem. Soc.* **81** (1959), 237–242.
86. Djerassi, C., Burrows, B. F., Overberger, C. G., Takekoshi, T., Gutsche, C. D., and Chang, C. T., *J. Am. Chem. Soc.* **85** (1963), 949–950.
87. Jones, R. B., Zander, J. M., and Price, P., *J. Am. Chem. Soc.* **89** (1967), 94–101.
88. Kirk, D. N., *J. Chem. Soc. Perkin Trans. 1* (1977), 2122–2148.
89. Lightner, D. A., and Docks, E. L., *Tetrahedron* **35** (1979), 713–720.

CHAPTER

7

Saturated Bicyclic Ketones

7.1. Decalins (Bicyclo[4.4.0]decanes)

Decahydronaphthalenes (decalins) have been known since the early 1890s. In 1918, Mohr[1] predicted that they would exist in two strain-free forms, now known as *cis*- and *trans*-decalin (Figure 7-1), and their rational syntheses were reported in 1925 by Hückel et al.[2] Mohr had predicted that *trans*-decalin would consist of two fused chair cyclohexanes and that *cis*-decalin would consist of two fused boats.[1] However, in the 1940s, Bastiansen and Hassel[3] challenged the latter prediction, asserting on the basis of electron diffraction analysis, that *cis*- as well as *trans*-decalin would exhibit all-chair cyclohexane substructures. These findings were reaffirmed in 1964 by Davis and Hassel,[4] who published a few of the molecular parameters (Table 7-1) for the all-chair structure found by gas-phase electron diffraction. The most recent refinement of electron diffraction data appeared in 1978,[5] and the various molecular parameters that were obtained compare favorably with those generated by molecular mechanics calculations[5,6] (e.g., PCMODEL).[7]

The relative stability of *cis*- and *trans*-decalin may be determined from data on heats of combustion (Table 7-2),[8,9,10] which indicate that the *trans* isomer is more stable than the *cis* by 2.69 ± 0.31 kcal/mole (liquid) and 3.09 ± 0.77 kcal/mole (gas). Probably the most accurate determination of the enthalpy difference was reported in 1959 by Allinger and Coke[11] at Wayne State University (*trans*-decalin ⇄ *cis*-decalin: $\Delta H_{313°C}$ = 2.72 ± 0.20 kcal/mole; $\Delta S_{313°C}$ = 0.55 ± 0.30 eu) by isomerization at 258–368°C over 5% Pd(C) and a hydrogen pressure of 35 atm.

Early attempts to derive the relative stability of *cis*- and *trans*-decalin by computational methods came from Barton in 1948,[12] who calculated that chair–chair *trans*-decalin was 0.52–8.23 kcal/mole more stable than chair–chair *cis*-decalin,

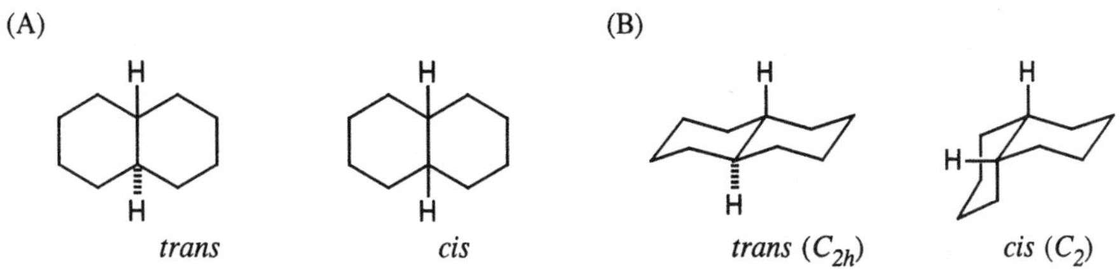

Figure 7-1. (A) Planar representations of *trans*- and *cis*-decalins. (B) Chair–chair conformations of *trans*- and *cis*-decalins.

Table 7-1. Comparison of Experimentally Derived and Computed Average Torsion Angles (ψ), Bond Angles (∠), and Bond Lengths (*r*) for *trans*- and *cis*-Decalins

trans

	Electron Diffraction		Molecular Mechanics			
trans	Geise[a] 1978	Hassel[b] 1964	PCMODEL[c]	Allinger[d] 1968	Lifson[e] 1968	Boyd[e] 1970
ψ (C–C–C–C)°	55.0(5)	—	56.5	—	56.2	56.0
∠ (C–C–C)°	111.4(2)	111.50	110.8	111.0–112.0	111.0	111.0
∠ (C–C–H)°	109.4(7)	—	109.4	—	109.2	109.6
r (C–C) Å	1.530(3)	1.537	1.537	1.523–1.526	1.530	1.541
r (C–H) Å	1.112(5)	—	1.116	—	1.106	1.092

cis

	Electron Diffraction		Molecular Mechanics			
cis	Geise[a] 1978	Hassel[b] 1964	PCMODEL[c]	Allinger[d] 1968	Lifson[e] 1968	Boyd[e] 1970
ψ (C–C–C–C)°	55.0(1.2)	—	54.9	—	54.7	54.2
∠ (C–C–C)°	111.4(4)	112.05	111.3	110.3–113.9	111.3	111.7
∠ (C–C–H)°	109.5(6)	—	109.1	—	109.4	109.5
r (C–C) Å	1.530(2)	1.536	1.537	1.523–1.526	1.530	1.542
r (C–H) Å	1.114(5)	—	1.115	—	1.106	1.092

[a] Data from ref. 5.
[b] Data from ref. 4.
[c] Ref. 7.
[d] Ref. 6.
[e] Cited in ref. 5.

which in turn was more stable than boat–boat *cis*-decalin by 2.87–7.28 kcal/mole. Subsequent molecular mechanics calculations in 1967 by Allinger et al.[13] used the Westheimer–Hendrickson–Wiberg method to predict that *trans*-decalin in the double-chair conformation is more stable than *cis*-decalin in the double chair by 2.87 kcal/mole. This work also found the double-chair *cis*-decalin conformation to be more stable than the boat–chair (5.57 kcal/mole), boat–boat (8.14 kcal/mole), and twist-boat combinations (6.63–7.34 kcal/mole). Thus, the all-nonchair and chair–nonchair conformations of *cis*-decalin are much higher in energy than the all-chair conformation, in contrast to both Barton's predictions[12] and those of Geneste and Lamaty.[14] The latter suggested a twisted double boat as the stable conformation,[14] but that hypothesis is not consistent with either the gas-phase electron diffraction measurements[4] or subsequent molecular mechanics calculations.[6,13,15–19]

A simple way to estimate the energy difference between *cis*- and *trans*-decalin[20] is to sum gauche butane interactions. One then finds that the all-chair *cis* conformation has three more gauche butane interactions than the all-chair *trans*. And if each gauche

Table 7-2. Comparison of Experimental Data on Heats of Combustion and Heats of Formation with Computed Heats of Formation for *trans*- and *cis*-Decalin

	Standard Heat of Combustion (liq.), kcal/mole			Standard Heat of Formation (kcal/mole)[c]		Computed ΔH_f (kcal/mole)			
Decalin	1925[a]	1941[b]	1960[c]	liquid	vapor	MM3[d]	Boyd[e]	Allinger[f]	PCMODEL[i]
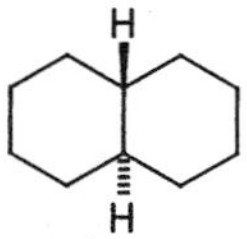 *trans*	1499.8	1500.3	1500.23 ± 0.23	–55.14 ± 0.22	–43.54 ± 0.55	—	–43.8	–43.50[g] –43.68[h]	–43.76
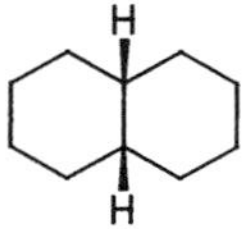 *cis*	1501.4	1502.4	1502.92 ± 0.22	–52.45 ± 0.22	–40.45 ± 0.55	—	–39.7	–40.77[g] –40.89[h]	–41.03
$\Delta\Delta H$ (*cis–trans*)	1.6	2.1	2.69 ± 0.31	2.69 ± 0.31	3.09 ± 0.77	2.83	4.1	2.87[f] 2.73[g] 2.79[h]	2.73

[a] Values from ref. 8.
[b] Values from ref. 9.
[c] Values from ref. 10.
[d] Value from ref. 17.
[e] Values from ref. 19.
[f] Value from ref. 13.
[g] Value from ref. 6.
[h] Value from ref. 16.
[i] Ref. 7.

butane interaction is worth approximately 0.8 kcal/mole of destabilization energy,[20] then the *trans* isomer is calculated to be about 2.4 kcal/mole more stable than the *cis*, a value identical to that computed by Turner.[21]

A decalin unit is found in many terpene natural products, often with an angular methyl group present. The presence of this group has been evaluated in terms of its influence on decalin configurational stability. Thus, from differences in heats of combustion, Dauben et al.[22] determined that *trans*-9-methyldecalin (Figure 7-2) is more stable than the *cis* by approximately 1.4 kcal/mole, which is about one-half the enthalpic difference found for the unmethylated parents[10,11] (Figure 7-2). Earlier, Turner had predicted an energy difference of 0.8 kcal/mole,[21] and, more recently, Allinger and Coke[23] determined energy differences of $\Delta H_{311°C} = 0.55 \pm 0.28$ kcal/mole and $\Delta S_{311°C} = -0.5 \pm 0.5$ eu, favoring the *trans* isomer following equilibration over 10% Pd(C) at 284–338°C. Simple analysis based on summarizing gauche butane interactions predicts an energy difference of $\Delta H \sim 0.8$ kcal/mole. The observation of a lower energy difference between *cis*- and *trans*-decalins with an added angular methyl might suggest that *cis*-9,10-dimethyldecalin should be more stable than its *trans* counterpart. Molecular mechanics calculations,[7] which reproduce the experimental $\Delta\Delta H$ values for *cis*- and *trans*-decalin, as well as *cis*- and *trans*-9-methyldecalin (Table 7-3) predict that *cis*-9,10-dimethyldecalin should be more stable than the corresponding *trans* isomer by about 0.9 kcal/mole. By way of comparison, the analysis of gauche butane interactions predicts that *cis* will be more stable than *trans* by 0.8 kcal/mole.

The interconversion of *cis*- and *trans*-decalins involves breaking and remaking bonds. Conformational changes such as the chair ⇄ chair cyclohexane ring inversion is impossible in *trans*-decalin. Ring deformations leading to boat and twist-boat conformations are possible, as in cyclohexane itself, but they are significantly higher in energy than the chair. The situation is different in *cis*-decalin, which can undergo

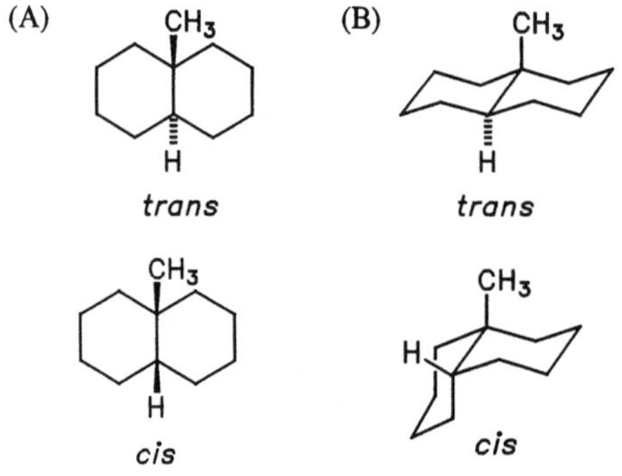

(C) ***Standard Heats of Combustion (liq)***

1658.04 ± 0.41

1659.43 ± 0.45

(D) $\Delta\Delta H° = 1.39 \pm 0.64$

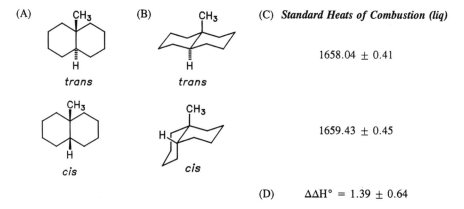

Figure 7-2. (A) Planar representations of *trans*- and *cis*-9-methyldecalins. (B) Stable (chair) conformations of *trans*- and *cis*-9-methyldecalins. (C) Standard heats of combustion (kcal/mole) at 25°C, from ref. 22. (D) Heat of isomerization (kcal/mole) calculated using the data of (C).

Table 7-3. Influence of Angular Methyls on the Enthalpy Differences ($\Delta\Delta H_f$, kcal/mole) for the *cis* ⇄ *trans* Isomerization of Decalins

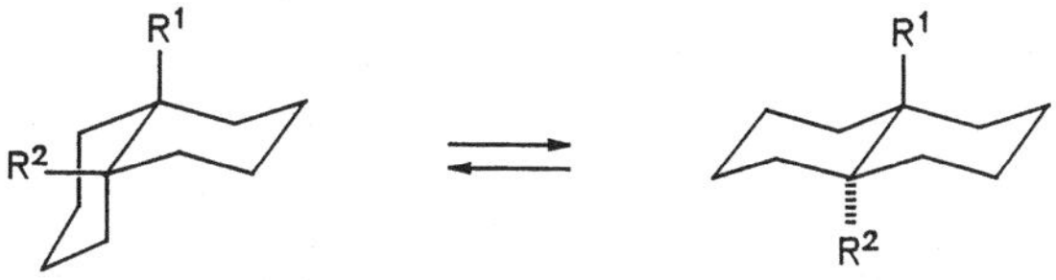

Substituent	Experimental $\Delta\Delta H_f$	Computed $\Delta\Delta H_f$ (PCMODEL)[c]	Computed Allinger
$R^1 = R^2 = H$	2.72[a]	2.73	2.73,[d] 2.79,[e] 2.83,[f] 2.87[g]
$R^1 = CH_3$, $R^2 = H$	0.55[b]	0.62	0.57,[d] 0.43[g]
$R^1 = R^2 = CH_3$	Not Reported	−0.88	—

[a] Ref. 11.
[b] Ref. 23.
[c] Ref. 7.
[d] Ref. 6.
[e] Ref. 16.
[f] Ref. 17.
[g] Ref. 13.

concerted chair ⇄ chair inversions in both rings. This process may be seen readily in the carbon (Figure 7-3) and proton NMR spectra of *cis*-decalin, which change as the temperature is lowered. In contrast, the NMR spectra of *trans*-decalin remain invariant. The barrier to conformational ring inversion in *cis*-decalin (Table 7-4) has been determined by variable-temperature NMR[24–27] to be a few kcal/mole higher than that of cyclohexane ($\Delta H^\ddagger = 10.71$ kcal/mole, $\Delta S^\ddagger = 2.2$ eu; Section 2.2) or even *cis*-1,2-dimethylcyclohexane ($\Delta H^\ddagger = 10.11 \pm 0.15$ kcal/mole, $\Delta S^\ddagger = 0.6 \pm 0.7$ eu).[26] Conformational analysis of the *cis*-decalin chair ⇄ chair double inversion has been analyzed by molecular mechanics calculations,[18] which predict a barrier of 12.0 kcal/mole, with chair–twist-boat and double twist-boat intermediates lying along the interconversion path.

7.2. Decalones

In addition to *cis* and *trans* configurational isomers, there are two regio-isomeric decahydronaphthalenones (decalones)—one with the carbonyl group at C(1), the other at C(2) (Table 7-5). As in the parent decalins, *trans*-1- and 2-decalones are more stable than the corresponding *cis* isomers[2c,28–33] and have been shown to favor the double-chair conformation by gas-phase electron diffraction studies,[34–36] lanthanide-induced shift (LIS)–NMR analysis,[37] X-ray crystallography,[38] and molecular mechanics calculations.[7,28,32–36] The experimental studies on ring conformation have been limited

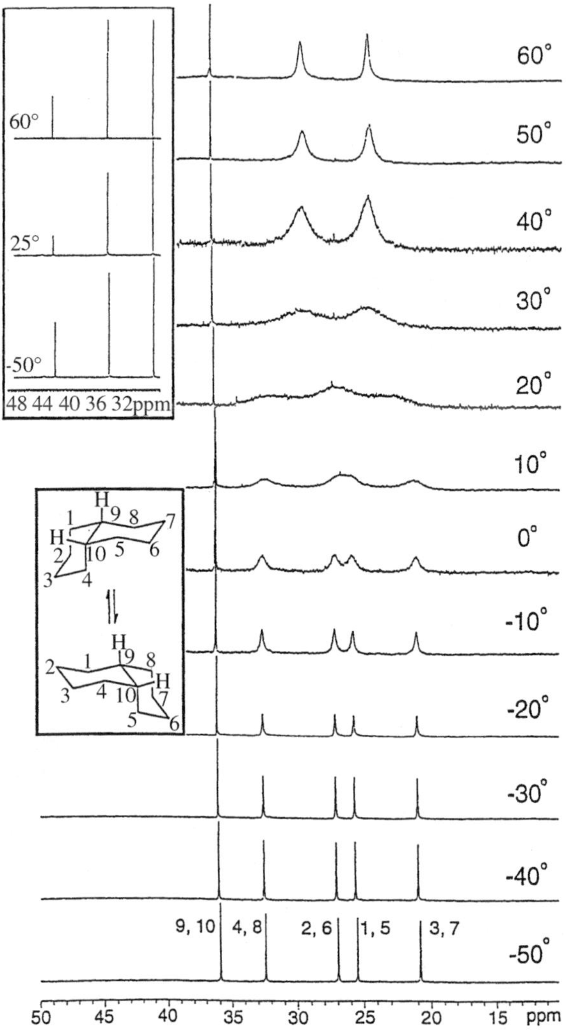

Figure 7-3. Variable-temperature 125-MHz ^{13}C–NMR spectra of *cis*-decalin, in $CDCl_3$ and reflecting the equilibrium shown (lower inset). For comparison, NMRs of *trans*-decalin are shown (upper inset). Temperatures are indicated in °C. We thank Mr. Michael Huggins for obtaining these spectra and creating this figure.

mainly to 2-decalones. For example, electron diffraction spectroscopy on *trans*-2-decalone vapor showed a double-chair conformation with the expected slight flattening near the carbonyl carbon (as in cyclohexanone) of one ring and also slight flattening in the other ring.[34] These conclusions are supported by LIS-NMR studies that indicate a ring-puckering angle (51°) identical to that in cyclohexanone.[37] Although an X-ray crystal study of the methyl ester of *trans*-2-decalone-10-carboxylic acid shows the

Table 7-4. Chair ⇄ Chair Conformational Equilibrium in *cis*-Decalin, *cis*-9-Methyldecalin, and *cis*-9,10-Dimethyldecalin and Corresponding Energies of Activation

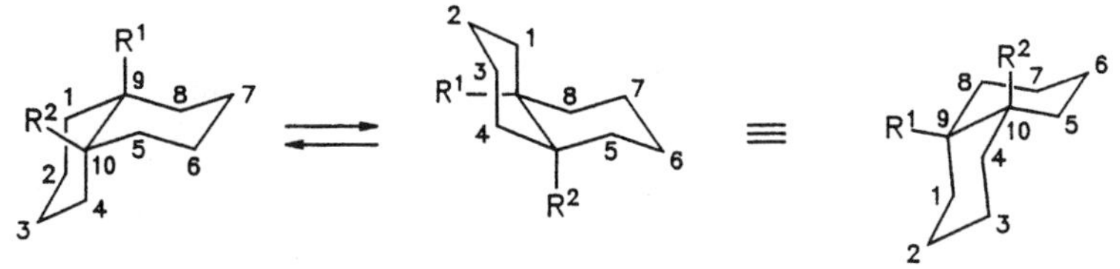

R^1	R^2	$\Delta G^{\ddagger}_{298°C}$ (kcal/mole)	$\Delta H^{\ddagger}$ (kcal/mole)	$\Delta S^{\ddagger}$ (cal/mole/deg)
H	H	12.7[a]	13.4	2.5
H	H	12.6[b]	13.6 ± 0.7	3.5 ± 3
H	H	12.30[c]	12.35 ± 0.11	+0.15 ± 0.44
H	H	12.59[d]	13.57 ± 0.27	+3.27 ± 0.96
H	H	12.65[e]	12.66 ± 0.36	+0.03 ± 0.82
H	H	12.45[f]	12.47 ± 0.13	+0.06 ± 0.52
CH_3	H	12.6[b]	12.4 ± 0.7	−0.7 ± 3
CH_3	CH_3	15.2[g]	—	—

[a] ^{1}H-NMR, −18°C in CS_2, ref. 24.
[b] ^{13}C-NMR, neat liquid, ref. 25.
[c] ^{13}C-NMR, 27°C in CD_2Cl_2, ref. 26.
[d] Ref. 26, recalculated using all data from ref. 25.
[e] Ref. 26, recalculated from ref. 25, but leaving out two extreme data values.
[f] Ref. 26, $CD_3C_6D_5$ solvent.
[g] ^{1}H-NMR, 25°C in CD_2Cl_2, ref. 27.

Table 7-5. Relative Stability of 1-Decalone and 2-Decalone as Determined by Equilibration ($\Delta G°$, kcal/mole)[a] and Heats of Combustion ($\Delta\Delta H_f°$, kcal/mole)[b]

	1-Decalones			2-Decalones	
	trans	*cis*		*trans*	*cis*
$\Delta G°$ (equil.)	0.0	2.2	$\Delta\Delta H_f°$[b]	0.0	2.2
$\Delta G°$ (equil.)	0.0	1.3			
$\Delta G°$ (equil.)	0.0	2.2			

[a] At 227°C, 90% *trans*, ref. 2c; at room temp (?), 90% *trans*, ref. 29; at 240–250°C, 10–11% *cis*, ref. 30.
[b] From heats of combustion, ref. 31.

double-chair conformation, it also indicates more ring flattening[38]—presumably due to crystal-packing forces. Although no similar experimental data exist on 1-decalones, the double-chair conformation is supported by molecular mechanics calculations as the most stable stereochemistry.[28]

There is only one all-chair conformation in *trans*-decalones, but *cis*-decalones have two: "steroid" and "nonsteroid" conformations (Table 7-6), so named because an *A/B cis* steroid can adopt only the "steroid" conformation (Chapter 8 and Table 8-3) and cannot undergo a chair $\rightleftarrows$ chair interconversion that would convert it into "nonsteroid." The steroid conformation is that found in *A/B cis* steroids. In decalones, steroid and nonsteroid conformations interconvert by a chair $\rightleftarrows$ chair inversion in each ring. Their relative stabilities have been calculated by molecular mechanics computations[7,28] which predict that (*i*) the steroid conformer is approximately 1 kcal/mole more stable than the nonsteroid in *cis*-1-decalone (Tables 7-5 and 7-6) and (*ii*) the nonsteroid conformer is slightly more stable (~0.25 kcal/mole) than the steroid in *cis*-2-decalone. LIS-NMR conformational analysis of the steroid $\rightleftarrows$ nonsteroid conformational equilibrium in cis-2-decalone indicates that when *cis*-2-decalone is complexed to $Yb(fod)_3$ in $CDCl_3$, the steroid conformer is favored by a 55:45 ratio at 25°C ($\Delta G° = 0.23$ kcal/mole).[37] At –87°C, however, a ^{13}C-NMR analysis of uncomplexed *cis*-2-decalone in CD_2Cl_2 shows a 65:35 ratio ($\Delta G° = 0.13$ kcal/mole) in favor of the nonsteroid conformer.[27] When combined, these data predict a $\Delta H° \sim 0.81$ kcal/mole and $\Delta S° =$ +3.4 eu for the nonsteroid $\rightleftarrows$ steroid equilibrium in *cis*-2-decalone. The data also support the predictions of molecular mechanics calculations that the nonsteroid conformer of 2-decalone is favored enthalpically over the steroid conformer. No equivalent NMR studies have yet been performed on *cis*-1-decalones. In principle, a variable low-temperature CD study of optically active *cis*-1- and 2-decalones should be able to provide experimentally derived conformational energies between steroid and nonsteroid, but this has not been reported. The steroid $\rightleftarrows$ nonsteroid inversion barrier has been estimated from temperature-dependent ^{19}F-NMR studies of

Table 7-6. Computed Relative Heats of Formation ($\Delta\Delta H_f$, kcal/mole) of Decalones

1-Decalones			2-Decalones		
trans	*cis*-steroid	*cis*-nonsteroid	*trans*	*cis*-steroid	*cis*-nonsteroid
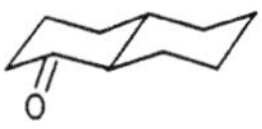	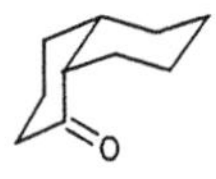	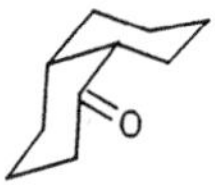	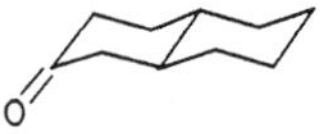	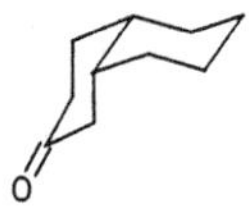	
0.17[a]	1.67[a]	2.68[a]	0.00[a]	2.00[a]	1.75[a]
0.00[b]	1.97[b]	2.87[b]	0.00[b]	2.20b	

[a] PCMODEL, ref. 7, relative to *trans*-2-decalone.
[b] Refs. 32, 38.

6,6-difluoro-*cis*-2-decalone: $\Delta H^{\ddagger} = 10.1–10.8$ kcal/mole, and $\Delta S^{\ddagger} = 4–6$ eu.[39,40] This conformational inversion barrier has not yet been determined for *cis*-1-decalone.

The presence of an angular methyl group has a profound influence on the relative stabilities of *cis*- and *trans*-decalones (Table 7-7). In 9-methyl-1-decalone, for example, the presence of the angular methyl leads to the interesting result that the *cis* isomer is more stable than the *trans*—a conclusion based on equilibration studies.[41] However, the equilibration was carried out at 250°C, and the equilibrium population at room temperature may not reflect a bias toward the *cis* isomer if the entropy favors the *cis* isomer at higher temperatures. Molecular mechanics calculations[28,32] predict only a small difference in enthalpy. 10-Methyl-1-decalone is easier to equilibrate, and acid- or base-catalyzed equilibration studies[28,42–45] indicate that the *trans* isomer is slightly favored over the *cis*. Molecular mechanics calculations[7] also indicate that the *trans* isomer is more stable than the *cis*. For 1-decalones, the presence of an angular methyl group clearly decreases the energy difference between *cis* and *trans* isomers, by nearly

Table 7-7. Relative Stability ($\Delta G°$, $\Delta\Delta H_f$, kcal/mole) of 1-Decalones and 2-Decalones with Angular Methyl Groups

9-Methyl-1-decalone			10-Methyl-1-decalone		
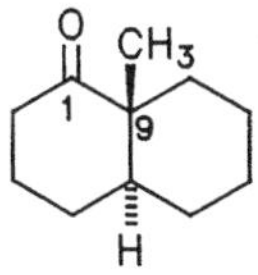		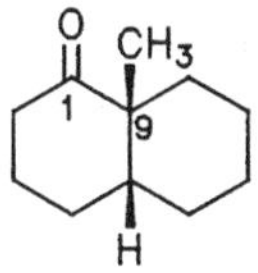	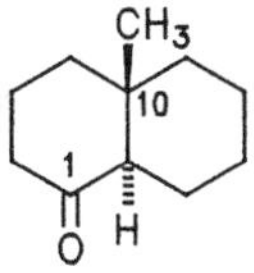		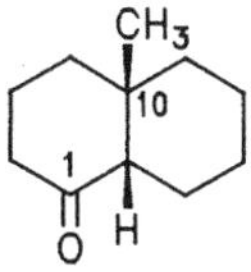
trans		cis	*trans*		*cis*
$\Delta G°$ (equil.)	0.38[a]	0.0[a]	$\Delta G°$ (equil.)	0.0[d]	~0.2[d]
$\Delta\Delta H_f$	0.07[b]	0.0[b]	$\Delta\Delta H_f$	0.0[e]	0.37/0.76[e]
$\Delta\Delta H_f$	0.0[c]	0.13/0.53[c]	$\Delta\Delta H_f$	0.0[b]	0.69[b]
			DDH_f	0.0[c]	0.78/0.89[c]

9-Methyl-2-decalone			10-Methyl-2-decalone		
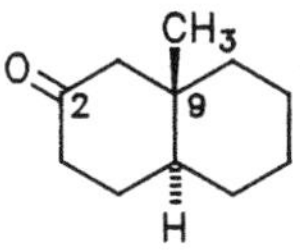		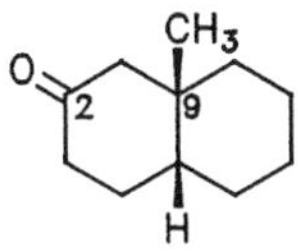	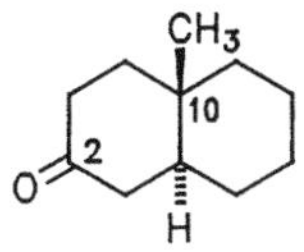		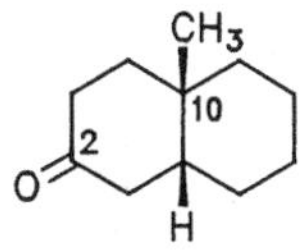
trans		cis	*trans*		*cis*
$\Delta\Delta H_f$	0.0[f]	1.11/0.65[f]	$\Delta\Delta H_f$	0.0[f]	0.98/0.30[f]

[a] Equilibrated over Pd(C) at 250°C, ref. 41.
[b] Ref. 32.
[c] Ref. 28, steroid/nonsteroid.
[d] Refs. 42–44, room temp to approximately 60°C.
[e] Ref. 45, steroid/nonsteroid.
[f] PCMODEL (ref. 7), steroid/nonsteroid.

two-thirds (Table 7-7). A similar sort of effect is seen in decalins. Qualitatively, the effect may be attributed to four additional gauche butane interactions introduced by an angular methyl on the *trans*-decalin skeleton vs. two additional gauche butane interactions from an angular methyl on the *cis*. Similarly, in 2-decalones, the introduction of an angular methyl should have a similar effect in reducing the energy difference between *cis* and *trans* isomers (Table 7-7). However, there is no experimental evidence to support this belief—only the predictions of molecular mechanics calculations.[32,33,45]

As in the parent decalones, *cis*-9- and 10-methyldecalones may adopt a steroid or a nonsteroid conformation. The steroid ⇄ nonsteroid conformational inversion barrier has been determined only for 10-methyl-2-decalone: $\Delta H^{\ddagger}$ = 9.8–10.4 kcal/mole, and $\Delta S^{\ddagger}$ = –5 eu, found by variable-temperature ^{19}F-NMR spectroscopy on the 6,6-difluoro analog.[39,40] The activation enthalpy is not very different from that of *cis*-2-decalone ($\Delta H^{\ddagger}$ = 10.1–10.8 kcal/mole)[39,40] or of cyclohexane ($\Delta H^{\ddagger}$ = 10.3 kcal/mole), a not unexpected result, since the steroid ⇄ nonsteroid conformational change involves a cyclohexane chair ⇄ chair inversion. The energy differences between steroid and nonsteroid conformations have been calculated by molecular mechanics computations and are not large (Table 7-8).[7] Interestingly, only in the case of 10-methyl-1-decalone is the steroid conformation predicted to be more stable than the nonsteroid. As with the parent nonmethylated decalones, conformational analyses would appear to be perfectly suited to investigation by CD spectroscopy, but this has not yet been done.

Table 7-8. Computed[a] Relative Heats of Formation ($\Delta\Delta H_f$, kcal/mole) of Decalones with an Angular Methyl Group

9-Methyl-1-decalone			10-Methyl-1-decalone		
trans	*cis*-steroid	*cis*-nonsteroid	*trans*	*cis*-steroid	*cis*-nonsteroid
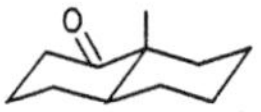	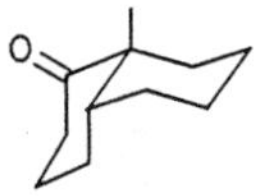		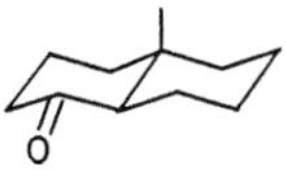	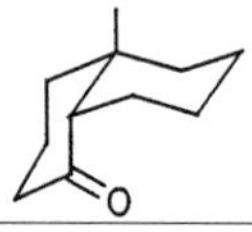	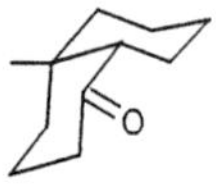
3.21	3.21	2.82	0.32	1.18	1.60
9-Methyl-2-decalone			**10-Methyl-2-decalone**		
trans	*cis*-steroid	*cis*-nonsteroid	*trans*	*cis*-steroid	*cis*-nonsteroid
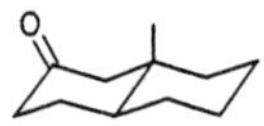	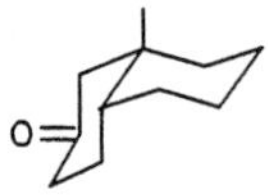		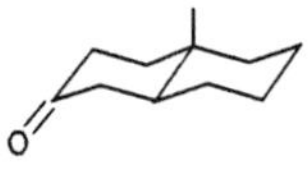	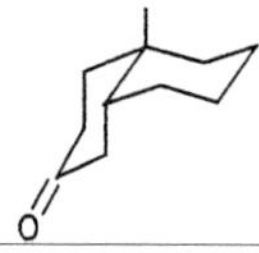	
0.00	1.11	0.65	0.62	1.60	0.92

[a] PCMODEL, ref. 7; all data are relative to ΔH_f of *trans*-9-methyl-2-decalone.

Rather early, Djerassi et al.[46,47] measured the ORD spectra of (+)-*cis*-10-methyl- and (+)-*cis*-9-methyl-1-decalone, finding a weak positive Cotton effect in the former (a = +11, $\Delta\varepsilon \sim +0.27$) and a stronger positive Cotton effect in the latter (a = +51, $\Delta\varepsilon \sim +1.27$).[47] These studies were carried out before the days of CD spectroscopy in the Djerassi lab and were not subsequently reinvestigated. The arguments in favor of a particular predominant conformer are relatively straightforward, however. If one (reasonably) assumes a double-chair conformation and applies the octant rule to the steroid and nonsteroid conformers of (+)-*cis*-9-methyl-1-decalone (Figure 7-4), the $n \rightarrow \pi^*$ Cotton effect of the steroid conformer is predicted to be controlled by an α-axial methylene group in a negative back octant. The angular methyl, C(2), C(4), C(6), and C(9) lie in octant symmetry planes; C(5) and C(7) lie equidistant from a symmetry plane, as do C(3) and C(10). Thus, the lone dissymmetric perturber is C(8), in an α-axial position. The Cotton effect of the steroid conformer should thus be strongly negative. The Cotton effect of the nonsteroid conformer is predicted to be dominated by the contributions of an α-axial methyl in a positive octant. C(2), C(4), C(8), and C(9) lie in octant symmetry planes and hence make no contribution. C(3) and C(10) lie equidistant from an octant symmetry plane, so their octant contributions cancel. C(5), C(6), and C(7) lie in front octants, where weak positive contributions are expected. The angular methyl lies in the α-axial orientation in a positive octant. The net Cotton effect of the nonsteroid conformer should be strongly positive. Since a strongly positive $n \rightarrow \pi^*$ Cotton effect ($\Delta\varepsilon \sim +1.27$)[47] is observed, (+)-*cis*-9-methyl-1-decalone is predicted by the octant rule to be largely in the nonsteroid conformation—in good agreement with the results of molecular mechanics calculations (Table 7-8).

Similarly, the octant rule was applied to the steroid and nonsteroid conformers of (+)-*cis*-10-methyl-1-decalone (Figure 7-4) to reach a qualitative conclusion regarding the "preferred" conformation. In the steroid conformer, the angular methyl lies in a positive octant as a β-equatorial position on the cyclohexanone ring. Ring atoms C(5), C(6), and C(7) lie in negative front octants, so their octant contributions should be weak and negative. Atoms C(2), C(4), C(9), and C(8) lie on octant symmetry planes and thus make no octant contributions. C(3) and C(10) lie equidistant from an octant symmetry plane, so their contributions cancel. Thus, the net predicted Cotton effect is moderately positive. In the nonsteroid conformer, C(2), C(4), C(9), and C(6) lie on octant symmetry planes and hence make no contribution to the Cotton effect. C(3) and C(10) lie equidistant from an octant symmetry plane, as do C(5) and C(7). Therefore, their octant contributions cancel. One is accordingly left with an α-axial C(8) CH_2 in a positive back octant and a β-axial angular methyl in a weakly positive front octant. The net Cotton effect is predicted to be strongly positive for the nonsteroid conformer. The observed moderately positive Cotton effect ($\Delta\varepsilon \sim +0.27$) indicates a predominance of the steroid conformer in (+)-*cis*-10-methyl-1-decalone. Again, these qualitative conclusions are supported by the predictions of molecular mechanics calculations (Table 7-8).

As described earlier (Chapter 5), variable-temperature CD spectra can be analyzed in order to probe conformational equilibria, to identify the most stable conformer, and

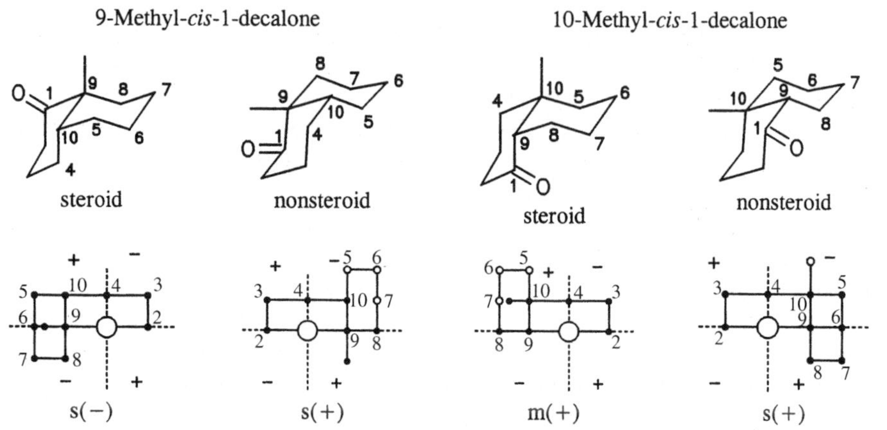

Figure 7-4. Double-chair conformers of 9- and 10-methyl-*cis*-1-decalones and (below) their octant projection diagrams. Open circles represent front octant perturbers; filled circles represent back octant perturbers. The summary $n \rightarrow \pi^*$ Cotton effect predictions are noted as s(−) (strongly negative), s(+) (strongly positive), and m(+) (moderately positive).

to predict Δε values for each conformer in a two-conformer equilibrium. No variable low-temperature CD studies have been carried out on the *cis*-1-decalones, however. Yet, it seems clear from applications in other cyclohexanone systems that such a study should lend itself to predicting the predominant conformer, as well as the difference in energy between the steroid and nonsteroid conformations.

As has already been mentioned, no experimental evidence is available to support the predictions of molecular mechanics calculations (Table 7-6) that the parent *cis*-1-decalone favors the steroid conformation. However, Djerassi and Staunton[48] showed by ORD that a bromination product of (−)-*cis*-1-decalone, namely, 2-bromo-*cis*-1-decalone (Figure 7-5), had a negative Cotton effect in isooctane (***a*** = −74, Δε ~ −1.8) and in methanol (***a*** = −51, Δε ~ 1.3). The authors argued in favor of the α-axial bromoketone with a nonsteroid conformation in isooctane and about 30% of the steroid conformer (with an α-equatorial bromine in methanol). Both conformers are predicted by the octant rule to give negative $n \rightarrow \pi^*$ Cotton effects, with the nonsteroid isomer

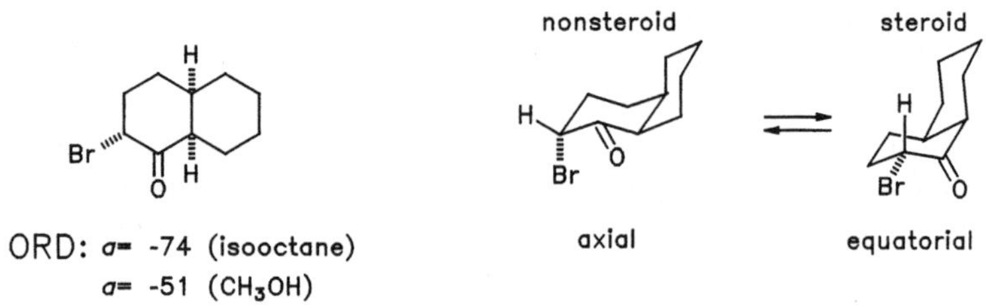

Figure 7-5. 2-Bromo-*cis*-1-decalones in its nonsteroid and steroid conformations.

Cotton effect dominated by the octant contribution from an α-axial bromine and the steroid isomer Cotton effect dominated by the octant contribution from an α-axial methylene group. Here again, variable low-temperature CD would be an excellent probe of the conformational equilibrium.

Attempts at conformational analysis of the *cis*-2-decalones with angular methyls proved to be much more complicated than analyzing the *cis*-1-decalones. 9-Methyl-*cis*-2-decalone is predicted by molecular mechanics calculations to favor the nonsteroid conformation (Table 7-8), and octant projection diagrams for the steroid and nonsteroid conformers (Figure 7-6) predict moderately strong negative and positive Cotton effects, respectively. This decalone should lend itself nicely to a simple CD analysis of its conformation and to a variable low-temperature CD analysis of its conformational energies; yet there are no available ORD or CD data. On the other hand, 10-methyl-*cis*-2-decalone (Figure 7-6) has been investigated carefully and extensively by ORD and CD. The observed negative $n \to \pi^*$ ORD Cotton effect (***a*** = −11, $\Delta\varepsilon \sim -0.27$) for (+)-10-methyl-*cis*-2-decalone[46,47] correlated best with a predominance of the steroid conformer, in which the dissymmetric perturbation comes mainly from a β-equatorial C(8) methylene group; hence, a moderate negative Cotton effect is predicted by the octant rule. The alternative, nonsteroid conformation is predicted to have a very weak Cotton effect, either positive or negative. Here the dissymmetric perturbers are a β-axial methylene C(8) in a negative front octant and a more distant C(7) methylene, possibly in a positive back octant.

At the time of the investigations,[46,47] the conclusion in favor of a steroid conformer for 10-methyl-*cis*-2-decalone seemed unusual, as it contradicted the earlier predictions of Klyne,[49] who favored a nonsteroid conformer on the basis of its fewer nonbonded 1,3-diaxial steric repulsions. However, Halsall and Thomas[50] suggested that, whereas in *cis*-2-decalone the nonsteroid conformer should be more stable than the steroid, in

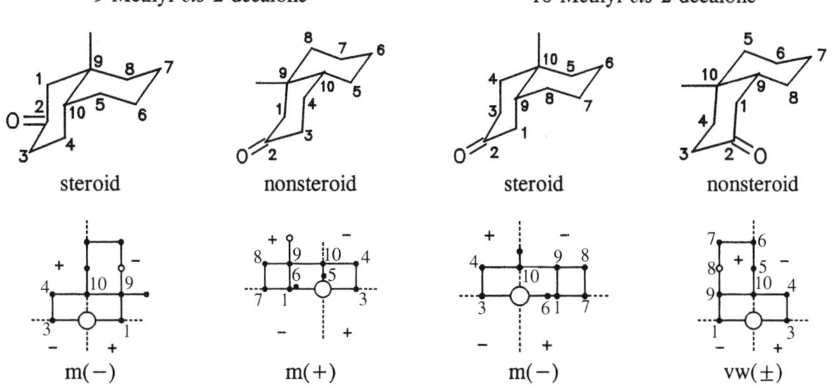

Figure 7-6. Double-chair conformers of 9- and 10-methyl-*cis*-2-decalones and (below) their octant projection diagrams. The summary $n \to \pi^*$ Cotton effect predictions are noted as m(−) (moderately negative), m(+) (moderately positive), and vw(±) (very weakly negative or positive).

10-methyl-*cis*-2-decalone the conformers should be of approximately equal energy. Molecular mechanics calculations (Table 7-8) indicate that the nonsteroid conformer will be the more stable, suggesting that its predicted weakly negative Cotton effect is overwhelmed by a stronger positive Cotton effect from the minor steroid conformer. A variable-temperature CD study should clarify the situation, but none is currently available.

A potentially less ambiguous example was conceived by Djerassi et al.[51] as 7,7,10-trimethyl-*cis*-2-decalone (Figure 7-7), which had previously been predicted to exist almost completely in the nonsteroid conformation.[51] In this decalone, the double-chair steroid conformer suffers from a severe nonbonded steric destabilization interaction between the C(1) methylene and C(7) *syn*-methyl, and this interaction was predicted to drive the equilibrium toward the nonsteroid isomer.[50,51] Unfortunately, the octant rule scarcely distinguishes between the two isomers, predicting a zero Cotton effect for the steroid conformer and a very weakly negative Cotton effect for the nonsteroid. Unexpectedly, a weakly positive $n \rightarrow \pi^*$ Cotton effect was observed, which correlates with neither conformer. Since the Cotton effect becomes increasingly positive at lower temperatures (Figure 7-8), it is clear that the most stable conformer has a positive Cotton effect. Since neither the steroid nor the nonsteroid conformer is predicted to have a positive Cotton effect, it was assumed that a small amount of twist conformer dominated the observed Cotton effect. However, this rationale does not seem entirely satisfactory, as molecular mechanics calculations[7] do not reveal a stable twist conformer and predict that the nonsteroid conformer should be more stable than the steroid by 3.3 kcal/mole.

In a related attempt to displace the steroid ⇄ nonsteroid conformational equilibrium clearly toward the nonsteroid in a 10-methyl-*cis*-2-decalone, Djerassi et al.[51] prepared an analog with a 7(*S*)-isopropyl group (Figure 7-9). The isopropyl group is axial in the

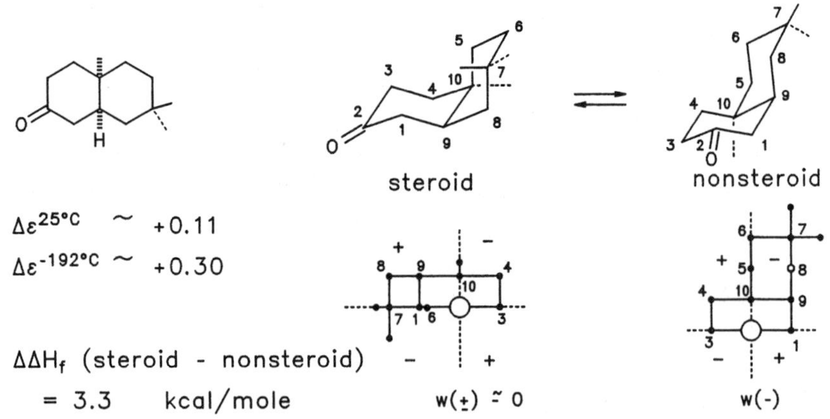

Figure 7-7. (+)-7,7,10-Trimethyl-*cis*-2-decalone and its steroid and nonsteroid conformers. CD values (ref. 51) were obtained in EPA solvent (Figure 7-8) and corrected to the values in the PhD dissertation of J.V. Burakevich, Stanford, 1964. $\Delta\Delta H_f$ is from PCMODEL (ref. 7).

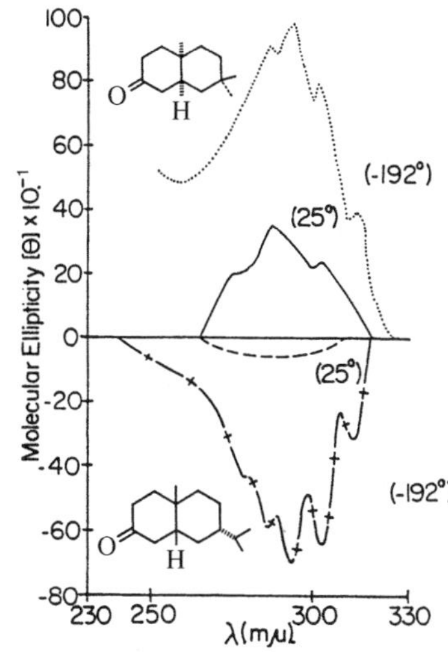

Figure 7-8. Circular dichroism curves for (+)-7,7,10-trimethyl-*cis*-2-decalone and for (–)-7(*S*)-isopropyl-10(*S*)-methyl-*cis*-2-decalone in ether–isopentane–ethanol (EPA:5:5:2 by vol.) at 25° and –192°C. [Reprinted with permission from ref. 51. Copyright © 1964 American Chemical Society.]

steroid conformation and equatorial in the nonsteroid; the latter should thus be more stable. The octant rule predicts moderately strong positive Cotton effects for both isomers; yet a weak negative Cotton effect is observed at room temperature and becomes much more strongly negative at low temperatures (Figure 7-8). These data indicate that the energetically most stable conformer has a negative Cotton effect—a conclusion that is inconsistent with both the nonsteroid and the steroid conformer and

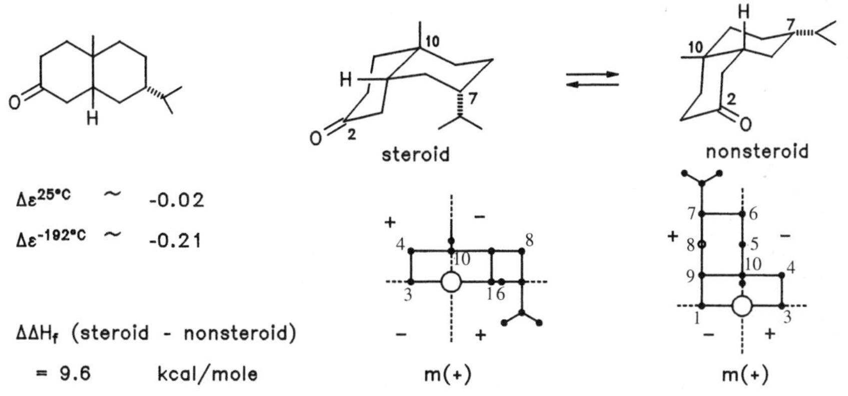

Figure 7-9. (–)-7(*S*)-Isopropyl-10(*S*)-methyl-*cis*-2-decalone and its steroid and nonsteroid conformers. CD (ref. 51) obtained in EPA solvent (Figure 7-8) corrected according to the ORD data. $\Delta\Delta H_f$ is from PCMODEL (ref. 7).

with what the octant rule predicts about them. These surprising results were interpreted in terms of the Cotton effect being dominated by a small amount of twist-boat conformer with a strongly negative Cotton effect. Yet, molecular mechanics calculations[7] indicate a perfectly ordinary stable nonsteroid conformer to be nearly 10 kcal/mole more stable than the steroid conformer. The cyclohexane ring of the steroid conformer is flattened to alleviate nonbonded steric interactions due to the axial isopropyl group. These results suggest that if a twist-boat conformer were present, it would be so only in very small amounts. And in order to exhibit such large Cotton effects, it would need to have an enormous negative Cotton effect, which seems improbable. This work bears repeating by variable low-temperature CD.

Evidence in support of a predominant nonsteroid conformation came from an analysis of the ^{1}H-NMR chemical shifts of the angular methyl group. Dauben et al.[52] noted that when the angular methyl group is axial in the cyclohexanone ring of 3-keto-steroids and decalones, the difference between the chemical shift of the angular methyl in the ketone and its parent hydrocarbon ($\Delta\delta$ of Table 7-9) is 0.20–0.24 ppm.[53,54] This phenomenon is best illustrated in the *trans*-decalones and in 5α-cholestan-3-one, wherein the ring conformations are relatively rigid. In the one conformationally fixed example in which the angular methyl is equatorial on the cyclohexanone ring (5β-cholestan-3-one), $\Delta\delta$ is much smaller, approximately 0.11. *cis*-Decalone steroid and nonsteroid conformations have angular methyls in the equatorial and axial sites, respectively, on the cyclohexanone rings. Since the observed $\Delta\delta$ values (Table 7-9) for the *cis*-decalones are all large (0.22–0.29), as best fits axial angular methyls, it was reasoned that these decalones favor the nonsteroid conformation.

Further evidence regarding the steroid $\rightleftarrows$ nonsteroid conformational equilibrium in 9-methyl- and 10-methyl-*cis*-2-decalones comes from angular methyl chemical shifts and line-shape analysis in the 60-MHz ^{1}H-NMR spectra of these compounds.[55a] For comparable pairs of derivatives of *cis*- and *trans*-9-methyl decalins, the spectral line widths (the widths at half the height, $w_{1/2}$, of the spectral lines) of the angular methyls were compared with that of tetramethylsilane. The difference in line width, $\Delta w_{1/2}$, was shown to be regularly greater for the *trans* than for the *cis* isomers, due to the greater number of 4-bond (*W* or *M* rule) coupling paths between the C–CH_3 bond and proximal C–H bonds in the former. The magnitude of $\Delta w_{1/2}$ could also be used to distinguish nonsteroid and steroid conformations in *cis*-decalones. In 10-methyl-*cis*-2-decalone and derivatives expected to adopt the nonsteroid conformation, the angular methyl resonance is relatively broad ($\Delta w_{1/2} = 0.4$–0.75), and a splitting ($J \sim 0.4$ Hz) is found in the 1α,3,3,8,8-pentadeuterated, 3α-methyl, and 3α-bromo analogs, which suggests coupling from the 4α-axial H in the nonsteroid conformation. In contrast, derivatives expected to adopt the steroid conformation (e.g., 3β-methyl and 3β-bromo) exhibit relatively narrow and unsplit resonances ($\Delta w_{1/2} = 0.4$–0.5 Hz). In derivatives of 9-methyl-*cis*-2-decalone that “prefer” the steroid conformation, namely, the 8α-methyl and 6α-*tert*-butyl analogs, the angular methyl is split ($J \sim 0.88$–0.95 Hz) from coupling to the 1α-axial H. The parent ketone, 9-methyl-*cis*-2-decalone, shows a smaller coupling ($J = 0.59$ Hz) reflecting a mixture of steroid and nonsteroid

Table 7-9. Chemical Shifts of C(19) Methyls in 5α and 5β-Cholestan-3-ones and Angular Methyls in *trans*- and *cis*-Decalones[a]

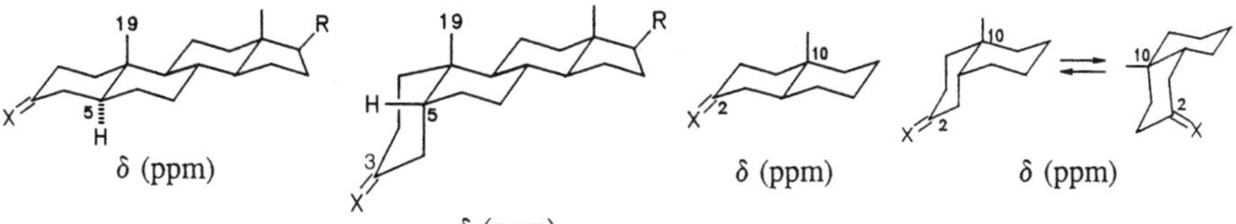

	δ (ppm)	δ (ppm)	δ (ppm)	δ (ppm)
X = H_2	0.77	0.92	0.83	0.95
X = O	0.99	1.03	1.03	1.17
Δδ	0.22	0.11	0.20	0.22

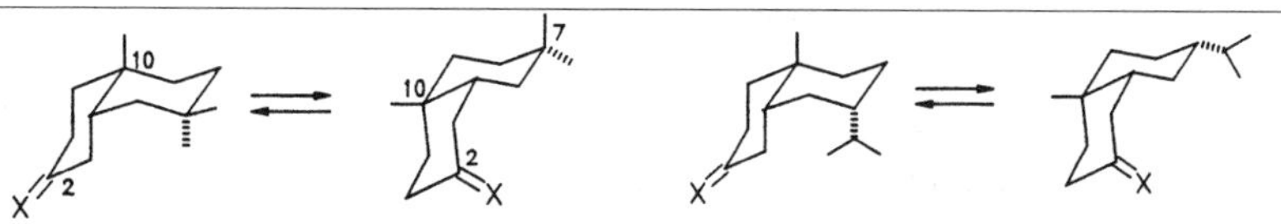

	δ (ppm)	δ (ppm)
X = H_2	0.95	0.95
X = O	1.24	1.22
Δδ	0.29	0.27

	δ (ppm)	δ (ppm)
X = H_2	0.83	0.95
X = O	1.07	1.17
Δδ	0.24	0.22

[a] Values from CCl_4 solvent in ppm downfield from $(CH_3)_4Si$; ref. 52. Δδ = δ(ketone) minus δ(parent hydrocarbon).

conformations. The data were interpreted in terms of a 1:1 mixture of nonsteroid and steroid conformers of 9-methyl-*cis*-2-decalone and a 1:2 mixture of a nonsteroid and steroid conformer of 10-methyl-*cis*-2-decalone in chloroform solvent.

On the other hand, an analysis of vicinal coupling constants suggested a predominant nonsteroid conformation in 10-methyl-*cis*-2-decalone.[55b] For example, the coupling constants, measured at 60 MHz, between the 1β-H and the 9-H is 5.1 Hz, and that between the 1α-H and the 9-H is 4.3 Hz, which is consistent with an equatorial 9-H on the ketone ring, as is found in the nonsteroid conformation. Consistent with the latter possibility, and on the basis of low-temperature ^{1}H-NMR shifts of its angular methyl, the nonsteroid conformation was thought to have a lower enthalpy than the steroid.[55b] These data also suggested that a twist conformer would not be present.[55b] The inconsistencies presented by conflicting CD and NMR data suggest that reinvestigating *cis*-decalone conformations should be a worthwhile endeavor.

One attempt at a resolution was provided in a study[57] of methyl *cis*-tetrahydro-α- and β-santoninates (Figure 7-10A), which differ only in the stereochemistry of the propionic ester side chain, which is remote from the ring fusion. Both ketones adopt a nonsteroid conformation in the crystal, as indicated by X-ray crystallography. This is at first surprising, because in the nonsteroid conformation the larger ester group is axial, whereas in the steroid conformer the methyl adjacent to the carbonyl is axial. However, in the nonsteroid conformer, the ketone carbonyl and the hydroxyl group are positioned to engage in conformation-stabilizing intramolecular hydrogen bonding. Despite this readiness, only *inter*molecular hydrogen bonding was found in the crystal.

Whether the nonsteroid conformation predominates in solution was addressed by CD and ^{1}H-NMR spectroscopy. In both protic and aprotic solvents, the CD spectra (Figure 7-10B) showed positive $n \rightarrow \pi^*$ Cotton effects for both ketones over the temperature range +20° to −150°C. Octant diagrams (Figure 7-10A) for the chair–chair conformations predict a positive Cotton effect for both the steroid and nonsteroid conformations. In the former, negative octant contributions coming from ring carbon 8 and its attached hydroxyl group are more than offset by the strong positive Cotton effect expected from the axial α-methyl perturber at carbon 1. In the latter, only a weak positive back-octant contribution is expected from the equatorial α-methyl perturber at carbon 1, but strong positive contributions are expected from ring carbon 7 and its attached propionate ester group, and a weak negative front-octant contribution is expected from ring carbon 8. The overall prediction is a positive Cotton effect for the nonsteroid conformation. It is unclear whether the steroid conformation will give a stronger positive Cotton effect than the nonsteroid; Ogura et al. predict the reverse.[57] Since the observed Cotton effects are more positive at lower temperatures (Figure

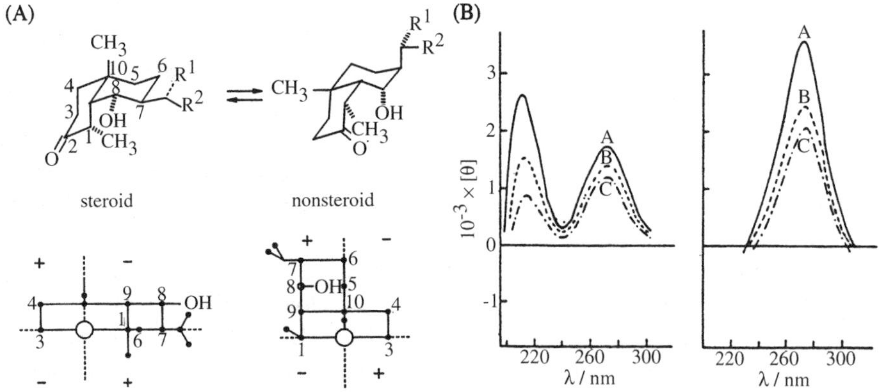

Figure 7-10. (A) Steroid and nonsteroid conformations of methyl *cis*-tetrahydro-α-santoninate and their octant diagrams (R^1=CH_3, R^2=CO_2CH_3). (B) CD Cotton effects of the α-santoninate (left) and β-santoninate (R^1=CO_2CH_3, R^2=CH_3) (right) in ethanol–methanol (4:1 by vol.) at −150°C (A), −50°C (B) and +20°C (C). The CD near 210 nm is due to the ester group. [(B) is reprinted with permission from ref. 57. Copyright © 1979 Royal Society of Chemistry.]

7-10B), the data confirm that both conformers are present at room temperature and suggest that the nonsteroid conformer is the more stable. In support of this hypothesis, PCMODEL[7] molecular mechanics calculations predict that the nonsteroid conformers are more stable than the steroid by 3–4 kcal/mole.

Unlike the interpretation of the $n \rightarrow \pi^*$ Cotton effects of *cis*-decalones, that of *trans*-decalones is usually rather straightforward and uncomplicated by ring conformational equilibria.[56] The octant rule correctly predicts the observed $n \rightarrow \pi^*$ Cotton effect signs for the *trans*-decalones of Table 7-10, and the observed $\Delta\varepsilon$ values show evidence of additivity. For example, the presence of the C(9) angular methyl acts as an α-axial methyl perturber (with estimated $\Delta\varepsilon$ contribution of +1.5 (Table 5-7)) and inverts the intrinsic Cotton effect of the parent *trans*-1-decalone ($\Delta\varepsilon = -0.95$) to give a net observed $\Delta\varepsilon$ of approximately +0.67. The predicted value, $\Delta\varepsilon = +1.5 - 0.95 = +0.55$, is in good agreement with the observed value.

Despite their inability to undergo chair–chair conformational ring inversion, *trans*-decalones may exhibit ring distortion in response to intramolecular nonbonded steric interactions. For example, there is a severe nonbonded steric 1,3-diaxial repulsion between the isopropyl and the angular methyl in *trans*-tetrahydroeremophilone[58,59] (Figure 7-11) when it adopts the chair–chair conformation. In this conformation, the main octant perturbers are ring carbons 5 and 6, which lie in a negative back octant; the methyl at carbon 5 that lies in a negative back octant; the angular methyl, which lies in a positive front octant; and the axial isopropyl, which lies partly in a negative front octant and partly in a positive back octant. While it is difficult to assess quantitatively, the octant contribution of the isopropyl is probably weakly positive, and with the other octant perturbers expected to weigh in with a moderately negative octant contribution (as in *trans*-10-methyl-1-decalone; see Table 7-10C), the net $n \rightarrow \pi^*$ Cotton effect may be predicted to be moderate and negative ($\Delta\varepsilon \sim -0.8$). However, an *intense negative* Cotton effect was observed ($a = -109$, $\Delta\varepsilon \sim -2.7$). This unexpected result is not consistent with the chair–chair decalone, but can be explained[58] if the

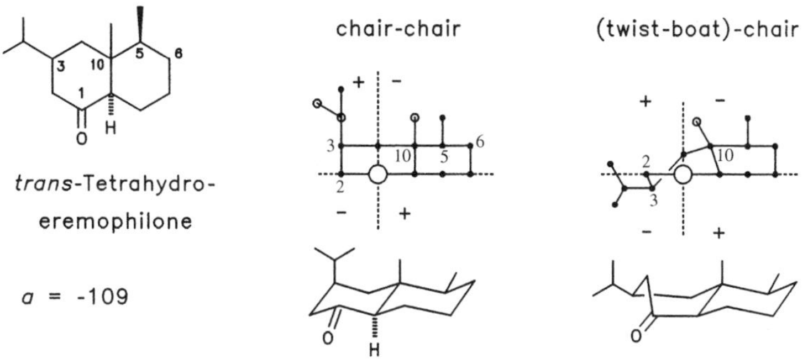

Figure 7-11. Octant diagrams for chair–chair and (twist-boat)-chair conformers of *trans*-tetrahydroeremophilone. A moderate negative $n \rightarrow \pi^*$ Cotton effect is predicted for the former, a strong negative Cotton effect is predicted for the latter. The observed Cotton effect from ORD studies is strong ($a = -109$). Open circles represent front-octant perturbers.

Table 7-10. Simple Decalone Conformations and Octant Projection Diagrams with Predicted and Observed $n \rightarrow \pi^*$ CD Cotton Effects[a]

Decalone	Conformation	Octant Projection Diagram[b]	CE	
			Pred.	Obs.[a]
(A) 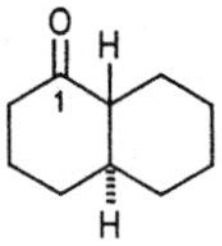	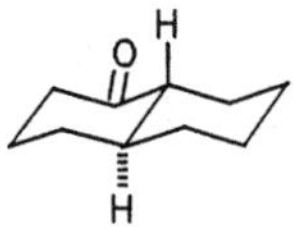	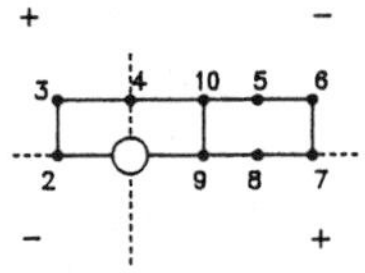	–	–0.95
(B) 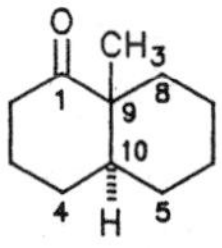	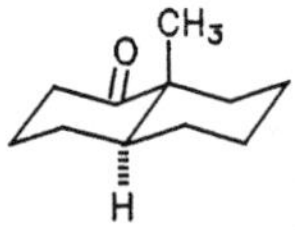	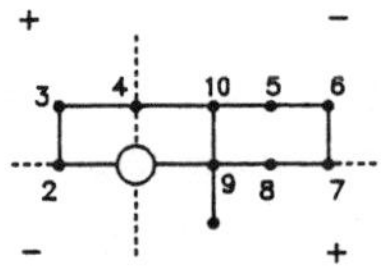	+	+0.67
(C) 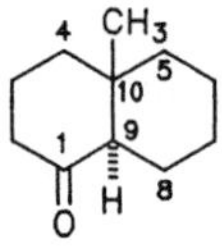	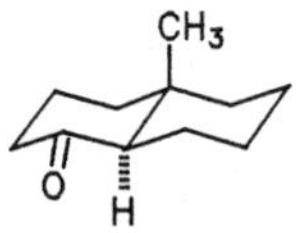	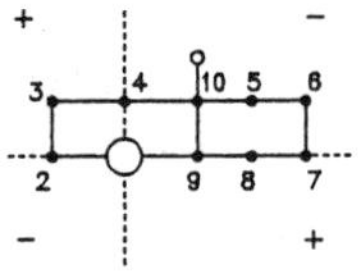	–	–0.79
(D) 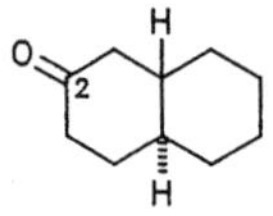	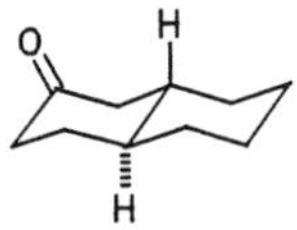	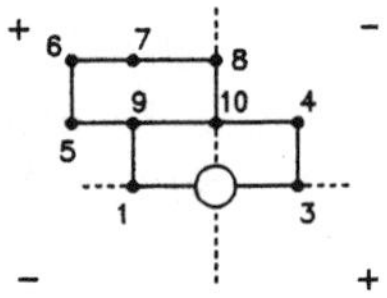	+	+1.4
(E) 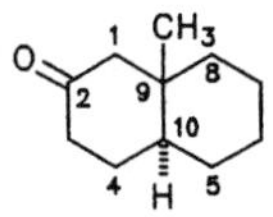	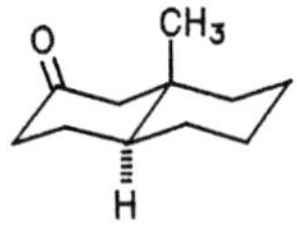	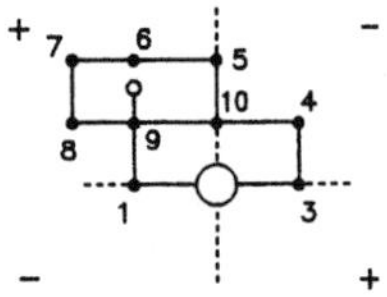	+	+1.8
(F) 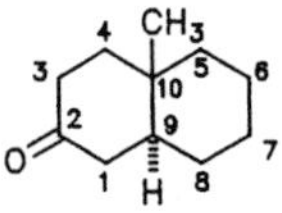	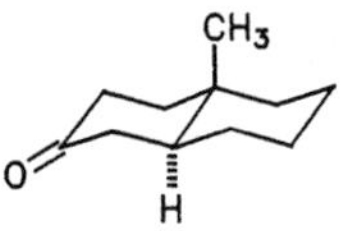	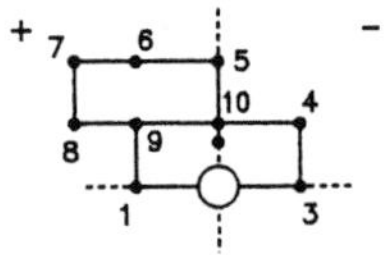	+	+1.1

[a] In CH_3OH; data from refs. 56 and 80.

[b] Open circles indicate front-octant perturbers.

ketone ring adopts a twist-boat conformation. In such a conformation, ring atoms 3 and 10, as well as the isopropyl group, lie in a negative back octant and add to the contributions of the other negative back-octant perturbers. Alternatively, Kirk[59c] has calculated a $\Delta\varepsilon$ value for *trans*-tetrahydroeremophilone in the all-chair conformation that agrees with experiment, but he calculates a value twice as large for the twist-boat isomer. On the other hand, PCMODEL molecular mechanics calculations predict that the twist boat is more stable by about 1.5 kcal/mole than the chair–chair. The energy difference lies largely in the strain energy due mainly to the 1,3-diaxial 10-methyl-isopropyl interaction in the latter.

The epimeric 3-iso-*trans*-tetrahydroeremophilone (Figure 7-12) has an equatorial isopropyl and thus adopts the chair–chair conformation. Its octant diagram suggests that the negative octant contributions of ring atoms 5 and 6 and the methyl at C(5) are essentially balanced by the octant contributions of the equatorial isopropyl—leaving other ring atoms lying on octant symmetry planes and thus making no contribution. The remaining octant perturber is the angular methyl, which is expected to make a very weak positive front-octant contribution. The overall octant prediction is for an extremely weak, possibly positive, $n \rightarrow \pi^*$ Cotton effect. Yet this ketone exhibits a large negative ORD $n \rightarrow \pi^*$ Cotton effect (a = –69, $\Delta\varepsilon \sim -1.7$). The discrepancy remains unresolved, and no CD or variable-temperature CD work has been carried out with 3-iso-*trans*-tetrahydroeremophilone or with *trans*-tetrahydroeremophilone.

Equilibration studies[59] have shown, as might be predicted, that *trans*-tetrahydroeremophilone is less stable than its *cis* counterpart. However, when the isopropyl is equatorial, as in the 3-iso isomer, *trans* is *more* stable than *cis*. These findings are confirmed by PCMODEL[7] molecular mechanics calculations, summarized in Table 7-11. The calculations predict that *cis*-tetrahydroeremophilone is most stable in the steroid conformation, which is approximately 1 kcal/mole more stable than the *trans* isomer (which is most stable in the twist-boat conformation). In contrast, in the 3-iso series, the *trans* isomer is about 3 kcal/mole more stable than the *cis* (nonsteroid).

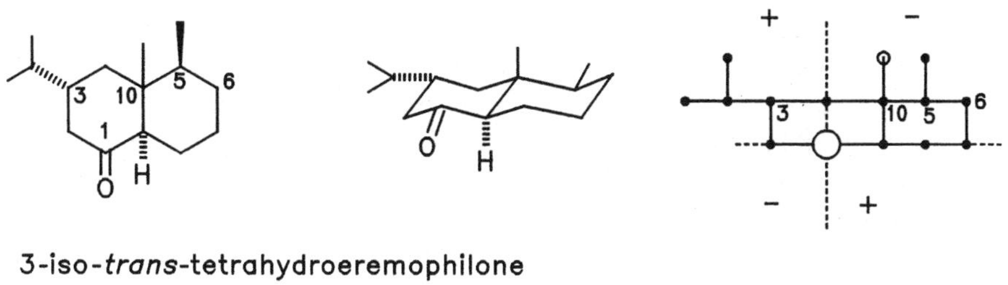

3-iso-*trans*-tetrahydroeremophilone
a = -69

Figure 7-12. Octant diagram (far right) for the chair–chair conformation of 3-iso-*trans*-tetrahydroeremophilone. A net very weak $n \rightarrow \pi^*$ Cotton effect is predicted, but the ORD Cotton effect is strongly negative.

Table 7-11. Computed[a] Relative Heats of Formation ($\Delta\Delta H_f$, kcal/mole) of Isomeric Tetrahydroeremophilones, Showing the Influence of the Isopropyl Configuration on Ring Conformation and Relative Stability

trans-tetrahydroeremophilone | *cis*-tetrahydroeremophilone

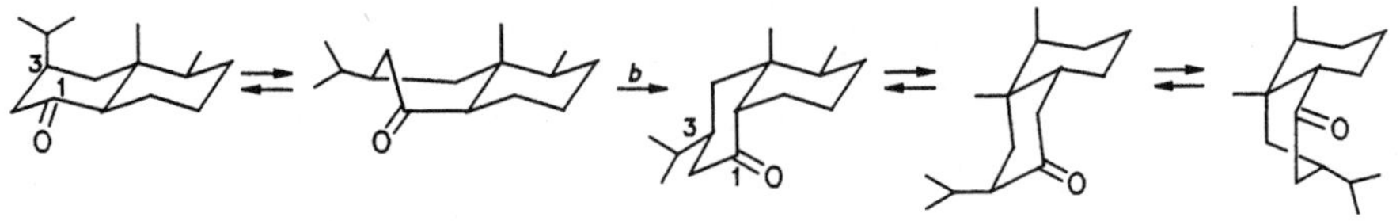

steroid | nonsteroid

$\Delta\Delta H_f$	4.5	2.0	1.0	8.0	6.0

3-iso-*trans*-tetrahydroeremophilone | 3-iso-*cis*-tetrahydroeremophilone

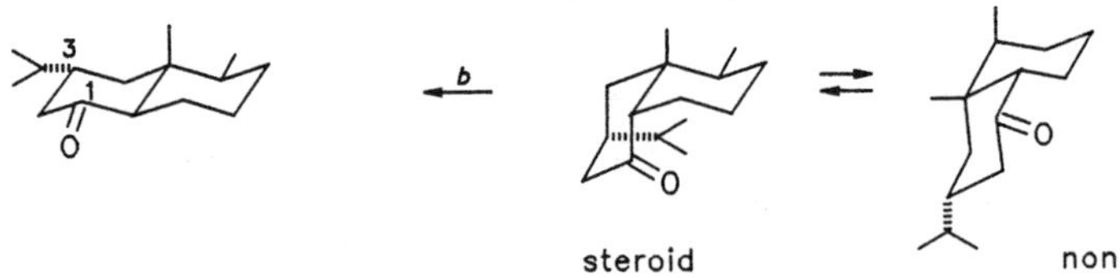

steroid | nonsteroid

$\Delta\Delta H_f$	0.0	5.7	2.9

[a] PCMODEL, ref. 7.

[b] Direction toward more stable isomer, from equilibration studies (ref. 59).

When the ketone group is moved closer to the isopropyl, as in the eremophilanones (Table 7-12), the *trans*-C8-eremophilanone is found by equilibration studies[59] to be more stable when the isopropyl is equatorial—as might be expected. No Cotton effects have been reported for the axial isopropyl isomers, but the observed positive ($a = +73$, $\Delta\varepsilon \sim +1.8$) ORD $n \rightarrow \pi^*$ Cotton effect for the more stable isomer matches the sign predicted by the octant rule and is similar in magnitude to that seen in a more naked decalone (Table 7-10F). In the *cis*-C8-eremophilanones, the 7-epi isomer is found[59] to be the more stable, which is opposite to that predicted by PCMODEL[7] molecular mechanics calculations: *cis*-C8-eremophilanone in the steroid conformation is calculated to be more stable than 7-epi-*cis*-C8-eremophilanone in the nonsteroid conformation by about 0.8 kcal/mole. In *cis*-C8-eremophilanone, the steroid conformation is predicted to be much more stable than the nonsteroid, but in the 7-epimer, the nonsteroid is predicted to be more stable than the steroid. No Cotton effects have been reported for the *cis*-C8, but a weak ORD $n \rightarrow \pi^*$ Cotton effect ($a = -9$, $\Delta\varepsilon \sim -0.2$) has been observed for the 7-epimer. However, according to the octant rule, a large positive Cotton effect might be expected for this isomer—with the magnitude depending on whether the isopropyl group projects into a front octant. As with the tetrahydroeremophilone series, no variable-temperature CD studies have been performed to attempt to sort out conformational questions.

Table 7-12. Computed[a] Relative Heats of Formation ($\Delta\Delta H_f$, kcal/mole) of Isomeric C8-Eremophilanones, Showing the Influence of the Isopropyl Configuration on Ring Conformation and Relative Stability

trans-C8-eremophilanone — 7-epi-*trans*-C8-eremophilanone

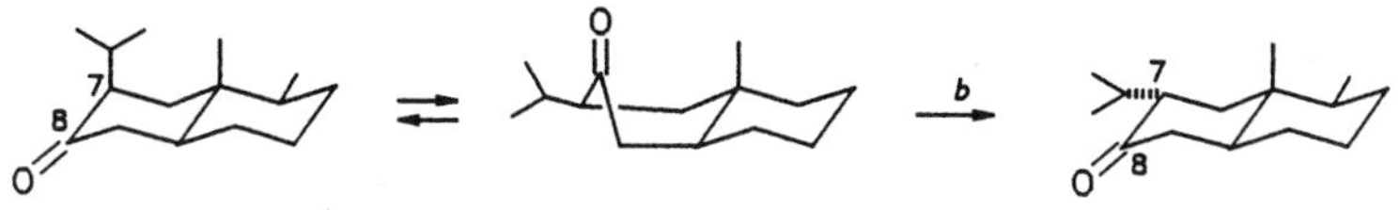

$\Delta\Delta H_f$	3.5	2.1	0.0

cis-C8-eremophilanone — 7-epi-*cis*-C8-eremophilanone

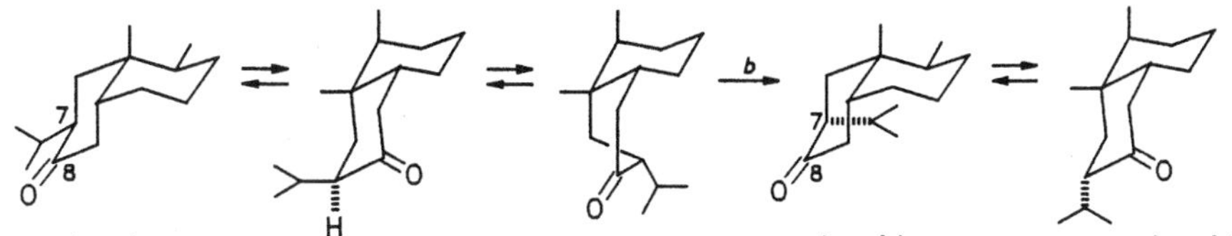

	steroid	nonsteroid	boat	steroid	nonsteroid
$\Delta\Delta H_f$	0.0	4.5	4.3	2.7	0.84

[a] PCMODEL, ref. 7.

[b] Direction toward more stable isomer, from equilibration studies (ref. 59).

The stereochemistry of an aza-decalone, (–)-1-oxoquinolizidine (Figure 7-13), presents an interesting example in which inversion at the bridgehead nitrogen interconverts the *cis* and *trans*-decalone types. PCMODEL[7] molecular mechanics calculations predict that the *trans* isomer is the most stable and that, between the two *cis* isomers, the nonsteroid should be more stable than the steroid. Octant diagrams predict a moderate negative $n \rightarrow \pi^*$ Cotton effect for the *trans* steroid, a strong positive Cotton effect for the *cis*, and a weak positive for the *cis* nonsteroid (due mainly to front-octant perturbers). A moderate negative CD Cotton effect is seen at +60°, suggesting a predominance of the *trans* isomer.[60] At –78°C, the Cotton effect becomes somewhat more negative, indicating that the *trans* isomer is more stable than the *cis*.

7.3. Hydrindanes (Bicyclo[4.3.0]nonanes)

Perhydroindenes (hydrindanes) have been known since the early 1900s. The *cis* and *trans* stereoisomers were first prepared as ketones, viz., β-hydrindanones, in 1926 by Hückel and Friedrich at Göttingen in a rational synthesis from *cis*- and *trans*-1,2-cyclohexanediacetic acid.[61] Subsequently, *cis* and *trans*-hydrindane were prepared by reduction of the ketone carbonyl.[61,62] At that time, it was thought that the cyclohexane ring in *trans*-hydrindane adopted a chairlike conformation, whereas in *cis*-hydrindane it was thought to adopt a boat.[61] Later, as ring conformation theory and experiment revealed that cyclohexane boat conformations were typically of higher energy than

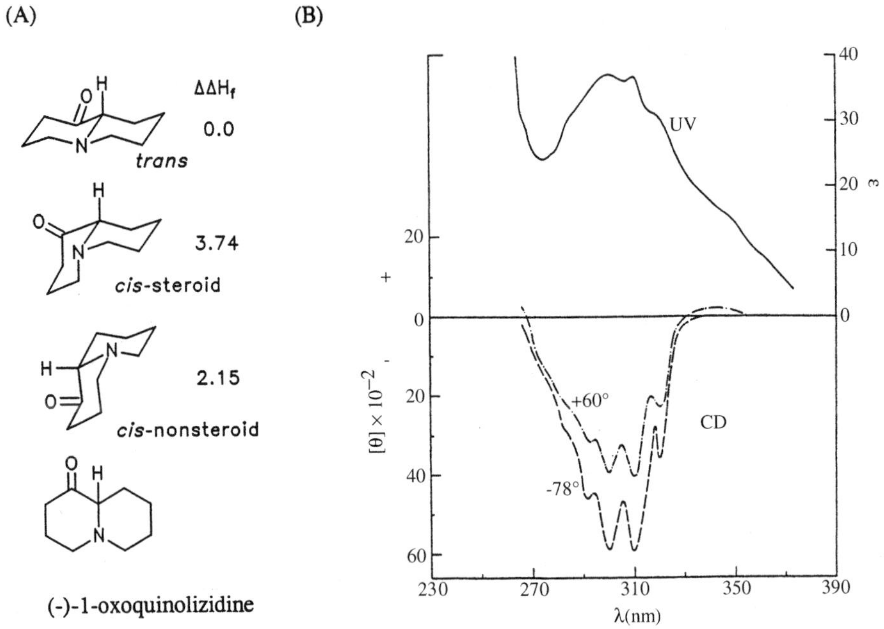

Figure 7-13. (A) (–)-1-Oxoquinolizidine and its chair–chair conformers. The relative heats of formation ($\Delta\Delta H_f$, kcal/mole) are from PCMODEL (ref. 7) molecular mechanics calculations. (B) UV and CD spectra of (–)-1-oxoquinolizidine in isooctane. [(B) is reproduced with permission from Crabbé, P., *Optical Rotatory Dispersion and Circular Dichroism in Organic Chemistry,* Holden-Day, Inc., Boca Raton, FL, 1965. Copyright © Holden-Day, Inc.]

chair, the six-ring boat conformation in *cis*-hydrindane became less attractive. Although molecular mechanics calculations in the 1960s[6] and 1970s[16,19] clearly revealed a chair cyclohexane conformation in both *cis*- and *trans*-hydrindane, direct experimental evidence did not become available until 1981,[63] through gas-phase electron diffraction spectroscopy. This study, as well as earlier[6,16,19] and subsequent molecular mechanics calculations[17,64] clearly revealed significant torsion angle strain in the six-membered ring of *trans*-hydrindane at the shared bond, which is opened to about 61° (from 55° in decalins) to accommodate the *trans*-fused cyclopentane ring (Table 7-13).

The relative stability of liquid *cis*- and *trans*-hydrindane was first addressed in 1935.[62] Determinations of heats of combustion showed that the *trans* isomer is 1.8 kcal/mole more stable than the *cis*. Better values for the heats of combustion were published in 1938,[65] with the energy difference lowered to 1.5 kcal/mole (Table 7-14). Subsequently, Browne and Rossini used more sensitive calorimetry to show that the *trans* isomer is favored over the *cis* by only 0.74 kcal/mole (in the liquid phase);[66] however, more recently, Finke et al.[67] found an enthalpy difference of 1.065 kcal/mole for the gas phase. The energy difference between *cis*- and *trans*-hydrindane isomers is thus only one-fourth to one-half the difference separating *trans*- from *cis*-de-

Table 7-13. Comparison of Experimentally Derived[a] and Computed[b] Ring Torsion Angles (ψ, °) in Hydrindanes with Data from Decalins and Cyclopentane

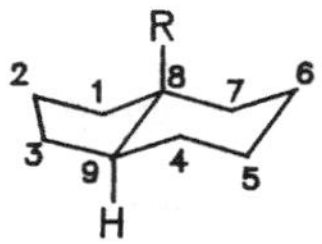

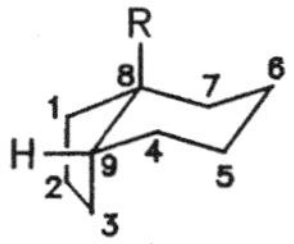

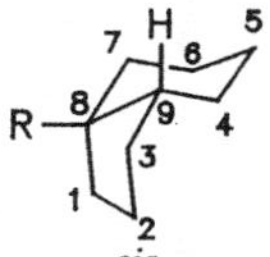

	METHOD (*trans*)			METHOD (*cis*)			METHOD (*cis*)		
	Electron Diffraction[a]	PCMODEL[b]		Electron Diffraction[a]	PCMODEL[b]		Electron Diffraction[a]	PCMODEL[b]	
	R=H	R=H	R=CH_3	R=H	R=H	R=CH_3	R=H	R=H	R=CH_3
ψ(1–8–9–3)		–46.5	–46.6		–40.9	–39.0		–40.7	–40.2
ψ(4–9–8–7)	61.1	62.1	62.4	46.6	47.4	47.8	–46.6	–47.3	–43.7
ψ(4–5–6–7)	–57.1	–54.3	–51.0	–58.2	–58.3	–56.1	58.2	58.4	59.2
ψ(2–1–8–9)		39.3	39.0		–36.3	–35.7		–29.3	34.1
ψ(2–3–9–8)		36.2	35.7		–30.0	–27.9		36.8	31.2
ψ(3–2–1–8)		–15.9	–18.1		18.1	19.1		–6.38	–15.2
ψ(1–2–3–9)		–12.6	–10.6		7.36	5.37		–18.8	–9.94
ψ(6–7–8–9)	–58.6	–58.8	–57.1	–51.5	–53.4	–52.1	47.4	47.2	44.8
ψ(5–4–9–8)	–58.6	–58.6	–52.8	–47.4	–47.3	–49.0	–51.5	53.4	52.0
ψ(5–6–7–8)	56.6	55.0	53.0	57.5	59.1	57.0	–52.8	–52.5	–52.8
ψ(6–5–4–9)	56.6	59.9	52.8	52.8	52.5	52.6	–57.5	–59.1	–59.6

 ψ[c]	55.0	56.5		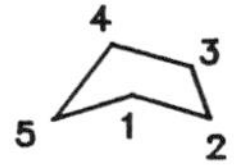			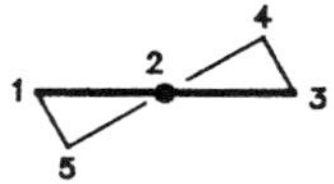		
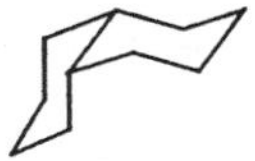 ψ[c]	55.0	54.9		ψ(1–2–3–4)	0.0	0.0	—	ψ(1–2–3–4) 13.2	13.2 —
				ψ(2–3–4–5)	25.0	25.1	—	ψ(2–3–4–5) 34.3	34.5 —
				ψ(3–4–5–1)	40.3	40.5	—	ψ(3–4–5–1) 42.3	42.5 —

[a] Ref. 63.

[b] Ref. 7.

[c] Average C–C–C–C torsion angle.

Table 7-14. Comparison of Data on Heats of Combustion and Experimental Heats of Formation with Computed Heats of Formation for *trans* and *cis*-Hydrindane

Hydrindane	Standard Heat of Combustion (kcal/mole)		Enthalpy Difference	Standard Heat of Formation (kcal/mole)		Computed ΔH_f (kcal/mole)			
	1938[a] (liq.)	1960[b] (liq.)	1972[c] (gas)	gas[b]	liq.[b]	MM3[d]	Boyd[e]	MM2	PCMODEL
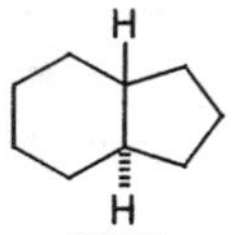 *trans*	1344.28 ±0.25	1350.86 ±0.40	—	–31.45 ±0.50	–42.15 ±0.40	–31.87	–30.5	–31.65[f] –31.07[g] –31.66[h]	–30.46
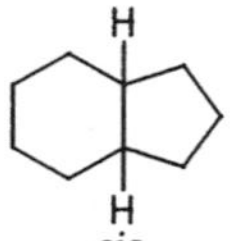 *cis*	1345.77 ±0.50	1351.60 ±0.36	—	–30.41 ±0.47	–41.41 ±0.36	–30.93	–31.0	–29.99[f] –29.86[g] –30.30[h]	–31.63
$\Delta\Delta H_f$(*cis* – *trans*)	1.49 ±0.9	0.74 ±0.52	1.0065	1.04	0.74	0.97	–0.5	0.66[f] 1.21[g] 0.63[h]	1.17

[a] Liquid, ref. 65.
[b] Liquid, ref. 66.
[c] Values from ref. 67
[d] Values from ref. 17.
[e] Values from ref. 19
[f] Values from ref. 6.
[g] Values from ref. 16.
[h] Values from ref. 64.

calin. The smaller difference has been attributed to a relatively more strained *trans*-hydrindane—to the ring strain imposed on a cyclohexane ring when it is fused *trans* with a cyclopentane ring.

Heats of isomerization for *cis* ⇄ *trans* equilibration of liquid hydrindane were determined at 193–365°C (with a Pd–C catalyst) in 1960 by Allinger and Coke at Wayne State University: $\Delta H_{297°C} = -1.070 \pm 0.09$ kcal/mole and $\Delta S_{297°C} = -2.3 \pm 0.1$ eu.[68] Thus, at 25°C, $\Delta G° = -0.38$ kcal/mole, but at approximately 200°C, $\Delta G° \sim 0$. A smaller heat of isomerization ($\Delta H° = -0.58 \pm 0.05$ kcal/mole) was obtained at Princeton in 1963 by Blanchard and Schleyer, who isomerized liquid hydrindane at –22 to 47°C over $AlBr_3$.[69] These heats of isomerization match reasonably well with those from calorimetry,[65–67] and, as with the calorimetric measurements, they are significantly lower than those for *cis*-decalin ⇄ *trans*-decalin isomerization: $\Delta H_{313°C} = -2.72$ kcal/mole and $\Delta S_{313°C} = -0.55$ eu.[10]

As in the decalins, the presence of an angular methyl on the hydrindanes is expected to reduce the difference between *cis* and *trans* isomers. Sokolova and Petrov[70] reported that *cis*-8-methylhydrindane was favored by 92:8 over *trans* during isomerization using a platinum catalyst at 200°C. This result gives a value for $\Delta G°_{200°C}$ (2.3 kcal/mole) that is larger than the *A*-value for methyl (1.74). Molecular mechanics calculations[7] predict that the presence of the angular methyl inverts the usual order of stability (i.e., that *trans* is more stable than *cis*) found for the parent hydrindanes (Table 7-15), with *cis*-8-methylhydrindane being 1.82 kcal/mole (enthalpy) more stable than *trans*. There are two *cis*-8-methylhydrindane conformers: one with an axial methyl (steroid conformation), and one with an equatorial methyl (nonsteroid conformation), on the cyclohexane ring. The former is predicted to be approximately 0.4 kcal/mole more stable than the latter (Table 7-15). Apparently, there is no major advantage in *cis*-8-methylhydrindane for an equatorial methyl, and the large difference in free energy ($\Delta G° = 2.3$ kcal/mole)[70] found in the isomerization of *cis*- and *trans*-8-methylhydrindanes is due to factors other than the presence of an axial vs. an equatorial angular methyl.

As in *trans* decalin, the *trans* fused hydrindane is structurally rigid: Chair ⇄ chair cyclohexane conformational inversion is not possible. However, in the *cis*-hydrindane, as in *cis*-decalin, the six-membered ring conformational inversion occurs relatively easily, with a concomitant inversion in the second ring. The activation barriers for this

Table 7-15. Computed[a] Relative Heats of Formation ($\Delta\Delta H_f$, kcal/mole) of Hydrindanes

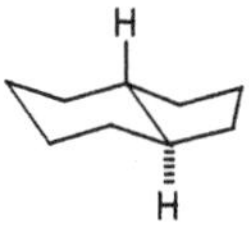	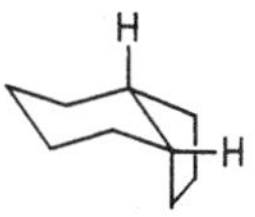	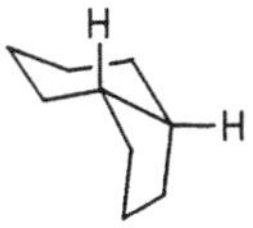	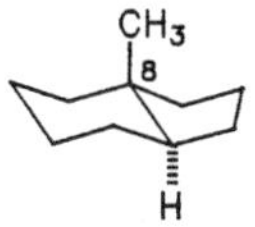	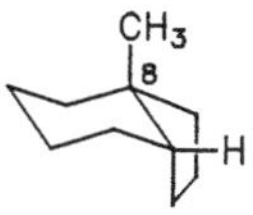	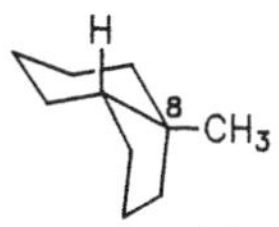
				steroid	nonsteroid
0.00	1.17[b]	1.17[b]	1.82[b]	0.00	0.43

[a] PCMODEL, ref. 7.

[b] $\Delta\Delta H_f$ (*cis-trans*) = 0.66, ref. 6.

process between the two isoenergetic conformers of *cis*-hydrindane were determined by Schneider and Nguyen-Ba in 1982 by dynamic ^{13}C-NMR spectroscopy: $\Delta H^{\ddagger} = 8.8 \pm 0.1$ kcal/mole and $\Delta S^{\ddagger} = 6.7 \pm 0.7$ eu.[71] Thus, at room temperature, $\Delta G^{\ddagger}_{298} = 6.9$ kcal/mole, which is only about one-half the value for *cis*-decalin (Table 7-4). This lower value of $\Delta G^{\ddagger}$ is apparently due to flattening in the cyclohexane ring to accommodate *cis* fusion with the cyclopentane ring. The activation energies just noted for the conformational inversion in *cis*-hydrindane[71] are somewhat greater than those determined earlier by Lack and Roberts,[39] who used dynamic ^{19}F-NMR to find an enthalpy of activation of $\Delta H^{\ddagger} = 6.5$ kcal/mole (and $\Delta S^{\ddagger} = -6$ eu) for the steroid → nonsteroid conformational change in 5,5-difluoro-*cis*-hydrindane and $\Delta H^{\ddagger} = 5.8$ kcal/mole (and $\Delta S^{\ddagger} = -7$ eu) for the nonsteroid → steroid change. The reasons for the lower $\Delta H^{\ddagger}$ values and the negative $\Delta S^{\ddagger}$ values here are not clear. The presence of an angular methyl group has been shown to increase $\Delta H^{\ddagger}$ and reduce the magnitude of $\Delta S^{\ddagger}$ to near zero. Thus, 7,7-difluoro-8-methyl-*cis*-hydrindane was shown[39] to have $\Delta H^{\ddagger} = 7.8$ kcal/mole (and $\Delta S^{\ddagger} = 0.0$ eu) for the steroid → nonsteroid conformational change and $\Delta H^{\ddagger} = 7.3$ kcal/mole (and $\Delta S^{\ddagger} = +0.5$ eu) for the nonsteroid → steroid conformational change. The energy difference $\Delta\Delta H^{\ddagger} = 0.5$ kcal/mole is quite close to the value of $\Delta\Delta H_f$ (0.43 kcal/mole) between the steroid and nonsteroid isomers of 8-methyl-*cis*-hydrindane, as determined by molecular mechanics calculations (Table 7-15). In each case, the steroid conformer is more stable.

7.4. Hydrindanones

There are four possible regioisomeric hydrindanones, in addition to *cis* and *trans* configurational isomers (Figure 7-14). In contrast, decalones have only two regioisomers, which have been studied extensively. Perhaps surprisingly, there have been fewer studies of hydrindanone stereochemistry. There are no gas-phase electron diffraction studies of hydrindanone structure; most information on the conformational stability and molecular structure of the compound comes from isomerization studies and molecular mechanics calculations. In the earliest work (1926), Hückel and Friedrich[61] reported that *trans*-2-hydrindanone had a smaller heat of combustion than *cis* by about 3.6 kcal/mole. Subsequently, Hückel et al.[62] showed that the energy difference was closer to 1.0 kcal/mole (Table 7-16). More recently, Sellers[72] demonstrated that *trans*-2-hydrindanone is more stable than *cis* on the basis of ΔH_f° values (Table 7-16). These data indicate that the presence of the 2-keto group can have a significant

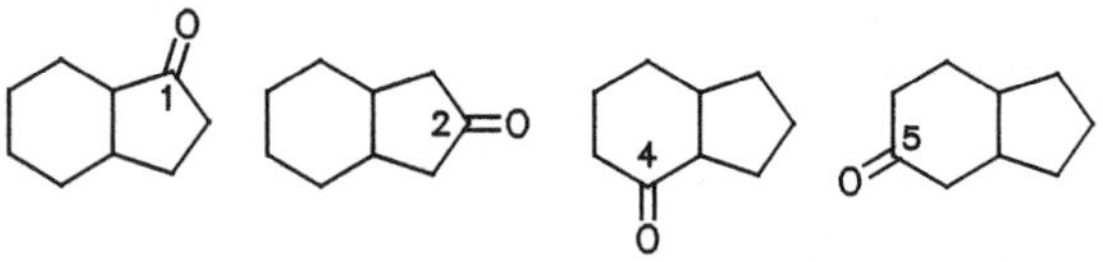

Figure 7-14. The four regioisomeric hydrindanones.

Table 7-16. Heats of Combustion and ΔH_f° (kcal/mole) of *trans*- and *cis*-2-Hydrindanone and Their 8-Methyl Analogs

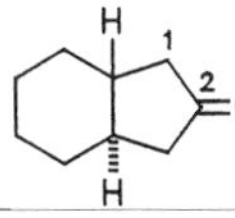	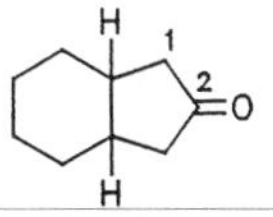	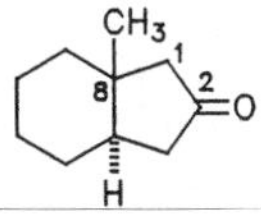	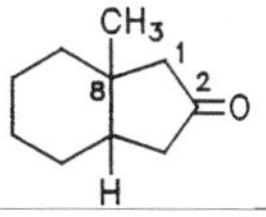
1243.9[a]	1244.9[b]	1403.8 ± 0.8[c]	1407.3 ± 0.5[c]
Δ (*cis* − *trans*) = 1.0[c]		Δ (*cis* − *trans*) = 3.5	
		$\Delta H_f^\circ = -79.71 \pm 0.53$ (l)[c]	$\Delta H_f^\circ = -83.15 \pm 0.81$ (l)[c]
$\Delta H_f^\circ = -59.7 \pm 0.4$[c,d]	$\Delta H_f^\circ = -59.3 \pm 0.4$[c,d]	$\Delta H_f^\circ = -65.77 \pm 0.55$ (g)[c]	$\Delta H_f^\circ = -68.59 \pm 0.81$ (g)[c]

[a] Average of 1243.8 and 1244.0, values from ref. 71; average replaces 1241.3 (value reported in ref. 61).

[b] Reported in ref. 61 and in Linstead, R. P., *Ann. Repts. Chem. Soc.* (1935), 305–330; see esp. p. 314.

[c] Values from ref. 72.

[d] Reported in ref. 37.

influence on the relative stability ($\Delta\Delta H_f^\circ$) of *cis*- and *trans*-hydrindane (Table 7-14). In the hydrocarbon, the *trans* isomer is more stable by $\Delta\Delta H_f^\circ \sim 1$ kcal/mole; in the 2-ketone, it is more stable by only about half that value. As in the parent hydrindanes (Table 7-15), the presence of an angular methyl dramatically inverts the order of stability, rendering *cis*-2-hydrindanone more stable than its *trans* counterpart (by 2.82 ± 0.91 kcal/mole)[72] (Table 7-16).

Other hydrindanones have been less well studied. Acid- or base-catalyzed epimerization of 1-hydrindanone indicates that the *cis* isomer is more stable than the *trans*.[73] Equilibration studies in triethylamine at 100°C (in a sealed tube) indicated that the *cis* isomer is more stable than the *trans* by $\Delta G_{100°C} = -0.82$ kcal/mole (Table 7-17).[74] The enthalpy difference ($\Delta\Delta H_f = -0.82$ kcal/mole), from molecular mechanics calculations,[32] is in good agreement with the experimental free-energy value. In 2-hydrindanones, the enthalpy difference is smaller ($\Delta\Delta H_f \sim 0.4$ kcal/mole) and favors the *trans* isomer as being the more stable (Table 7-17). This conclusion, opposite to that concerning 1-hydrindanones, is supported by molecular mechanics calculations.[32] PCMODEL calculations[7] indicate that *trans*-1-hydrindanone and *trans*-2-hydrindanone are more stable than the corresponding *cis* conformers (Table 7-18) and that the steroid conformation of *cis*-1-hydrindanone is more stable than the nonsteroid.

There is no available information on the influence of an angular methyl on the energy difference between *cis* and *trans* isomers of 8-methyl-1-hydrindanone and 8-methyl-3-hydrindanone, but molecular mechanics calculations predict a large enthalpy difference, with the *cis* isomer being favored by 2–3 kcal/mole.[32] This energy difference is similar to that found in 8-methyl-2-hydrindanone by experiment[72] and calculation.[32,75,76] The $\Delta\Delta H_f$ value (−2.35 kcal/mole) computed by the MM3 force field[76] for 8-methyl-2-hydrindanone is essentially the same as that from a modified MM2 force field.[75]

Hydrindanones with the carbonyl group in the six-membered ring have been investigated to a lesser extent. *cis*-4-Hydrindanone was reported to be more stable than the *trans* isomer in equilibration studies[62] in which mainly the *cis* isomer was found.[77]

Table 7-17. Comparison of Experimental[a] and Computed[b] Relative Enthalpies of Formation ($\Delta\Delta H_f$ (*cis* – *trans*), kcal/mole) for Hydrindanones[c]

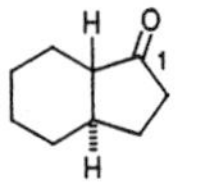	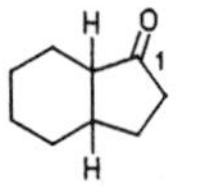	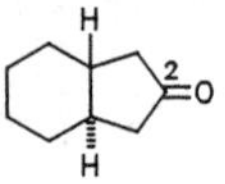	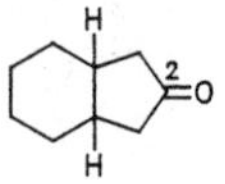	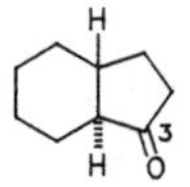	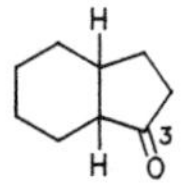
	−0.82[a,d] −0.82[b,e]		+0.4[a,f] −0.67[b]		−0.82[a,d] −0.10[b,i]
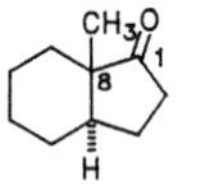	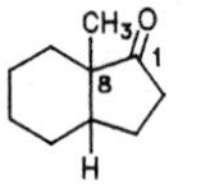	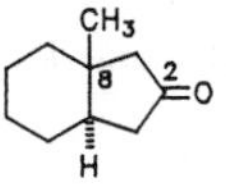	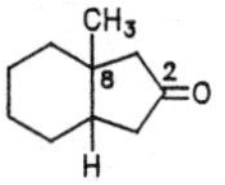	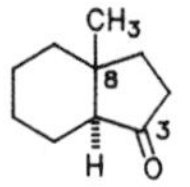	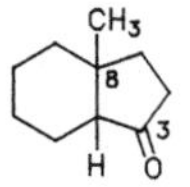
	— −2.35[b,e] — —		−2.82[a,f] −2.10[b] −2.34[g] −2.35[h]		— −2.73[b,i] — —

[a] Experimental value is top entry.
[b] Values from ref. 32.
[c] Numerical data under the more stable isomer.
[d] ΔG (100°C) from ref. 74.
[e] Carbonyl next to axial bridgehead hydrogen.
[f] Ref. 72.
[g] MM2 value, ref. 75.
[h] MM3 value, ref. 76.
[i] Carbonyl next to equatorial bridgehead hydrogen.

Table 7-18. Computed[a] Relative Heats of Formation ($\Delta\Delta H_f$, kcal/mole) of Hydrindanones

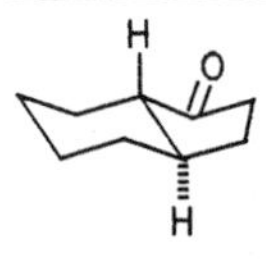	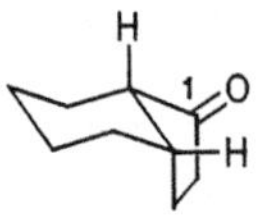	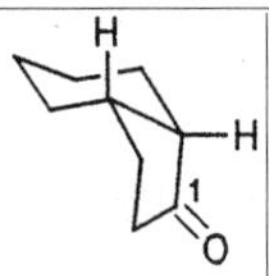	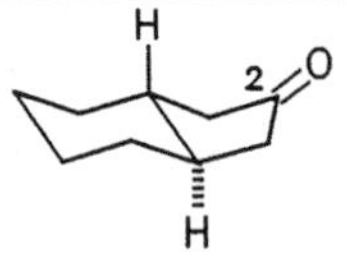	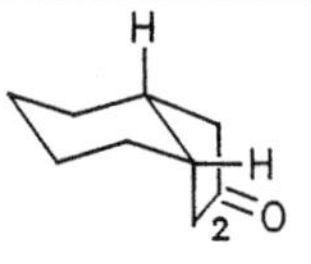	
0.00 (0.70)[b]	0.33	0.91	0.00 (0.00)[b]	0.84	
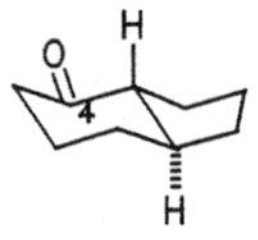	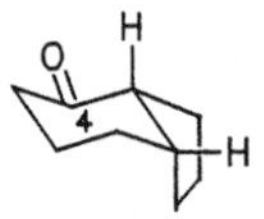	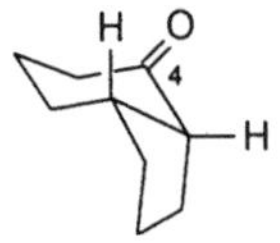	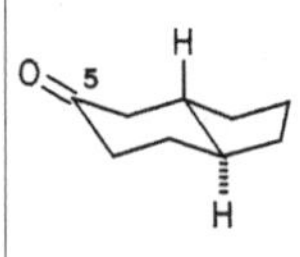	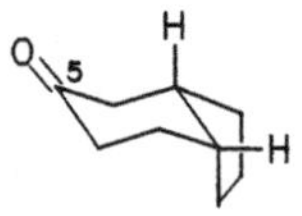	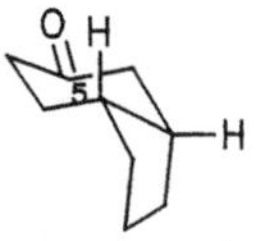
0.00 (3.31)[b]	0.91	1.41	0.00 (3.04)[b]	1.42	0.96

[a] PCMODEL, ref. 7.
[b] Relative to hydrindan-2-one, which is more stable than hydrindan-l-one, hydrindan-4-one, and hydrindan-5-one by 0.70, 3.31, and 3.04 kcal/mole, respectively.

More recently, 4-hydrindanone was observed to give a 76:24 mixture of *cis*:*trans* isomers following equilibration at reflux in tetrahydrofuran–water–ethanol (10:2:2, by vol.) with sodium hydroxide as a catalyst.[78] In this reaction, $\Delta G_{100°C} \sim -0.76$ kcal/mole (*cis* ⇄ *trans*), a value rather similar to that found in 1-hydrindanone. Unexpectedly, molecular mechanics calculations[7] (Table 7-18) predict an enthalpy difference ($\Delta\Delta H_f$) of 0.91 kcal/mole favoring the *trans* isomer. The introduction of an angular methyl at C(8) leads to a 92:8 mixture of *cis*:*trans* isomers under the same equilibration conditions,[78] with $\Delta G_{100°C} \sim 1.6$ kcal/mole. Again, the presence of an angular methyl drives the equilibrium toward the *cis* isomer. No experimental evidence is available on the relative stability of the other 6-ring ketone, 5-hydrindanone. As with the 4-hydrindanone, molecular mechanics calculations[7] predict a greater stability for the *trans* isomer. Here, however, the nonsteroid *cis* conformer is predicted to be more stable than the steroid (Table 7-18), whereas in *cis*-4-hydrindanone, the steroid isomer is predicted to be more stable than the nonsteroid.

Table 7-19. Computed[a] Relative Heats of Formation ($\Delta\Delta H_f$, kcal/mole) of 8-Methylhydrindanones

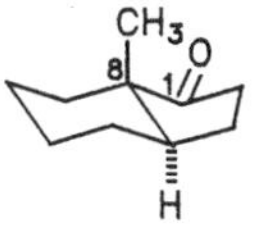	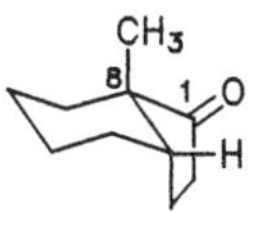	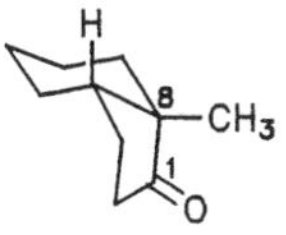	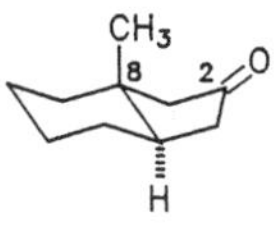	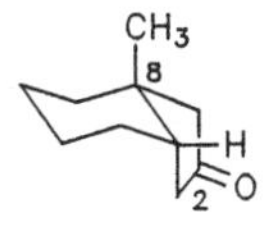	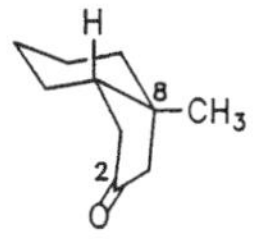
3.11	0.00 (0.18)[b]	0.92	2.33	0.00 (0.02)[b]	0.45
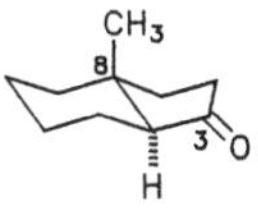	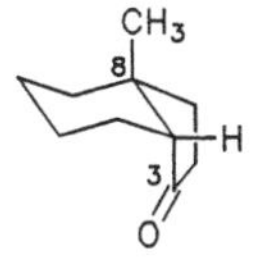	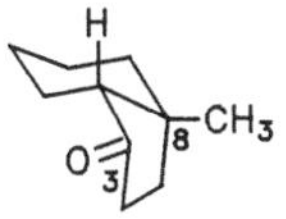	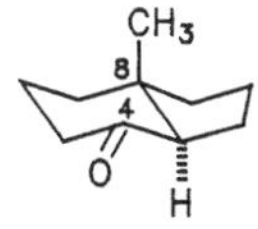	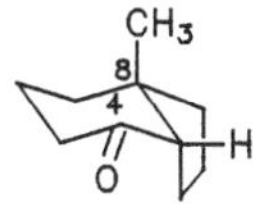	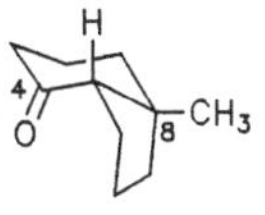
2.98	0.00 (0.00)[b]	1.36	0.76	0.00 (4.08)[b]	0.00 (4.08)[b]
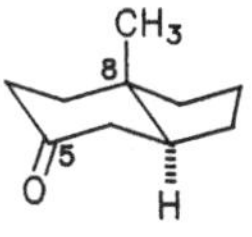	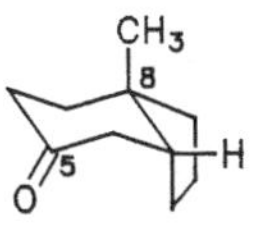	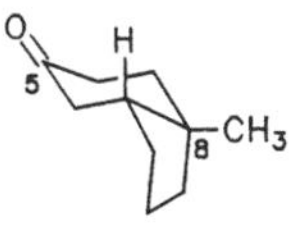	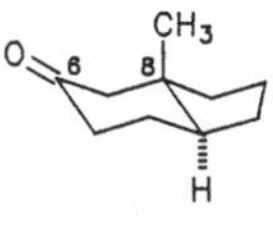	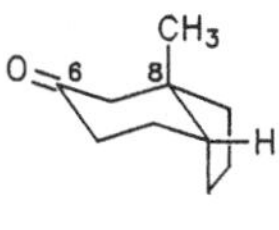	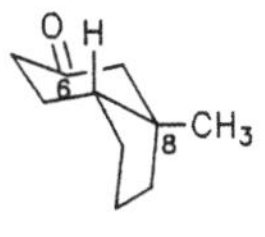
1.82	0.00 (3.10)[b]	1.16	1.19	0.00 (3.15)[b]	0.49
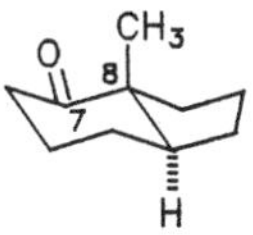	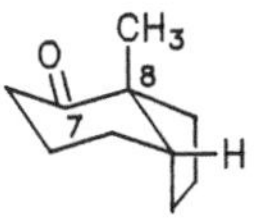	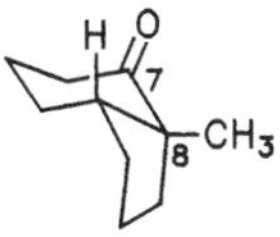			
2.04	0.00 (5.05)[b]	0.74			

[a] Values computed by PCMODEL molecular mechanics calculations (ref. 7); values are relative to the lowest energy isomer in a set of three.

[b] Relative to 8-methyl-*cis*-hydrindan-3-one in the steroid conformation.

The influence of an angular methyl on hydrindanone *cis/trans* and steroid/nonsteroid stability may be assessed by the molecular mechanics calculations[7] summarized in Table 7-19. For every regioisomer, the *cis* stereoisomer is predicted to be more stable than the *trans*. And in each *cis* regioisomer, except the 4-keto, the steroid conformer is predicted to be more stable than the nonsteroid. In the 8-methyl-*cis*-4-hydrindanone, the steroid and nonsteroid isomers are predicted to be isoenergetic. Unfortunately, there is no experimental evidence relating to the relative stabilities of the steroid and nonsteroid isomers of the *cis*-hydrindanones of Tables 7-18 and 7-19. Variable-temperature CD should be an excellent way to analyze and determine the energy differences. Dynamic NMR might offer a useful way to estimate conformational energies and determine activation energies for the steroid ⇄ nonsteroid conformational interconversion. However, such measurements have not yet been carried out.

The relatively few simple hydrindanones studied by ORD or CD spectroscopy have generally been limited to the conformationally less flexible *trans* isomers (Table 7-20),[79–85] and all of them follow the octant rule. The presence of an angular methyl

Table 7-20. Comparison of $n \rightarrow \pi^*$ Ketone Cotton Effect Signs and Magnitudes for Simple Hydrindanones and Their Analogs[a,b]

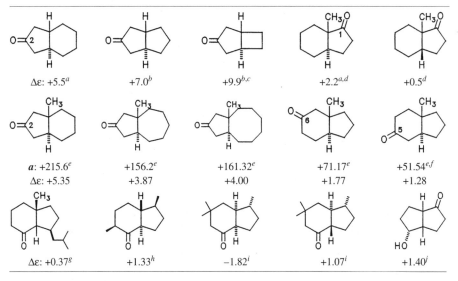

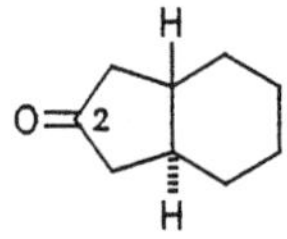	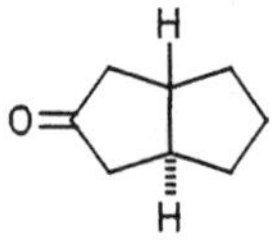	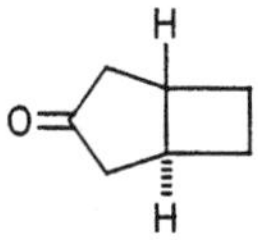	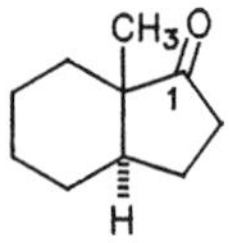	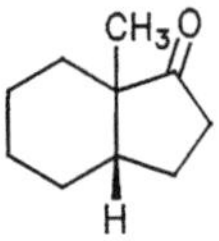
Δε: +5.5[a]	+7.0[b]	+9.9[b,c]	+2.2[a,d]	+0.5[d]
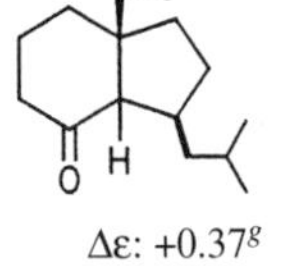	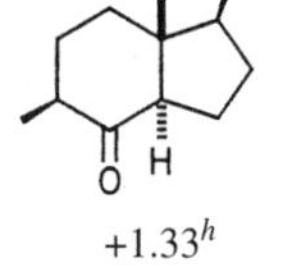	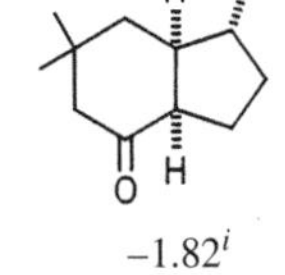	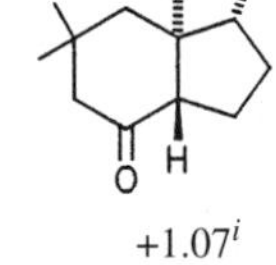	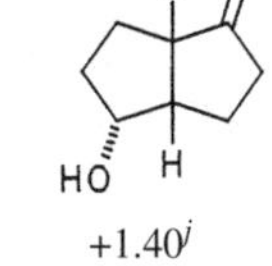
a: +215.6[e]	+156.2[e]	+161.32[e]	+71.17[e]	+51.54[e,f]
Δε: +5.35	+3.87	+4.00	+1.77	+1.28
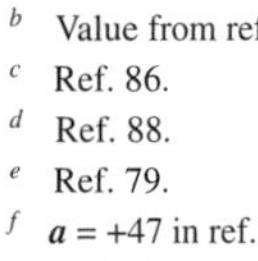	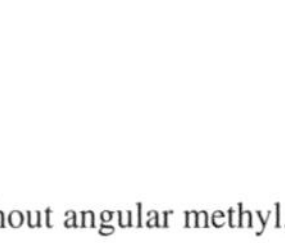			
Δε: +0.37[g]	+1.33[h]	−1.82[i]	+1.07[i]	+1.40[j]

[a] Value from ref. 80; hydrindanone numbering.
[b] Value from ref. 81.
[c] Ref. 86.
[d] Ref. 88.
[e] Ref. 79.
[f] ***a*** = +47 in ref. 88; without angular methyl, ***a*** = +53.
[g] Value from ref. 82.
[h] Value from ref. 83.
[i] Value from ref. 84.
[j] Value from ref. 85.

group on 2-hydrindanone makes only a very small negative contribution to the $n \rightarrow \pi^*$ Cotton effect. As in cyclopentanones (Chapter 6), the magnitude is dominated by the β-ring atom contributions from the cyclopentanone component (in 3(*R*)-methylcyclopentanone, $\Delta\varepsilon \sim +2.1$). This domination persists in hydrindanone analogs with larger[79] and smaller[81,86] rings. One of the largest CD Cotton effects observed for a simple ketone is that found in bicyclo[3.2.0]heptan-3-one ($\Delta\varepsilon = +9.9$),[86] and this has been analyzed and explained theoretically by Bouman at Southern Illinois University.[87] When the keto group is moved to C(1) in hydrindane, the $n \rightarrow \pi^*$ Cotton effect magnitude is reduced somewhat, presumably due to a change in conformation in the cyclopentanone ring.[80] For hydrindanones with the keto group in the six-membered ring, the Cotton effect magnitudes are reduced considerably over those with the keto group in the five-membered ring,[79] and, as might be expected, the values are comparable to those of the corresponding decalones. *cis*-Hydrindanones have not been investigated by CD spectroscopy; still, variable-temperature CD offers a useful way to analyze the conformations (steroid $\rightleftarrows$ nonsteroid) of these conformationally mobile systems.

7.5. Other Bicyclo[m.n.0]alkanones

There are very few reports of CD or ORD spectra of other simple unconjugated bicyclo[m.n.0]alkanones, except for [m.1.0]alkanones (cyclopropyl ketones), which are discussed in Chapter 11. A search of *Chemical Abstracts On-line* of the various ring systems of Table 7-21 indicated that no CD or ORD spectra had been obtained between 1967 and 1999 for the parent ketones or for parent ketones with angular methyl groups, except for the decalones ([4.4.0]) and hydrindanones ([4.3.0]) reported in Sections 7.2 and 7.4, respectively, and the perhydropentalenone ([3.3.0]) and bicyclo[3.2.0]heptanone of Table 7-20.

Most CD or ORD spectra of bicyclo[m.n.0]alkanones were reported prior to 1967 and include a wealth of data on decalones, hydrindanones, and perhydroazulenones. Conformational studies of the first two have been reported in this chapter; but, surprisingly, there are no corresponding ORD or CD data on the parent perhydroazulenones or parents with angular methyl groups. CD or ORD studies of perhydroazulenones have been concerned mainly with sesquiterpene natural products and their derivatives.[80] The only examples of chiroptical studies from the [6.4.0], [5.4.0], [6.3.0], and [5.3.0] series were reported in 1964 by Djerassi and Gurst at Stanford[79] in a detailed study of the influence of ring size on the ORD amplitudes of *trans*-bicyclo[m.n.0]alkanones, in which the ketone is in a five- or six-membered ring. There are no reports on the *cis* analogs. The data for cyclopentanones are summarized in Table 7-20; those for cyclohexanones may be found in Table 7-22. As the nonketone ring size decreases from eight to six membered, the intensity of the $n \rightarrow \pi^*$ Cotton effect was generally found to increase. No low-temperature CD measurements were carried out to help provide an explanation for the observed increase. With five- or six-membered rings fused *trans*, the systems are stereochemically rigid, and no

Table 7-21. Bicyclo[m.n.0]alkane Ring Systems Searched for CD or ORD Data on Parent Ketones and Parents with Angular Methyl Groups

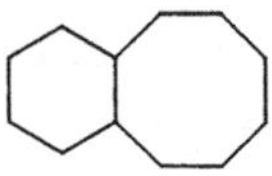	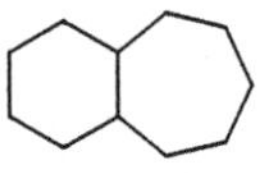	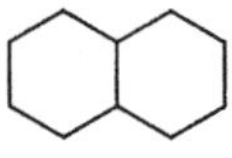	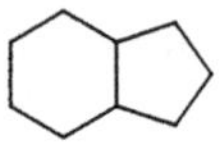	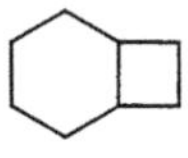
[6.4.0][a]	[5.4.0][a]	[4.4.0][b]	[4.3.0][c]	[4.2.0][a]
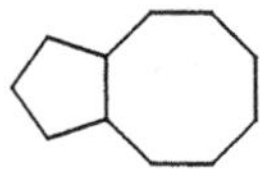	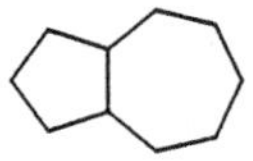	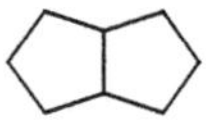	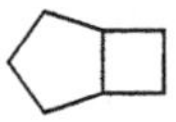	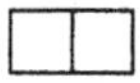
[6.3.0][d]	[5.3.0][d]	[3.3.0][d]	[3.2.0][d]	[2.2.0][a]

[a] No CD or ORD.
[b] See Section 7.2.
[c] See Section 7.4.
[d] See Table 7-20.

significant changes in low-temperature CD are expected; however, with conformationally mobile seven- and eight-membered rings, low-temperature CD measurements can be expected to limit the number of different conformations contributing to the $n \rightarrow \pi^*$ Cotton effects.

There have been no reports of CD or ORD data on the parent bicyclo[6.4.0], [5.4.0], [6.3.0], and [5.3.0]alkanones devoid of angular methyls. And save for the two examples in Table 7-20, there have also been no reports of CD or ORD of bicycloalkanones from the [4.2.0], [3.3.0], [3.2.0], and [2.2.0] systems. Collectively, except possibly for the already well-studied decalones and hydrindanones, the bicyclic systems of Table 7-21 offer attractive targets for future ketone stereochemical studies using circular dichroism spectroscopy.

Table 7-22. Comparison of $n \rightarrow \pi^*$ Ketone Cotton Magnitudes for Bicyclo[m.n.0]alkanones[a]

	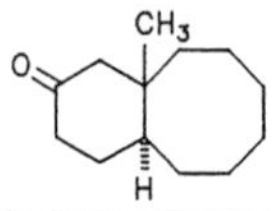	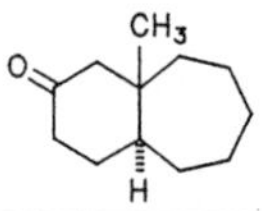	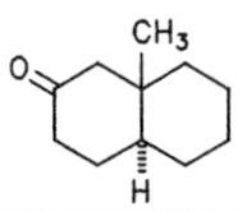	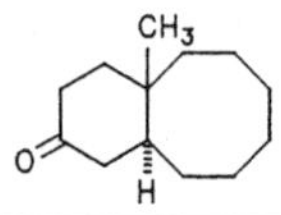	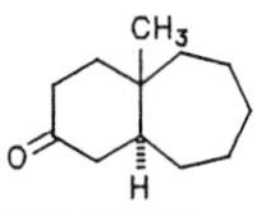	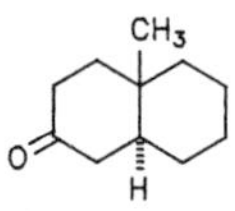
a	+47.6	+59.1	+71.2	+33.2	+41.1	+51.5
$\Delta\varepsilon$:	+1.18	+1.47	+1.77	+0.82	+1.02	+1.29

[a]Data from ref. 79.

7.6. Spiro[m.n]alkanones

Except for spiro[m.2]alkanones, which have a cyclopropane ring and are discussed in Chapter 11, few simple spiroalkanones have been investigated by ORD or CD spectroscopy. Most investigations have focused on spiro[4.4]nonanones. The parent spiro[4.4]nonanone is nominally achiral. However, when a methyl group is present, as in *cis*- and *trans*-6-methylspiro[4.4]nonan-1-ones (Figure 7-15), the ketone may exhibit optical activity.[89] The CD spectra (Figure 7-15) of these epimeric ketones were determined at room temperature and at –160°C or –165°C. Both ketones exhibit

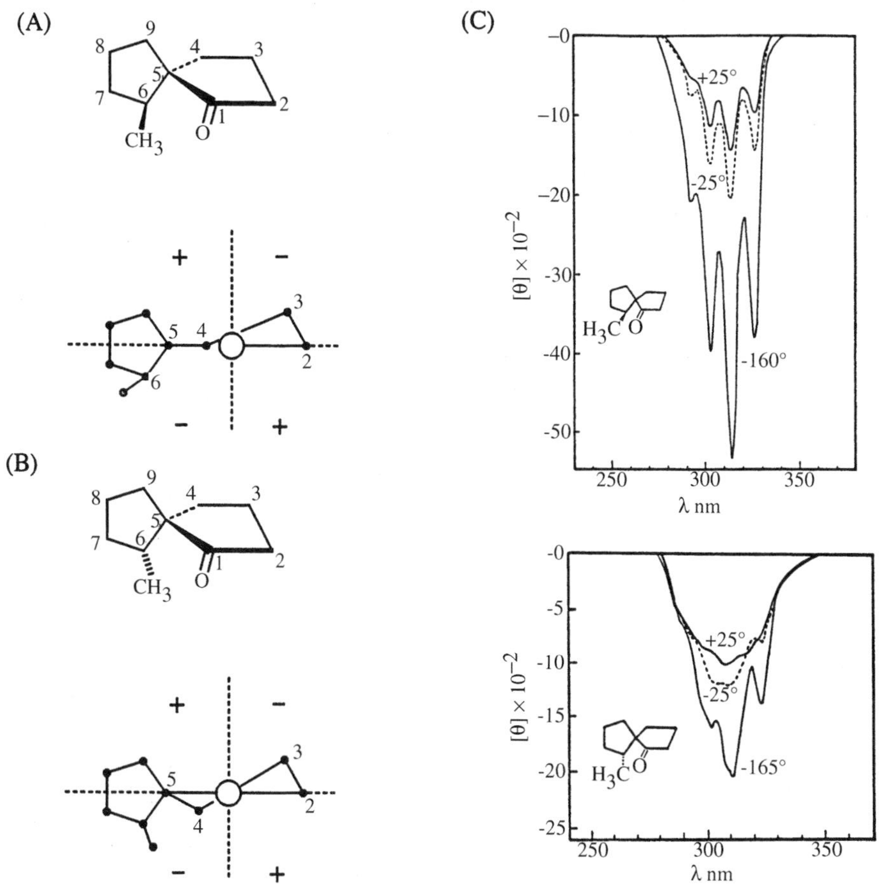

Figure 7-15. (A) (–)-*cis*-6-methylspiro[4.4]nonan-1-one and its octant diagram. (B) (–)-*trans*-6-methylspiro[4.4]nonan-1-one and its octant diagram. (C) Variable-temperature CD spectra of the *cis* (upper) and *trans* ketones in methylcyclohexane–isopentane (1:5 by vol.). Temperatures are given on the curves. The reduced rotatory strengths [*R*] for the ketone are, for *cis*, –1.0 at 25°C and –2.8 at –165°C, and for *trans*, –1.2 at 25°C and –1.7 at –165°C. [(C) is reprinted from ref. 89, copyright © 1972, with permission from Elsevier Science.]

negative $n \rightarrow \pi^*$ Cotton effects, as predicted by octant diagrams for the most stable conformers determined by PCMODEL[7] molecular mechanics calculations. In both compounds, to a first approximation, the cyclopentane ring atoms would appear to make no octant contributions. The main octant perturbers are from cyclopentanone ring atoms 3 and 4, which make negative contributions, and from the methyl groups. In the *trans* isomer, the methyl group lies in or near the curved third nodal surface, making little or no octant contribution. In the *cis* isomer, the methyl projects toward oxygen, but apparently falls behind the curved third nodal surface and thus into a negative back octant. The orientation of the methyl with respect to the carbonyl group is interesting. In the original octant rule (Chapter 4), according to which the third nodal surface is assumed to be a plane, the *cis* methyl lies in a front octant. But in the revised octant rule, which assumes that the third nodal surface is curved, the *cis* methyl lies behind it and in a back octant. The net result is a more negative $n \rightarrow \pi^*$ CD Cotton effect for the *cis* isomer (Figure 7-15C).

Somewhat more complicated spiroketones (Table 7-23) were used to provide a clear experimental proof of the existence of front octants in accordance with the octant rule (see Section 4.4),[90] and more recently the octant rule was used to assign the configurations of spiroketones with cyclobutanones connected (spiro) to steroids (Table 7-23).[91] In the latter, the signs and magnitudes of the ORD $n \rightarrow \pi^*$ Cotton effect were utilized to distinguish the two possible orientations of the spirocyclobutanone ketone

Table 7-23. Chiroptical Data for the $n \rightarrow \pi^*$ Transition of Spirocyclobutanones

	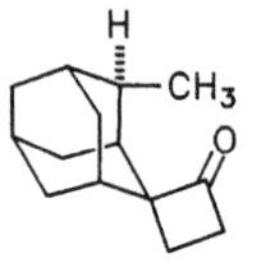	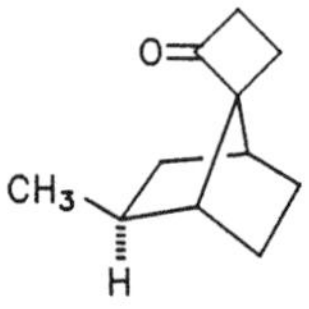	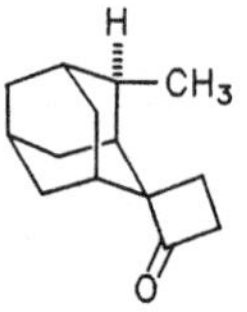
Δε:	-0.12^a	$+0.91^b$	-0.056^a

	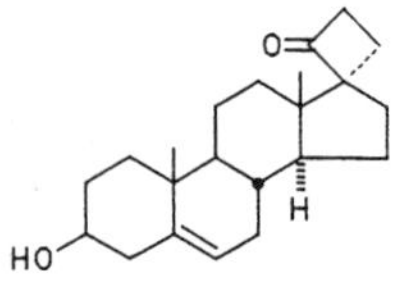	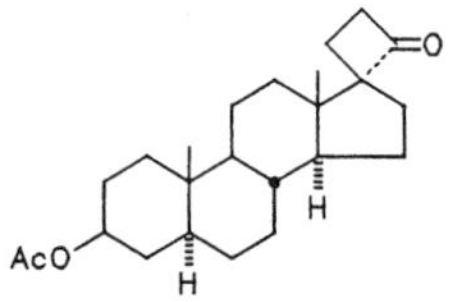	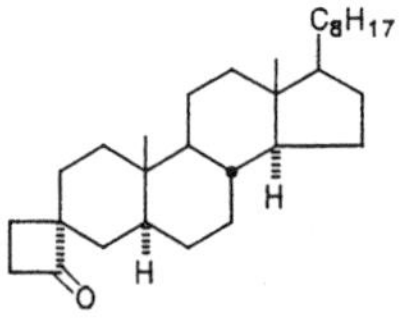	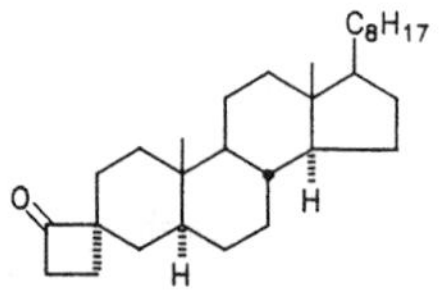
α:	$+124.59^c$	—	$+13.53^c$	$+24.06^c$
Δε:	+3.1	—	+0.33	+0.60
Prediction:	$(+)^d$	$(-)^d$	$w(+)^{c,d}$	$s(+)^{c,d}$

[a] In methylcyclohexane–isopentane (4:1, by vol.), ref. 90c.
[b] In isopentane, ref. 90b.
[c] In CH_2Cl_2, ref. 91.
[d] Using the octant rule.

with respect to the steroid unit. The assignments were confirmed by ^{1}H-NMR of the angular methyls.

7.7. Bridged Bicyclo[m.n.o]alkanones

Most of the bridged bicyclic ketones studied by ORD or CD spectroscopy belong to the bicyclo[2.2.1]heptane series and thus have fairly rigid carbocyclic skeletons. This was an advantage when one was interested in sorting out the origin of "antioctant" effects, defining the existence of front octants, and clarifying the location of the third nodal surface. (See Chapter 4.) It was also an advantage in revealing the angular dependence of the octant contribution of an α-methyl group in α-methylcyclohexanones. The symmetric bicyclo[2.2.1]heptan-7-one framework was especially useful in the former situation, and the positive $n \rightarrow \pi^*$ CD Cotton effect seen for the *exo*-methyl derivative (Table 7-24)[92] could arise only if the lone dissymmetric perturber lay in a front octant. The *exo* methyl was found to be an even clearer example of a front-octant perturber than a β-axial methyl perturber (adamantanone, Table 7-24) at room temperature. Consistent with these observations, an *exo*-chloro was also found to give an $n \rightarrow \pi^*$ Cotton effect opposite to that predicted by the original octant rule,[93] but in agreement with the chlorine lying in a front octant. In contrast, and as expected, *endo* (and β-equatorial) substituents project into a back octant and give the $n \rightarrow \pi^*$ Cotton effect signs predicted by the octant rule. When the size of the perturber is increased to that of ethyl or *n*-hexyl, the perturber projects even farther into a front (or back) octant, and the magnitude of the Cotton effect increases (Figure 7-16).

An unsuccessful attempt was made to kick the *exo* methyl group into a back octant by tilting the ketone carbonyl away, as would be expected in the symmetrical bicyclo[3.2.1]octan-8-one framework (Table 7-24).[92b] Although the one-carbon bridge

Table 7-24. Comparison of Reduced Rotatory Strengths[a] for Bicyclic Ketones and Adamantanones with Methyl Perturbers in Front (*Exo* or Axial) and Back (*Endo* or Equatorial) Octants

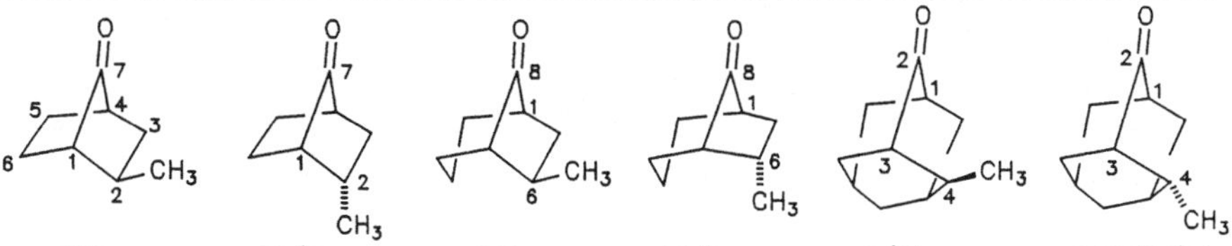

	exo	*endo*	*exo*	*endo*	axial	equatorial
$[R]^{25\pm2}$	+0.38	−1.9	+0.23[b]	−6.3[b]	−0.078	−1.5
$[R]^{-100\pm2}$	+0.50	−1.8	+0.46[b,c]	−6.1[b,c]	+0.44	−1.8
$[R]^{-175\pm2}$	+0.58	−1.9			+0.76	−1.9

[a] $[R]$ = rotatory strength × 1.08 × 10^{40}; superscript temperatures in °C, in methylcyclohexane–isopentane, 4:1 by vol.
[b] Pentane solvent. [c] −112°C; data from ref. 92.

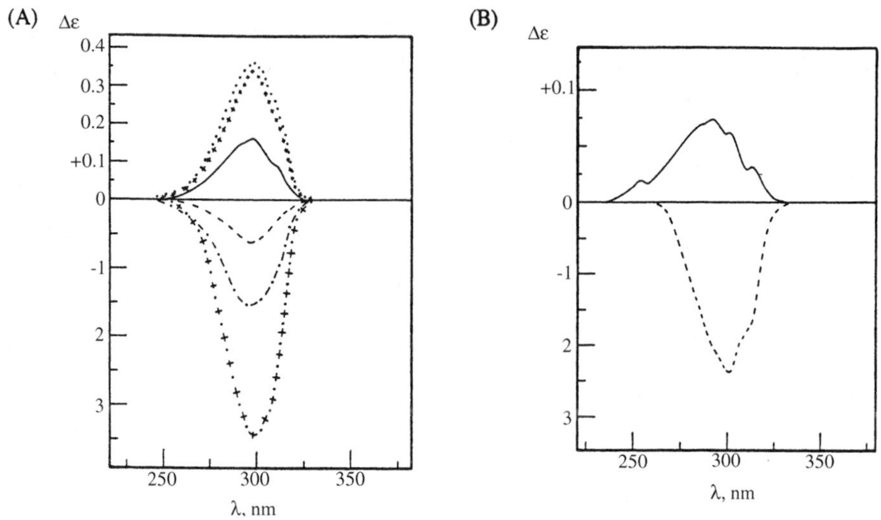

Figure 7-16. (A) Circular dichroism spectra of (1*S*,4*R*)-*exo*-2(*R*)-alkylbicyclo[2.2.1]heptan-7-ones: methyl (——), ethyl (· · ·), and *n*-hexyl (+ + +); and (1*S*,4*R*)-*endo*-2(*S*)-alkylbicyclo[2.2.1]heptan-7-ones: methyl (– - –), ethyl (– ·), and *n*-hexyl (+ ·), both compounds in methylcyclohexane–isopentane (4:1, v/v) at 25°C. (B) CD spectra of (1*R*,5*S*)-*exo*-6(*R*)-methylbicyclo[3.2.1]-octan-8-one (——) and (1*R*,5*S*)-*endo*-6(*S*)-methylbicyclo[3.2.1]octan-8-one (----) in pentane at 20°C. (Note the different scales for + and –Δε.) Sample concentrations in (A) and (B) were 0.01–0.001 M, and the data are corrected to 100% ee. [Reprinted with permission from ref. 92b. Copyright © 1985 American Chemical Society.]

tilts away from the two-carbon bridge, thus moving the *exo*-methyl perturber farther behind the carbonyl group, the perturber still lies in a front octant and gives an $n \rightarrow \pi^*$ Cotton effect opposite in sign to that found in the *endo* epimer, but quite comparable to that seen in *exo*-2-methylbicyclo[2.2.1]heptan-7-one. This study served to locate the curved third nodal octant surface in accordance with the octant rule (Chapter 4). However, the unexpectedly large Cotton effect (Table 7-24) found for *endo*-6-methylbicyclo[3.2.1]octan-7-one suggests ring distortion.

Other bicyclo[2.2.1]heptanones related to camphor with a ketone carbonyl at C(2) and methyl substitutions variously at C(1), (3), (4), (6), and (7) were first studied by Jacob, Ourisson, and Rassat in 1959 by ORD,[94] and a rationalization in terms of the octant rule was attempted by Klyne at Westfield College.[88a] Later, the compounds were examined by Coulombeau and Rassat in Grenoble,[95] using CD spectroscopy. These data, along with data obtained from many other cyclohexanones, decalones, and steroid ketones, led to Coulombeau and Rassat's formulation of a curved (convex) third nodal surface. This work stimulated the CNDO/S calculations[96] that led to the current octant rule (Chapter 4) and stimulated a comprehensive study by Kirk[81] to explain the CD of strained rings.

A report of octant-dissignate or "antioctant" contributions from the α-methyl groups of α and β-pinanones[97] led to a systematic investigation of the octant contributions of the α-methyl group on cyclohexanones.[98] In this study, important use was made of the symmetry and well-defined conformations of bicyclo[2.2.1]heptan-3-ones and bicyclo[3.2.1]octan-3-ones. (See Section 5.2.) Thus, from the conventional O=C–C_α–CH_3 torsion angles found in α-equatorial and α-axial methylcyclohexanones (–5° and +102°, respectively), a full range of torsion angles was found in the ketones (Table 5-7). The data showed that, for angles close to equatorial and axial, the octant rule is obeyed, but for bisected angles it is not. Ab initio calculations using random-phase approximation (RPA, Chapter 4) indicated that, as the conformation changes from equatorial to axial, changes also occur in the intensity mechanism and in the relative importance of "perturber" and framework couplings.[98]

These results may provide a way to understand the "antioctant" effects ascribed to α-axial oxygen-containing substituents found in bornane-2-one and 3-one derivatives and in selected steroids (Table 7-25).[99] In those examples, whether α-OH or α-OAc, the axial-like *endo* group makes an oppositely signed (dissignate) octant contribution (see ΔΔε) from that of ordinary alkyl perturbers; however, the equatorial-like *exo* group makes an octant-consignate contribution. Moreover, the CD absorption in the *endo* examples is bathochromically shifted, compared with the *exo* analog and the parent. This shift suggests that the α-oxygen *p*-orbitals and the *n* and π orbitals of the carbonyl chromophore interact electronically. Although bathochromically shifted, an *endo* ethyl

Table 7-25. Comparison of Bornane-2-one and Bornane-3-one $n \rightarrow \pi^*$ CD Cotton Effects for α-Hydroxy and α-Acetoxy Groups[a]

	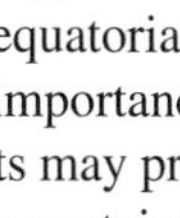	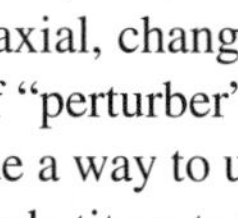	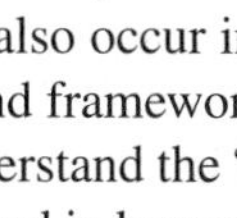	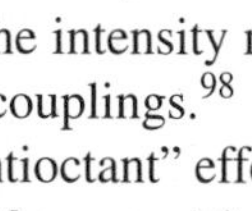	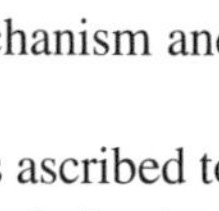
$\Delta\varepsilon(\lambda^{max})$:[b]	+1.50 (303)	+0.31 (320)	+1.85 (304)	+0.36 (325)	+2.18 (304)
ΔΔε:	0.00	−1.19 (*d*)[c]	+0.35 (*c*)[c]	−1.14 (*d*)[c]	+0.68 (*c*)[c]
	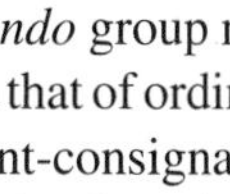	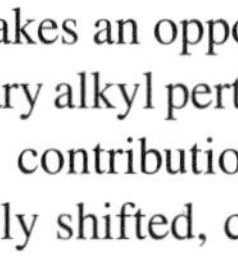	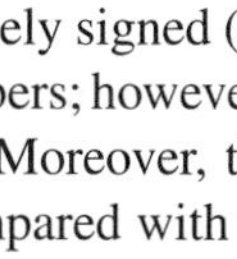	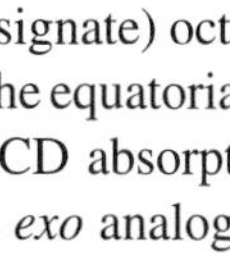	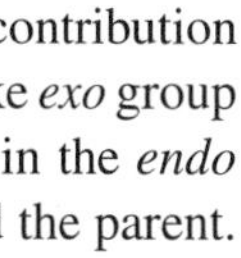
$\Delta\varepsilon(\lambda^{max})$:[b]	−1.46 (306)	−0.76 (317)	−2.07 (304)	−0.23 (327)	−2.34 (308)
ΔΔε:	0.00	+0.70 (*d*)[c]	−0.61 (*c*)[c]	+1.23 (*d*)[c]	−0.88 (*c*)[c]
	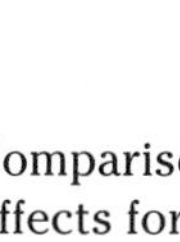	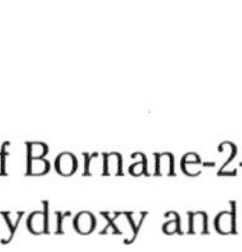	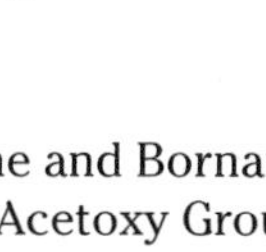	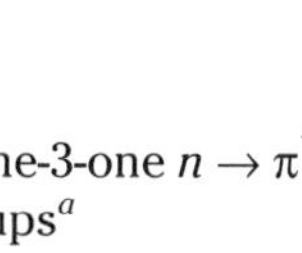	
$\Delta\varepsilon(\lambda^{max})$:[d]	+1.69 (302)	+1.90 (315)	+0.84 (313)	+0.71 (324)	+2.01 (310)
ΔΔε:	0.00	+0.21 (*c*)[c]	−0.85 (*d*)[c]	−0.98 (*d*)[c]	+0.31 (*c*)[c]

[a] In *n*-hexane. $\Delta\Delta\varepsilon = \Delta\varepsilon_{derivative} - \Delta\varepsilon_{parent\ ketone}$

[b] Data from ref. 99.

[c] *d* = octant dissignate; *c* = octant consignate.

[d] Data from ref. 100.

group is octant consignate, as is an *endo* α-amino group (Table 7-25).[100] However, an *endo* α-methylamino or dimethylamino makes an octant-dissignate contribution. *Endo*-α-amino groups also cause a bathochromic shift of the $n \rightarrow \pi^*$ CD band. Apparently, the relative orientation of the *p*-orbital(s) on *endo* α-N or O is of considerable importance in causing this shift and in determining whether the octant contribution will be dissignate or consignate.[101] Similar octant-dissignate contributions were found in 16-acetoxy-17-keto steroids and 17-acetoxy-16-keto steroids.[102]

Relatively few simple and flexible bicyclo[m.n.o]alkanones have been studied by CD spectroscopy.[81] Among bicyclo[2.2.2]octanones (Table 7-26), a negative $n \rightarrow \pi^*$ CD Cotton effect was reported for (1*R*,2*S*,4*S*)-2 carbomethoxybicyclo[2.2.2]octan-6-one, and a positive Cotton effect was found for (1*S*,2*S*,4*R*)-2-carboxylmethylbicyclo[2.2.2]octan-6-one[81,103] both in agreement with predictions of the octant rule. Klyne[88] suggested that (1*R*,5*S*)-bicyclo[3.2.1]octan-6-one should have a positive $n \rightarrow \pi^*$ Cotton effect, based on an analysis of natural product derivatives. This was subsequently shown to be correct,[104] although the measured ORD amplitude falls short of that calculated (Δε +2.3).[81] The isomeric (1*R*,5*R*)-bicyclo[3.2.1]octan-2-one (Table 7-26) of questioned[81] optical purity was found to exhibit a negative $n \rightarrow \pi^*$ Cotton effect with magnitude only about one-third of that calculated (Δε −2.0).[81] ORD Cotton

Table 7-26. Comparison of ORD and CD Data for the $n \rightarrow \pi^*$ Cotton Effects of Bicyclo[m.n.o]-alkanones

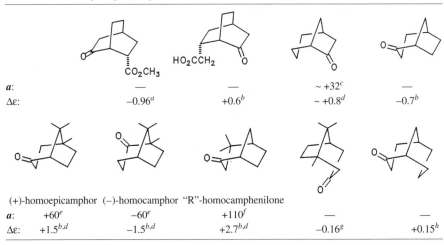

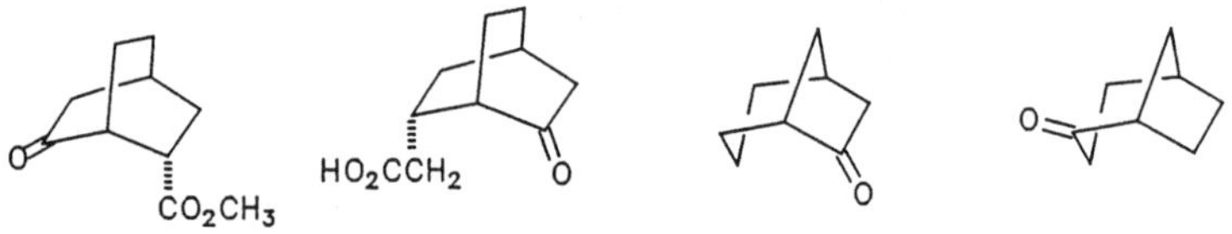

a:	—	—	~ +32[c]	—
Δε:	−0.96[a]	+0.6[b]	~ +0.8[d]	−0.7[b]

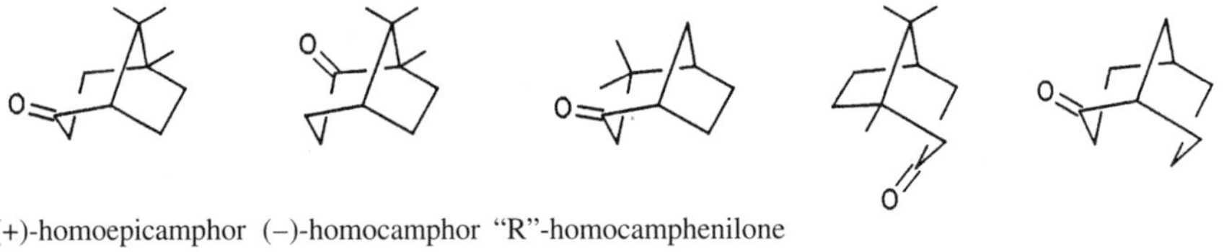

	(+)-homoepicamphor	(−)-homocamphor	"R"-homocamphenilone		
a:	+60[e]	−60[e]	+110[f]	—	—
Δε:	+1.5[b,d]	−1.5[b,d]	+2.7[b,d]	−0.16[g]	+0.15[h]

[a] Ref. 103.
[b] Ref. 81.
[c] Ref. 104.
[d] Calculated from ORD *a* value (÷ 40·28).
[e] From refs. 81 and 88a, estimated from incomplete ORD curves.
[f] Refs. 81 and 88a.
[g] Ref. 105.
[h] Ref. 107.

effects of methyl-substituted bicyclo[3.2.1]octan-2-ones have been measured (Table 7-26).[81,88a] but no further studies have been carried out by CD or variable-temperature CD spectroscopy.

In the more symmetric (+)-(1*S*,5*R*)-1,8,8-trimethylbicyclo[3.2.1]octan-3-one (Table 7-26), a weakly negative $n \rightarrow \pi^*$ CD Cotton effect was found in methanol,[105] as predicted by the octant rule. However, in hydrocarbon solvent, the compound exhibited a mainly positive CD at room temperature that became negative at lower temperatures (Figure 7-17). This behavior was interpreted in terms of a chair $\rightleftarrows$ boat equilibrium, which was later discounted by extensive molecular mechanics and ab initio molecular orbital calculations.[106] The boat isomer is much too high in energy to be present in a meaningful way, and the observed changes in the temperature-dependent CD spectra are probably due to solvation effects.

The simplest chiral bicyclo[3.3.1]nonanone (Table 7-26) shows a weak positive $n \rightarrow \pi^*$ CD Cotton effect,[107] as predicted by the octant rule. As with most other flexible bridged bicyclic ketones, no variable temperature CD studies have been carried out to aid in sorting out conformational equilibria.

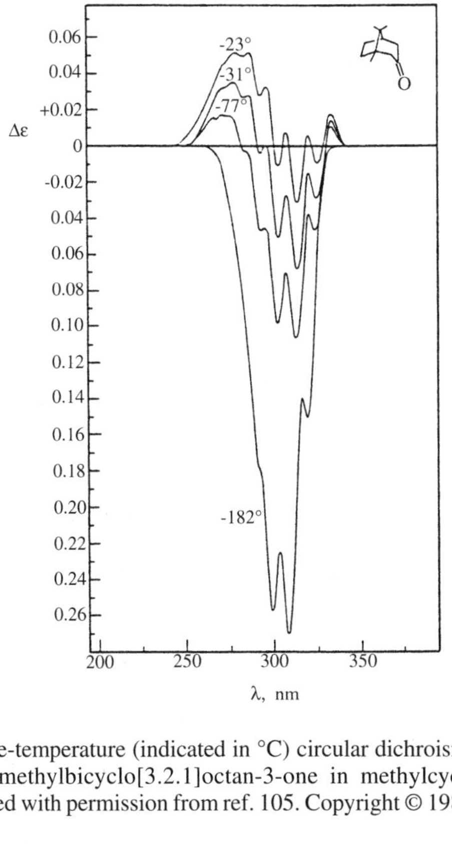

Figure 7-17. Variable-temperature (indicated in °C) circular dichroism spectra of 0.0050 M (+)-1(*S*),5(*R*)-1,8,8-trimethylbicyclo[3.2.1]octan-3-one in methylcyclohexane–isopentane (4:1, by vol.). [Reprinted with permission from ref. 105. Copyright © 1982 American Chemical Society.]

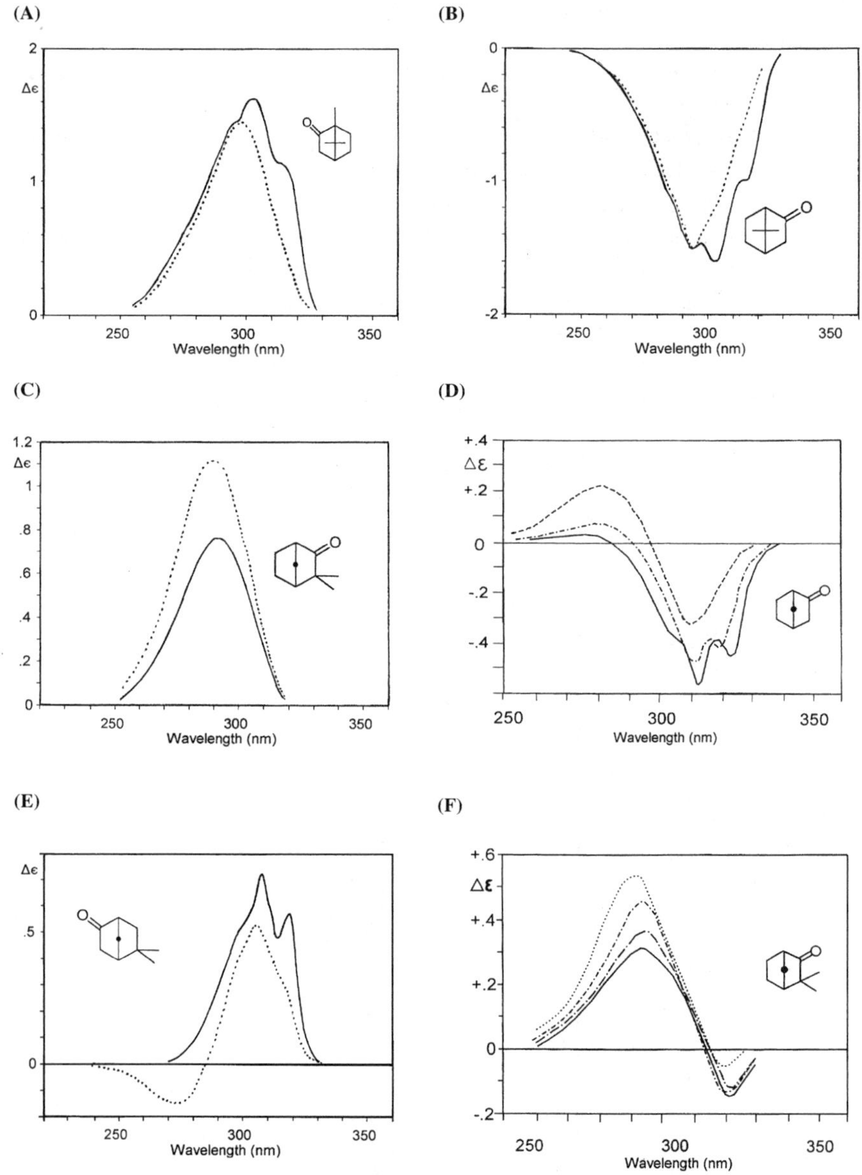

Figure 7-18. Solvent effects on the circular dichroism spectra of (A) (+)-camphor; (B) (–)-α-fenchocamphorone; (C) (+)-fenchone; (D) (+)-norcamphor; (E) (–)-β-fenchocamphorone; (F) (+)-camphenilone. (———) and ethanol (- - - -). In (D), methanol (· · · ·) and dioxane (-·-·-·). In (F), diethyl ether (—·—·) and trichloroethylene (- - · - - ·). [Redrawn from ref. 108a.]

Analyses of the CD of bicyclo[m.n.o]alkanones found in polycyclic systems have been summarized by Kirk.[81] With these compounds, an attempt was made to calculate and predict the sign, and especially the magnitude, of the ketone $n \rightarrow \pi^*$ Cotton effects based on C_α–H and C_α–C_β bond contributions; on the other hand, the special role of solvation effects was not taken into account.

Unusual solvation effects were first detected in studies of conformationally rigid bicyclo[2.2.1]heptanones by ORD and then CD spectroscopy. In 1966, Coulombeau and Rassat published a comprehensive ORD and CD study of various norcamphor derivatives in polar and nonpolar solvents (Figure 7-18).[108] Although the ketones generally obeyed the octant rule, an unusual phenomenon was noted: bisignate CD curves, attributed to solvation effects. In view of the structural rigidity of the bicyclic systems, such solvent-dependent CD could not be attributed to conformational changes. Rather, a solvation equilibrium had been proposed earlier[108,109] to account for the phenomenon. More recently, a theoretical approach has shown that hydrogen-bonding asymmetric solvation effects are small and that the observed solvent effects are due to the influence of solvent on (1) the electric transition moment and the angle between the electric and magnetic moment vectors and (2) vibronic coupling effects.[110]

References

1. Mohr, E., *J. Prakt. Chem.* **98** (2) (1918), 315–353.
2. (a) Hückel, W., *J. Liebigs Ann. Chem.* **441** (1925), 1–8.
   (b) Hückel, W., and Mentzel, R., *J. Liebigs Ann. Chem.* **441** (1925), 8–21.
   (c) Hückel, W., and Brinkman, E., *J. Liebigs Ann. Chem.* **441** (1925), 21–34.
   (d) Hückel, W., and Goth, E., *J. Liebigs Ann. Chem.* **441** (1925), 34–48.
3. (a) Bastiansen, O., and Hassel, O., *Nature* **157** (1946), 765.
   (b) Bastiansen, O., and Hassel, O., *Tids. Kjemi Bergvesen Met.* **6** (1946), 70–71.
4. Davis, M. I., and Hassel, O., *Acta Chem. Scand.* **18** (1964), 813–814.
5. Van den Enden, L., Geise, H. J., and Spelbas, A., *J. Mol. Struct.* **44** (1978), 177–185.
6. Allinger, N. L., Hirsch, J. A., Miller, M. A., Tyminski, I. J., and Van Catledge, F. A., *J. Am. Chem. Soc.* **90** (1968), 1199–1210.
7. PCMODEL versions 4.0–7.0, Serena Software, Inc., Bloomington, IN 47402–3076. PCMODEL uses the MMX modification of Allinger's MM2 force field.
8. Roth, W. A., and Lassé, R., *J. Liebigs Ann. Chem.* **441** (1925), 48–53.
9. Davies, G. F., and Gilbert, E. C., *J. Am. Chem. Soc.* **63** (1941), 1585.
10. Speros, D. M., and Rossini, R. D., *J. Phys. Chem.* **64** (1960), 1723–1727.
11. Allinger, N. L., and Coke, J. L., *J. Am. Chem. Soc.* **81** (1959), 4080–4082.
12. Barton, D. H. R., *J. Chem. Soc.* (1948), 340–342.
13. Allinger, N. L., Miller, M. A., VanCatledge, F. A., and Hirsch, J. A., *J. Am. Chem. Soc.* **89** (1967), 4345–4357.
14. (a) Geneste, P., and Lamaty, G., *Bull. Soc. Chim. France* (1964), 2439–2447.
    (b) Geneste, P., and Lamaty, G., *Tetrahedron Lett.* (1964), 3545–3550.
15. Gerig, J. T., and Roberts, J. D., *J. Am. Chem. Soc.* **88** (1966), 2791–2799.
16. Allinger, N. L., Tribble, M. T., Miller, M. A., and Wertz, D. H., *J. Am. Chem. Soc.* **93** (1971), 1637–1648.
17. Allinger, N. L., Yuh, Y. H., and Lii, J.-H., *J. Am. Chem. Soc.* **111** (1989), 8551–8566.
18. Baas, J. M. A., van de Graaf, B., Tavernier, D., and Vanhee, P., *J. Am. Chem. Soc.* **103** (1981), 5014–5021.

19. Chang, S., McNally, D., Shary-Therany, S., Hickey, M. J., Sr., and Boyd, R. H., *J. Am. Chem. Soc.* **92** (1970), 3109–3118.
20. Dauben, W. G., and Pitzer, K. S., in *Steric Effects in Organic Chemistry* (Newman, M. S., ed.), J. Wiley, New York, 1956.
21. Turner, R. B., *J. Am. Chem. Soc.* **74** (1952), 2118–2119.
22. Dauben, W. G., Rohr, O., Labbauf, A., and Rossini, F. D., *J. Phys. Chem.* **64** (1960), 283–284.
23. Allinger, N. L., and Coke, J. L., *J. Org. Chem.* **26** (1961), 2096–2097.
24. Jensen, F. R., and Beck, B. H., *Tetrahedron Lett.* (1966), 4523–4528.
25. Dalling, D. K., Grant, D. M., and Johnson, L. F., *J. Am. Chem. Soc.* **73** (1971), 3678–3682.
26. Mann, B. E., *J. Magn. Reson.* **21** (1976), 17–23.
27. Brown, L. M., Klinck, R. E., and Stothers, J. B., *Can. J. Chem.* **57** (1979), 803–806.
28. Allinger, N. L., Lane, G. A., and Wang, G. L., *J. Org. Chem.* **39** (1974), 704–708.
29. Heathcock, C. H., Ratcliffe, R., and Van, J., *J. Org. Chem.* **37** (1972), 1796–1807.
30. Zimmerman, H. E., and Mais, A., *J. Am. Chem. Soc.* **81** (1959), 3644–3651.
31. Hückel, W., Mentzel, R., Brinkmann, E., and Kamenz, I., *J. Liebigs Ann. Chem.* **451** (1927), 109–132.
32. Allinger, N. L., Tribble, M. T., and Miller, M. A., *Tetrahedron* **28** (1972), 1173–1190.
33. Burkert, U., and Allinger, N. L., *Molecular Mechanics* (ACS Monograph 177), American Chemical Society, Washington, DC, 1982.
34. Schubert, W., Schafer, L., and Pauli, G. H., *J. Mol. Struct.* **21** (1974), 53–60.
35. Pauli, G. H., Askari, M., Schubert, W., and Schäfer, L., *J. Mol. Struct.* **32** (1976), 145–152.
36. Askari, M., Pauli, G. H., Schubert, W., and Schäfer, L., *J. Mol. Struct.* **37** (1977), 275–281.
37. Abraham, R. J., Bergen, H. A., and Chadwick, D. J., *Tetrahedron* **38** (1982), 3271–3275.
38. Chadwick, D. J., and Dunitz, J. D., *J. Chem. Soc. Perkin Trans. 2* (1979), 276–284.
39. Lack, R. E., and Roberts, J. D., *J. Am. Chem. Soc.* **90** (1968), 6997–7001.
40. Lack, R. E., Ganter, C., and Roberts, J. D., *J. Am. Chem. Soc.* **90** (1968), 7001–7007.
41. Ross, A., Smith, P. A. S., and Dreiding, A. S., *J. Org. Chem.* **20** (1955), 905–908.
42. Rao, B., and Weiler, L., *Tetrahedron Lett.* (1971), 927–930.
43. Marshall, J. A., and Hochstetler, A. R., *J. Am. Chem. Soc.* **91** (1969), 648–657.
44. Sondheimer, F., and Rosenthal, D., *J. Am. Chem. Soc.* **80** (1958), 3995–4001.
45. Allinger, N. L., Hirsch, J. A., Miller, M. A., and Tyminski, I. J., *J. Am. Chem. Soc.* **91** (1969), 337–343.
46. Djerassi, C., and Marshall, D., *J. Am. Chem. Soc.* **80** (1958), 3986–3995.
47. Moffitt, W., Woodward, R. B., Moscowitz, A., Klyne, W., and Djerassi, C., *J. Am. Chem. Soc.* **83** (1961), 4013–4018.
48. Djerassi, C., and Staunton, J., *J. Am. Chem. Soc.* **83** (1961), 736–743.
49. Klyne, W., *Experientia* **12** (1956), 119–124.
50. Halsall, T. G., and Thomas, D. B., *J. Chem. Soc.* (1956), 2431–2443.
51. Djerassi, C., Burakevich, J., Chamberlin, J. W., Elad, D., Toda, T., and Stork, G., *J. Am. Chem. Soc.* **86** (1964), 465–471.
52. Dauben, W. G., Coates, R. M., Vietmeyer, N. D., Durham, L. J., and Djerassi, C., *Experientia* **21** (1965), 565–566.
53. Bhacca, N. S., and Williams, D. H., *Applications of NMR Spectroscopy in Organic Chemistry*, Holden-Day, Inc., San Francisco, 1964.
54. Zürcher, R. F., *Helv. Chim. Acta* **46** (1963), 2054–2088.
55. (a) Robinson, M. J. T., *Tetrahedron Lett.* (1965), 1685–1692.
(b) Elliott, D. R., Robinson, M. J. T., and Riddell, F. G., *Tetrahedron Lett.* (1965), 1693–1701.
56. Kirk, D. N., and Klyne, W., *J. Chem. Soc. Perkin Trans. 1* (1974), 1076–1103.
57. Ogura, H., Takayanagi, H., Harada, Y., and Iitaka, Y., *J. Chem. Soc. Perkin Trans. 1* (1979), 1142–1146.
58. Djerassi, C., and Klyne, W., *Proc. Natl. Acad. Sci. U.S.* **48** (1962), 1093–1098.
59. (a) Djerassi, C., Mauli, R., and Zalkow, L. H., *J. Am. Chem. Soc.* **81** (1959), 3424–3429.
(b) Zalkow, L. H., Shaligram, A. M., Hu, S-H., and Djerassi, C., *Tetrahedron* **22** (1966), 337–350.

(c) Kirk, D. N., *J. Chem. Soc. Perkin Trans. 1* (1976), 2171–2177.
60. Mason, S. F., Schofield, K., and Wells, J. R., *Proc. Chem. Soc.* (1963), 337.
61. Hückel, W., and Friedrich, H., *J. Liebigs Ann. Chem.* **451** (1926), 132–160.
62. Hückel, W., Sachs, M., Yantschulewitz, J., and Nerdel, F., *J. Liebigs Ann. Chem.* **518** (1935), 155–183.
63. Van den Enden, L., and Geise, H. J., *J. Mol. Struct.* **74** (1981), 309–320.
64. Maime, C., and Osawa, E., *Tetrahedron* **39** (1983), 2769–2778.
65. Hückel, W., *J. Liebigs Ann. Chem.* **533** (1938), 1–45.
66. Browne, C. C., and Rossini, F. D., *J. Phys. Chem.* **64** (1960), 927–931.
67. Finke, H. L., McCullough, J. P., Messerly, J. F., Osborn, A., and Douslin, D. R., *J. Chem. Thermodyn.* **4** (1972), 477–494.
68. Allinger, N. L., and Coke, J. L., *J. Am. Chem. Soc.* **82** (1960), 2553–2556.
69. Blanchard, K. R., and Schleyer, P. v. R., *J. Org. Chem.* **28** (1963), 247–248.
70. Sokolova, I. M., and Petrov, A. A., *Neftekhimiya* **17** (1977), 498–502, CA **87**: 184059b.
71. Schneider, H. J., and Nguyen-Ba, N., *Org. Magn. Reson.* **18** (1982), 38–41.
72. Sellers, P., *Acta Chem. Scand.* **24** (1970), 2453–2458.
73. Hückel, W., and Egerer, S., *J. Liebigs Ann. Chem.* **645** (1961), 162–176.
74. House, H. O., and Rasmussen, G. H., *J. Org. Chem.* **28** (1963), 31–38.
75. Bowen, J. P., Pathiaseril, A., Profeta, S., Jr., and Allinger, N. L., *J. Org. Chem.* **52** (1987), 5162–5166.
76. Allinger, N. L., Chen, K., Rahman, M., and Pathiaseril, A., *J. Am. Chem. Soc.* **133** (1991), 4505–4517.
77. Allinger, N. L., and Tribble, M. T., *Tetrahedron* **28** (1972), 1191–1202.
78. Lo Cicero, B., Weisbuch, F., and Dana, J., *J. Org. Chem.* **46** (1981), 914–919.
79. Djerassi, C., and Gurst, J. E., *J. Am. Chem. Soc.* **86** (1964), 1755–1761.
80. For leading references, see Crabbé, P., *Applications de la Dispersion Rotatoire Optique et du Dichroïsme Circulaire Optique en Chimie Organique*, Gauthier–Villars, Paris, 1968.
81. Kirk, D. N., *J. Chem. Soc. Perkin Trans. 1* (1977), 2122–2148.
82. Adinolfi, M., DeNapoli, L., DiBlasio, B., Iengo, A., Pedone, C., and Santacroce, C., *Tetrahedron Lett.* (1977), 2815–2818.
83. Tori, M., Sono, M., and Asakawa, Y., *J. Chem. Soc., Perkin Trans.* **1** (1990), 2849–2850.
84. Tori, M., Nakashima, K., Asakawa, Y., Connolly, J. D., Harrison, L. J., Rycroft, D. S., Singh, J., and Woods, N., *J. Chem. Soc., Perkin Trans. 1* (1995), 593–597.
85. Perard-Viret, J., and Rassat, A., *Tetrahedron Asymmetry* **5** (1994), 1–4.
86. Windhorst, J. C. A., *J. Chem. Soc. Chem. Commun.* (1976), 331–332.
87. Bouman, T. D., *J. Chem. Soc. Chem. Commun.* (1976), 665–666.
88. (a) Klyne, W., *Tetrahedron* **13** (1961), 29–47.
(b) Klyne, W., *Bull. Soc. Chim. France* (1960), 1396–1406.
89. Lightner, D. A., and Christiansen, G. D., *Tetrahedron Lett.* (1972), 883–886.
90. (a) Lightner, D. A., and Chang, T. C., *J. Am. Chem. Soc.* **96** (1974), 3015–3016.
(b) Lightner, D. A., *J. Chem. Soc. Chem. Commun.* (1974), 344–345.
(c) Lightner, D. A., Chang, T. C., Hefelfinger, D. T., Jackman, D. E., Wijekoon, W. M. D., and Givens, J. W., III., *J. Am. Chem. Soc.* **107** (1985), 7499–7508.
91. Trost, B. M., and Scudder, P. H., *J. Am. Chem. Soc.* **99** (1977), 7601–7610.
92. (a) Lightner, D. A., and Jackman, D. E., *J. Am.Chem. Soc.* **96** (1974), 1938–1939.
(b) Lightner, D. A., Crist, B. V., Kalyanam, N., May, L. M., and Jackman, D. E., *J. Org. Chem.* **50** (1985), 3867–3878.
93. McDonald, R. N., and Steppel, R. N., *J. Am. Chem. Soc.* **92** (1970), 5664–5670.
94. Jacob, G., Ourisson, G., and Rassat, A., *Bull. Soc. Chim. France* (1959), 1374–1377.
95. Coulombeau, C., and Rassat, A., *Bull. Soc. Chim. France* (1971), 516–526.
96. Bouman, T. D., and Lightner, D. A., *J. Am. Chem. Soc.* **98** (1976), 3145–3154.
97. Hirata, T., *Bull. Chem. Soc. Jpn.* **45** (1972), 3458–3464.
98. Lightner, D. A., Bouman, T. D., Crist, B. V., Rodgers, S. L., Knobeloch, M. A., and Jones, A. M., *J. Am. Chem. Soc.* **109** (1987), 6248–6259.

99. Bartlett, L., Kirk, D. N., Klyne, W., Wallis, S. R., Erdtman, H., and Thorén, S., *J. Chem. Soc. (C)* (1970), 2678–2682.
100. Beckett, A. H., Khokhar, A. Q., Powell, G. P., and Hudec, J., *J. Chem. Soc. Chem. Commun.* (1971), 326–327.
101. Levin, C. C., Hoffmann, R., Hehre, W. J., and Hudec, J., *J. Chem. Soc. Perkin Trans. 2* (1973), 210–220.
102. Djerassi, C., Fishman, J., and Nambara, K., *Experientia* **17** (1961), 565–566.
103. Varech, D., and Jacques, J., *Bull. Soc. Chim. France* (1972), 5671–5679.
104. Numata, A., Suzuki, T., Ohoro, K., and Uyeo, S., *Yakugaku Zasshi* **88** (1968), 1298–1305.
105. Crist, B. V., Rodgers, S. L., and Lightner, D. A., *J. Am. Chem. Soc.* **104** (1982), 6040–6045.
106. Jaime, C., Buda, A. B., and Osawa, E., *Tetrahedron Lett.* **25** (1984), 3883–3886.
107. Gerlach, H., *Helv. Chim. Acta* **61** (1978), 2773–2776.
108. (a) Coulombeau, C., and Rassat, A., *Bull. Soc. Chim. France* (1966), 3752–3762.
(b) Coulombeau, C., and Rassat, A., *Bull. Soc. Chim. France* (1963), 2673–2674.
(c) Gervais, H. P., and Rassat, A., *Bull. Soc. Chim. France* (1961), 743–747.
109. (a) Moscowitz, A., Wellman, K. M., and Djerassi, C., *Proc. Natl. Acad. Sci. (U.S.)* **50** (1963), 799–804.
(b) Moscowitz, A., in *Optical Rotatory Dispersion and Circular Dichroism in Organic Chemistry* (Snatzke, G., ed.), Heyden & Son, Ltd., London, 1967.
110. Ruiz-Lopez, M. F., Rinaldi, D., and Rivail, J. L., *Chem. Phys.* **110** (1986), 403–414.

CHAPTER

8

Saturated Polycyclic Ketones

Most ORD and CD studies of ketones have been carried out on polycyclics, particularly diterpenes, triterpenes and (especially) steroids.[1–11] The latter contain the perhydrophenanthrene nucleus (Figure 8-1), which, along with perhydroanthracene, has been the subject of considerable interest in conformational analysis.[12] However, very few simple perhydrophenanthrenones and perhydroanthracenones have been studied by ORD or CD spectroscopy. Those which have been examined belong to the more stable and least conformationally mobile *trans-anti-trans* perhydrophenanthrenone and *trans-syn-trans* perhydroanthracenone stereochemistry.

In perhydrophenanthrenes, there are six different isomers—four chiral and two meso—based on stereochemistry at the ring junctions (Table 8-1). These compounds were synthesized and their stereochemistry elucidated some 50 years ago by Linstead et al.;[13] the relative stabilities of the isomers have since been studied on a regular basis. Of these isomers, the *trans-anti-trans* found in many steroids and terpene natural products is most stable. In the 1950s, W. S. Johnson, then at Wisconsin, found it important to know the relative energies of the isomers for use in steroid total syntheses and calculated the energies on the basis of summations of gauche butane interactions.[14] Shortly after, Dauben and Pitzer, at Berkeley, found the same values for all but the *tst* and *csc* isomers (for notations, see Table 8-1), which were computed to be slightly

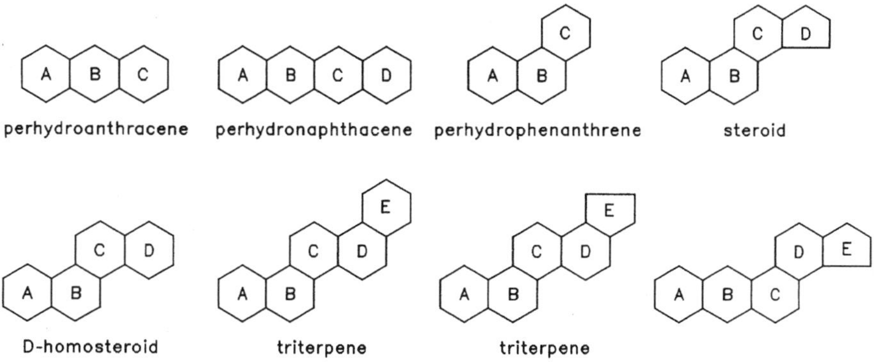

Figure 8-1. Ring systems found in numerous polycyclic ketones studied by ORD and CD spectroscopy.

Table 8-1. Stable Conformations and Relative Enthalpies of Formation, $\Delta\Delta H_f$ (kcal/mole), of Isomeric Perhydrophenanthrenes[a,b] and Perhydroanthracenes[a,b]

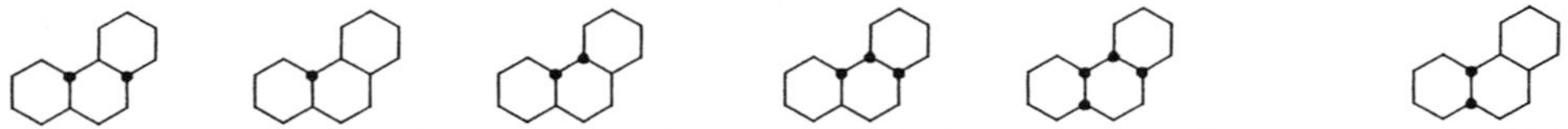

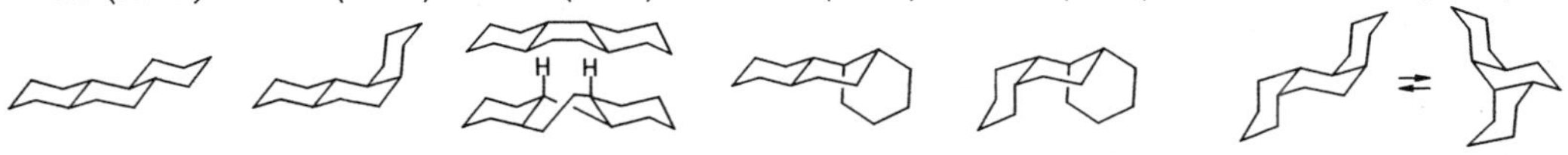

	tat (chiral)	*tac* (chiral)	*tst* (meso)	*tsc* (chiral)	*csc* (meso)	*cac* (chiral)
$\Delta\Delta H_f$ (expt):[c]	0.00	2.66 ± 0.22	8.98 ± 0.70	2.25 ± 0.22	7.43 ± 0.56	4.60 ± 0.36
$\Delta\Delta H_f$ (calc):	0.00	2.4,[d] 2.63,[e] 2.79[f]	6.4,[d] 8.04[f] (upper) 7.49,[e] 9.44[f] (lower)	2.4,[d] 2.53,[e] 2.81[f]	>8.2,[d] 8.46,[e] 10.09[f]	5.54,[e] 5.78[f]; 0.4[d] 4.53[f]; 4.27[e]

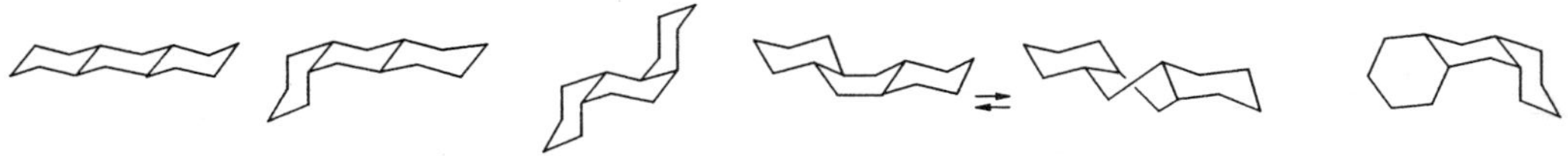

	tst (meso)	*cst* (chiral)	*cac* (meso)	*tat* (chiral)	*csc* (meso)
$\Delta\Delta H_f$ (expt):[g]	0.00	2.76 ± 0.28	5.58 ± 0.28	4.15 (5.02 ± 0.28)	8.74 ± 0.61
$\Delta\Delta H_f$ (calc):	0.00	2.4,[h] 2.62,[i] 2.83[f]	4.8,[h] 5.56,[i] 5.82[f]	≥5.6[h] 5.86,[i] 7.11[f]	≥9[h] 8.13,[i] 8.50[f]

[a] Perhydrophenanthrenes in upper half of table, perhydroanthracenes in lower half.

[b] • = H up; no • = H down; *t* = trans, *c* = cis, *a* = anti, *s* = syn.

[c] Values from ref. 16.

[d] Ref. 15 and 14.

[e] Value from MM2, ref. 16.

[f] Value from MM3, ref. 17.

[g] Value form ref. 18.

[h] Value from ref. 15.

[i] Value from MM1 molecular mechanics calculations, ref. 18.

higher (6.4 and >8.2 kcal/mole, respectively).[15] During the decades that followed, there appeared a succession of calculations,[16] culminating in the more recent molecular mechanics computations based on the MM2 and MM3 force fields (Table 8-1).[16,17] The calculated values (including those from additivity)[14,15] for the *tac, tsc*, and *cac* isomers are generally found to be in good agreement with the experimental values.[16] However, where the B-ring is not a chair (*tst*), the computed values match the experimental values less well, and in the *csc* isomer the match is poorest, especially when the more recent MM3 force field is employed (Table 8-1). The reasons for the mismatches are as yet unclear.

In perhydroanthracenes there are five different isomers—three meso forms and two chiral—based on stereochemistry at the ring junctions (Table 8-1). The experimentally determined relative energies[18] of the isomers clearly shows that the *trans-syn-trans* isomer is much more stable than all the others. The relative stabilities were predicted by Dauben and Pitzer,[15] and more recent molecular mechanics calculations[17,18] have given more refined values. One of these, however, an MM3 calculation for the *tat* isomer in which ring B adopts a twist-boat conformation, is rather far off, which suggests a need for further analysis.[17]

8.1. Tricyclic Ketones: Perhydrophenanthrenones and Perhydroanthracenones

Only a few simple perhydroanthracenones and perhydrophenanthrenones have been studied by CD spectroscopy (Figure 8-2).[19] Studies of the latter have been limited only to the *trans-anti-trans* stereochemistry, the former only to the *trans-syn-trans* (Table 8-2), and these compounds are found to obey the octant rule (Chapter 4). Note that in the perhydrophenanthren-1-one (steroid numbering; 4-keto in phenanthrene numbering), three of the ring-C carbons lie in a (negative) front octant. These carbons augment the negative back-octant contributions from ring B and lead to an $n \rightarrow \pi^*$ Cotton effect of large magnitude.

Other, substituted perhydrophenanthrenones were investigated earlier by ORD.[3] The $n \rightarrow \pi^*$ Cotton effect intensities, translated into $\Delta\varepsilon$ values from ORD amplitudes, are presented in Table 8-2. The observed Cotton effect signs are predicted by the octant rule. No unusual effects were found by changing the B/C ring fusion [cf. (D)–(G)], or by contracting ring C [cf. (A) and (I)]. However, flattening ring B by the introduction of a carbon–carbon double bond (H) produced a much-diminished Cotton effect compared with G. The ketones of Figure 8-2 and Table 8-2 cover most of the sites for locating a carbonyl group, but no other isomers due to changes in ring junction stereochemistry have been investigated by CD, thus leaving a potentially rich source for future research involving CD and variable-temperature CD.

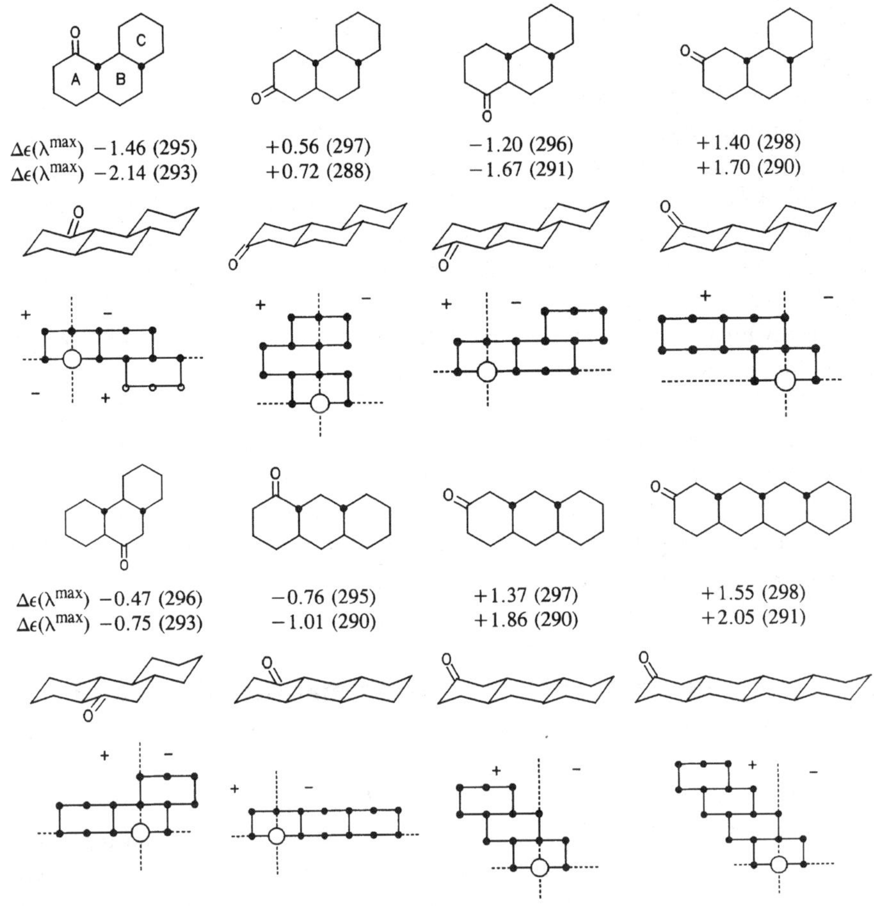

Figure 8-2. Perhydrophenanthrenones and perhydroanthracenones with $n \rightarrow \pi^*$ CD data (upper values: in *n*-hexane; lower value: in methanol), conformational drawings, and octant projection diagrams. Empty circles represent carbons in front octants. Data for mirror-image structures are from ref. 19.

8.2. Tetracyclic Ketones: Steroids

The perhydrophenanthrene skeleton is found in most steroids, and the *trans-anti-trans* and *trans-anti-cis* ring junctures are most common. Hundreds of ORD and CD spectra have been run on steroid ketones since the early days of ORD.[1–11] Those erstwhile studies formed the experimental basis of the octant rule[1–4] and the consignate–dissignate zigzag rules.[9] The work of Kirk et al.[9] contains extensive summaries of CD data on steroid and terpene ketones. In most of these studies, the carbocyclic skeleton either was fairly rigid (as in the all-*trans* stereochemistry examples) or possessed limited flexibility—important features when one is attempting to explain experimental data

Table 8-2. Perhydrophenanthrenones and their $n \rightarrow \pi^*$ Cotton Effects[a]

	(A)	(B)	(C)
	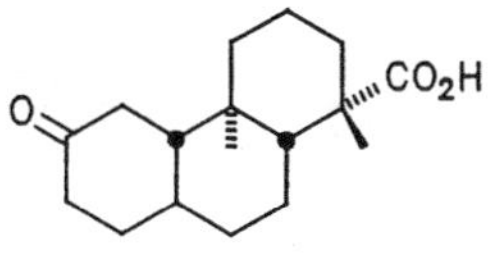	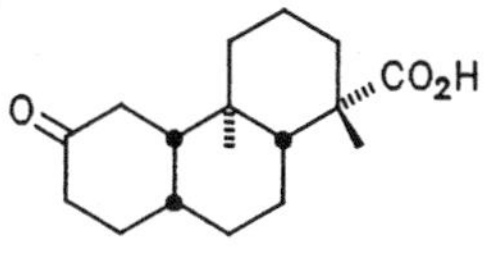	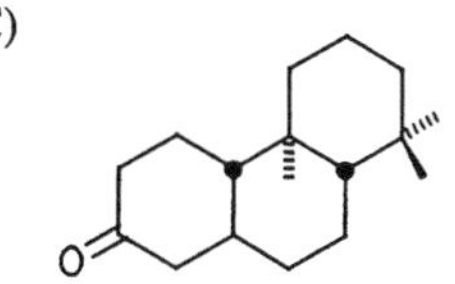
Δε:	+1.76	−0.17	+1.19
	(D)	**(E)**	**(F)**
	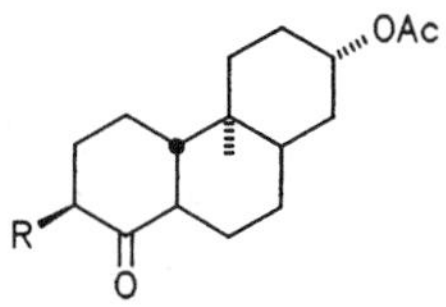	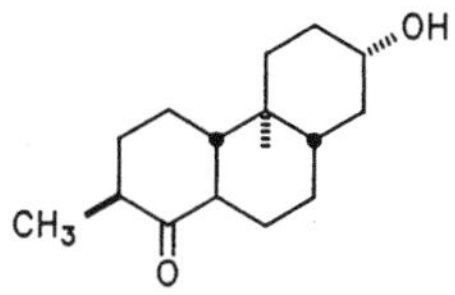	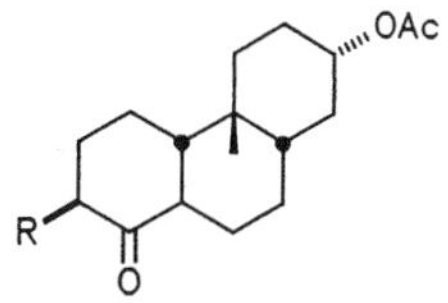
Δε:	−1.34 (R=CH_3)[b] −1.49 (R=H)	−1.42 (R=CH_3)[b]	−1.54 (R=H) −1.34 (R=CH_3)
	(G)	**(H)**	**(I)**
	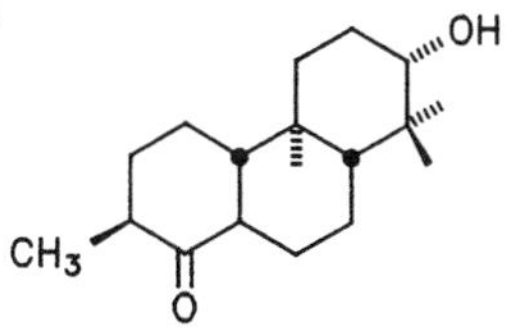	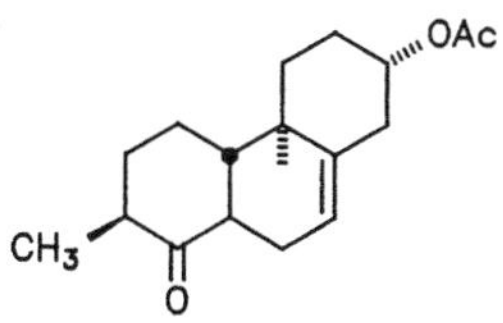	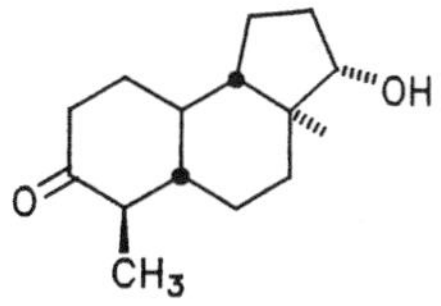
Δε:	−1.44	−0.32	−1.96

[a] ORD amplitudes from ref. 3, converted to Δε ($a = 40.28 \times \Delta\varepsilon$) in methanol.
[b] In dioxane.

with an eye toward putting the octant rule or the zigzag rule on a quantitative basis. Typical $n \rightarrow \pi^*$ CD curves for selected steroid ketones are shown in Figure 8-3.

From the many available steroid ketone $n \rightarrow \pi^*$ ORD Cotton effect data compiled by Djerassi and Klyne[3,4] and by Crabbé,[5,7] and from the extensive compilation by Kirk and Klyne,[9a] approximate expected Δε values can be correlated with each possible site for locating the ketone (Table 8-3). As might be expected, the predicted Cotton effect signs of the A/B *trans* ketones are the same as those seen in the corresponding perhydrophenanthrenones (Figure 8-2). The differences in observed Δε values reflect octant contributions from angular methyl groups (not present in the perhydrophenanthrenone analogs), from the D-ring atoms, and from potential inaccuracies in early ORD measurements.

In the 1-keto steroid, rings A and B contribute the same octant perturbers as found in the corresponding decalone (Table 7-10); these rings lie in a negative back octant. Rings C and D are thought to be negative front-octant perturbers. The collection of negative front-octant perturbers is sufficient to overcome the strong positive back-octant contribution from the C(19) angular methyl and results in a net weakly negative $n \rightarrow \pi^*$ Cotton effect. The magnitude of Δε (−0.36) may be contrasted with

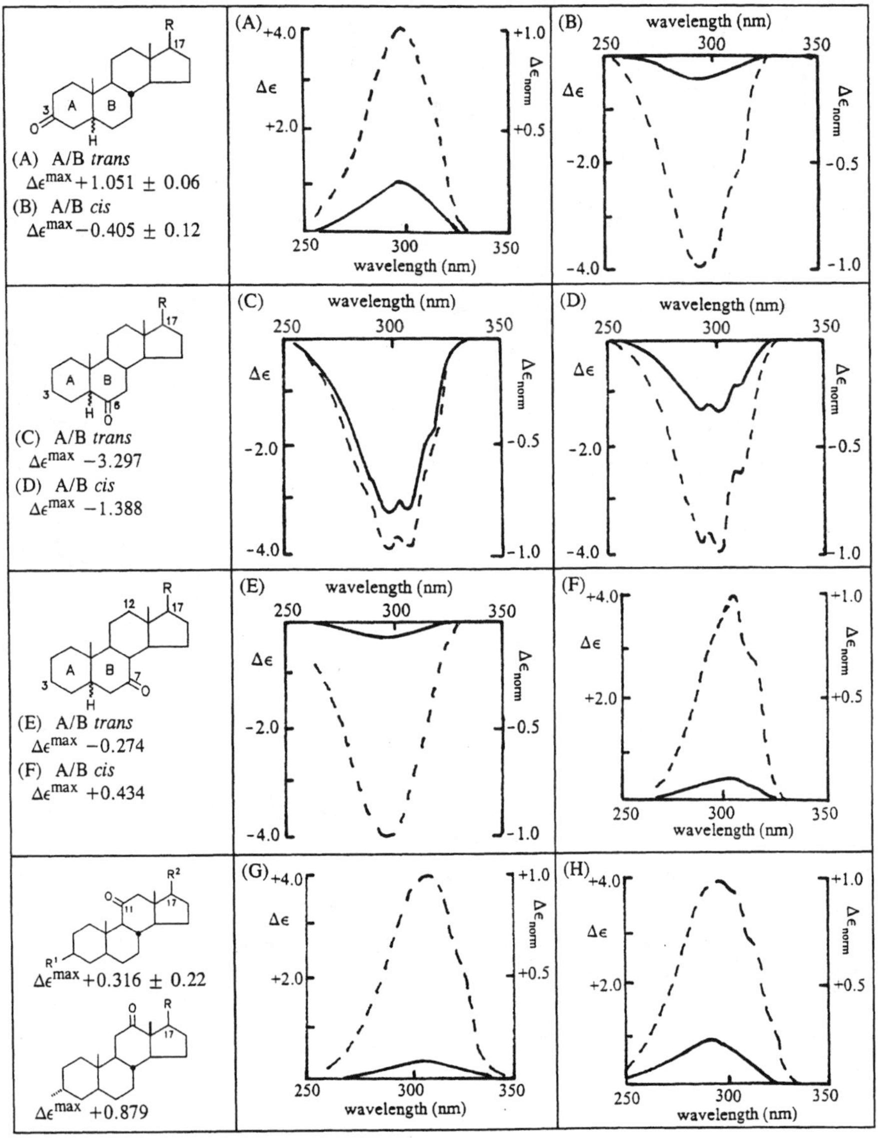

Figure 8-3. Examples of steroid ketone $n \rightarrow \pi^*$ CD Cotton effects in dioxane and of the influence of ring junction stereochemistry and the location of the carbonyl site. Data, replotted from ref. 6, were obtained in dioxane solutions: (———) is the actual curve, (- - -) shows curves normalized to the same height. The R group varies.

Table 8-3. Experimentally Determined Steroid Ketone $n \rightarrow \pi^*$ Cotton Effects from ORD and CD Spectroscopy[a] and the Predicted Cotton Effects Based on Octant Projection Diagrams[b]

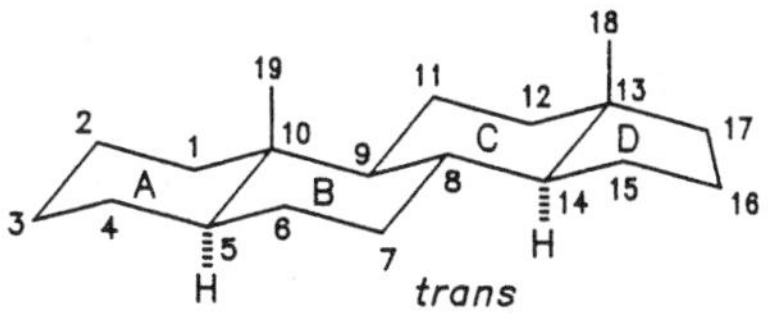

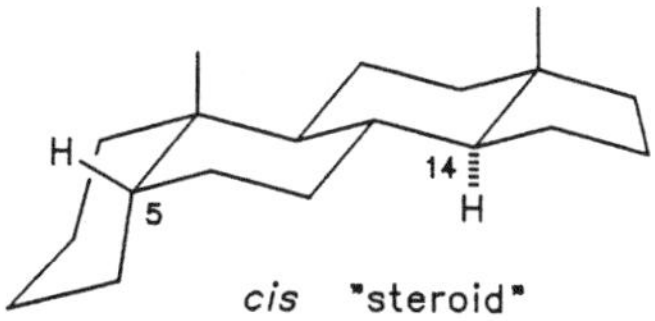

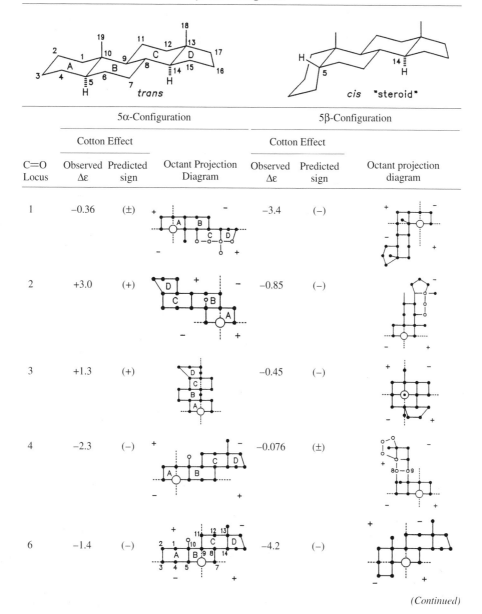

	5α-Configuration			5β-Configuration		
	Cotton Effect			Cotton Effect		
C=O Locus	Observed Δε	Predicted sign	Octant Projection Diagram	Observed Δε	Predicted sign	Octant projection diagram
1	–0.36	(±)	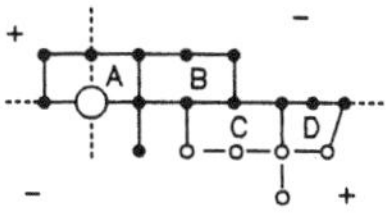	–3.4	(–)	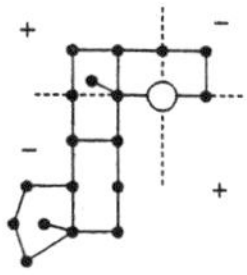
2	+3.0	(+)	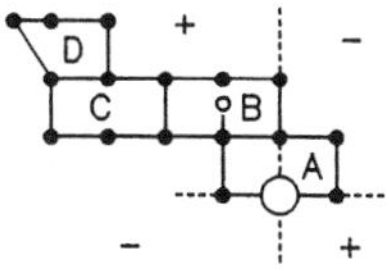	–0.85	(–)	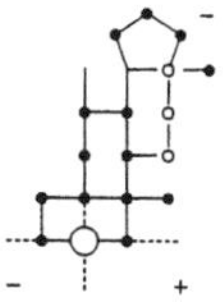
3	+1.3	(+)	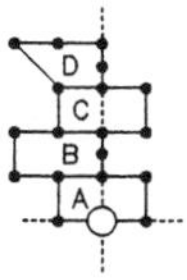	–0.45	(–)	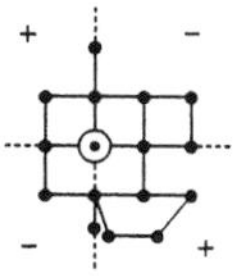
4	–2.3	(–)	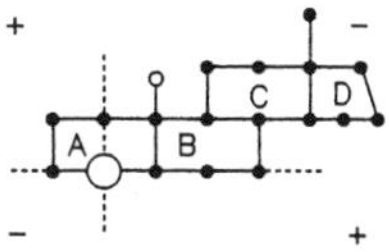	–0.076	(±)	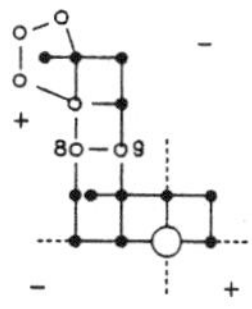
6	–1.4	(–)	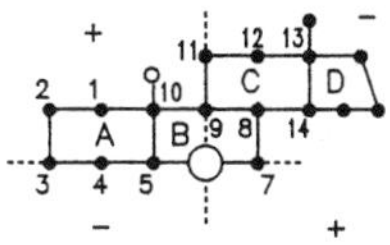	–4.2	(–)	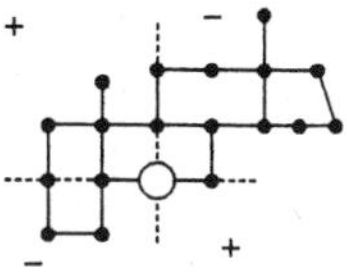

(Continued)

Table 8-3. *(Continued)*

	5α-Configuration			5β-Configuration		
	Cotton Effect			Cotton Effect		
C=O Locus	Observed Δε	Predicted sign	Octant Projection Diagram	Observed Δε	Predicted sign	Octant projection diagram
7	−0.67	(−)	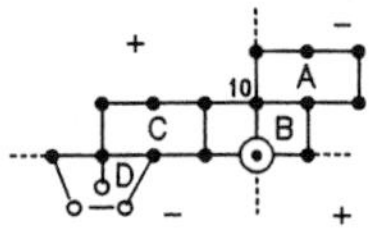	+0.73	(+)	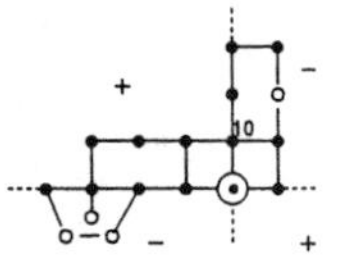
11	+0.33	(+)	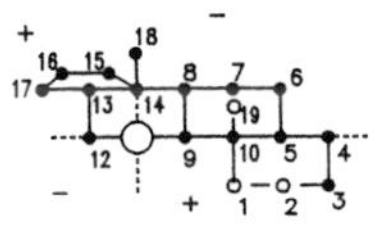	+0.37	(+)	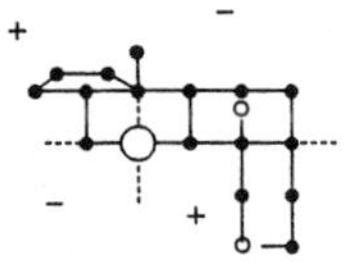
12	+1.0	(+)	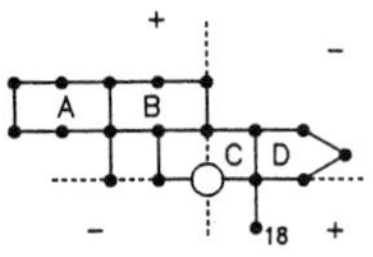	+0.58	(+)	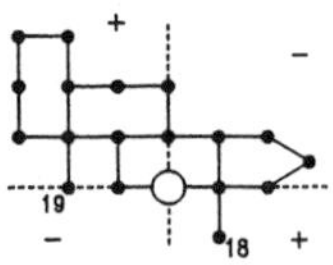
15[b]	+3.3	(+)	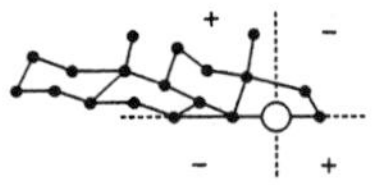	−2.5	(−)	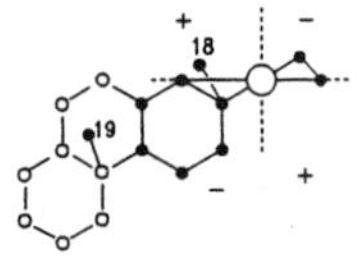
16[b]	−5.8	(−)	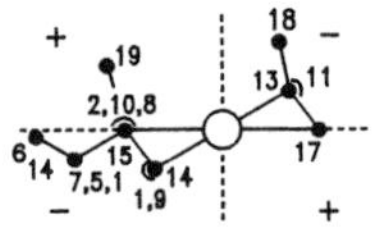	+3.4	(+)	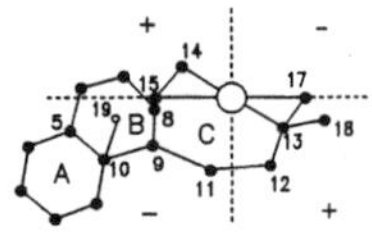
17[b]	+3.5	(+)	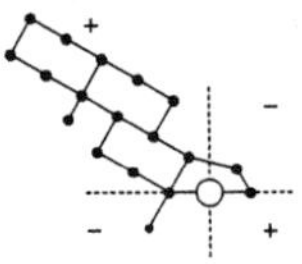	+1.1	(±)	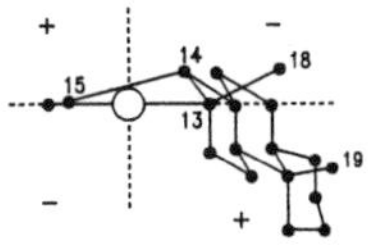

[a] Data from ref. 5.

[b] 14β-Configuration for 15, 16, and 17 C/D *cis* ketones. Open circles are perturbers in front octants.

the strong negative $n \rightarrow \pi^*$ Cotton effect (Δε −2.14) from the corresponding perhydrophenanthrenone, which lacks the angular methyl group (Figure 8-2).

The octant rule accurately predicts the signs of the 6- and 7-keto- and the 11- and 12-keto-steroids. The 6-, 7- and 12-keto-steroid octant projection diagrams show no complicating front-octant contributions. In contrast, the 11-keto steroid, like the 1-keto steroid, has front-octant contributions, mainly from the C(19) angular methyl and carbons 1 and 2 of ring A. Since most of the remaining skeletal carbons lie on octant planes or are canceling, only a few carbons (3, 6, 7, 15, 16, 17, and 18) remain as back-octant contributors to the $n \rightarrow \pi^*$ Cotton effect. Their contributions are probably weak, given their remoteness from the C=O group. Carbon 19 lies in a positive front octant, carbon 6 in a negative back octant, and carbons 1 and 2 in a negative front octant. The rough summation would predict a weak positive Cotton effect, which was observed experimentally.

The change from A/B *trans* to A/B *cis* stereochemistry (5β-hydrogen) results in a predictable change in the $n \rightarrow \pi^*$ Cotton effects, in accordance with the octant rule. There are no complications from a nonsteroid conformation. Note that the weak negative Cotton effect seen for the 4-keto-5β-steroid is probably due to an accumulation of front-octant contributions (mainly from ring carbons 7, 8, and 9) overcoming the positive octant contribution of the C(19) angular methyl. As with the 5α-1-keto-steroid, it is somewhat more difficult to predict the Cotton effect sign for the 4-keto-5β-steroid from the octant diagram, as one must include estimates of the magnitudes of the individual contributions.

Ring D ketones are cyclopentanones and as such (Chapter 6), are predicted to exhibit more intense Cotton effects than ring A, B, or C ketones. Larger Δε values are reported, except for the 17-keto-14β-steroid, in which the strong negative octant contributions from the twisted cyclopentanone ring β-carbons are canceled somewhat by the positive octant contributions from most of the carbons of rings A, B, and C. However, as we shall see in the next section, evidence for twisted rings and conformational distortion is not limited to the five-membered ring.

8.3. Conformational Distortions in Steroid and Triterpene Ketones

In the relatively few reported ORD or CD studies of steroid ketones with *cis* ring fusions, the majority have dealt with A/B *cis*-fused examples (Table 8-3).[2,5–7] These *cis*-fused ketones adopt chair conformations in rings A, B, and C and have been found to follow the predictions of the octant rule. Even fewer studies have been carried out on B/C or C/D *cis*-fused steroid ketones with the ketone carbonyl on a six-membered ring, but these ketones were found to exhibit unusual ORD spectra that were best interpreted by assuming that either the B or the C ring adopts a boat or twist-boat conformation.[3] For example, the 3-keto steroid of Table 8-4 is thought to have a boat B ring or C ring. Both are predicted by the octant rule to give a positive $n \rightarrow \pi^*$ Cotton effect, but the two possible conformations could not be distinguished by ORD.[3]

Unlike the 3-keto steroid of Table 8-4, whose Δε value is not grossly different from that of an all-*trans* 3-keto steroid (Table 8-3), the 11-keto steroid Δε values (Table 8-4) are very different from the Δε value (Δε +0.33) for an average all-*trans* 11-keto steroid (Table 8-3) or even for an average A/B *cis* 11-keto steroid (Δε +0.37). The enhanced Δε values and inverted signs seen in Table 8-4 are an indicator of gross conformational alterations from the chair. Thus, the 8β,9β-11-keto steroid is thought to have either its B or C ring in a boat or twist-boat conformation. But at least three twist boats are thought to be possible—all predicted to have a positive $n \rightarrow \pi^*$ Cotton effect.[3] In the

Table 8-4. Nonnatural Steroid Ketones and Their $n \rightarrow \pi^*$ Cotton Effects,[a] Ring B and C Torsion Angles from Molecular Mechanics Calculations,[b] and Octant Diagrams

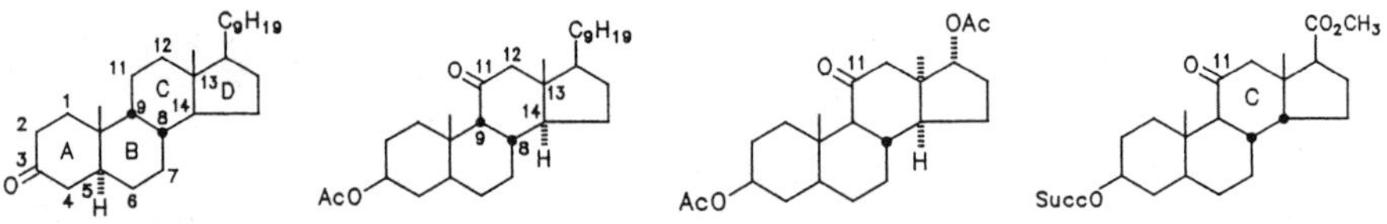

Δε	+1.07[c]	+2.28[d]	−4.64[c]	−0.92[d]
Torsion Angle	(°)	(°)	(°)	(°)
9–11–12–13	34.19	38.45	53.63	55.71
8–9–11–12	−69.37	−68.22	−0.26	−9.66
11–12–13–14	26.29	24.61	−54.12	−44.58
11–9–8–14	28.49	28.78	−50.95	−44.54
8–14–13–12	−63.16	−63.40	3.95	−7.90
9–8–14–13	33.74	33.82	49.73	54.53
5–10–9–8	−47.27	−50.11	−53.11	−54.79
6–5–10–9	56.59	57.48	56.44	56.03
7–8–9–20	42.72	44.96	53.39	56.27
7–6–5–10	−60.80	−60.21	−59.29	−58.17
6–7–8–9	−47.25	−47.13	−54.05	−55.48
5–6–7–8	55.88	54.88	56.53	56.12
$\Delta\Delta H_f$	10.4	8.60[e]	0.0[e]	0.45[e]

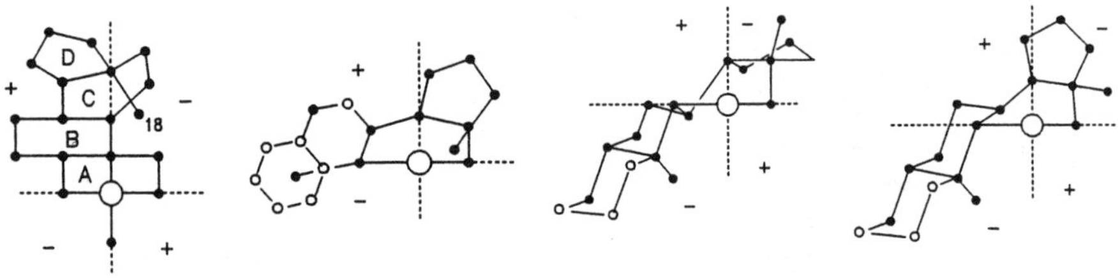

[a] Δε Values from ORD amplitudes of ref. 3; $a = 40.28 \times \Delta\varepsilon$. Open circles are in front octants.
[b] From PCMODEL (ref. 20).
[c] In CH_3OH.
[d] In dioxane. [e] $\Delta\Delta H_f$ in kcal/mole.

13α-11-keto steroid, the very strong negative Cotton effect cannot be rationalized in terms of an all-chair conformation, but appears to be more compatible with a twist-boat C ring. Similarly, the 14β-11-keto steroid is thought to have an abnormal C-ring conformation.[3]

Molecular mechanics calculations[20] on the four ketones of Table 8-4, modeled as androstanones, confirms a twist-boat conformation of the C ring for the B/C *cis* 3-keto steroid, while the A and B rings remain in the chair conformation. This configuration causes the C(18) angular methyl to move from an octant symmetry plane into a negative octant, thereby decreasing the positive $\Delta\varepsilon$ value (~ +1.3) found in all-*trans* 3-keto steroids.

In the 11-keto steroids with B/C or C/D *cis* ring fusions, the C ring is computed[20] to favor a boatlike conformation in each example. In the B/C *cis*, three twist boats were thought to be possible,[3] but the C-ring twist boat described in Table 8-4 is found by molecular mechanics computations to be the energy-minimum conformation, while the B ring remains in the chair. The presence of the carbonyl group in ring C (at C(11)) lowers the enthalpy of formation by approximately 2 kcal/mole relative to carbonyl in ring A (at C(3)). Unlike the all-*trans* 11-keto steroid, the B/C *cis* isomer has a much larger $n \rightarrow \pi^*$ Cotton effect that is attributable to a completely altered octant diagram (cf. Table 8-4 with Table 8-3) in which most of the ring A carbons are thrust into a positive *front* octant, while ring B atoms and the C(19) methyl lie in a positive back octant.

Even greater changes in the $n \rightarrow \pi^*$ Cotton effects of 11-keto steroids attend the creation of a C/D *cis* ring juncture. In both examples (13α-CH_3 and 14β), the Cotton effect signs are inverted, and in the former the magnitude is quite large ($\Delta\varepsilon$ = −4.64 vs. $\Delta\varepsilon$ = +0.33 for the all-trans isomer; cf. Tables 8-4 and 8-3). In the 13α-methyl 11-keto steroid, ring C adopts an almost perfect chair in its lowest energy conformation. This leads to an octant diagram in which almost all of the most important ring atom octant perturbers lie in negative back octants, thereby explaining the large negative Cotton effect. Other conformations of ring A are less stable. For example, the twist-boat ring C has 1.3 kcal/mole higher energy. In the 11-keto-14β steroid also, the C ring adopts a boatlike conformation, which leads to an octant diagram with more ring atoms in positive octants than in the octant diagram of 13α-CH_3 and a resulting less strongly negative Cotton effect. However, no further investigations of the conformations of any of the ketones of Table 8-4 have been reported, and no CD studies have been carried out since the early ORD work.

Indications of ring deformation from chair to boat are not limited to those attributed to B/C *cis* or C/D *cis* steroid ketones, however. In the 1950s, Djerassi et al.[21] investigated the influence of alkyl substituents and carbon–carbon double bonds on the ORD spectra of all-*trans* steroid and triterpene ketones and found a few examples where the observed ORD was very different from that predicted by the octant rule for the all-chair conformation. While excellent correlations between $n \rightarrow \pi^*$ Cotton effects that were predicted by the octant rule and those that were actually observed were found for most all-*trans*, all-chair ketones,[1–8,9a] no good correlations were discovered in 2,2 and 4,4-dimethyl ketones,[21–23] in triterpenes with a *gem*-dimethyl group at C(4) and an axial methyl at C(8), or in a few other methylated steroid ketones.[3,24]

The influence of a *gem*-dimethyl group is dramatically illustrated in the CD curves of Figure 8-4A, which shows that the strongly positive $n \rightarrow \pi^*$ Cotton effect of 5α-cholestan-3-one becomes inverted to a moderate negative Cotton effect upon the introduction of a *gem*-dimethyl at C(4).[22] For ring A in a chair conformation, 5α-cholestan-3-one is predicted by the octant rule to exhibit a strong positive Cotton effect

(Figure 8-4B), which is, in fact, observed. The introduction of a *gem*-dimethyl at ring carbon 4 on the all-chair conformation is predicted by the octant rule to add a strong negative increment (Δε ~ −1.5; see Table 5-7), due to the axial 4β-methyl, and a much weaker positive increment (Δε ~ +0.3; see Table 5-7) from the equatorial 4α-methyl. With these increments added to the base (Δε ~ +1.3) from the parent unsubstituted ketone, a net very weakly positive Cotton effect may be predicted for 4,4-dimethyl-5α-cholestan-3-one, rather than the moderately negative Cotton effect (Δε ~ −0.3) that is actually observed. This discrepancy was explained in terms of a conformational deformation of the chair ring A that relieves a severe 1,3-diaxial methyl–methyl interaction between the axial 4β-methyl and the angular 19-methyl. That the source of the inverted Cotton effect comes from the presence of a 4β-methyl, and not the 4α-methyl, could be seen from the Cotton effects of the corresponding monomethyl

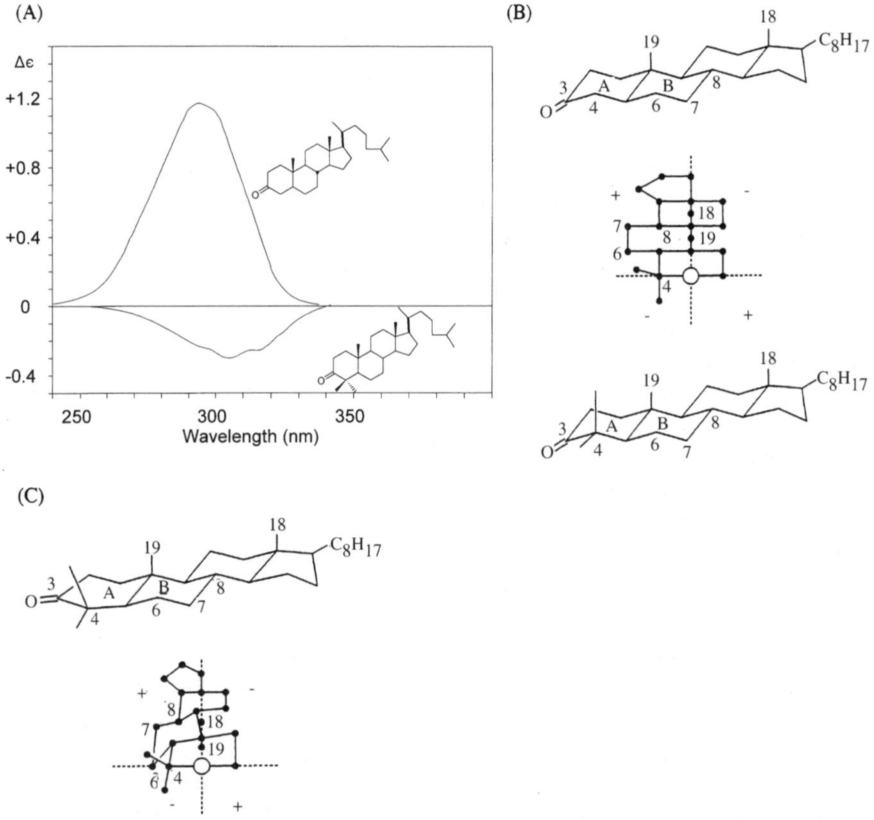

Figure 8-4. (A) CD curves of 5α-cholestan-3-one (Δε = +1.2) and 4,4-dimethyl-5α-cholestan-3-one (Δε = −0.30) in dioxane. (B) All-chair conformations of the same ketones and their octant diagram. (C) Flattened chair conformation of 4,4-dimethyl-5α-cholestan-3-one and its octant diagram. [(A) is redrawn from ref. 22.]

ketones (Table 8-5). A 4α-methyl group (equatorial) does not especially perturb the Cotton effect of the parent ketone, but an axial 4β-methyl group leads to a significant diminution in magnitude of the Cotton effect, although without inverting the sign. Again, this state of affairs was taken as an indicator of a conformational change associated with the severe nonbonded 1,3-diaxial methyl–methyl buttressing that is present when ring A adopts the chair conformation.

This nonbonded steric compression may be relieved by tilting C(3) toward C(10), flattening the chair conformation in the vicinity of the carbonyl group (Figure 8-4C), as suggested by Allinger and DaRooge,[23] who argued that a flattened chair was preferred over either of two extreme boat conformations for ring A on the basis of dipole measurements. Such a conformation would also accommodate the sign inversion in the Cotton effect by moving the major positive octant contributors (C(6) and C(7)) near to a symmetry plane, while shifting the 4β methyl away from a fully axial position (with a strongly negative octant contribution) and closer to a bisected orientation where its octant contribution was still expected to be negative, but less strongly so than in the fully axial position. Although it was not known then, we now know that even small changes in the O=C–C_α–CH_3 (axial) dihedral angle can result in strong decreases in the magnitude of an axial methyl group's contribution to the Cotton effect. In the bisected orientation, the Cotton effect contribution would be both weak and *positive*, for a methyl in a negative back octant (Table 5-7).

Various approaches directed toward understanding the *4,4-dimethyl effect* have been summarized by Tsuda and Kiuchi at Kanazawa University in Japan,[25] with a consensus emerging to the effect that the A-ring deformed chair is the most probable conformation in most 4,4-dimethyl-3-keto steroids, and the A-ring deformed twist boat is the most probable conformation in 4,4,8-trimethyl-3-keto steroids. Independent experimental evidence for a slightly flattened chair in 4,4-dimethyl-3-keto steroids

Table 8-5. Influence of α-Methyl Substitution on the $n \rightarrow \pi^*$ Cotton Effect of 3-Keto Steroids[a]

Ketone	Parent	4α-CH_3	4β-CH_3	4,4-$(CH_3)_2$	2α-CH_3	2β-CH_3	2,2-$(CH_3)_2$
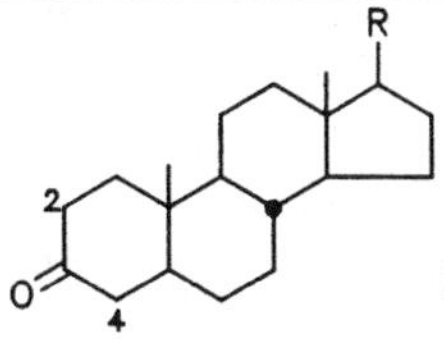	+1.3 +1.6	+1.3	+0.27	−0.27	+1.6	+1.8	+2.0
R =	OAc C_8H_{17}	C_8H_{17}	C_8H_{17}	C_8H_{17}	C_8H_{17}	C_8H_{17}	OH
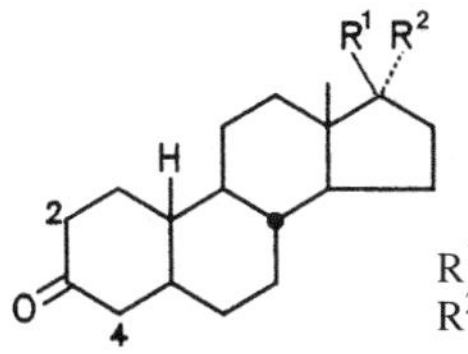	+1.4	—	—	−0.50	+1.4	—	+3.5
R^1 = R^2 =	OH H			OH CH_3	OAc H		OH CH_3

[a] Values in Δε; Data from ORD *a* values of ref. 23; *a* = 40.28 × Δε.

comes from an analysis of the ^{1}H-NMR chemical shift of the C(19) angular methyl[26] and an analysis of the vicinal coupling constants of the C(2) hydrogens (from the C(1) hydrogens).[27a] The former analysis provides evidence for a chair rather than a twist boat because the chemical shift of the C(19) methyl indicates that it does not fall into a diamagnetic anisotropy cone of the 3-keto carbonyl and thus falls into the normal region for 3-keto steroids.[26] The latter analysis indicates H–H coupling constants with values $^3J_{1\alpha,2\beta} = 13.7$ Hz, $^3J_{1\alpha,2\alpha} = 5.1$ Hz, $^3J_{1\beta,2\alpha} = 6.4$ Hz, $^3J_{1\beta,2\beta} = 3.1$ Hz in CCl_4 (all ±0.2 Hz) and nearly identical values in d_6-benzene for 4,4-dimethyl-5α-cholestan-3-one[27a]—data consistent with a chairlike, and not a boatlike, conformation in ring A. These conclusions were supported by NMR analysis of the C(19) angular methyl chemical shifts in 4,4-dimethyl-5α-cholestan-3-one and 5α-cholestan-3-one using lanthanide-shift reagents.[28] The shifts with added Eu(fod)$_3$ were remarkably alike, which was taken as an indication that the 4,4-dimethyl group did not greatly alter the A-ring chair conformation. Other evidence regarding the conformation of 4,4-dimethyl-3-keto steroids comes from X-ray crystallography,[29,30] which shows a slightly deformed chair, and not a twist-boat conformation, for ring A.

Given this qualitative understanding of the 4,4-dimethyl effect,[23] it was surprising to note that the presence of a remote 8β-axial methyl reestablishes a positive $n \rightarrow \pi^*$ Cotton effect (Figure 8-5). The presence of both the C(4) *gem*-dimethyl group and an 8β-methyl is typical of many triterpenes, which have been a rich source for investigating the 4,4,8-trimethyl effect.[22] To explain the positive CD, the 8β-methyl is thought to distort further an A ring already flattened somewhat by the *gem*-dimethyl group at C(4). The increased twisting of ring A is believed to reorient the C(19) angular methyl into a positive octant (Figure 8-5B), while lifting C(6) away from an octant symmetry plane.[22] Independent evidence comes from the X-ray crystallography of 3-keto triterpenes with a 4,4-dimethyl and an 8β-methyl group showing a distorted (twist) chair conformation of the A-ring.

A situation similar to the 4,4-dimethyl effect is found for a *gem*-dimethyl group placed opposite C(4)—that is, at C(2). Thus, in 2,2-dimethyl-5α-cholestan-3-one, the $n \rightarrow \pi^*$ Cotton effect is significantly larger than that of the parent (Figure 8-6 and Table 8-5). And the observed Cotton effect ($\Delta\varepsilon \sim +2.0$) is smaller than expected for simply adding the increments from a 2β-axial methyl ($\Delta\varepsilon \sim +1.5$) and a 2α-equatorial methyl ($\Delta\varepsilon \sim -0.3$) to the parent ketone ($\Delta\varepsilon \sim +1.3$)—suggesting a reorientation of octant perturbers (Figure 8-6B) attending an all-chair conformation. As with the addition of *gem*-dimethyls to C(4), the presence of *gem*-dimethyls at C(2) is thought to lead to a flattening of ring A. Again, it is the β-axial methyl that forces the flattening in order to relieve a severe nonbonded steric compression with the C(19) angular methyl.[23] However, the shift in the magnitude of the Cotton effect for a 2β-methyl is much less than that of a 4β-methyl, suggesting differing A-ring conformations.

Molecular mechanics calculations have been employed to analyze the 4,4-dimethyl effect on the A-ring chair conformation of 4,4-dimethyl-3-keto steroids.[27b,31] Using the 1973 (MM1) force field, Burkert and Allinger computed the flattened chair of androstan-3,17-dione to be only 0.2 kcal/mole lower in energy than the twist, suggesting an equilibrium in solution.[27a] Subsequently, the MM2 force field indicated an

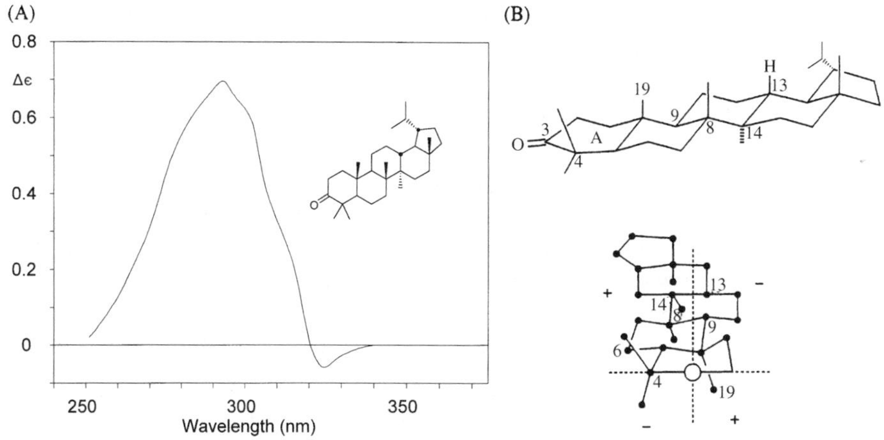

Figure 8-5. (A) CD spectrum of 3-lupanone (Δε = +0.70) in methanol. (B) Conformational drawing of 3-lupanone with a flattened A ring. The octant digram of this compound moves the C(19) angular methyl into a positive octant and left C(6) from an octant symmetry plane. (Cf. Figure 8-4C.) [(A) is redrawn from ref. 22.]

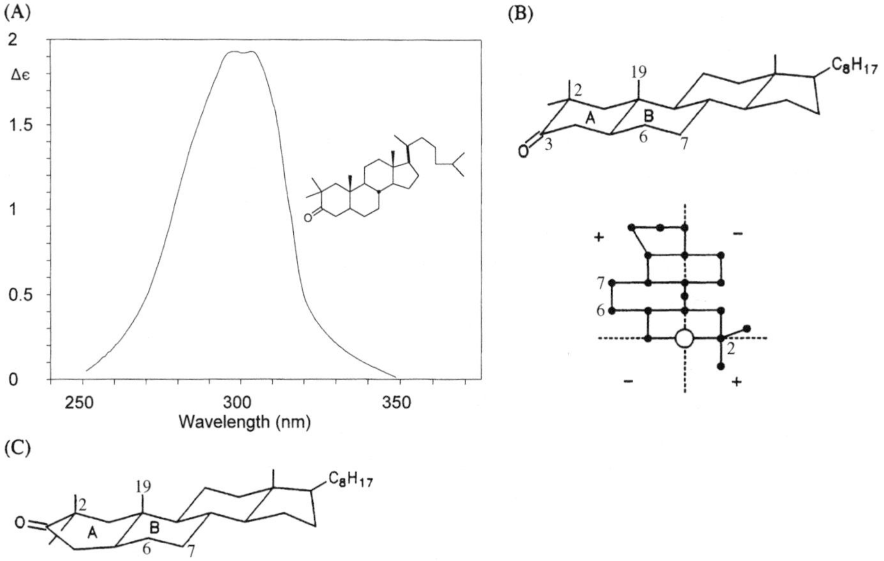

Figure 8-6. (A) $n \rightarrow \pi^*$ Cotton effect of 2,2-dimethyl-5α-cholestan-3-one in dioxane. (B) Conformation of all-chair 2,2-dimethyl-5α-cholestan-3-one and its octant diagram. (C) Flattened chair in ring A of 2,2-dimethyl-5α-cholestan-3-one. [(A) is redrawn from ref. 22.]

energy difference of 0.7 kcal/mole.[32] The computed A-ring torsion angles match well with those found in X-ray crystal structures[29] and clearly indicate a slightly flattened chair conformation.

When the possible conformations of the 2,2-dimethyl, the 4,4-dimethyl, and the 4,4,8-trimethyl ketones are investigated by PCMODEL molecular mechanics calculations,[20] one finds that the A ring does indeed suffer distortion from the chair cyclohexanone conformation (Table 8-6). With a *gem*-dimethyl group at C(4) or C(2), the distortion is not large: Only a slight flattening is observed in the region of the ketone carbonyl. However, when an 8β-methyl is added to the 4,4-dimethyl ketone, a much more severe distortion of the chair is seen, approaching the sofa conformation shown in Figure 8-5B. That the source of the distortion is a 1,3-diaxial methyl–methyl repulsion between the C(19) angular methyl and the 2β or 4β methyl can also be seen in the data of Table 8-6, which show that the ring A deformations are comparable to those seen with *gem*-dimethyls at C(2) or C(4). In contrast, the equatorial 2α-methyl induces no appreciable ring conformational change relative to the parent, and an equatorial 4α methyl causes a puckering of the carbonyl end of the ring—presumably to accommodate a nonbonded *peri* steric interaction between the 4α-methyl and the C(6)α-hydrogen (a gauche butane interaction between the C(6) CH_2 and the 4α-CH_3). There is much less steric strain in the 2α and 4α-methyl ketones than in the 2β and 4β, the 2,2 and 4,4-dimethyl, or (especially) the 4,4,8β-trimethyl-substituted 5α-androstane-3-ones.

While the molecular mechanics results of Table 8-6 may be seen to support the earlier[21,22] explanations for the Cotton effects shown in Figures 8-4, 8-5, and 8-6, they also indicate that boatlike conformations should be considered. Although the computed difference in enthalpy of formation ($\Delta\Delta H_f$) between the chair and boat of the ring A strongly favors the chair in the parent ketone and its 4α and 2α methyl derivatives, for the other conformations of Table 8-6, the energy difference is far smaller. The data suggest that boatlike conformations of ring A should be considered to be in equilibrium with the flattened chair for *gem*-dimethyl ketones, as Tsuda and Kiuchi concluded,[25] and that a reinvestigation of their conformations using variable low-temperature CD might be useful.

The importance of the C(19) angular methyl to the *gem*-dimethyl effect on conformation and CD or ORD was investigated by determining ORD spectra of selected 19-norsteroids.[22] As might be expected from their octant diagrams, which place the 19-methyl group on an octant symmetry plane (Figure 8-4B), the observed $n \rightarrow \pi^*$ Cotton effects of 5α-dihydrotestosterone acetate and 19-nor-dihydro-5α-testosterone differ little (Table 8-5). The addition of *gem*-dimethyl groups to C(2) or C(4) of a 19-norsteroid does not introduce a 1,3-diaxial methyl–methyl steric compression, and one might assume that the A ring would "prefer" to adopt the chair conformation. In such conformations, the 4,4-dimethyl ketone might be expected to give a weakly positive $n \rightarrow \pi^*$ Cotton effect, the 2,2-dimethyl ketone a strongly positive Cotton effect. However, in the former, the Cotton effect is negative and more intense than the counterpart possessing a C(19) angular methyl. And in the latter, the observed Cotton effect ($\Delta\varepsilon \sim +3.5$) is much more intense than that expected ($\Delta\varepsilon \sim +2.5$). The data again

Table 8-6. Deformation of Ring-A Chair Cyclohexanone by 2-, 4-, and 8-Methyl Substitution on 5α-Androstan-3-one, as Measured by Torsion Angles from Energy-Minimized Structures Computed by Molecular Mechanics[a]

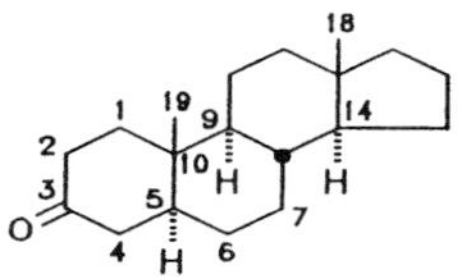

	Torsion Angle (°)	Parent X-Ray[b]	Parent	4,4 X-Ray[c]	4,4	4,4,8β	18-nor 4,4,8β	4β	4α	2,2	2β	2α
CHAIR A-RING	1–2–3–4	44.5	44.30	47.8	46.74	36.63	34.33	43.01	51.04	32.99	31.24	46.13
	2–3–4–5	–46.1	–45.47	–37.6	–36.99	–31.24	–29.22	–37.50	–51.24	–42.73	–39.38	–48.31
	3–2–1–10	–50.3	–50.46	–57.5	–56.16	–49.78	–48.75	–52.78	–52.77	–38.16	–38.10	–50.62
	3–4–5–10	53.1	52.96	39.2	38.66	40.89	40.32	42.78	53.58	55.75	53.90	54.38
	2–1–10–5	56.5	55.72	56.3	55.02	55.72	55.95	54.96	54.26	49.88	50.40	55.14
	1–10–5–4	–58.1	–57.07	–48.7	–47.70	–52.72	–53.14	–50.76	–54.95	–57.19	–57.74	–56.64
	O=C–C_4–βCH_3	—	—	—	–88.71	–82.85	–80.99	–82.37	—	—	—	—
	O=C–C_2–βCH_3	—	—	—	—	—	—	—	—	86.07	78.51	—
	O=C–C_4–αCH_3	—	—	—	25.20	30.98	32.99	—	2.92	—	—	—
	O=C–C_2–αCH_3	—	—	—	—	—	—	—	—	–29.38	—	–10.04
	Relative Strain Energy[a,d]		0.0		8.6	20	14	6.0	2.4	5.3	4.5	0.5

(Continued)

Table 8-6. *(Continued)*

Torsion Angle (°)		Parent	4,4 X-Ray[c]	4,4,8β	18-nor 4,4,8β	4β	4α	2,2	2β	2α
TWIST-BOAT A-RING	1–2–3–4	–54.28	–60.94	–59.20	–59.05	–61.36	–56.12	–44.76	–54.76	
	2–3–4–5	19.86	31.30	30.99	30.30	30.15	24.19	21.68	21.34	
	3–2–1–10	27.72	26.74	23.53	23.90	28.90	26.02	12.21	27.20	
	3–4–5–10	38.96	28.36	29.05	29.78	30.07	34.88	33.74	37.88	
	2–1–10–5	27.74	28.31	31.68	31.25	27.37	29.72	38.82	28.39	
	1–10–5–4	–62.98	–58.93	–60.99	–61.22	–59.37	–62.30	–62.86	–62.86	
	O=C–C_4–βCH_3	—	–23.77	–24.93	–25.11	–19.07	—	—	—	
	O=C–C_2–βCH_3	—	—	—	—	—	—	14.80	–0.74	
	O=C–C_4–αCH_3	—	90.89	90.28	89.64	—	76.67	—	—	
	O=C–C_2–αCH_3	—	—	—	—	—		–82.48	—	
	Relative Strain Energy[a,d]	0.0	6.0	17	11	3.9	2.2	3.2	0.3	2.7
	$\Delta\Delta H_f^{d,e}$	2.8	0.26	0.49	0.81	0.70	2.6	0.74	1.3	5.1

[a] PCMODEL, ref. 20.
[b] From X-ray crystallography, ref. 29.
[c] From X-ray crystallography of 17β-benzoyloxy-4,4-dimethyl-5α-androstan-3-one, ref. 27b.
[d] kcal/mole, steroid-parent.
[e] ΔH_f (twist boat)-ΔH_f (chair).

hint at distorted chair conformations in ring A, but different from those suggested earlier for 2,2 and 4,4-dimethyl 3-ketones with C(19) angular methyls.

Molecular mechanics calculations using PCMODEL[20] indicate very little deformation of the A-ring chair cyclohexanone in 19-nor-3-keto steroids with substitution of a *gem*-dimethyl group at either C(2) or C(4) (Table 8-7). The O=C–C–CH_3 torsion angles suggest fully axial and equatorial methyls. The addition of an 8β-methyl does not perturb the chair conformation of the A ring in the 19-norsteroid, and the twist-boat conformations are of much higher energy in the 4,4,8β-trimethyl derivative of the normal steroid than in the 19-nor steroid. The situation is much different from that seen when the C(19) angular methyl is present (Table 8-6), and with the 19-nor ketones, one should assume normal octant projection diagrams based on the all-chair geometry. In the 2,2-dimethyl ketone, the 2β-axial methyl lies in a positive octant, and its octant contribution, expected to be about +0.5 (Table 8-5), adds to the summed contributions from ring atoms ($\Delta\varepsilon \sim +1.3$) and the equatorial 2α-methyl ($\Delta\varepsilon \sim +0.3$) , yielding a net $\Delta\varepsilon$ of +2.1, close to the value that is observed. In the 4,4-dimethyl ketone, the 4β-axial methyl lies in a negative octant, and its octant contribution, expected to be $\Delta\varepsilon \sim -0.5$ (Table 8-5), adds to the mainly positive octant contributions of ring atoms ($\Delta\varepsilon \sim +1.3$) and the $\Delta\varepsilon \sim 0$ contribution of the 4α-equatorial methyl group to yield a net positive $\Delta\varepsilon \sim +0.8$. The observation that the sums of the individual contributions do not match the observed Cotton effects suggests modest ring deformations or inaccuracies in recording ORD data and translating ORD amplitudes to CD $\Delta\varepsilon$ values.

Related to the 4,4,8-trimethyl effect in tetra- and pentacyclic triterpene and steroid ketones is a similar steric effect in the tetracyclic triterpene α-onoceran-3,21-dione, a structure in which two *trans*-decalones are linked by an ethylene unit. The strong positive $n \rightarrow \pi^*$ Cotton effect (Figure 8-7) seen for the 8αH, 14βH diastereomer[33] has about twice the intensity of that of 3-lupanone (Figure 8-5A), a 4,4,8β-methyl 3-keto triterpene with four contiguous rings, thus suggesting a similarly distorted (twist) A ring in both ketones. This conclusion is supported by an X-ray crystallographic study of 8α*H*, 14β*H*-onoceran-3,21-dione which indicates that ring A adopts a twist- or flattened-boat conformation.[34] The Cotton effect sign inversion seen (Figure 8-7) when the 8-methyl is inverted from β to α was unexpected and is believed to alter the A-ring conformation from twist boat to chairlike.[33] Here again, a variable low-temperature CD study is suggested as a way to sort out possible conformation equilibria and identify the minimum-energy conformation of ring A in the onocerandiones.

Other examples of the influence of the *gem*-dimethyl group on CD and ring stereochemistry are found in D-homosteroids. 3β-Acetoxy-D-homo-5α-androstan-17a-one exhibits a very weak positive $n \rightarrow \pi^*$ Cotton effect (Table 8-8A), with a near-zero magnitude resulting from a balancing of the expected strong negative octant contributions from the C(18) axial angular methyl by the positive octant contributions from ring carbons in the A, B, and C rings.[3] The introduction of a 17α-equatorial methyl easily dominates the near-zero Cotton effect of the parent ketone, inverting the sign to give a net weak negative Cotton effect, as predicted by the octant rule. The introduction of a 17β-axial methyl has a more profound influence and leads to a very intense positive $n \rightarrow \pi^*$ Cotton effect. This relationship was taken as persuasive

Table 8-7. Deformation of Ring-A Chair Cyclohexanone by Methyl Substitution on 19-nor-5α-Androstan-3-one, as Measured by Torsion Angles from Energy-Minimized Structures Computed by Molecular Mechanics[a]

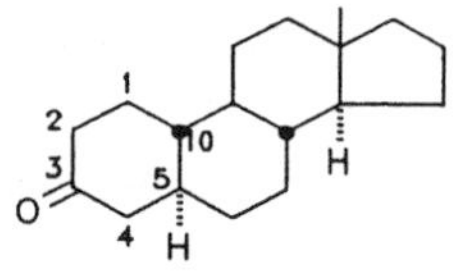

Torsion Angle (°)	Parent	2,2	4,4	4,4, X Ray[b]	4,4,8β
CHAIR A-RING					
1–2–3–4	51.32	46.59	55.20	57	53.79
2–3–4–5	–52.11	–51.02	–50.73	–55	–51.39
3–2–1–10	–52.90	–49.01	–55.28	–55	–53.04
3–4–5–10	54.04	54.79	48.25	50	51.31
2–1–10–5	55.24	54.95	53.97	53	54.76
1–10–5–4	–55.59	–56.14	–51.38	–52	–54.49
O=C–C_4–βCH_3	—	—	–106.89	—	–107.6
O=C–C_2–βCH_3	—	103.49	—	—	—
O=C–C_4–αCH_3	—	—	9.25	—	8.55
O=C–C_2–αCH_3	—	–14.73	—	—	—
Relative Strain E^c	0.0	2.6	5.6	—	15
TWIST-BOAT A-RING					
1–2–3–4	–54.27	–47.89	–60.64	—	–59.70
2–3–4–5	16.65	21.50	23.36	—	22.85
3–2–1–10	31.21	17.85	34.15	—	31.84
3–4–5–10	41.10	34.13	36.23	—	37.61
2–1–10–5	25.12	35.15	22.90	—	25.93
1–10–5–4	–62.73	–62.01	–60.74	—	–63.41
O=C–C_4–βCH_3	—	—	–33.95	—	–34.57
O=C–C_2–βCH_3	—	12.07	—	—	—
O=C–C_4–αCH_3	—	—	81.77	—	81.14
O=C–C_2–αCH_3	—	–104.52	—	—	—
Relative Strain E^c	0.0	3.5	5.0	—	14
$\Delta\Delta H_f^{c,d}$	3.2	4.1	2.6		2.8

[a] PCMODEL, ref. 20.
[b] From ref. 25.
[c] kcal/mole.
[d] ΔH_f (twist boat) – ΔH_f (chair).

evidence that the chair cyclohexanone D-ring was distorted to a twist boat, which relieves a severe 1,3-diaxial methyl–methyl interaction between the C(18) angular methyl and the 17β-methyl.[3,22] The introduction of a *gem*-dimethyl at C(17) does not lead to a simple sum of Cotton effects from the 17α-methyl and 17β-methyl ketones, suggesting the presence of a different twist conformation.

In contrast, a related D-homo steroid ketone, 5α-D-homoandrostan-3β-ol-17-one (Table 8-8B) exhibits a strong negative $n \rightarrow \pi^*$ Cotton effect, comparable in magnitude

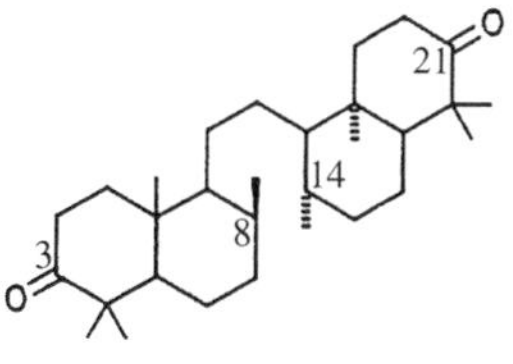

(———)
$8\alpha H,14\beta H$-Onoceran-3,21-dione

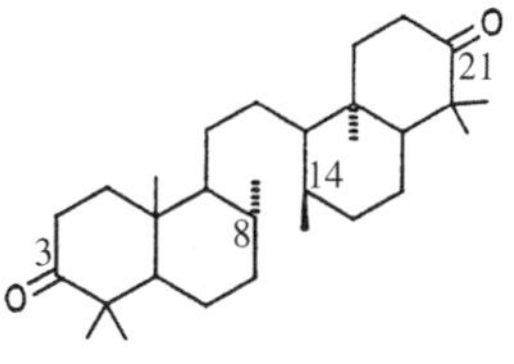

(- - - -)
$8\beta H,14\alpha H$-Onoceran-3,21-dione

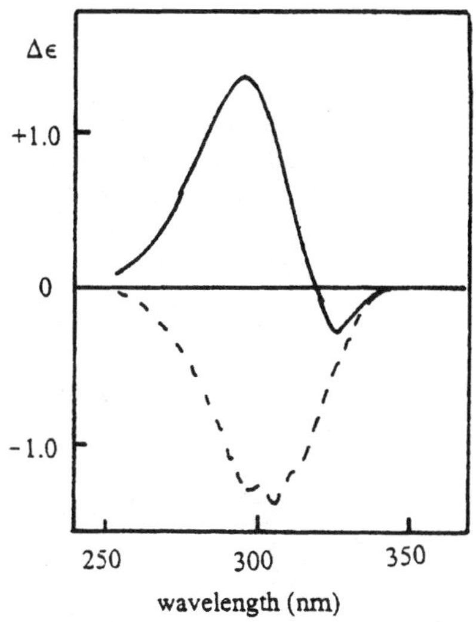

Figure 8-7. CD spectra (right) of diastereomeric onoceran-3,21-diones: $8\alpha H$, $14\beta H$ (———), $8\beta H$, $14\alpha H$ (- - - -) in dioxane. CD data are redrawn from ref. 33.

to that of a 3-keto steroid (Table 8-3) and in accord with predictions of the octant rule. The introduction of an equatorial 17a β-methyl makes the expected weak negative octant contribution. Unlike the situation with an isomeric 17a keto D-homo steroid, the introduction of an axial 17a α-methyl makes the expected strong positive octant contribution, yielding a net weakly negative Cotton effect. Similarly, there is no detectable deformation from the *gem*-dimethyl at 17a.

Molecular mechanics calculations[20] indicate that the D-homo steroid parent ketones (Tables 8-8A and B) adopt a fairly ordinary chair cyclohexanone conformation. In the 17,17-dimethyl-17a-ketone, the introduction of a *gem*-dimethyl group strongly perturbs (flattens) the chair conformation, a consequence of relieving a 1,3-diaxial methyl–methyl steric repulsion. A similar result attends the 17β-methyl ketone. However, no such ring deformation is evident in the 17-ketone, either in the parent or following the introduction of a *gem*-dimethyl group at 17a.

In other examples of the influence of methyl substitution on the $n \rightarrow \pi^*$ Cotton effect of steroid ketones, octant rule correlations were poor when equatorial methyl substituents were present at position 1β of 3-keto steroids[24] and at positions 8β and 14α of 12-keto steroids.[3] Thus, the larger-than-expected $\Delta\varepsilon$ value (+1.8, Table 8-8C) of 1β-methyl-5α-dihydrotestosterone acetate has been attributed to a twisted A-ring conformation adopted in order to relieve the nonbonded steric interaction between the 1β-equatorial methyl and the 11α-equatorial hydrogen. (The nonbonded distance $H_3C \cdots H{-}C_{11}$ is 2.78Å.) And, indeed, PCMODEL[20] molecular mechanics calculations indicate a mild distortion of the A ring, but whether this distortion provides a sufficient rationale for the observed Cotton effect is unclear. Ordinary 12-keto steroids have an $n \rightarrow \pi^*$ $\Delta\varepsilon \sim +1.0$, but with 8β and 14α methyls present, the sign

Table 8-8. Influence of Methyl Substitution on the $n \rightarrow \pi^*$ Cotton Effects[a,b] of Skeletally Deformed All-*trans* Steroid Ketones. Deformation in the Ketone-Bearing Ring from Chair Cyclohexanone Is Measured by the Ring Torsion Angles from the Energy-Minimized Structure Found from Molecular Mechanics Calculations[c]

(A)

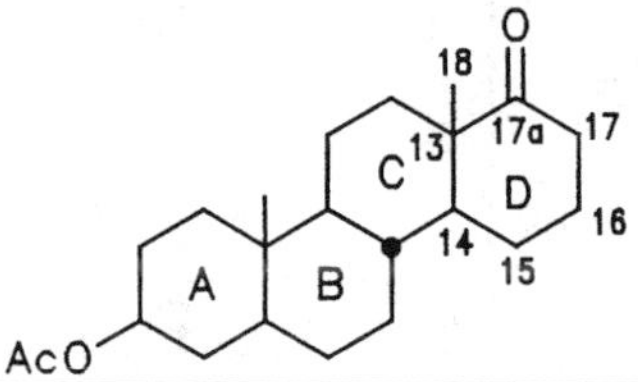

(B)

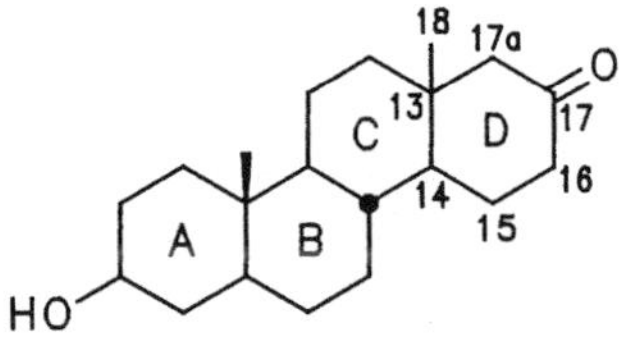

Δε	+0.06[d]	−1.2	+4.2[d]	−0.4[d]	−2.0[d]	−0.45[d]	−2.4	−0.52
Torsion Angle (°)	Parent	17,17-$(CH_3)_2$	17-βCH_3	17-αCH_3	Parent	17a,17a-$(CH_3)_2$	17a-βCH_3	17a-αCH_3
13–17a–17–16	50.18	31.38	33.99	50.13	53.54	52.27	55.57	51.87
14–13–17a–17	−51.04	−37.09	−38.49	−51.34	−56.64	−54.56	−57.05	−54.74
15–16–17–17a	−50.57	−40.67	−41.94	−50.64	−48.96	−50.72	−51.46	−49.06
15–14–13–17a	53.23	50.99	51.13	53.25	57.36	57.29	57.74	56.29
14–15–16–17	55.88	58.06	57.16	56.35	49.06	50.00	50.30	49.07
13–14–15–16	−57.81	−62.79	−62.79	−57.84	−54.37	−54.19	−55.01	−53.44

(C)

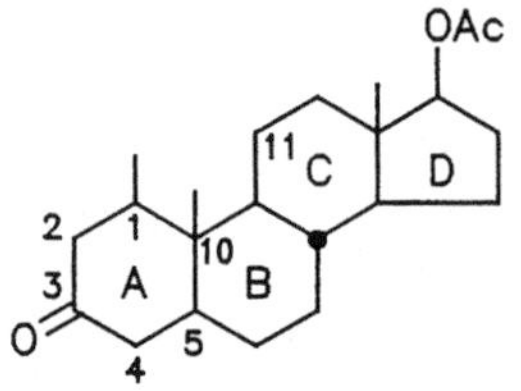

(D)

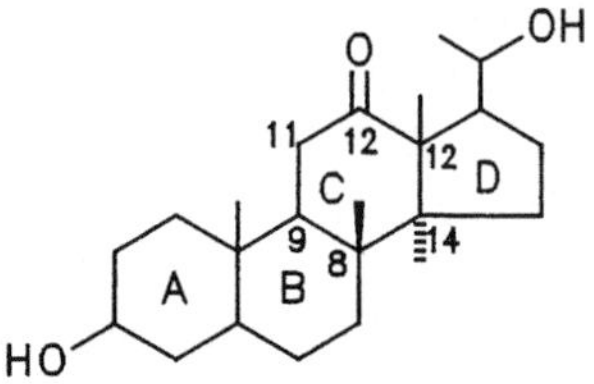

Δε	+1.8		−0.92
Torsion Angle	(°)	Torsion Angle	(°)
1–2–3–4	49.27	11–12–13–14	55.20
2–3–4–5	−55.54	9–11–12–13	−39.90
3–2–1–10	−43.35	8–14–13–12	−69.86
3–4–5–10	59.65	8–9–11–12	30.86
2–1–10–5	43.92	9–8–14–13	67.03
1–10–5–4	−53.27	11–9–8–14	−42.50

[a] In CH_3OH or C_2H_5OH solvent.

[b] ORD data from ref. 3 converted to Δε ($a = 40.28 \times \Delta\varepsilon$).

[c] PCMODEL, ref. 20.

[d] Data from ref. 9a.

inverts (Table 8-8D).[3] The reason for this inversion is not clear, but is probably tied to a distortion of the C ring, as is confirmed by PCMODEL[20] molecular mechanics calculations. No further CD studies were carried out on the 12-ketones to reach firm conclusions on the ring geometry, but variable-temperature CD studies were conducted on the 1-methyl-3-keto steroid, as is discussed in the next section.

8.4. Variable-Temperature CD Studies of Steroid and Triterpene Ketones

Relatively few variable-temperature CD studies of steroid or triterpene ketones have been carried out, and almost all of them were conducted in Djerassi's laboratory at Stanford in the 1960s. Most of the other variable-temperature CD studies were directed toward understanding conformational equilibria in monocyclic ketones (Chapters 5 and 6) and bicyclic ketones (Chapter 7).

The simplest situation in solution is when a single molecular species is present. In this case, the rotatory strength of the species is effectively independent of the temperature of the solution. As examples, all-*trans* 5α-cholestan-2-one and 5α-cholestan-3-one are thought to exist as a stable polycyclic array of chair cyclohexanes and chair cyclohexanones. Not surprisingly, therefore, the intensity of the former's $n \to \pi^*$ transition remains essentially invariant between 25°C and −192°C (Figure 8-8A),[35] and that of the latter remains invariant: $[R]^{25°} = +3.57$ and $[R]^{-192°} = +3.66$ in EPA; and $[R]^{25°C} = +2.68$ and $[R]^{132°C} = +2.57$ in decalin.[36,37] Similarly, the reduced rotatory strengths of 5α-antrostane-16-one, in which the ketone group is on the five-membered D ring, remain essentially invariant between 25°C and −192°C ($[R]^{25°C} = +18.7$ and $[R]^{-192°C} = +18.1$ in EPA),[36] as do those of 5α-androstan-3β-ol-17-one $[R]^{25°C} = +10.5$ and $[R]^{164°C} = +10.2$ in decalin).[37] As might be expected, the substitution of a methyl at 5α does not alter the essential temperature independence of the 5α-cholestan-3-one low-temperature CD (Figure 8-8B), but the reduced rotatory strengths are enhanced: $[R]^{25°C} = +4.28$ and $[R]^{-192°C} = +4.43$ in EPA relative to the parent by approximately +0.7 reduced-rotatory-strength units. Overall, these data signify conformational invariance in the wide temperature range cited; that is, the ketones are already in their most stable conformations at room temperature. In the special case of the 5α-methyl-cholestan-3-one, the data would seem to indicate that a 3-axial methyl makes a moderately large (0.7) back octant contribution rather than the expected front-octant contribution. This result is probably due to a displacement of the 5α-axial methyl into a positive back octant when the ring-A chair distorts slightly, as is predicted by molecular mechanics calculations (Table 8-9).

In view of the temperature-invariant CD seen with 2, 3, 16, and 17-keto steroids, it was surprising to find that the CD spectrum of 5α-androstane-11-one, with the ketone in the C ring, underwent a sign inversion between 25°C and −192°C (Figure 8-8C).[36] These data were interpreted in terms of the existence of at least a second, hitherto unsuspected, conformation of ring C at room temperature, with the most stable conformer exhibiting a weak negative $n \to \pi^*$ Cotton effect and a less stable conformer,

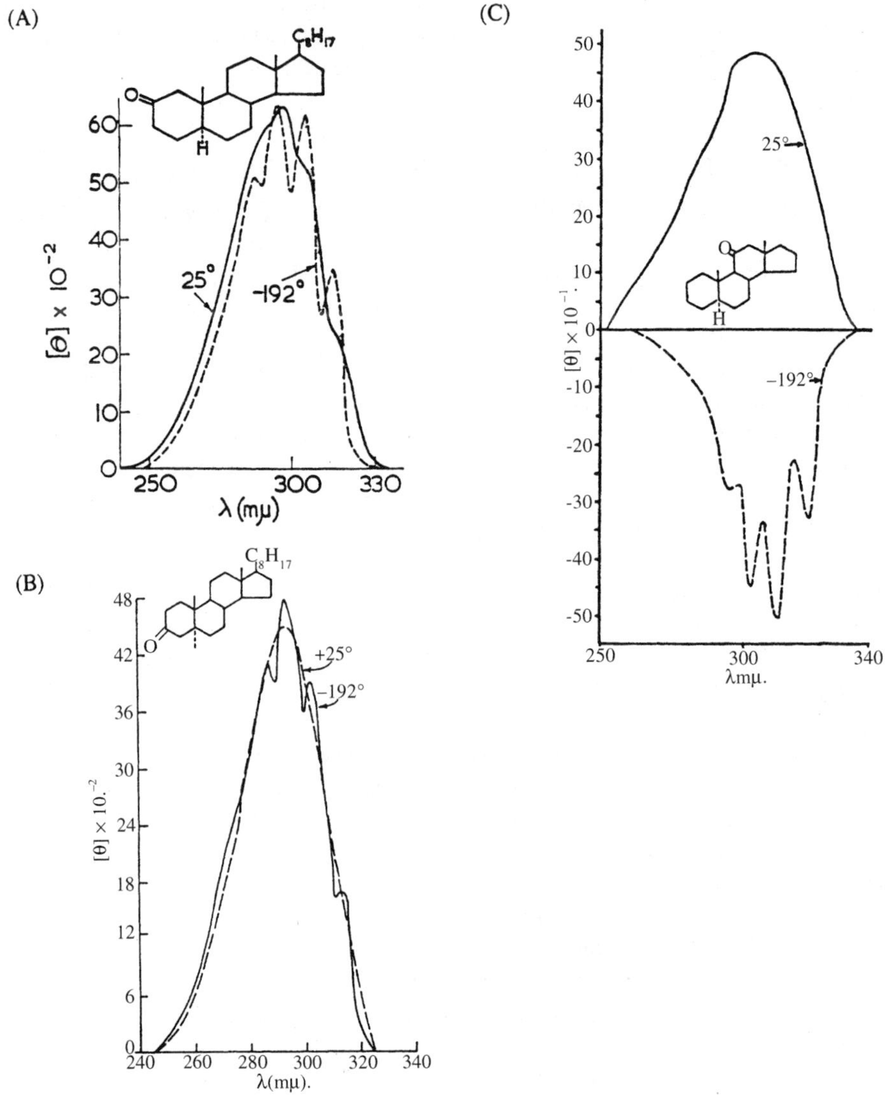

Figure 8-8. Circular dichroism spectra of (A) 5α-cholestan-2-one, (B) 5α-methylcholestan-3-one, and (C) 5α-androstan-11-one in EPA solvent at 25°C and −192°C. EPA is ether–isopentane–ethanol, 5:5:2 by vol., a glass-forming solvent. [(A) is reprinted with permission from ref. 35. Copyright© 1964 Royal Society of Chemistry. (B) is reprinted with permission from ref. 37 and (C) from ref. 36. Copyright © 1965, 1963 American Chemical Society.]

possibly present to a very small extent, exhibiting a more intense, positive Cotton effect. However, the most stable conformer found by molecular mechanics calculations[20] has a relatively normal chair cyclohexanone. Its octant diagram (Table 8-3) reveals a balance of positive and negative perturbers, making it difficult to predict what the net sign should be and thus indicating that small deformations of ring C may lead to an upsetting of this balance in favor of either a weak net positive or a net negative Cotton effect. As is expected, the molecular mechanics calculations do not indicate a conformation so severely deformed that it would yield a large positive Cotton effect (Table 8-9). Yet it appears that the Cotton effect sign inversion that is seen (Figure 8-8C) is not due to solvation effects. (See Section 8.5 and Chapter 7.)

Another case that defied easy interpretation[38] is the variable-temperature CD (Figure 8-9) of 3-lupanone, a pentacyclic 3-keto triterpene with a 4,4-dimethyl group and an 8β-methyl. The $n \rightarrow \pi^*$ Cotton effect of this compound in EPA is virtually independent of temperature from +25 to –41°C, indicating a fairly static population distribution over that temperature range—a conclusion that is reinforced by ^{1}H-NMR measurements.[39] Unexpectedly, below –41°C, the Cotton effect magnitude drops, and the CD becomes bisignate, with considerable vibrational fine structure at –192°C. The data would seem to suggest a conformational equilibrium dominated by a strong positive Cotton effect from a minor isomer.

The influence of methyl substitution at C(1) in 3-keto steroids was discussed in Section 8.3 in connection with a severe nonbonded steric interaction between C(11) (11α-H) and C(1) (1β-CH_3) that led to a larger-than-predicted $n \rightarrow \pi^*$ Cotton effect. As may be seen from the low-temperature CD of 1β-methyl-5α-dihydrotestosterone

Table 8-9. Influence of Methyl Substitution on the Ring-A Chair Conformation Torsion Angles of 5α-Androstan-3-ones, and the Ring-C Torsion Angles for 5α-Androstan-11-one[a]

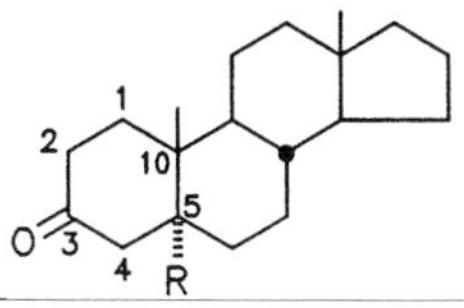

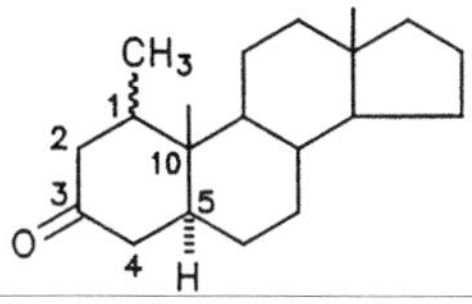

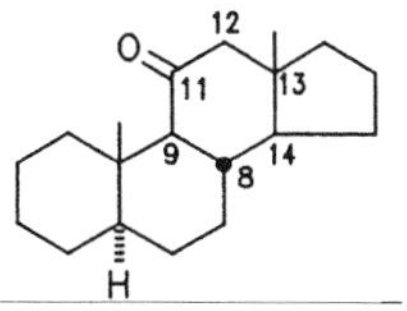

Torsion Angle (°)	R = H	R = CH_3	1β	1α	Torsion Angle (°)	
1–2–3–4	44.82	40.55	49.61	50.72	8–9–11–12	50.25
2–3–4–5	–46.03	–46.66	–55.85	–49.02	11–9–8–14	–48.49
3–2–1–10	–50.59	–45.01	–43.35	–53.74	9–11–12–13	–55.64
3–4–5–10	53.28	56.56	59.56	51.43	9–8–14–13	57.75
2–1–10–5	55.51	54.56	43.25	54.87	11–12–13–14	58.73
1–10–5–4	–57.00	–60.10	–53.02	–54.27	8–14–13–12	–63.19
$\Delta\Delta H_f$ (kcal/mole)		1.9	4.0	0.0	0.20 (relative to 3-one)	

[a] Data from PCMODEL molecular mechanics calculations (ref. 20). Values of $\Delta\Delta H_f$ are calculated for the methylated steroids and for 5α-androstan-11-one relative to the 3-one.

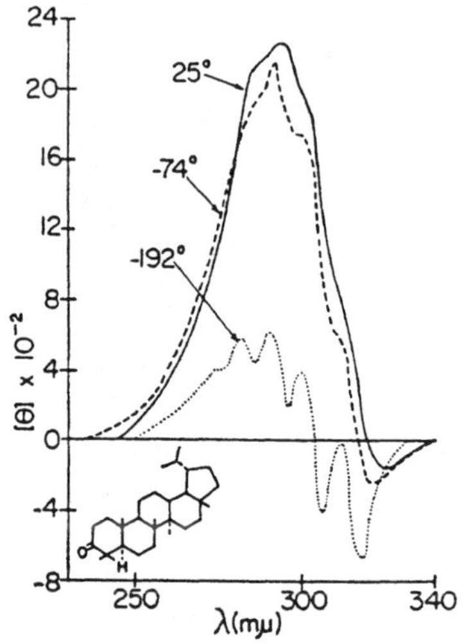

Figure 8-9. Variable-temperature CD curves of 3-lupanone in ether–isopentane–ethanol (EPA, 5:5:2 by vol.). Reprinted from ref. 38 with permission of the authors.

(Figure 8-10A), this ketone is already in its most stable conformation: $[R]^{25°C} = +5.96$ and $[R]^{-192°C} = +6.21$.[37] The equatorial 1β-methyl is expected to make a negative octant contribution (Δε ~ –0.6) to the 3-keto steroid base (Δε ~ +1.3, Table 8-3), but the net predicted Δε (~ +0.7) falls far short of that observed (Δε ~ +1.8). Apparently, the slight distortion of the ring-A chair conformation predicted by molecular mechanics calculations (Table 8-8) is sufficient to make the sum of the contributions from the ring atoms and the C(19) angular methyl more positive, while decreasing the magnitude of the contribution from the 1β-methyl. Yet, the intensity of the observed Cotton effect could not have been anticipated, and the –192°C CD curve indicates that it cannot be explained by the presence of more than one stable conformation. Molecular mechanics calculations[20] indicate that ring A is a slightly deformed chair, suggesting that subtle conformational changes in this ring can have large effects on the CD.

Unlike the 1β-methyl isomer, 1α-methyl-5α-dihydrotestosterone undergoes modest changes in $n \rightarrow \pi^*$ Cotton effect intensity (Figure 8-10B) in going from +25°C to –192°C: $[R]^{25°C} = +3.00$, $[R]^{-74°C} = +3.38$, and $[R]^{-192°C} = +3.91$.[37] The axial 1α-methyl lies near the interface of a negative back and positive front octant—as in a 3β-axial methylcyclohexanone. And as seen with the corresponding methyl adamantanone, the octant contribution of the axial methyl is markedly enhanced at low temperatures, compared with a near-zero value at room temperature. (See Figure 4-12.) The data shown in Figure 8-10B are consistent with this observation. The 1α-axial methyl makes a much larger positive octant contribution at lower temperatures than at room temperature. One need not invoke a conformational equilibrium, and molecular mechanics calculations[20] do not indicate serious deformation of the A-ring chair conformation.

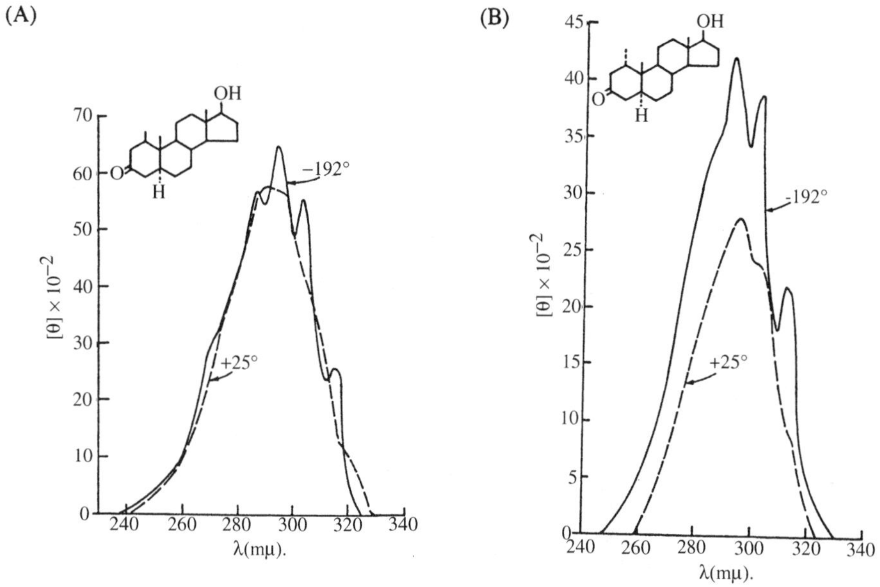

Figure 8-10. Circular dichroism $n \rightarrow \pi^*$ Cotton effects for (A) 1β-methyl-5α-dihydrotestosterone and (B) 1α-methyl-5α-dihydrotestosterone in EPA at +25°C and −192°C. EPA is ether–isopentane–ethanol, 5:5:2 by vol. [Reprinted with permission from ref. 37. Copyright © 1965 American Chemical Society.]

Variable-temperature CD measurements (Figure 8-11) were used to explore rotameric conformational equilibria in 20-keto steroids, in which the ketone is not directly attached to a ring, but comes from an acetyl group attached to C(17) of the cyclopentane D ring, as is commonly found in pregnanolones. One might imagine free rotation about the C(17)–C(20) bond, but in practice, free rotation of the 17β-acetyl is restricted by the C(18) angular methyl, and free rotation of the 17α-acetyl is restricted by axial hydrogens at C(12), C(14), and, possibly C(16) in 5α-pregnan-20-one. In the 17α-acetyl, a large rotatory strength is found for the $n \rightarrow \pi^*$ Cotton effect at +25°C and a slightly larger one at −192°C—with no change of sign.[35,40] Thus, in the 17α series, 5α-pregnan-3β-ol-20-one exhibits $[R]^{+25°C} = -12.1$ and $[R]^{-192°C} = -14.7$,[40] and 5α-pregnan-20-one exhibits $[R]^{+25°C} = -7.45$ and $[R]^{-192°C} = -8.0$[40] (Figure 8-11A). The temperature dependence of the Cotton effect is small, suggesting one very stable conformation or several isoenergetic conformations. In fact, molecular mechanics calculations predict that one rotamer (Table 8-10D) is much more stable than any of the others, as suggested earlier.[40] This most stable rotamer is predicted by the octant rule to give a negative Cotton effect, as is observed, with no temperature dependence (Figure 8-11A).[40] In contrast, the 17β epimer is predicted by molecular mechanics calculations to have two isoenergetic stable rotamers (Tables 8-10D and E) and one rotamer of only slightly higher energy (Table 8-10B). In a comprehensive

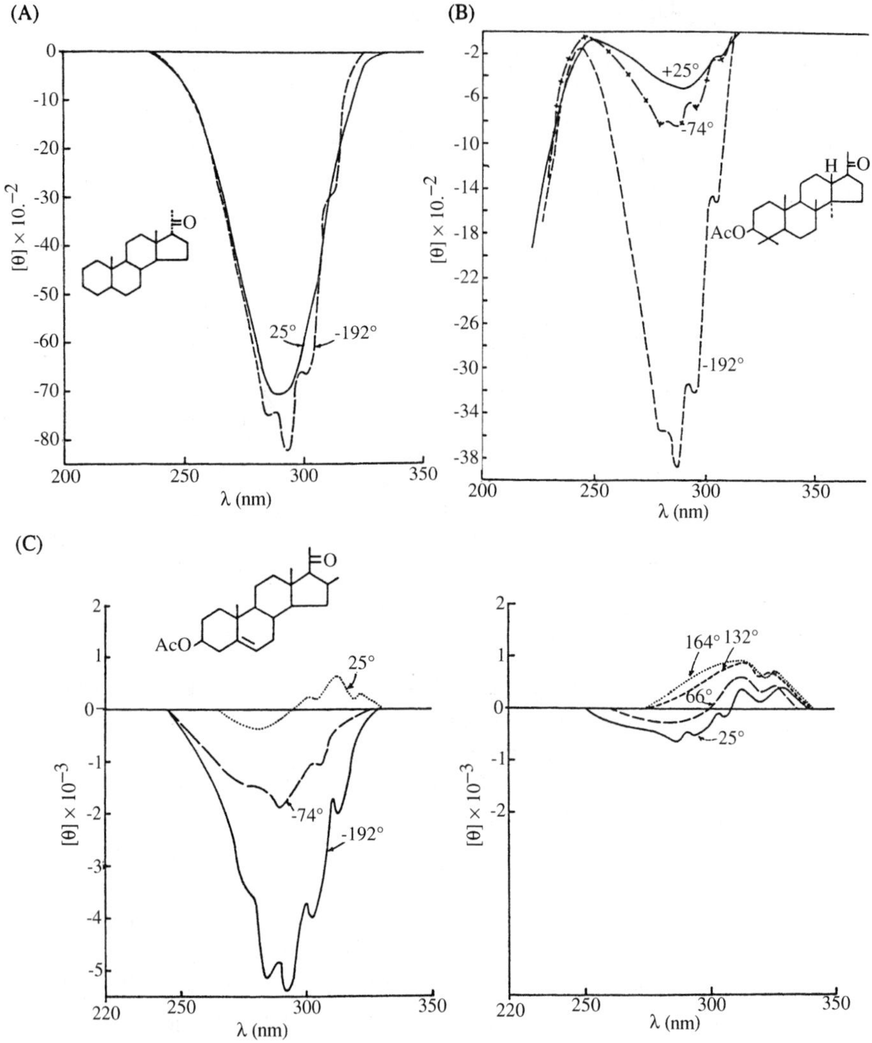

Figure 8-11. Variable-temperature CD spectra for the $n \rightarrow \pi^*$ transition of (A) 5α-pregnan-20-one in EPA, (B) 3β-acetoxyhexanordammar-20-one in EPA, and (C) 3β-acetoxy-16β-methyl-5-pregnen-20-one in EPA (left) and in decalin (right). EPA is ether–isopentane–ethanol (5:5:2 by vol.). [Reprinted with permission from ref. 40. Copyright © 1965 American Chemical Society.]

collection of X-ray crystallographic data[29] for 17β-acetyl steroids, the average rotameric conformation, with an average 16–17–20–O torsion angle of 0 to –46°, is closer to D of Table 8-10 than to any of the other conformations shown. D and E are predicted by the octant rule to exhibit strong positive Cotton effects; B is predicted to exhibit a strong negative Cotton effect. The small temperature effect on the CD, in which the Cotton effect is positive at 25° and becomes slightly more positive at –192°, is consistent with the relative stabilities of the rotamers predicted by molecular mechanics calculations.

The removal of the C(18) angular methyl, as in 3β-acetoxyhexanordammar-20-one (Figure 8-11B), leads to a weak negative $n \rightarrow \pi^*$ Cotton effect at room temperature, which becomes very strongly negative at –192°C: $[R]^{+25°C} = -0.51$ and $[R]^{-192°C} = -3.67$.[39] Molecular mechanics calculations[20] predict that the most stable rotamer

Table 8-10. Conformational Energy Differences and Torsion Angles (ψ) Calculated form Molecular Mechanics[a] for 20-Keto Steroids

Predicted Cotton Effect for 17β		–	–	–	+	+
Ketone/Conformation		(A) 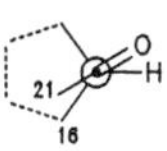	(B) 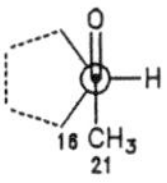	(C) 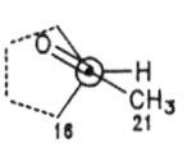	(D) 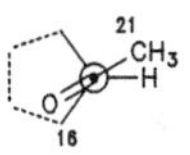	(E) 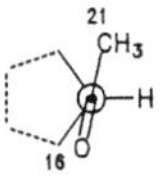
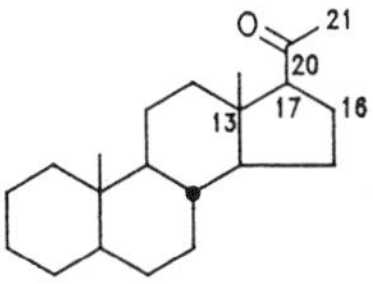	$\Delta\Delta H_f$	1.8	0.23	1.7	0.0	0.0
	ψ(H–17–20–O)	38.40	101.1	163.7	–154.3	–125.9
	ψ(H–17–20–21)	–142.7	–76.75	–17.23	25.33	52.33
	ψ(16–17–20–21)	–24.33	40.36	99.46	143.7	171.0
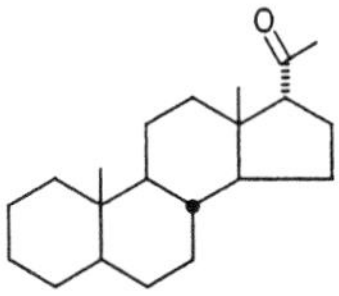	$\Delta\Delta H_f$	2.3	4.1	1.3	0.0	1.8
	ψ(H–17–20–O)	–40.32	–83.81	–158.9	158.7	136.1
	ψ(H–17–20–21)	140.1	92.48	22.30	–20.33	–42.35
	ψ(16–17–20–21)	21.44	–23.42	–96.92	–139.8	–159.7
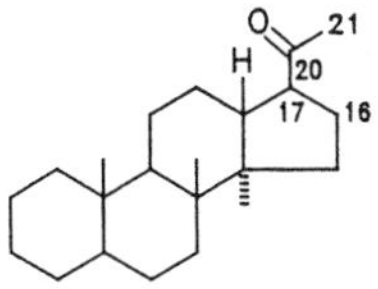	$\Delta\Delta H_f$	1.5	0.85	0.0	0.4	1.2
	ψ(H–17–20–O)	23.02	101.1	120.15	–173.0	–142.6
	ψ(H–17–20–21)	–158.1	–77.96	–59.00	7.33	37.83
	ψ(16–17–20–21)	–36.25	43.04	62.00	129.3	159.0
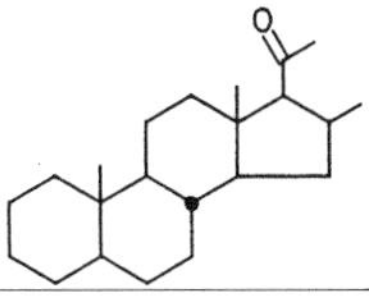	$\Delta\Delta H_f$	3.4	2.4	0.77	0.0	1.1
	ψ(H–17–20–O)	27.35	103.8	150.0	–167.6	–138.1
	ψ(H–17–20–21)	–154.3	–72.41	–30.77	12.83	42.76
	ψ(16–17–20–21)	–38.41	43.55	85.43	130.6	160.5

[a]Using PCMODEL, ref. 20; $\Delta\Delta H_f$ in kcal/mole.

(Table 8-10C) lies some 1.5 kcal/mole below a local minimum (Table 8-10A). The other rotameric conformations (Tables 8-10B, D, and E) are not minima. Since the minimum-energy conformation is expected to exhibit a negative $n \rightarrow \pi^*$ Cotton effect, but lie in a broad well in which a positively contributing rotamer (Table 8-10D) may also contribute, it is not surprising that the Cotton effect at room temperature is only weakly negative and that the low-temperature Cotton effect is more strongly negative.

When a 16β-methyl group is added to 17β-pregnenolone acetate, a weak bisignate $n \rightarrow \pi^*$ Cotton effect is found at room temperature and a strong negative Cotton effect is encountered at −192°C (Figure 8-11C). Molecular mechanics calculations[20] predict that the most stable rotamer (Table 8-10D) is one predicted by the octant rule to give a positive $n \rightarrow \pi^*$ Cotton effect and that it lies in a broad potential-energy well in which negatively contributing rotamers (Table 8-10C) are found. In this example, the variable-temperature CD spectra do not match up well with the rotamer stabilities predicted by molecular mechanics.

Variable-temperature CD measurements have also been carried out to examine possible effects of rotameric contributions of 2-isopropyl substituents on the CD of 5α-cholestan-3-one.[37] Thus, in 2α-isopropyl-5α-cholestan-3-one when an equatorial isopropyl is present, the $n \rightarrow \pi^*$ Cotton effect magnitudes increase when the temperature is lowered: $[R]^{+25°C} = +1.65$, $[R]^{-74°C} = +1.73$, and $[R]^{-192°C} = +2.30$. A similar result was seen (Figure 8-12A) with a 19-nor analog, 2α-isopropyl-19-nor-5α-andro-

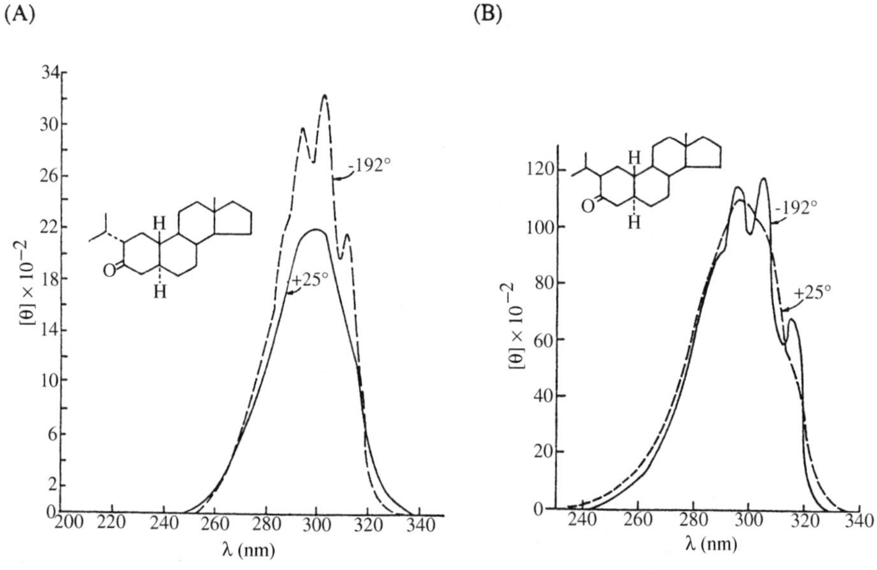

Figure 8-12. Circular dichroism $n \rightarrow \pi^*$ Cotton effect of (A) 2α-isopropyl-19-nor-5α-androstan-3-one and (B) 2β-isopropyl-19-nor-5α-androstan-3-one at +25°C and −192°C in ether–isopentane–ethanol (EPA, 5:5:2 by vol.). [Reprinted with permission from ref. 37. Copyright © 1965 American Chemical Society.]

stan-3-one: $[R]^{+25°C} = +2.05$, $[R]^{-74°C} = +2.15$, and $[R]^{-192°C} = +2.67$. (The slightly elevated rotatory strength of the 19-nor analog is consistent with the fact that the parent ketone, 5α-androstan-3-one, has $[R]^{+25°C} = +3.93$, which is slightly higher than that of 5α-cholestan-3-one, $[R]^{+25°C} = +3.57$.)[37] In contrast, the epimeric 2β-isopropyl-19-nor-5α-androstan-3-one exhibits very little temperature dependence of its $n \rightarrow \pi^*$ CD (Figure 8-12B), suggesting little conformational variation over the temperature range. Analysis of the various staggered rotamers by molecular mechanics computations[20] (Table 8-11) indicates that, for the 2β-axial isopropyl, one rotamer is substantially lower in energy (by about 2 kcal/mole; see Table 8-11D) than the other two staggered rotamers. In contrast, the three staggered rotamers of the 2α-equatorial isopropyl differ

Table 8-11. Comparison of Relative Energies ($\Delta\Delta H_f$) and Torsion Angles of Staggered Isopropyl Rotamers in 2α-Isopropyl and 2β-Isopropyl-19-nor-5α-Androstan-3-ones[a]

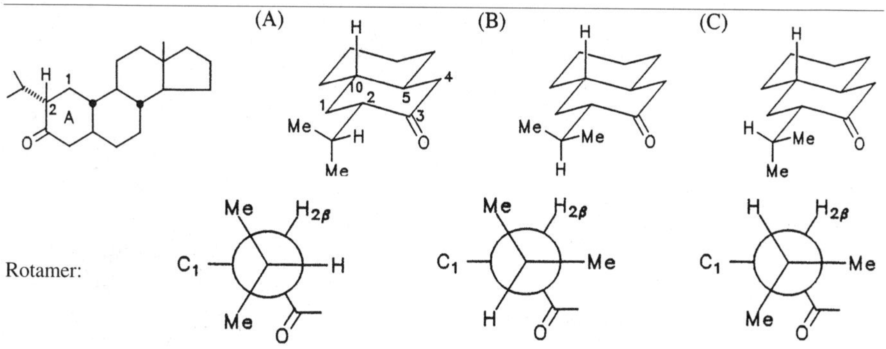

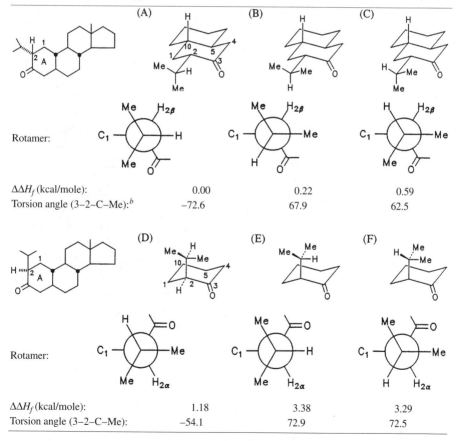

	(A)	(B)	(C)
$\Delta\Delta H_f$ (kcal/mole):	0.00	0.22	0.59
Torsion angle (3–2–C–Me):[b]	–72.6	67.9	62.5

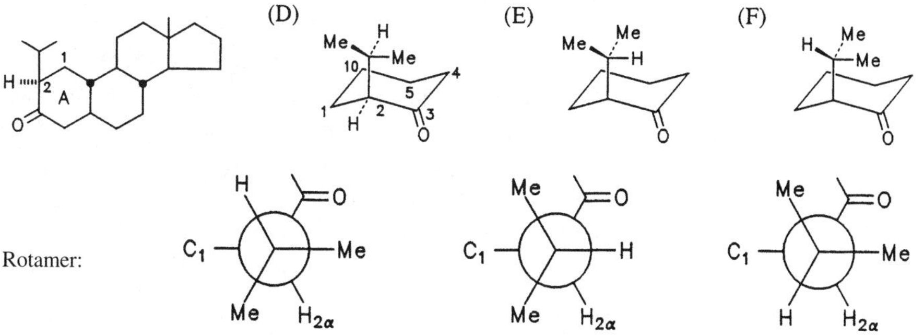

	(D)	(E)	(F)
$\Delta\Delta H_f$ (kcal/mole):	1.18	3.38	3.29
Torsion angle (3–2–C–Me):	–54.1	72.9	72.5

[a] Relevant values from PCMODEL molecular mechanics calculations, ref. 20.

[b] Me nearest to carbonyl or, when ambiguous, the *pro-S* methyl.

little in energy (Tables 8-11A–C), with the most stable being only 0.22 or 0.59 kcal/mole more stable than the other two. The computed[20] order of rotamer stability (A > B > C) differs from that indicated earlier (B > A > C)[41] and is supported by the CD analysis proposed.[37] Since the rotatory strength increases upon going to lower temperatures, conformer (A) is expected to exhibit a more positive Cotton effect than (B) or (C). This expectation is supported by an octant rule analysis. The difference between (A) and (B) lies in the proximity to the carbonyl of a *pro-S* methyl in (A) and a *pro-R* methyl in (B). The *pro-R* methyl lies in a negative front octant, the *pro-S* methyl in a positive front octant. Consequently, the *pro-S* methyl is expected to make a more positive octant contribution, and, assuming that all other contributions are equal, conformer (A) should have a more positive net Cotton effect than conformer (B).

8.5. Influence of Solvent on the Cotton Effects of Steroid Ketones

In the previous section, we noted that the $n \rightarrow \pi^*$ Cotton effects of 5α-androstan-16-one, 5α-cholestan-2-one and 5α-cholestan-3-one were essentially temperature invariant in the glass-forming solvent ether–isopentane–ethanol (EPA, 5:5:2 by volume). Over the temperature range +25°C to −192°C, [*R*] changes by less than 3%. Such data support the presence of but a single molecular species in solution. On the other hand, large changes in rotatory strength have been measured for 5α-cholestan-3-one at room temperature with only a change in solvent (Table 8-12). And unlike the variable-temperature CD in EPA, in a hydrocarbon solvent the positive $n \rightarrow \pi^*$ Cotton effects of 5α-cholestan-3-one are observed to undergo substantial change (Figure 8-13A). The rotatory strength at −192°C is approximately double that at 25°C. The −192°C CD curve is blueshifted, and does not exhibit the expected vibrational fine structure typically seen at low temperatures. (See, for example, Figures 8-8 to 8-11.) The latter two features are indicative of an increase in solvation compared with that existing at room temperature.[38]

A similar behavior is exhibited by 2α-*tert*-butyl-5α-cholestan-3-one, whose $n \rightarrow \pi^*$ Cotton effect in isopentane–methylcyclohexane is bisignate at 25°C (Figure 8-13B). The positive component shrinks in going from 25°C to −74°C and then grows from

Table 8-12. Solvent-Dependence of the Rotatory Strength of 5α-Cholestan-3-one[a]

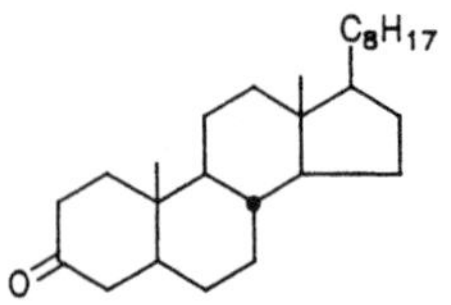

Rotatory Strength (× 10^{40} cgs) in:					
i-octane	benzene	dioxane	CH_3CN	CH_3OH	$CHCl_3$
+2.52	+2.73	+3.17	+3.41	+3.94	+4.32

[a]Data from ref. 40.

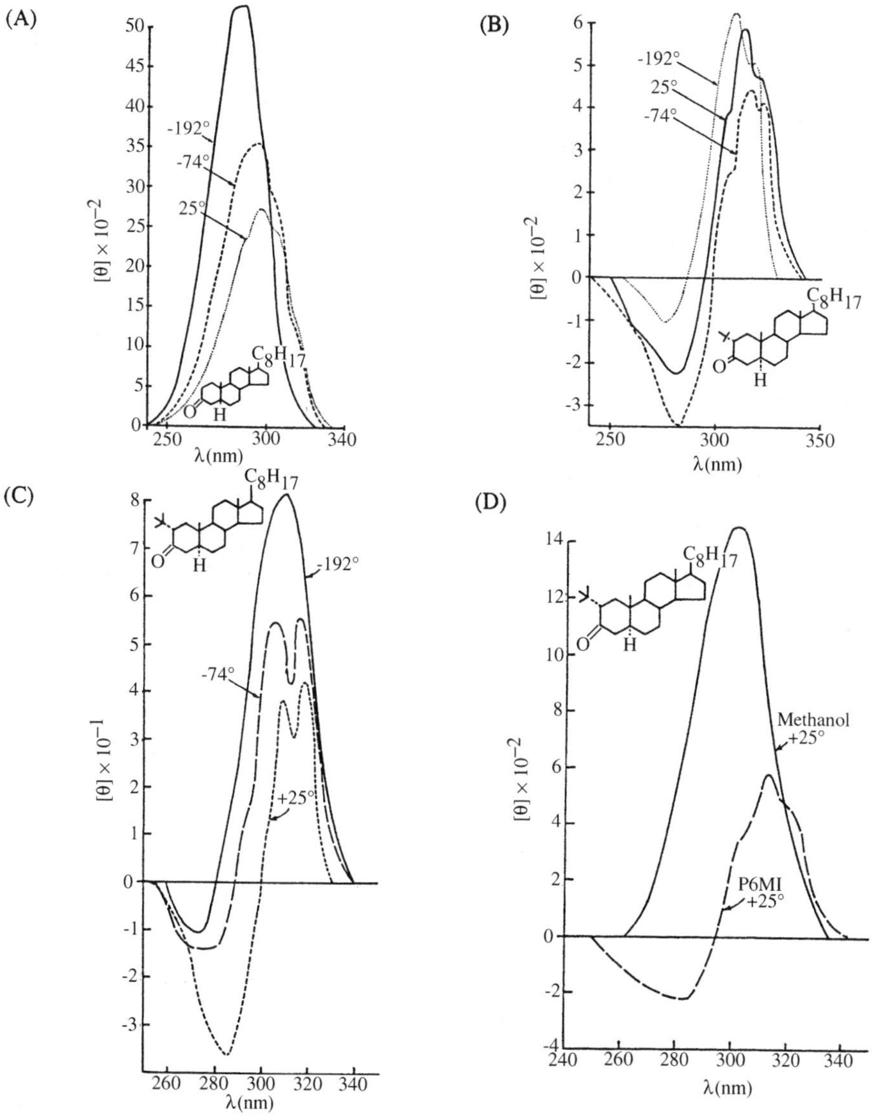

Figure 8-13. (A) Variable-temperature CD spectra of 5α-cholestan-3-one in isopentane-methylcyclohexane, 5:1 by vol. (B) Variable-temperature CD spectra of 2α-*tert*-butyl-5α-cholestan-3-one in the same solvent as (A). (C) Variable-temperature CD spectra of 2α-*tert*-butyl-5α-cholestan-3-one from +25°C to −192°C in ether–isopentane–ethanol (EPA, 5:5:2 by vol.). (D) Influence of solvent on the CD of 2α-*tert*-butyl-5α-cholestan-3-one. (P6MI is isopentane–methylcyclohexane, 6:1 by vol.) [(A) and (B) are reprinted from ref. 38 with permission of the authors. (C) and (D) are reprinted with permission from ref. 37. Copyright© 1965 American Chemical Society.]

−74°C to −192°C, while the negative component grows and then shrinks concomitantly. There is less vibrational fine structure at −192°C than at 25°C. Unlike 5α-cholestan-2-one, whose $n \rightarrow \pi^*$ Cotton effects remain invariant in EPA solvent[37] and show increased vibrational fine structure at −192°C (Figure 8-8A), 2α-*tert*-butyl-5α-cholestan-3-one in EPA (Figure 8-13C) gives bisignate CDs with no vibrational fine structure at −192°C. Unlike the compound's behavior in the hydrocarbon solvent, in EPA the positive Cotton effect steadily increases and the negative Cotton effect steadily decreases in going from 25°C to −74°C to −192°C.

Such low-temperature behavior *cannot* be associated exclusively with, or taken superficially as evidence for, conformational equilibrium, such as that seen in the 2α-isopropyl analogs of the previous section. In fact, since all staggered rotamers of *tert*-butyl are equivalent in energy, a rotameric conformational equilibrium akin to that of the isopropyl seems rather unlikely. Yet, the variable-temperature CD (Figures 8-13B, C) clearly indicates the presence of at least two species in equilibrium in EPA and more than two forms in equilibrium in isopentane–methylcyclohexane. The general behavior may be taken as evidence for a solvation equilibrium between solvated and unsolvated species. Although solvent perturbs the Cotton effect in a major way (Figure 8-13D), specific structures for solvated and unsolvated species remain unclear at the present time. Molecular mechanics calculations[20] on the unsolvated molecule indicate a normal staggered *tert*-butyl rotamer and only a very slightly deformed A ring (Table 8-13).

Other evidence of solvent effects on the $n \rightarrow \pi^*$ Cotton effect of ketones may be found in a comprehensive 1970 study by Kirk, Klyne, and Wallis.[42] In this study, trifluoroethanol was found to yield the largest effects on the $n \rightarrow \pi^*$ ORD Cotton effect amplitudes, attributed to H-bonding to the ketone oxygen.[43] Chloroform was placed

Table 8-13. Stable Rotameric Conformation of 2α-*tert*-Butyl-5α-cholestan-3-one and Comparison of its A-Ring Torsion Angles with Those of 5α-cholestan-3-one and Cyclohexanone[a]

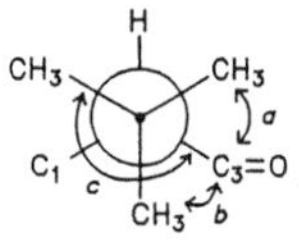

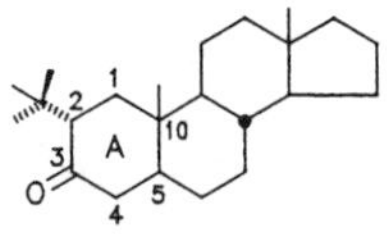

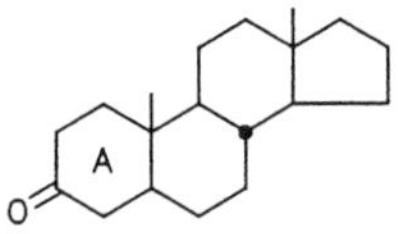

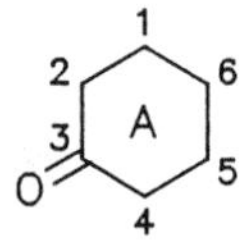

CH_3–C–C_2–C_3	Torsion Angle	(°)	(°)	(°)
a: 61.38	1–2–3–4	48.85	46.03	49.8
b: −61.38	2–3–4–5	−52.56	−47.44	−49.8
c: −178.38	3–2–1–10	−51.03	−50.68	−52.8
	3–4–5–10	56.17	53.75	52.8
	2–1–10–5	54.77	54.88	57.7
	1–10–5–4	−56.07	−56.54	−57.7

[a]Calculated using PCMODEL molecular mechanics, ref. 20.

next to trifluoroethanol in the solvent-effect order. Like trifluoroethanol, chloroform strongly hydrogen bonds to ketones. Since the work of Kirk et al.,[42] there have relatively few reports of solvent effects on ketone CD spectra and no further studies at variable temperatures of "asymmetric" solvation.

8.6. Bridged-ring Polycyclic Ketones

Only a modest number of polycyclic bridged-ring ketones have been investigated by ORD or CD spectroscopy.[9d] However, these typically rigid, conformationally fixed structures have provided some of the most intriguing and fundamentally important information relating stereochemistry to chiroptical properties. Among the earliest important studies[44] were those conducted on ketones derived from tricyclic adamantane, which is highly symmetric (T_d), is achiral, and consists of four unstrained chair cyclohexane rings (Figure 8-14). Adamantanone, also symmetric and achiral, is among the most studied, since it provides a conformationally immobile chair cyclohexanone while retaining the qualitative completeness[45,46] of the octant rule (Chapter 4). Because adamantanone is achiral, it is predicted by the octant rule to exhibit no $n \rightarrow \pi^*$ Cotton effect; yet even a minor modification, as in [4-^{13}C]adamantan-2-one[47] or homoadamantan-2-one[48] (Figure 8-15) can lead to a weak Cotton effect with considerable vibrational fine structure. However, earlier ORD and CD investigations of 4-substituted adamantanones were among the most extensive and important. Studies of adamantanones with C(4) axial substituents were the first unambiguous examples of "antioctant" effects[44] and subsequently led to a clarification of the octant rule boundaries.[49] And studies of adamantanones with *equatorial* C(4) substituents have provided standards for quantifying octant contributions of β-equatorial groups on cyclohexanones (Chapter 4).[44,49] More recently, isotopically substituted (^{2}H,^{13}C,^{17}O) adamantanones served to ignite an interest in understanding the effects of isotopes on the octant rule and on conformation (Chapter 9).

Although adamantane is highly symmetric and consists entirely of (four) chair cyclohexanes, its isomer, twistane (Figure 8-14), consists only of (five) twist-boat cyclohexanes and is dissymmetric (D_2). Twistane is chiral, with all twist boats having the same helical sense. Its two distinct ketones (Figure 8-14) are also chiral, and each has the cyclohexanone constrained to a twist-boat conformation. The absolute configuration of twistan-2-one was the subject of early controversy[50,51,52] resulting from its assignment based on the octant rule and, specifically, what became known as the Djerassi–Klyne rule[24] for twist-boat cyclohexanones. An examination of the octant projection diagram for twist-boat cyclohexanone (Table 8-14A) places the only octant contributors (C(3) and C(5)) in octants having the same sign, and the contributions were thought to dominate the sign of the Cotton effect in the examples studied.[24] When the Djerassi–Klyne rule applied to (+)-twistan-2-one, more weight was given to these atoms than to the other two octant contributors (denoted by asterisks (*) in Table 8-14B), with the effect that the observed positive $n \rightarrow \pi^*$ Cotton effect was thought to predict an absolute configuration opposite to that shown.[50] Subsequently, it was

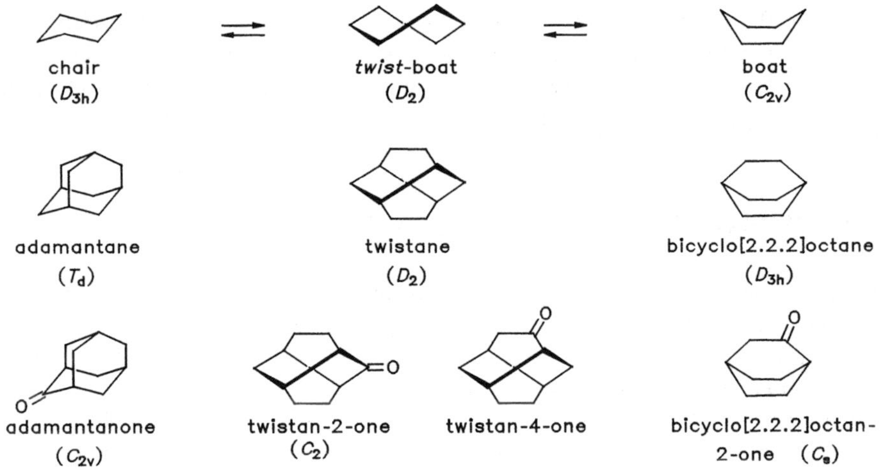

Figure 8-14. (Upper) Conformational isomers of cyclohexane. (Middle) The same conformational isomers locked into polycyclic analogs. (Lower) Corresponding conformationally locked ketones. Only twistane and its ketones are chiral. The symmetry group is noted below each name.

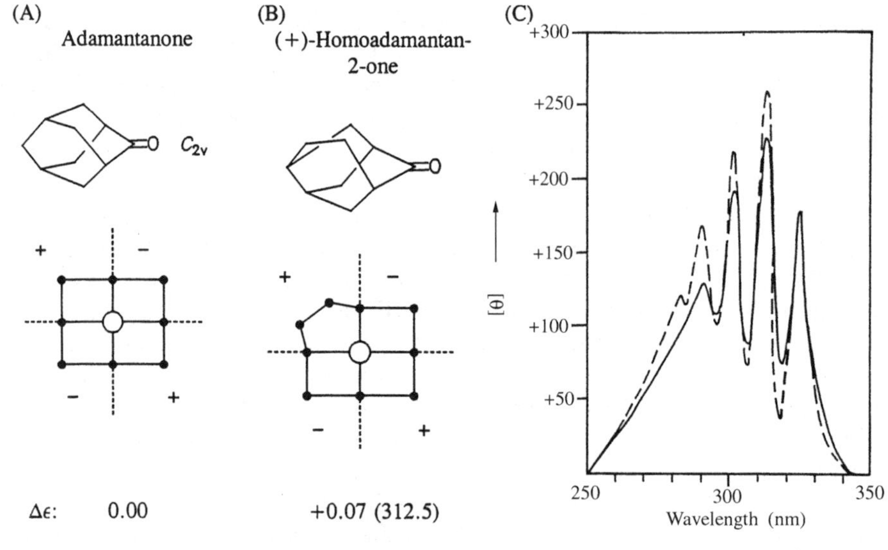

Figure 8-15. (A) Adamantanone (achiral) and its octant diagram. (B) (+)-Homoadamantan-2-one and its octant diagram. (C) CD spectrum of (+)-homoadamantan-2-one at 25°C (———) and –68°C (— — —) in methylcyclohexane–isopentane, 5:1 by vol. [(C) is reprinted with permission from ref. 48. Copyright© 1977 Chemical Society of Japan.]

demonstrated by independent experiments that the absolute configuration should be reversed (corresponding to that in Table 8-14B).[51,52,53] These findings also indicated that the contributions of C(3) and C(5) in twist-boat cyclohexanone were not dominant, according to the octant rule. They prompted the following theoretical explanation by T. D. Bouman[54]:

> In 1962, Djerassi and Klyne[24] suggested that a twisted boat conformation of the 6-membered ring containing the carbonyl group might explain the abnormally large rotatory amplitudes of certain ketones. In such a conformation, the plane formed by the three carbons farthest from the carbonyl makes an angle of as much as 45° with the plane of the latter. [A projection looking from O to C for twist-boat cyclohexanone or its enantiomer is shown in Table 8-14A.] Djerassi and Klyne predict a strong contribution from C(3) and C(5), in a sense obeying the octant rule, based on a similar effect observed in derivatives of cyclopentanone.
>
> On the basis of the Djerassi–Klyne rule and on applying the octant rule to a carbonyl substituent, absolute configurations were assigned[50] to (+)-2-twistanone and (+)-4-twistanone which were subsequently shown by stereospecific synthesis to refer to their enantiomers [Tables 8-14B and C].[51,53] This was cited as a failure of the two rules, although an *a posteriori* reconciliation of the observed structures with those rules was attempted.[52]
>
> To examine the role of the twisted-boat ring in determining the observed chiroptical properties, we calculated [CNDO/S][55] the rotatory strengths for the two isomeric twistanones mentioned, and also for cyclohexanone itself in the twist-boat conformation. [The results are given in Table 8-14.] The $<n|\partial/\partial z|\pi^*>$ matrix element was analyzed as above into one- and two-center contributions, weighted by the product of MO coefficients.[55] It was seen that the contribution from the twisted

Table 8-14. Twistanones and their Octant Diagrams and $n \rightarrow \pi^*$ Cotton Effects

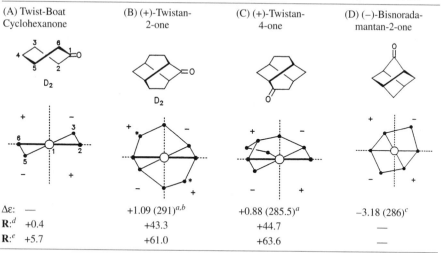

	(A) Twist-Boat Cyclohexanone	(B) (+)-Twistan-2-one	(C) (+)-Twistan-4-one	(D) (−)-Bisnoradamantan-2-one
	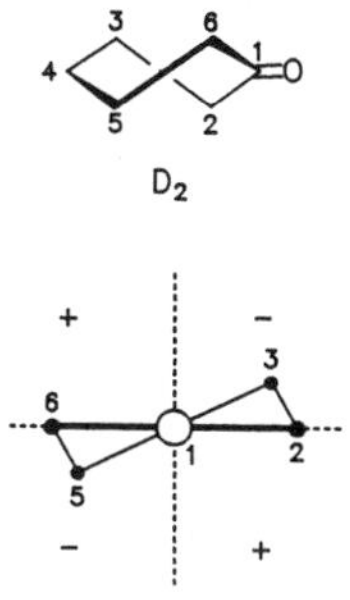	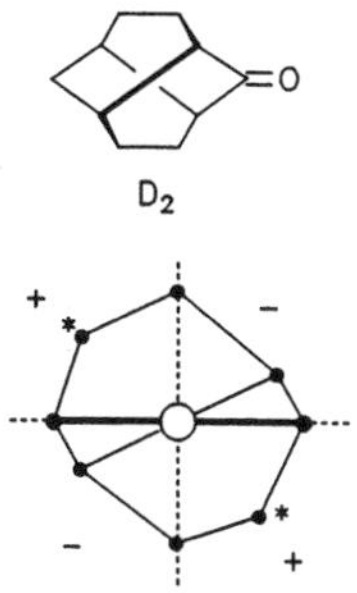	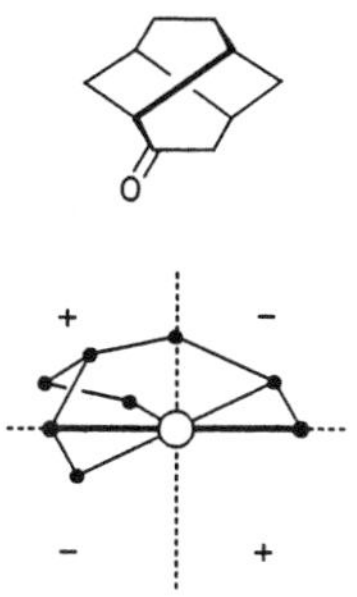	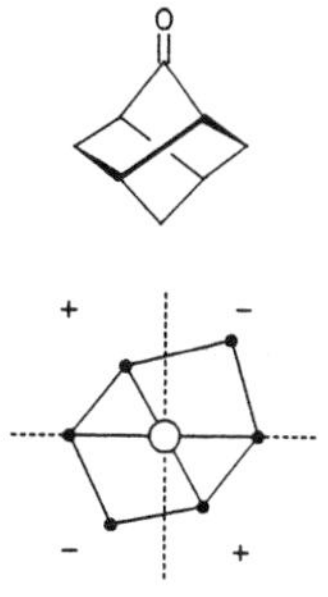
Δε:	—	+1.09 (291)[a,b]	+0.88 (285.5)[a]	−3.18 (286)[c]
R:[d]	+0.4	+43.3	+44.7	—
R:[e]	+5.7	+61.0	+63.6	—

[a] Data from ref. 53, in ethanol.

[b] Given in ref. 50 as +10.6 erroneously for the enantiomeric twistan-2-one.

[c] Data from ref. 57 in isooctane.

[d] Rotatory strength (dipole length form), **R** = val × 10^{-40} cgs.

[e] Rotatory strength (dipole velocity form), **R** = val × 10^{-40} cgs. These values were calculated by Prof. Thomas D. Bouman (ref. 54) using CNDO/S methds (ref. 55).

ring consists mostly of C–C bond terms largely canceled by oppositely signed C–H and nonbonded terms. Only a small moment is induced in the C–O bond itself by the dissymmetry of the molecule as a whole. By contrast, in the twistanones, a very large term appears in the C–O bond, augmented by large values from atoms lying along a primary zigzag![9d]

These results suggest that large rotatory amplitudes associated with a twisted 6-membered ring conformation result not from the inherent chirality of the ring itself, but from either (i) many atoms being brought into the same octant by the twist, as in the examples originally adduced by Djerassi and Klyne, or (ii) a primary zigzag[9a,d] being extended by the bridging atoms holding the central ring in the twisted conformation, as in 2-twistanone. The twisted ring itself contributes only weakly to the CD, and in an anti-octant sense. This result is to be contrasted with the behavior of a twisted cyclopentanone fragment, as demonstrated by Richardson.[56] In their calculations, the chirality of the ring was the dominant factor for the CD of β-substituted derivatives, while for α-substituents the substituent itself determined the chiroptical properties.

Subsequent investigations of other polycyclics with twist-boat cyclohexanes indicated that octant contributors lying near the symmetry planes make much weaker contributions than those lying at equivalent distances midway between the planes, as in the moderately strong $n \rightarrow \pi^*$ Cotton effect seen in (–)-bis-noradamantan-2-one[57] (Table 8-14D).

Following the resolution of the absolute configuration of the twistanones, Nakazaki's group at Osaka University initiated a comprehensive study of gyrochiral hydrocarbons and ketones.[58] The core structure in this work is bicyclo[2.2.2]octane, which, when twisted slightly may adopt either of two enantiomeric propellerlike shapes with D_3 symmetry (Figure 8-16). By adding new bridges, the core may be held in either of two gyrochiral shapes, as illustrated for (–)-twistane by adding an ethano bridge to carbons 2 and 5 (or for (+)-twistane, by connecting carbons 3 and 6 with an

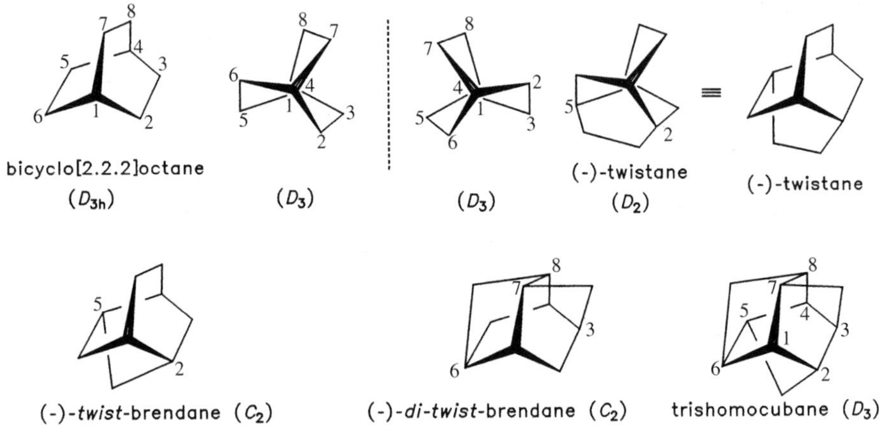

Figure 8-16. (Upper) Distortion of bicyclo[2.2.2]octane D_{3h} symmetry into twisted, chiral conformations with D_3 symmetry and held in a dissymmetric conformation by an ethano bridge between C(2) and C(5). (Lower) Examples of twist bicyclo[2.2.2]octanes in dissymmetric polycyclic hydrocarbons formed with one, two, and three methano bridges in *twist*-brendane, di-*twist*-brendane, and trishomocubane, respectively.

ethano bridge). A host of examples follow simply from adding methano bridges connecting bicyclo[2.2.2]octane carbons 2 and 5, 3 and 7, or 6 and 8. Thus, (–)-*twist*-brendane comes from connecting carbon 2 to carbon 5, 3 to 7 or 6 to 8 with a single methano bridge; (–)-di-*twist*-brendane comes from connecting any two of the pairs, each with a methano bridge; and D_3-trishomocubane comes from connecting each of the three with a methano bridge (Figure 8-16). The enantiomers would be reached by connecting carbon 2 to carbon 8, 3 to 6, or 5 to 7. The various gyrochiral bridged-ring polycyclics prepared mainly by Nakazaki et al.[58] are collected in Table 8-15.

The extensive synthetic and stereochemical studies of gyrochiral compounds by Nakazaki et al.,[48,50a,57–60] carried out over more than a decade, led to an impressive, almost bewildering, array of bridged-ring polycyclic ketones whose $n \rightarrow \pi^*$ Cotton effects were measured (Table 8-16). In all cases, the dominant octant contributions came not from the off-symmetry-plane atoms of twist-boat cyclohexanone, but from

Table 8-15. Gyrochiral Tricyclic, Tetracyclic and Pentacyclic Hydrocarbons from 2,5 and 3,7 and 6,8 Diagonal Bridging of D_3 Twisted Bicyclo-[2.2.2]octane[a]

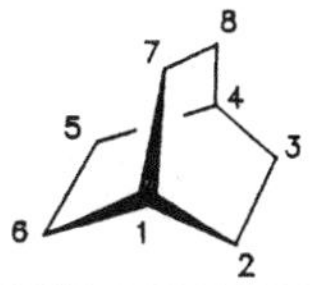

Gyrochiral	k	l	m	Trivial Name	Symmetry
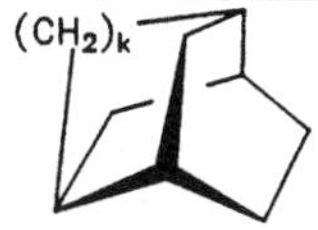	1	—	—	*twist*-brendane	C_2
	2	—	—	twistane	D_2
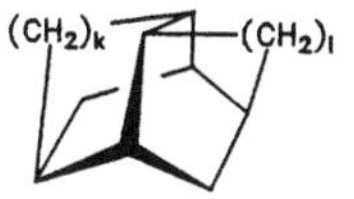	1	1	—	di-*twist*-brendane	C_2
	1	2	—	C_1-methanotwistane	C_1
	2	2	—	ditwistane	C_2
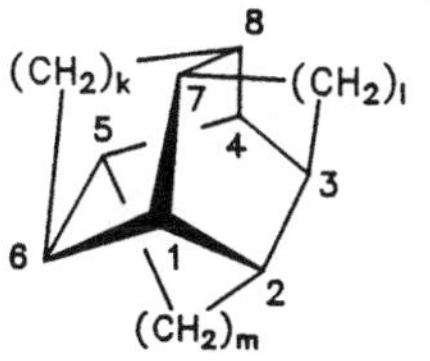	0	0	0	cubane	O_h
	0	0	1	homocubane	C_{2v}
	0	0	2	basketane	C_{2v}
	0	1	1	C_2-bishomocubane	C_2
	0	1	2	C_1-homobasketane	C_1
	0	2	2	C_2-dehydroditwistane	C_2
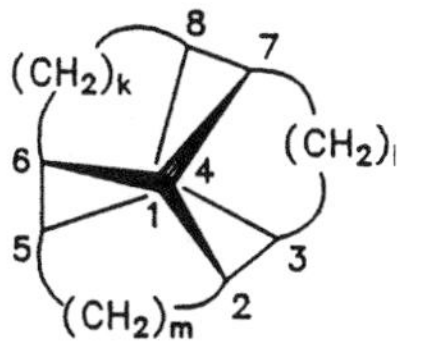	1	1	1	D_3-trishomocubane	D_3
	1	1	2	C_2-bismethanotwistane	C_2
	1	2	2	C_2-methanoditwistane	C_2
	2	2	2	D_3-tritwistane	D_3

[a]See ref. 58 for a summary of gyrochiral cage compounds and key references.

Table 8-16. Gyrochiral Polycyclic Ketones and their $n \rightarrow \pi^*$ Cotton Effects[a]

	(A) (+)-*Twist*-brendan-2-one	(B) (−)-*Twist*-brendan-9-one	(C) (−)-Brexan-2-one	(D) Norbrexan-2-one
		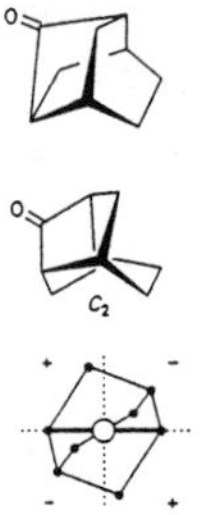	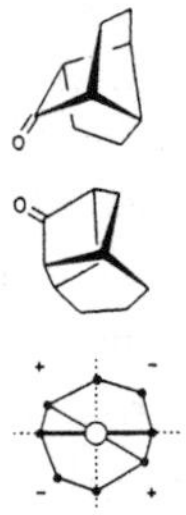	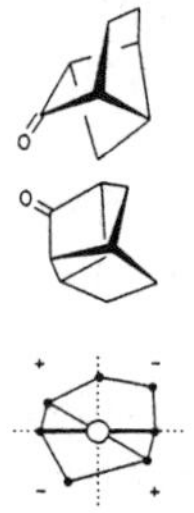
Δε:	−4.5 (289)[b,j]	+0.61 (285)[c,j]	−2.23 (300)[d]	−1.21 (296)[d,j]
	(E) (−)-Di-*twist*-brendan-3-one	(F) (−)-Di-*twist*-brendan-5-one	(G) (−)-C_1-Methano-twistan-1-one	(H) (+)-C_2-Bishomocuban-6-one
		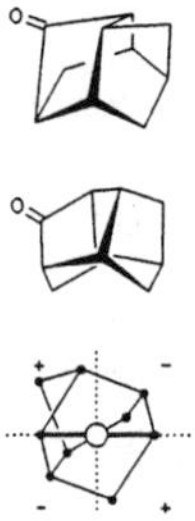	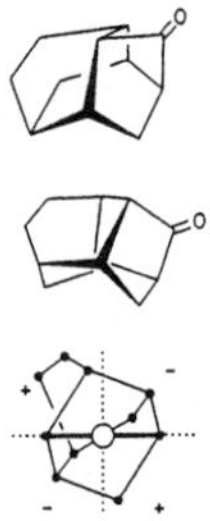	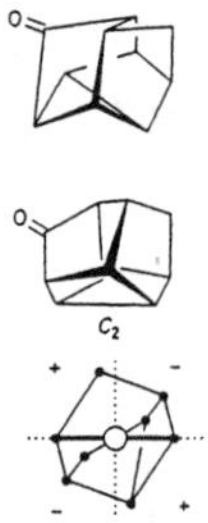
Δε:	−3.58 (295.5)[e]	+2.13 (293)[f]	+3.61 (290.5)[g]	+0.39 (301.5)[h]
	(I) (−)-C_1-Homobasketan-4-one	(J) (−)-D_3-Tris-homocuban-4-one	(K) (−)-C_2-Methano-ditwistan-4-one	(L) (−)-D_3-Tri-twistan-4-one
	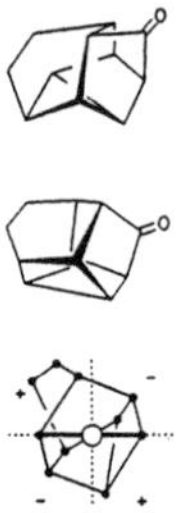	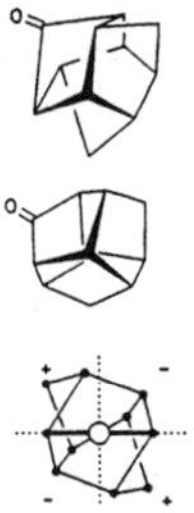	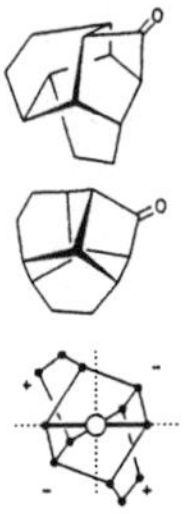	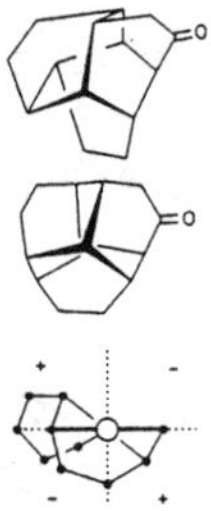
Δε:	+1.61 (301.5)[g]	+1.70 (292.5)[f]	+2.91 (291)[i]	+0.25 (303.5)[i]

[a] Measured in Isooctane, unless noted otherwise.
[b] Ref. 59a.
[c] Refs. 58c and 59b.
[d] Ref. 59c.
[e] Ref. 59f.
[f] Ref. 58c and 59e.
[g] Ref. 58b.
[h] Refs. 59d, 59e, and 61.
[i] Ref. 58c.
[j] Enantiomer of that reported in the literature.

ring atoms situated farther away from the octant planes, as in Tables 8-16A, C, D, and E. In many of the remaining entries, even the greater number of negative contributors is more than offset by the few positive contributors, though sometimes just barely, as in Tables 8-16H and L.

Nakazaki has reviewed a wide range of high-symmetry chiral organic compounds, including his own extensive studies of cage-shaped molecules.[62] He has proposed a generic name, "triblattane," for the parent hydrocarbons with diagonal bridging (Table 8-15). The tri-, tetra-, and pentacyclic cages are to be called [*n*]-, [*m.n*]-, and [*m.n.p.*]-triblattanes, where *m*, *n*, and *p* may be 0, 1, and 2, corresponding respectively to a single bond or to $-CH_2-$ or $-CH_2-CH_2-$ bridges. Thus, D_2-twistane, C_2-ditwistane, D_3-trishomocubane, and C_1-homo-basketane of Table 8-15 are called, simply, [2]-, [2.2]-, [1.1.1]-, and [2.1.0] triblattanes, respectively. The ketones would be called triblattanones.

Circular dichroism data for other bridged-ring polycyclic ketones—largely those derivatives of steroids and terpenes—may be found in a compilation of Kirk[9d] and in a more recent one by Boiadjiev and Lightner.[11]

References

1. Djerassi, C., *Optical Rotatory Dispersion*, McGraw-Hill, New York, 1960.
2. Moffitt, W., Woodward, W. B., Moscowitz, A., Klyne, W., and Djerassi, C., *J. Am. Chem. Soc.* **83** (1961), 4013–4018.
3. Djerassi, C., and Klyne, W., *J. Chem. Soc.* (1962), 4929–4944.
4. Djerassi, C., and Klyne, W., *J. Chem. Soc.* (1963), 2390–2402.
5. Crabbé, P., *Optical Rotatory Dispersion and Circular Dichroism in Organic Chemistry*, Holden-Day, San Francisco, 1965.
6. Velluz, L., Legrand, M., and Grosjean, M., *Optical Circular Dichroism*, Verlag Chemie, Weinheim, 1965.
7. Crabbé, P., *Applications de la Dispersion Rotatoire Optique et du Dichroisme Circulaire Optique en Chimie Organique*, Gauthier-Villars, Paris, 1968.
8. Legrand, M., and Rougier, M. J., in *Stereochemistry, Volume 2: Dipole Moments, CD or ORD* (Kagan, H. B., ed.), Geo. Thieme Publ., Stuttgart, 1977.
9. (a) Kirk, D. N., and Klyne, W., *J. Chem. Soc., Perkin Trans.* **1** (1974), 1076–1103.
   (b) Fernandez, F., Kirk, D. N., and Scopes, M., *J. Chem. Soc., Perkin Trans.* **1** (1974), 18–21.
   (c) Kirk, D. N., *J. Chem. Soc. Perkin Trans.* **1** (1976), 2171–2177.
   (d) Kirk, D. N., *J. Chem. Soc. Perkin Trans.* **1** (1977), 2122–2148.
   (e) Kirk, D. N., *J. Chem. Soc., Perkin Trans.* **1** (1980), 787–803.
   (f) Kirk, D. N., *J. Chem. Soc., Perkin Trans.* **1** (1980), 1810–1819.
10. For a review of the literature from 1977 to mid-1995, see Boiadjiev, S., and Lightner, D. A., "Chiroptical Properties of Compounds Containing C=O Groups," in *Supplement A3: The Chemistry of Double-Bonded Functional Groups* (Patai, S., ed.), John Wiley & Sons, Ltd., Chichester, U.K., 1997, Chap. 5. For earlier work, see Crabbé, P., *ORD and CD in Chemistry and Biochemistry*, Academic Press, New York, 1972.
11. Boiadjiev, S. E., and Lightner, D. A., in *The Chemistry of Double Bond Functional Groups* (Patai S., and Rappoport Z., eds.), J. Wiley, Ltd., Chichester, Sussex, U. K., 1997.
12. For leading references, see Eliel, E. L., and Wilen, S. H., *Stereochemistry of Organic Compounds*, J. Wiley & Sons, Inc., New York, 1994.

13. (a) Linstead, R. P., von Doering, W. E., Davis, S. B., Levine, P., and Whetstone, R. R., *J. Am. Chem. Soc.* **64** (1942), 1985–1991.
(b) Linstead, R. P., and Whetstone, R. R., *J. Chem. Soc.* (1950), 1428–1432.
14. Johnson, W. S., *J. Am. Chem. Soc.* **75** (1953), 1498–1500.
15. Dauben, W. G., and Pitzer, K. S., in *Steric Effects in Organic Chemistry* (Newman, M. S., ed.), John Wiley & Sons, New York, (1956), p. 33.
16. For leading references, see Hönig, H., and Allinger, N. L., *J. Org. Chem.* **50** (1985), 4630–4632.
17. Allinger, N. L., Yuh, Y. H., and Lii, J. -H., *J. Am. Chem. Soc.* **111** (1989), 8551–8566 (perhydrophenanthrene values in supplementary material).
18. Allinger, N. L., and Wuesthoff, M. T., *J. Org. Chem.* **36** (1971), 2051–2053.
19. Alcaide, B., and Ferandez, F., *J. Chem. Soc., Perkin Trans.* **1** (1983), 1665–1671.
20. PCMODEL, versions 5.0–7.0, Serena Software, Inc., Bloomington, IN 47402-3076.
21. Djerassi, C., Halpern, O., Halpern, V., and Riniker, B., *J. Am. Chem. Soc.* **80** (1958), 4001–4015.
22. Witz, P., Herrmann, H., Lehn, J- M., and Ourisson, G., *Bull. Soc. Chim. France* (1963), 1101–1112.
23. Allinger, N. L., and DaRooge, M. A., *J. Am. Chem. Soc.* **84** (1962), 4562–4567.
24. Djerassi, C., and Klyne, W., *Proc. Natl. Acad. Sci.* **48** (1962), 1093–1098.
25. For leading references, see Tsuda, T., and Kiuchi, F., *Chem. Pharm. Bull. Jpn.* **32** (1984), 4806–4819.
26. (a) Zürcher, R. F., *Helv. Chim. Acta* **46** (1963), 2054–2088.
(b) Zürcher, R. F., *Helv. Chim. Acta* **44** (1961), 1380–1395.
27. (a) Burkert, U., and Allinger, N. L., *Tetrahedron* **34** (1978), 807–809.
(b) Allinger, N. L., Burkert, U., and De Camp, W. H., *Tetrahedron* **33** (1977), 1891–1895.
28. Dougherty, D. A., Mislow, K., Huffman, J. W., and Jacobus, J., *J. Org. Chem.* **44** (1979), 1585–1589.
29. Duax, W. L., Weeks, C. M., and Rohrer, D. C., "Crystal Structures of Steroids," in *Topics in Stereochemistry* Vol. 9 (Allinger, N. L., and Eliel, E. L., eds.), John Wiley & Sons, New York, 1976.
30. Ferguson, G., Macauley, E. W., Midgley, J. M., Robertson, J. M., and Whalley, W. B., *J. Chem. Soc., Chem. Commun.* (1970), 954–955.
31. Wertz, D., and Allinger, N. L., *Tetrahedron* **30** (1974), 1579–1586.
32. Burkert, U., and Allinger, N. L., *Molecular Mechanics* ACS Monograph 177, American Chemical Society, Washington, DC, 1982.
33. Tsuda, Y., Yamashita, T., and Sano, T., *Chem. Pharm. Bull. Jpn.* **32** (1984), 4820–4832.
34. Tsuda, Y., Yamaguchi, K., and Sakai, S., *Chem. Pharm. Bull. Jpn.* **32** (1984), 313–321.
35. Djerassi, C., *Proc. Chem. Soc.* (1964), 314–330.
36. Wellman, K. M., Bunnenberg, E., and Djerassi, C., *J. Am. Chem. Soc.* **85** (1963), 1870–1872.
37. Wellman, K. M., Briggs, W. S., and Djerassi, C., *J. Am. Chem. Soc.* **87** (1965), 73–81.
38. Moscowitz, A., Wellman, K. M., and Djerassi, C., *Proc. Natl. Acad. Sci. (US)* **50** (1963), 799–804.
39. Lehn, J -M., Levisalles, J., and Ourisson, G., *Bull. Soc. Chim. France* (1963), 1096–1101.
40. Wellman, K. M., and Djerassi, C., *J. Am. Chem. Soc.* **87** (1965), 60–66.
41. Cotterill, W. D., and Robinson, M. J. T., *Tetrahedron* **20** (1964), 777–790.
42. Kirk, D. N., Klyne, W., and Wallis, S. R., *J. Chem. Soc. (C)* (1970), 350–360.
43. Beecham, A. F., and Hurley, A. C., *Austral. J. Chem.* **32** (1979), 1643–1648.
44. (a) Snatzke, G., Ehrig, B., and Klein, H., *Tetrahedron* **25** (1969), 5601–5609, and references therein.
(b) Snatzke, G., and Eckhardt, G., *Tetrahedron* **26** (1970), 1143–1155.
45. Yeh, C -Y., and Richardson, F. S., *Theor. Chim. Acta* **43** (1977), 253–260.
46. (a) Lightner, D. A., and Toan, V. V., *J. Chem. Soc. Chem. Commun.* (1987), 210–211.
(b) Lightner, D. A., and Toan, V. V., *Tetrahedron* **43** (1987), 4905–4916.
47. Sing, Y. L., Numan, H., Wynberg, H., and Djerassi, C., *J. Am. Chem. Soc.* **101** (1979), 5155–5158. Errata. *Ibid.* **101** (1979), 7439.
48. Nakazaki, M., and Naemura, K., *J. Org. Chem.* **42** (1977), 4108–4113.
49. (a) Lightner, D. A., Chang, T. C., Hefelfinger, D. T., Jackman, D. E., Wijekoon, W. M. D., and Givens, J. W., III, *J. Am. Chem. Soc.* **107** (1985), 7499–7508.
(b) Lightner, D. A., Bouman, T. D., Wijekoon, W. M. D., and Hansen, Aa. E., *J. Am. Chem. Soc.* **108** (1986), 4484–4497.
50. (a) Adachi, K., Naemura, K., and Nakazaki, M., *Tetrahedron Lett.* (1968), 5467–5470.

(b) Tichý, M., and Sicher, J., *Tetrahedron Lett.* (1969), 4609–4613.
51. Tichý, M., *Tetrahedron Lett.* (1972), 2001–2004.
52. Snatzke, G., and Werner-Zamojska, F., *Tetrahedron Lett.* (1972), 4275–4278.
53. Tichý, M., *Coll. Czech. Chem. Commun.* **39** (1974), 2673–2684.
54. Bouman, T. D., personal communication (late 1970s), edited.
55. Bouman, T. D., and Lightner, D. A., *J. Am. Chem. Soc.* **98** (1976), 3145–3154.
56. Richardson, F. S., Shillady, D. D., and Bloor, J. E., *J. Phys. Chem.* **75** (1971), 2466–2479.
57. (a)Nakazaki, M., Naemura, K., and Arashiba, N., *J. Chem. Soc. Chem. Commun.* (1978), 678–679.
(b)Nakazaki, M., Naemura, K., and Arashiba, N., *J. Org. Chem.* **43** (1978), 888–891.
58. For leading references, see:
(a)Nakazaki, M., Naemura, K., Arashiba, N., and Iwasaki, M., *J. Org. Chem.* **44** (1979), 2433–2438.
(b)Nakazaki, M., Naemura, K., Kondo, Y., Nakahara, S., and Hashimoto, M., *J. Org. Chem.* **45** (1980), 4440–4444.
(c)Nakazaki, M., Naemura, K., Chikamatsu, G., Iwasaki, M., and Hashimoto, M., *J. Org. Chem.* **46** (1981), 2300–2310.
59. (a)Naemura, K., and Nakazaki, M., *Bull. Chem. Soc. Jpn.* **46** (1973), 888–892.
(b)Nakazaki, M., Naemura, K., and Harita, S., *Bull. Chem. Soc. Jpn.* **48** (1975), 1907–1913.
(c)Nakazaki, M., Naemura, K., and Kadowaki, H., *J. Org. Chem.* **41** (1976), 3725–3730.
(d)Nakazaki, M., and Naemura, K., *J. Org. Chem.* **42** (1977), 2985–2988.
(e)Nakazaki, M., Naemura, K., and Arashiba, N., *J. Org. Chem.* **43** (1978), 689–692.
(f)Nakazaki, M., Naemura, K., and Nakahara, S., *J. Org. Chem.* **43** (1978), 4745–4750.
60. Nakazaki, M., Naemura, K., and Kondo, Y., *J. Org. Chem.* **41** (1976), 1229–1233.
61. Eaton, P., and Leipzig, B., *J. Org. Chem.* **43** (1978), 2483–2484.
62. Nakazaki, M. In *Topics in Stereochemistry*, Volume 15 (Eds. Eliel, E.L.; Wilen, S.H.; Allinger, N.L.) Wiley-Interscience, New York 1984, pp 199–251.

CHAPTER

9

Isotopically Perturbed Ketones

In 1949, Eliel[1] showed that 1(*R*)-deuterio-1-phenylethane (Figure 9-1A), a molecule whose chirality is due solely to isotopic substitution, exhibited weak, but measurable, optical activity: $[\alpha]_D = -0.30°$. Subsequently, numerous other examples were reported in a review by Verbit[2] in 1970. In most cases the $[\alpha]_D$ values were less than 1°, and at the time, only three examples were known of attempts to measure the influence of isotopic perturbation by ORD spectroscopy. Results of the first attempt to study the influence of deuterium on the ketone $n \rightarrow \pi^*$ Cotton effect were published by Djerassi et al. in 1956.[3] In this investigation, the researchers could detect no difference between the ORD spectrum of 3β-acetoxy-6β-deuterio-5α-cholestan-7-one (Figure 9-1B) and that of the parent undeuterated ketone. Since the major contributions to the Cotton effect come largely from dissymmetric perturbers other than deuterium, their success would depend on knowing the exact concentrations and purity of the samples they used, as well as on maintaining identical operating conditions. These difficulties and ORD instrument sensitivity levels probably contributed to the failure of the investigation. Some 10 years later, Meyer and Lobo[4] prepared (+)-camphor-9,9,9-d_3 (Figure 9-1C) and measured an ORD amplitude (a = +60.92°) that was 3% smaller than that of the undeuterated parent. From this result, the authors suggested that deuterium makes an octant-dissignate (or "antioctant") contribution to the octant rule.

A better model is one in which the isotope is the only chiral perturber, as in 3(*S*)-deuteriocyclopentanone (Figure 9-1D). Djerassi and Tursch[5] prepared and examined this compound by ORD spectroscopy. Since it was known in 1961 that 3-methylcyclopentanone exhibited a strong ORD Cotton effect (Section 6.3), it was

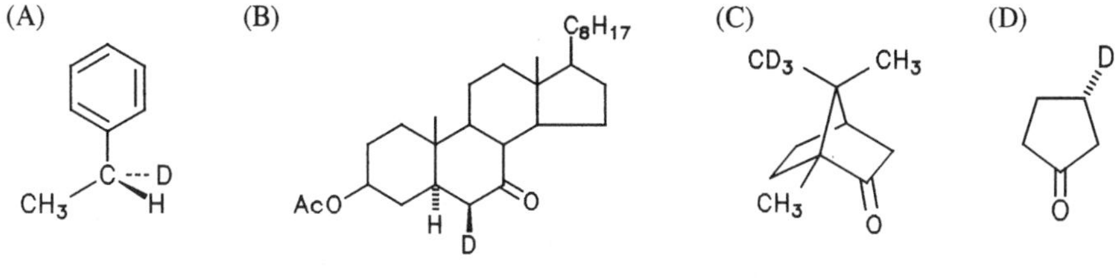

Figure 9-1. Early examples of optically active molecules with deuterium as an isotopic perturber. From (A) Eliel, 1949 (ref. 1), (B) Djerassi et al., 1956 (ref. 3), (C) Meyer and Lobo, 1966 (ref. 4), and (D) Djerassi and Tursch, 1961 (ref. 5).

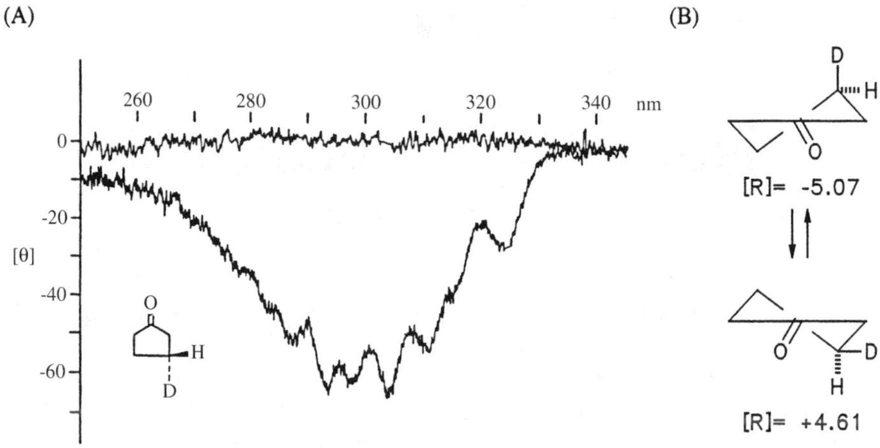

Figure 9-2. (A) Circular dichroism spectrum of 3(*R*)-deuteriocyclopentanone in EPA at 25°C. (EPA is ether–isopentane–ethanol, 5:5:2, by vol.) (B) Interconverting half-chair conformers of 3(*R*)-deuteriocyclopentanone with axial (upper) and equatorial (lower) deuterium and computed reduced rotatory strengths [*R*]. [Reprinted from ref. 7 Copyright© 1975, with permission from Elsevier Science.]

expected that 3-deuteriocyclopentanone would be the best example in which to detect an $n \rightarrow \pi^*$ Cotton effect. However, no such Cotton effect was detected, and neither was any rotation between 589 and 280 nm under conditions wherein a rotation greater than 42° was detectable. For nearly 15 years, little, if any, further work was carried out to observe isotopically induced ketone $n \rightarrow \pi^*$ Cotton effects.

The mid-1970s ushered in an explosion of publications on CD due to isotopic substitution.[6] The first observation of an $n \rightarrow \pi^*$ CD Cotton effect in a monoketone with chirality due solely to deuterium (Figure 9-2) was observed in 3(*R*)-deuteriocyclopentanone (the enantiomer shown in Figure 9-1D) and reported by Simek et al.:[7] $\Delta\varepsilon_{304}^{max} = -0.019$ (25°C) and $\Delta\varepsilon_{302}^{max} = -0.021$ (−196°C). (By way of comparison, $\Delta\varepsilon = +2.1$ in the corresponding 3(*R*)-methylcyclopentanone; see Chapter 6.) At nearly the same time, an ORD spectrum of a partially α-deuterated cyclopentanone of unknown absolute configuration was reported by Hine and Li.[8] Calculated[9] and experimental data for 3(*R*)-deuteriocyclopentanone indicated that deuterium is a weakly dissignate octant perturber. These studies launched a series of investigations of isotopically substituted, conformationally mobile and rigid cyclic ketones to probe the influence of isotopic perturbers on conformational equilibria and to learn their octant contributions and signs.[6]

9.1. Deuterated Cyclopentanones

The octant contributions of a β-deuterium can be more easily assessed when the cyclopentanone is held in an envelope conformation, as in 2-deuterio-7-norbornanones and their $n \rightarrow \pi^*$ CD.[10] In 7-norbornanone, which is achiral, the cyclopentanone moiety (Figure 9-3A) is constrained to adopt the envelope conformation. The substitution of a deuterium (for hydrogen) at an *exo* or *endo* site (Figures 9-3B and C) lifts the symmetry, creating chiral ketones, in which deuterium is the lone dissymmetric element. Consequently, these model compounds of known absolute configuration offer ideal structures for detecting and evaluating CD due to isotopic substitution.

Both deuterioketones were found to exhibit *positive* $n \rightarrow \pi^*$ Cotton effects of weak to moderate intensity (Figure 9-4), with $\Delta\varepsilon$ for the *endo* isomer being about an order of magnitude greater than that of the *exo*. In contrast, the short-wavelength (~200 nm) Cotton effects were oppositely signed, negative for both *exo* and *endo*. These data may be compared with those of the corresponding 2-methyl-7-norbornanones (enantiomers shown in Figure 4-9), in which methyl is the lone dissymmetric perturber: for *exo*-CH_3, $\Delta\varepsilon_{296}^{max} = +0.15$ and $\Delta\varepsilon_{187}^{max} = -2.8$; for *endo*-$CH_3$, $\Delta\varepsilon_{305}^{max} = +0.6$ and $\Delta\varepsilon_{189}^{max} = -1.2$ (isopentane).[11] An *endo*-deuterium thus behaves as a weak octant-dissignate perturber, whereas an *exo*-deuterium is weak and consignate. The octant location of the 2-*endo*-deuterium on 7-norbornanone is similar to that of a β-equatorial deuterium on cyclohexanone; the octant location of the 2-*exo*-deuterium is similar to that of a β-axial deuterium. If the 2-*exo*-deuterium were to lie on a front octant, as does a 2-*exo*-methyl, then it, too, would be labeled octant dissignate. These conclusions were supported by CNDO/S calculations of rotatory strengths,[9,10] which predicted that $R^r = +0.02$ and $R^\nabla = +0.07 \times 10^{-40}$cgs for the *exo*-deuterio ketone (Figure 9-3B) and $R^r = +0.28$ and $R^\nabla = +0.51 \times 10^{-40}$cgs for the *endo*, where R^r is the length form and R^∇ is the velocity form of the rotatory strength. (See Section 4.7.)

In contrast to the envelope conformation, the more stable half-chair conformation of cyclopentanone is dissymmetric and consists of two enantiomeric conformers. In 3(*R*)-deuteriocyclopentanone,[7] there are two distinct, nearly isoenergetic, half-chairs (Figure 9-2B) in equilibrium. In one of these, the deuterium occupies an axial site; in the other, it occupies an equatorial site—and the ring conformations are enantiomeric. If a deuterium substituent were to perturb the equilibrium of Figure 9-2B, the rotatory strength might be expected to vary with lowered temperature; yet, the reduced rotatory strengths $[R]^{25°C} = -0.0605$, $[R]^{-100°C} = -0.0543$, and $[R]^{-192°C} = -0.0769$ in EPA were

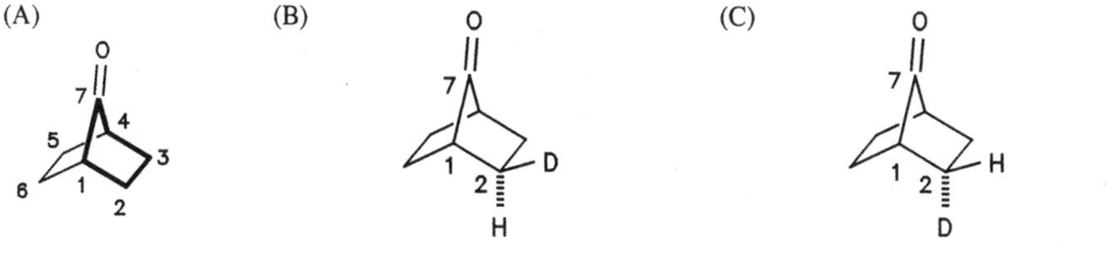

$\Delta\epsilon_{296}^{max} = +0.033$ $\quad R^{25°C} = +0.073$ $\qquad \Delta\epsilon_{292}^{max} = +0.132$ $\quad R^{25°C} = +0.338$

Figure 9-3. (A) 7-Norbornanone, with cyclopentanone (heavy lines) held in the envelope conformation. (B) *exo*-2(*R*)-Deuterio-7-norbornanone. (C) *endo*-2(*S*)-Deuterio-7-norbornanone. ***R*** is the rotatory strength, with value $\times 10^{-40}$ cgs. Data from ref. 10.

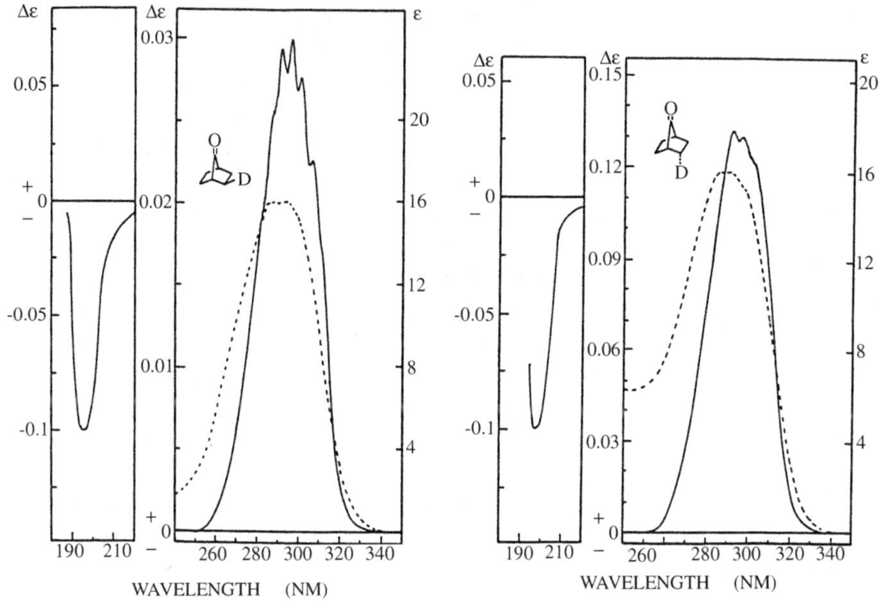

Figure 9-4. Circular dichroism (———) and ultraviolet (- - -) spectra of *exo*-2(*R*)-deuterio-bicyclo-[2.2.1]heptan-7-one (left) and *endo*-2(*S*)-deuteriobicyclo[2.2.1]heptan-7-one (right) in *n*-heptane at 25°C. Corrections are made to 100% enantiomeric excess and 100% d_1. Baselines were obtained by using equivalent concentrations of (achiral) 7-norbornanone with no difference from pure *n*-heptane. [Reprinted with permission from ref. 10. Copyright© 1980 American Chemical Society.]

initially thought to suggest no significant shift in the equilibrium.[7] According to the octant rule, the ring conformation with the axial deuterium is predicted to give a negative Cotton effect, and that with the equatorial is predicted to give a positive Cotton effect. The octant contribution of the deuterium perturber is thought to be small. Using a CNDO calculation, Bouman[9] predicted reduced rotatory strengths [*R*] = –5.07 for the axial, and [*R*] = +4.61 for the equatorial, conformers (Figure 9-2). Assuming a 50:50 mixture of conformers, one would predict an observed [*R*] = –0.23, which has the correct sign and a magnitude acceptably close to the experimental value.[7]

Subsequently, by means of variable low-temperature CD, the conformational equilibrium of 3(*R*)-deuteriocyclopentanone was reinvestigated more carefully, along with that of two other analogs: 3(*S*),4(*S*)-dideuteriocyclopentanone and 3(*S*),4(*S*)-2,2,3,4,5,5-hexadeuteriocyclopentanone (Figure 9-5A).[12] In all three examples, the magnitude of the $n \rightarrow \pi^*$ Cotton effect was observed to increase upon lowering the temperature from +20°C to –196°C in EPA solvent, with the main change in intensity occurring below –120°C. Virtually identical increases in intensity were found in a hydrocarbon solvent glass (isopentane–methylcyclohexane, 4:1, by vol.)—an indication that the CD changes which were observed were due to conformational, rather than solvational, effects. The data of Figure 9-5A suggest that the population of the

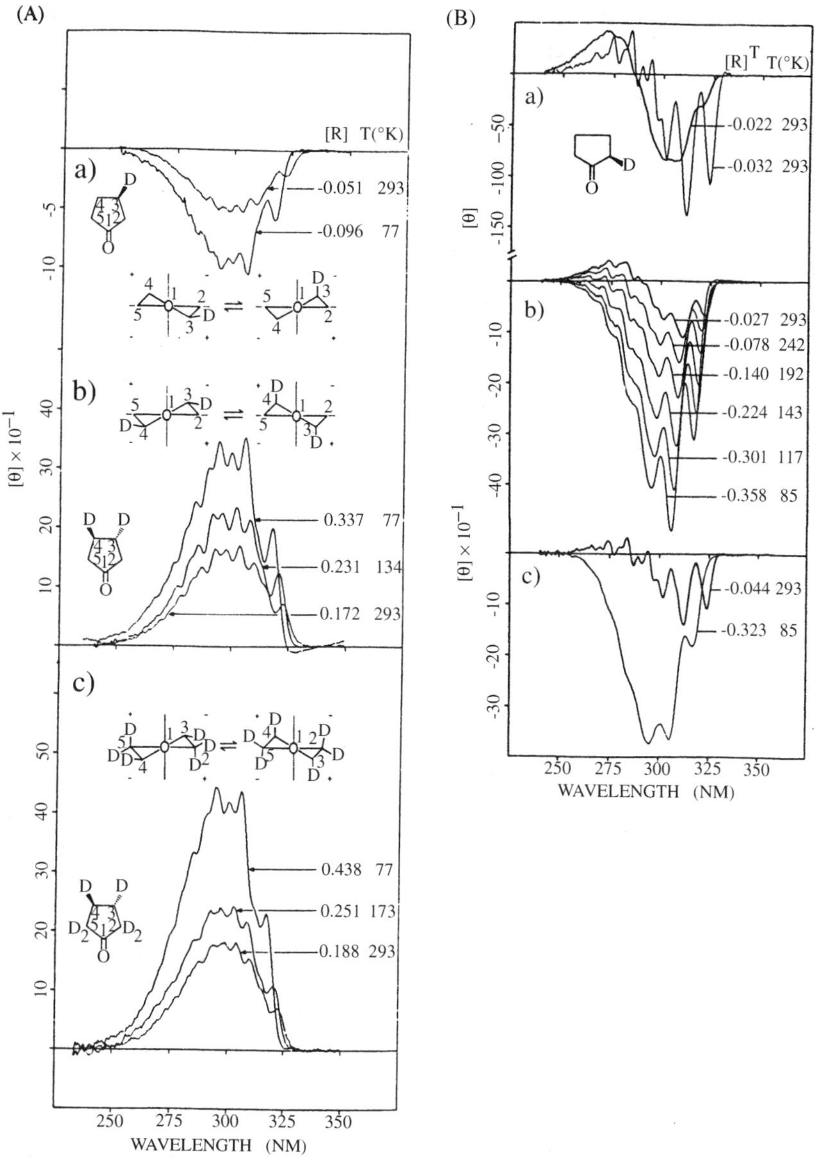

Figure 9-5. (A) CD spectra at various temperatures and octant diagrams of both twist conformations of (a) 3(*R*)-deuteriocyclopentanone, (b) 3(*S*),4(*S*)-3,4-dideuteriocyclopentanone, and (c) 3(*S*),4(*S*)-2,2,3,4,5,5-hexadeuteriocyclopentanone in ether–isopentane–ethanol (EPA, 5:5:2 by vol.). (B) CD spectra of 2(*R*)-deuteriocyclopentanone (a) in methanol (heavy line) and isooctane (thin line); (b) in EPA; and (c) in isopentane–methylcyclohexane (IPM, 4:1), at various temperatures between 293 and 85°K. [(A) is reprinted from ref. 12 with permission from Elsevier Science. (B) is reprinted with permission from ref. 16. Copyright© 1981 American Chemical Society.]

half-chair cyclopentanone conformation with one or more β-axial deuterium substituents increases as the temperature is lowered. In each example, this means that β-axial deuterium is more stable than β-equatorial deuterium—that is, that the isotope of "smaller size" (deuterium) "prefers" to occupy the position of larger, nonbonded steric strain—namely, the axial position.

Assuming a reduced rotatory strength $[R] = \pm 17$ for the enantiomeric half-chair conformers of cyclopentanone, as estimated by Kirk,[13] Sing et al. calculated the enthalpy differences between the β-axial and β-equatorial deuterio conformers (Figure 9-5A) from the variable low-temperature CD data using the method outlined by Moscowitz et al.[14] (Section 5.6). For 3(*R*)-deuteriocyclopentanone, $\Delta H^\circ = -1.11$ cal/mole; for 3(*S*),4(*S*)-dideuteriocyclopentanone, $\Delta H^\circ = -4.04$ cal/mole; and for the hexadeuterio isomer, $\Delta H^\circ = -6.15$ cal/mole—all with the β-axial deuterio conformers predicted to be more stable than the β-equatorials, conclusions supported by force-field calculations.[12]

Cyclopentanones with an α-deuterium substituent behave differently from those with β-deuteriums and, apparently, anomalously. The earliest report of a Cotton effect from an α-deuteriocyclopentanone was that of a 2,5-dideuteriocyclopentanone prepared by the chiral base-catalyzed α-deuterium exchange of 2,2,5,5-tetradeuteriocyclopentanone, conducted by Hine et al.[8] A positive $n \rightarrow \pi^*$ Cotton effect was found for the dideuterioketone, whose configuration was posited to be (2*S*,5*S*) on the basis of the reaction mechanism. The data suggested that an α-deuterium makes a *consignate* octant contribution (Figure 9-6), which was surprising in light of the dissignate behavior of β-deuterium. Other anomalous behavior came from the work of Dauphin et al.,[15] who obtained an optically active 2α-deuteriocyclopentanone by the enzymic reduction of 2-deuterio-2-cyclopentenone and reported a negative $n \rightarrow \pi^*$ Cotton effect in the CD spectrum. The absolute stereochemistry was taken to be 2(*R*) by analogy to

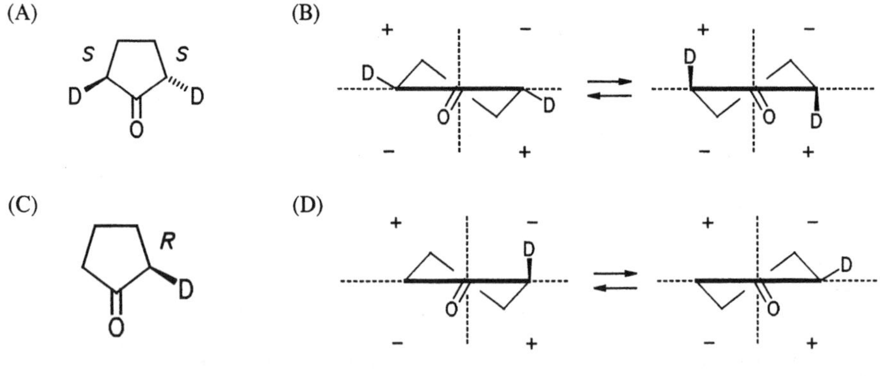

Figure 9-6. (A) 2(*S*),5(*S*)-Dideuteriocyclopentanone and (B) its half-chair ⇄ half-chair conformational equilibrium. (C) 2(*R*)-Deuteriocyclopentanone and (D) its half-chair ⇄ half-chair conformational equilibrium. Conformations on the left in (B) and the right in (D) have pseudoequatorial deuteriums, while those on the right in (B) and the left in (D) have pseudoaxial deuteriums. Octant diagrams and signs are superimposed on the conformers.

reductions of 2-methyl-2-cyclopentenone and 2-methyl-2-cyclohexenone. The authors noted that the *R* absolute configuration and negative Cotton effect implied that an α-deuterium makes a *consignate* octant contribution (Figure 9-6), which was unexpected. Shortly thereafter, Djerassi's group at Stanford[16] reported an asymmetric synthesis of 2(*R*)-deuteriocyclopentanone and its variable low-temperature CD spectra (Figure 9-5B), wherein the observed predominantly negative $n \rightarrow \pi^*$ Cotton effect at room temperature becomes increasingly negative as the temperature is lowered to 85°K. The data may be interpreted as a displacement of the conformational equilibrium of Figure 9-6D toward the right—toward the half-chair conformer with a pseudoequatorial deuterium and a negative $n \rightarrow \pi^*$ Cotton effect. (The major (–)-octant contributions come from the cyclopentanone ring β-CH_2 groups.) From this analysis, one may conclude, surprisingly, that in cyclopentanone an α-deuterium "prefers" to occupy the α-pseudoequatorial position of lower nonbonded steric strain. Typically, deuterium "prefers" the more sterically hindered position, as was shown in the case of 3β-deuteriocyclopentanone. By analyzing the variable low-temperature CD spectra (Figure 9-5B) using Moscowitz's method[14] (Section 5.6) and Kirk's estimate of $[R] = \pm 17$[13] for a twisted cyclopentanone, the energy difference between the conformers of Figure 9-6D is calculated to be 10 ±2 cal/mole. The physical origin of this unusual conformational isotope effect, in which deuterium appears to be larger than hydrogen, is as yet unclear.

9.2. Deuterated Cyclohexanones

The first report of CD spectra in a cyclohexanone, in which deuterium is the lone dissymmetric perturber, came from the University of Nevada in 1977.[17] This initial report of a weak positive Cotton effect [$\Delta\varepsilon_{295}^{max} = +0.082$ for 1(*R*),3(*S*)-4^e-deuterioadamantan-2-one (Figure 9-7A)] indicated that a β-equatorial deuterium on a conformationally immobile chair cyclohexanone makes an octant-dissignate contribution. Shortly thereafter, Numan and Wynberg[18] reported CD spectra of the same ketone and its β-axial deuterio epimer (Figure 9-7B), both of which exhibited weak positive $n \rightarrow \pi^*$ Cotton effects. Unlike β-axial alkyl perturbers, which project into front octants, a β-axial deuterium is thought to reside in a back octant. Consequently, both the β-axial and β-equatorial deuterium may be viewed as an octant-dissignate perturber, with the axial isomer being weaker than the equatorial. CD data for the $n \rightarrow \pi^*$ transition of β-deuterio-adamantanones[18] and the corresponding 7-norbornanones[10] are summarized in Table 9-1.[19]

At about the same time that the preceding studies were carried out, Sundararaman and Djerassi reported on the CD of β-deuterio and α-deuterio chair cyclohexanones that were conformationally stabilized by *tert*-butyl and isopropyl groups—viz., β-deuterio-4-*tert*-butylcyclohexanones and α-deuterio-4-isopropylcyclohexanones (Figure 9-8).[20] The CD data from the equatorial β-deuteriocyclohexanone and adamantanone are consistent: A β-equatorial deuterium is an octant-dissignate perturber. But the CD data from the β–axial deuteriocyclohexanone and adamantanone differ (Table 9-1).

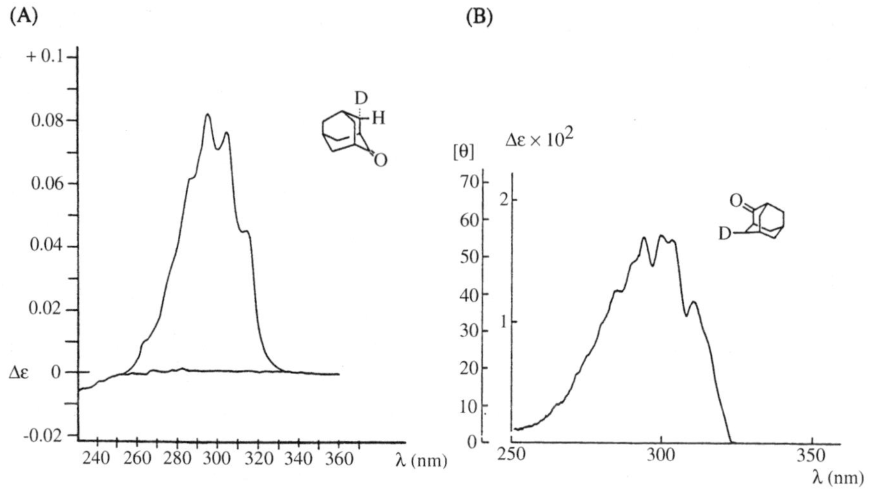

Figure 9-7. (A) Circular dichroism spectrum of $(4S)^e$-deuterioadamantan-2-one in isopentane at 23°C. The line near $\Delta\varepsilon = 0$ is the baseline. (B) Circular dichroism spectrum of $4(R)^a$-deuterioadamantan-2-one in isooctane. [(A) is reprinted from ref. 17, Copyright© 1977, with permission from Elsevier Science. (B) is reprinted with permission from ref. 18. Copyright© 1978 American Chemical Society.]

Whether a β-axial deuterium is an octant-dissignate or octant-consignate perturber depends on whether it is located in a front or back octant. And since a β-axial deuterium lies close to the third nodal surface (Chapter 4), which varies somewhat for each ketone, this may be an example in which the β-axial deuterium lies in a front octant in the 4-*tert*-butylcyclohexanone and in the back octant in the corresponding adamantanone. If so, then it is octant dissignate in both instances.

The data from the α-deuteriocyclohexanones (Figure 9-8) predict that an α-axial deuterium makes an octant contribution opposite to that predicted by the octant rule, whereas an α-equatorial deuterium is octant consignate. These results, which were confirmed independently by ORD spectroscopy[21] and by CD spectroscopy of α-deuterio cholestan-2-ones and 7-ones,[22] are summarized in Table 9-2. From those limited data, one may conclude that, although deuterium is more often an octant-dissignate perturber, as determined from one example of a conformationally fixed cyclohexanone, it is apparently octant consignate when it is α-equatorial on cyclohexanone.

These general conclusions were supported and refined by investigations of the CD of α-deuteriocyclohexanones held in a chair or deformed chair conformation by one- or two-carbon belts connecting β-carbons of the cyclohexanone (Figure 9-9).[23] Thus, α-axial deuterium in Figure 9-9A is found to make an octant-dissignate contribution, whereas α-equatorial deuterium (Figure 9-9B) is found to make an octant-consignate contribution. When the cyclohexanone ring is forced to adopt a sofa conformation, as in Figures 9-9C and D, deuterium makes a very weak octant-dissignate contribution. Here, the $D{-}C_\alpha{-}C{=}O$ torsion angle is approximately ±60°, whereas in Figure 9-9A it is about +96°, and in Figure 9-9B it is around −20°. So it would appear that on

Table 9-1. Structures, Octant Diagrams, and Observed $n \rightarrow \pi^*$ CD Cotton Effects for β-Deuterioadamantanones and 2-Deuterio-7-norbornanones

Ketone	Octant Projection Diagram	Observed $n \rightarrow \pi^*$ Cotton Effect[a,b]
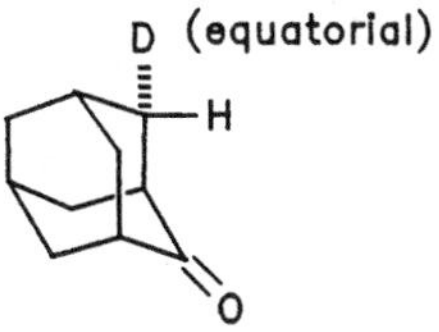	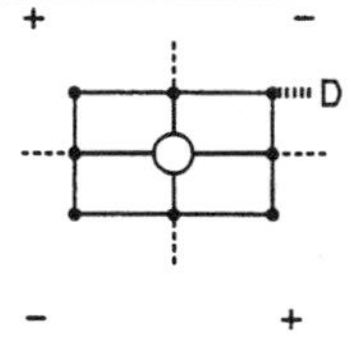	$\Delta\varepsilon_{295}^{max} \sim +0.11$ $[R]^{+25°} = +0.27$
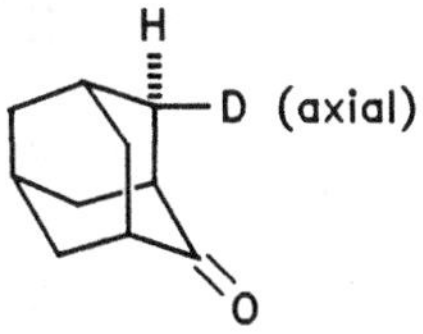	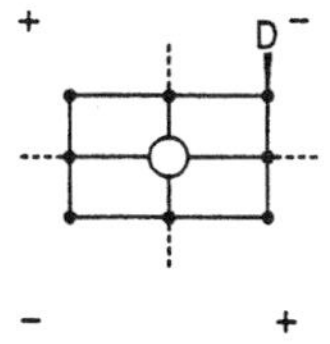	$\Delta\varepsilon_{291}^{max} \sim +0.009$ $[R]^{+25°} = +0.021$
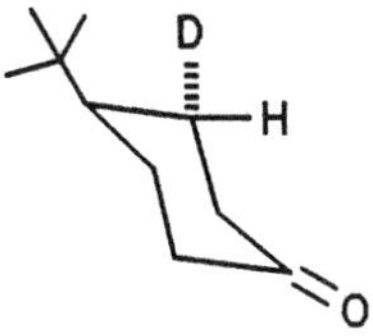	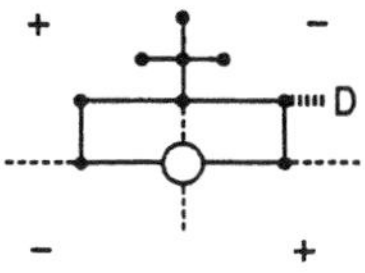	$\Delta\varepsilon_{298}^{max} \sim +0.088$ $[R]^{+25°} = +0.34$
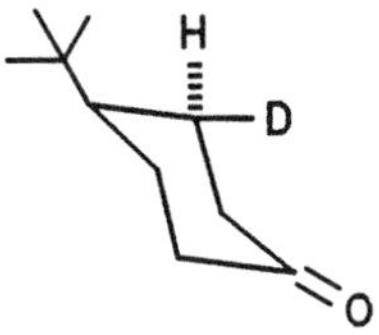	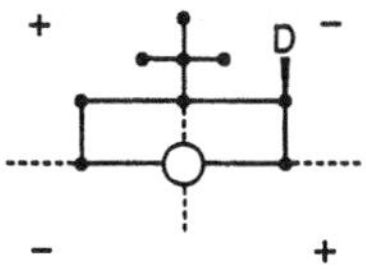	$\Delta\varepsilon_{297}^{max} \sim -0.0085$ $[R]^{+25°} = -0.029$
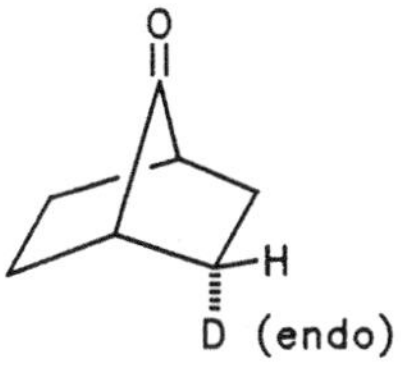	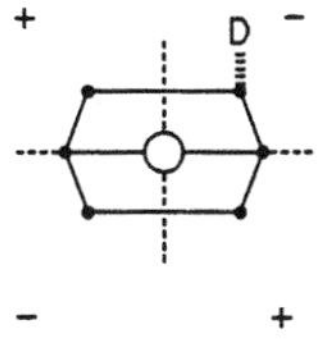	$\Delta\varepsilon^{max} \sim +0.13$[c] $[R]^{25°} = +0.36$[c]
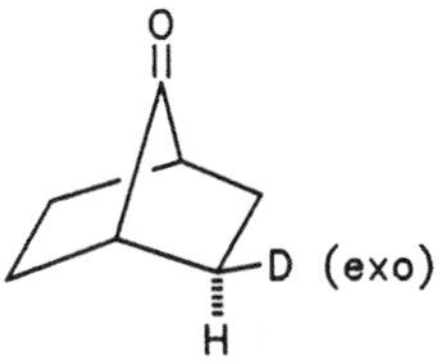	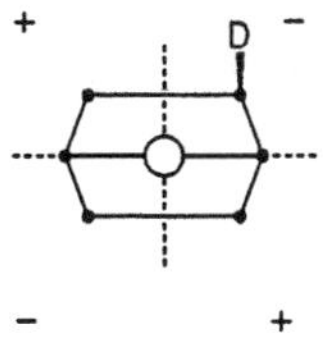	$\Delta\varepsilon^{max} \sim +0.33$[c] $[R]^{25°} = +0.079$[c]

[a] Ether–isopentane–ethanol, 5:5:2, by vol.

[b] Data for the first two ketones are from ref. 19. Data for the next two ketones are from ref. 29.Data for the last two ketones are from ref. 10.

[c] Data from *n*-heptane solvent.

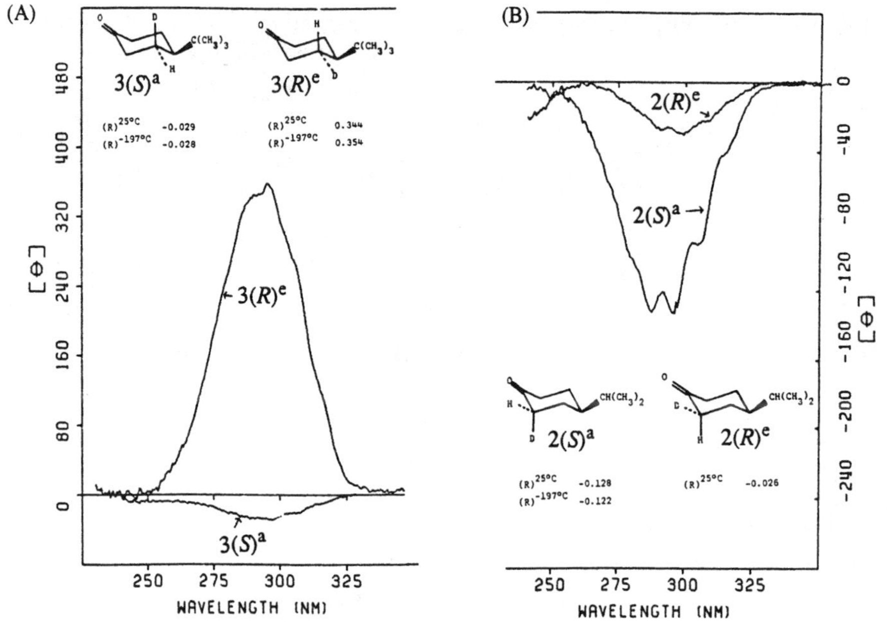

Figure 9-8. Circular dichroism curves and reduced rotatory strengths for (A) 3(*S*)^a and 3(*R*)^e-deuterio-4(*R*)-*tert*-butylcyclohexanone and (B) 2(*S*)^a and 2(*R*)^e-deuterio-4(*R*)-isopropylcyclohexanone at 25°C in EPA solvent. [Reprinted from ref. 20, Copyright© 1978, with permission from Elsevier Science.]

Table 9-2. Chair Conformations, Octant Projection Diagrams, and CD Cotton Effects of (2*R*, 4*R*) and (2*S*, 4*R*)-2-Deuterio-4-isopropylcyclohexanone (Upper and Lower, Respectively)[a]

Ketone	Octant Projection Diagram	Observed $n \rightarrow \pi^*$ Cotton Effect
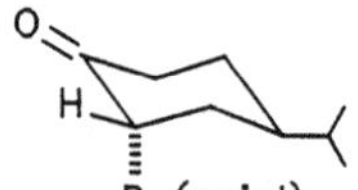 D (axial)	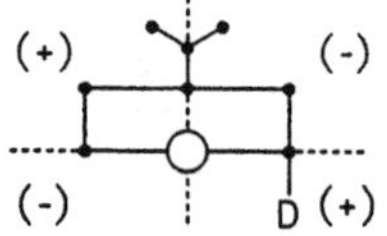	$\Delta\varepsilon \sim -0.090$
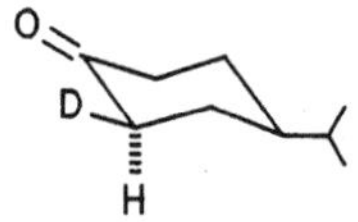 (equatorial)	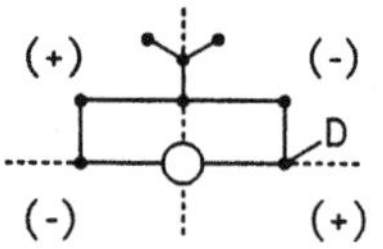	$\Delta\varepsilon \sim -0.0091$

[a] CD data from ref. 20.

cyclohexanones, α-equatorial deuterium is octant consignate. But as the chair deforms toward the sofa, with a concomitant increase in the D–C_α–C=O torsion angle from about 9° (equatorial) to approximately 60° (sofa), the contribution becomes octant dissignate and remains so through larger torsion angles, including the axial (~108°).

Interestingly, CD spectra determined in *n*-heptane for the deuterated nopinones of Figures 9-9C and D exhibit exceptional vibrational fine structure (Figure 9-10). (The well-structured curves of Figure 9-10 collapse to the typical bell-shaped curves of Figures 9-9E and F when the deuterionopinones are studied in trifluoroethanol solvent.) Similar, but less developed, fine structure has been found previously, for example, with the epimeric pinanones,[24] and very similar fine structure was observed with spiro-2-cyclopropane-isopinanone.[25] The CD curves with rotatory strengths of mixed sign in the vibrational substructure of a single electronic transition represent almost perfect examples in which the rotatory strength originates from second-order (vibrational) contributions. The ketone carbonyl C=O stretching vibration is centered near 1700 cm^{-1} in the ground state and near 1200 cm^{-1} in the $n \rightarrow \pi^*$ excited state. In the top spectrum of Figure 9-10, a (+) "allowed" vibrational progression of the 1200 cm^{-1} C=O stretch in the $n \rightarrow \pi^*$ excited state based on a totally symmetric mode is superimposed on a (–) combination that takes its origin in odd quanta of a nontotally symmetric mode and quanta of a totally symmetric mode. In the bottom spectrum of

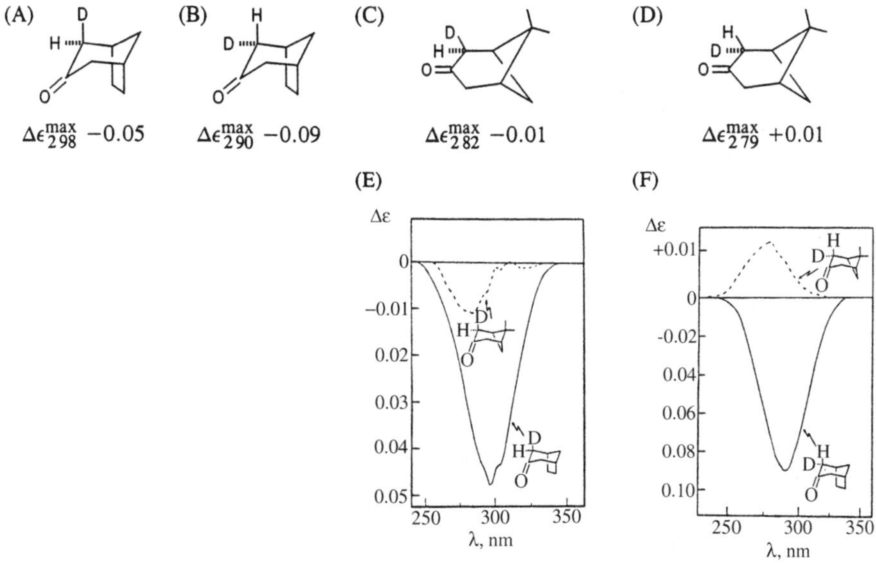

Figure 9-9. α-Deuteriocyclohexanones constrained in a bicyclic framework to adopt chair conformations (A) and (B) and sofa conformations (C) and (D). Circular dichroism Δε data are reported for the $n \rightarrow \pi^*$ transition. (E) Circular dichroism spectra of (A) and (C) at 23°C in CF_3CH_2OH solvent. (F) Circular dichroism spectra of (B) and (D) at 23°C in CF_3CH_2OH solvent. The CD spectra and data are corrected to 100% e.e. and 100% d_1. [(E) and (F) are reprinted from ref. 23, Copyright© 1985, with permission from Elsevier Science.]

Figure 9-10, the "allowed" and "forbidden" vibrational progressions are of opposite sign, as is expected from molecules that are essentially mirror images of each other. The origin of vibrational structuring in electronic CD transitions has been explained by Weigang.[26] The spectra of Figure 9-10 show that overlapping bands of mixed sign may originate from factors other than conformational and solvation effects.[27]

Conformationally mobile α-deuterio and β-deuteriocyclohexanones were studied by the Stanford group. Djerassi et al.[28] reported the $n \rightarrow \pi^*$ CD Cotton effects ($\Delta\varepsilon_{297}^{\max} = +0.046$, $[R]^{25°} = +0.15$) of 3(*S*)-deuteriocyclohexanone at +25°C and −192°C (Figure 9-11). The observed positive Cotton effect at +25°C increases in magnitude by only about 10% in going to −192°C. If one assumes that the equilibrium distribution between both chair conformers is unaffected by deuterium substitution, the observed Cotton effects represent an average value, 50% equatorial and 50% axial. Since the

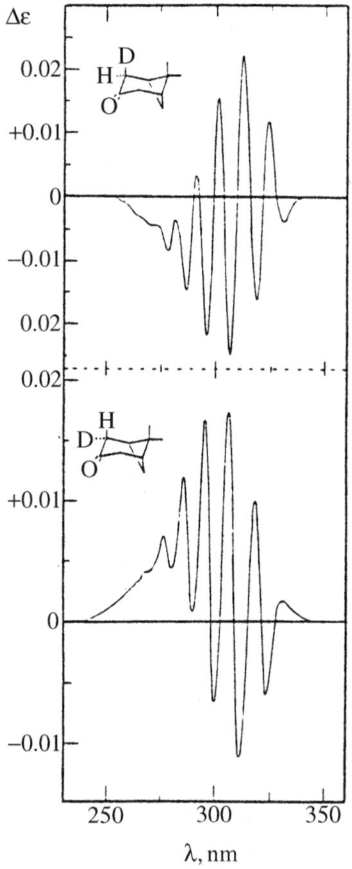

Figure 9-10. CD spectra of α-deuterio-pinanones (Figures 9-9C and D) in *n*-heptane at 24°C, corrected to 100% e.e. and 100% d_1. [Reprinted from ref. 23, Copyright© 1985, with permission from Elsevier Science.]

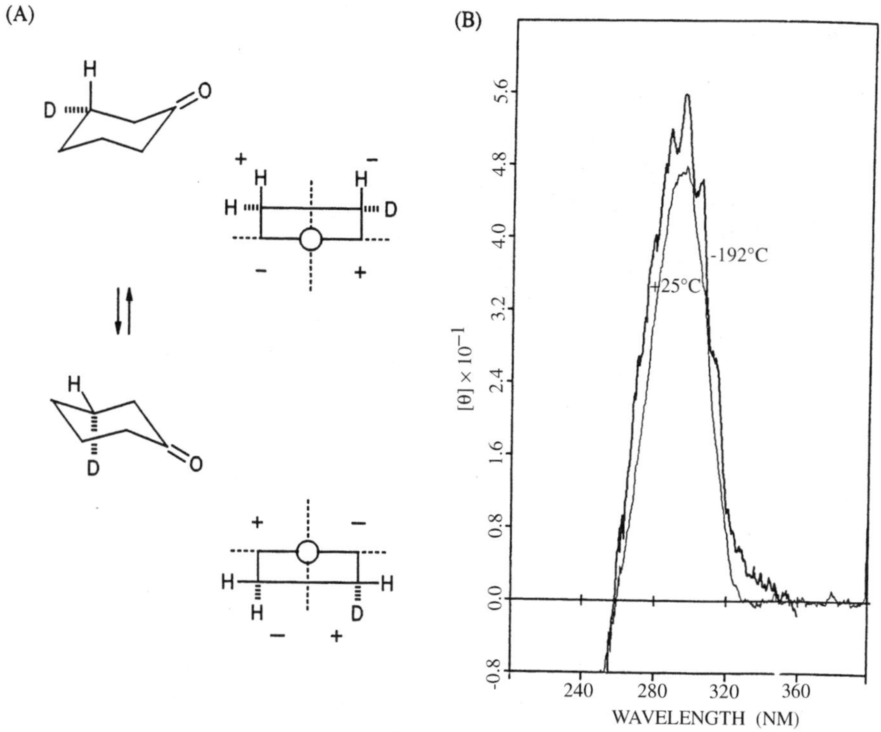

Figure 9-11. (A) Chair ⇄ chair cyclohexanone equilibrium interconverting equatorial (upper) and axial (lower) 3(*S*)-deuteriocyclohexanone. The corresponding octant projection diagrams are shown at the right. (B) Circular dichroism spectra of 3(*S*)-deuteriocyclohexanone at +25°C and −192°C in ether–isopentane–ethanol (EPA, 5:5:2, by vol.). [(B) is reprinted from ref. 28, Copyright© 1978, with permission from Elsevier Science.]

octant contribution from a 3-equatorial deuterium is about an order of magnitude larger than that from a 3-axial deuterium, the positive Cotton effect that is observed (Figure 9-11B) is dominated by the octant dissignate contribution of the equatorial deuterium, and the magnitude is approximately one-half the value of the corresponding conformationally immobile analogs (Table 9-1).

More recently, variable low-temperature CD spectra were reported for 2(*R*)-deuteriocyclohexanone,[29] and here again, the changes in rotatory strength of the positive Cotton effect were miniscule in going from +23°C to −196°C (Figure 9-12). In both chair conformers (Figure 9-12B), the deuterium perturbers lie in a negative octant, and the α-axial deuterium is expected to make the dominant octant-dissignate contribution. Unlike the conformationally immobile examples discussed earlier (Table 9-2), it is difficult here to determine whether the α-equatorial deuterium makes an octant-consignate contribution. However, this question and the conformational isotope effect on the equilibrium were explored by Djerassi et al.[29] in selected cyclohexanones.

In 2(*R*)-deuterio and 3(*S*)-deuteriocyclohexanones, careful CD measurements at variable low temperatures indicated, within experimental error, an invariant rotatory strength over a range of 220° (Figures 9-11 and 9-12)[29]—this despite the fact that the deuterium substituents lie in quite different environments with respect to nonbonded steric interactions (Figures 9-11A and 9-12B), and one might thus expect an unbalancing of the chair ⇄ chair conformational equilibrium. However, the energy difference is probably only a few cal/mole, which would translate into only very small equilibrium shifts. And since the CD intensities are inherently low, equilibrium shifts on the order of a percent or two exceed the reliability of the measurement. It thus became clear that in order to investigate the influence of deuterium as a conformational perturber by CD, the participating conformers should have large Cotton effects of opposite sign.

The system chosen by Djerassi et al.[29] for such an investigation was 2,2-dimethylcyclohexanone, which is achiral, but exists in a degenerate equilibrium between two chair conformational enantiomers (Figure 9-13). Most important, the intensity of the $n \rightarrow \pi^*$ Cotton effect is expected to be dominated by the α-axial methyl in each conformer, so the inherent magnitudes are expected to be large and comparable to that observed for the 4(*R*)-*tert*-butyl analog (Figure 9-13B)—*but oppositely signed* in each conformer of Figure 9-13A. Thus, any unbalancing of the conformational equilibrium by a deuterium perturber can be expected to result in a much larger change in the CD intensity than in cyclohexanone itself. Using this chiral probe, Djerassi et al.[30] prepared and studied three different deuterium-substituted 2,2-dimethylcyclohexanones:[6,30] 3(*R*)-deuterio, 5(*S*)-deuterio, and 4(*S*)-deuterio. Variable low-temperature CD spectra were measured for each of these (Figure 9-14) and interpreted in terms of chair ⇄

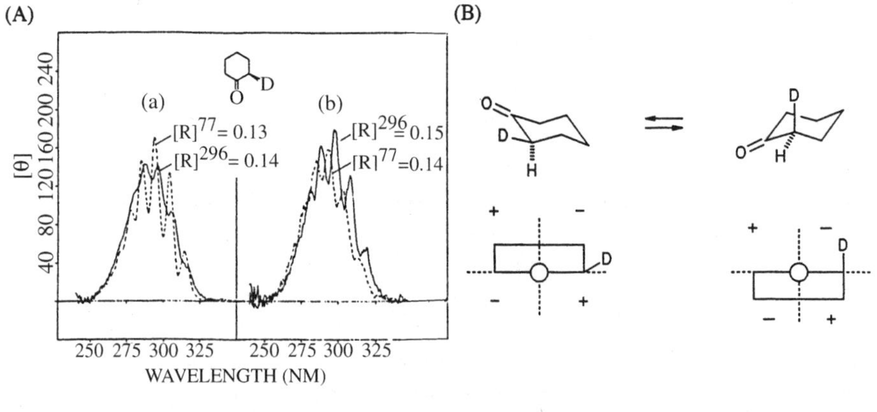

Figure 9-12. (A) Circular dichroism spectra of 2(*R*)-deuteriocyclohexanone at room temperature (solid line) and at 77°K (dashed line) in (a) ether–isopentane–ethanol (EPA, 5:5:2, by vol.) and in (b) isopentane–methylcyclohexane, 4:1 by vol. (B) Chair ⇄ chair conformational equilibrium for 2(*R*)-deuteriocyclohexanone (upper drawing) with corresponding octant projection diagrams (lower drawing). [(A) is reprinted with permission from ref. 29. Copyright© 1980 American Chemical Society.]

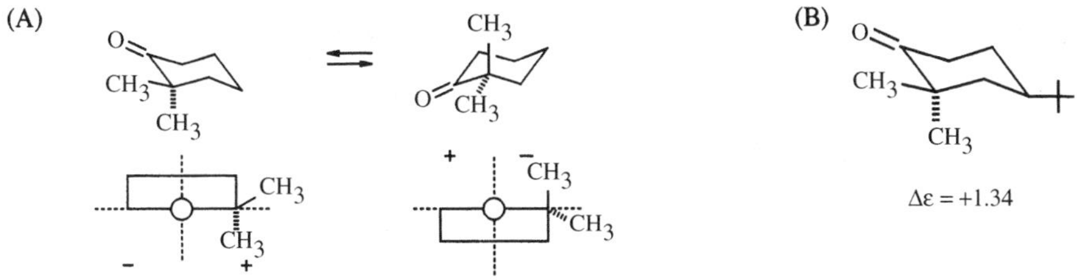

Figure 9-13. (A) Degenerate chair ⇄ chair conformational equilibrium in 2,2-dimethylcyclohexanone, giving conformational enantiomers. The corresponding octant projection diagrams are shown for each conformer. (B) 2,2-Dimethyl-4(*R*)-*tert*-butylcyclohexanone and its $n \rightarrow \pi^*$ Cotton effect, as reported in ref. 6.

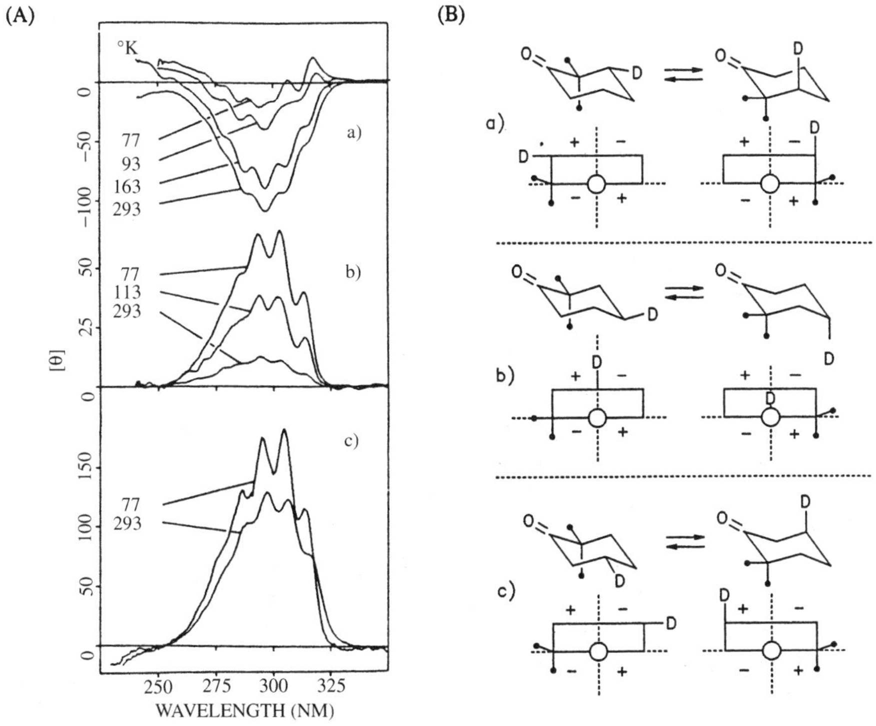

Figure 9-14. (A) Variable low-temperature circular dichroism spectra of (a) 3(*R*)-deuterio-2,2-dimethylcyclohexanone, (b) 4(*S*)-deuterio-2,2-dimethylcyclohexanone, and (c) 5(*S*)-deuterio-2,2-dimethylcyclohexanone. Spectra (a) and (c) were obtained from the compound in isopentane–methylcyclohexane 4:1, by vol, spectra (b) in ether–isopentane–ethanol (EPA, 5:5:2, by vol.). (B) Chair ⇄ chair conformational equilibria for the corresponding ketones and their octant projection diagrams. • = CH_3 [(A) is reprinted from ref. 6, Copyright© 1981, with permission from Elsevier Science.]

Table 9-3. Variable Low-Temperature Circular Dichroism Reduced Rotatory Strengths $[R]^T$, Calculated Reduced Rotatory Strengths $[R]$, % Axial Conformer, and Conformational Enthalpies for Deuteriocyclohexanones[a]

Ketone	$[R]^T$	(T, °K)	Conformer	$[R]$	% Axial (+20°, −196°C)	$\Delta H°$ (cal/mol)
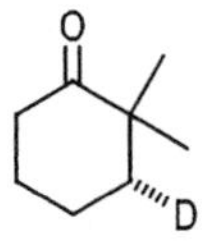	−1.03	(293)	 ax			
	−1.00	(253)		+6.04	50.41, 51.55	
	−0.072	(163)				−9.5
	−0.020	(93)	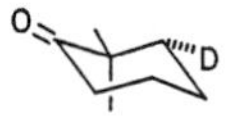 eq			
	+0.003	(77)		−6.31		
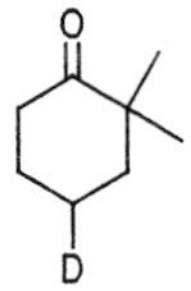	+0.011	(293)	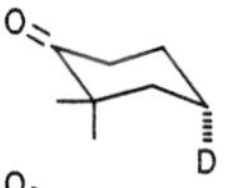 ax			
	+0.015	(228)		+5.97	50.18, 50.69	
	+0.018	(180)				−4.2
	+0.026	(128)	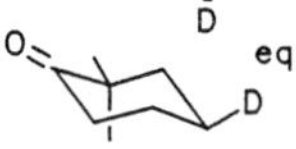 eq	−5.97		
	+0.034	(113)				
	+0.058	(77)				
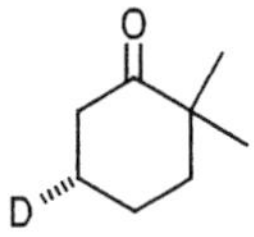	+0.134	(293)	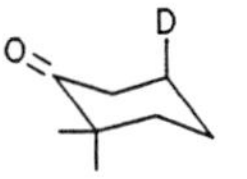 ax	+5.90	50.12, 50.44	
	+0.133	(203)				−2.7
	+0.141	(133)				
	+0.161	(77)	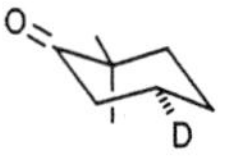 eq	−5.65		

[a] Experimental rotatory strengths $[R]^T$ and reduced rotatory strengths $[R]$ are extrapolated for each conformer from the experimental values; $\Delta H°$ values from temperature-dependent CD data are from ref. 6.

chair conformational equilibria. In the β- and γ-deuteriocyclohexanones, the observed $n \rightarrow \pi^*$ CD Cotton effects change in the direction of the more stable isomer. In each case they become increasingly positive, signifying an increasing population of the conformer with an α-axial methyl in a positive octant (i.e., the conformers with β- or γ-axial deuterium). In other words, *the conformer with a β or γ axial deuterium is energetically preferred over the conformer with a β or γ equatorial deuterium.* Since the β-axial site is also the more sterically hindered, this conclusion is consistent with that drawn earlier (Section 9.1) for β-deuteriocyclopentanones.

The qualitative conclusions just described were put on a more quantitative basis by analyzing the variable low-temperature CD data using the method described in Section 5.6. The results (Table 9-3) indicate that the axial deuterio conformer is slightly favored over the equatorial. Thus, the isotope of "smaller size" (deuterium) preferentially occupies the position of larger nonbonded strain, the axial site in the examples, by enthalpies ranging from 2.7 to 9.5 cal/mole. The fact that the $\Delta H°$ values are not the same for the isomeric 3-deuterio and 5-deuterio-2,2-dimethylcyclohexanones has been explained in terms of a small distortion of the chair caused by the *gem*-dimethyl group.[30a] In particular, the dihedral angle between the α-equatorial methyl and β-axial

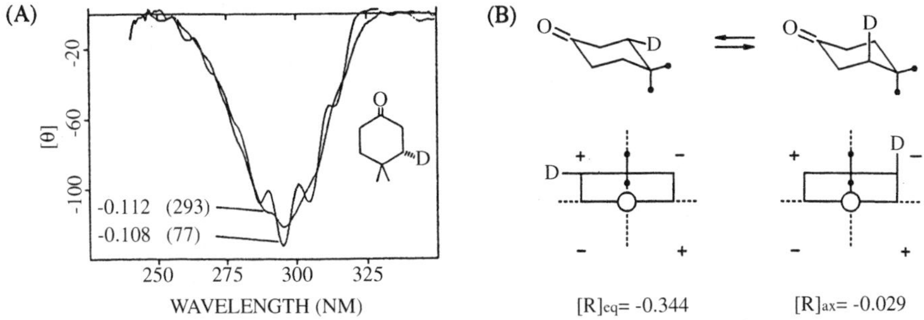

Figure 9-15. Circular dichroism spectra of 3(*S*)-deuterio-4,4-dimethylcyclohexanone at 293°K and 77°K and their associated reduced rotatory strengths. (B) Chair ⇄ chair conformational equilibria of 3(*S*)-deuterio-4,4-dimethylcyclohexanone, with octant projection diagrams and estimated reduced rotatory strengths for each conformer. • = CH_3 [(A) is reprinted with permission from ref. 30b. Copyright© 1978 American Chemical Society.]

hydrogen is narrowed, and the 3-axial site receives additional nonbonded steric strain not found at the 5-axial site.

In support of the general conclusion, a fourth 3(*S*)-deuteriocyclohexanone was prepared, with the *gem*-dimethyl group located at C(4).[30b] The *gem*-dimethyl group lies in an octant symmetry plane and is thus not expected to act as a chiral probe. Although a small equilibrium shift might be expected on the basis of the $\Delta H°$ value of –4.2 cal/mole (Table 9-3), the observed CD rotatory strengths are nearly invariant between 293°K and 77°K (Figure 9-15), a result similar to those of Figures 9-11 and 9-12.

9.3. Deuterated Cyclobutanones

There have been only two reports of CD on optically active cyclobutanones in which the chirality is due solely to isotopic substitution.[31,32] In one example,[31] a C(2) *gem*-dimethyl was used as a conformational probe, and variable low-temperature CD spectra were reported for 3(*S*) and 3(*R*)-deuterio-2,2-dimethylcyclobutanone (Figure 9-16). If the cyclobutanone ring is planar (Figure 9-16A), the octant rule predicts no CD. If the ring is puckered, two different conformations are available, one with a pseudoaxial deuterium (Figure 9-16B, upper drawing), the other with a pseudoequatorial one (Figure 9-16B, lower diagram). If the former predominates, a positive $n \rightarrow \pi^*$ Cotton effect is predicted; if the latter predominates, a negative Cotton effect is predicted. In fact, a weak negative CD is observed at room temperature (Figure 9-16C). Clearly, the molecule cannot be planar. Although the observed $n \rightarrow \pi^*$ Cotton effect is negative at room temperature, it becomes less negative as the temperature is lowered, essentially completely inverting its sign at 77°K. These results indicate that as the conformational equilibrium (Figure 9-16B) is shifted toward the thermodynamically more stable isomer, according to the octant rule, this is the conformer with a pseudoaxial deuterium. That is, the heavier isotope adopts the more sterically hindered position.

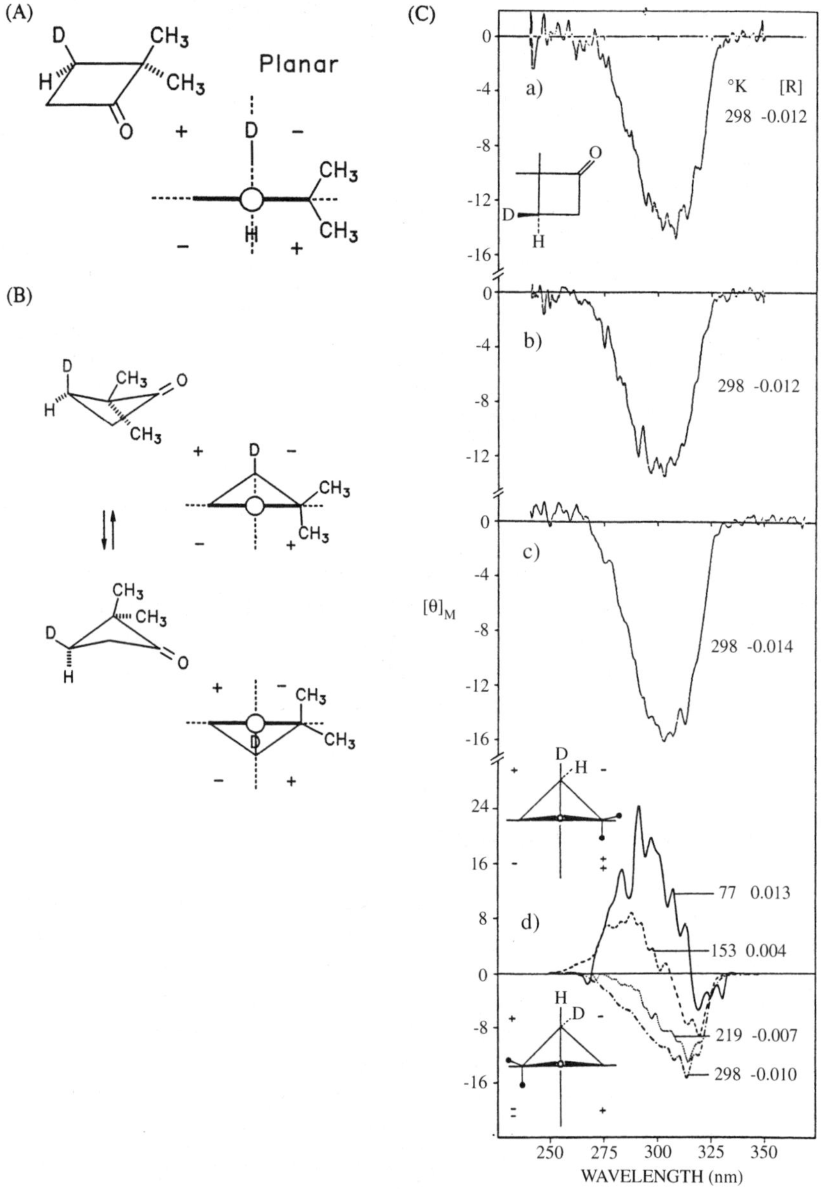

Figure 9-16. (A) Planar 3(*R*)-deuterio-2,2-dimethylcyclobutanone and its octant diagram. (B) Interconverting puckered cyclobutanone conformers and their octant diagrams. (C) Circular dichroism spectra in (a) isooctane at 298°K, (b) methanol at 298°K, (c) EPA at 298°K, and (d) isopentane–methylcyclohexane (4:1, by vol.) at 298 to 77°K. Temperatures and reduced rotatory strength are indicated next to the curves. EPA is ether–isopentane–ethanol (5:5:2, by vol.). [(B) is reprinted with permission from ref. 31. Copyright© 1983 American Chemical Society.]

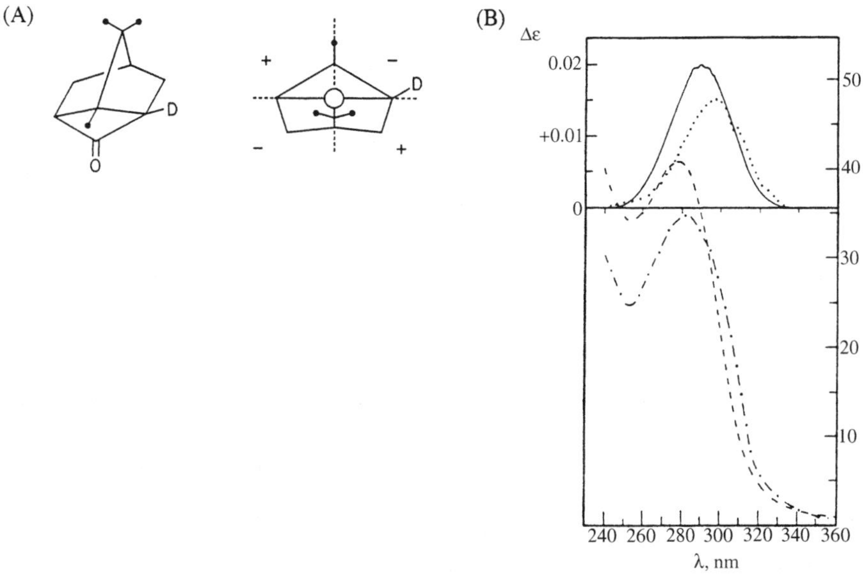

Figure 9-17. (A) (1*R*, 6*S*)-3(*S*)-Deuterio-6,7,7-trimethyltricyclo[3.2.1.0^{3,6}]octan-4-one and its octant diagram. (B) Circular dichroism spectra in methanol (———) and *n*-heptane (· · · · ·), and ultraviolet spectra measured in methanol (- - - - -) and *n*-heptane (— · — ·) at room temperature and concentration 8–9 × 10^{-3} M. • = CH_3 [(B) is reprinted from ref. 32 with permission of the author. Copyright© R.J.R. Paré Establishment for Chemistry.]

If one assumes a 50:50 mixture of conformers (Figure 9-16B) at room temperature, it would appear that the conformer with the pseudoequatorial deuterium makes a greater octant contribution than the conformer with the pseudoaxial deuterium. This is unusual, as the deuterium perturber is thought to lie in an octant symmetry plane, thus making no contribution to the Cotton effect. Yet, the results imply that the C(3) deuterium/hydrogen either makes a direct octant contribution or that it influences the ring puckering such that the *gem*-dimethyls of the two conformers do not lie in a strictly enantiomeric relationship.

In a second example, the cyclobutanone is held conformationally fixed by fusion to the bornane skeleton (Figure 9-17A).[32] The positive $n \rightarrow \pi^*$ Cotton effect that is observed (Figure 9-17B) is consistent with the lone deuterium perturber making an octant-dissignate contribution. The results are consistent with those seen by Djerassi et al.[16] for an α-quasi-equatorial deuterium on cyclopentanone and support their conclusion that "in cyclopentanones the α-quasi-equatorial position is of lower non-bonded strain energy."[16]

9.4. Trideuteriomethyl Perturbers

cis-3,5-Dimethylcyclohexanone is thought to exist predominantly in the chair conformation, with both methyls equatorial. This symmetric, achiral conformer becomes

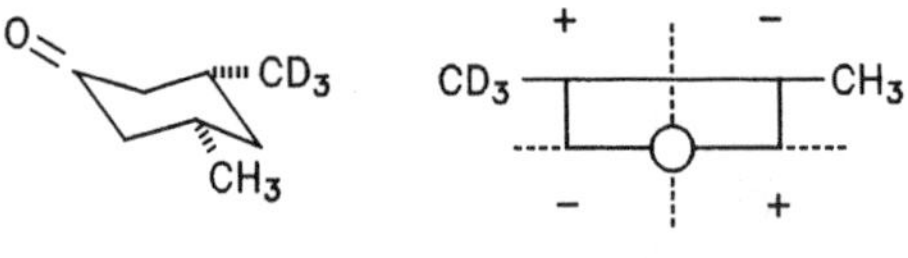

Figure 9-18. Stable chair conformation of 3(*S*)-methyl-5(*R*)-trideuteriomethylcyclohexanone (left) and its octant projection diagram (right).

chiral, however, when one methyl is replaced by trideuteriomethyl. This model thus serves as an excellent example for determining the octant contribution of a β-equatorial CD_3 group on cyclohexanone or, more exactly, the contribution of deuterium on CD_3. Pak and Djerassi[33] reported on the synthesis and circular dichroism of 3(*S*)-methyl-5(*R*)-trideuteriomethylcyclohexanone, which was found to give a negative $n \rightarrow \pi^*$ Cotton effect ($\Delta\varepsilon = -0.025$). These results, consistent with the earlier generalization that deuterium usually makes octant-dissignate contributions (Figure 9-18), are supported by CD data comparing β-methyladamantanones and β-trideuteriomethyladamantanones.[34] In that study, a β-equatorial CD_3 group in a positive back octant was found to exhibit a slightly smaller Cotton effect than a β-equatorial CH_3 perturber, and a β-axial CD_3 group was found to make a less octant-dissignate contribution than a β-axial CH_3. The data support the notion that *the heavier isotope (D) is the weaker octant perturber.*

The conformational preference of CD_3 vs. CH_3 was examined in an α,α- and a β,β-dimethylcyclohexanone (Figure 9-19A).[35] A negative $n \rightarrow \pi^*$ Cotton effect is observed in both compounds at 293°K and becomes increasingly negative at 77°K (Figure 9-19B). In the former, the chair ⇄ chair conformational equilibrium of Figure 9-19A surprisingly favors the "smaller" CD_3 group in the less sterically hindered equatorial position, whereas in the latter, the CD_3 group is favored to lie in the more sterically crowded axial position. The reason for this differing behavior is unclear. From an analysis of the variable low-temperature data and the CD data from the conformationally locked 4-*tert*-butyl analogs (Table 9-4), one can calculate that the cyclohexanone chair conformer with an α-axial CH_3 (Figure 9-19A) is 3.4 ± 0.7 cal/mole more stable than the conformer with an α-axial CD_3.[35] These findings are, however, not consistent with the general observation that the heavier isotope (CD_3) "prefers" to adopt the sterically more crowded (axial) site, but they are consistent with

Table 9-4. Circular Dichroism Reduced Rotatory Strengths [*R*] for Deuterated 2,2-Dimethyl-4(*R*)-*tert*-butylcyclohexanone[a]

Ketone	[*R*]	Ketone	[*R*]	Ketone	[*R*]
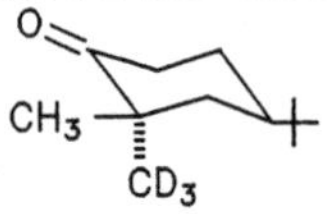	+4.30	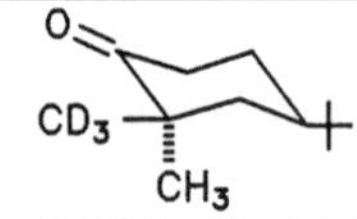	+4.57	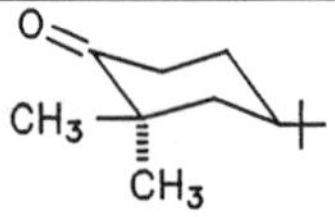	+4.59

[a] In methylcyclohexane–isopentane (1:4, by vol.), ref. 35.

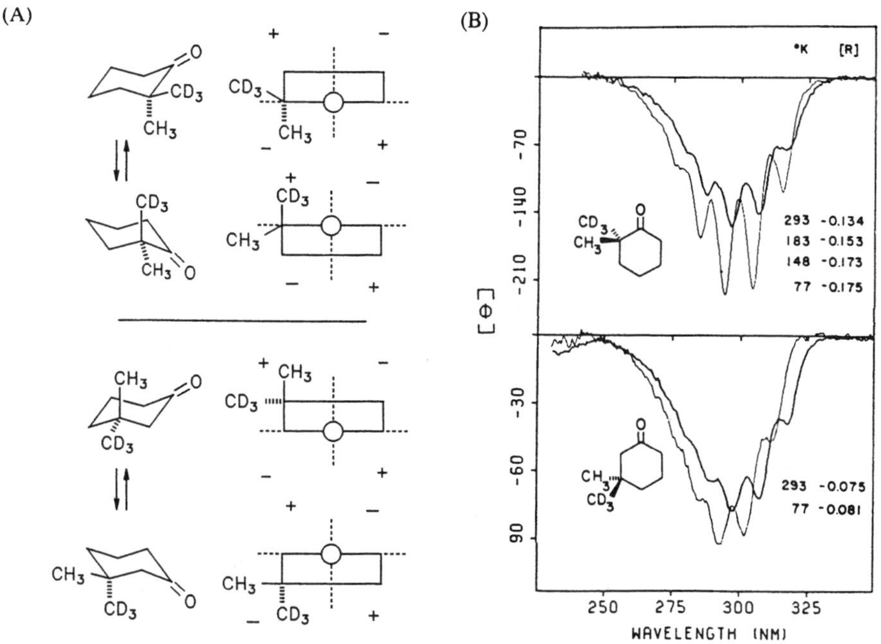

Figure 9-19. (A) Chair ⇄ chair conformational equilibria in (*S*)-2-methyl-2-trideuteriomethylcyclohexanone (upper left drawing) and (*R*)-3-methyl-3-trideuteriomethylcyclohexanone (lower left drawing). Corresponding octant projection diagrams are shown to the right. (B) Variable low-temperature circular dichroism spectra of 2(*S*)-methyl-2-trideuteriomethylcyclohexanone and (*R*)-3-methyl-3-trideuteriomethylcyclohexanone in isopentane–methylcyclohexane (4:1, by vol.) at 293°K (heavy line) and 77°K (thin line). Reduced rotatory strength [*R*] data are provided to the right of the indicated temperatures (°K). [(B) is reprinted with permission from ref. 35. Copyright© 1980 American Chemical Society.]

earlier indications that conformational preferences involving deuterium are only a few cal/mole. In fact, the conformational energies are so small that the CD of (*S*)-2-methyl-2-trideuteriomethylcyclohexanone is nearly identical to the average value of the CD curves of 2(*R*),4(*R*)-2-methyl-2-trideuteriomethyl-4-*tert*-butylcyclohexanone and 2(*S*),4(*R*)-2-methyl-2-trideuteriomethyl-4-*tert*-butylcyclohexanone.[35]

When the CD_3 perturber is moved to the β-position in cyclohexanone, as in 3(*R*)-methyl-3-trideuteriomethylcyclohexanone (Figure 9-19), the intensity of the negative $n \rightarrow \pi^*$ Cotton effect is smaller compared with that of (*S*)-2-methyl-2-trideuteriomethylcyclohexanone. The temperature effect is much reduced, suggesting that the conformational energy difference is less than ±3 cal/mole, based on $[R] = \pm 1.3$.[35]

9.5. ^{13}C as an Octant Perturber

In 1978, Pak and Djerassi[33] published the first example in which ^{13}C acted as an octant perturber. In 3(*S*),5(*R*)-3-methyl-[5-^{13}C]-methylcyclohexanone, a β-equatorial $^{13}CH_3$

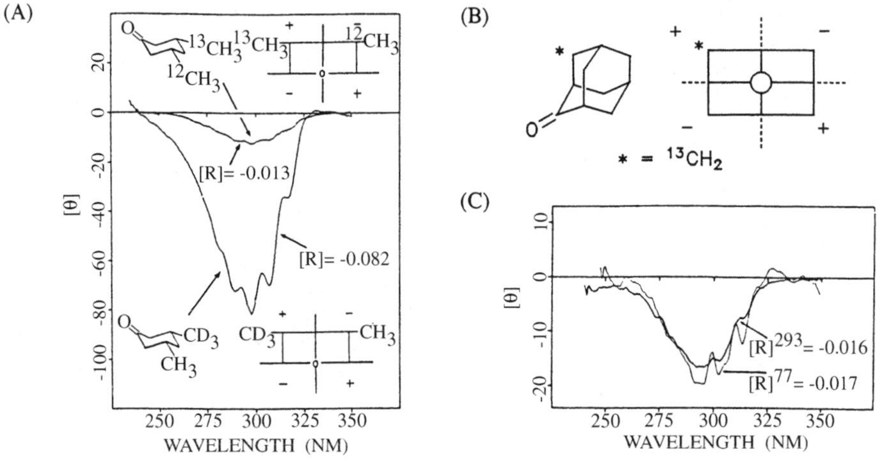

Figure 9-20. (A) Circular dichroism spectra of (3*S*,5*R*)-3-methyl-5-^{13}C-methylcyclohexanone and (3*S*, 5*R*)-3-methyl-5-trideuteriomethylcyclohexanone at 293°K in isopentane–methylcyclohexane (4:1, by vol.). (B) (1*S*)-2-adamantanone-4-^{13}C and (C) its circular dichroism spectra in ether–isopentane–ethanol (EPA, 5:5:2, by vol.) at 293°K and 77°K. Reduced rotatory strengths [*R*] are indicated beside the curves. The corresponding octant projection diagrams are shown to the right of the structures. [(A) is reprinted from ref. 33 with permission, copyright© 1978 from Elsevier Science. (B) is reprinted with permission from ref. 36. Copyright© 1979 American Chemical Society.]

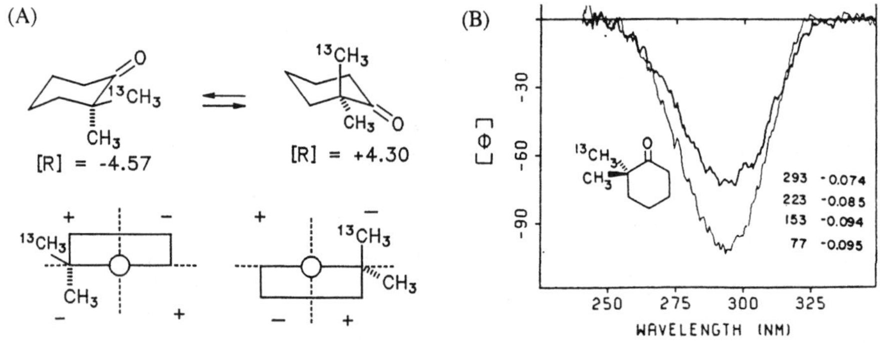

Figure 9-21. (A) Chair ⇄ chair conformational equilibrium of 2(*S*)-methyl-2-^{13}C-methylcyclohexanone and their corresponding octant diagrams. Reduced rotatory strengths [*R*] are given below each conformer. (B) Circular dichroism spectra at 293°K and 77°K in isopentane–methylcyclohexane (4:1, by vol.). The reduced rotatory strengths [*R*] at variable low temperatures are indicated to the right of the curves. [(B) is reprinted with permission from ref. 35. Copyright© 1980 American Chemical Society.]

is paired against a β′-equatorial CH_3 (Figure 9-20A), and the ^{13}C is the lone dissymmetric element. This model is similar to that discussed before, wherein β-equatorial CD_3 is paired with β′-equatorial CH_3 (Figure 9-18). And, as with the CD_3 group, the $^{13}CH_3$ is found to give a very weak, but negative, $n \rightarrow \pi^*$ Cotton effect (Figure 9-20A). Again, the heavier isotope is found to be the weaker perturber. That is, when paired with $^{13}CH_3$, $^{12}CH_3$ controls the sign of the Cotton effect.

The same conclusions can be reached with a different model in which a β-carbon of adamantanone is replaced with ^{13}C (Figure 9-20B,C).[36] In this conformationally rigid ketone, a β-^{13}C is paired against a β-^{12}C, and a negative $n \rightarrow \pi^*$ Cotton effect is observed when the ^{13}C lies in a positive octant.

The conformational preference of a $^{13}CH_3$ group for the more hindered axial position in cyclohexanone has been shown in the variable low-temperature CD spectra of 2(*S*)-methyl-2-^{13}C-methylcyclohexanone (Figure 9-21).[35] Analysis of the CD spectra indicates a conformational preference of 1.5 cal/mole for the axial α-$^{13}CH_3$.[35] Consistent with earlier findings, the heavier isotope "prefers" the more sterically hindered site.

9.6. Isotopic Perturbation of Diketones

In one of the first examples of circular dichroism due to isotopic perturbation, Kokke and Oosterhoff made the symmetric α-diketone, α-fenchocamphorone, dissymmetric and optically active by replacing one oxygen with ^{18}O.[37,38] The resulting (1*R*)-2-^{18}O-α-fenchocamphorone paired ^{18}O vs. ^{16}O and exhibited a highly vibrationally struc-

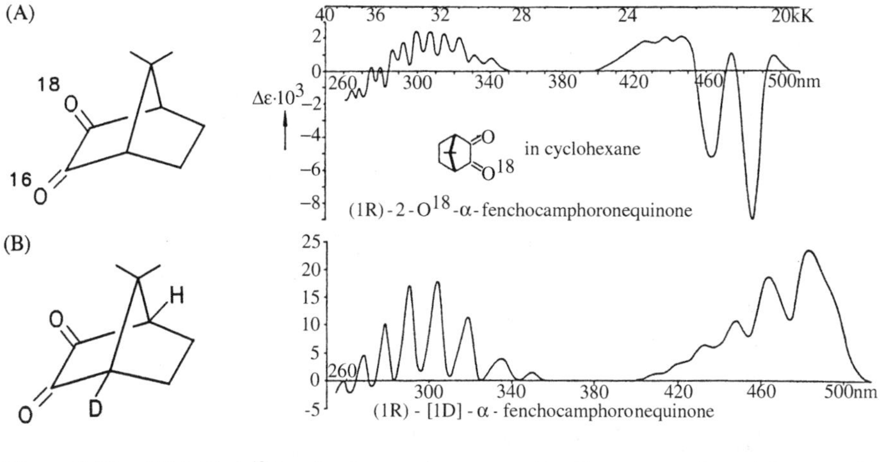

Figure 9-22. (A) 1(*R*)-2-^{18}O-α-fenchocamphor-one and its CD spectrum in cyclohexane. (B) 1(*R*)-deuterio-α-fenchocamphorone and its CD spectrum in cyclohexane. [CD curves in (A) are reprinted with permission from refs. 37 and 38, respectively. Copyright© 1972, 1973 American Chemical Society.]

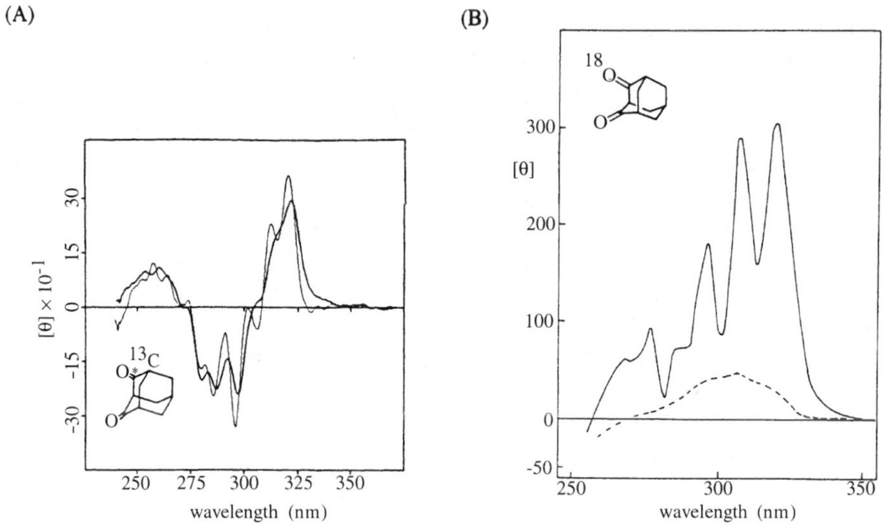

Figure 9-23. (A) Circular dichroism spectra of 1(*S*)-2,4-adamantanedione-4-^{13}C in ether–isopentane–ethanol (EPA, 5:5:2, by vol.) at 293°K (heavy line) and 77°K (thin line). (B) Circular dichroism of 1(*S*)-2,4-adamantanedione-4-^{18}O in cyclohexane at 293°K (——), baseline (----). [(A) and (B) are reprinted from refs. 36 and 42, respectively. Copyright© 1979, 1982 American Chemical Society.]

tured CD (Figure 9-22A). Similarly, 1(*R*)-deuterio-α-fenchocamphorone pairs off D vs. H, and this dione exhibits a well-structured CD (Figure 9-22B).[38] Shortly after Kokke and Oosterhoff's work, Dezentje and Dekkers[39] calculated the rotatory strengths of these isotopically substituted α-diketones using a polarizability model. More recently, Polonski[40] showed that the oxidation product of the deuterio-α-fenchocamphorone, viz., 1(*R*)-deuterio-apocamphoric anhydride, its imide, and the N-nitroso derivative of the imide all exhibited CD.

Using the adamantane skeleton, Djerassi et al.[36] converted an otherwise symmetric 1,3-diketone, adamantan-2,4-dione, into an optically active derivative with ^{13}C in one of the carbonyl carbons. The optically active (1*S*)-2,4-adamantanedione-4-^{13}C was found to exhibit a polysignate $n \rightarrow \pi^*$ Cotton effect (Figure 9-23A). Except for the appearance of vibrational fine structure, the Cotton effects were invariant down to 77°K, and they were independent of the solvent. Consequently, it was concluded that three of probably four $n \rightarrow \pi^*$ transitions[41] were being detected by CD. In a related diketone, in which ^{18}O is substituted for ^{16}O, only one band is apparently seen (Figure 9-23B).[42]

9.7. Isotopically Perturbed β,γ-Unsaturated Ketones

The only example of CD from a deuterium-substituted β,γ-unsaturated ketone comes from the structurally rigid 7-norbornenone (Figure 9-24), which is achiral in the

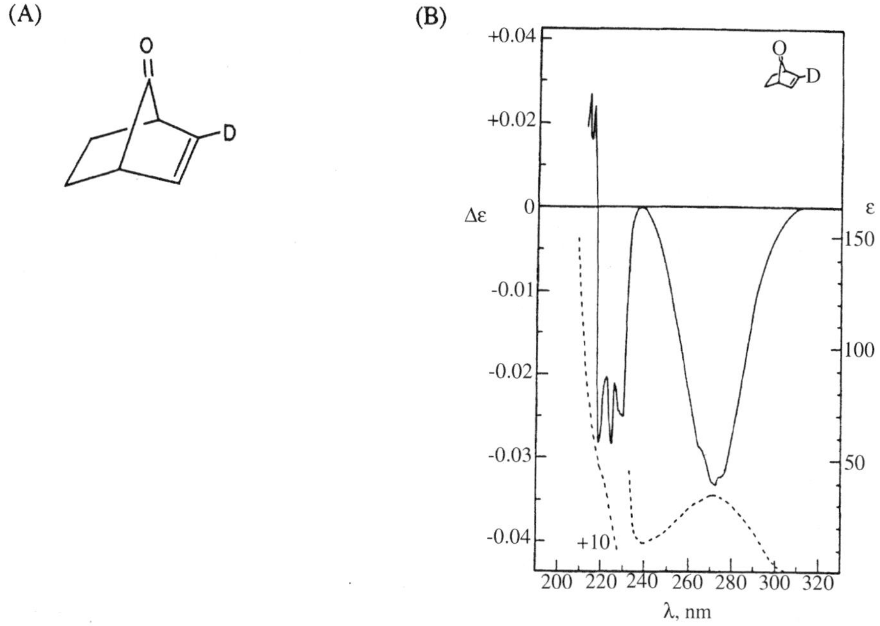

Figure 9-24. (A) 1(*R*)-2-deuteriobicyclo-[2.2.1]hepten-7-one and (B) its circular dichroism (—) and ultraviolet (- - - -) spectra in *n*-heptane at 25°C. The CD curves are corrected to 100% e.e. and 100% d_1. [(B) is reprinted with permission from ref. 43. Copyright© 1981 American Chemical Society.]

absence of the deuterium perturber.[43] More interesting than the negative $n \rightarrow \pi^*$ Cotton effect, which is again consistent with the heavier isotope (D) making the weaker octant contribution, is the observation of a Cotton effect near 225 nm corresponding to the "mystery band" due to the interaction of the carbonyl and olefin chromophores. Theoretical analysis[43] correlates the mystery band with a mixed transition containing a substantial (~15%) $\pi_{C=C} \rightarrow \pi_{C=O}$ component and major contributions coming from $\sigma_{C-C_\alpha} \rightarrow \pi^*_{C=O}$ (~15%), $n(\sigma)_{C=O} \rightarrow \pi^*_{C=O}$ (~30%), and $\sigma_{C=O} \rightarrow \pi^*_{C=O}$ (~30%).

9.8. Isotopic Perturbers and the Octant Rule

In accordance with the octant rule (Chapter 4), the net signed contribution to the Cotton effect of a given perturber (*a*) is evaluated relative to that perturber (*b*) located in a position reflected across the carbonyl symmetry plane. The important contrasting features are (1) differences in the chemical nature of the perturbers and (2) differences in bond lengths. For most perturbers studied heretofore, differences in the *chemical*

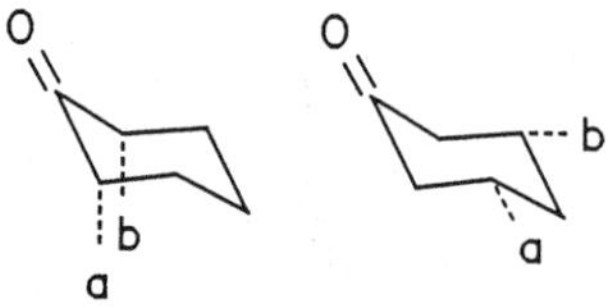

nature of the perturbers were sufficiently large to be sign determining (i.e., *a* and *b* ordinarily need not be in exactly comparable positions in their respective octants). Thus, considerations of differences in bond length were rendered inconsequential. For example, in *i* in the figure shown, with a = H and b = CH_3, the chemical difference between CH_3 and H is sufficient to allow a correct prediction of the Cotton effect without resorting to a consideration of the different C–CH_3 and C–H bond lengths. This approximation and that of neglecting H contributions form the basis for the qualitative application of the classical octant rule. The chemical nature of a deuterium perturber, however, is virtually indistinguishable from that of H (e.g., as in *ii* in the figure), and one must therefore resort to a consideration of differences in bond length. When the small difference between the C–D and C–H bond lengths is incorporated into our theoretical treatment,[9,10] we conclude that the H perturber makes the sign-determining contribution to the observed Cotton effect because the C–D bond is shorter and D is closer to the carbonyl symmetry plane. (The reason for the latter observation can be discerned through an examination of the contribution of the C–D bond to the z component of the dipole velocity transition moment matrix element, as explained in Chapter 4.) In our computations,[9,10] the C–D bond contributes *less* than its mirror-image C–H counterpart. *Thus, the net contribution to the sign of the rotatory strength is determined by the comparatively more consignate*[44] *contribution of the C–H bond that is located across the carbonyl mirror plane from the C–D bond.*

9.9. Other Isotopically Perturbed Chromophores

Circular dichroism has been detected and measured for a small number of other chromophores with deuterium as the lone dissymmetric perturber. The structures are shown in Table 9-5. A discussion of these and other isotopically perturbed chromophores found in dissymmetric molecules appears in the excellent summary of Barth and Djerassi.[6]

Table 9-5. Circular Dichroism Data for Nonketone Chromophores with Deuterium and other Isotopes as the Lone Dissymmetric Perturber

Chromophore	Structure	Δε
Olefin[45]		$\Delta\epsilon_{200}^{max}$ −0.20
		+0.14
		+0.15
		+0.23
Diene[46]	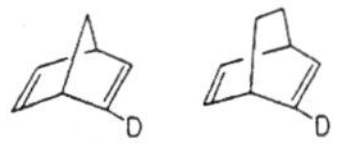	$\Delta\epsilon_{237}^{max}$ −0.032, $\Delta\epsilon_{216}^{max}$ +0.14, $\Delta\epsilon_{205}^{max}$ −0.37; $\Delta\epsilon_{222}^{max}$ −0.027
Triene[47]	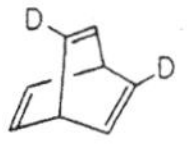	$\Delta\epsilon_{246}^{max}$ +0.008, $\Delta\epsilon_{215}^{max}$ −0.07
Paracyclophane[48]	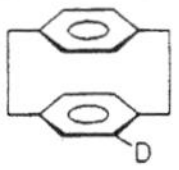	$\Delta\epsilon_{\sim 300}$ −0.02/+0.004, $\Delta\epsilon_{255}$ −0.19
Benzene[49,50,51]	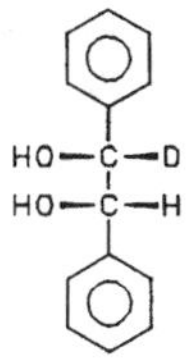	$\Delta\epsilon_{268}^{max}$ −0.006, $\Delta\epsilon_{262}^{max}$ −0.007, $\Delta\epsilon_{255}^{max}$ −0.006
	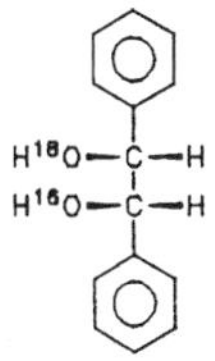	$\Delta\epsilon_{268}^{max}$ −0.003, $\Delta\epsilon_{259}^{max}$ −0.006, $\Delta\epsilon_{253}^{max}$ −0.005
	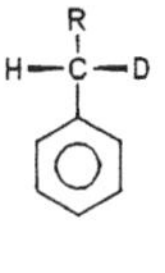 R=CH_3, CD_2H, D-C_3H_7, CH_2OH, CH_2Br	$\Delta\epsilon_{240\text{-}270}^{max}$ +0.004 to +0.02, $\Delta\epsilon_{219}^{max}$ −0.07 to −0.2
	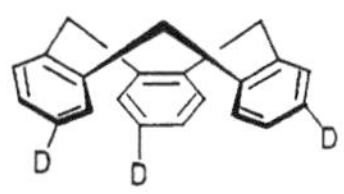	$\Delta\epsilon_{265}^{max}$ −0.060, $\Delta\epsilon_{258}^{max}$ +0.015, $\Delta\epsilon_{267}^{max}$ +0.012, $\Delta\epsilon_{252}^{max}$ +0.006, $\Delta\epsilon_{263}^{max}$ +0.007
Acid[52]	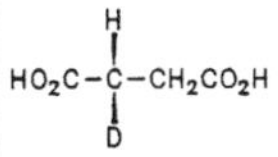	$\Delta\epsilon_{210}^{max}$ +0.058
	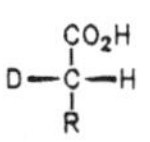 R=CH_3, NH_2, OH	$\Delta\epsilon_{210}^{max}$ +0.04 to +0.02
Azide[53]	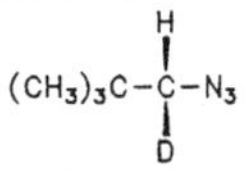	$\Delta\epsilon_{290}^{max}$ +0.009
Thione[54,55]	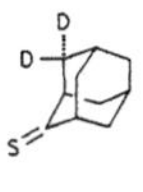	$\Delta\epsilon_{507}^{max}$ −0.041, $\Delta\epsilon_{252}^{max}$ −0.23, $\Delta\epsilon_{220}^{max}$ +0.53
Sulfoxide[56]	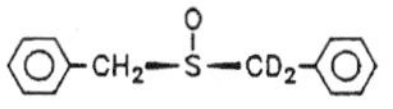	$\Delta\epsilon_{270}^{max}$ +0.017, $\Delta\epsilon_{265}^{max}$ +0.044, $\Delta\epsilon_{250}^{max}$ +0.062
Phosphate[57]	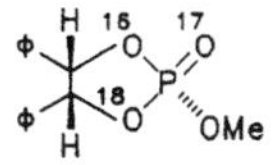	$\Delta\epsilon_{208}^{max}$ +0.0012
Epoxide[58]	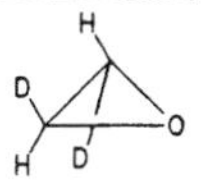	$\Delta\epsilon_{170}^{max}$ +1.47

[a] Literature references cited are found at the end of this section in References.

References

1. Eliel, E. L., *J. Am. Chem. Soc.* **71** (1949), 3970–3972.
2. Verbit, L., *Progr. Phys. Org. Chem.* **7** (1970), 51–127.
3. Djerassi, C., Closson, W., and Lipman, A. E., *J. Am. Chem. Soc.* **78** (1956), 3163–3166.
4. Meyer, W. L., and Lobo, A. P., *J. Am. Chem. Soc.* **88** (1966), 3181–3182.
5. Djerassi, C., and Tursch, B., *J. Am. Chem. Soc.* **83** (1961), 4609–4612.
6. For leading references, see Barth, G., and Djerassi, C., *Tetrahedron* **37** (1981), 4123–4142.
7. Simek, J. W., Mattern, D. L., and Djerassi, C., *Tetrahedron Lett.* (1975), 3671–3674.
8. (a) Hine, J., and Li, W-S., *J. Am. Chem. Soc.* **97** (1975), 3550–3551. (b) See also Hine, J., Li, W-S., and Ziegler, J. P., *J. Am. Chem. Soc.* **102** (1980), 4403–4409.
9. Thomas D. Bouman provided the calculated data mentioned in ref. 7. His calculation assumed a slightly shorter C–D vs. C–H bond and used the CNDO method described in Bouman, T. D., and Lightner, D. A., *J. Am. Chem. Soc.* **98** (1976), 3145–3154 and in ref. 10.
10. Lightner, D. A., Gawroński, J. K., and Bouman, T. D., *J. Am. Chem. Soc.* **102** (1980), 1893–1990.
11. Lightner, D. A., Crist, B. V., Kalyanam, N., May, L. M., and Jackman, D. E., *J. Org. Chem.* **50** (1985), 3867–3878.
12. Sing, L. Y., Lindley, M., Sundararaman, P., Barth, G., and Djerassi, C., *Tetrahedron* **37** (Supplement No. 1) (1981), 181–189.
13. Kirk, D. N., *J. Chem. Soc. Perkin Trans.* **1** (1976), 2171–2177.
14. Moscowitz, A., Wellman, K. M., and Djerassi, C., *J. Am. Chem. Soc.* **85** (1963), 3515–3516.
15. Dauphin, G., Gramain, J. C., Kergomard, A., Renard, A. F., and Veschambre, H., *Tetrahedron Lett.* **21** (1980), 4275–4278.
16. Sundararaman, P., Barth, G., and Djerassi, C., *J. Am. Chem. Soc.* **103** (1981), 5004–5007.
17. Lightner, D. A., Chang, T. C., and Horwitz, J., *Tetrahedron Lett.* (1977), 3019–3020. Erratum: *Tetrahedron Lett.* (1978), 696.
18. Numan, H., and Wynberg, H., *J. Org. Chem.* **43** (1978), 2232–2236.
19. Lightner, D. A., Bouman, T. D., Wijekoon, W. M. D., and Hansen, Aa. E., *J. Am. Chem. Soc.* **108** (1986), 4484–4497.
20. Sundararaman, P., and Djerassi, C., *Tetrahedron Lett.* (1978), 2457–2460.
21. Levine, S. G., and Gopalakristinan, B., *Tetrahedron Lett.* (1979), 699–702.
22. Meyer, W. L., Lobo, A., Ernstbrunner, E. E., Giddings, M. R., and Hudec, J., *Tetrahedron Lett.* (1978), 1771–1774.
23. Lightner, D. A., and Crist, B. V., *Tetrahedron* **41** (1985), 3021–3028.
24. Hirata, T., *Bull. Chem. Soc. Jpn.* **45** (1972), 3458–3464.
25. Bessière-Chrétien, Y., and El Gaied, M. M., *Bull. Soc. Chim. France* (1971), 2189–2194.
26. (a) Weigang, O. E., Jr., *J. Chem. Phys.* **43** (1965), 3609–3618. (b) Weigang, O. E., Jr., and Ong, E. C., *Tetrahedron* **30** (1974), 1783–1793.
27. Wellman, K. M., Laur, P. H. A., Briggs, W. S., Moscowitz, A., and Djerassi, C., *J. Am. Chem. Soc.* **87** (1965), 66–72.
28. Djerassi, C., Van Antwerp, C. L., and Sundararaman, P., *Tetrahedron Lett.* (1978), 535–538.
29. Sundararaman, P., Barth, G., and Djerassi, C., *J. Org. Chem.* **45** (1980), 5232–5236.
30. (a) Lee, S.-F., Barth, G., and Djerassi, C., *J. Am. Chem. Soc.* **103** (1981), 295–301. (b) Lee, S.-F., Barth, G., and Djerassi, C., *J. Am. Chem. Soc.* **100** (1978), 8010–8012.
31. Harris, R. N., III, Sundararaman, P., and Djerassi, C., *J. Am. Chem. Soc.* **105** (1983), 2408–2418.
32. Lightner, D. A., Wijekoon, W. M. D., and Crist, B. V., *Spectroscopy: An Intl. J.* **2** (1983), 255–259.
33. Pak, C. S., and Djerassi, C., *Tetrahedron Lett.* (1978), 4377–4378.
34. Lightner, D. A., and Wijekoon, W. M. D., *J. Org. Chem.* **47** (1982), 306–310.
35. Lee, S-F., Edgar, M., Pak, C-S., Barth, G., and Djerassi, C., *J. Am. Chem. Soc.* **102** (1980), 4784–4790.
36. Sing, Y. L., Numan, H., Wynberg, H., and Djerassi, C., *J. Am. Chem. Soc.* **101** (1979), 5155–5158. Errata: *J. Am. Chem. Soc.* **101** (1979), 7439.

37. Kokke, W. C. M. C., and Oosterhoff, L. J., *J. Am. Chem. Soc.* **94** (1972), 7583–7584.
38. Kokke, W. C. M. C., and Oosterhoff, L. J., *J. Am. Chem. Soc.* **95** (1973), 7159–7160.
39. Dezentje, R. F. R., and Dekkers, H. P. J. M., *Chem. Phys.* **18** (1976), 189–197.
40. Poloński, T., *J. Chem. Soc. Perkin Trans. 1* (1983), 305–309.
41. Spafford, R., Baiardo, J., Wrobel, J., and Vala, M., *J. Am. Chem. Soc.* **98** (1976), 5217–5225.
42. Meijer, E. W., and Wynberg, H., *J. Am. Chem. Soc.* **104** (1982), 1145–1146.
43. Lightner, D. A., Gawroński, J. K., Hansen, Aa. E., and Bouman, T. D., *J. Am. Chem. Soc.* **103** (1981), 4291–4296.
44. In the phrase "comparatively more consignate," it is not implied that the C–H bond by itself makes a consignate contribution to the rotatory strength. It has been argued on the basis of experimental data from chiral cyclopentanones that α C–H bonds make a dissignate contribution to the observed Cotton effect; see Kirk, D. N., *J. Chem. Soc. Perkin Trans. 1* (1976), 2171.
45. Paquette, L. A., Doecke, C. W., Kearney, F. R., Drake, A. F., and Mason, S. F., *J. Am. Chem. Soc.* **102** (1980), 7228–7233.
46. Lightner, D. A., Gawroński, J. K., and Bouman, T. D., *J. Am. Chem. Soc.* **102** (1980), 5749–5754.
47. Lightner, D. A., Paquette, L. A., Chayangkoon, P., Lin, H-S., and Peterson, J. R., *J. Org. Chem.* **53** (1988), 1969–1973.
48. Hoffman, P. H., Ong, E. C., Weigang, O. E., and Nugent, M. J., *J. Am. Chem. Soc.* **96** (1974), 2619–2620.
49. Cullis, P. M., Lowe, G., and Bayley, P. M., *J. Chem. Soc. Chem. Commun.* (1978), 850–852.
50. Ringdahl, B., Craig, J. C., and Mosher, H. S., *Tetrahedron* **37** (1981), 859–862.
51. Canceill, J., Collet, A., and Gottarelli, G., *J. Am. Chem. Soc.* **106** (1984), 5997–6003.
52. Ringdahl, B., Craig, J. C., Keck, R., and Retey, J., *Tetrahedron Lett.* (1980), 3965–3968.
53. Anderson, P. H., Stephenson, B., and Mosher, H. S., *J. Am. Chem. Soc.* **96** (1974), 3171–3177.
54. Numan, H., Meuwese, F., and Wynberg, H., *Tetrahedron Lett.* (1978), 4857–4858.
55. Lightner, D. A., Bouman, T. D., Wijekoon, W. M. D., and Hansen, Aa. E., *J. Am. Chem. Soc.* **106** (1984), 934–944.
56. Anderson, K. K., Cinquini, M., Colonna, S., and Pilar, F. L., *J. Org. Chem.* **40** (1975), 3780–3782.
57. Cullis, P. M., and Lowe, G., *J. Chem. Soc. Chem. Commun.* (1978), 512–514.
58. Ben-Tzur, S., Basil, A., Gedanken, A., Moore, J. A., and Schwab, J. M., *J. Am. Chem. Soc.* **114** (1992), 5751–5753.

CHAPTER 10

Diketones

In chiroptical spectroscopy, the concept of additivity lies at the heart of applications of the octant rule. According to this concept, for the ketone carbonyl chromophore the octant contributions of each vicinal dissymmetric perturber are summed to give the predicted net sign and magnitude of the $n \rightarrow \pi^*$ Cotton effect (Chapter 4). Additivity was investigated early in the history of modern optical rotatory dispersion and circular dichroism spectroscopy[1–5] and has been repeatedly examined since,[5–9] but it was not limited to monoketones. An entirely related question was posed early for diketones:[1,10] Is the observed $n \rightarrow \pi^*$ Cotton effect the simple sum of the Cotton effects of the monoketones in which the absorption bands are situated more or less at the same wavelength? Failure of the additivity rule[1,3] might be interpreted as vicinal interaction between the two carbonyl chromophores, and such interaction would invalidate or render tenuous any application of the octant rule to diketones. At the time of the early studies,[1,10] the answer was clear for certain ketones: Strong vicinal interaction was known for α-diketones, which are conjugated, and weaker interaction for a few

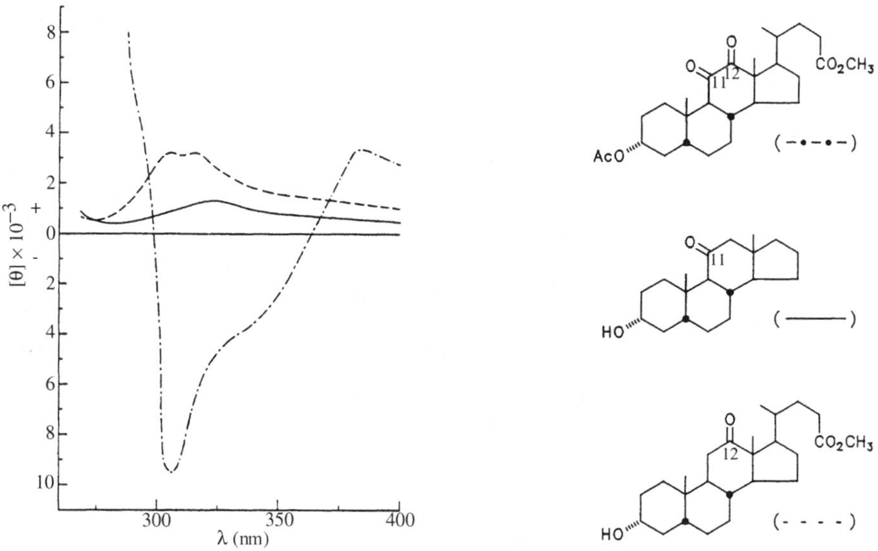

Figure 10-1. Optical rotatory dispersion of the α-diketone methyl 3α-acetoxy-11,12-diketo-cholanate (— · — · —) and its component monoketones: 3α-hydroxy-5β-androstan-11-one (———) and methyl 3α-hydroxy-12-ketocholanate (– – – –). [ORD curves are reprinted from ref. 3 with permission from Holden-Day, Inc.].

homoconjugated β-diketones. (Here, one assumes no enol forms—that the spectra represent only the diketone.) In these special cases, it was anticipated that the ORD or CD spectrum would *not* be the sum of the corresponding spectra of the component monoketones. Thus, the ORD Cotton effect of the 11,12-dione of Figure 10-1 bears scant relation to those of the parent 11- and 12-mono-ketones. However, for more widely separated diketones, could the spectra be assumed to obey the additivity rule? This issue was explored by Djerassi and Closson in the early 1950s,[10] using ORD, which was then believed to offer the most precise means of determining the presence and extent of additivity in diketones.[1]

10.1. Additivity of Octant Contributions vs. Vicinal Interaction

The additivity of ketone $n \rightarrow \pi^*$ octant contributions in diketones is tested very simply by obtaining ORD or CD spectra of the two component monoketones and comparing the summed spectral curves with the observed ORD or CD curve of the diketone. Since ORD curves have values over the entire spectral range, ORD is inherently more sensitive than CD for this comparison; nevertheless, in the region of absorption (e.g., of the $n \rightarrow \pi^*$ transition), CD comparisons are perfectly satisfactory. One important caveat in this approach to additivity is that the carbocyclic skeletons and their nonketone appendages should be identical and have identical conformations. Of course, this is usually not possible, so additivity is seldom in perfect agreement with experimental observations.

An early example of additivity was found in *trans*-8-methylhydrindan-2,5-dione, whose ORD curve was discovered to be nearly identical to the sum of the oppositely signed ORD curves from the two monoketones *trans*-8-methylhydrindan-2-one and -5-one (Figure 10-2).[1,11] From this finding, it was concluded that there existed "no perceptible vicinal interaction between the two carbonyl groups of the diketone."[1] Thus, no interactions were found in a 1,5 or δ-diketone, in which the carbonyl carbons lie approximately 4Å apart (nonbonded distance). In fact, the monoketone ORD curves and amplitudes (Figure 10-2) do not sum exactly to reproduce those observed for the diketone. The data differ by about 8.5%, probably an acceptable difference, given the errors incurred in ORD measurements in the 1950s and the fact that the monoketone carbocyclic skeletons have only one sp^2 carbon, so their conformations cannot be exactly the same as that of the diketone. The match is surprisingly good, considering the approximations.

As one might expect, when the diketone carbonyl groups are even more widely separated, as is possible in steroid diketones, the additivity rule is satisfactorily met. Thus, some of the earliest ORD investigations of additivity indicated that there was *no* vicinal interaction in 5α-androstan-3,17-dione and methyl 3,12-diketocholanate, in which the ketone carbonyl carbons lie roughly 9 and 6.5 Å apart, respectively (Figure 10-3). In one study,[10] the calculated ORD curves were generated by using only one-half of the molar rotation [ϕ] values of the component monoketone curves. This is not

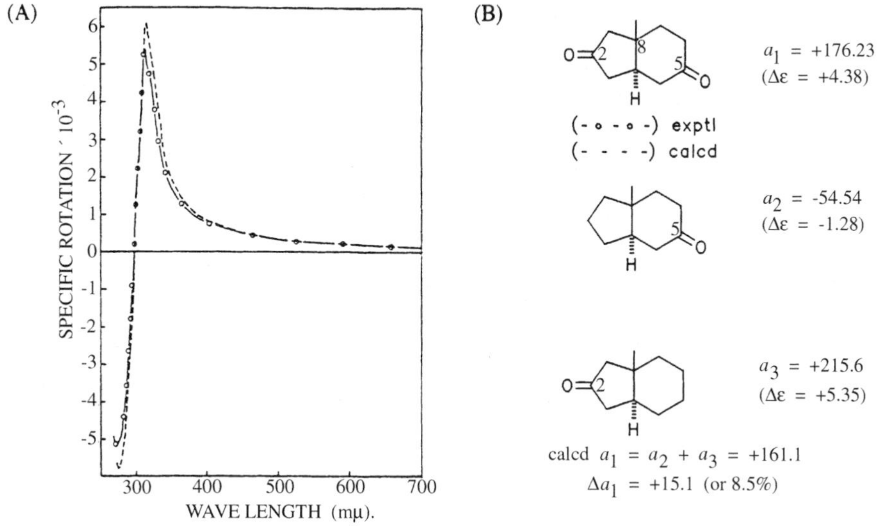

Figure 10-2. Optical rotatory dispersion of (+)-*trans*-8-methylhydrindan-2,5-dione in CH_3OH: observed (○ — ○ — ○), and calculated (- - - -) by summing the ORD curves of the component monoketones. (B) Cotton effect amplitudes (*a*) for the diketone and its component monoketones. The *a* values are from ref. 5 and Table 7-20. Δε values are computed by dividing the *a* values by 40.28. [(A) is reprinted with permission from ref. 11. Copyright© 1958 American Chemical Society.]

correct; the full [ϕ] values should have been used, as was recognized shortly thereafter by the authors.[1] Unfortunately, the full ORD curves were not available, due to limitations of instrumentation at the time; however, when the full values are used, the agreement between calculated and observed data is much more satisfactory[1] and indicates that the additivity rule is followed. Even when the carbonyl groups lie closer together, as in the steroid 3,11-dione and the 7,12-dione of Figure 10-3, in which the ketone carbon lies some 5 and 4.5 Å apart, respectively, the calculated curves (after correction) are in fairly good agreement with the observed ones. Here, too, vicinal interaction was thought to be essentially absent[10]—although subsequently, the 3,11-dione was believed not to follow the additivity rule.[1]

From the early ORD study,[10] two clear examples of nonadditivity were suggested: 5α-cholestan-3,6-dione and methyl 3,7-diketocholanate. The calculated ORD curves (Figure 10-4) were a poor match to those that were observed. This is more evident for the 3,6-dione, but less obvious for the 3,7-dione after the calculated curves are corrected to their full values. The 3,6-dione, whose carbonyl carbons are only approximately 4 Å apart, was a clear and often-cited example of vicinal interaction (and nonadditivity).[1,3–5] However, the nonadditivity of this γ-diketone contrasts sharply with the additivity observed in the bicyclic diketone of Figure 10-2, raising the spectre that for vicinal interaction, both the spatial orientation of the two chromophores and the distance between them play important roles. Further evidence for this notion was

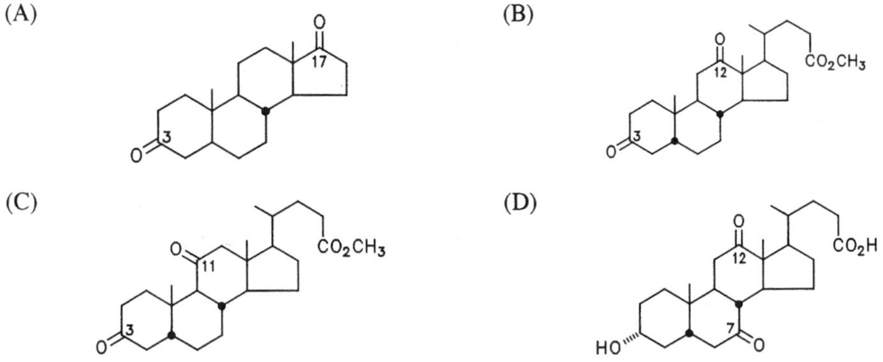

Figure 10-3. Steroid diketones that obey additivity in the octant rule. (A) 5α-androstan-3,17-dione; (B) methyl 3,12-diketocholanate; (C) methyl 3,11-diketocholanate; (D) 3α-hydroxy-7,12-diketocholanic acid.

provided at Shionogi in Japan by Takeda and Minato's ORD study[12] of yet a different δ-diketone, the sapogenin 25D,5β,20β,22α-spirostan-2,11-dione. Here, too, vicinal interaction was concluded to have taken place, but no vicinal interaction was detected in the 5α-analog. Similarly, no vicinal interaction was detected in 25D,5α,20β, 22α-spirostan-3,7-dione, also a δ-diketone, and vicinal interaction, if any, was concluded to be weak in the ε-diketones 25D,5β,20β,22α-spirostan-3,11-dione and its 5α-isomer.

When circular dichroism instrumentation became available, Velluz and Legrand at Roussel-UCLAF in Paris and Grosjean in Nancy reported on the CD spectra of a number of steroid di- and monoketones.[4] CD data on diketones for which data on the component monoketone are also available are shown in Table 10-1. Not surprisingly, in view of the ORD studies,[1,10] diketones with widely separated carbonyl chromophores (e.g., 5α-androstane-3,17-dione and 5β-pregnane-3,20-dione, in which the carbonyl carbons are spaced roughly 9 and 10 Å apart, respectively) are found by CD to follow the additivity rule. At a somewhat closer separation (7–8 Å), a 6,20-diketo *i*-steroid (16β-carbomethoxy-3,5-cyclo-17α-progestan-6,20-dione) was reported by Crabbé et al.[13] to follow the additivity rule.[3] Here, the inexact match between calculated and observed Δε values was attributed to the uncertainty in the ORD amplitude (an incomplete ORD curve) of the 6-monoketone from which the Δε value was obtained.[3] Unexpectedly, CD data suggest that there is no interaction in the often-touted example of vicinal interaction, 5α-cholestan-3,6-dione (Table 10-1), in which the carbonyl carbons are much closer together (~4 Å) than in the 3,17- and 3,20-diones cited above. This finding is unusual, because the sum of the ORD curves of 5α-cholestan-3-one and 5α-cholestan-6-one was reported[11] to be quite different from the observed ORD curve of 5α-cholestan-3,6-dione (Figure 10-4A). Where vicinal interaction *is* clearly indicated by CD is in the 11,17-diketone, which, like the 3,6-dione and the bicyclic diketone of Figure 10-2, is also a γ-diketone. Intermediate

Table 10-1. Comparison of Diketone CD Δε Values with the Sum (Σ) of the Δε Values for the Component Monoketones at λ^{max} (nm)[a]

			Σ[b]	ΔΔε[c]
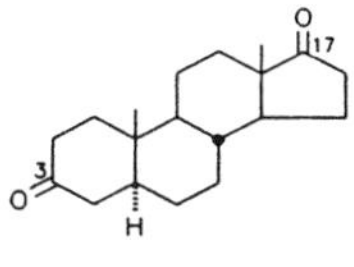 Δε: +4.27 (305)	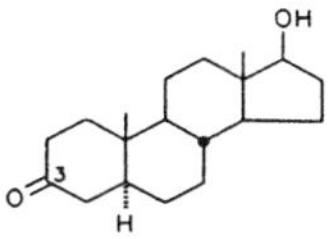 +0.90 (295)	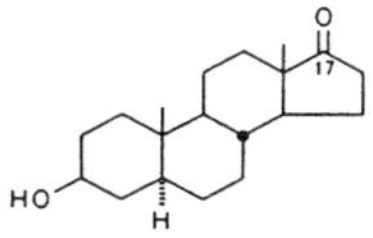 +3.28 (304)	+4.18	+0.09
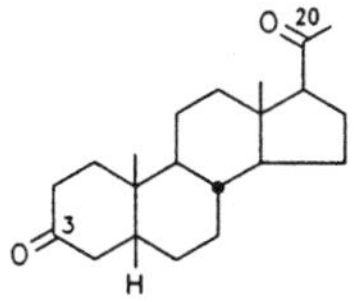 Δε: +2.9 (294)	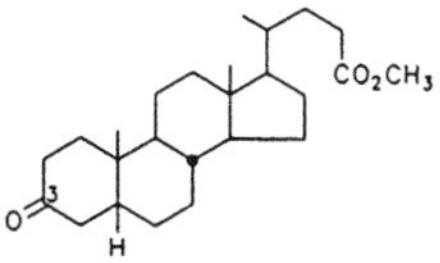 −0.44 (295)	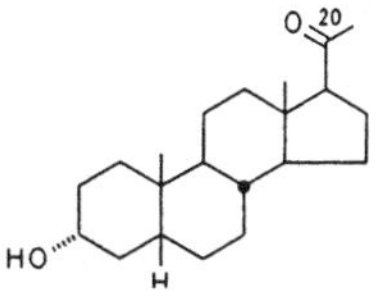 +3.45 (294)	+3.0	−0.1
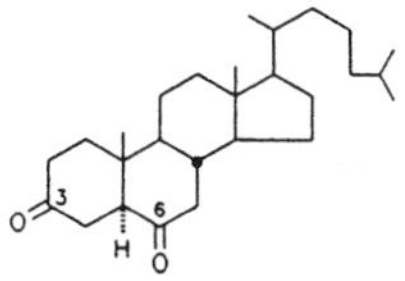 Δε: −0.30 (302)	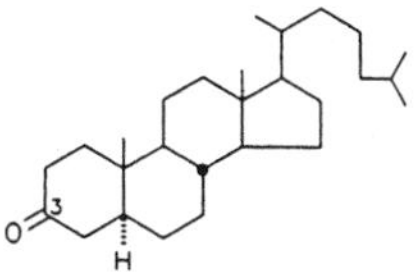 +1.13 (295)	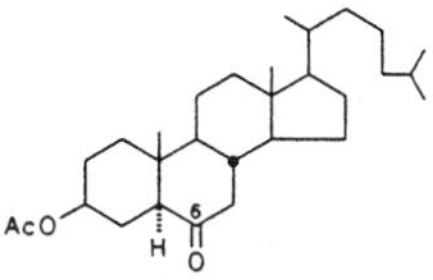 −1.39 (302)	−0.26	−0.04
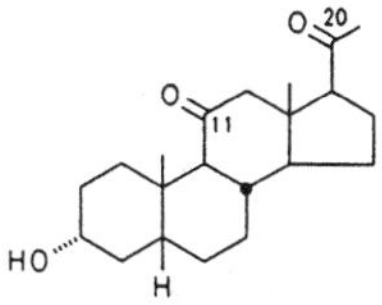 Δε: +4.89 (304)	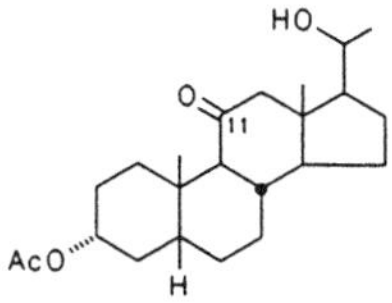 +0.25 (307)	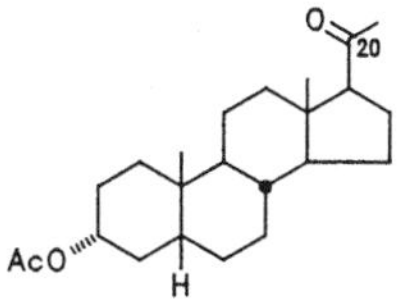 +3.33 (304)	+3.58	+1.31
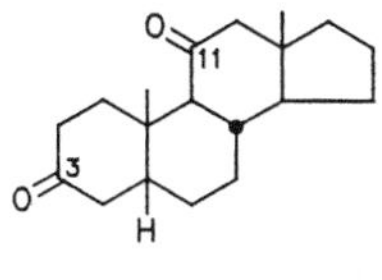 Δε: +0.27 (312)	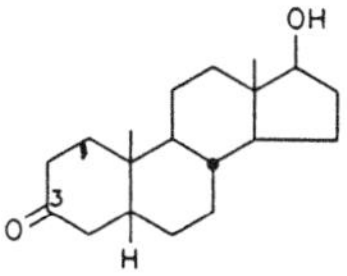 −0.30 (294)	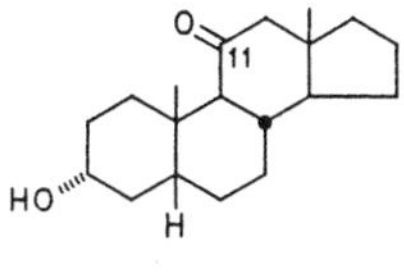 +0.25 (307)	−0.05	+0.32
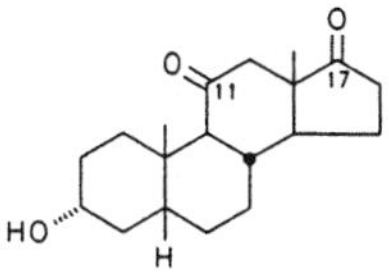 Δε: +3.32 (296)	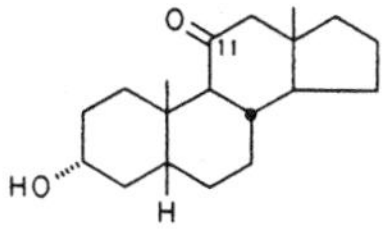 +0.45 (305)	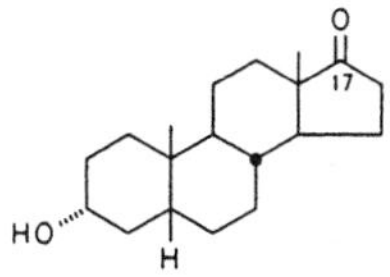 +3.45 (294)	+3.90	−0.58

[a] Data in dioxane solvent from ref. 6.

[b] Σ = sum of monoketone Δε values.

[c] Δε (diketone) − Σ.

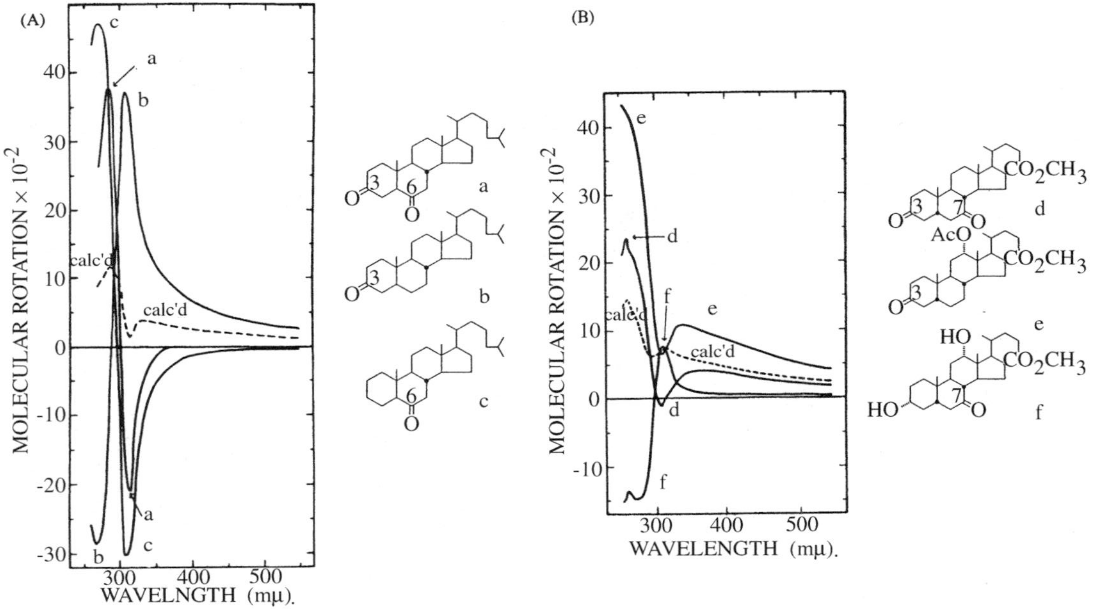

Figure 10-4. (A) Optical rotatory dispersion of 5α-cholestan-3,6-dione (**a**) and the calculated composite curve (- - - -) from adding [ϕ] values (times one-half) from the curves of the monoketones 5α-cholestan-3-one (**b**) and 5α-cholestan-6-one (**c**). (B) ORD of methyl 3,7-diketocholanate (**d**) and the calculated composite curve (- - - -) from adding [ϕ] values (times one-half) from the curves of the monoketones methyl 3-keto-12α-acetoxycholanate (**e**) and methyl 3α,12α-dihydroxy-7-ketocholanate (**f**). [The ORD curves are reprinted with permission from ref. 10. Copyright © 1956 American Chemical Society.]

cases of questionable vicinal interaction are found in 3,11 and 11,20-diketones (with carbonyl carbon separations of approximately 5 and 4 Å, respectively).

It is difficult to assess whether *modest* differences between observed and predicted $\Delta\varepsilon$ values for the ketones $n \rightarrow \pi^*$ transition (Table 10-1) indicate vicinal interaction or simply the imperfections of the analysis and the models used. Although cited as examples wherein the additivity rule[3] works, the 6,20-, 3,11- and 11,20-diketones are worthy of careful reinvestigation using potentially more accurate ORD and CD instrumentation in conjunction with analyses of ^{13}C-NMR carbonyl chemical shifts[14] to determine whether vicinal interaction between ketone chromophores can occur over modestly long distances.

The data for the diketones cited[3] as examples of vicinal interaction (the 2,11-, 11,17-, and 3,6-diones), but where the additivity rule breaks down, might also be reinvestigated. The contrasting CD and ORD data for the 3,6-dione particularly need to be reconciled. And the clearer nonadditivity of the 2,11- and 11,17-diones contrasts with the additivity behavior of *trans*-8-methyl-hydrindan-2,5-dione (Figure 10-2), suggesting that the relative orientation of the carbonyl groups plays an important role. The importance of orientation and distance between ketone carbonyl groups is examined further in the rest of the chapter.

10.2. Transannular Orbital Interaction in Nonconjugated Diketones

The early investigations of the additivity rule[3] in diketone ORD and CD spectra were the forerunners of more general studies of electronic interactions between chromophores. Some 10 years later, Roald Hoffmann described the phenomenon of electron delocalization through homoconjugation in terms of direct ("through-bond") and indirect ("through-space") interactions from localized chromophores.[14] In addition to being examined by chiroptical spectroscopy, orbital interactions have been studied by IR spectroscopy,[15] UV spectroscopy,[16] photoelectron spectroscopy (PES)[17] of ground-state orbitals, electron transmission spectroscopy (ETS)[17c,18] of excited-state orbitals, ^{13}C-NMR,[19] and even the kinetics of solvolysis reactions.[20] From such research, new insights were forthcoming on the spatial determinants of vicinal interaction for diketones (and for homoconjugated ketones and dienes—see Chapters 11 and 12).

Using PES, for example, investigators detected interchromophoric interactions between ketone carbonyl chromophores of certain diketones as splittings of the *n*-orbitals. Understandably, the largest *n*-orbital splittings (~2 eV) occur from α-diketones, which are conjugated (Table 10-2). However, homoconjugated β-diketones also show *n*-orbital splittings (Figure 10-5 and Table 10-2) on the order of 0.5 eV, thought to derive from interactions through σ-orbitals.[21a] With optimal alignment, as in the tetramethylcyclobutan-1,3-dione,[21b,c] where the carbonyl carbon-to-carbonyl carbon nonbonded distance is only 2 Å, the splitting is larger (0.73 eV), suggesting some through-space interaction.

Splitting generally falls off with distance: As the orbital interaction through bonds is extended to γ-diketones, *n*-orbital splitting is much reduced. For example, the splitting in cyclohexane-1,4-dione is only 0.15 eV, and in bicyclo[2.2.1]heptane-2,5-dione, in which the overlap between *n*-orbitals and the relevant σ-*p* ring orbitals is poor, the splitting is also small (0.16 eV).[17d] However, in γ-diketones, in which the ketone carbonyls are aligned for favorable interaction, the splitting can become large.[21d]

With even larger through-bond separations of the two ketone carbonyls, as in the δ-diones of Table 10-2, the splittings are quite small. In the adamantan-2,6-dione, the resolution was insufficient to detect a verifiable splitting, but the measured orbital energy is higher than that found in adamantan-2-one, indicating that the "nonbonding" MO of the dione is quite delocalized and the ionic state is destabilized inductively by the presence of a second carbonyl group.[21e] The situation is not very different in the bicyclo[3.3.1]nonane-2,6-dione (Figure 10-5), in which the splitting is computed to be small (0.16 eV) if it even exists.[19a] Again, an exception is found in the bicyclo[3.3.0]octane skeleton, where carbonyl groups are held close together. (The O-to-O nonbonding distance is 6.2 Å.)[21d]

Although not as extensively investigated, transannular orbital interactions have also been detected by ^{13}C-NMR spectroscopy.[19] Typically, the carbonyl carbon chemical shift ($\delta_{C=O}$) of the dione is found upfield from that of the corresponding monoketone, and the difference in chemical shift (Δδ) between di- and monoketones decreases with increasing separation. The largest shieldings are found when the carbonyl carbons lie close together, as in homoconjugated β-diketones (Table 10-3A), in which the O=C-to-C=O distances are shortest. However, even the γ- and δ-diketones of Table 10-3B exhibit moderately large Δδ values, consistent with transannular orbital interaction

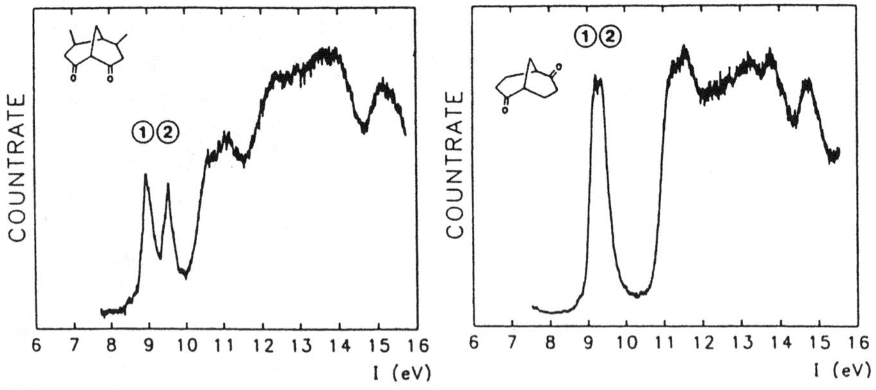

Figure 10-5. He(I) photoelectron spectra of the β-dione 4,6-dimethylbicyclo[3.3.1]nonane-2,8-dione (left) and the δ-dione bicyclo[3.3.1]nonane-2,6-dione (right). ① and ② are the orbital energies for the n_- and n_+ states, respectively, with a much larger splitting (0.53 eV) in the β-dione than in the δ-dione. (See Table 10-2.) [Reprinted with permission from ref. 19a. Copyright© 1992 American Chemical Society.]

Table 10-2. Carbonyl *n*-Orbital Interactions in Diketones, with Splittings Measured by Photoelectron Spectroscopy[a]

α-Diketone[b]	n_+	n_-	Δn	β-Diketone[b]	n_+	n_-	Δn
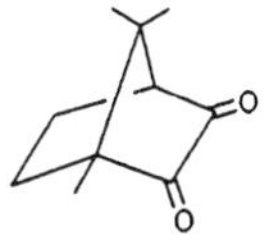	10.46	8.71	1.75	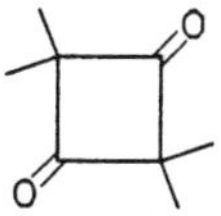	9.53[c]	8.80[c]	0.73
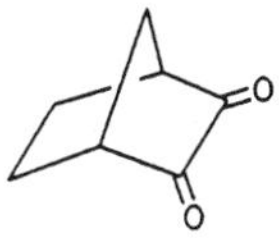	10.85	9.00	1.85	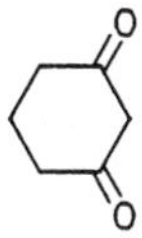	10.04	9.60	0.44
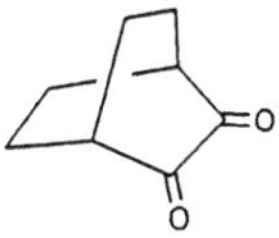	10.5	9.0	1.5		10.40	9.48	0.92
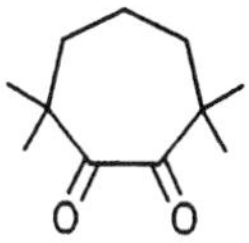	10.60	8.70	1.90	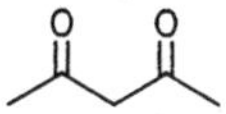	10.15	~9.6	~0.55
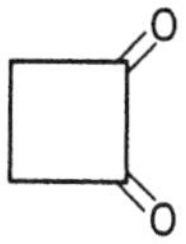	11.71	9.61	2.10	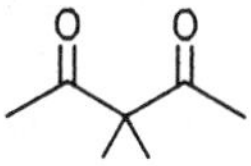	9.86	9.30	0.56
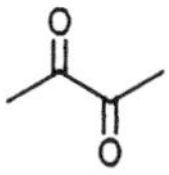	11.46	9.55	1.91	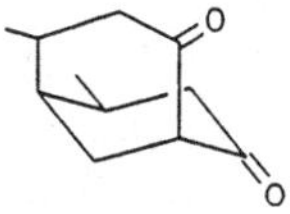	9.58	9.05	0.53
γ-Diketone[b,d]	n_+	n_-	Δn	**δ-Diketone**	n_+	n_-	Δn
	9.80	9.65	0.15	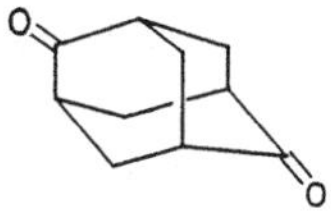	9.06[d]	9.06[d]	small
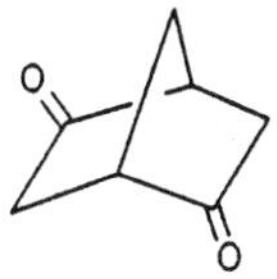	9.63	9.47	0.16	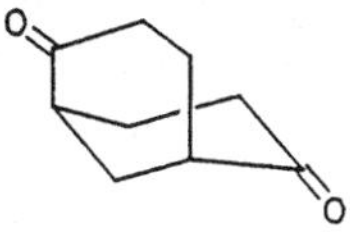	9.33[e]	9.33[e]	small
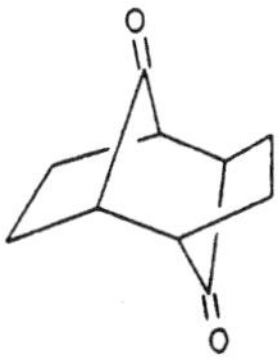	9.45	9.05	0.41	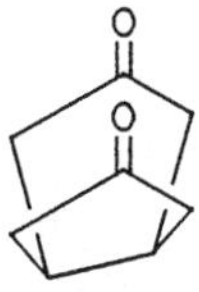	9.58[f]	9.31[f]	0.29 ($\Delta\pi^*$ 0.65)[g]

Table 10-2. *(Continued)*

[a] Values in eV. $n_+ = (1/\sqrt{2})(n_1 + n_2)$ and $n_- = (1/\sqrt{2})(n_1 - n_2)$.

[b] Data from refs. 17d and 20a.

[c] Data from refs. 21b and c.

[d] Data from ref. 21e; splitting is very small; the n-orbital energy in adamantan-2-one is 8.59 eV, indicating that there is some orbital interaction in the dione.

[e] Data from ref. 19a; see Figure 10-5. The n-orbital energy of the monoketone is 8.94 eV, suggesting a weak interaction in the dione.

[f] Data from ref. 21d.

[g] ETS data from ref. 18a, where a positive splitting indicates that the π_-^* orbital is less stable than the π_+^* orbital, and a negative splitting indicates the reverse.

over nonbonded distances of 2.8–4.0 Å. When the chromophores are separated by 4 σ-bonds, but the nonbonded distance is short (3.0 Å), as in the γ-diketone bicyclo[3.3.1]nonane-3,7-dione, Δδ is –3.9, and the origin of this effect is thought to come mainly from through-space interactions.[22] The orientation of the carbonyl groups seems to play an important role, as is evidenced by the larger Δδ values for the γ-diketone bicyclo[2.2.1]heptane-2,5-dione (Δδ = –5.8), the δ-diketone adamantane-2,6-dione (Δδ = –4.9), and the cage ε-diketone (Δδ = –3.8). These data indicate that, with a favorable alignment, moderately large Δδ shifts can be seen even when the interchromophoric distances are large. However, if the distances *are* large, as in the ε-diketone decalindione, as well as the acyclic δ-diketone heptane-2,5-dione, and the nonbonded orientation is favorable, the difference in chemical shift (Δδ) falls to very low values.

Some while ago, a variety of diketones was treated by CNDO-SCF-CI MO methods to analyze the importance of spatial orientation on n and π^* orbital splittings and on the $n \rightarrow \pi^*$ transition.[23] The work included strongly interacting α-diketones, β-diketones, which interact less strongly, and remote γ- and δ-diketones. The last included, *inter alia*, cyclohexane-1,4-dione (γ-dione, with a calculated Δn = 0.15 eV for the twist-boat conformation) and adamantane-2,6-dione (δ-dione) of Table 10-2. β-Diketones included cyclobutane-1,3-dione (calculated Δn = 0.91 eV) and spiro[4.4]nonane-1,6-dione; α-diketones included cyclohexane-1,2-dione (computed Δn = 1.8 eV) and camphorquinone. Splitting between two n-orbitals (and two π^* orbitals) was found to depend on the distance between the two carbonyl groups, their spatial orientation, and the nature of the molecular fragment connecting them (e.g., whether it was saturated or unsaturated). As expected, the n-orbital splitting is large (~1.8 eV) in α-diketones and smaller in the others. The splitting is significantly lower when the carbonyl groups lie in-line; then the $2p_\pi$ nodal planes found in some γ- and δ-diones, such as the adamantane-2,6-dione, are orthogonal.

Recent MO calculations on diketones using the Gaussian 92 HF/3-21G basis set show that (i) the electron density is greater around the carbonyl carbons in many of the diketones of Table 10-3, compared with the corresponding monoketones, and (ii) the variation in the magnitude of the electron density reorganization correlates qualitatively well with the variation of Δδ.[19c] Thus, although the two major causes of the

Table 10-3A. Influence of Homoconjugation on the ^{13}C-NMR Chemical Shifts of Diketones[a]

β-Diketone	$\delta_{C=O}$	Monoketone	$\delta_{C=O}$	$\Delta\delta^{b}$	r(Å)[c]
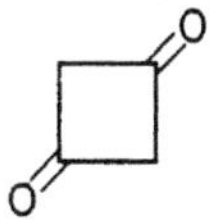	197.9		208.6[d]	−10.7	2.0
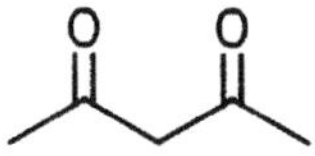	202.2	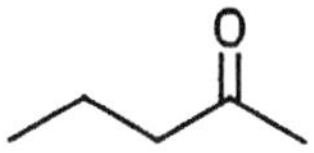	209.0	−6.8	2.4
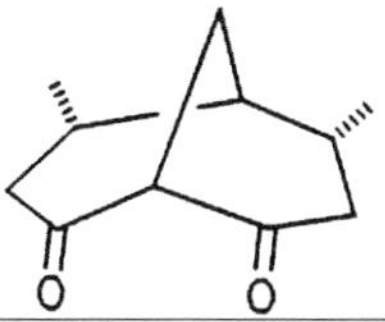	203.3[e]	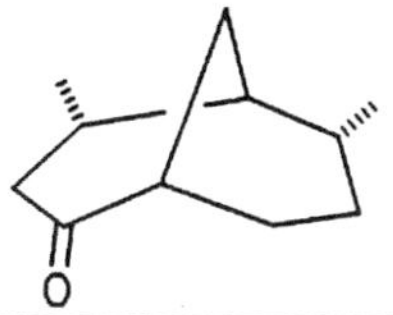	213.3[e]	−10.0	2.5[f]
γ-Diketone	$\delta_{C=O}$	Monoketone	$\delta_{C=O}$	$\Delta\delta^{b}$	r(Å)[c]
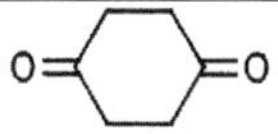	208.4	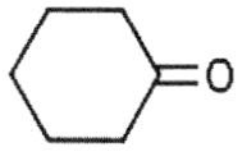	212.1	−3.7	2.9
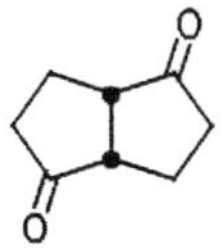	220.3[g]	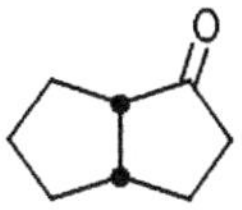	223.2[g]	−2.9	3.1
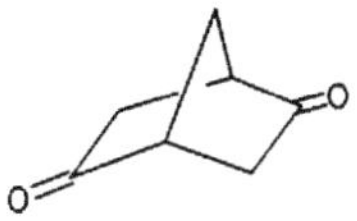	212.3[f]	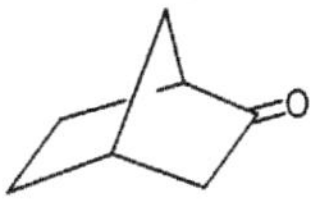	218.1[f]	−5.8	2.8
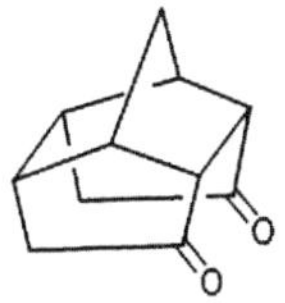	Not given[h]	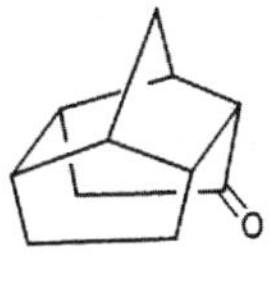	Not given[h]	−9.5[h]	2.6[h]
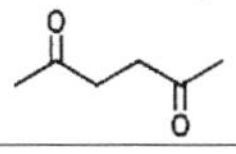	207.1	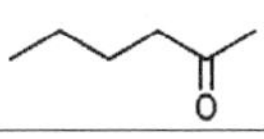	209.4	−2.3	3.8

[a] Measured in $CDCl_3$. δ in ppm downfield from $(CH_3)_4Si$ (ref. 19b, unless otherwise specified).

[b] Dione minus ketone.

[c] Interchromophoric nonbonded (O = C ··· C = O), distances from PCMODEL.

[d] Cerichelli, G., Frachey, G., and Galli, C., *Gazz. Chim. Ital.* **116** (1986), 683–686.

[e] Data from ref. 19a. The methyl configuration is *endo* (Bishop, R., private communication). Order of stability from PCMODEL, kcal/mole (distance, Å): chair–chair, −92.72 (2.48); chair–boat, −88.94 (2.44); boat–boat,did not minimize.

[f] Data from ref. 22.

[g] Whitesell, J. K., and Matthews, R. S., *J. Org. Chem.* **42** (1977), 3878–3882.

[h] Data from ref. 19c.

Table 10-3B. Influence of Homoconjugation on the ^{13}C-NMR Chemical Shifts of Diketones[a]

δ-Diketone	$\delta_{C=O}$	Monoketone	$\delta_{C=O}$	Δδ[b]	r(Å)[c]
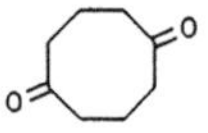	213.2[i]	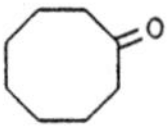	217.7[i]	−4.5	2.9
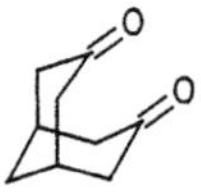	208.6[f]	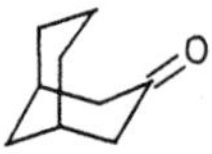	212.5[f]	−3.9	3.0[i]
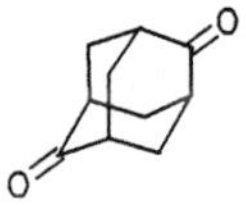	213.4[f]	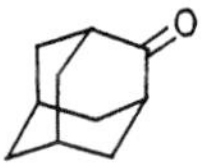	218.3[f]	−4.9	3.5
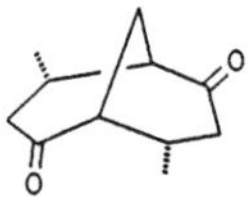	208.9[e]	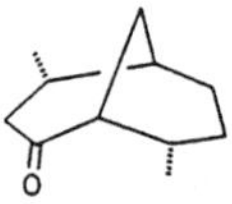	213.3[e]	−4.4	3.5[k]
	218.0	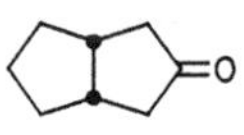	220.9[g]	−2.9	4.0
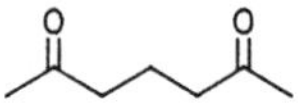	208.4	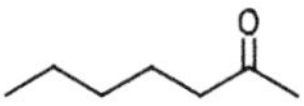	209.4	−1.0	5.0
ε-Diketone	$\delta_{C=O}$	**Monoketone**	$\delta_{C=O}$[b]	Δδ	r(Å)[c]
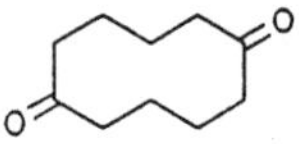	213.6[l]	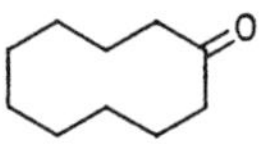	214.4[d]	−0.8	4.8
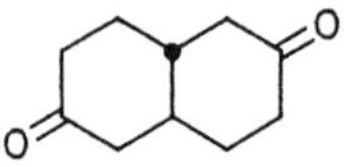	209.6[m]	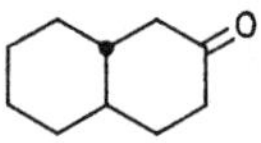	209.2[m]	+0.1	5.3
	Not given[h]		Not given[h]	−3.8[h]	5.3[h]
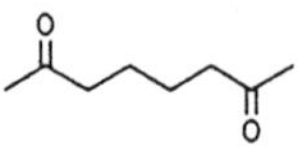	208.7	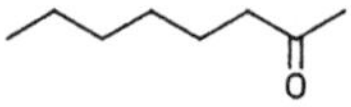	209.4	−0.7	6.2

[a] Measured $CDCl_3$. δ in ppm downfield from $(CH_3)_4$ Si. (Data from ref. 19b, unless otherwise specified.)
[b] Dione minus ketone.
[c] Interchromophoric nonbonded (O=C · ··C=O) distances from PCMODEL.
[d] Cerichelli, G., Frachey, G., and Galli, C., *Gazz. Chim. Ital.* **116** (1986), 683–686.

(Continued)

Table 10-3B. Continued

[e] Data from ref. 19a. The methyl configuration is *endo* (Bishop, R., private communication). Order of stability from PCMODEL, kcal/mole (distance, Å): chair–chair, −92.72 (2.48); chair–boat, −88.94 (2.44); boat–boat did not minimize.

[f] Data from ref. 22.

[g] Whitesell, J. K., and Matthews, R. S., *J. Org. Chem.* **42** (1977), 3878–3882.

[h] Data from ref. 19c.

[i] Data from ref. 22c.

[j] Order of stability from PCMODEL, kcal/mole (distance, Å): chair–chair, −82.45 (2.97); chair–boat, −78.69 (3.58); boat–boat, −75.53 (4.05).

[k] Order of stability from PCMODEL, kcal/mole (distance, Å): chair–chair, −92.53 (3.50); chair–boat, −89.91 (3.39); boat–boat, −85.22 (3.59).

[l] House, H. O., Lee, J. H. C., VanDerveer, D., and Wissinger, J. E., *J. Org. Chem.* **48** (1983), 5285–5288.

[m] Jones, J. B., and Dodds, D. R., *Can. J. Chem.* **65** (1987), 2397–2404.

$\Delta\delta$ shifts in diketones are thought to be orbital interactions (through-bond and through-space) and electrostatic dipole–dipole interactions, the calculations of Paddon-Row favor the latter.[19c] Vicinal interaction over long distances in diketones has been explored further by ORD and CD spectroscopy, as outlined in the next section.

10.3 Remote Diketones: γ- and δ-Diones

Aside from the steroid diketones of Section 10.1, there are relatively few examples of more recent studies of vicinal interaction in diketones by ORD or CD spectroscopy. These are collected in Table 10-4 and consist only of data on γ- and δ-diketones, most for conformationally rigid or restricted bicyclic ketones. The presence or absence of vicinal interaction may be detected by comparing the $\Delta\varepsilon$ value of the diketone with the sum of the component monoketone $\Delta\varepsilon$ values, $\Delta\Delta\varepsilon = \Delta\varepsilon$ (diketone) $- \Sigma\Delta\varepsilon$ (monoketones). For bicyclo[2.2.1]-heptane-2,5-dione, $\Delta\Delta\varepsilon$ is very modest, suggesting little, if any, vicinal interaction where the two carbonyl groups lie essentially in-line. This relative inactivity may be contrasted with the large effect seen (Table 10-3) on the ^{13}C-NMR chemical shift of the carbonyl carbons, but is consistent with the findings on the camphane-2,5-dione of Table 10-4, which shows a $\Delta\Delta\varepsilon$ of only 0.1.[24] The somewhat larger and negative $\Delta\Delta\varepsilon$ value (−0.81) found for the fenchane-2,5-dione may be ascribed to the inaccuracies of measurement, where the value of $\Delta\varepsilon = +0.65$ comes from the conversion of ORD data.[24] In a related γ-diketone, bicyclo[2.2.2]octane-2,5-dione (Figure 10-6), $\Delta\varepsilon = -0.78$,[25] but the monoketone is achiral, thus making it difficult to assess vicinal interaction. A better estimate of $\Delta\Delta\varepsilon$ might have been made if the CD data of 5-hydroxybicyclo[2.2.2]octan-2-one had been available.

In the different orientation of the two carbonyl groups provided by *cis*-bicyclo-[3.3.0]octane-2,6-dione,[26] the $n \rightarrow \pi^*$ CD Cotton effect is large and positive—and some 60% larger than that of the component monoketone (Table 10-4, Figure 10-7). Yet, the UV spectrum (Figure 10-7) of the diketone has nearly the same absorbance as twice the concentration of the component monoketone. Both the CD and UV λ^{max} of the diketone are also shifted (a bathochromic shift) from the corresponding λ^{max} of the

Table 10-4. Vicinal Interaction in Diketones and a Triketone Relative to that in Monoketone Components, as Detected by CD Δε Values

γ-Diketone	Monoketones		ΔΔε
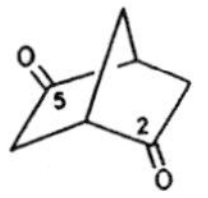			+0.21
Δε(nm): +1.75 (307)[a]	+0.77(307)[a]		
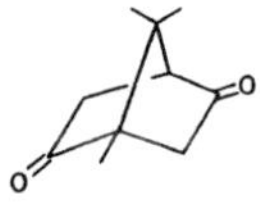	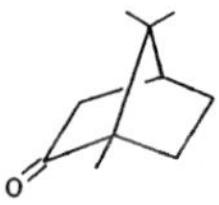	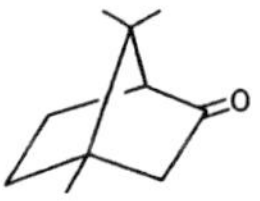	+0.10
Δε(nm): −3.22 (294)[b]	−1.78(296)[b]	−1.54(296)[b]	
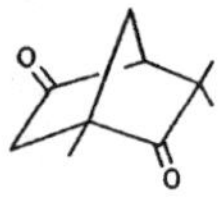	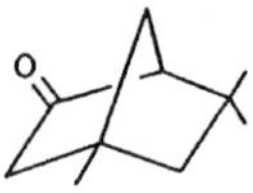	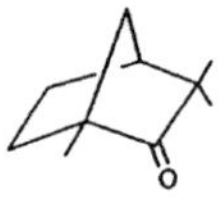	−0.81
Δε(nm): −0.30 (303)[b]	+0.65(304)[b]	−0.14(294)[b]	
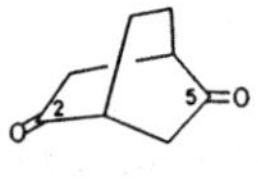	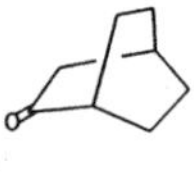		−0.78
Δε(nm): −0.78(298)[c]	achiral		
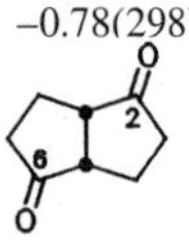	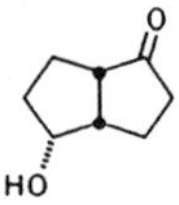		+1.87
Δε(nm): +4.67(300)[d]	+1.40(296)[d]		

δ-Diketone	Monoketone	ΔΔε
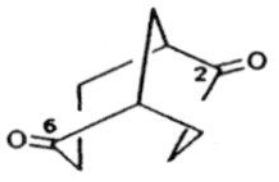	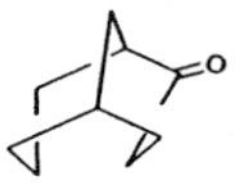	+2.9
Δε(nm): +3.2 (297, 302)[e] +1.2 (295)[f]	+0.15(292)e	(+0.9)[f]
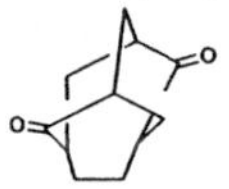	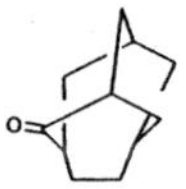	+0.26
Δε(nm): +0.40(302)[g]	+0.07(312)[g]	

γ-Triketone	Monoketone	ΔΔε
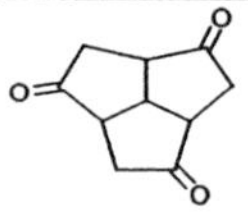	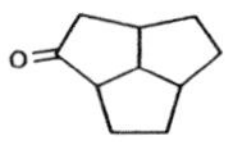	+1.6
Δε(nm): +4.0(302)[h]	+0.8(294)[i]	

γ-Diketone	Monoketone
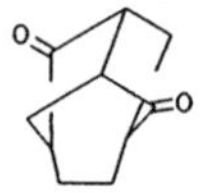	
Δε(nm): −2.19(301)[g]	

[a] Data from Pak, C. S., Ph.D. dissertation, *Circular Dichroism Studies of Bicyclo[3.1.0]hexanones and Bicyclo[2.2.1]heptanones*, University of Nevada, Reno (1977).

[b] Data from ref. 24, in 95% ethanol.

[c] Data from ref. 25, in cyclohexane.

[d] Data from ref. 26, in CH_2Cl_2.

[e] Data from ref. 30, in dioxane.

[f] Data from ref. 31, in ethanol.

[g] Data from ref. 27, in isooctane.

[h] Data from ref. 28.

[i] Data from ref. 29.

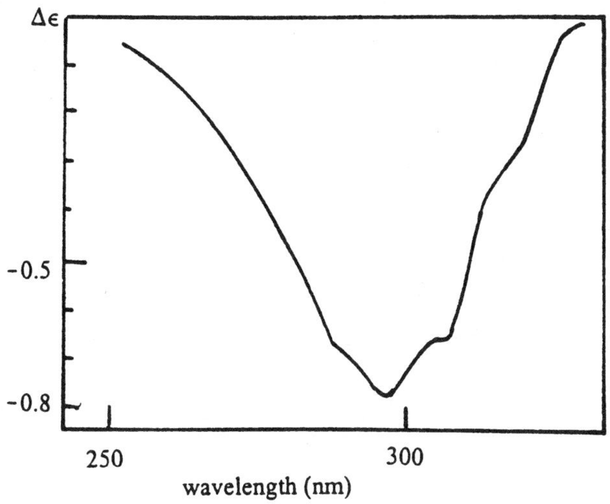

Figure 10-6. CD spectrum of bicyclo[2.2.2]octane-2,5-dione in cyclohexane. $\Delta\varepsilon^{max} = -0.78$. [Redrawn from ref. 25.]

component monoketone. This γ-diketone example appears to offer a clear-cut case for vicinal interaction of the carbonyl groups. When the compound is examined by ^{13}C-NMR carbonyl chemical shifts (Table 10-3), here, too, $\Delta\delta$ is consistent with vicinal interaction. In a related γ-dione,[27] in which carbons 3 and 7 are tied to an ethano bridge (Table 10-4, bottom right), the magnitude of the Cotton effect (Figure 10-8) is considerably reduced (from 4.67 to 2.19),[27] indicating that the vicinal interaction is sensitive to changes in conformation of the bicyclo[3.3.0]octane framework. Unfortunately, with no tricyclic monoketone CD or ORD data available, estimates of vicinal interaction in the tricyclic diketone cannot yet be made.

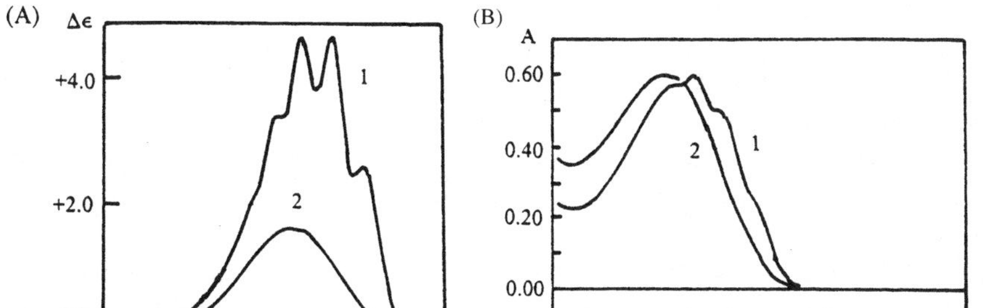

Figure 10-7. (A) CD and (B) UV spectra of (1*S*,5*S*)-bicyclo[3.3.0]octane-2,6-dione (curve 1) and (1*S*,2*R*,5*S*)-endo-2-hydroxybicyclo[3.3.0]octan-6-one (curve 2) in CH_2Cl_2. In (B), the absorbance of curve 2 is doubled for comparison purposes. [Redrawn from ref. 26.]

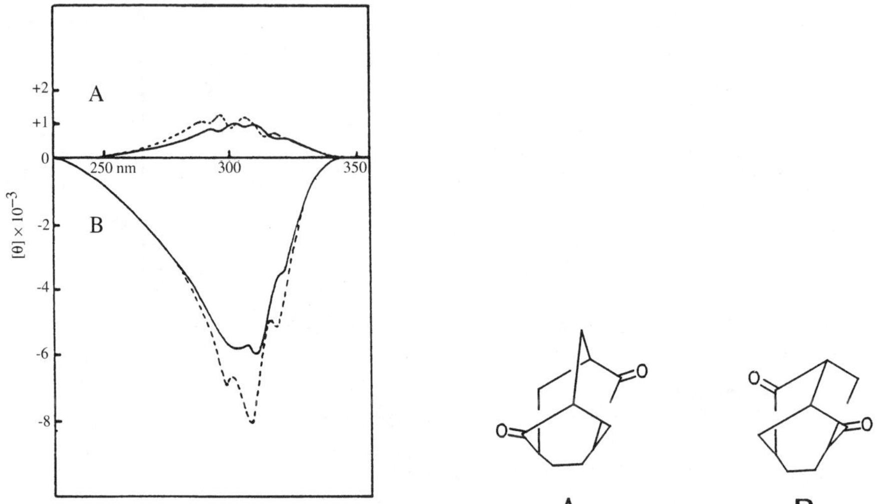

A B

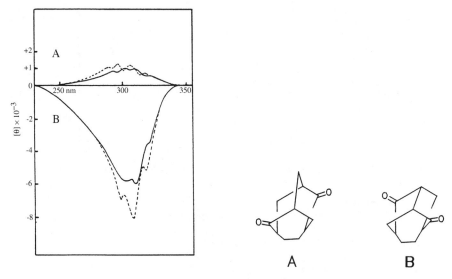

Figure 10-8. CD spectra at 25°C (———) and −190°C (- - - -) of (A) (+)-homoadamantane-2,7-dione and (B) (−)-tricyclo[3.3.2.0^{3,7}]-decane-2,6-dione in ether–isopentane–ethanol (EPA, 5:5:2 by vol.). [Spectral data are reprinted with permission from ref. 27. Copyright© 1980 American Chemical Society.]

The only example of a rigid triketone—in this case, a γ-trione (Table 10-4)—indicates strong vicinal interaction[28] when its Δε value is compared with three times the Δε value of the component monoketone.[29]

Strong vicinal interaction can apparently also be found in a δ-diketone (Table 10-4), bicyclo[3.3.1]nonane-2,6-dione[30,31]; however, there is some concern that the earlier Δε value[30] may be high by a factor of about 3. For this diketone, strong vicinal interaction was also detected by ^{13}C-NMR Δδ for the carbonyl carbons[19a] (Table 10-3). However, as with the related γ-dione, when carbons 3 and 7 are tied together by an ethano bridge to make a more rigid tricylic dione,[27] the Cotton effect magnitude (Figure 10-8) is substantially reduced, again indicating that the vicinal interaction is considerably sensitive to changes in the relative orientation of the two carbonyl chromophores. Weaker vicinal interaction is found in the tricyclic dione.

Most CD or ORD studies of long-range interchromophoric interaction in diketones have been carried out with conformationally rigid systems. However, the largest vicinal effects have been seen where the carbocyclic framework is the most mobile, as in bicyclo[3.3.0]octane-2,6-dione and bicyclo[3.3.1]nonane-2,6-dione (Table 10-4). Conformational analysis of the latter by molecular mechanics (MM2, 1991 force field)[31b] indicates that it exists as a mixture of two conformers: a chair–chair (**cc**) and a chair–boat (**cb**) (Figure 10-9), which are predicted to have oppositely signed $n \rightarrow \pi^*$ (and higher energy) Cotton effects.[31b] PCMODEL calculations[32] predict that the **cc** conformer predominates by only 0.21 kcal/mole of enthalpy. In the **cc** conformer, the chairs are somewhat flattened to minimize the 3,7-nonbonded interaction. In the **cb**

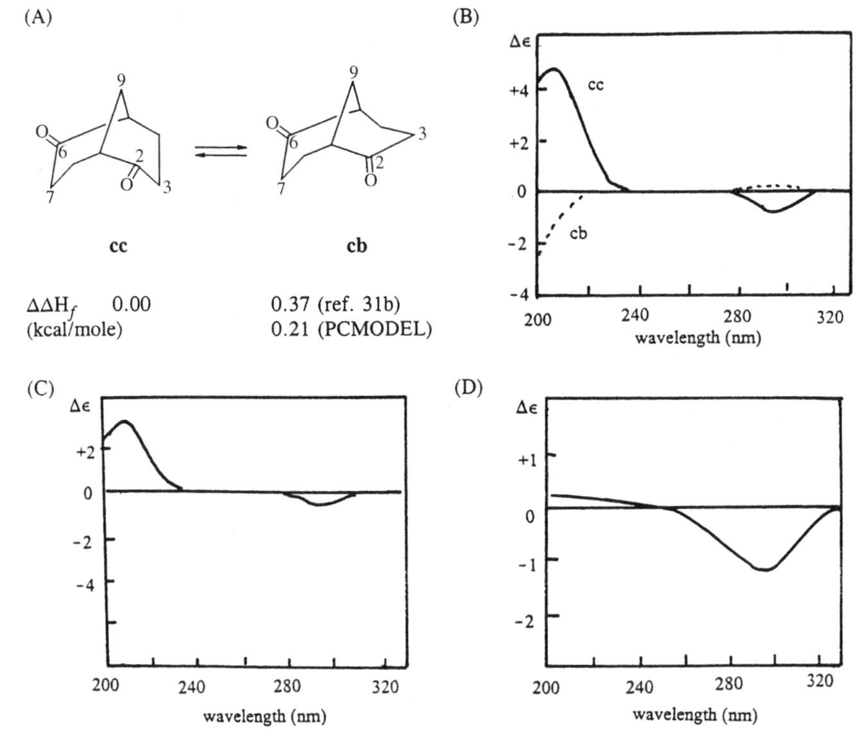

Figure 10-9. (A) Interconverting chair–chair (**cc**) and chair–boat (**cb**) conformations of (–)-(1*R*,5*R*)-bicyclo[3.3.1]nonane-2,6-dione and the calculated differences in their heats of formation. (B) Calculated (ref. 31b) CD spectra of the **cc** and **cb** conformers. (C) Calculated net CD spectrum of the dione, based upon a 75% **cc**, 25% **cb**, conformer population. (D) The observed CD spectrum of (–)-(1*R*,5*R*)-bicyclo[3.3.1]nonane-2,6-dione in 95% ethanol. [Data are redrawn from ref. 31b.]

conformer, the boat is somewhat flattened to minimize the 3,9-nonbonded steric interaction. No similar conformational analysis has been carried out on bicyclo[3.3.0]octane-2,6-dione, and in neither case have there been follow-up conformational studies involving variable low-temperature CD.

Chiroptical spectroscopic investigations of longer range vicinal interaction in diketones (e.g., ε-diones) reside with the steroids of Section 10-1. No follow-up studies have been conducted, but it would appear from those data and from the data of Table 10-4 that vicinal interaction becomes very weak with interchromophoric distances exceeding those of δ-diketones.

10.4. Proximal Diketones: β-Diones

With decreasing separation between the carbonyl groups of a diketone, increasing vicinal interaction is expected. As will be seen, spectral changes are most pronounced

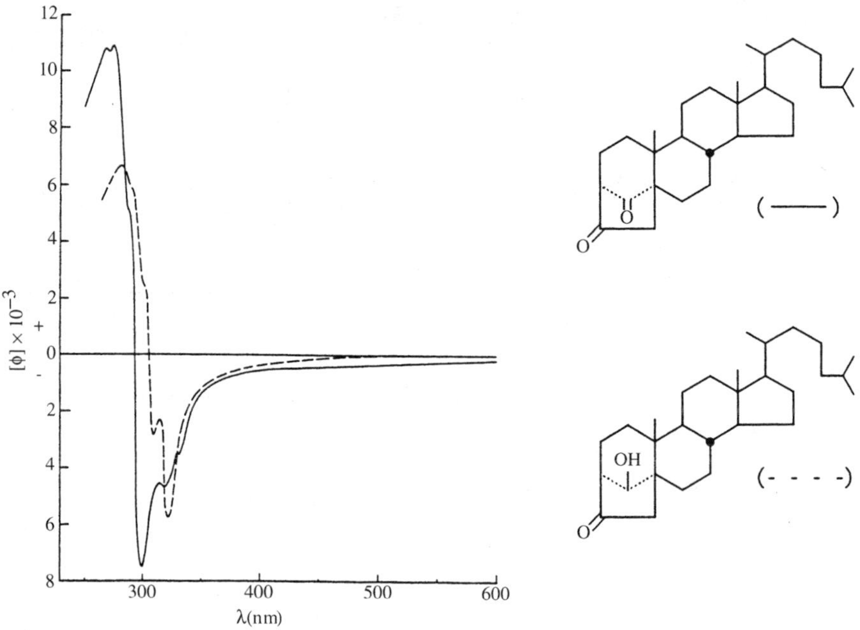

Figure 10-10. ORD curves for the $n \rightarrow \pi^*$ transition of the steroid β-diketone (———) and one of its ketols (- - - -). [ORD curves are reprinted from ref. 3 with permission from Holden-Day, Inc.]

when the carbonyl groups are conjugated, as in α-diones, but strong vicinal interaction has been detected by ORD and CD spectroscopy in β-diketones, of which there are surprisingly few examples. One of the earliest reported[3] examples of interchromophoric interaction detected by ORD in β-diketones showed (Figure 10-10) that the overall $n \rightarrow \pi^*$ Cotton effect was shifted to shorter wavelengths. However, it is not entirely clear whether the component ketone Cotton effects are additive, because data are not available for both components.

At about the same time, Gerlach[33] reported CD data for the $n \rightarrow \pi^*$ Cotton effects of (+)-5(*R*)-spiro[4.4]nonane-1,6-dione (Table 10-5), which was of special interest because Cram and Steinberg[34] had shown much earlier that the dione UV ($\varepsilon_{305}^{max} = 126$) was not simply twice that of the parent monoketone ($\varepsilon_{295}^{max} = 25$). Rather, it was bathochromically shifted and considerably enhanced. The CD data for the dione (Table 10-5) show two Cotton effects in the $n \rightarrow \pi^*$ region and a more intense short-wavelength Cotton effect near 222 nm, which are presumably due to short-range vicinal interaction (homoconjugative coupling). Variable low-temperature CD studies (Figure 10-11)[35] showed relatively minor changes in the $n \rightarrow \pi^*$ bands in the hydrocarbon solvent and larger changes in the more polar EPA solvent. Although at that time it was not possible to distinguish between conformational changes in those very flexible rings (see Sections 6.3 and 7.6) and asymmetric solvation as probable

Table 10-5. Circular Dichroism Data, Δε, and λ^{max}(nm), for Bicyclic and Polycyclic β-Diketones

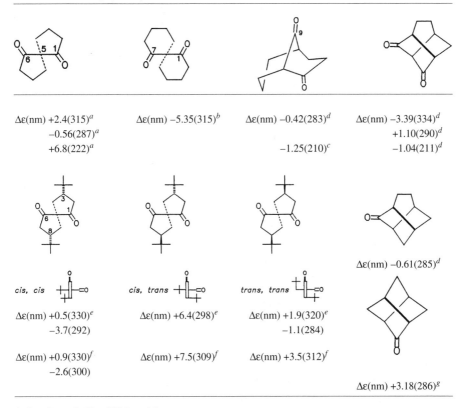

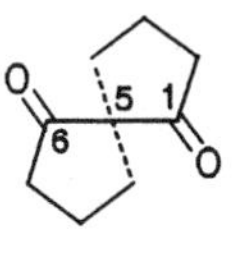	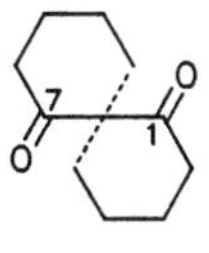	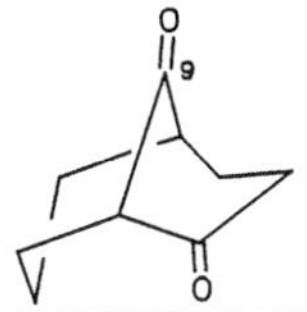	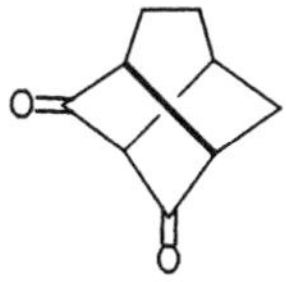
Δε(nm) +2.4(315)[a] −0.56(287)[a] +6.8(222)[a]	Δε(nm) −5.35(315)[b]	Δε(nm) −0.42(283)[d] −1.25(210)[c]	Δε(nm) −3.39(334)[d] +1.10(290)[d] −1.04(211)[d]
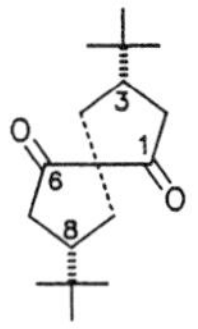	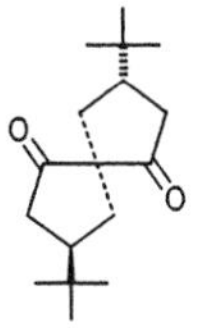	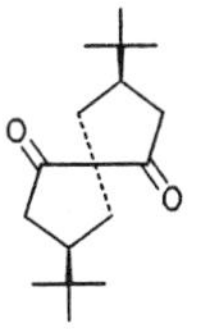	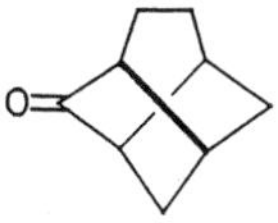
			Δε(nm) −0.61(285)[d]
cis, cis 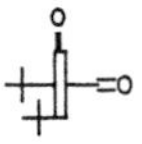	*cis, trans* 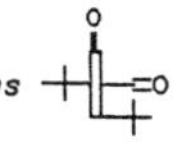	*trans, trans* 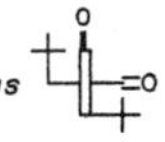	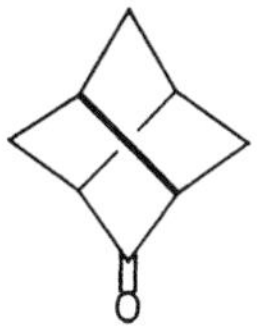
Δε(nm) +0.5(330)[e] −3.7(292)	Δε(nm) +6.4(298)[e]	Δε(nm) +1.9(320)[e] −1.1(284)	
Δε(nm) +0.9(330)[f] −2.6(300)	Δε(nm) +7.5(309)[f]	Δε(nm) +3.5(312)[f]	Δε(nm) +3.18(286)[g]

[a] Data from refs. 33 and 39, in cyclohexane.
[b] Data from ref. 41, in cyclohexane.
[c] Data from ref. 31, in ethanol.
[d] Data from ref. 42, in isooctane.
[e] Data from ref. 40a, in methanol.
[f] Data from ref. 40b, in isooctane.
[g] Data from ref. 43, in isooctane.

causes of the temperature and solvent effects on the CD curves, the data supported vicinal interaction between the two carbonyl groups.

An X-ray crystallographic image of the structure of the dione[36] indicates a conformation intermediate between envelope and half-chair forms for the two rings, albeit closer to the latter, with an angle of 84° between the two carbonyl C=O vectors. Using these data, Longuet-Higgins and Murrell[37] attempted to apply coupled chromophore theory to calculate the CD of the dione. According to these researchers, when two chromophores are brought sufficiently close together in a molecule, electrostatic coupling of locally excited states with or without a transfer of charge leads to an

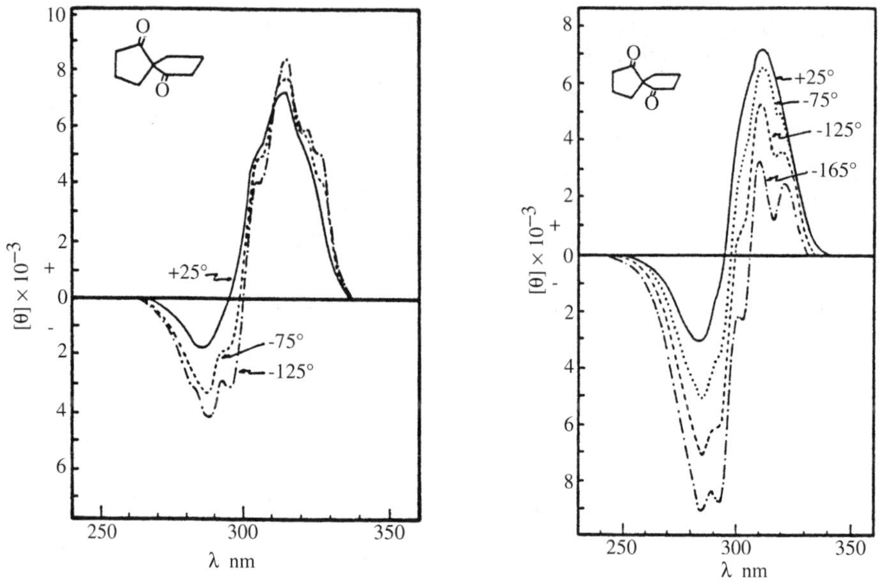

Figure 10-11. Variable low-temperature CD spectra of (+)-5(*R*)-spiro[4.4]nonane-1,6-dione in (left) methylcyclohexane–isopentane (1:5 by vol.) and (right) ether–isopentane–ethanol (EPA, 5:5:2 by vol.). [Reprinted from ref. 35, Copyright© 1972, with permission from Elsevier Science.]

absorption spectrum that is not merely the sum of the spectra of the separate chromophores. Aage Hansen at the H.C. Ørsted Institute applied the theory to calculate chiroptical properties of (+)-(1*R*)-5-methylenebicyclo-[2.2.1]hept-2-ene, (+)-1(*R*)-bicyclo[2.2.2]oct-5-ene-2-one, and bicyclo[2.2.1]heptan-2,7-dione.[38] Using well-known semiempirical techniques, he predicted accurately the ordinary absorption spectra and optical activities of the first two compounds, but calculations on the last compound still await experimental verification. Applied to the spirodione of Figure 10-11, Hansen's method[38a] correctly predicted the sign of the long-wavelength transition.[35] Coupling of the two locally excited carbonyl $n \rightarrow \pi^*$ transitions is apparent in the computed transitions at 296 and 290 nm (Table 10-6).

A different theoretical treatment[23] using CNDO–SCF–CI MO calculations predicts large n and π^*-orbital splittings for the approximately 90° orientation of the two carbonyl chromophores and a 0.086-eV (~700 cm^{-1}) splitting of the two $n \rightarrow \pi^*$ transitions. The lower energy $n \rightarrow \pi^*$ transition was computed to have a positive Cotton effect, and the higher $n \rightarrow \pi^*$ transition was computed to have a negative Cotton effect, for the homoconjugated ketone carbonyl system with right-handed chirality.[39]

Subsequently, Shingu et al.[40] at Osaka University measured and analyzed the CD spectra of three diastereomeric 3,8-di-*tert*-butylspiro[4.4]nonane-1,6-diones (Table 10-5, Figure 10-12). Figure 10-12 indicates the exquisite sensitivity of the CD curves to small low-energy changes in the spatial orientation of the two carbonyl chromo-

Table 10-6. Calculated[a] and Experimental Optical Properties[b] of (+)-(5*R*)-Spiro[4.4]nonane-1,6-dione

Calculated Results			Experimental Results	
λ^{max}(nm)	Oscillator Strength	Reduced Rotational Strength (cgs)	λ^{max}(nm)	Reduced Rotational Strength (cgs)
295	1.19×10^{-4}	+10.6	314	+4.13
290	2.00×10^{-6}	−1.13	287	−0.51
164	0.26	−200	—	—
163	0.22	+202	—	—

[a]Coupled chromophore calculations by Hansen's method (ref. 38a) and reported in ref. 35.
[b]Measured in isopentane–methylcyclohexane (5:1) at +25°C, as reported in ref. 35.

phores, in addition to strong homoconjugation. The authors attribute the $n \rightarrow \pi^*$ Cotton effects principally to contributions from the chirality of the rings and a pronounced vicinal interaction between the two carbonyls, but they were not able to assess the relative contributions of each.

More recently, the CD spectra of two additional β-diketones (Table 10-5) were measured: *S*-(−)-spiro[5.5]undecane-1,7-dione[41] and (−)-1(*R*),5(*S*)-bicyclo[3.3.1]-nonane-2,9-dione.[31] Although the former does not show the strong UV enhancement ($\varepsilon_{310}^{max} = 73$) seen in the spiro[4.4]nonane-1,6-dione ($\varepsilon_{305}^{max} = 125$), the large Δε values have been taken as evidence of an interaction between the two ketone carbonyl groups.[41] No conformational analyses or variable low-temperature CD studies have yet been carried out. In contrast, the bicyclo[3.3.1]nonane-2,9-dione exhibits a very weak $n \rightarrow \pi^*$ Cotton effect and no bathochromic shift—an indication perhaps of a dicarbonyl spatial orientation that is unfavorable for vicinal interaction.[31] Conformational analysis by MM2[31b] and PCMODEL[32] molecular mechanics calculations indicates that the chair–boat (**cb**) conformation is more stable than the chair–chair (**cc**) (Figure 10-13A). The CD spectra of these two conformers have been predicted[31b] (Figure 10-13B), and from these spectra, a composite CD spectrum (Figure 10-13C) may be calculated, assuming a 3:1 population of **cb** to **cc**. (The **cb** conformer is computed (Δ*H*) to be about 0.5 kcal/mole more stable than the **cc**.) For comparison, the observed CD is presented in Figure 10-13D. Variable-temperature CD studies have not yet been carried out on this diketone.

The conformationally more complicated β-diketone bicyclo[5.3.1]undecane-8,11-dione (Table 10-7) was studied by Berg et al.[44] using dynamic NMR spectroscopy, MM3 molecular mechanics calculations, and CD spectroscopy. Dynamic NMR spectroscopy indicated the existence of two conformers (or sets of conformation families), separated by a barrier of 8.2 ± 0.4 kcal/mole. At −110°C, only one conformer was appreciably populated. Conformational analysis by molecular mechanics using the MM3 force field gave six conformational isomers lying within $\Delta\Delta H_f = 2$ kcal/mole.

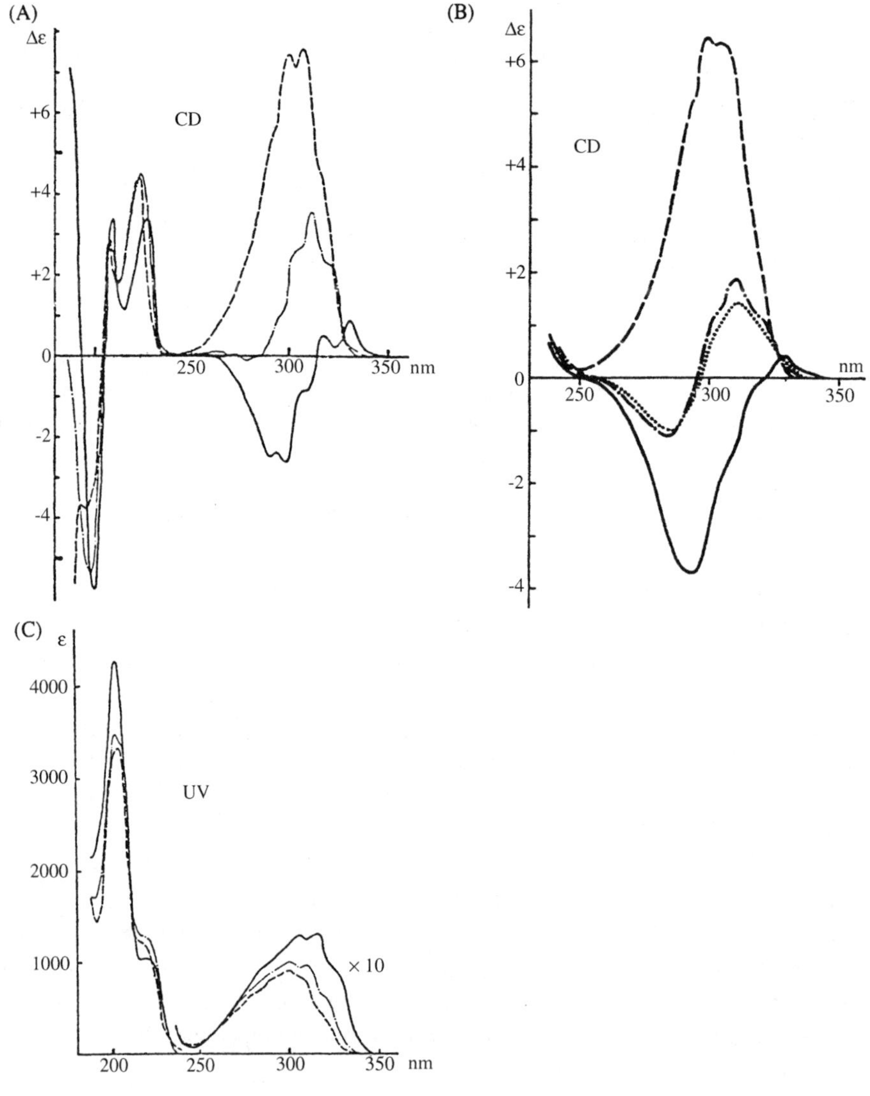

Figure 10-12. Circular dichroism (A) and (B) and ultraviolet (C) spectra of 3(*S*),5(*R*),8(*S*)-(−)-*cis,cis*-di-tert-butylspiro-[4.4]nonane-1,6-dione (——— in (A) and (C), - - - - in (B)) and its (+)-3(*R*),5(*R*),8(*S*)-*cis,trans* (— · — · — ·) and (+)-3(*R*),5(*R*),8(*R*)-*trans,trans* (- - - -, in (B), —— in (A) and (C)) isomers. In (B), · · · · is for the parent dione, (+)-5(*R*)-spiro[4.4]nonane-1,6-dione. (A) and (C) in isooctane, (B) in methanol. [Reprinted from ref. 40, Copyright© 1980, with permission from Elsevier Science.]

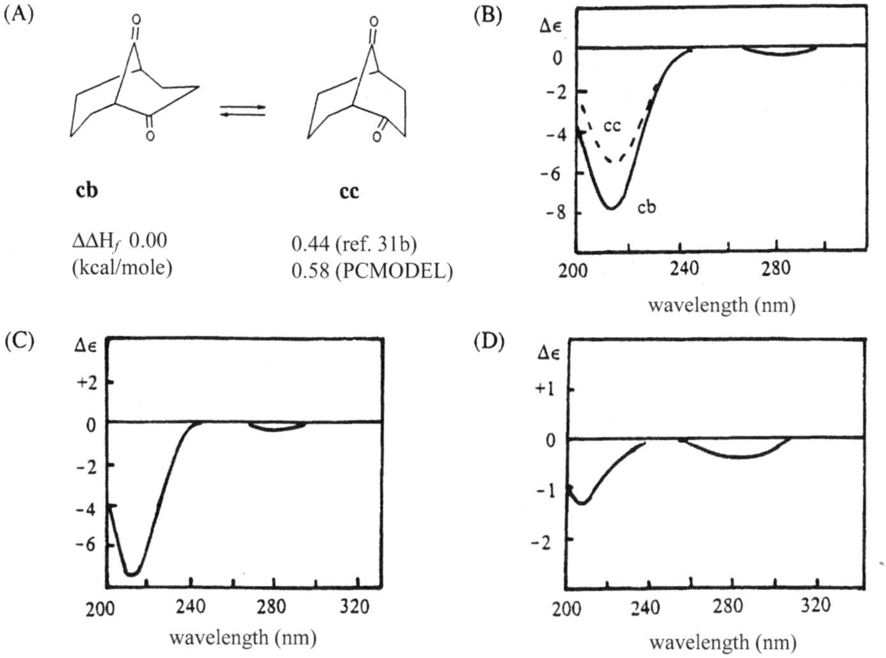

Figure 10-13. (A) Interconverting chair–chair (**cc**) and chair–boat (**cb**) conformations of (−)-1(*R*),5(*S*)-bicyclo[3.3.1]nonane-2,9-dione and the calculated differences in their heats of formation. (B) Calculated (ref. 31b) CD spectra of the **cc** and **cb** conformers. (C) Calculated net CD spectrum of the dione, based on a 75% **cb**–25% **cc** conformer population. (D) The observed CD spectrum of the (−)-1(*R*),5(*S*)-bicyclo[3.3.1]nonane-2,9-dione in 95% ethanol. [Data are redrawn from ref. 31b.]

Table 10-7. Computed[a] Low-Energy Conformers of Bicyclo[5.3.1]undecane-8,11-dione

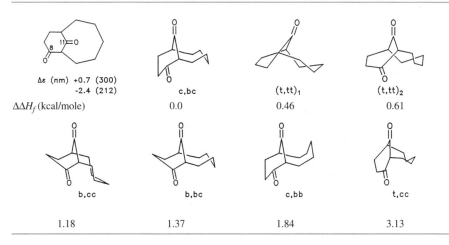

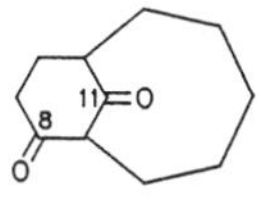 Δε (nm) +0.7 (300) −2.4 (212)	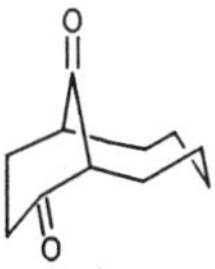 c,bc	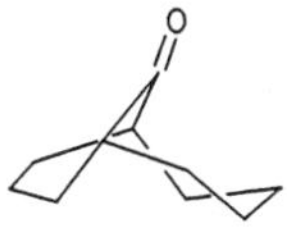 $(t,tt)_1$	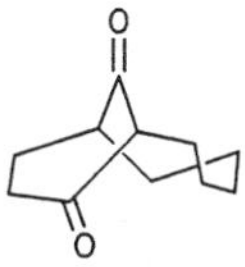 $(t,tt)_2$
$\Delta\Delta H_f$ (kcal/mole)	0.0	0.46	0.61
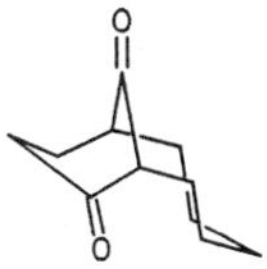 b,cc	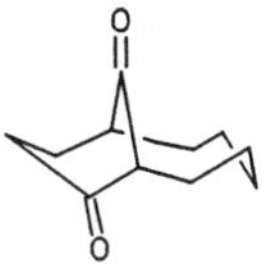 b,bc	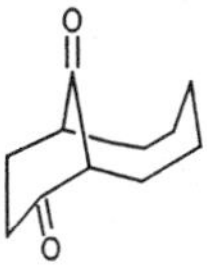 c,bb	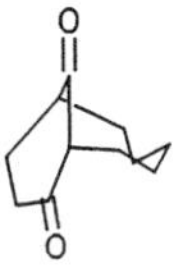 t,cc
1.18	1.37	1.84	3.13

[a]Ref. 44. $\Delta\Delta H_f$ relative to c,bc (c = chair, b = boat, t = twist).

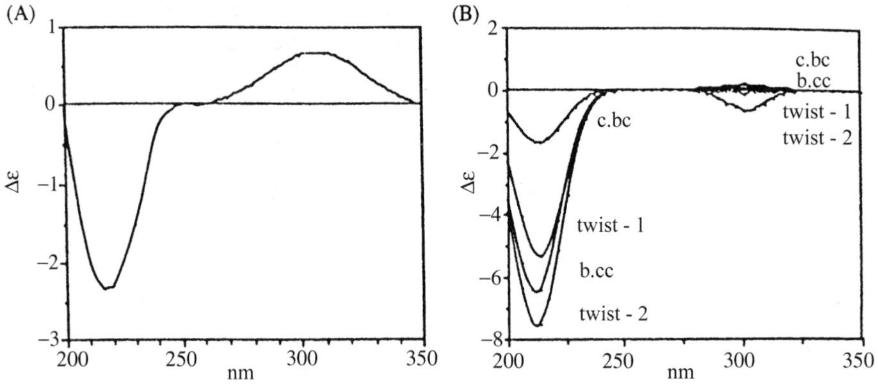

Figure 10-14. (A) Observed circular dichroism spectrum of (+)-bicyclo[5.3.1]undecane-8,11-dione in ethanol. (B) Computed CD spectra of each of the conformers of Table 10-7. [Reprinted from ref. 44, copyright© 1997, with permission from Elsevier Science.]

The next highest (t,cc) lies approximately 3.1 kcal/mole above the global minimum at c,bc.

The diketone was resolved by chromatography through a column packed with swollen microcrystalline triacetylcellulose, giving enantiomerically enriched diones, first positive and then negative.[44] The (+) isomer gave the CD spectrum shown in Figure 10-14A, and its absolute configuration was assigned 1(*S*),7(*R*) using the octant rule—an assignment that would be rendered tenuous in the event of interchromophoric coupling between the two ketone carbonyl chromophores. Circular dichroism spectra were computed for each conformer of Table 10-7 (Figure 14B). Only the MM3-computed lowest energy conformer (c,bc) and one other (b,cc) were calculated to exhibit

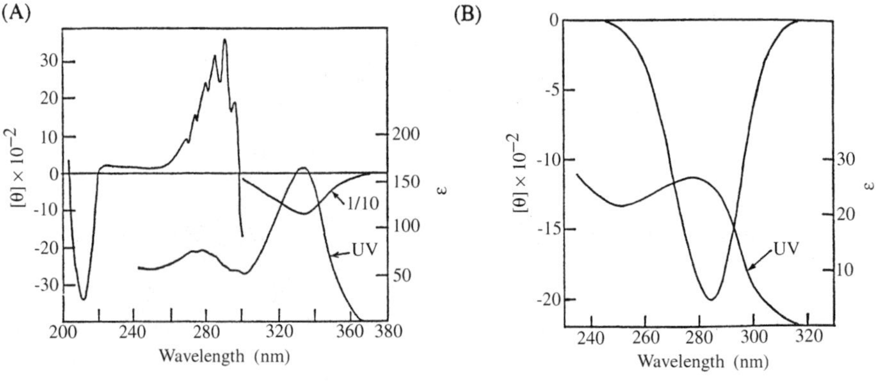

Figure 10-15. (A) CD and UV spectra of (+)-2,9-*twist*-brendanedione and (B) CD and UV spectra of (+)-9-*twist*-brendanone, in isooctane. [Reprinted from ref. 42 copyright© 1975, with permission from the Japanese Chemical Society.]

positive long-wavelength Cotton effects. Here again, conformational analysis using variable low-temperature CD could be useful.

(+)-*Twist*-brendane-2,9-dione[42] is the only conformationally rigid diketone shown in Table 10-5. Its UV spectrum exhibits (*i*) a strong $n \rightarrow \pi^*$ band ($\varepsilon_{334}^{max} = 167$), bathochromically shifted relative to the monoketone component, (+)-9-*twist*-brendanone ($\varepsilon_{279}^{max} = 27$), and (*ii*) a second, weaker band ($\varepsilon_{279}^{max} = 77$). The CD absorptions of the two $n \rightarrow \pi^*$ dione transitions are intense and oppositely signed (Table 10-5, Figure 10-15), suggesting strong vicinal interaction or homoconjugation. One component monoketone, (+)-9-*twist*-brendanone, is known and has a weakly negative $n \rightarrow \pi^*$ CD Cotton effect. Data for the corresponding 2-*twist*-brendanone are not available, but using (–)-bisnoradamantan-2-one as an analog, one finds a strong positive $n \rightarrow \pi^*$ CD Cotton effect (Table 10-5) and a typical λ^{max}. The dione CD spectrum is far from the spectrum obtained by adding the spectra of these two monoketones. Although no computations of the chiroptical properties have been reported, the data indicate that this β-diketone has its two carbonyl chromophores optimally disposed for strong coupling of locally excited $n \rightarrow \pi^*$ states.[37]

10.5. Adjacent Diketones: α-Diones

The ORD spectrum of a steroid α-diketone was reported long ago by Djerassi and Closson[10] and, as anticipated, was found to bear no resemblance to the ORD curves

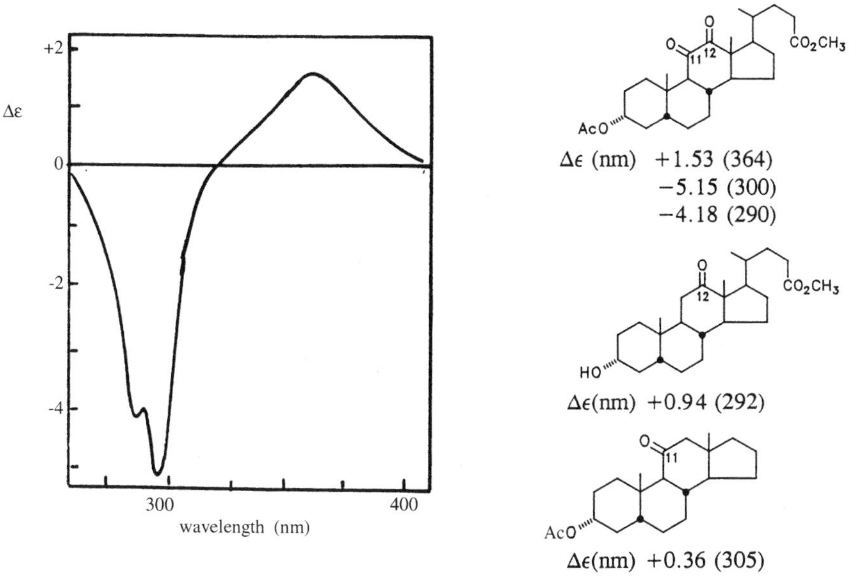

Figure 10-16. Circular dichroism spectrum of methyl 3α-acetoxy-11,12-diketocholanate and CD data for the 11,12-dione may be contrasted with those of its monoketone analogs 3α-hydroxy-12-ketocholanic acid and 3α-acetoxy-5β-androstane-11-one. (See Figure 10-1.) [Data redrawn from ref. 4.]

of the component monoketones, as shown in Figure 10-16. The differences become even clearer in CD spectroscopy: The α-diketone exhibits multiple Cotton effects, one near 360 nm and two others, much stronger, near 300 nm. The component monoketones, on the other hand, exhibit only weaker positive Cotton effects near 290 nm (Figure 10-16 and Table 8-3). The strongly shifted $n \rightarrow \pi^*$ absorption is typical of conjugation. The $n \rightarrow \pi^*$ transitions from each carbonyl interact and split. The lowest energy transition has been assigned to a nearly pure $n_+ \rightarrow \pi^*_-$ state, the second to an admixture of $n_- \rightarrow \pi^*_-$ and $n_+ \rightarrow \pi^*_+$ states.[45] The subscripts denote symmetric (+) and antisymmetric (–) combinations of interacting carbonyl orbitals, and the ordering of the highest occupied molecular orbitals is $n_+ > \pi_+ > n_- > \pi_-$, all localized on the α-diketone moiety. As noted in Section 10.2, the energy separation between the highest occupied molecular orbitals, n_+ and n_-, is approximately 2 eV.[21] Thus, at least two well-separated $n \rightarrow \pi^*$ Cotton effects are observed in α-diketone UV and CD spectra.

Table 10-8. α-Diketones, Their CD Data, and ψ(O=C–C=O) Torsion Angles[a]

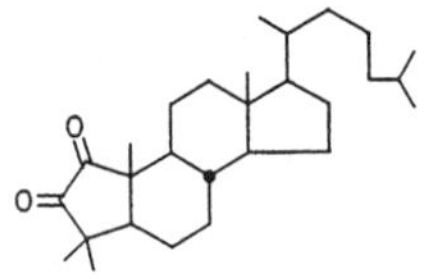	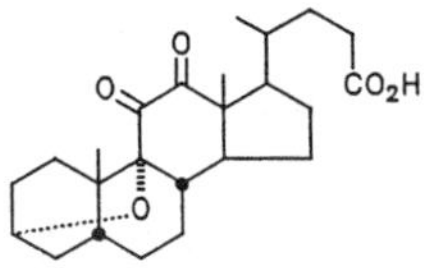	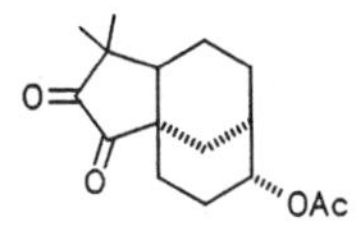	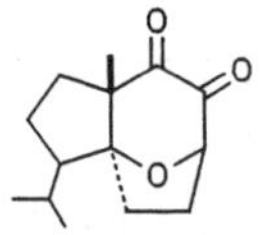
Δε(nm)[b] +2.1 (480) –0.30 (305)	Δε(nm)[c] +2.7 (~390) –4.2 (~300) +1.3 (~245)	Δε(nm)[b] –0.57 (~520) +0.57 (~460) +4.57 (~290)	Δε(nm)[d] +0.04 (464 –1.60 (406) +6.40 (300) +3.15 (281)
ψ(O=C–C=O) –7.7°	–2.0°	+0.8°	+7.1°
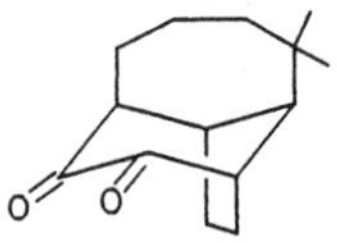	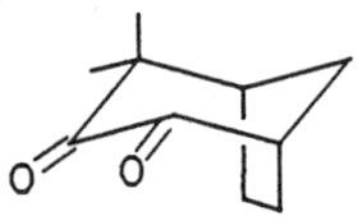	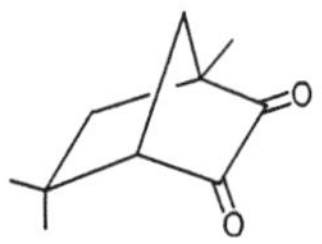	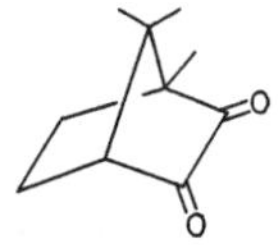
Δε(nm)[c,e] +0.82 (425) –0.77 (298)	Δε(nm)[c,e] + (> 400) – (< 400)	Δε(nm)[c] +0.38 (478) – (300)	Δε(nm)[c] +0.45 (484) –0.29 (293) +1.27 (222) –0.83 (204) +1.47 (193)
ψ(O=C–C=O) –2°	+1.2° (+33.9°)[f]	–1.7°	+0.1° (0°)[g]

[a] Torsion angles from PCMODEL calculations (ref. 32) and from X-ray crystallography (values in parentheses).
[b] Data from ref. 3.
[c] Data from ref. 39.
[d] Data from Levisalles, J., and Rudler, H., *Bull. Soc. Chim. France* (1967), 2059–2066.
[e] ORD data from Jacob, G., Ourisson, G., and Rassat, A., *Bull. Soc. Chim. France* (1959), 1374–1377.
[f] Data from ref. 48.
[g] Data from ref. 49.

The first attempt to correlate α-diketone $n \rightarrow \pi^*$ Cotton effects with molecular chirality came from a theoretical study by Hug and Wagnière[39] in Zürich, who proposed that a positive long-wavelength Cotton effect correlates with a right-handed (***P***) α-dione helicity. However, the correlation is not always found, quite possibly because extrachromophoric vicinal perturbers dominate.[46] The helicity rule can have practical value only when Cotton effect contributions from dione helicity dominate all others,[47] and that was not evident in the examples studied (Table 10-8). A subsequent theoretical treatment of α-diketones by Richardson and Caliga[45] at Virginia attributed a negative long-wavelength Cotton effect to a ***P***-helical dione and significant contributions from vicinal effects of substituents.

Actually, relatively few α-diketones in which the ketone carbonyls are not conjugated with other π-chromophores have been studied by chiroptical spectroscopy. Those are summarized in Table 10-8, along with the ψ(O=C–C=O) torsion angles computed by PCMODEL molecular mechanics calculations[32] and a few from X-ray crystallographic determinations.[48,49] It has been suggested that the second $n \rightarrow \pi^*$ band of α-diketones correlates better than the long-wavelength transition with the dione helicity,[46] and that correlation seems to hold up reasonably well for the diones of Table 10-8.

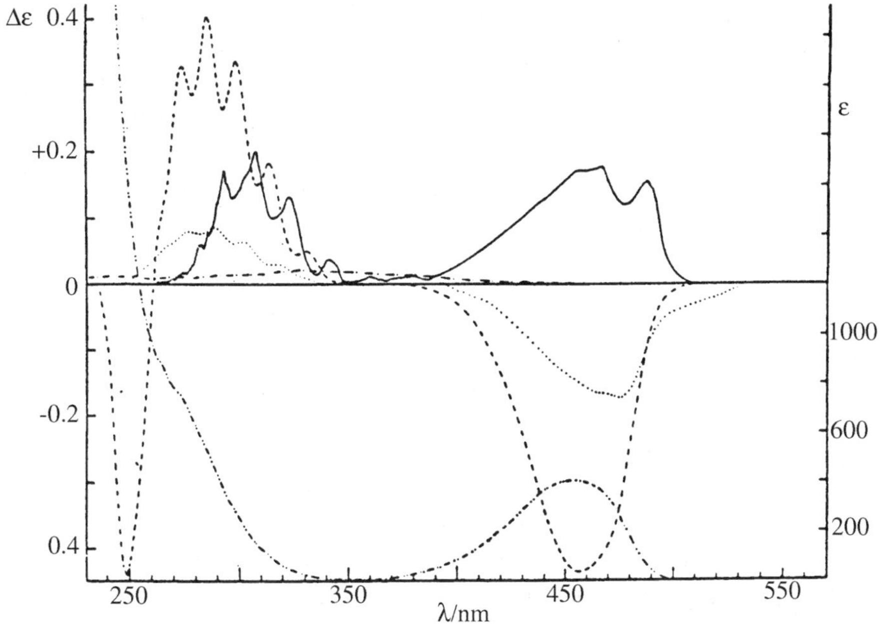

Figure 10-17. CD spectra of (1*R*)-camphorquinone(bornane-2,3-dione): (———) 1.50 mg/ 200 mg KBr (density of KBr = 2.75 g cm^{-3}) at 25°C; (- - - -) 7.34 × 10^{-3} M solution in CF_3CH_2OH at 124°C; (· · · ·) at 124°C in the gas phase; (— · — · — ·) baseline CD spectrum of (±)-bornane-2,3-dione; (- ·· - ·· -): UV spectrum in CF_3CH_2OH. [Reprinted from ref. 51, copyright© 1980, with permission from the Royal Society of Chemistry.]

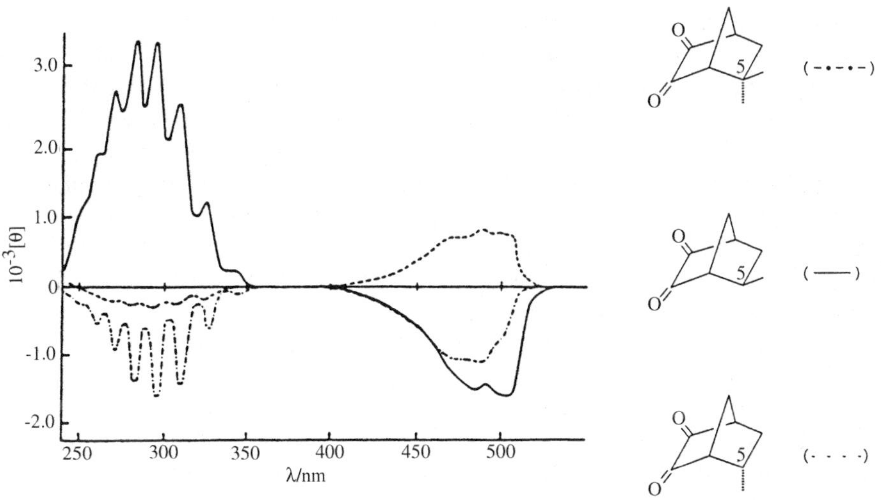

Figure 10-18. Circular dichroism spectra of (1*S*)-5,5-dimethylbicyclo[2.2.1]heptane-2,3-dione (— · — · — ·), (1*R*)-5-*exo*-methylbicyclo[2.2.1]heptane-2,3-dione (———), and (1*R*)-5-*endo*-methylbicyclo[2.2.1]heptane-2,3-dione (- - - -) in cyclohexane. [Reprinted from ref. 50, copyright© 1986, with permission from the Royal Society of Chemistry.]

The most studied α-diketone is camphorquinone,[49–53] which has a ψ(O=C–C=O) dihedral angle of approximately 0°.[49,50] Its CD has been measured in solution, in the solid phase (in KBr), and in the gas phase (Figure 10-17).[51] Although the solution and gas-phase spectra have the same signed Cotton effects, the long-wavelength Cotton effect measured in the solid is oppositely signed. The change in sign has been attributed to molecular association in the crystal.[50] The sign of the long-wavelength Cotton effect found in solution and in the gas phase is thought to be determined largely by vicinal substituents acting on a planar dione.[50]

The sensitivity of planar diones to vicinal substituent effects has been explored by Poloński and Dauter[50] in Gdansk using various camphorquinone analogs and other bicyclo[2.2.1]heptane-2,3-diones. The sensitivity of the CD to slight changes in geometry was shown in the CD spectra of 1(*R*)-*exo*-5-methylbicyclo[2.2.1]heptane-2,3-dione and its *endo* epimer, which do not sum to the CD of the 5,5-dimethyl analog (Figure 10-18).

In sum, two effects seem to control the CD of α-diketones: (i) the intrinsic chirality of the dione[39,47,50] and (ii) vicinal contributions from α-axial substituents.[46,50] The Cotton effect is difficult to predict when these two mechanisms work in opposition, and no clear rule is yet available for such situations.[50]

References

1. Djerassi, C., *Optical Rotatory Dispersion*, McGraw-Hill, New York, 1960.

2. Moffitt, W., Woodward, W. B., Moscowitz, A., Klyne, W., and Djerassi, C., *J. Am. Chem. Soc.* **83** (1961), 4013.
3. (a) Crabbé, P., *Optical Rotatory Dispersion and Circular Dichroism in Organic Chemistry*, Holden-Day, San Francisco, 1965.
   (b) Crabbé, P., *Applications de la Dispersion Rotatoire Optique et du Dichroisme Circulaire Optique en Chimie Organique*, Gauthier-Villars, Paris, 1968.
4. Velluz, L., Legrand, M., and Grosjean, M., *Optical Circular Dichroism*, Verlag Chemie, Weinheim, Germany, 1965.
5. For a review of the literature from 1977 to mid-1995, see Boiadjiev, S., and Lightner, D. A., "Chiroptical Properties of Compounds Containing C=O Groups," in *Supplement A3: The Chemistry of Double-Bonded Functional Groups* (Patai, S., ed.), John Wiley & Sons, Ltd., Chichester, U.K., 1997, Chap. 5. For earlier work, see Crabbé, P., *ORD and CD in Chemistry and Biochemistry*, Academic Press, New York, 1972.
6. Legrand, M., and Rougier, M. J., *Stereochemistry, Volume 2: Dipole Moments, CD or ORD* (Kagan, H. B., ed.), Geo. Thieme Publ., Stuttgart, 1977.
7. Kirk, D. N., *Tetrahedron* **42** (1986), 777–818.
8. Ripperger, H., *Z. Chem.* **17** (1977), 250–258.
9. (a) Lightner, D. A., Bouman, T. D., Wijekoon, W. M. D., and Hansen, Aa. E., *J. Am. Chem. Soc.* **108** (1986), 4484–4497.
   (b) Lightner, D. A., and Toan, V. V., *J. Chem. Soc. Chem. Commun.* (1987), 210–211.
10. Djerassi, C., and Closson, W., *J. Am. Chem. Soc.* **78** (1956), 3761–3769.
11. Djerassi, C., Marshall, D., and Nakano, T., *J. Am. Chem. Soc.* **80** (1958), 4853–4857.
12. Takeda, K., and Minato, H., *Steroids* **1** (1963), 345–359.
13. (a) Crabbé, P., Pérez, M., and Vera, G., *Can. J. Chem.* **41** (1963), 156–164.
    (b) Crabbé, P., McCapra, F., Comer, F., and Scott, A. I., *Tetrahedron* **20** (1964), 2455–2465.
14. (a) Hoffmann, R., Imamura, A., and Hehre, W. J., *J. Am. Chem. Soc.* **90** (1968), 1499–1509.
    (b) Hoffmann, R., *Acc. Chem. Res.* **4** (1971), 1–9.
15. Orloski, R. F., Ph.D. dissertation (*Interactions between Nonadjacent Chromophores in Rigid Systems*) University of California at Los Angeles, 1964.
16. (a) Houk, K. N., *Chem. Rev.* **76** (1976), 1–74.
    (b) Robbins, T. A., Toan, V. V., Givens, J. W., III, and Lightner, D. A., *J. Am. Chem. Soc.* **114** (1992), 10799–10810.
    (c) Winstein, S., DeVries, L., and Orloski, R., *J. Am. Chem. Soc.* **83** (1961), 2020–2021.
    (d) Lightner, D. A., Gawroski, J. K., Hansen, Aa. E., and Bouman, T. D., *J. Am. Chem. Soc.* **103** (1981), 4291–4296, and references therein.
17. Reviews:
    (a) Gleiter, R., *Angew. Chem. Int. Ed. Engl.* **13** (1974), 696–701.
    (b) Paddon-Row, M. N., *Acc. Chem. Res.* **15** (1982), 245–251.
    (c) Paddon-Row, M. N., and Jordan, K. D., in *Modern Models of Bonding and Delocalization* (Liebman, J. F., and Greenberg, A., eds.), Verlag Chemie, Weinheim, Germany, 1988, 115–194.
    (d) Martin, H.-D., and Mayer, B., *Angew. Chem. Int. Ed. Engl.* **22** (1983), 283–310.
    (e) Gleiter, R., and Schäfer, W., *Acc. Chem. Res.* **23** (1990), 369–375.
18. (a) Balaji, V., Jordan, K. D., Gleiter, R., Jähne, G., and Müller, G., *J. Am. Chem. Soc.* **107** (1985), 7321–7323.
    (b) Balaji, V., Ng, L., Jordan, K. D., Paddon-Row, M. N., and Patney, H. K., *J. Am. Chem. Soc.* **109** (1987), 6957–6969.
19. (a) Doerner, T., Gleiter, R., Robbins, T. A., Chayangkoon, P., and Lightner, D. A., *J. Am. Chem. Soc.* **114** (1992), 3235–3241, and references therein.
    (b) Gurst, J. E., Schubert, E. M., Boiadjiev, S. E., and Lightner, D. A., *Tetrahedron* **49** (1993), 9191–9196.
    (c) For leading references, see Paddon-Row, M. N., *Tetrahedron* **50** (1994), 10813–10828.
20. (a) Bruch, P., Thompson, D., and Winstein, S., *Chem. and Ind.* (London) (1960), 590–591; Winstein, S., Hansen, R., *Tetrahedron Lett.* **25** (1960), 4–8, and references therein.

(b) See Olah, G. A., and Schleyer, P. v. R., eds., *Carbonium Ions*, Vol. III, J. Wiley, New York, 1972.
(c) See, for example, Sargent, G. D., "The 2-Norbornyl Carbon," in *Carbonium Ions*, Vol. III (Olah, G. A., and Schleyer, P. v. R., eds.) J. Wiley, New York, 1972, 1099–1200, and references therein.
21. (a) Dougherty, D., Brint, P., and McGlynn, S. P., *J. Am. Chem. Soc.* **100** (1978), 5597–5603.
(b) Cowan, D. O., Gleiter, R., Hashmall, J. A., Heilbronner, E., and Hornung, V., *Angew. Chem. Int. Ed. Engl.* **10** (1971), 401–402.
(c) Klasinc, L., and McGlynn, S. P., in *The Chemistry of Quinoid Compounds*, Vol. II (Patai, S., and Rappoport, Z., eds.), J. Wiley, New York, 1988, 155–201, and references therein.
(d) Jähne, G., and Gleiter, R., *Angew. Chem. Int. Ed. Engl.* **22** (1983), 488–489.
(e) Worley, S. D., Mateescu, G. D., McFarland, C. W., Fort, R. C., Jr., and Sheley, C. F., *J. Am. Chem. Soc.* **95** (1973), 7580–7586.
22. (a) Bishop, R., and Lee, G-H., *Austral. J. Chem.* **40** (1987), 249–255.
(b) Senda, Y., Ishiyama, J., and Imaizumi, S., *J. Chem. Soc., Perkin Trans.* **2** (1981), 90–93.
(c) Bishop, R., *Austral. J. Chem.* **37** (1984), 319–325.
23. Hug, W., Kuhn, J., Seibold, K. J., and Labhart, H., Wagnière, G., *Helv. Chim. Acta* **54** (1971), 1451–1466.
24. Bays, D. E., Cannon, G. W., and Cookson, R. C., *J. Chem. Soc. (B)* (1966), 885–892.
25. Hill, R. K., Morton, G. H., Peterson, J. R., Walsh, J. A., Paquette, L. A., *J. Org. Chem.* **50** (1985), 5528–5533.
26. Pérard-Viret, J., and Rassat, A., *Tetrahedron: Asymmetry* **5** (1994), 1–4.
27. Nakazaki, M., Naemura, K., Sugano, Y., and Kataoka, Y., *J. Org. Chem.* **45** (1980), 3232–3236.
28. Almansa, C., Moyano, A., and Serratosa, F., *Tetrahedron* **44** (1988), 2657–2662.
29. Paquette, L. A., Farnham, W. B., and Ley, S. V., *J. Am. Chem. Soc.* **97** (1975), 7273–7279.
30. Gerlach, H., *Helv. Chim. Acta* **61** (1978), 2773–2776.
31. (a) Berg, J., and Butkus, E., *J. Chem. Res.* **5** (1993), 116–117.
(b) Berg, J., and Butkus, E., *J. Chem. Res.* **5** (1994), 356–357.
32. PCMODEL, versions 5.0–7.0. Serena Software, Inc., Bloomington, IN 47402–3076. PCMODEL uses the MMX force field, a variation of Allinger's MM2 force field.
33. Gerlach, H., *Helv. Chim. Acta* **51** (1968), 1587–1593.
34. Cram, D. J., and Steinberg, H., *J. Am. Chem. Soc.* **76** (1954), 2753–2757.
35. Lightner, D. A., Christiansen, G. D., and Melquist, J. L., *Tetrahedron Lett.* (1972), 2045–2048.
36. Altona, C., DeGraaff, R. A. G., Leeuwestein, C. H., and Romers, C., *J. Chem. Soc., Chem. Commun.* (1971), 1305–1307.
37. Longuet-Higgins, H. C., and Murrell, J. N., *Proc. Phys. Soc. (London)* **A68** (1955), 601–611.
38. (a) Hansen, Aa. E., *On the Optical Activity of Compounds Containing Coupled Chromophores*, licentiate thesis H.C. Ørsted Institute, Copenhagen, 1964.
(b) See also Moscowitz, A., Hansen, Aa. E., Forster, L. S., and Rosenheck, K., *Biopolymers (Symposium No. 1)*, (1964), 75–89.
(c) See also Moscowitz, A., *Proc. Roy. Soc. (London)* **A297** (1967), 40–42.
39. Hug, W., and Wagnière, G., *Helv. Chim. Acta* **54** (1971), 633–649.
40. (a) Sumiyoshi, M., Kuritani, H., Shingu, K., and Nakagawa, M., *Tetrahedron Lett.* **21** (1980), 2855–2856.
(b) Sumiyoshi, M., Kuritani, H., Shingu, K., and Nakagawa, M., *Tetrahedron Lett.* **21** (1980), 1243–1246.
41. Brünner, R., and Gerlach, H., *Tetrahedron: Asymmetry* **5** (1994), 1613–1620.
42. Nakazaki, M., Naemura, K., and Harita, S., *Bull. Chem. Soc. Jpn.* **48** (1975), 1907–1913.
43. Nakazaki, M., Naemura, K., and Arashiba, N., *J. Chem. Soc., Chem. Commun.* (1976), 678–679.
44. Berg, U., Butkus, E., Trejd, T., and Bromander, S., *Tetrahedron* **53** (1997), 5339–5348.
45. Richardson, F. S., and Čaliga, D., *Theor. Chim. Acta* **36** (1974), 49–66.
46. Burgstahler, A. W., and Naik, N. C., *Helv. Chim. Acta* **54** (1971), 312–313.
47. Hug, W., and Wagnière, G., *Helv. Chim. Acta* **55** (1972), 67–68.
48. (a) Lee, B., Seymour, J. P., and Burgstahler, A. W., *J. Chem. Soc., Chem. Commun.* (1974), 235–236.
(b) Seymour, J. P., Lee, B., and Burgstahler, A. W., *Acta Crystallogr.* **B33** (1977), 2667–2669.

49. Bright, W. M., Cannon, J. F., Langs, D. A., and Silverton, J. V., *Cryst. Struct. Commun.* **9** (1980), 251–256.
50. Polonński, T., and Dauter, Z., *J. Chem. Soc., Perkin Trans.* **1** (1986), 1781–1788.
51. Lightner, D. A., Crist, B. V., and Flores, M. J., *J. Chem. Soc., Chem. Commun.* (1980), 273–275.
52. Charney, E., and Tsai, L., *J. Am. Chem. Soc.* **93** (1971), 7123–7132.
53. Luk, C. K., and Richardson, F. S., *J. Am. Chem. Soc.* **96** (1974), 2006–2009.

CHAPTER

11

Unsaturated and Cyclopropyl Ketones

Conjugation alters the ketone or aldehyde carbonyl electronic transitions in a significant way.[1] Simplistically, in an α,β-unsaturated ketone, the component carbonyl and olefin π-systems interact, and the energy levels split to lower the energy level of the lowest unoccupied π^* molecular orbital (LUMO) and raise the energy level of the highest occupied π molecular orbital (HOMO). The $n \rightarrow \pi^*$ transition energy is thus lower, and the wavelength is bathochromically shifted to around 320–350 nm in the electronic spectrum of an α,β-unsaturated ketone, compared with the saturated ketone counterpart (Table 11-1).[1] The transition intensity is typically increased, by a factor of two to three. In addition, an even more intense $\pi \rightarrow \pi^*$ type absorption is found at higher energy, near 220–260 nm—a region where neither of the component chromophores absorbs light.

Homoconjugated ketones may also exhibit bathochromically shifted $n \rightarrow \pi^*$ transitions, often with very large increases in intensity (Table 11-1).[2,3] The intensity and wavelength shift are dependent on the relative orientation of the carbonyl and olefinic chromophores. Here and in α,β-unsaturated ketones, conjugation or coupling of the carbonyl and olefin chromophores is expressed through coulombic mixing of the local states of the separated chromophores and through the intrusion of charge-transfer states associated with the transfer of electronic charge between the chromophores.[4] The relative percent of charge transfer diminishes with increasing interchromophoric distance. Thus, for β,γ-unsaturated ketones, charge transfer is of small importance, and the observed bathochromic shift and enhanced absorption of the erstwhile $n \rightarrow \pi^*$ transition can be understood in terms of the coulombic coupling of locally excited states: The carbonyl $n \rightarrow \pi^*$ transition borrows electric dipole transition moment from the olefinic $\pi \rightarrow \pi^*$ transition.[4] In α,β-unsaturated ketones, however, both coulombic and charge-transfer interactions are important for coupling the carbonyl and olefin chromophores.

There are far fewer examples of interchromophoric interaction occurring over longer through-bond distances. One example of a transannular interaction is the one that occurs in the δ,ε-ketone, *trans*-5-cyclodecenone[5] (Table 11-1), wherein the olefin chromophore lies rather close (in nonbonded distance) to the carbonyl chromophore and causes the appearance of a photodesmotic transition near 215 nm.[5b]

Cyclopropane has long been recognized to exhibit olefinic properties and an olefinic character in conjugation with carbonyl groups.[6] Thus, cyclopropyl conjugation or homoconjugation in ketones is found to influence the $n \rightarrow \pi^*$ transition and lower lying transitions, albeit less significantly than olefin conjugation does

Table 11-1. Influence of Conjugation on the Electronic Absorption Spectra of Ketones

Saturated			Unsaturated or Cyclopropyl		
Ketone	ϵ^{max}	λ^{max}(nm)	Ketone	ϵ^{max}	λ^{max}(nm)
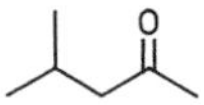	20[a]	283	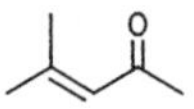	47[a] 11,500[a]	328 231
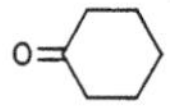	15[a]	291	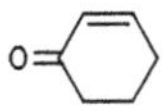	36[b] 13,800[b]	320 225
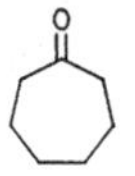	17[a]	292	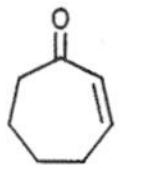	52[c] 9,600[b]	317 228
			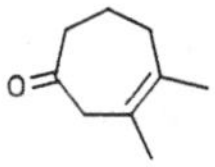	450[c]	295
	23[c]	295[c]		290[c] 2,800[c]	304 215
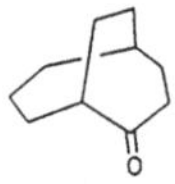	22[c]	300	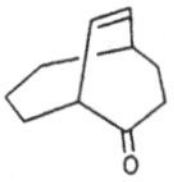	110[c]	307
	14[e] 33[e]	290 (215)[f]	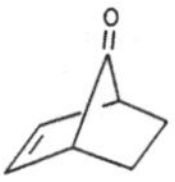	36[e] 500[e]	274 (215)[f]
			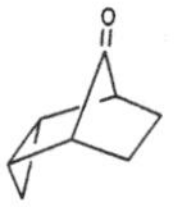	44[e] 215[e]	(276) (215)[f]
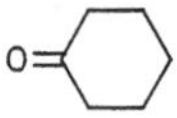	15[a]	291	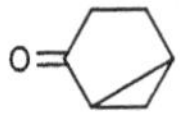	20[d] 5,400[g]	285 197
			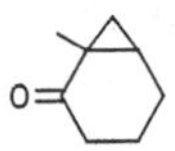	33[h] 15,264[h]	275 207.5
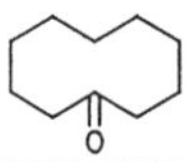	15	288[j]	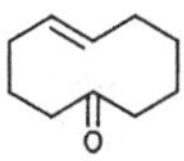	73[i] 2,300[j]	302 215

[a] Ref. 1, isooctane. [b] Ref. 1, ethanol. [c] Ref. 2. [d] Ref. 6c. [e] Ref. 6d. [f] Not max. [g] Ref. 6a. [h] Ref. 6e, H_2O. [i] Ref. 5a. [j] Ref. 5b.

(Table 11-1).[6e,f] As with perturbation by olefins, the extent of perturbation by a cyclopropane on the carbonyl chromophore depends on the relative orientation of the two chromophores.

11.1 α,β-Unsaturated Ketones

As indicated in the first paragraph of this chapter, α,β-unsaturated ketones typically exhibit UV absorption bands, one near 330 nm, which is due to an $n \rightarrow \pi^*$ transition (and is called the *R*-band), and one near 230 nm, which has considerable $\pi \rightarrow \pi^*$ character (and is called the *K*-band).[1] Most chiroptical studies of α,β-unsaturated ketones have focused on steroids and other natural products, convenient sources of optically active compounds. In the 1950s, the carbonyl $n \rightarrow \pi^*$ transition was studied by ORD; however, the high intensity of the $\pi \rightarrow \pi^*$ absorption precluded ORD measurements in the 220–260-nm region.[7a,b] A systematic investigation revealed that structural alterations remote from the chromophore did not influence the ORD $n \rightarrow \pi^*$ Cotton effect, nor did α-substituents such as alkyl or halogen or α′ substituents such as methyl and acetoxy. However, allylic axial substituents did.[7a] With improved signal detection capabilities in ORD instruments, and with the advent of CD in the 1960s, both the $n \rightarrow \pi^*$ and $\pi \rightarrow \pi^*$ Cotton effects were studied and correlated with stereochemistry. Jacek Gawroński at Adam Mickiewicz University, Poznan and the late David Kirk (of Queen Mary College, London) provided the most comprehensive CD measurements of α,β-unsaturated ketones and also interpreted their behavior.[8,9]

Circular dichroism spectra of α,β-unsaturated ketones typically exhibit not just two, but three and sometimes four, Cotton effects between about 185–360 nm,[7c,8,9] as may be seen in 5α-cholest-4-en-3-one (Figure 11-1). As noted earlier, the weak transition between approximately and 320–350 nm corresponds to the redshifted ketone carbonyl $n \rightarrow \pi^*$ excitation, which often shows considerable vibrational fine structure in the CD spectrum. This excitation is followed (Fig. 11-1) by one or more $\pi \rightarrow \pi^*$ transitions, with one corresponding to the typical UV band between 230 and 260 nm and a second often overlapping with the first and lying in the 200–220-nm range.[10]

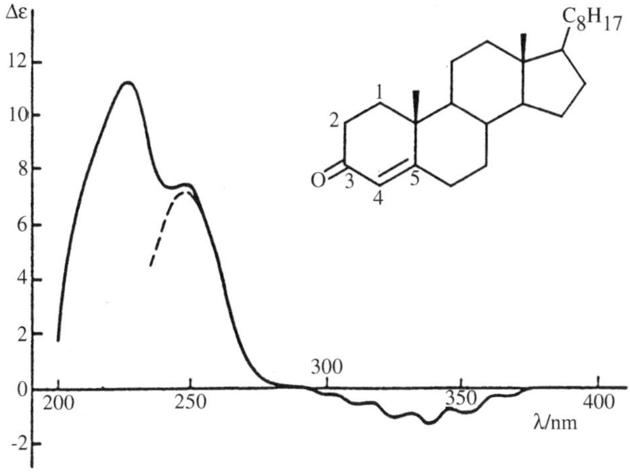

Figure 11-1. Circular dichroism spectrum of cholest-4-en-3-one in *n*-hexane. [Reprinted from ref. 9, copyright© 1986 with permission from Elsevier Science.]

Although the rotatory strength of the second band is typically high, it is not normally as apparent in the UV spectrum.[8,9] Yet another, more energetic, band has been found to lie near 185 nm and is thought to involve mainly an $n \rightarrow \sigma^*$ transition.[8,9]

The spectral properties of α,β-unsaturated ketones derive from the relative orientation of the carbonyl and olefin chromophores and the influence of that orientation on the extrachromophoric components of the molecule. The enone conformation can be characterized by the torsion angle (ω) about the bond connecting the C=C and C=O units (Figure 11-2). The two limiting cases are planar *s-cis* and planar *s-trans*, in which $\omega = 0°$ and 180°, respectively. The intermediate conformations are nonplanar. One of the complicating factors in interpreting ORD or CD Cotton effects of cyclic α,β-unsaturated ketones is the greater flexibility than that of their saturated analogs. For example, the cyclohexenone system of cholest-4-en-3-one is very sensitive to substitution and can adopt a wide variety of conformations, as revealed by X-ray crystallography.[8a,9] Although the usual conformation is the half-chair in ring A, substituents elsewhere in the steroid may cause it to adopt sofa conformations, and bulky 2β-substituents result in an inverted half-chair.[9]

Various chirality rules have been proposed for correlating the Cotton effects of α,β-unsaturated ketones with their stereochemistry. Two fundamental considerations had to be taken into account: (i) contributions from the intrinsic helicity of the enone chromophore, which, when nonplanar, is an inherently dissymmetric chromophore, and (ii) contributions from extrachromophoric perturbation. Given the success of the octant rule for saturated alkyl ketones and aldehydes, at the time of the early investigations of enone ORD, it was attractive to formulate modified "octant rules" based on extrachromophoric perturbation of the then only accessible ($n \rightarrow \pi^*$) transition.[11] However, the situation for an α,β-unsaturated ketone can very often be that of an inherently dissymmetric chromophore; and the octant rule, whose derivation rests on the concept of an inherently symmetric carbonyl chromophore, does not apply in general. As it turns out, the ORD and CD spectra of α,β-unsaturated ketones and aldehydes are very sensitive to extrachromophoric perturbers *and* to the intrinsic

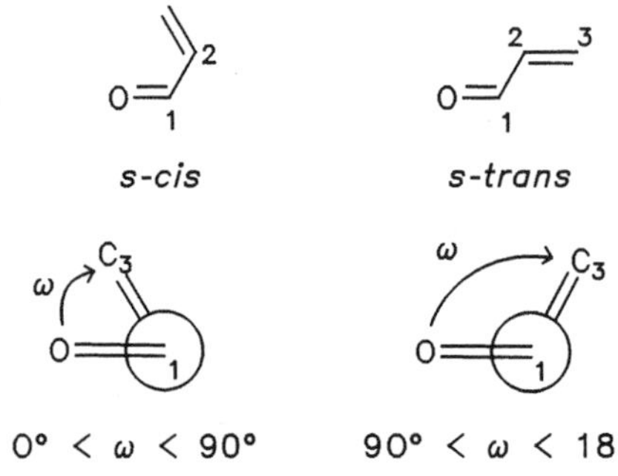

Figure 11-2. Enone conformation and helicity (positive helicity and torsion angle ω indicated). According to the earliest proposed helicity rules (refs. 10,11), the positive helicity shown predicts positive $\pi \rightarrow \pi^*$ and negative $n \rightarrow \pi^*$ Cotton effects.

chirality of the enone, a chirality that may be dictated by vicinal substituents, especially when the enone is conformationally flexible.[8–12,13,14] Thus, contributions from the helicity of the chromophore, as well as from the extrachromophoric perturbers, both weigh in on the Cotton effects, and it is not always easy to sort out the dominant contribution.

The first enone helicity rule was proposed in 1962 by Djerassi et al.[12a] and Whalley[12b] for the 230–260-nm $\pi \rightarrow \pi^*$ Cotton effect: A positive enone helicity (Figure 11-2) correlates with a positive Cotton effect. Shortly thereafter, Snatzke[11] extended the helicity rule to the $n \rightarrow \pi^*$ transition, predicting a negative Cotton effect for positive enone helicity. For cyclopentenones, however, an inverse rule was required.[8,11b] Kirk subsequently summarized the enone helicity rules by relating the helicity of the π–p orbitals at enone carbons C(2) and C(3) to the signs of the $n \rightarrow \pi^*$ and $\pi \rightarrow \pi^*$ Cotton effects (Figure 11-3) for *s-cis* and *s-trans* enones unperturbed by polar substituents. A number of semiempirical calculations were undertaken to correlate enone chromophore helicity with the Cotton effect, ranging from simple Hückel-type wave functions for the $\pi \rightarrow \pi^*$ transition[12a] to SCF–CNDO–CI calculations for $n \rightarrow \pi^*$ and $\pi \rightarrow \pi^*$.[15] With the acrolein chromophore as a model, the latter predict same-signed Cotton effects for the *s-cis*, and oppositely signed effects for the *s-trans*, enone (Figure 11-4). A satisfactory correlation between the helicity of *s-trans* cyclohexenones (but not cyclopentenones) and the $n \rightarrow \pi^*$ Cotton effect sign was achieved using "qualitative MO theory,"[11e,f] but a fuller understanding of enone Cotton effects awaits the application of more sophisticated theoretical calculations. Clearly, however, as the (helical) torsion angle ω (Figure 11-2) approaches 0° or 180°, the Cotton effect contribution from enone helicity drops to zero. As the enone assumes planarity, Cotton effect contributions from extrachromophoric perturbers must dominate.

For $n \rightarrow \pi^*$ transitions of planar enones, Snatzke[11a,f] proposed a sector rule (Figure 11-5) similar to the octant rule (Chapter 4), but with opposite signs for the back octants. The rule seems to be applicable as long as the *s-trans* enone is planar, as illustrated by

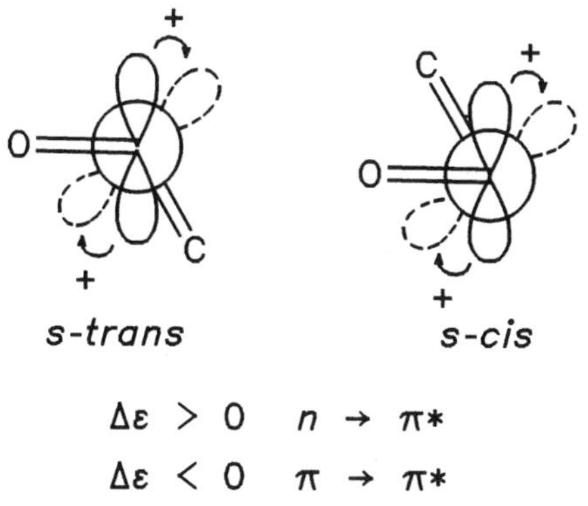

Figure 11-3. Kirk's enone orbital helicity rule correlating the positive π p-orbital helicity at C(1) and C(2) with a positive $n \rightarrow \pi^*$ Cotton effect and a negative $\pi \rightarrow \pi^*$ Cotton effect. See refs. 8a and 9.

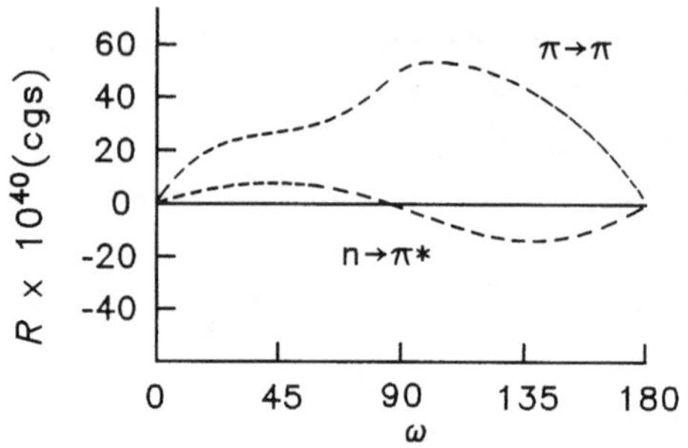

Figure 11-4. Computed dependence of the $n \to \pi^*$ and $\pi \to \pi^*$ rotatory strengths (R) with enone torsion angle (ω). Data from ref. 15.

the Δ^3-octalin-2-one (Figure 11-6A) in the "preferred" sofa conformation of ring A.[8a] The A/B ring steroid analog of Δ^3-octalin-2-one, 5α-cholest-1-en-3-one (Figure 11-6B), exhibits the same signed $n \to \pi^*$ Cotton effect, as predicted, and the shorter wavelength Cotton effects also have corresponding Cotton effects with the same sign.[8b] The rule seems applicable as well to the $n \to \pi^*$ Cotton effect of $\Delta^{1(9)}$-octalin-2-one (Figure 11-6C).[8b] Note here, however, that while the $n \to \pi^*$ Cotton effect is negative, as predicted, the $\pi \to \pi^*$ Cotton effect has the same sign as the $n \to \pi^*$ effect, unlike that of Figure 11-6A. The corresponding steroid analog, cholest-4-en-3-one (Figure 11-6D) also exhibits the predicted negative $n \to \pi^*$ Cotton effect, but here the $\pi \to \pi^*$ is oppositely signed.[8b] When an axial 6β-methyl group is added to ring B, both the $n \to \pi^*$ and $\pi \to \pi^*$ Cotton effect signs invert. This inversion is thought to be due to A/B ring skeletal distortion and a change in enone helicity accommodating the relief of a severe 1,3-diaxial methyl–methyl interaction between the 6β-methyl and the C(19) angular methyl group.[8,9,13a] In this enone system, π-donor substituents at the C(6) allylic site also seem to have an especially strong influence on the Cotton effects—but only if those substituents are β (axial). There is little or no effect when they are α (equatorial), irrespective of whether the C(19) methyl is present. Interest-

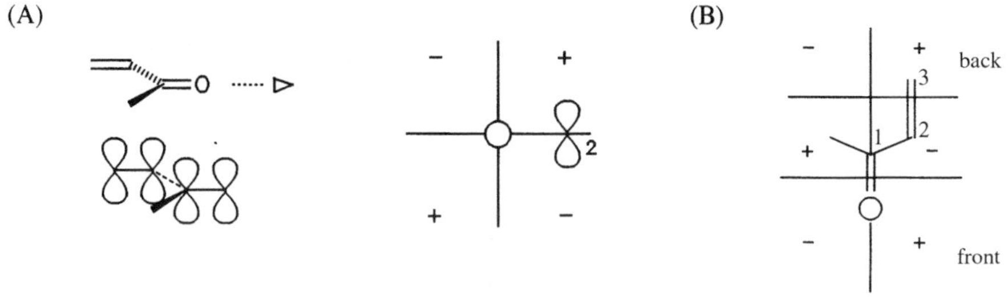

Figure 11-5. Sector rule for the $n \to \pi^*$ transition of an *s-trans* planar enone (A) looking down the C=O bond from oxygen to C(1) with signs given for the back sectors and (B) from the top view. According to Snatzke (refs. 11a and 11f).

(A)

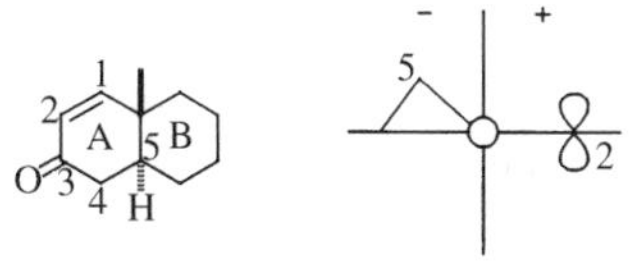

$(n \rightarrow \pi^*)$	$\Delta\epsilon_{349}^{max}$	−0.91
$(\pi \rightarrow \pi^*)$	$\Delta\epsilon_{227}^{max}$	+4.2
	$\Delta\epsilon_{202}^{max}$	−6.2

(heptane)

(B)

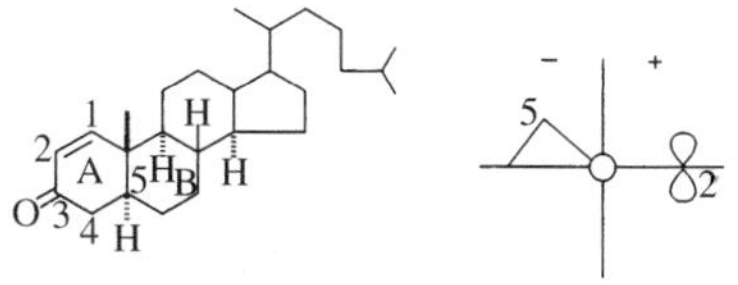

$(n \rightarrow \pi^*)$	$\Delta\epsilon_{349}^{max}$	−0.94
$(\pi \rightarrow \pi^*)$	$\Delta\epsilon_{226}^{max}$	+11.6
	$\Delta\epsilon_{199}^{max}$	−10.2

(hexane)

(C)

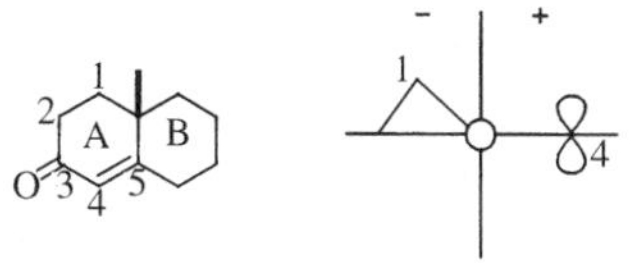

$(n \rightarrow \pi^*)$	$\Delta\epsilon_{335}^{max}$	−0.98
$(\pi \rightarrow \pi^*)$	$\Delta\epsilon_{249}^{max}$	−1.6
	$\Delta\epsilon_{220}^{max}$	+5.4

(hexane)

(D)

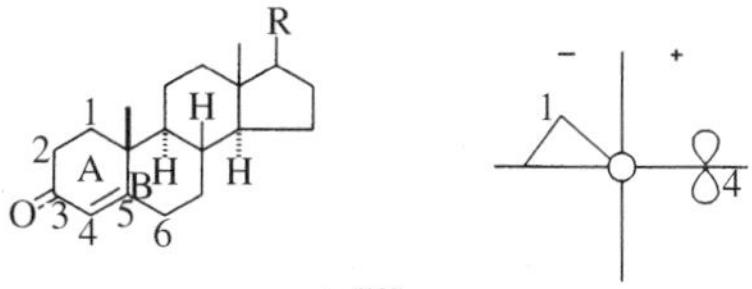

	$\Delta\epsilon^{max}$ $n \rightarrow \pi^*$	$\pi \rightarrow \pi^*$		solvent
parent	−1.41(337)	+5.5(230)^{sh}	+11.0(207)	hexane
6β-CH_3	+0.70(333)	−1.8(248)	+5.8(221)	methanol
6α-OH	−1.50(319)	+7.9(242)	+9.0(203)	methanol
6β-OH	−0.70(321)	−5.3(250)	+10.7(215)	methanol
6α-Br	−1.05(337)	+6.6(230)^{sh}	+13.8(210)	hexane
6β-Br	+2.00(353)	−19.9(249)	+9.8(218)	hexane

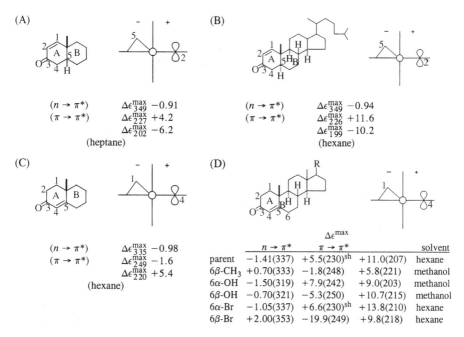

Figure 11-6. Correlation between the $n \rightarrow \pi^*$ Cotton effect sign of planar *s-trans* enones in a sofa conformation with Snatzke's sector rule (ref. 11). Data from ref. 8.

ingly, the second "$\pi \rightarrow \pi^*$" transition, near 215 nm, is uniformly positive in all the examples of Figures 11-6C and D (but not B).

Subsequently, Kuball and Neubrech in Kaiserslautern, together with Schönhofer in Berlin,[14a] used polarized spectroscopy to analyze some 18 *transoid* Δ^4-3-one systems related to those of Figure 11-6D. They concluded that, while the observed changes in the sign of the $n \rightarrow \pi^*$ CD transition might be attributed to ring conformational changes (in the helicity of the O=C–C=C chromophore), a more likely explanation is based on two different vibrational progressions: an "allowed" transition with a typically positive CD and a "forbidden" transition that is negative. In general, $\Delta\varepsilon$ is dominated by the "forbidden" progression (Figures 11-1, 11-6C, and 11-6D), but γ-allylic substituents may alter the relative intensities of the two vibrational progressions, permitting the "allowed" progression to dominate the net $n \rightarrow \pi^*$ CD and thus lead to a change in sign. The authors[14a] cautioned that helicity or sector rules can be applied only when the CD spectrum of the compound studied belongs to the same vibronic transitions as those compounds for which the rule was derived.

This interpretation served to explain the bisignate $n \rightarrow \pi^*$ Cotton effect occasionally seen in *s-cis* enones. Frelek and Szczepek in Warsaw and Weiss in Bochum[14b,c] surveyed over 20 *s-cis* enones (polycyclic α,β-unsaturated cyclohexenones) and found that, for bisignate $n \rightarrow \pi^*$ Cotton effects, the long-wavelength component correlated with the enone helicity rule. Frelek, Szczepek, and Weiss, in collaboration with Kuball et al.,[14d] used polarized spectroscopy to show that the apparent breakdown of the helicity rule is a consequence of an intensification of the contribution of the allowed vibronic progression to the CD so as to dominate the usually more evident (or dominant) "forbidden" progression that does correlate with the helicity rule (Figure 11-3). Thus, (i) for symmetric or nearly symmetric cyclohexenone units, the sign of the long-wavelength part of the bisignate band (or the sign of the monosignate band) is governed by the enone helicity rule (Figure 11-3), but (ii) if the cyclohexenone unit adopts a nonsymmetrical conformation, the sign of the $n \rightarrow \pi^*$ Cotton effect correlates with the chirality of the cyclohexenone ring, and the C=C behaves as an ordinary (octant) α-substituent—as in planar *s-cis* enones (see below).[8]

The importance of allylic axial substituents was first shown by Burgstahler et al.,[16] who reported a breakdown of the enone helicity rule[12] for the $\pi \rightarrow \pi^*$ transition of certain *s-cis* steroid enones. To give but a few examples,[8,10,16] 5α-cholest-8(14)-en-15-one, which has a negative enone torsion angle (ω = −12°) and negative orbital helicity (Figure 11-3), is found to exhibit a positive $n \rightarrow \pi^*$ Cotton effect and a positive $\pi \rightarrow \pi^*$ Cotton effect (Figure 11-7). The latter, a violation of the enone helicity rules, is thought to arise from net positive allylic axial contributions—positive from the C(18) angular methyl and 9α-hydrogen and negative from the 7α-hydrogen.[16] Similarly, the *s-cis* A-norcholest-5-en-3-one exhibits a negative $\pi \rightarrow \pi^*$ Cotton effect, which is opposite in sign to that predicted by the helicity rules for the compound's positive helicity (ω = +21°)[17] (Figure 11-7).[16] Here[16] and elsewhere,[18] there is a balancing of the two types of octant contributions: that from the intrinsic dissymmetry of the enone chromophore and that from vicinal perturbers. When the latter are allylic and axial, there is proper alignment for overlap between the enone π-system and the sp^3 orbital on the allylic carbon pointing toward the axial substituent. This situation, seen also in *s-trans* enones[8,9] (Figure 11-6), has been treated theoretically in dienes, where allylic axial substituents can also dominate and control the sign of the $\pi \rightarrow \pi^*$ Cotton effect.[19]

The importance of γ-allylic and α′ substituents was investigated more recently[14b,c] in polycyclic α,β-unsaturated cyclohexenones including over 20 *s-cis* enones. It was concluded that, for *transoid* enones, axial and equatorial γ-allylic substituents, including O and Cl, strongly affect the intensity and sign of the $\pi \rightarrow \pi^*$ Cotton effect by giving a contribution whose sign depends on the helicity of the substituent C_γ–C_β=C_α system. In contrast, for nonfunctionalized *cisoid* enones, neither the geometry (conformation) of the C=C-containing ring nor the chiral interaction of the γ-allylic axial bond is a controlling factor for the $\pi \rightarrow \pi^*$ Cotton effect. An axial bromine at the α′-carbon, however, has an overriding influence on this (and the $n \rightarrow \pi^*$) Cotton effect, exerting an octant-like contribution in accord with the Br–C_α–C=O helicity.

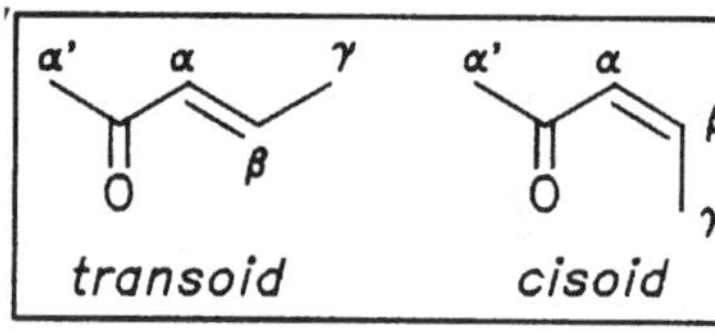

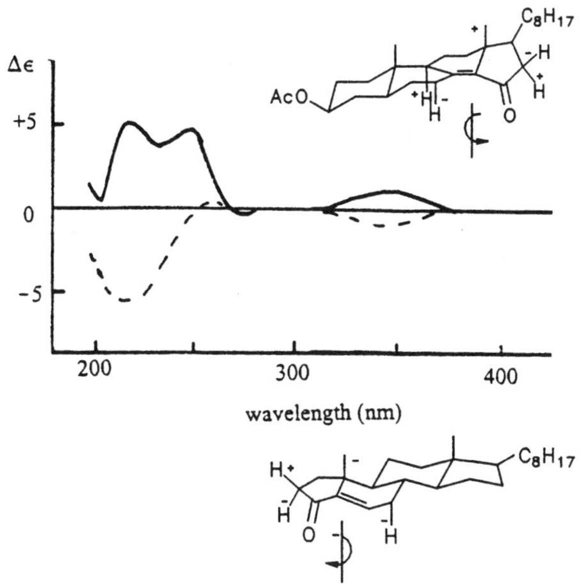

Figure 11-7. CD Cotton effects of 3β-acetoxy-5α-cholest-8(14)-en-15-one (—) and A-nor-cholest-5-en-3-one (– – – –) in *n*-hexane. The signed contributions of the allylic axial extrachromophoric perturbers are noted on the structures. [Redrawn from ref. 16b.]

When the *s-cis* enone is essentially planar (ω ~ 0°), as in the bornane derivatives of Table 11-2, extrachromophoric perturbers must dominate the Cotton effect signs.[20] Here, octantlike contributions may be expected to dominate the $n \rightarrow \pi^*$ Cotton effects, so it is not surprising that the α,β-unsaturated ketones shown in the table have the same signs as those of the parent camphor and epicamphor. The $\pi \rightarrow \pi^*$ band is apparently also dominated by ring atom contributions, probably of the allylic axial type suggested by Burgstahler et al.[16,21]—that is, the positive C(4)–C(5) "axial" allylic bond contribution of α-methylene and α-isopropylidene camphor and the negative C(1)–C(6) allylic bond contribution of α-methylene epicamphor (Table 11-2).

Most of the preceding discussion of enones has focused on the $n \rightarrow \pi^*$ transition lying between approximately 320 and 350 nm and the $\pi \rightarrow \pi^*$ transition lying between about 230 and 260 nm. A third transition, near 200–220 nm, has been detected by CD spectroscopy.[8] This transition has a low oscillator strength, making it difficult to detect in the UV spectrum, but its high rotatory strength makes it far easier to detect by CD, in which the intensity is typically comparable to that of the approximately 230–260-nm $\pi \rightarrow \pi^*$ transition (Figures 11-1, 11-6, and 11-7). Although it was thought at first that the sign of the Cotton effect for the transition could be correlated with the chirality contribution of the axial bond at the α′-carbon adjacent to the enone carbonyl carbon,[16] it now seems more likely that the olefinic substituents are more dominating.[8,9,11] Gawroński proposed a rule for correlating the sign of the approximately 210-nm Cotton effect with the helicity of the $C_\alpha{=}C_\beta{-}C_\gamma{-}R$ torsion angle (Figure 11-8): A

Table 11-2. CD Cotton Effects of *s-cis* α,β-Unsaturated Ketones from Camphor Epicamphor in Methylcyclohexane[a]

α,β-Unsaturated Ketone	Δε^max (nm)			
	$n \rightarrow \pi^*$	$\pi \rightarrow \pi^*$	Third transition	ω^b
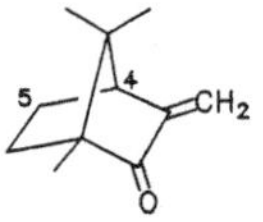	+0.81 (344)	+4.99 (228)	−1.47 (196)	−1.3°
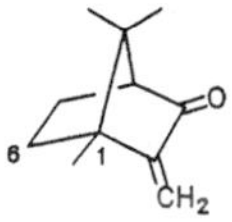	−1.04 (346)	−5.54 (228)	+0.92 (196)	+1.0°
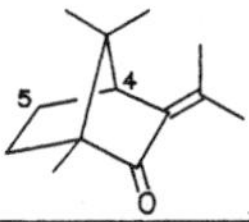	+0.60 (340)	+9.1 (240)	−0.87 (205)	−1.0°

[a] Data from ref. 20. [b] See Fig. 11-3; calculated using PCMODEL (ref. 17).

right-handed (positive) helicity predicts a positive Cotton effect, and a left-handed (negative) helicity predicts a negative Cotton effect.[8] This correlation holds up quite well for the ketones of Figure 11-6, as well as many other ketones.[8b] Recently, it was shown that in *s-cis* enones the band is always of opposite sign to that of the $n \rightarrow \pi^*$ transition and that the sign of the Cotton effect is governed by the helicity of the $C_\beta{=}C_\alpha$–C–X unit.[14] Yet a fourth electronic transition in α,β-unsaturated ketones with α′ or β′ axial substituents has been detected by CD spectroscopy, between 185 and 195 nm. This transition is thought to involve an $n \rightarrow \sigma^*$ excitation,[8,9] and the signed contributions of the Cotton effect-dominating axial α′ and β′ substituents on 2-cyclohexenone are given in Figure 11-9.

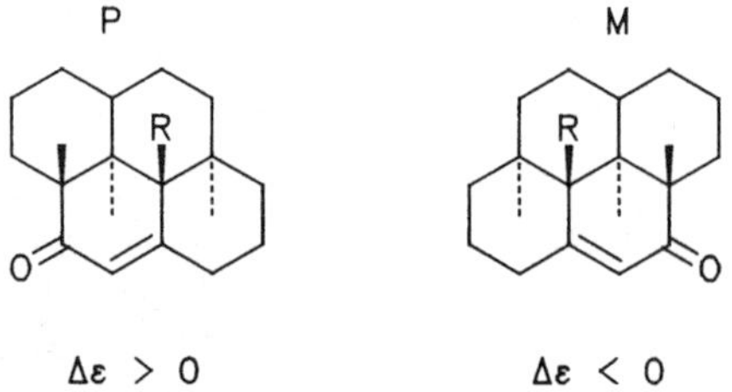

Figure 11-8. Correlation rule for the sign of the approximately 200–220-nm Cotton effect of *s-trans* α,β-unsaturated ketones and the helicity of the C=C–C–R torsion angle. A positive (*P*) helicity gives a positive Cotton effect, and a negative (*M*) helicity predicts a negative Cotton effect. According to ref. 8b.

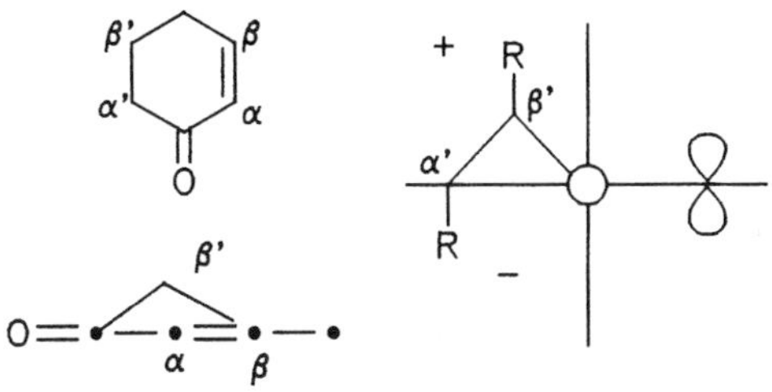

Figure 11-9. Correlation rule for the sign of the 185–195-nm Cotton effect in 2-cyclohexenones. The sign of the Cotton effect is determined by α' and β'-axial substituents as shown. According to ref. 8b.

Correlating ORD and CD spectra with the stereochemistry of α,β-unsaturated ketones and aldehydes is complicated by difficulties in assessing the relative importance of contributions from the helicity of the enone chromophore and extrachromophoric perturbers. A variety of sector rules have been proposed for predicting the Cotton effect, when the enone is planar (and helicity contributions are unimportant) assuming that the conformations can be ascertained. One of the difficulties in assessing the conformations is the small energy differences between them,[8] a problem not typical of saturated ketones. In addition, it is often difficult to sort out the relative contributions of extrachromophoric perturbers, because, unlike the $n \rightarrow \pi^*$ transition of saturated ketones, the $n \rightarrow \pi^*$ and (especially) the $\pi \rightarrow \pi^*$ and $n \rightarrow \sigma^*$ transitions of α,β-unsaturated ketones (and aldehydes) are not pure. Although mainly the indicated transitions, they are typically a mixture of transitions[8a]—and the relative mix of the components depends on the intrinsic stereochemistry of the enone, as well as the extrachromophoric parts.

Attempts to analyze the conformation of α,β-unsaturated ketones using variable low-temperature CD have been reported, but there are few examples. In one of the simplest,[8b] (–)-5(*R*)-methyl-2-cyclohexenone is believed to adopt either of two equilibrating sofa or half-chair conformations—one with an equatorial methyl, the other with an axial. Upon lowering the temperature from +20° to –150°C in hydrocarbon solvent (Figure 11-10) the CD Cotton effects intensify while maintaining their sign, and the weak Cotton effect (+0.08) near 237 nm disappears. It may be assumed that the conformer with an equatorial methyl is more stable than that with an axial. PCMODEL molecular mechanics calculations[17] predict the equatorial conformer to be 1.2 kcal/mole more stable than the axial. Analysis of the variable-temperature CD predicts 80–85% of the equatorial conformer at room temperature, corresponding to $\Delta G° \sim 0.8$–1.0 kcal/mole.[8b] Earlier,[22] variable low-temperature CD spectra of this ketone in ether–isopentane–ethanol solvent (EPA, 5:5:2 by vol.) for the $n \rightarrow \pi^*$ transition only were measured and analyzed; the researchers concluded that the equatorial conformation gave a positive Cotton effect and was more stable than the axial conformation by 0.8 kcal/mole. The axial conformation was predicted to have a positive $n \rightarrow \pi^*$ Cotton effect. Surprisingly, however, the equatorial conformer is

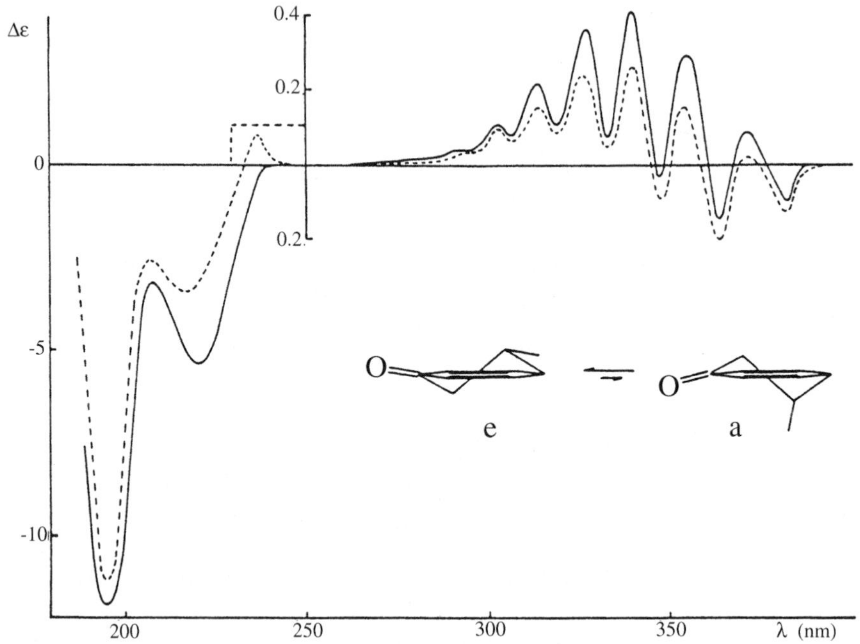

Figure 11-10. Variable low-temperature CD spectra of (–)-5(*R*)-methyl-2-cyclohexenone, shown in two preferred equilibrating sofa or half-chair conformations, at +20 °C (– – – –) and –150 °C (—) in methylcyclohexane–isopentane (4:1 by vol). [Reprinted from ref. 8b copyright© 1982, with permission from Elsevier Science.]

predicted by Snatzke's sector rule for planar enones[11] (Figure 11-5) to have a negative $n \rightarrow \pi^*$ Cotton effect, to go along with aforementioned prediction of a positive $n \rightarrow \pi^*$ Cotton effect.[8b,22] These unexpected sign reversals were explained in terms of a dominant effect from enone helicity.[8b] PCMODEL calculations[17] predict nearly planar enones with oppositely signed torsion angles ($\omega = -178°$ for the equatorial and $\omega = +177°$ for the axial), and the signs of the torsion angles predict the observed $n \rightarrow \pi^*$ and $\pi \rightarrow \pi^*$ Cotton effect signs using Kirk's orbital helicity rule[9] (Figure 11-3). Whether such small deviations from enone planarity are real and whether they can in fact dominate the Cotton effect signs is unclear and awaits a more detailed examination.

The earlier study[22] also analyzed variable low-temperature CD spectra of 4(*R*)-methyl-2-cyclohexenone and concluded that (*i*) the equatorial conformer is more stable than the axial one in EPA by $\Delta G° = 1.1–1.25$ kcal/mole and (*ii*) the equatorial conformer has a negative Cotton effect, whereas the axial conformer has a positive Cotton effect. PCMODEL calculations[17] indicate that the enone is nearly planar and that the equatorial conformer is more stable than the axial (Figure 11-11). The predicted $n \rightarrow \pi^*$ Cotton effect signs correspond to those determined

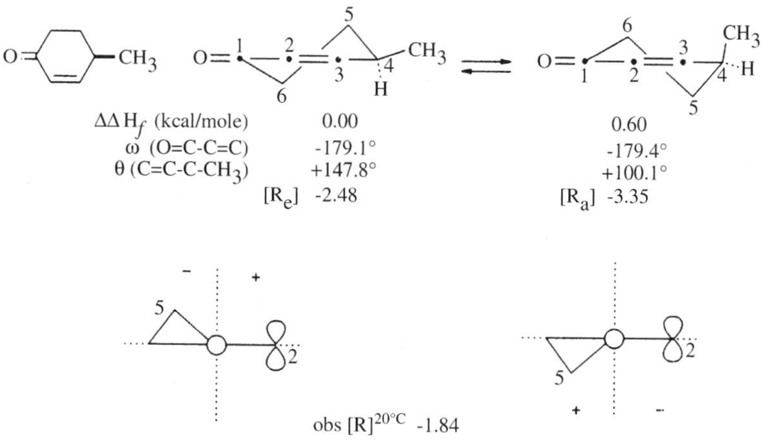

Figure 11-11. Stable conformations of 4(*R*)-methyl-2-cyclohexenone and their $n \rightarrow \pi^*$ sector diagrams [according to Snatzke (ref. 11) and Figures 11-5 and 11-6]. The $\Delta\Delta H_f$ and ϕ values are from PCMODEL (ref. 17).

by applying Snatzke's sector rule[11] (Figures 11-5, 11-6, and 11-11), and the net observed negative Cotton effect ($R_0^{+25°} = -1.7$) fits the analysis.

Several additional 2-cyclohexenones, all natural products based on the *p*-menthane skeleton (Figure 11-12), have been studied by variable low-temperature CD[23] (Figure 11-13). Using Snatzke's sector rules for planar enones (Figures 11-5 and 11-6) and neglecting contributions from the methyl and isopropyl or isopropenyl groups, investigators predict negative $n \rightarrow \pi^*$ Cotton effects for the equatorial methyl conformer of (−)-*p*-menthenone (Figure 11-12A) and the equatorial isopropyl conformer of (−)-piperitone (Figure 11-12B), whereas they predict positive Cotton effects for the equatorial methyl conformer of (−)-carvenone (Figure 11-12D) and the equatorial isopropenyl conformer of (+)-carvone (Figure 11-12C). However, vibronically structured, weakly positive $n \rightarrow \pi^*$ Cotton effects are observed (Figure 11-13) for all but (+)-carvone (Figures 11-12C and 11-13C), which has a negative Cotton effect, suggesting that the octantlike contributions of the methyl, isopropenyl, and isopropyl groups weigh in or that enone helicity dominates the Cotton effects. PCMODEL calculations[17] predict that the equatorial conformer will be more stable in the four α,β-unsaturated ketones (Figure 11-12), although barely so in (−)-piperitone (Figure 11-12B). On the basis of enone helicity (ω, Figure 11-12), net positive $n \rightarrow \pi^*$ Cotton effects are predicted for *p*-menthenone (Figure 11-12A), (−)-piperitone (Figure 11-12B) and (−)-carvenone (Figure 11-12D), but a net negative Cotton effect is predicted for (+)-carvone (Figure 11-12C). Conformational analysis of these ketones by variable low-temperature CD should be reexamined using the $\pi \rightarrow \pi^*$ transitions.

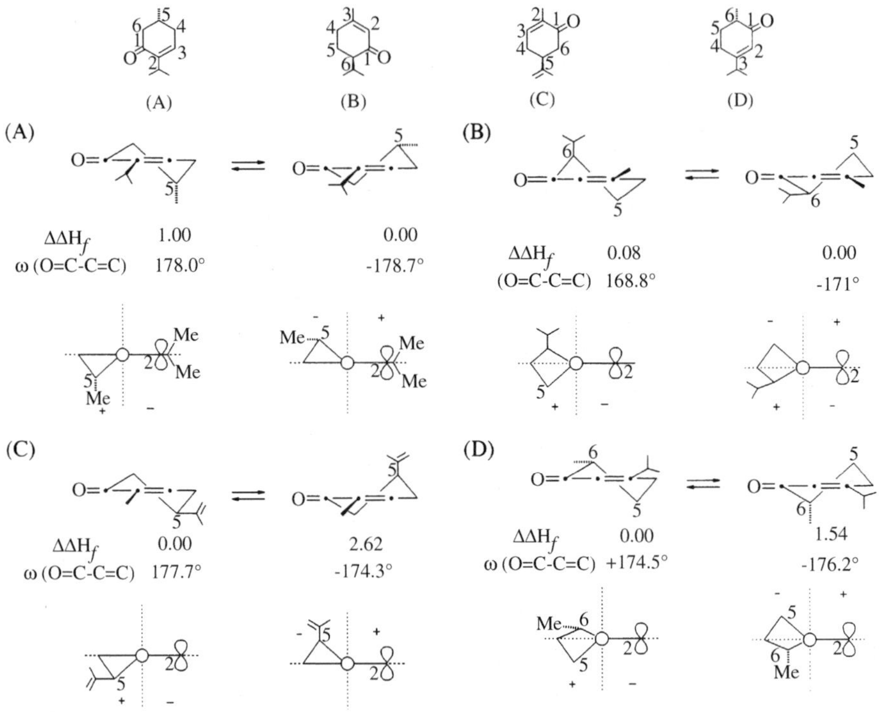

Figure 11-12. Absolute stereochemistry and standard numbering for α,β-unsaturated ketone natural products and their conformations, relative heats of formation ($\Delta\Delta H_f$, kcal/mole), enone torsion angle (ω, °), and sector diagrams (see Figures 11-5 and 11-6) of (A) (−)-*p*-menthene-2-enone, (B) (−)-piperitone, (C) (+)-carvone and (D) (−)-carvenone. (The numbering system is not the natural-product numbering system.) $\Delta\Delta H_f$ and ω values are from PCMODEL (ref. 17).

Conformational analysis of even fewer *s-cis* α,β-unsaturated ketones has been investigated by variable low-temperature CD. (+)-Pulegone (Figure 11-14) was shown to have a positive $n \rightarrow \pi^*$ Cotton effect[8b,24] and a negative $\pi \rightarrow \pi^*$ Cotton effect.[8b] The intensity of each increases with decreasing temperature. Two chairlike conformations were considered to be the most stable, with the equatorial methyl conformer thought to be more stable by $\Delta G° = 1.1$ kcal/mole[24] at 20°C. PCMODEL calculations[17] also predict it to be more stable, by approximately 2.5 enthalpy units over the axial methyl conformer. Unlike the endocyclic enones, these exocyclic enones have large torsion angles (ω): +57° for the equatorial methyl conformer and about −61° for the axial methyl conformer, as computed by PCMODEL. The positive ω correlates well with the observed positive $n \rightarrow \pi^*$ and negative $\pi \rightarrow \pi^*$ Cotton effects, as predicted by the orbital helicity rule (Figure 11-3).[9]

A related bicyclic *s-cis* enone, fukinone (Figure 11-15), was found to exhibit negative $n \rightarrow \pi^*$ and positive $\pi \rightarrow \pi^*$ Cotton effects that become more intense without

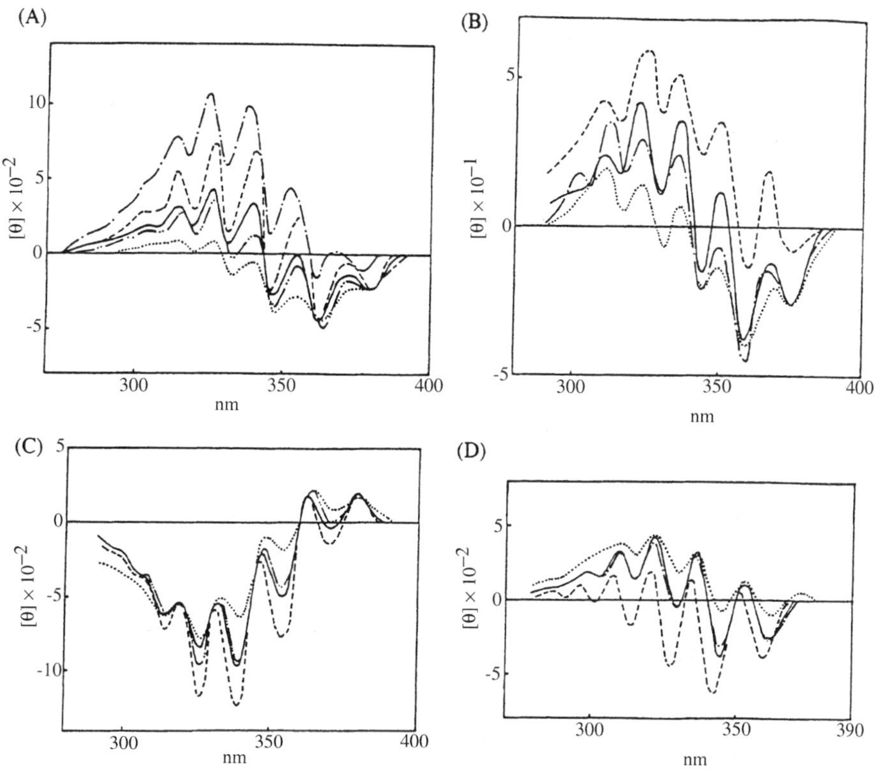

Figure 11-13. Variable low-temperature CD spectra for the $n \rightarrow \pi^*$ transition of α,β-unsaturated ketones in methylcyclohexane–isopentane (MI 1:3 by vol) at 25 °C (—), −91 °C (– – – –), and −186 °C (– · – · – ·) and in decalin at +25 °C (– ·· –) and approximately +147 °C (····), unless noted otherwise, for (A) (−)-*p*-menth-3-en-5-one; (B) (−)-piperitone; (C) (+)-carvone; and (D) (−)-carvenone. In (C) (– ·· – ··) is for decalin at +25 °C; in (D) (—) is for MI at −127 °C. See Figure 11-12 for structures. [Reprinted from ref. 23 with permission from the Chemical Society of Japan.]

changing signs as the temperature is lowered.[25] Like pulegone, fukinone has a large enone torsion angle (ω), whose sign depends upon whether the *cis*-decalone adopts a steroid or nonsteroid conformation (Figure 11-15). The observed Cotton effect signs and the enone orbital helicity rule (Figure 11-3) predict a predominance of the steroid conformer, in concert with the predicted energy difference: Molecular mechanics $\Delta\Delta H_f$[17] predicts that the steroid conformer should be more stable by about 2.9 kcal/mole. Analysis of the temperature dependence of the $n \rightarrow \pi^*$ Cotton effect indicates that $\Delta G° = 0.9$ kcal/ mole, and temperature-dependent NMR analysis gives $\Delta G° \sim 1.3$ kcal/mole, favoring the steroid conformer.

A recent comprehensive survey of data on α,β-unsaturated ketone circular dichroism is found in ref. 7c.

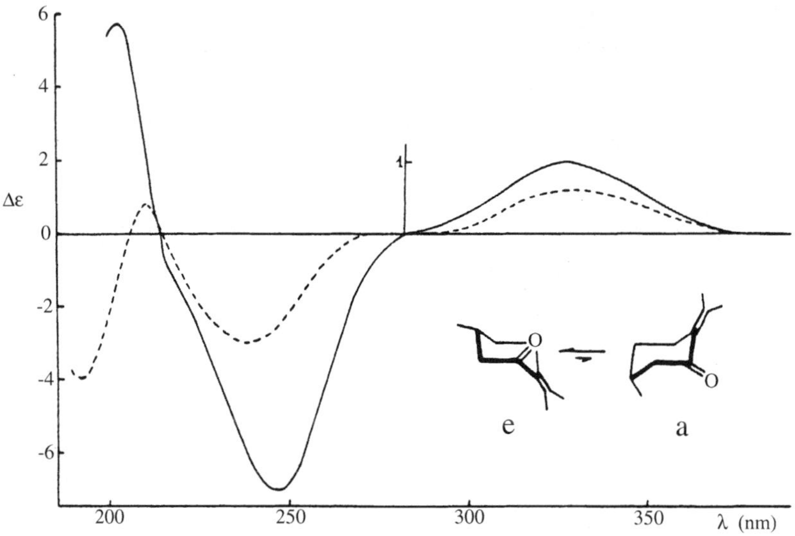

Figure 11-14. Variable low-temperature CD spectra of (+)-pulegone at +20 °C (– – – –) and –150 °C (—) in methylcyclohexane–isopentane (4:1 by vol). The equatorial methyl conformer on the left is more stable than the axial conformer by about 2.5 kcal/mole ($\Delta\Delta H_f$). [Reprinted from ref. 8b, copyright© 1982, with permission from Elsevier Science.]

(A)

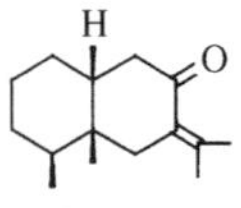

fukinone

T(°C)	Δε (325 nm)	Δε (243 nm)
+24	−0.38	+3.42
−68	−0.53	+5.12
−190	−0.46	+7.21

(B)

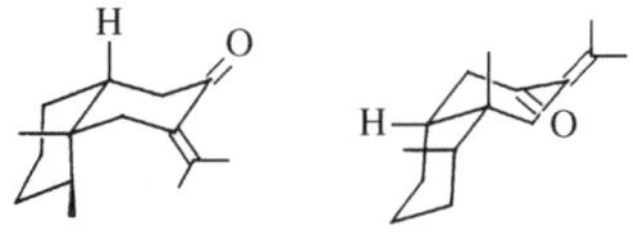

	"steroid"	"non-steroid"
$\Delta\Delta H_f$ (kcal/mole)	0.00	2.86
ω(O=C–C=C)	−56.7°	+62.6°

Figure 11-15. (A) Fukinone and its variable low-temperature CD data for the $n \rightarrow \pi^*$ and $\pi \rightarrow \pi^*$ transitions near 325 and 243 nm, respectively (data from ref. 25). (B) Fukinone steroid and nonsteroid conformations, with relative heats of formation ($\Delta\Delta H_f$, kcal/mole) and enone torsion angle (ω) determined by PCMODEL (ref. 17).

11.2. β,γ-Unsaturated Ketones and the Extended Octant Rule

An interaction between homoconjugated carbonyl and olefin chromophores has been recognized for many years by rotatory dispersion and UV spectroscopy. For example, in 1936, Asahina et al.[26] at Tokyo University reported that dehydrocamphor had a very

Table 11-3. Comparison of Specific Rotation, UV $n \rightarrow \pi^*$ Absorption, and CD Data for (–)-Dehydrocamphor and (+)-Camphor[a]

		$[\alpha]_D$	ϵ^{max} (λ,nm)	$\Delta\epsilon^{max}$ (λ,nm)
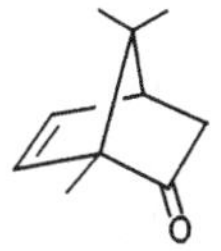	(–)-Dehydrocamphor	–731°	308 (304)	–17.85 (306)
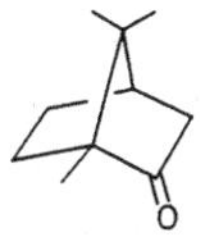	(+)-Camphor	+44°	32 (291)	+1.54 (296)

[a] Values in ethanol from refs. 47b and 47c.

large specific rotation compared with that of camphor (Table 11-3). UV and chiroptical data were not available at that time. Much earlier, however, in 1881, the β,γ-unsaturated ketones santonide and parasantonide (Table 11-4), whose structures were then unknown, were reported by Nasini et al. to give remarkably large specific rotations.[27] Subsequently, in 1939, unusually intense UV absorption coefficients, abnormally large CD maxima, and ORD Cotton effects orders of magnitude greater than most saturated ketones were reported for santonide and parasantonide by Mitchell and Schwarzwald.[28] Although at the time the structures of the two ketones had not been fully established, by 1950 they were solved by Woodward and Kovach at Harvard,[29] who, noting their unusual optical characteristics (Table 11-4), pointed out that (i) the ketone carbonyl electric dipole moment and the very high electric dipole moment of the enol lactone group are held rigidly in a particular orientation to each other and (ii) "the high

Table 11-4. Optical Rotation, ORD, UV, and CD Data for Santonide, Parasantonide, and Parasantonic Acid

		$[\alpha]_D{}^a$	a^b	$\epsilon^{max}(\lambda,nm)^b$	$\Delta\epsilon^{max}(\lambda,nm)^b$
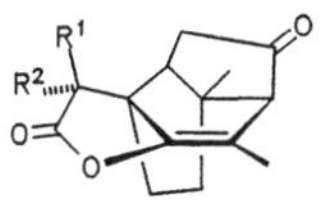	Santonide (R^1=H, R=CH_3)	+758°	+1390	970(300)	+26.7(300)
	Parasantonide (R^1=CH_3,R^2=H)	+897°	+1648	1190(300)	+36.7(300)
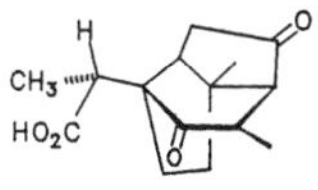	Parasantonic Acid	–98°	—	~40(300)	—

[a] $[\alpha]_D$ values from refs. 27 and 29. [b] ORD a value, UV ϵ and CD $\Delta\epsilon$ values in ethanol at 17°C from the data of ref. 29.

rotation and abnormally high carbonyl absorption intensity have a common basis."[29a] At about the same time, the exaltation of the $n \rightarrow \pi^*$ UV absorption found in phenylacetones (α-phenylketones) was attributed by Alpen et al. to a "through-space" interaction between carbonyl and phenyl chromophores.[30] Shortly thereafter, Cookson and Wariyar,[31a] then at Birkbeck College, London, surveyed the UV spectra of a large number and variety of β,γ-unsaturated ketones and α-phenyl ketones, noting that abnormally intense absorption of the carbonyl $n \rightarrow \pi^*$ transition was associated with a special geometric relationship between the carbonyl and olefin chromophores (i.e., when the *p*-orbitals of C(1) and C(3) point to one another in $O{=}C_1{-}C_2{-}C_3{=}C_4$ (Table 11-5). These investigators indicated that the transition responsible for the increase in intensity in the carbonyl absorption might come from charge-transfer from the olefin group to the carbonyl group, but they preferred to assign it to a normal $n \rightarrow \pi^*$ carbonyl transition influenced by the adjacent π-electrons. And they, too, suggested that high optical activity in β,γ-unsaturated ketones would be found when an intense carbonyl UV absorption was also present.

The earliest meaningful theoretical treatment of the phenomenon was proposed in 1959 by Labhart and Wagnière,[32] who introduced a simple LCAO Hückel-type MO delocalized ground-state model to explain the intensification of the carbonyl absorption and the appearance of a new absorption band near 220 nm in some β,γ-unsaturated ketones. In α,β-unsaturated ketones, in which the carbonyl and olefin π-systems overlap strongly, but the *n* and π-systems do not (except when the enone is twisted), the $n \rightarrow \pi^*$ transition is bathochromically shifted, but not intensified, and a $\pi_{C=C} \rightarrow \pi^*_{C=O}$ charge-transfer transition is intensified and bathochromically shifted. By treating β,γ-unsaturated ketones as stretched α,β-unsaturated ketones, the intensifica-

Table 11-5. Influence of Relative Orientation of Chromophores on the Carbonyl UV Absorption of β,γ-Unsaturated Ketones[a]

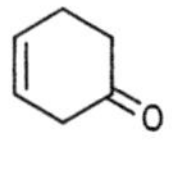	19 (280)	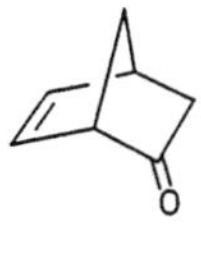	315 (295)
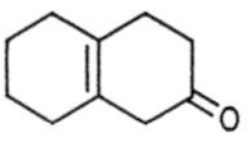	38 (281)	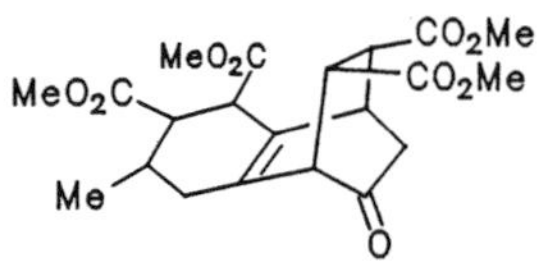	300 (297)
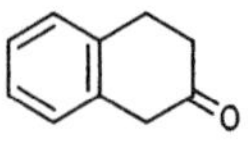	18 (321) 89 (294)	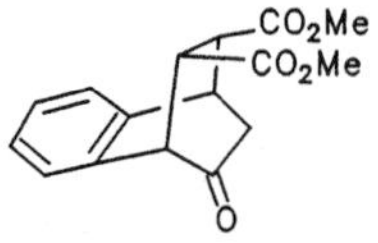	500 (295)

[a] UV data in ethanol from ref. 31. Values reported as ϵ^{max} (λ, nm).

tion of the carbonyl transition in bicyclo[2.2.1]hept-5-en-2-one and bicyclo[2.2.2]oct-5-en-2-one was explained as being due to overlap of the *n*-orbital on oxygen with the olefin π-orbital. In this case, the classification of orbitals into *n* and π thus no longer strictly holds, because the *n*-orbital has some $\pi_{C=C}$ character mixed in, and the symmetry forbiddenness of the $n \to \pi^*$ transition is removed. The erstwhile $n \to \pi^*$ carbonyl transition becomes a transition from an $(n + \lambda\pi')$ nonbonding orbital (where λ is a mixing parameter determined by perturbation theory and π′ is the highest bonding π-orbital) to a π^*-orbital formed from a linear combination of $\pi^*_{C=C}$ and $\pi^*_{C=O}$. Since the relevant nonbonding and antibonding orbitals are now not orthogonal by symmetry, the transition takes on an allowed $(\pi \to \pi^*)$ character, and the intensity is considerably enhanced. The degree of mixing (indicated by the coefficient λ) is, of course, dependent on the relative orientation of the carbonyl and olefin chromophores. Some orientations favor *n* and $\pi_{C=C}$ overlap and thus lead to strongly intensified $n \to \pi^*$ transitions, as in the orientations in the β,γ-unsaturated ketones of Tables 11-3 and 11-4. Other orientations do not, however, and the intensities are weaker, as may be seen in the β,γ-unsaturated ketones of Table 11-1. The model also makes predictions on $\pi_{C=C} \to \pi^*_{C=O}$ charge transfer and localized $\pi \to \pi^*$ transitions.

The foregoing considerations were important to later studies relating enhanced $n \to \pi^*$ intensity, specific rotation, ORD, CD, and stereochemistry of β,γ-unsaturated ketones. However, no attempt was made to explain or correlate the high optical rotations with the stereochemistry of β,γ-unsaturated ketones having exalted $n \to \pi^*$ absorptions until 1960, when Mislow et al., then at New York University, and Djerassi, at Stanford, published a collaborative study of the ORD spectra of atropisomeric biaryls.[33,34] On the basis of their early investigations that successfully correlated biaryl absolute configuration and ORD Cotton effect signs, Mislow et al.[33a] discovered that (−) and (+)-2,2′-dimethyldibenzsuberone (Figure 11-16A) exhibited an extraordinarily large ORD amplitude for the carbonyl transition, which they attributed to homoconjugation between carbonyl and benzene π-electrons. Almost immediately thereafter, Mislow and Djerassi[33b] reported on the striking coincidence between the long-wavelength ORD Cotton effect of (+)-2,2′-dimethyldibenzsuberone and that of parasantonide (Figure 11-16B). Within a year, Moscowitz, Mislow, Glass, and Djerassi[35a] explained that (*i*) the unusually intense UV long-wavelength carbonyl transitions previously noted must also give rise to a marked increase in the rotatory strength for the same electronic excitation and (*ii*) the β,γ-unsaturated ketone Cotton effect is due mainly to the chiral orientation of the unsaturated group with respect to the carbonyl group (i.e., the helicity of the composite chromophore from carbonyl and ethylenic (or other π-electron) chromophores may dominate contributions from extrachromophoric octant-like asymmetrically disposed perturbers and may thus determine the sign and magnitude of the Cotton effect for the erstwhile $n \to \pi^*$ ketone transition). The latter was called a generalization of the octant rule.[35a]

A generalized or extended octant rule for β,γ-unsaturated ketones was proposed and justified by Moscowitz et al.[35b] in early 1962, following analyses of the ORD and UV spectra of various homoconjugated ketones, including, *inter alia*, the previously studied 2,2′-dimethyldibenzsuberone (Figure 11-16), santonide (Table 11-4), and

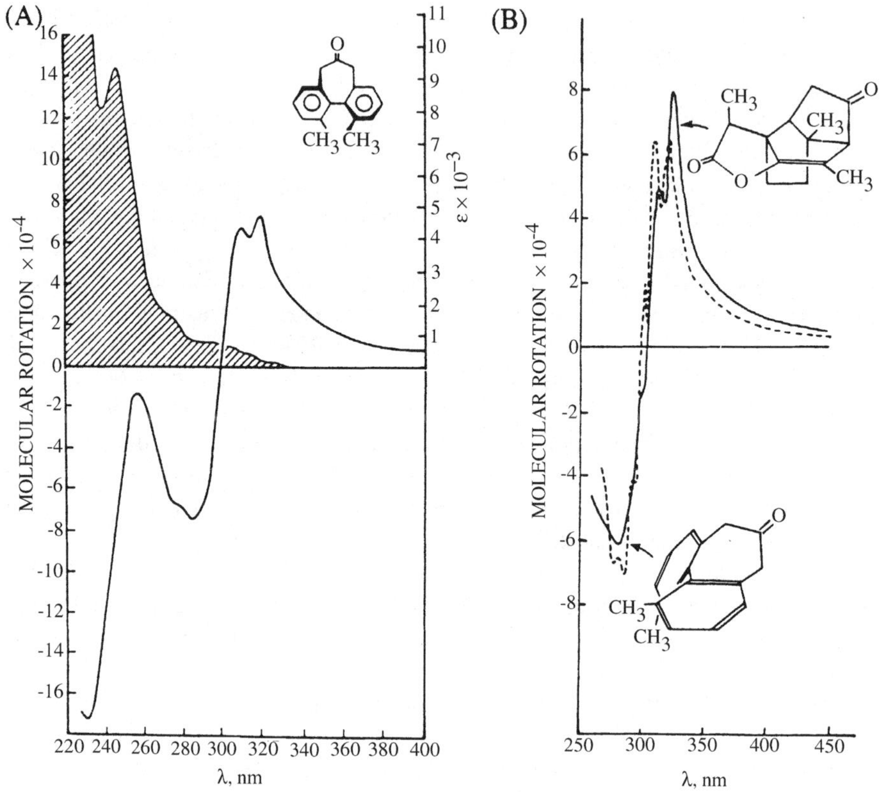

Figure 11-16. (A) ORD spectra (—) and UV (shaded) of (+)-dimethyldibenzsuberone in isooctane. (B) Comparison of ORD spectra of (+)-dimethyldibenzsuberone and parasantonide in isooctane. [Reprinted from ref. 34b with permission of the New York Academy of Sciences.]

parasantonide (Table 11-4, Figure 11-16B); the contrasting epimeric 3-phenyl-5α-cholestan-2-ones (Figure 11-17A) and chrysanthenone and verbenalin (Figure 11-17B); and the prototypic β,γ-unsaturated ketones—bicyclo[2.2.1]hept-5-en-2-one (Figure 11-17C) and bicyclo[2.2.2]oct-5-en-2-one (Figure 11-17D)—which had then just become available in optically active form.[36] Within six months, CD spectra of many of these unsaturated ketones had been determined and showed the expected, unusually strong carbonyl $n \rightarrow \pi^*$ Cotton effects,[36,37] as illustrated for bicyclo[2.2.1]hept-5-en-2-one in Figure 11-17C. The influence of the olefin chromophore on the ketone carbonyl Cotton effect is strikingly demonstrated in comparing the ORD spectra of Figure 11-17C, in which the weak negative $n \rightarrow \pi^*$ Cotton effect of the saturated ketone undergoes a reversal of sign and tremendous intensification when the β,γ-olefinic chromophore is present. The importance of the orientation of the C=O and olefinic or aromatic chromophores in influencing the magnitude (and sign) of the carbonyl Cotton effect is strikingly apparent in the contrasting ORD curves (Figure

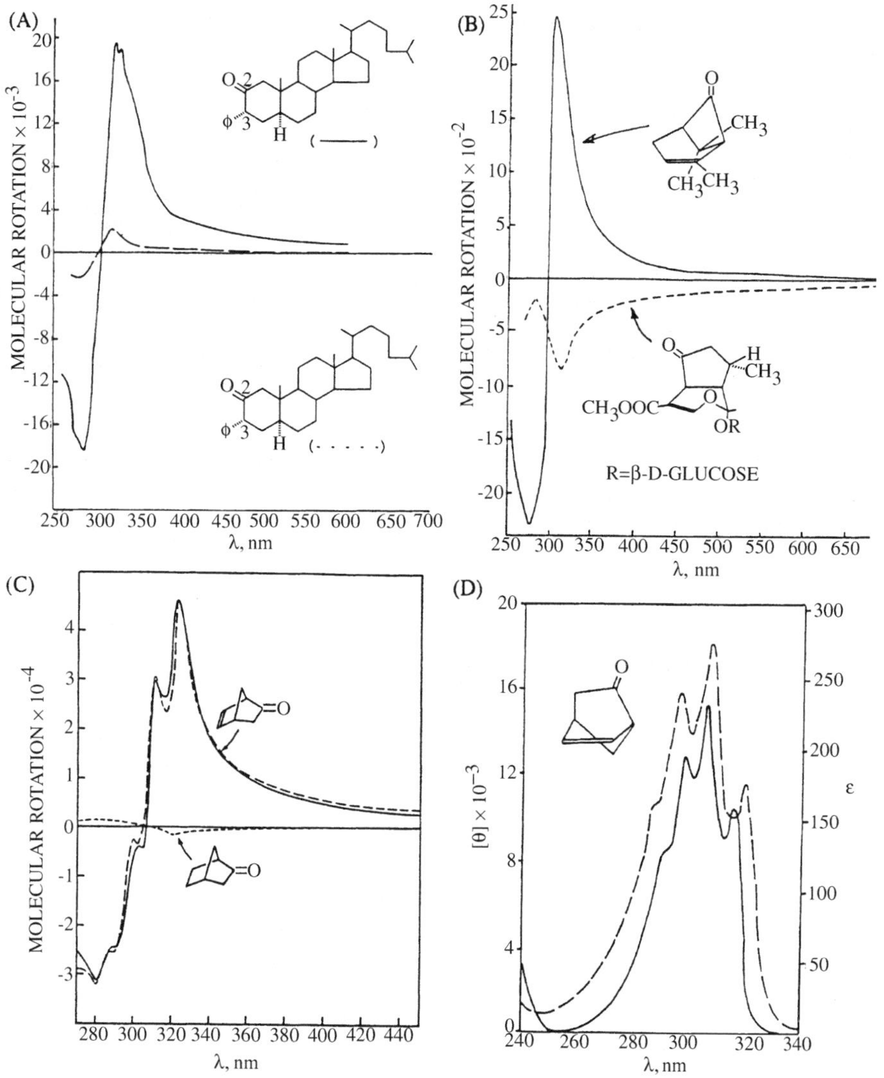

Figure 11-17. ORD spectra of (A) 3α-phenyl-5α-cholestan-2-one (—) and 3β-phenyl-5α-cholestan-2-one (– – –) in dioxane, (B) chrysanthenone (—) and verbenalin (– – –) in isooctane, and (C) (+)-1(*R*)-bicyclo[2.2.1]hept-5-en-2-one (—) and (+)-1(*S*)-bicyclo[2.2.1]heptan-2-one (– – –) in isooctane. (D) CD (—) and UV (– – –) spectra of (+)-bicyclo[2.2.2]oct-5-en-2-one in isooctane. In (C), the computed ORD curve is given by (– – –). [(A) and (B) are reprinted with permission from ref. 35b, (C) is reprinted with permission from ref. 36, and (D) is reprinted from ref. 37. Copyright© 1962 American Chemical Society.]

11-17A) of the epimeric α-phenyl ketones and in the comparatively rather weak Cotton effects of 3β-phenyl-5α-cholestan-2-one (Figure 11-17A) and verbenalin (Figure 11-17B). The various important relative orientations of carbonyl and olefin chromophores (in $O{=}C_1{-}C_2{-}C_3{=}C_4$) of the β,γ-unsaturated ketones studied by Moscowitz et al.[35b] are shown in Figure 11-18. The most well represented orientation is shown in Figure 11-18A. Figure 11-18B represents the slightly altered geometry found in chrysanthenone, which has an $n \rightarrow \pi^*$ ORD Cotton effect amplitude (***a*** ~ +48) an order of magnitude lower than the ***a*** values typically associated with the geometry shown in Figure 11-18A (e.g., (+)-bicyclo[2.2.1]hept-5-en-2-one has ***a*** ~ +770). Figure 11-18C represents the greatly altered geometry found in verbenalin, whose Cotton effect is comparatively much weaker (***a*** ~ 6.5) and oppositely signed. Thus, the Cotton effect intensities of β,γ-unsaturated ketones was discovered to depend critically on the relative disposition of the carbonyl and olefin groups. Yet, from these limited data, a chirality rule could be proposed.

Using the formalism of Labhart and Wagnière,[32] Moscowitz et al.[35b] were able to reproduce the ORD curves of various β,γ-unsaturated ketones from their UV curves, thereby justifying quantitatively the earlier suggestion[29] of a qualitative relationship between exalted rotations or rotatory dispersion amplitudes and exalted UV absorption. More importantly, Moscowitz et al. provided a theoretical basis for extracting information from Cotton effects of inherently dissymmetric chromophores,[38] which aided in understanding the relationship between the Cotton effect sign of a β,γ-unsaturated ketone and the relative orientation of its olefinic and carbonyl chromophores. This work led to a generalization of the octant rule, which is not really an octant rule, but a chirality rule that correlates the sign of the β,γ-unsaturated ketone $n \rightarrow \pi^*$ Cotton effect to the handedness of the inherently dissymmetric homoconjugated $O{=}C_1{-}C_2{-}C_3{=}C_4$ chromophore (Figure 11-19).[34b,35b] Figure 11-19 shows the various relative geometries of the carbonyl and olefin chromophores of the β,γ-unsaturated ketones studied,[35b] which have positive carbonyl $n \rightarrow \pi^*$ Cotton effects. The mirror-image geometries would have negative carbonyl Cotton effects.

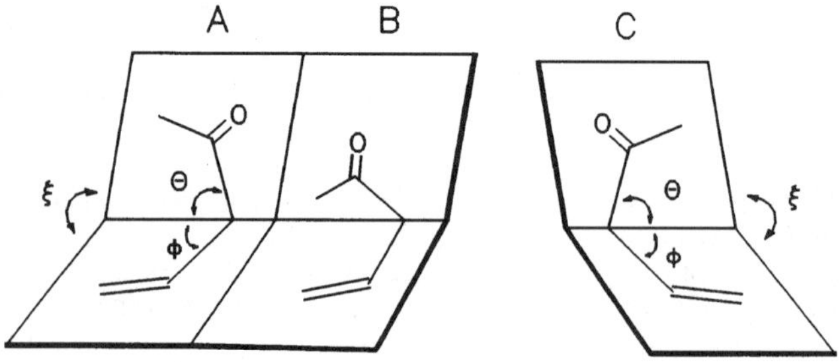

Figure 11-18. Orientation of carbonyl and olefin chromophores of β,γ-unsaturated ketones: (A) (+)-2,2′-dimethyldibenzsuberones, santonide, parasantonide, (+)-(1*R*)-bicyclo[2.2.1]hept-5-en-2-one, (+)-1(*R*)-bicyclo[2.2.2]oct-5-en-2-one, (+)-dehydrocamphor, and other ketones of ref. 34b; (B) chrysanthenone and (C) verbenalin. (See ref. 35b).

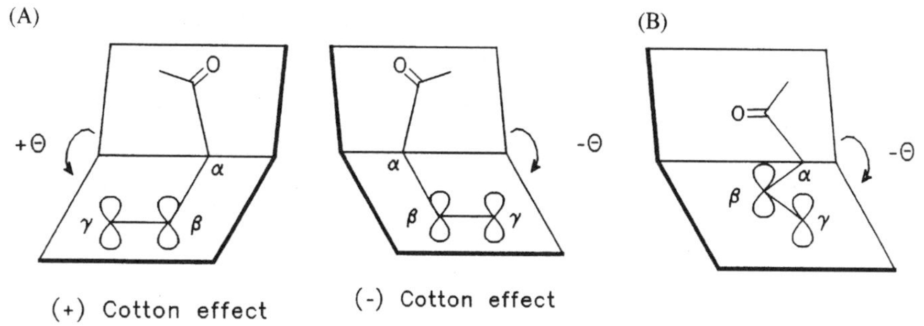

Figure 11-19. Chirality rule for β,γ-unsaturated ketones, where the absolute value of the interplanar angle $\theta > 90°$ for (A) enantiomeric *s-cis* and (B) *s-trans* geometry. β,γ-Unsaturated ketones with the geometry shown in (B) often have only weak Cotton effects (ref. 35b).

While these studies were being carried out, Cookson et al. continued to find examples of strong homoconjugation, as seen in the exalted carbonyl UV absorption in β,γ-unsaturated ketones, as well as in ORD spectra.[39] Following analyses of α-phenylketone and β,γ-unsaturated ketone UV and ORD spectra, Cookson and Hudec[40] noted that it was not quite correct to assume that maximum intensification of the carbonyl UV absorption would occur when the *p*-orbitals at C_1 and C_3 of $O{=}C_1{-}C_2{-}C_3{=}C_4$ were in the same plane.[31] However, from their analysis of the work of Labhart and Wagnière,[32] it was thought that such an orientation would be favorable for maximum $\pi \rightarrow \pi^*$ charge transfer and that the intensity borrowed by the $n \rightarrow \pi^*$ transition would come from the charge-transfer transition. Cookson and Hudec envisaged two extreme situations in which β,γ-unsaturated ketones would *not* show intensified $n \rightarrow \pi^*$ absorption: (i) when the overlap integral between the appropriate carbonyl *p* and olefin π-orbitals was zero and (ii) when there was no charge transfer. They cited examples wherein the carbonyl and olefin groups face one another and, due to lack of overlap, show the charge-transfer band but not the intensified $n \rightarrow \pi^*$ absorption. It was suggested that in the symmetric ketone 7-norbornenone (Table 11-6) there was no carbonyl-*p*-orbital-to-olefin-π-orbital overlap; thus, while a charge-transfer absorption occurred near 225 nm, there was only a weak $n \rightarrow \pi^*$ absorption near 275 nm. In contrast, the nonsymmetrical orientation of the carbonyl and olefin chromophores in 2-norbornenone was conducive to both the charge transfer (at 215 nm) and an enhanced $n \rightarrow \pi^*$ absorption (at 290 nm). Enhanced carbonyl $n \rightarrow \pi^*$ UV absorption of β,γ-unsaturated ketones was thus explained as being due to borrowing intensity from a $\pi \rightarrow \pi^*$ charge-transfer transition. This explanation was proposed as the basis for the previously observed exaltation of optical rotation and rotatory dispersion Cotton effects. A special case of the octant rule was proposed: If the olefin π-orbital is in an upper left or lower right quadrant, a positive $n \rightarrow \pi^*$ Cotton effect is predicted; if it is in an upper right or lower left quadrant, a negative $n \rightarrow \pi^*$ Cotton effect is predicted.[40]

Table 11-6. Comparison of the UV Spectral Data of 7-Norbornenone and 2-Norbornenone in Isooctane Solvent[a]

		ϵ^{max} (λ,nm)	ϵ^{max} (λ,nm)
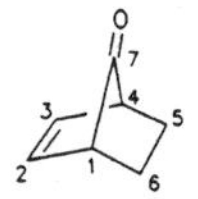	7-Norbornenone (Bicyclo[2.2.1]hept-5-en-7-one)	33 (275)	1200 (225)
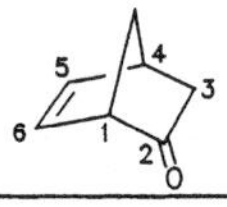	2-Norbornenone (Bicyclo[2.2.1]hept-5-en-2-one)	304 (290)	2800 (215)

[a] Data from ref. 40.

A more recent analysis of 7-norbornenone[41] indicates that, while the lowest energy transition is essentially an isolated carbonyl $n \rightarrow \pi^*$ excitation, the 225-nm band is not a pure charge-transfer transition. Rather, it is predicted to be a mixed transition with a substantial (~15%) charge-transfer component of the form $\pi_{C=C} \rightarrow \pi^*_{C=O}$, but also with major components originating in three different carbonyl in-plane $\sigma \rightarrow \pi^*$ excitations: approximately 15% $\sigma_{C-C_\alpha} \rightarrow \pi^*_{C=O}$, about 30% $n(\sigma)_{C=O} \rightarrow \pi^*_{C=O}$, and around 30% $\sigma_{C=O} \rightarrow \pi^*_{C=O}$.[41]

As the earlier studies were developing, Mason[42] showed that the rotatory power of β,γ-unsaturated ketones was proportional to the square root of the enhancement in UV intensity, effectively linking rotatory power to (UV) intensity borrowing in the Labhart and Wagnière mechanism.[32] Further studies[43] linked the enhanced UV absorption and ORD or CD Cotton effects to the mixing of a $\pi \rightarrow \pi^*$ charge-transfer transition with the $n \rightarrow \pi^*$ carbonyl transition. The model formulated by Moscowitz et al.[35b] for explaining the enhanced UV and chiroptical properties of the $n \rightarrow \pi^*$ transition was criticized[43c] as implying that enantiomers should be absorptive at different wavelengths, but the criticism was retracted[44a] when it was learned that it was unfounded.[44b]

In the 1960s, even more sophisticated MO calculations, based on the coupled-oscillator approach of Longuet-Higgins and Murrell,[45] were being carried out by Hansen et al.[4] on the prototypic β,γ-unsaturated ketones bicyclo[2.2.1]hept-5-en-2-one (2-norbornenone) and bicyclo[2.2.2]oct-5-en-2-one. The results indicated that the Labhart and Wagnière model,[32,35b] which used the Hückel MO approach without configuration interaction, gave an incomplete picture of the situation—although it seemed adequate for assigning absolute configurations and (in some cases) for making order-of-magnitude estimates. The major inadequacies derive from the fact that the Hückel method made no inquiry into the origin and nature of the coupling between the carbonyl and olefin chromophores, nor did it take into account configuration interaction, which turns out to be quite important in the coupled-chromophore approach. This early semiempirical calculation[4] was appealing because it imported considerable experimental data, including (i) $n \rightarrow \pi^*$ and $\pi \rightarrow \pi^*$ transition energies of the ketone carbonyl chromophore and the $\pi \rightarrow \pi^*$ olefin chromophore of appropriate models such as camphor, norcamphor, norbornene, etc.; and (ii) ionization potentials of the same chromophores, also from suitable model compounds. Interestingly, and counter to suppositions derived from the Labhart and Wagnière model,[39–43] the calculation showed that *overlap and charge-transfer effects were relatively unimportant*—that the sign and magnitude of the approximately 300-nm transition of β,γ-unsaturated ketones could be accounted for correctly by the electrostatic coupling or mixing of locally excited ketone carbonyl $n \rightarrow \pi^*$ and olefin $\pi \rightarrow \pi^*$ states from the μ-m exciton treatment.[4]

In steroids and triterpenes, examples of unusually strong Cotton effects were also noted,[43c,46] as in cholest-5-en-3-one and 17β-hydroxy-10α-androst-5-en-3-one (Figure 11-20A).[7b] Although the enhanced magnitudes were thought to arise from homoconjugative effects, the Cotton effect magnitudes are somewhat weaker than those seen with the β,γ-unsaturated ketones reported earlier[34–37,39,40] (Tables 11-3, 11-4, and 11-5 and Figures 11-16 and 11-17). Interestingly, the pentacyclic β,γ-unsaturated steroid ketone (related to 7-norbornenone) with olefin and carbonyl chromophores symmetrically disposed unexpectedly gives a very strong ORD Cotton effect magnitude (***a*** = –500, Δε = –12.4)—much stronger than the ordinary Cotton effect magnitude of its saturated analog [***a*** = –39, Δε ~ –0.98 (Figure 11-20B)]. The large magnitude of the enone was unexpected, in view of the analysis of Cookson and Hudec,[39,40] and the discrepancy remains unresolved.

Further studies of the $n \rightarrow \pi^*$ and $\pi \rightarrow \pi^*$ charge-transfer excitations of β,γ-unsaturated ketones such as dehydrocamphor and dehydroepicamphor (Table 11-7) were

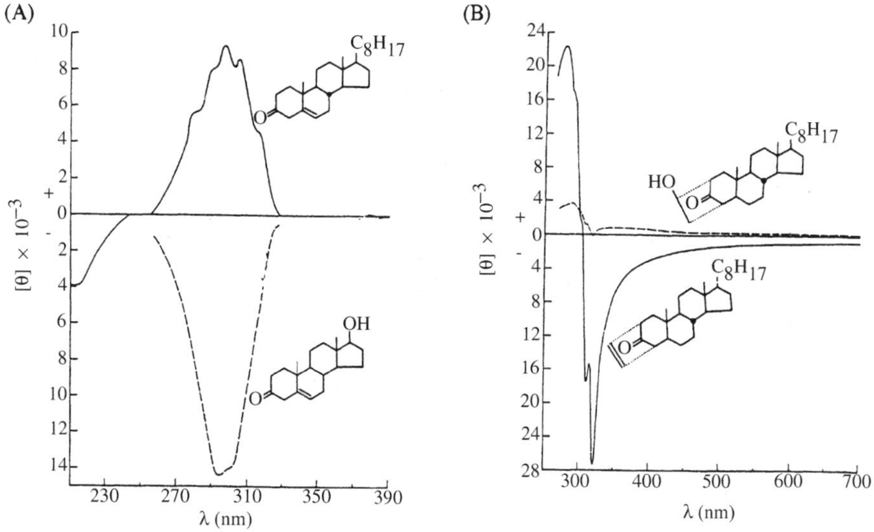

Figure 11-20. (A) CD spectra of cholest-5-en-3-one (—) and 17β-hydroxy-10α-androst-5-en-3-one. (B) ORD spectra of pentacyclic steroid ketones; β,γ-unsaturated (—) and β-ketol (– – – –). [Reprinted from ref. 7b with permission from Holden-Day, Inc.]

Table 11-7. Influence of Phenyl Substitution on the $n \rightarrow \pi^*$ CD and UV of Dehydrocamphor and Dehydroepicamphor in Ethanol[a]

	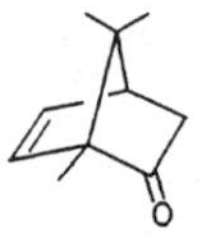	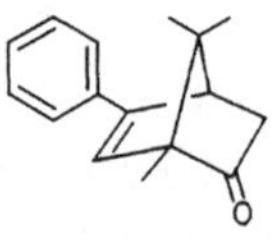	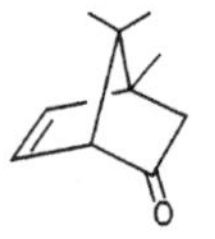	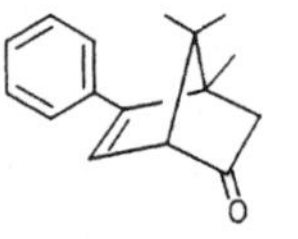
$\Delta\epsilon$ (λ,nm)	−17.9 (306)	−48.9 (315)	−18.2 (305)	−39.2 (309)
ϵ (λ,nm)	308 (304)	6070 (314)	330 (303)	2830 (312)

[a] Data from ref. 47b.

carried out by Cookson et al.,[47] who showed that extremely large $n \rightarrow \pi^*$ absorption and CD exaltations could be achieved with phenyl substituents on the enone carbon–carbon double bond.[47b] Their work was extended to other examples of β,γ-unsaturated ketones (Table 11-8, #1–6).[47b,c] At that time, it was recognized[47d] that the $n \rightarrow \pi^*$ intensity came more likely from mixing the locally excited olefin $\pi \rightarrow \pi^*$ and carbonyl $n \rightarrow \pi^*$ states, from electrostatic coupling rather than charge transfer. It was thought that the description of the short-wavelength band in β,γ-unsaturated ketones as a charge-transfer transition might also be misleading and should in some cases be dropped. However, that suggestion now seems overly cautious in light of subsequent theoretical analysis of 7-norbornenone[41] and later work.

Mislow et al. at Princeton corrected[48a] the previously reported (low) $\Delta\varepsilon$ value of 2-norbornenone (reported on less than optically pure material)[47b] and showed that 2-benzobornenone (Table 11-8, #7) exhibited an extraordinarily large $n \rightarrow \pi^*$ CD Cotton effect and UV absorption.[48b,49] The influence of an α-phenyl on the CD spectra of 2-norbornanone was investigated by Thomas and Mislow,[50] who showed that the *exo* isomer was much more effective than the *endo* in causing exaltation of the ketone carbonyl $n \rightarrow \pi^*$ UV band (Figure 11-21A). However, in the CD spectrum (Figure 11-21B), the parent, 2-norbornanone, has a stronger $n \rightarrow \pi^*$ Cotton effect than the *endo*-α-phenyl isomer and a weaker $n \rightarrow \pi^*$ Cotton effect than the *exo*-α-phenyl epimer. Here again, the orientation of the two chromophores is paramount, but ordinary octant effects also play an important role. Thus, front-octant or antioctant effects appear to contribute to the weaker-than-expected $n \rightarrow \pi^*$ Cotton effect of the *exo*-phenyl epimer (cf. Figure 11-17A), and back octant contributions weigh into the weak Cotton effect of the *endo*-phenyl epimer (cf. Figure 11-17A). The rotational orientation of the phenyl chromophore can be expected to have an influence on both the homoconjugation and octant contributions to the $n \rightarrow \pi^*$ Cotton effect, but no variable low-temperature CD studies have been carried out to assess rotamer contributions.

Other examples of β,γ-unsaturated ketone CD came from Erman et al. (Table 11-8, #8–10).[51] The smaller-than-expected $\Delta\varepsilon$ values might be attributed to low enantiomeric enrichments, but the observed Cotton effects correlate well with the generalized

Table 11-8. β,γ-Unsaturated Ketones with Strong $n \rightarrow \pi^*$ CD Cotton Effects and UV Absorption

#	Ketone	CD $\Delta\epsilon^{max}$ (λ,nm)	UV ϵ^{max} (λ,nm)	#	Ketone	CD $\Delta\epsilon^{max}$ (λ,nm)	UV ϵ^{max} (λ,nm)
1[a]	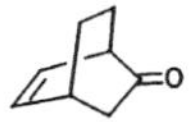	+10.1 (298)	110 (298) 3000 (202)	8[e]	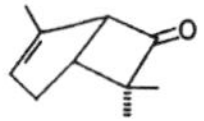	+3.75 (313)	260 (310)
2[a,b]	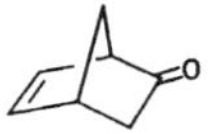	+9.6 (306)[a] +20.6 (306.5)[b] −7.6 (215)[b]	290 (308) 3000 (210)	9[e]	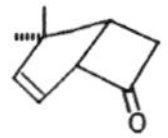	−6.83 (304)	289 (307)
3[a]	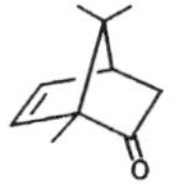	−17.9 (306)[a]	308 (304) 2425 (216)	10[e]		+1.35 (293)	102 (293)
4[a]	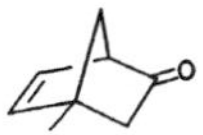	+18.2 (305)	330 (303) 2660 (216)	11[f]		−1.16 (296)	126 (295)
5[c]	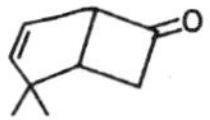	+14.4 (304) −6.6 (224)	285 (302) 2100 (224)[sh]	12[f]	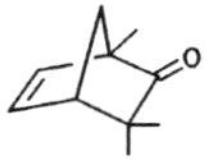	+19.4 (306)	
6[c]	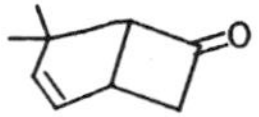	−14.6 (307)	190 (307)	13[g]	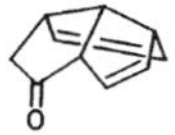	+25 (295)	250 (298)
7[d]	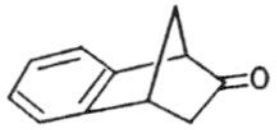	+187.9 (307.5)	796 (309)	14[h]	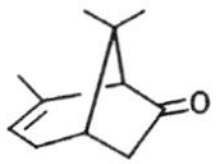	+127 (292)	267 (292)
				15[h]	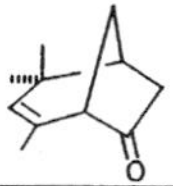	+130 (292)	314 (292)

[a] In ethanol, ref. 47b. [b] In isooctane, ref. 48a. [c] In ethanol, ref. 47c. [d] In isooctane, ref. 48b; $R = +31.1 \times 10^{-40}$cgs in ref. 49. [e] Ref. 51, not 100% e.e. [f] In ethanol, ref. 52. [g] In isooctane, ref. 53. [h] UV in cyclohexane, CD in CH_3OH, ref. 54; the $\Delta\epsilon$ values are probably too high by a factor of 10 (scaling error?).

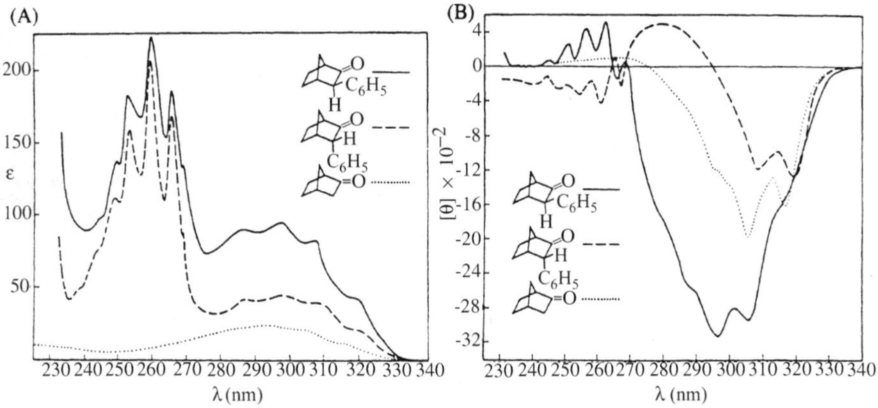

Figure 11-21. (A) UV and (B) CD spectra of (+)-3-*exo*-phenyl-2-norbornanone (—), (–)-3-*endo*-phenyl-2-norbornanone (– – – –), and 2-norbornanone (·····) in isooctane. [Reprinted with permission from ref. 50. Copyright© 1970 American Chemical Society.]

octant rule. (+)-Dehydrofenchone (Table 11-8, #12) was shown to give the expected large Δε values and positive Cotton effect.[52] Paquette et al.[53] at Ohio State noted the strong CD Cotton effect of (+)-2,3-dihydrotriquinacen-2-one (Table 11-8, #13) and found a reasonably good linear correlation between the long-wavelength log Δε and log ε for the $n \rightarrow \pi^*$ transition of various β,γ-unsaturated ketones known in 1975 (Figure 11-22). Extremely large CD Δε values were reported by Weissberger et al. for

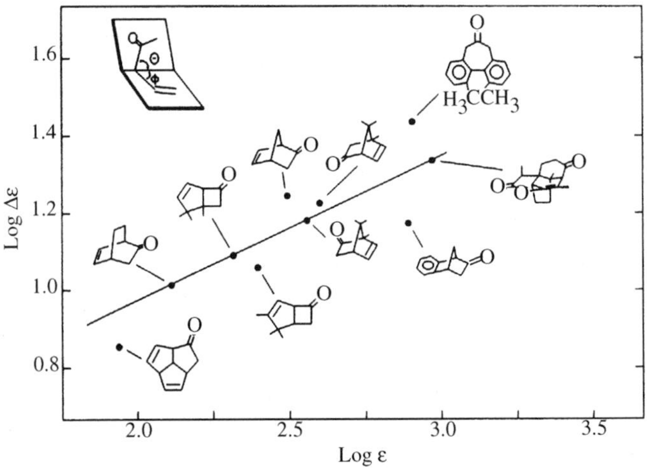

Figure 11-22. Correlation of the $n \rightarrow \pi^*$ CD Cotton effect log Δε and UV log ε of β,γ-unsaturated ketones. [Redrawn from ref. 53.]

a pair of bicyclo[3.2.1]oct-2-en-7-ones (Table 11-8, #14, 15), but the values were later thought to be high by an order of magnitude due to a scaling error.[54]

The bulk of the β,γ-unsaturated ketones studied through the early 1970s gave large carbonyl $n \rightarrow \pi^*$ Cotton effects and enhanced UV absorption (Tables 11-3, 11-4, 11-6, 11-7, 11-8; Figures 11-16, 11-17 and 11-21), and all of them had the carbonyl and olefin chromophore orientation shown in Figures 11-19A and B. Verbenalin (Figure 11-17B (- - -)) remained the only exception, and it possessed the rare interchromophore geometry shown in Figure 11-19B. By the mid-1970s, however, new β,γ-unsaturated ketones with weak $n \rightarrow \pi^*$ Cotton effects began to be studied (Table 11-9). Lightner et al.[55] at the University of Nevada noted that a spiroketone (Table 11-9, #1) with nearly the same $O{=}C_1{-}C_2{-}C_3{=}C_4$ orientation as that of verbenalin (Figure 11-17B) gave a positive carbonyl $n \rightarrow \pi^*$ Cotton effect that had the same sign and nearly the same magnitude as that of its saturated ketone analogs, in which a methyl group replaced the exocyclic methylene. As mentioned earlier,[35,38] caution was urged in using the generalized octant rule in cases where ordinary octant contributions were comparable in magnitude to Cotton effect contributions due to the inherent dissymmetry of the enone chromophore—that is, in cases where unfavorable geometries lead to small $\Delta\varepsilon$ values and small rotatory strengths.

Working on a rigid system, Nakazaki et al. at Osaka examined the CD and UV spectra of two enones of the same geometry as that expressed in Figure 11-18C.[56] These authors reported on the CD of (–)-9-methylene-*twist*-brendan-2-one (Table 11-9, #2) which exhibited a moderately strong, bathochromically shifted $n \rightarrow \pi^*$ Cotton effect magnitude ($\Delta\varepsilon_{302.5}^{max} = -5.76$) and intensity-enhanced UV ($\varepsilon_{303}^{max} = 278$), compared with its saturated ketone analog, (–)-9-methyl-*twist*-brendan-2-one ($\Delta\varepsilon_{292}^{max}$ –2.36, $\varepsilon_{292.5}^{max} = 37$).[56a] Here, too, the enone geometry is not ideal; yet the interchromophoric interaction dominates the UV and CD spectra. Shortly thereafter, Nakazaki et al.[56b] reported on the CD of the analogs 4-methylene and 4-isopropylidene-D_{2d}-dinoradamantan-2-one (Table 11-9, #3 and 4), whose bathochromically shifted, moderately strong $n \rightarrow \pi^*$ Cotton effects and exalted UV absorption ($\varepsilon_{300.5}^{max} = 445$),[56c] compared with the parent ketone, (–)-bisnoradamantan-2-one, ($\Delta\varepsilon_{286}^{max} = -3.06$, $\varepsilon_{282}^{max} = 25$),[56b,d,e] indicate again the dominance of interchromophoric interaction in the less favorable enone geometry, although the CD magnitudes are only modest. In both examples, the Cotton effect contributions from coupling of the dissymmetrically oriented carbonyl and olefin chromophores have the same sign (negative) as those attributable to ordinary octant perturbers, and the generalized octant rule is followed.

Working with the more flexible spiro[4.4]nonane system, Shingu et al.[57] at Osaka found weak $n \rightarrow \pi^*$ Cotton effects from β,γ-unsaturated ketones (Table 11-9, #1 and 5), in which the $O{=}C_1{-}C_2{-}C_3{=}C_4$ orientation is more comparable to that of Figure 11-18C than 11-18A or B. There is evidence in the UV spectrum for a relatively small exaltation of absorption, but the Cotton effect magnitudes are comparable to those perturbations due to dissymmetric vicinal action. Indeed, as pointed out,[55b] the magnitudes of the exocyclic enone are comparable to those found in the dihydro (methyl analogs). At low temperatures, the enone rotatory strength may even be less

Table 11-9. β,γ-Unsaturated Ketones with Strong UV Absorption and Weak-Moderate CD Cotton Effects

#	Ketone	CD $\Delta\epsilon^{max}$ (λ,nm)	UV ϵ^{max} (λ,nm)	#	Ketone	CD $\Delta\epsilon^{max}$ (λ,nm)	UV ϵ^{max} (λ,nm)
1[a,b]	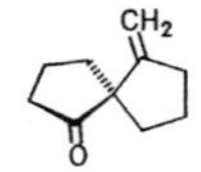	−0.88 (308)[a] +1.47 (301)[b]	66 (301)[b]	7[h]	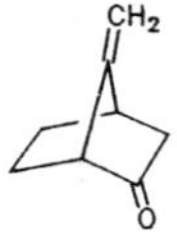	+7.6 (307) −45 (197)	180 (307) 7000 (197)
2[c]	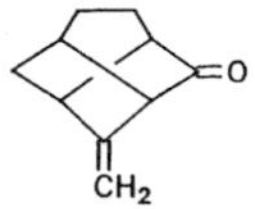	−5.76 (302.5)	278 (303)	8[i]	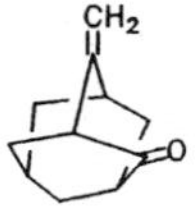	+2.55 (305) −10 (200)	110 (295) 7500 (200)
3[d,e]	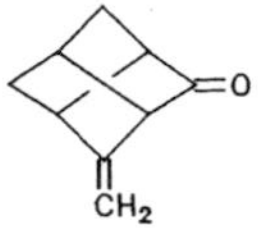	−5.15 (301) +29.7 (199.5)	445 (300.5)[e]	9[i]	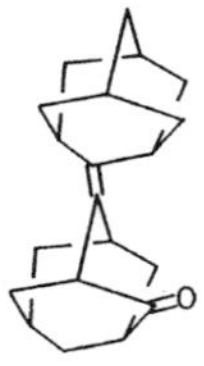	+2.58 (300) −14 (215)	401 (306.5) 20,000 (200)
4[d]	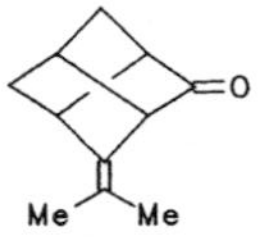	−4.15 (298.5) +9.88 (214)					
5[f]	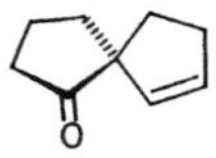	−3.14 (300)	87 (306)	10[j]	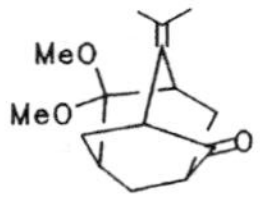	+3.25 (302)	380 (298)
6[g,h]	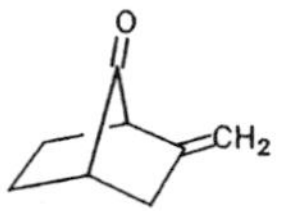	+2.0 (298) −6.4 (200) −5.7 (303)[h] +24 (207)[h]	84 (300) 135 (298) 8100 (210)	11[k]	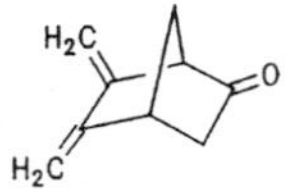	+0.59 (318) −1.20 (287.5)	470 (309)

[a] Enantiomer in isopentane, ref. 55. [b] In isooctane, ref. 57. [c] In isooctane, ref. 56a. [d] In isooctane, ref. 56b. [e] In isooctane, ref. 56c. [f] In isooctane, ref. 57a. [g] In isopentane, ref. 55b. [h] For the enantiomer in n-heptane, ref. 59. [i] In n-heptane, ref. 59a. [j] In ethanol, ref. 59a. [k] In isooctane, ref. 58a.

Table 11-10. Comparison of Reduced Rotatory Strengths for the $n \rightarrow \pi^*$ Transitions of (–)-5(*S*)-6-Methylenespiro[4,4]nonan-1-one and Its Methyl Analogs.[a]

T (°C)	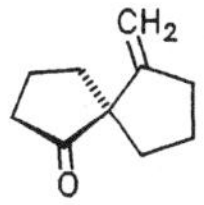	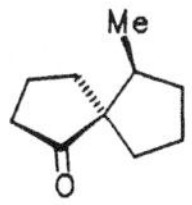	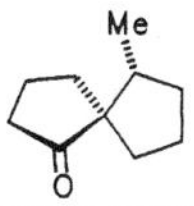
+25°	−2.93	−1.01	−1.17
−25°	−2.28	−1.25	−1.41
−125°	—	−1.97	−1.75
−160°	−1.81	−2.67	−1.75

[a] Data in isopentane-methylcyclohexane (5:1 by vol) from ref. 55b.

than that of a saturated ketone analog (Table 11-10).[55b] It was thought that the generalized octant rule was not applicable to the endocyclic enone, which showed dissignate behavior,[57] and a new symmetry rule was proposed to account for this discrepancy.[57b] Lightner et al.[55b] pointed to the weak $n \rightarrow \pi^*$ Cotton effect of (+)-1(*R*)-2-methylenebicyclo[2.2.1]heptan-7-one (Table 11-9, #6), noting that the Cotton effect magnitude was comparable to that of the saturated analog of the compound.

Subsequently, Sonney and Vogel found that the dienone (Table 11-9, #11) gave only a very weak $n \rightarrow \pi^*$ Cotton effect and not the sign predicted by the generalized octant rule; yet the UV spectrum showed a strongly enhanced carbonyl absorption.[58a] This ketone, too, has the relative geometry of Figure 11-18C, but the lack of agreement between its Cotton effect sign and stereochemistry remains unexplained. The dienone and other ketones of Table 11-11, were shown to exhibit a second broad band, suggesting a "delocalized" charge transfer from a diene π-orbital to the carbonyl π^*, as predicted by CNDO/S molecular orbital calculations.[58b] However, it is currently unclear what stereochemical information can be extracted from the suggested charge-transfer band, and no attempts have been made to extract any. Later, Vogel et al.[58c]

Table 11-11. Experimental and Calculated (CNDO/S) Transitions (λ, nm) for Seven β,γ-Unsaturated Ketones.[a]

	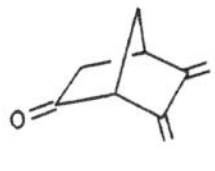	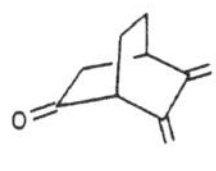		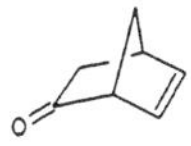	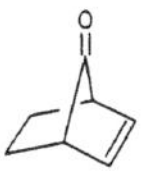	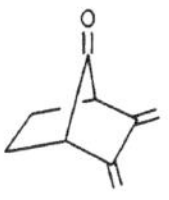	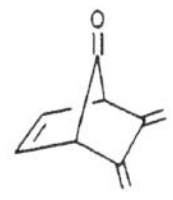
$n \rightarrow \pi^*$ (C=O)	318 (276)	320 (293)	295 (267)	306 (265)	275 (253)	(271)	(285)
C-T band	288 (238)	280 (237)	213 (189)	<200 (198)	233 (209)	(242)	(235)
$\pi \rightarrow \pi^*$ (C=C)	248 (196)	253 (203)	203 (179)	214 (180)	(178)	(197)	(211)

[a] Data from ref. 58b; C-T charge-transfer. Calculated values are in parentheses.

found that a single chlorine or bromine on the carbon–carbon double bond of 2-norbornenone causes the $n \rightarrow \pi^*$ transition to increase in magnitude, as seen in both the UV and CD spectra.

Schippers and Dekkers[59a] at the University of Leiden proposed a quantitative chirality rule for the $n \rightarrow \pi^*$ transition of β,γ-unsaturated ketones and applied the rule to all of the enones of Tables 11-8 and 11-9, incorporating new data for the enantiomer of #6 of Table 11-9 and data for new enones.[59b] For the enone $O{=}C_1{-}C_2{-}C_3{=}C_4$ system, two angles are defined: ξ, the angle of intersection of axes drawn along and through the O=C and $C_3{=}C_4$ bonds, and θ, the angle between the electric and magnetic transition moments of the erstwhile $n \rightarrow \pi^*$ transition (Figure 11-23). Here, $\cos\theta = -(\text{sign } xy)\cos\xi$. The sign of the Cotton effect is predicted by the angle ξ: When $-90° < \xi < 90°$ (or when cos ξ is positive), a (+) Cotton effect is predicted; when $90° < \xi < 270°$ (or when cos ξ is negative), a (–) Cotton effect is predicted. The agreement between predicted and observed Cotton effect signs is very good (Table 11-12), although at present only one example has been reported that gives a *negative* value of cos ξ.

In what was (1989) then the most definitive theoretical analysis to date on the $n \rightarrow \pi^*$ transition of β,γ-unsaturated ketones, Bouman at Southern Illinois University and Hansen at the H.C. Ørsted Institute applied the *ab initio* Random Phase Approximation to 1(*S*),4(*S*)-bicyclo[2.2.1]hept-5-en-2-one (2-norbornenone, entry #1 of Table 11-12).[60a] The calculated transition energy (4.51 eV, ~275 nm) compares quite satisfactorily to the experimental value (4.1 eV, 302 nm). From the data of Table 11-13, the

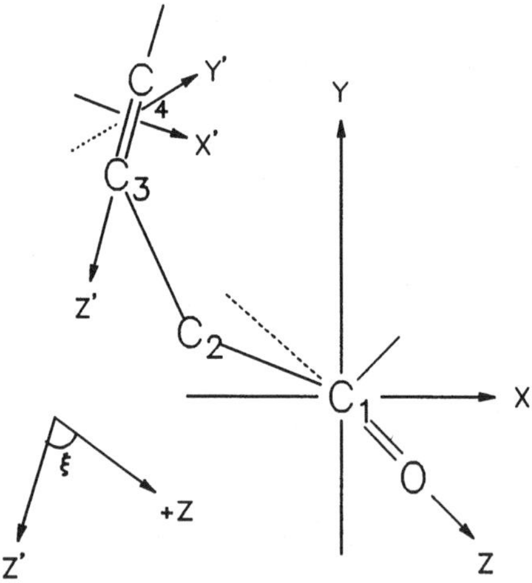

Figure 11-23. Schippers and Dekkers' reference frame for the carbonyl group (*x*, *y*, *z*) and ethylenic group (*x*′, *y*′, *z*′) in β,γ-unsaturated ketones (ref. 59a). The C_2 atom connecting the moieties lies in the *xz* plane; the *y*′ axis is perpendicular to the plane of the ethylenic group. ξ denotes the angle between the *z* and *z*′ axes. In this chirality rule, the sign of the $n \rightarrow \pi^*$ Cotton effect is predicted by the sign of cos ξ, and $\cos\theta = -(\text{sign } xy)\cdot(\cos\xi)$, where θ (not shown) is the angle between the electric and magnetic moments of the erstwhile $n \rightarrow \pi^*$ transition. For the C=O and C=C chromophores, θ is taken to be equal to ξ.

Table 11-12. Correlation between the Observed Sign of β,γ-Unsaturated Ketone $n \rightarrow \pi^*$ CD Cotton Effects and the Sign of cos ξ.[a] [ξ is the angle of intersection of the z and z' axes (Figure 11-23), which align along the C=O and C=C bonds, respectively. θ is the angle between the C=O axis and $\vec{r}$, the electric dipole moment associated with the $n \rightarrow \pi^*$ transition. cos θ = −(sign xy) cos ξ.]

Ketone	ξ(°)	cos ξ	Obs. Δε	θ(°)	cos θ
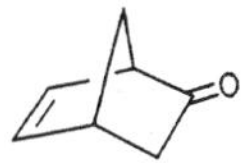	55	+0.57	+18.8	48	+0.67
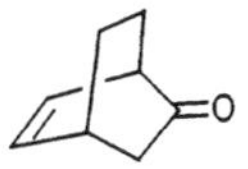	60	+0.50	+12.0	55	+0.57
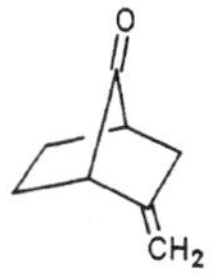	70	+0.34	+5.69	74	+0.28
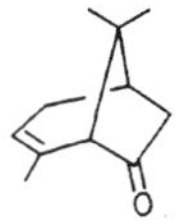	75	+0.26	+4.71	76	+0.24
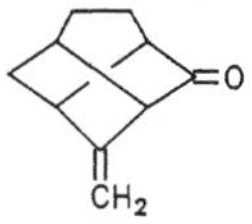	85	+0.09	+5.7	79	+0.19
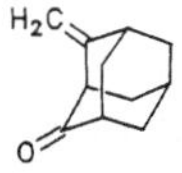	90	0	+2.55	83	+0.13
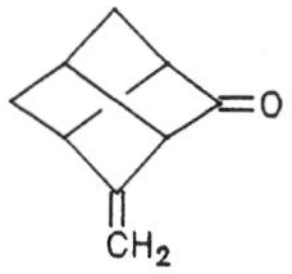	90	0	+5.4	82	+0.14
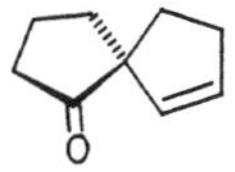	105	−0.26	−3.14	101	−0.19

[a] Data from ref. 59b.

Table 11-13. Compound Intensities and Leading Decomposition for the $n \rightarrow \pi^*$ Transition of 1(*S*),4(*S*)-Bicyclo[2.2.1]hept-5-en-2-one: UV Oscillator Strength *f* and CD Rotatory Strength (*R*). The Leading Contributions to the Total *R* are Expressed in Terms of μ–μ Components.[a]

	UV (*f*)	CD ($R/10^{40}$ cgs)		
		total	μ-m	μ-μ
expt.	0.006	−51		
∇^a	0.002	−30	−25	−6
r^a	0.003	−32		
$r\nabla^a$	0.002			
$n/n^{b,c}$	0.000	5.2	2.1	1.0
n/1-6		−11.6	−9.6	−1.9
n/1-7		23.6	19.4	4.1
n/5-6		−70.7	−58.8	−11.9
n/3-4		5.4	4.3	1.1
n/3-H		5.9	4.6	1.3
n/7-H		8.4	7.0	1.4
n/others		8.4	7.3	1.1
5-6/5-6	0.007	−1.5	−1.5	.0
5-6/others[d]	−0.006	−76.4	−60.7	−15.7

[a] ∇, r, and r∇ indicate velocity, length and mixed forms of the intensity expressions. [b] Bond-bond couplings $R_{\alpha\beta}{}^{q}$ and $f_{\alpha\beta}{}^{q}$. [c] *n* stands for the non-bonding oxygen orbital plus bonds 1-2 and 2-3 and these contributions are summed accordingly. See Figure 11-24 for atom numbering. [d] Includes the 5-6/*n* = *n*/5-6 coupling. Data from ref. 60a.

intensity of the $n \rightarrow \pi^*$ UV transition (oscillator strength *f*) is seen to come from a large (0.007) contribution from coupling to the locally excited (5–6/5–6) olefin chromophore, which is nearly counterbalanced (–0.006) by bond–bond couplings (5–6/others) to the rest of the molecule. The intensity of the Cotton effect, measured by its rotatory strength (*R*), is dominated by the μ–m contribution containing a C=O bond magnetic transition moment and a C=C bond electric transition moment. (See Chapter 4.) Although the μ–m mechanism dominates, μ–μ contributions are important and are seen to follow the μ–m contributions in sign and general magnitude—consistent with the trends noted in saturated ketones (Section 4.7).[61] In the β,γ-unsaturated ketone, however, the enhancement is due to the dominant *n*/5–6 coupling between the extended nonbonding orbital and the olefin double bond. The olefin chromophore is seen to couple almost exclusively to the nonbonding system, whereas the latter couples to a number of bonds, with signs and expected magnitudes corresponding to the predictions of the octant rule—which reduces the overall rotatory strength value

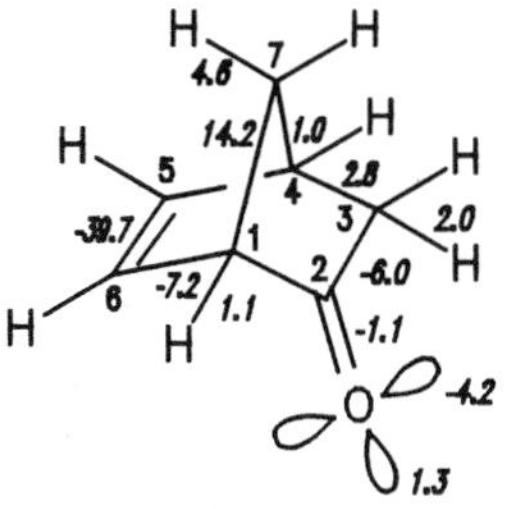

Figure 11-24. 1(*S*),4(*S*)-Bicyclo[2.2.1]hept-5-en-2-one (2-norbornenone) and its numbering system. Italicized numbers are the leading effective bond contributions to the $n \rightarrow \pi^*$ R, $\times 10^{-40}$ cgs. Bond contributions not shown sum to less than 1×10^{-40} cgs. Data from ref. 60a.

considerably. This analysis reinforces the previously stated qualifications concerning the dominance of the Cotton effect intensity from contributions intrinsic to the enone, compared with those from extrachromophoric perturbations.[35,38,55] Such asymmetry in the extrachromophoric rotatory strength couplings for the olefinic group and the nonbonding system is evident in the gross bond contribution terms of Figure 11-24. Note the octant-consignate zigzag behavior[9] of the C_1–C_7–H bonds.

The nature of the coupling between the C=O and C=C chromophores has been analyzed.[60a] The erstwhile carbonyl $n \rightarrow \pi^*$ transition is primarily $n \rightarrow \pi^*$, with 65% of the normalization of the excitation coming out of the oxygen nonbonding orbital into the local carbonyl π^* orbital and 25% coming from the extended nonbonding system (primarily the C_1–C_2 and C_2–C_3 bonds of the enone $O{=}C_1{-}C_2{-}C_3{=}C_4$) into the local carbonyl π^* orbital. Excitations out of the olefin π orbital amount to 5.6%, with 3.4% from a $\pi_{C=C} \rightarrow \pi^*_{C=O}$ charge-transfer excitation and 0.6% from the local $\pi_{C=C} \rightarrow \pi^*_{C=C}$ excitation. The earlier assumption that a local olefin $\pi_{C=C} \rightarrow \pi^*_{C=C}$ excitation was essentially the sole contributor to the total electric dipole transition moment is not supported. It would appear that the charge-transfer excitation plays an important role in twisting the transition moment away from the C=C direction, for the calculations show that neither the C=C electric dipole transition moment nor the total electric dipole transition moment for the excitation lies along the C=C bond direction. The bond moment forms an angle of 42° with the C=C bond; the total moment forms an angle of 36–50°. Charge transfer of the type $n_{C=O} \rightarrow \pi^*_{C=C}$ amounts to only 1.2% and is thus thought not to be significant. In terms of overall results, the approach of Bouman and Hansen[60a] agrees with the model proposed by Schippers and Dekkers.[59a] Areas of disagreement concern the direction of the electric dipole transition moment and the involvement of the extrachromophoric perturbers and do not support the assumptions behind the model.

11.3. γ,δ- and δ,ε-Unsaturated Ketones and Beyond

Interchromophoric interaction in nonconjugated systems involving remote chromophores was detected in 1958–1961 by Leonard and Owens[5a] at Illinois and Kosower et al.[5b] at Wisconsin in the δ,ε-unsaturated ketone *trans*-5-cyclodecenone (Table 11-1). Although the two chromophores lie some distance apart, as measured through bonds, they have a small nonbonded distance (~2.7 Å).[17] Apparently, "through-space" interaction of the chromophores leads to a bathochromically shifted and enhanced $n \rightarrow \pi^*$ carbonyl absorption band, as well as the appearance of a photodesmotic band near 215 nm,[5b] which might be assumed to be a charge-transfer transition from C=C to C=O. At about the same time, Cookson et al. reported on the enhancement of the carbonyl $n \rightarrow \pi^*$ transition and on what was suspected to be a charge-transfer band near 225 nm in the UV spectra of certain isomeric pentacyclic nonconjugated dienones (Table 11-14, top row) obtained by the reaction of norbornadiene with $Fe(CO)_5$.[62] In subsequent studies, UV evidence was presented for a modest enhancement of the carbonyl $n \rightarrow \pi^*$ absorption of γ,δ-unsaturated ketones (Table 11-14). Cookson et al. found evidence for a charge-transfer transition near 224 nm in polycyclic γ,δ-unsaturated ketones (Table 11-14, bottom), dihydro analogs of earlier work[62] (Table 11-14, top) that also exhibited an enhanced $n \rightarrow \pi^*$ absorption.[63] The $n \rightarrow \pi^*$ enhancement and the appearance of the charge-transfer transition seem to depend on the orientation of the carbonyl and olefin chromophores. At about the same time, evidence for very strongly enhanced $n \rightarrow \pi^*$ UV absorption was presented for gibberellic acid derivatives of γ,δ-unsaturated ketones in which the olefin group is conjugated with a phenyl and is thus part of a styrene chromophore (Table 11-14).[39b] Shortly thereafter, UV investigations of long-range charge-transfer homoconjugative interactions found in 17-keto steroids with Δ^2, Δ^5, or $\Delta^{9,11}$ olefin groups[43a] determined that the interactions were artifacts.[43c] Since the early 1960s, there have been relatively few investigations of long-range carbonyl olefin interactions in γ,δ-unsaturated ketones by UV spectroscopy (Table 11-14, middle).[64,65]

The first detection of homoconjugative effects in γ,δ-unsaturated ketones by chiroptical methods appears to be from Hudec et al.[66] who reported that both 5α-cholest-6-en-3-one and 5α-androst-6-en-3-one exhibited the usual weak $n \rightarrow \pi^*$ UV absorption with weak *negative* CD Cotton effects at approximately 296 nm (Figure 11-25). This finding was very striking, because the corresponding saturated ketones gave *positive* $n \rightarrow \pi^*$ CD Cotton effects near 296 nm, in accordance with the octant rule. The results indicated that the octant rule breaks down for certain γ,δ-unsaturated ketones, even though there is little evidence of homoconjugative interaction in the $n \rightarrow \pi^*$ UV intensity. There seems to be little distortion in the A-ring of the unsaturated steroid—only slight flattening (as judged by the ^{1}H–NMR chemical shift of the C(19) methyl).[66b] Apparently, the relative orientation of the carbonyl and olefin chromophores is very important; in contrast, the steroid and tricyclic γ,δ-unsaturated ketones of Table 11-15 all obey the octant rule (as do their saturated ketone analogs).[66b]

Other examples of bis-homoconjugation come from the 7-methylene-11-keto steroid of Table 11-15 (row 1), which has the same relative disposition of olefin and carbonyl chromophores and also exhibits a negative $n \rightarrow \pi^*$ Cotton effect, whereas the saturated analog has a (weaker) positive Cotton effect. The tricyclic γ,δ-unsaturated ketone (Table 11-15, row 1) has the same disposition of the two chromophores and exhibits a negative Cotton effect, but so does its saturated analog, suggesting that, in

Table 11-14. Influence of Remote Homoconjugation on the $n \rightarrow \pi^*$ UV Band of γ,δ-Unsaturated Ketones

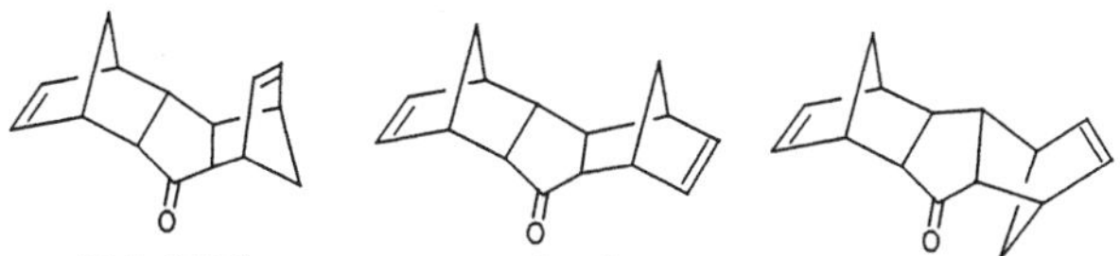

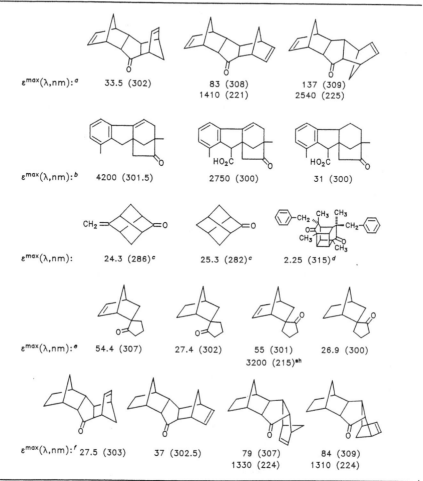

$\varepsilon^{max}(\lambda,nm)$:[a]	33.5 (302)	83 (308) 1410 (221)	137 (309) 2540 (225)

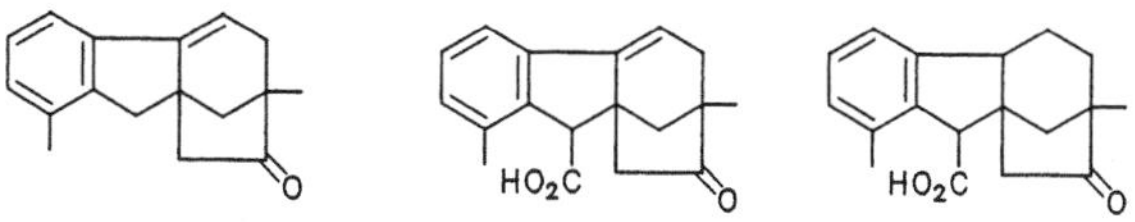

$\varepsilon^{max}(\lambda,nm)$:[b]	4200 (301.5)	2750 (300)	31 (300)

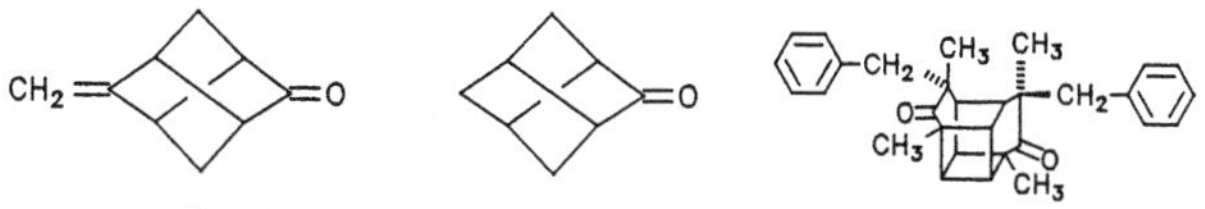

$\varepsilon^{max}(\lambda,nm)$:	24.3 (286)[c]	25.3 (282)[c]	2.25 (315)[d]

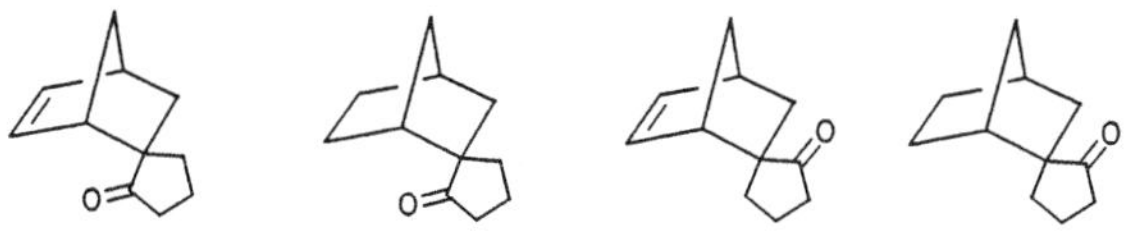

$\varepsilon^{max}(\lambda,nm)$:[e]	54.4 (307)	27.4 (302)	55 (301) 3200 (215)[sh]	26.9 (300)

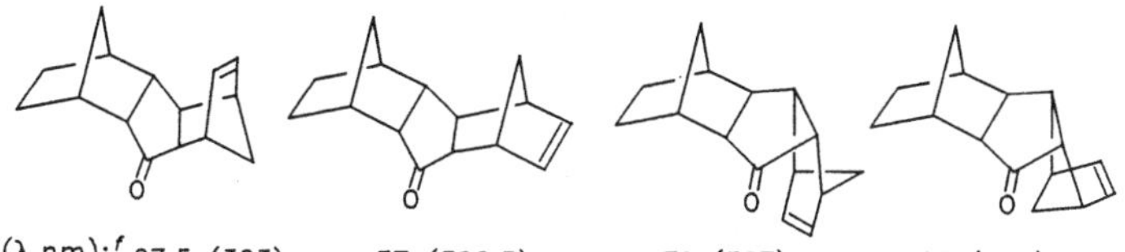

$\varepsilon^{max}(\lambda,nm)$:[f]	27.5 (303)	37 (302.5)	79 (307) 1330 (224)	84 (309) 1310 (224)

[a] In ethanol, ref. 62. [b] In ethanol, ref. 39b. [c] In isooctane, ref. 56b. [d] In dioxane, ref. 65. [e] In cyclohexane, ref. 64. [f] In ethanol, ref. 63.

terms of the octant rule, the C=C bond gives a strong negative antioctant contribution. An antioctant contribution is seen again in the 6-methylene-3-keto steroid of row 3 of Table 11-15, but in the remaining steroid γ,δ-unsaturated ketones listed in the table, the C=C bond appears to make an unexpectedly strong octant contribution. Of course, the relative orientation of the carbonyl and olefin chromophores differs substantially from that of the 7-methylene-11-keto steroid and the Δ^6-3-keto steroid of Figure 11-25. Just how the remote C=C bond influences the carbonyl $n \rightarrow \pi^*$ CD transition is at present unclear and awaits a detailed theoretical treatment. Note that the C=C bond

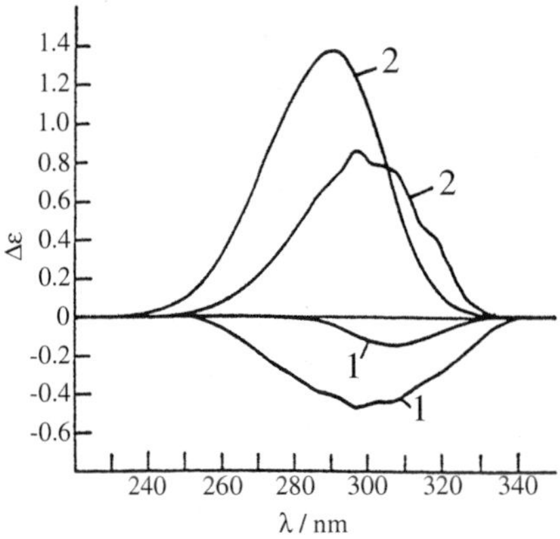

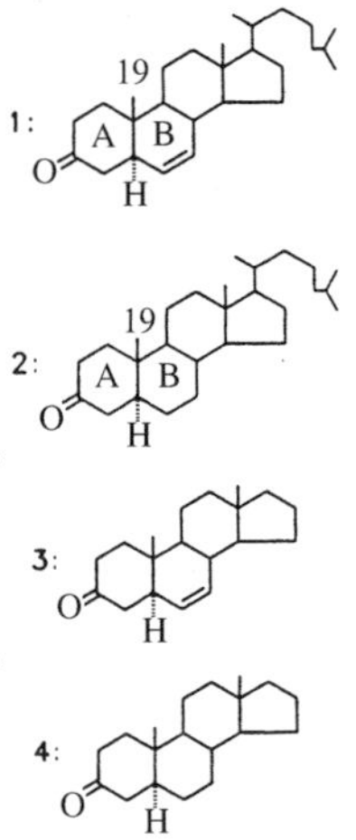

	$\Delta\varepsilon^{max}$ (λ,nm)	ε^{max} (λ,nm)
1:	−0.47 (297)	21 (287)
2:	+0.84 (297)	17 (288)
3:	−0.41 (296)	23 (287)
4:	+0.83 (296)	18 (289)

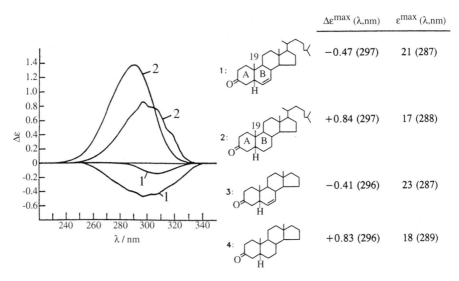

Figure 11-25. CD and UV spectra and data for 5α-cholest-6-en-3-one (**1**) and 5α-cholestan-3-one (**2**), and data for 5α-androst-6-en-3-one (**3**) and 5α-androstan-3-one (**4**). (Left) The CD readings from top to bottom are **2** in ethanol, **2** in *n*-hexane, **1** in ethanol, and **1** in *n*-hexane. (Right) All of the numerical data to the right ($\Delta\varepsilon^{max}$ and ε^{max}) are in *n*-hexane. [Spectra reprinted from ref. 66b with permission of the Royal Society of Chemistry.]

in the smallest γ,δ-unsaturated ketone, 5-methylene-2-norbornanone, scarcely alters the $n \rightarrow \pi^*$ CD Cotton effect seen of the parent, 2-norbornanone (Table 11-15, last row).[67]

More recently, CD and UV spectra of the epimeric γ,δ-unsaturated ketones, α-benzyl derivatives of camphor, have been studied and found to contrast with those of the corresponding β,γ-unsaturated α-phenyl analogs (Table 11-16).[68a] The UV spectra of the benzyl ketones show unexpectedly intense $n \rightarrow \pi^*$ transitions and relatively unshifted band centers compared with the UV spectrum of camphor. The $n \rightarrow \pi^*$ UV transitions of the former are more intense than normal carbonyl absorptions, and the UV intensities of the *endo* epimer approach the values (Tables 11-8 and 11-9) of β,γ-unsaturated ketones in which strong interchromophoric interaction has been noted. In the orbital formalism, the markedly enhanced $n \rightarrow \pi^*$ UV transitions provide evidence for a mixing in of the aromatic $\pi \rightarrow \pi^*$ transition with the C=O $n \rightarrow \pi^*$ transition or for a coupling of the locally excited states of each component of the extended chromophore. The effectiveness of the coupling is expected to be highly dependent on geometry, as has been found in other γ,δ-unsaturated ketones (Table 11-14). With unfavorable orientations of the chromophores, no enhanced ε for the $n \rightarrow \pi^*$ transition is observed. As seen in Dreiding models, a conformation is apparently available for the benzyl ketones in which the C=O and aromatic ring transition dipoles line up in a parallel orientation. Unlike the benzylcamphors, however, the

Table 11-15. CD and UV Data for the $n \rightarrow \pi^*$ Transition of γ,δ-Unsaturated Ketones

	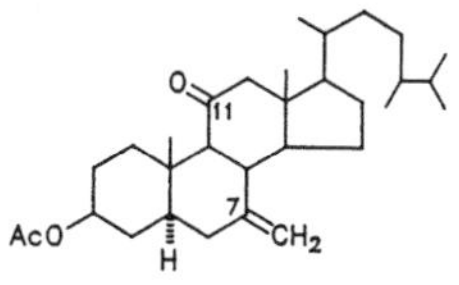	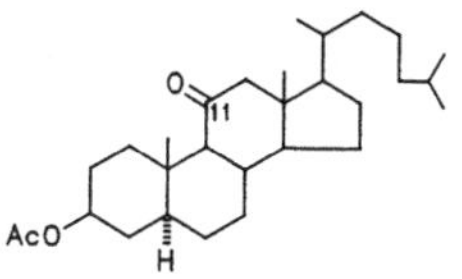	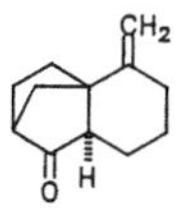	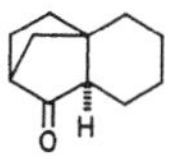
$\Delta\varepsilon^{max}(\lambda, nm)$:[a]	−1.87 (302)	+0.26 (299)	−3.14 (296)	−1.66 (294)
$\varepsilon^{max}(\lambda, nm)$:[a]	46 (300)	38 (297)	48 (295)	28 (293)
	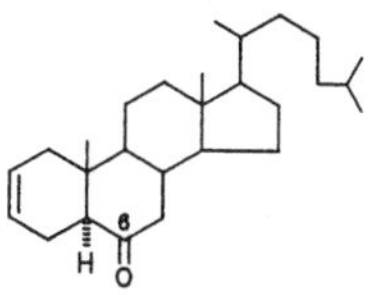	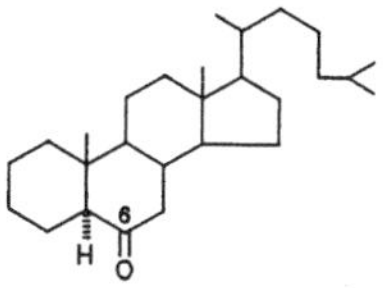	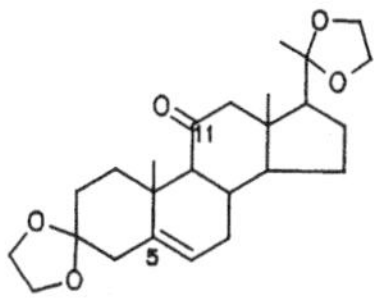	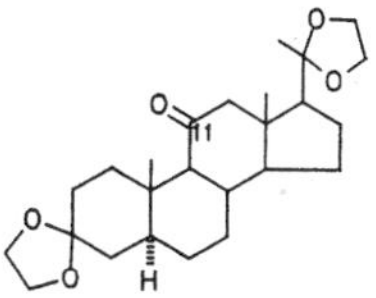
$\Delta\varepsilon^{max}(\lambda, nm)$:	−2.72 (293)[a]	−1.55 (293)[a]	+1.50[b]	+0.24[b]
$\varepsilon^{max}(\lambda, nm)$:	46 (289)[a]	39 (290)[a]		
	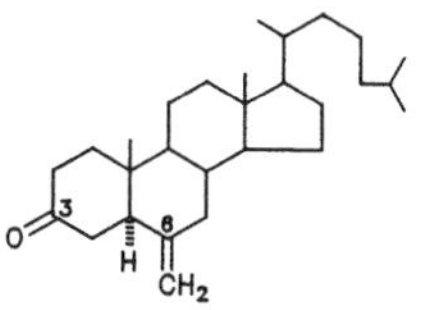	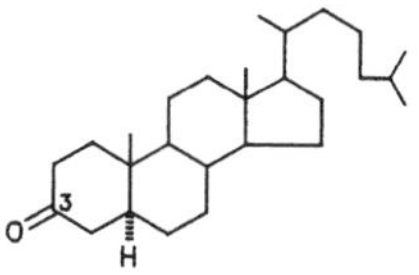	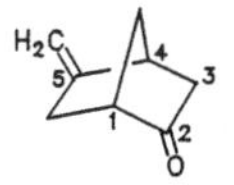	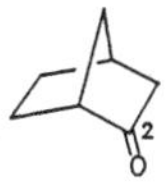
$\Delta\varepsilon^{max}(\lambda, nm)$:	+0.28 (296)[b]	+0.84 (297)[b]	+0.89 (308)[c]	+0.77 (307)[c]
$\varepsilon^{max}(\lambda, nm)$:	16 (286)[b]	17 (288)[b]	23 (294)[c]	26 (293)[c]

[a] In ethanol, ref. 66b. [b] In *n*-hexane, ref. 66b. [c] In isooctane, ref. 67.

β,γ-unsaturated exo- and endo-phenyl epicamphors show much less intense $n \rightarrow \pi^*$ transitions, with ε values not much larger than that of the parent ketone.

The CD spectra of α-benzyl ketones indicate a powerful influence of the benzyl group. From the enhanced UV $n \rightarrow \pi^*$ transitions, one should expect a contribution to the Cotton effect coming from orbital interaction as well as from octant contributions. The *exo*-benzyl group is expected to lie in a front octant, adding its positive front-octant contribution to the positive back-octant contributions inherent in the camphor skeleton. If these were the only contributions to the Cotton effect, *exo*-benzylcamphor would be expected to have a $\Delta\varepsilon$ greater than +1.8, not the extremely weak negative value observed ($\Delta\varepsilon \sim -0.4$). The dominating influence of the benzyl group is clearly evident through the mechanism of an inherently dissymmetric *extended* chromophore—similar, apparently, to that seen in 5α-cholest-6-en-3-one and other γ,δ-unsaturated steroid ketones of Table 11-15. The octant contribution of the *endo*-benzyl group depends considerably on its orientation. If the group is in a lower right or upper left back octant, it will make a positive contribution to the $n \rightarrow \pi^*$ CD Cotton effect. If the phenyl ring juts into a front octant, however, as seems probable from models, the contribution could be weakly negative or weakly positive. Considering the very large $n \rightarrow \pi^*$ $\Delta\varepsilon$ values

Table 11-16. Comparison of CD and UV Data for γ,δ-Unsaturated Ketones (Epimeric α-Benzylcamphors) and β,γ-Unsaturated Ketones (α-Phenylepicamphors) in *n*-Heptane[a]

Ketone	$\Delta\epsilon^{max}(\lambda, nm)$	$\epsilon^{max}(\lambda, nm)$	Ketone	$\Delta\epsilon^{max}(\lambda, nm)$	$\epsilon^{max}(\lambda, nm)$
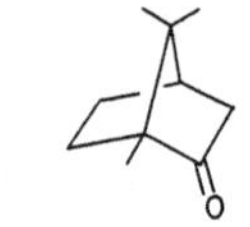	+1.85 (302)	27 (290)	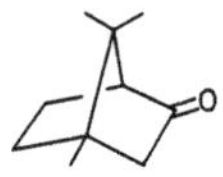	−1.45 (307)	26 (294)
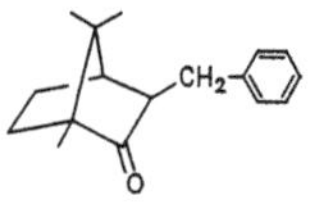	+0.030(324) −0.247(312) −0.362(301) −0.298(293)	82 (298)^{sh} 84 (291)	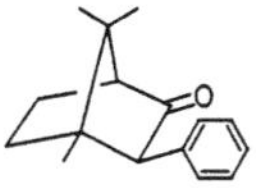	−1.39 (299)	34 (294)
	−0.238(267) −0.192(262) −0.102(255)	159 (268) 218 (259) 182 (254)			178 (266) 224 (259) 190 (244)
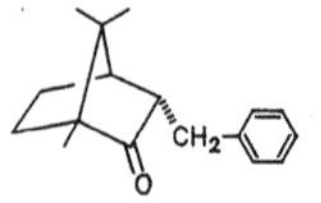	+2.27(313)^{sh} +3.32(302) +2.81(295)	303 (285)	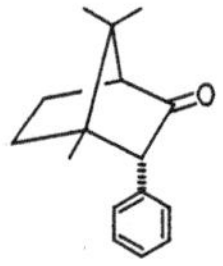	−1.20 (322) −1.46 (310) −1.06 (300)	23 (321) 43 (310) 46 (300)
	+0.73(268)^{sh} +0.41(261)^{sh}	335 (268) 318 (264) 318 (261) 317 (259) 246(253)^{sh}		−0.35 (267) −0.41 (260) −0.30 (253)	153 (264) 198 (258) 155 (252) 113 (248)

[a] Data from ref. 68a.

for the *endo*-benzyl epimer, it appears likely that the normal octant contributions of the camphor system are being augmented by the sort of homoconjugation effects described for the *exo*-benzyl group and noted for the ketones of Table 11-15. In contrast, neither the *exo* nor the *endo* α-phenyl group shows such a profound influence on the CD. Here, however, as with the des-methyl analogs (Figure 11-21),[50] a weaker transition moment coupling due to an unfavorable interchromophoric geometry has a far smaller influence on the $n \rightarrow \pi^*$ CD $\Delta\varepsilon$ values.

The epimeric γ,δ-unsaturated 4-phenyladamantan-2-ones were investigated and compared with the corresponding δ,ε-unsaturated 4-benzyladamantan-2-ones.[68b] Although the UV and CD $n \rightarrow \pi^*$ transitions of the phenyl ketones are unexceptional, the UV spectra of the benzyl ketones show moderately strong enhancement of the carbonyl $n \rightarrow \pi^*$ and aromatic $\pi \rightarrow \pi^*$ transitions (Figure 11-26). There are no special enhancements of the $n \rightarrow \pi^*$ Cotton effects. Variable low-temperature CD measurements show the same effect in both axial ketones, viz., an approximately 40%

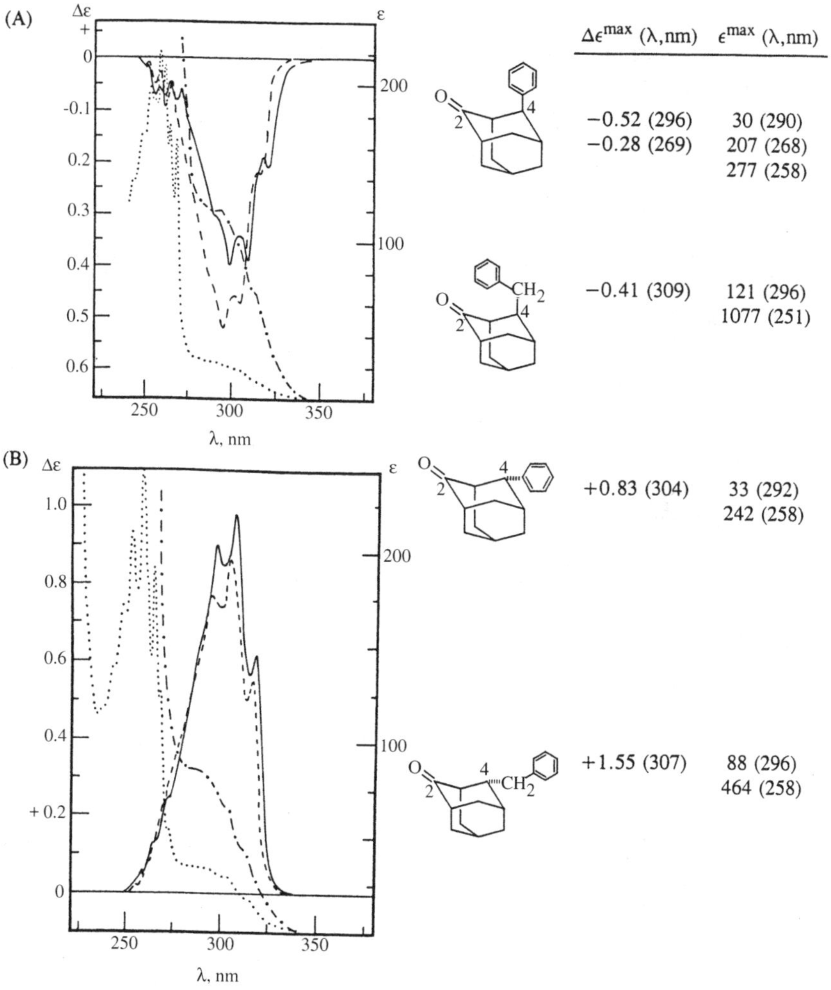

Figure 11-26. (A) CD and UV spectra of δ,ε-unsaturated and γ,δ-unsaturated 4-axial-adamantan-2-ones: phenyl (CD – – –) and (UV ·····), and benzyl (CD —) and (UV – · – · –). (B) CD and UV spectra of 4-equatorial-adamantan-2-ones: phenyl (CD – – –) and (UV ·····), and benzyl (CD —) and (UV – · – · – ·). [Spectra were obtained in methylcyclohexane–isopentane (4:1 by vol.) and are reprinted with permission from ref. 68b. Copyright© 1987 American Chemical Society.]

increase in rotatory strengths in going from +25°C to −175°C, whereas in the equatorial phenyl, the rotatory strength of the phenyladamantanone decreases by about 50% and that of the benzyladamantanone increases by approximately 50%, in going from +25°C to −150°C.[68b] The unusual exaltations of the $n \rightarrow \pi^*$ UV and CD transitions of the δ,ε-unsaturated ketones are thought to be due to "through-space" interactions.[68b] One other δ,ε-unsaturated ketone, 7-methylene-5α-

Table 11-17. Comparison of CD Intensity of a δ,ε-Unsaturated Steroid Ketone with That of Its Saturated Parent Ketone[a]

	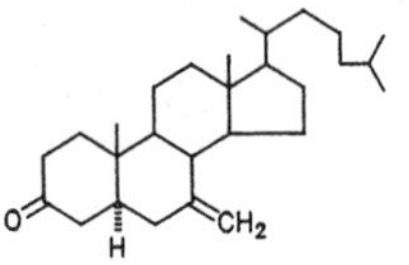	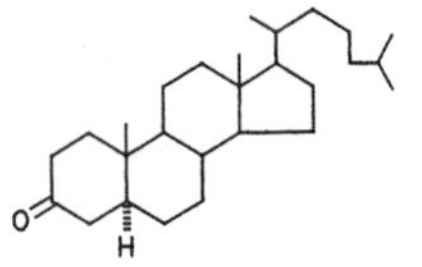
$\Delta\epsilon^{max}$ (λ, nm)	+1.26 (296)	+0.84 (297)

[a] Data from ref. 66.

cholestan-3-one (Table 11-17), has been shown to exhibit a larger $n \rightarrow \pi^*$ Cotton effect than its parent, and this has been taken as evidence of a long-range electronic interaction.[66]

An even longer range homoconjugation interaction was reportedly detected by CD spectroscopy in the δ,ε-unsaturated ketone of Table 11-18.[69] At first, the unsaturated ketone was reported to have a larger negative $n \rightarrow \pi^*$ Cotton effect,[69b] but this was corrected later.[69a] Still, the ε,ζ olefinic double bonds apparently reverse the normal octant rule Cotton effect contributions seen in the saturated analog.

Most recently, the influence of δ,ε-unsaturation (Table 11-19, #1,2,3), and even more distant ζ,η-unsaturation (Table 11-19, #4,5,6), on the ketone carbonyl absorption may be seen from comparison with UV and CD data on their saturated analogs (dihydro-**1** through **6** of Table 11-19).[70] Evidence for long-range interchromophoric interaction may be seen in the significantly more intense $n \rightarrow \pi^*$ UV intensities (Table 11-19, #2, 3, 5, 6) of the unsaturated ketones, compared with their dihydro analogs. The exaltations, which are found for surprisingly large interchromophoric distances, are comparable to those seen in β,γ-unsaturated ketones (Tables 11-8 and 11-9), in which such interactions have been well studied. In contrast, the exaltation is not noticeable in the δ,ε-unsaturated ketone with the smallest interchromophoric distance (Table 11-19, #1). In the $n \rightarrow \pi^*$ CD Cotton effects, however, enhancements are seen only for the δ,ε-unsaturated ketones (Table 11-19, #1,2,3) and seem to fall off with distance (Table 11-19, #4, 5, 6).

In addition to UV exaltations for the $n \rightarrow \pi^*$ UV and CD absorptions, a new band appears near 220–230 nm in the CD spectra of the unsaturated ketones of Table 11-19. The band is seen in the δ,ε-unsaturated ketone, whose chromophores are ideally oriented for σ-*pp* orbital overlap (Table 11-19, #1) over a short (2.9 Å) distance, but it is not seen in the compound's dihydro analog. The new transition persists in the norbornylog (Table 11-19, #4), in which a moderately intense transition is seen at 221 nm—a transition that is absent in the dihydro analog. This new band, appearing in an unusual spectral region (220–230 nm), bears a strong similarity to that seen in a chiral 7-norbornenone[41] and likewise probably contains a $\pi_{C=O} \rightarrow \pi^*_{C=O}$ charge-transfer component. Keeping the same orbital alignment but increasing the distance between

Table 11-18. Comparison of CD and UV Data Pertaining to a Polycyclic ε,ζ-Unsaturated Ketone and its Saturated Analog[a]

		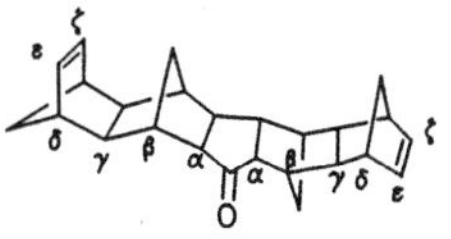	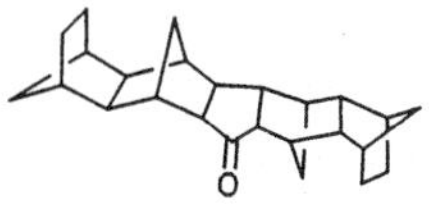
UV	ϵ^{max} (λ,nm)	45.8 (312)	49 (312)
CD	$\Delta\epsilon^{max}$ (λ,nm)	− 8.17 (323)[b]	+6.39 (324)

[a] In cyclohexane; data from ref. 69a. [b] Reported as −30.3 in ref. 69b.

the C=C and C=O chromophores, decreases the intensity of the band. However, when the relative orientation is altered to disfavor σ-*pp* overlap (as in #3 and 6 of Table 11-19), the approximately 230-nm charge-transfer band is absent and is replaced by a band near 206 nm.

For an interchromophoric geometry lying somewhere between that found in #1 and that found in #3 of Table 11-19, #2 does not share the same favorable orientation for σ-*pp* overlap found in #1, but still exhibits a very intense band near 218 nm, again probably due to a charge-transfer component. Increasing the distance between the chromophores drastically weakens the 218-nm band and even changes its sign (Table 11-19, #5). The intense band near 206 nm in #3 finds a counterpart in its norbornylog (#6) and also in the 205-nm band of 2-norbornenone of the same absolute configuration (Table 11-19).

These observations, related to excited states, are consistent with the conclusions drawn from an analysis of ^{13}C=O chemical shifts for the ground state, viz., that orbital interaction through space is an important component of transannular orbital interaction in #1 and probably #2, but is less important in #3, in which orbital interaction through bonds may dominate. On the basis of CD charge-transfer bands, one might also conclude that orbital interaction through space is more important in #4 than in #5 or #6.

Homoconjugation thus appears to operate over surprisingly long distances, as detected by UV and CD spectroscopy. The influence of the distance between, and the relative orientation of, ketone carbonyl and olefin chromophores can be detected by the intensity of the $n \rightarrow \pi^*$ transitions and also by the presence of shorter wavelength transitions near 220–230 nm. The latter are probably of mixed origin, containing a $\pi_{C=C} \rightarrow \pi^*_{C=O}$ charge transfer component. Long-range transannular molecular orbital interaction in the ground state may be found in the relatively more shielded ^{13}C=O NMR resonances of δ,ε- and ζ,η-unsaturated ketones, compared with their saturated analogs.[70]

Table 11-19. Influence of Remote Homoconjugation (δ,ε- and ζ,η-Unsaturated) on the UV and CD Spectral Data and on the ^{13}C–NMR Carbonyl Carbon Chemical Shift of Dimethanonaphthalene and Trimethanoanthracene Ketones[a]

	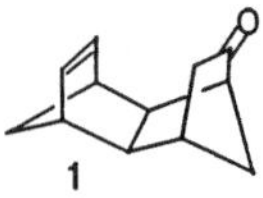 1	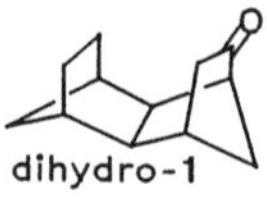 dihydro-1	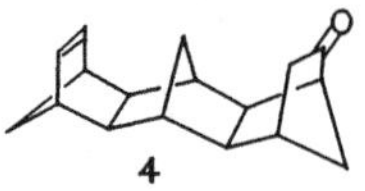 4	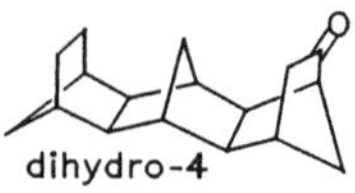 dihydro-4
ϵ^{max} (λ,nm)[b]	30 (277) 321 (227)^{sh}	11 (280)	40 (281) 840 (217)^{sh}	41 (279)
$\Delta\epsilon^{max}$ (λ,nm)[b]	−3.3 (288) +10 (233)	−0.83 (280)	< −0.01 (284) +2.8 (221)	< −0.01 (284)
$\delta_{C=O}$[c]	217.1	219.6	217.6	217.8
	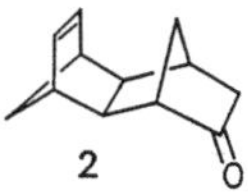 2	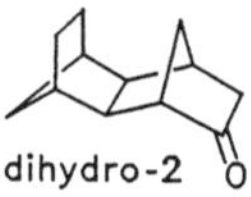 dihydro-2	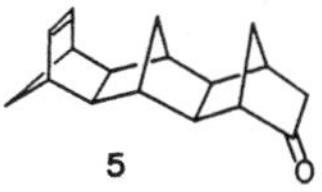 5	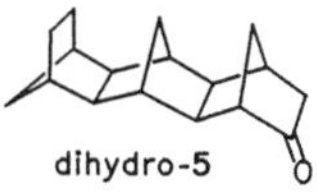 dihydro-5
ϵ^{max} (λ,nm)[b]	118 (283)	71 (281)	296 (284) 1265 (219)^{sh}	119 (285)
$\Delta\epsilon^{max}$ (λ,nm)[b]	−7.5 (284) +21 (218)	−4.7 (281)	−6.7 (285) −0.68 (215)	−6.8 (284)
$\delta_{C=O}$[c]	216.4	217.2	214.8	216.0
	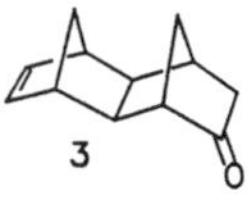 3	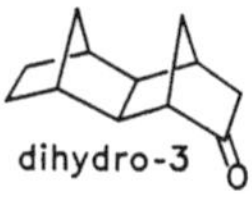 dihydro-3	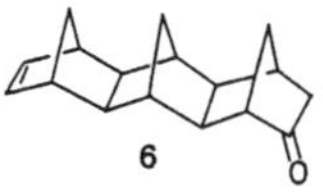 6	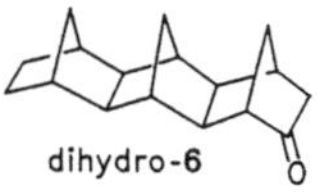 dihydro-6
ϵ^{max} (λ,nm)[b]	150 (283)	85 (282)	154 (279)	165 (279)
$\Delta\epsilon^{max}$ (λ,nm)[b]	−8.9 (283) −16 (206)	−5.8 (282)	−8.5 (284) −4.0 (206)	−8.0 (284)
$\delta_{C=O}$[c]	215.6	216.2	215.5	216.0
	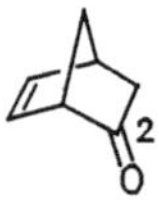	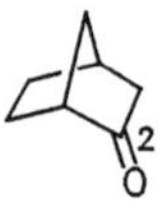	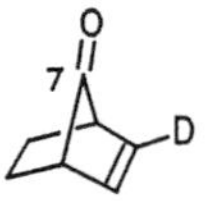	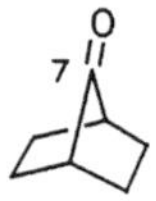
ϵ^{max} (λ,nm)[b]	300 (294) 2000 (214)^{sh}	23 (293)	35 (272) 450 (220)^{sh}	18 (293)
$\Delta\epsilon^{max}$ (λ,nm)[b]	−19 (294) +4.4 (224) −5.4 (205)	+0.76 (304)	−0.033 (273) −0.028 (224) +0.029 (219)	0
$\delta_{C=O}$[c]	212.9	215.0	205.1	216.2

[a] Data from ref. 70. [b] In CF_3CH_2OH, 2×10^{-3} M, at 22°C. CD values corrected to 100% ee.
[c] In $CDCl_3$, 2×10^{-2} M, at 22°C.

11.4. Cyclopropyl Ketones

Djerassi et al.[71] first noted that the ORD curves of ketones were strongly perturbed by conjugation with α,β-cyclopropane rings. Subsequently, Norin[72] observed that the octant rule failed to correctly assign the cyclopropyl configuration in thujane ketones. This discovery and that of Legrand et al.[73] that certain steroid α,β-epoxy ketones did not follow the octant rule led to further investigations of the ORD and CD spectra of α,β-cyclopropyl and α,β-epoxy ketones.[74,75,76] Djerassi et al.[74] and Snatzke and Schaffner[75] applied the octant rule to ORD data obtained from a large number of α,β-cyclopropyl ketones and α,β-epoxy ketones and concluded that cyclopropyl rings adjacent to the carbonyl group make Cotton effect contributions *opposite* to those made by ordinary alkyl perturbers—and these oppositely directed contributions often control the net Cotton effect sign (as in Figure 11-27). An "inverse" or "reversed" octant rule was thus proposed to account for this unusual behavior.

Among the many cyclopropyl ketones examined, two general α,β-conjugated types were noted—one with the cyclopropane as part of an [n.1.0]bicyclic unit (Figure 11-28A), the other with a spiro-cyclopropane[77] (Figure 11-28B). In the former, the cyclopropane is thought to make a dominant "inverse" octant contribution to the $n \rightarrow \pi^*$ Cotton effect. In the latter, the effect is thought to be very weak. Only a few exceptions to the reversed octant rule have been noted: the 3,5-cyclosteroid 6-ketones of Figure 11-28B, but not the 5α,7α-cyclosteroid 4-ketone; and (+)-carone of Figure 11-28A, which has a *gem*-dimethyl on the cyclopropane.[74] The unexpected behavior of (+)-carone may be attributed to the influence of the *gem*-dimethyl group, since the

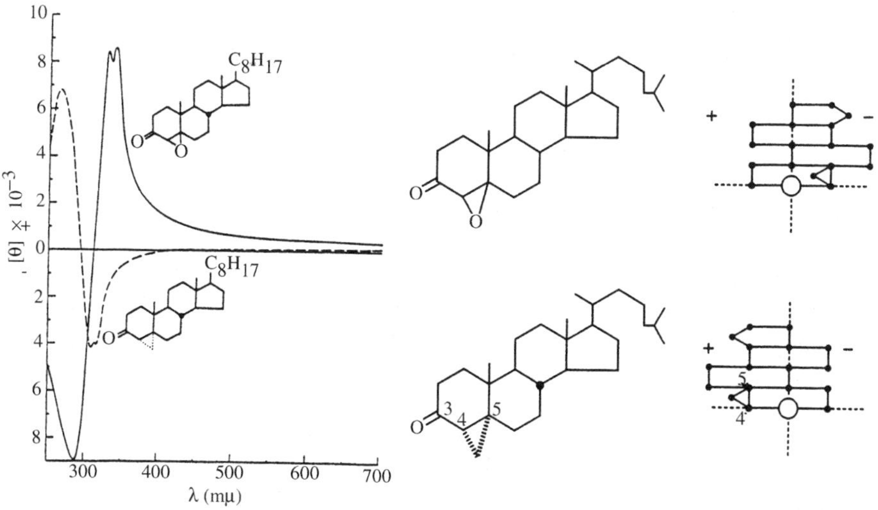

Figure 11-27. ORD Cotton effects for the $n \rightarrow \pi^*$ carbonyl transitions of 4α,5α-cyclopropylcholestan-3-one (– – –) and 4β,5β-epoxycholestan-3-one (—). [Spectra reprinted from ref. 74, copyright© 1965, with permission from Elsevier Science.]

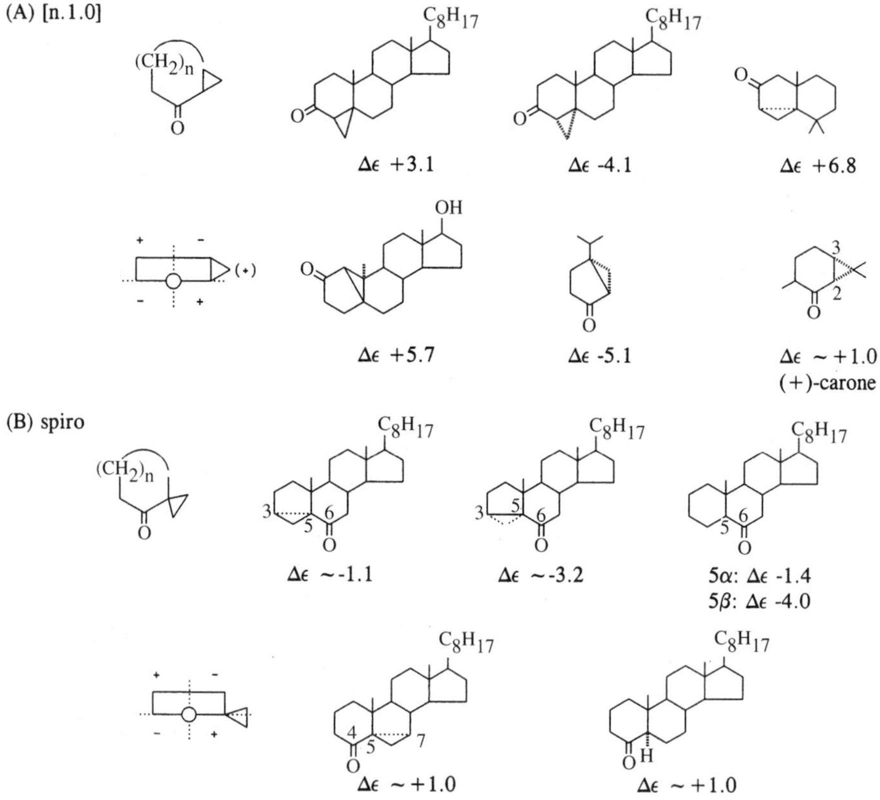

Figure 11-28. Examples of α,β-cyclopropyl ketones and their $n \rightarrow \pi^*$ Cotton effect intensities for (A) the [n.1.0]bicyclic ketone and (B) the spiro ketone. In the former, the cyclopropane ring is thought to make a dominant "inverse" octant contribution. In the latter, the cyclopropane is nearly bisected by an octant symmetry plane. Data from ref. 74.

parent ketone, (–)-2(*S*),3(*R*)-bicyclo[4.1.0]heptan-1-one, of the same absolute configuration, exhibits a negative $n \rightarrow \pi^*$ Cotton effect,[76] as predicted by the "inverse" octant rule (Table 11-20). In like accord, (–)-2(*S*),3(*R*)-oxidocyclohexanone (Table 11-20) gives a negative $n \rightarrow \pi^*$ Cotton effect.[78] The *gem*-dimethyl group apparently alters the cyclohexanone conformation. In contrast, the presence of the *gem*-dimethyl group in the more rigid 1(*S*)-6,6-dimethyl[3.1.0]hexan-2-one gives the positive $n \rightarrow \pi^*$ Cotton effect predicted by the "inverse" octant rule and seen in the parent α,β-cyclopropylketone, 1(*R*)-bicyclo[3.1.0]hexan-2-one (Figure 11-29).[79] However, the failure of the cyclopropylsteroid ketones of Figure 11-28B to obey the "inverse" octant rule is also found in two much simpler cases: (*i*) (–)-6-methylspiro[2.5]octan-4-one (Table 11-20), whose positive $n \rightarrow \pi^*$ Cotton effect appears to be dominated by the methyl group contribution, and (*ii*) the structurally rigid cyclocamphor, which exhibits a positive $n \rightarrow \pi^*$ Cotton effect of intensity comparable to that of the parent camphor.

Table 11-20. CD Data for α,β-Cyclopropyl and α,β-Epoxy Cyclohexanones[a]

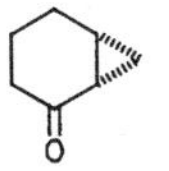	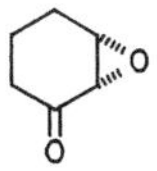	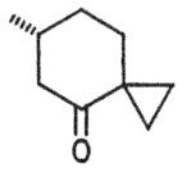	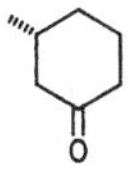	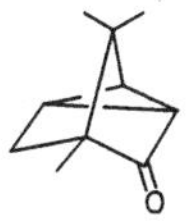	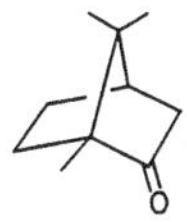
-0.61 (313)[b]	-1.88 (317)[c]	+0.3[d]	+0.6[d]	+2.08 (283)[e] 30 (282)	+1.54 (296)[e] 32 (291)

[a] Data are Δε (λ,nm). [b] In ref. 76b only the Cotton effect *sign* was reported. The reported Δε (λ, nm) in *n*-heptane is from Dr. P. Chayangkoon. [c] Ref. 78. [d] Ref. 77. [e] In ethanol, ref. 47c.

Taken collectively, the data illustrate the sensitivity of CD spectroscopy to the relative orientation of the carbonyl and cyclopropyl chromophores of α,β-cyclopropyl ketones.

The dependence on orientation of the "inverse" octant rule, as well as the rule's applicability, was explored by Tocanne,[80] using both CD and ^{1}H–NMR with many conformationally mobile α,β-cyclopropyl acyclic ketones. Tocanne proposed a new revised octant rule that retains the two local symmetry planes of the carbonyl group and assumes a third nodal surface, passing through the middle of the carbonyl bond and sharply convex toward the C=O oxygen bond (Figure 11-30). Variable low-tem-

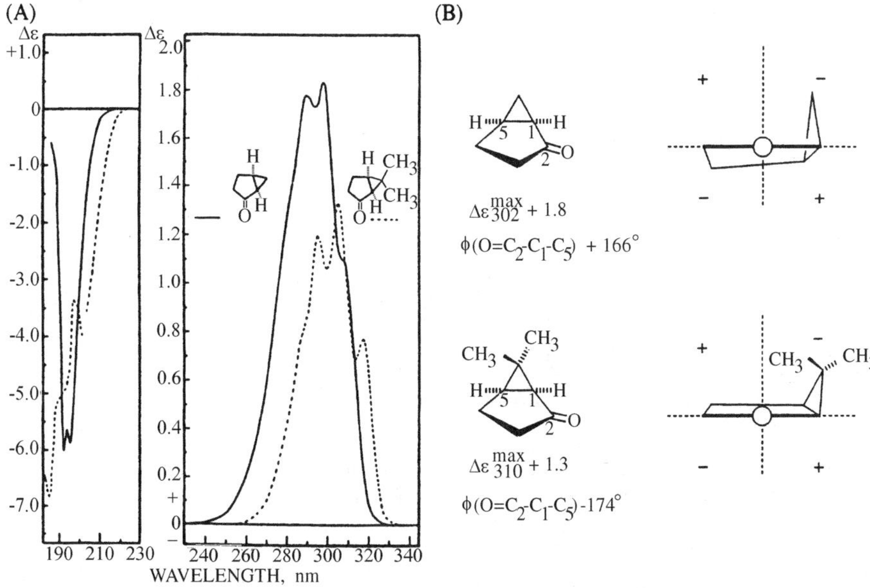

Figure 11-29. (A) CD spectra of 1(*R*)-bicyclo[3.1.0]hexan-2-one (—) and 1(*S*)-6,6-dimethyl-bicyclo[3.1.0]hexan-2-one in isopentane. (B) Structures, octant diagrams, and torsion angles. [(A) is reprinted from ref. 79, copyright© 1975, with permission from Elsevier Science.]

perature CD studies of several simple α,β-cyclopropyl ketones show a small temperature effect with increasingly positive Cotton effects as more of the *s-cis* conformer of the *trans* ketone is selected (Figure 11-30A) and as more of the *s-trans* conformer of the *cis* ketone is selected (Figure 11-30B). Molecular mechanics calculations predict two stable conformers for the *trans* and two for the *cis*, with only small energy differences relative to those proposed earlier.[80] Recent molecular mechanics calculations on methyl cyclopropyl ketone and other alicyclic α,β-cyclopropyl ketones indicate that the former adopts the *s-cis* conformation as the global minimum, some 2.8 kcal/mole below the *s-trans* conformation.[81] The ordinary octant rule predicts a positive $n \rightarrow \pi^*$ Cotton effect in each case for the more stable conformer[80] and a

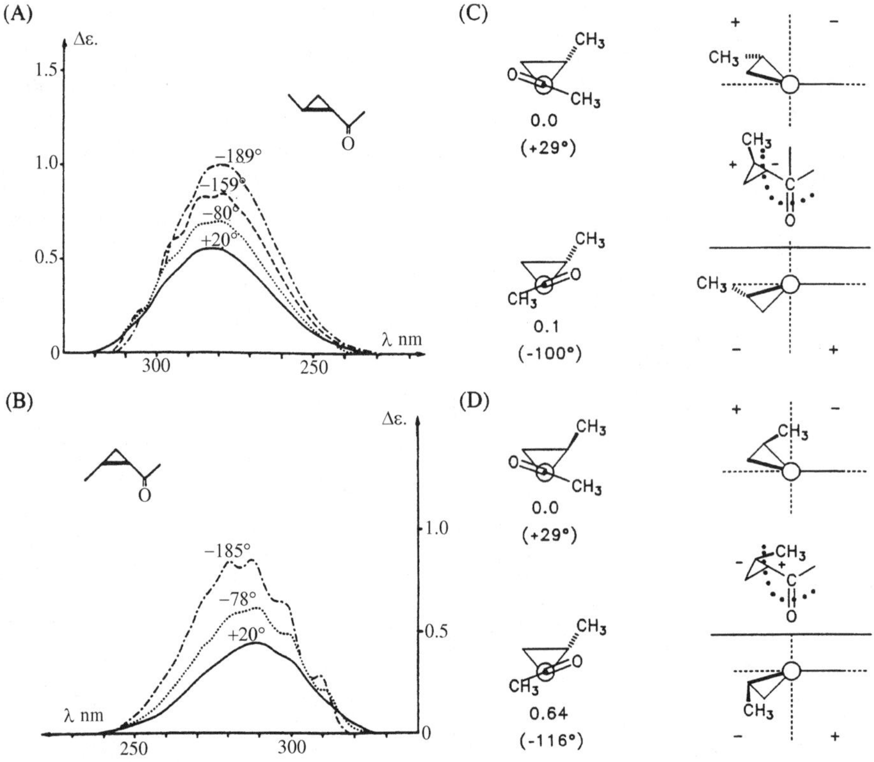

Figure 11-30. Variable low-temperature CD spectra in methylcyclohexane–isopentane (1:3 by vol.) and conformations of (A) *trans*-2-methylcyclopropyl methyl ketone and (B) *cis*-2-methylcyclopropyl methyl ketone. The two minimum energy conformers shown for *trans* and for *cis* were determined by molecular mechanics calculations.[17] The computed relative enthalpies ($\Delta\Delta H_f$ kcal/mole), in addition to the torsion angle O=C–C–CH_2, are given below each structure. (C) Modified octant rule applied to the most stable conformations of the *trans*-ketone conformer of (A). (D) Octant rule applied to the most stable *cis*-ketone conformer of (B). The suggested third nodal convex surface is shown by (·····). [(A) and (B) are reprinted from ref. 80, copyright© 1980, with permission from Elsevier Science.]

negative for the other. It is not entirely clear that these conformers obey an inverse octant rule; however, the $O{=}C_1{-}C_2{-}C_3$ ring torsion angles are much larger than the $O{=}C{-}C{-}CH_2$ torsion angles of Figure 11-30, as in the bicyclo[4.1.0]heptan-1-one of Table 11-20 (+169°) and the bicyclo[3.1.0]hexan-2-ones (+166°, −176°) of Figure 11-29. The important relative orientation, as indicated by Djerassi et al.,[74] appears to be one in which the carbonyl carbon $p{-}\pi$ and $p{-}\pi^*$ orbitals overlap with the adjacent cyclopropane carbon perimeter-in-place $p{-}sp^{4.12}$ Walsh orbital,[82] a delocalization (Figure 11-31) described by Ingraham.[83] For effective overlap, the cyclopropane ring should be perpendicular to the plane of the carbonyl carbon and its adjacent α and α′ carbons.[74] This, indeed, is the relative orientation found in the cyclopropyl ketones of Figure 11-29 and others of Figures 11-27 and 11-28A.

The "inverse" octant rule for α,β-cyclopropyl ketones was extended to the two β,γ-cyclopropyl ketones 4-isocaranone and 4-caranone by Brown and Suzuki,[84] who suggested that the ordinary octant rule (Chapter 4) could not account for the observed negative $n \rightarrow \pi^*$ Cotton effects seen in for these homoconjugated cyclopropyl ketones (Figure 11-32). These were the first examples of "antioctant" effects in β,γ-cyclopropyl ketones. Shortly thereafter, Lightner and Beavers,[85] then at UCLA, reported on the CD of rigid β,γ-cyclopropyl ketones. Particularly attractive examples were based on the achiral bicyclo[2.2.2]octanone skeleton, made chiral by introduction of the cyclopropane $-CH_2-$ in either a *syn* or *anti* configuration (Figure 11-33). The *syn*-cyclopropyl ketone gave a weak negative $n \rightarrow \pi^*$ CD Cotton effect; the *anti* gave a strong positive Cotton effect, comparable in magnitude to that of the β,γ-unsaturated ketone, bicyclo[2.2.2]oct-5-en-2-one. Yet in both examples, only a weak positive Cotton effect is predicted by the octant rule for the lone dissymmetric cyclopropyl CH_2 perturber, and thus, the observed data appear to be associated mainly with the chirality and orientation of the homoconjugated carbonyl and cyclopropane chromophores behaving as inherently dissymmetric chromophores. Analogous β,γ-unsaturated cyclopropyl ketones based on the 2-norbornanone skeleton behaved similarly (Figure 11-33C). Here, however, the 2-norbornanone is chiral and gives a weak negative $n \rightarrow \pi^*$ Cotton effect, as predicted by the octant rule (Figure 11-33D). Only weak $n \rightarrow \pi^*$ Cotton

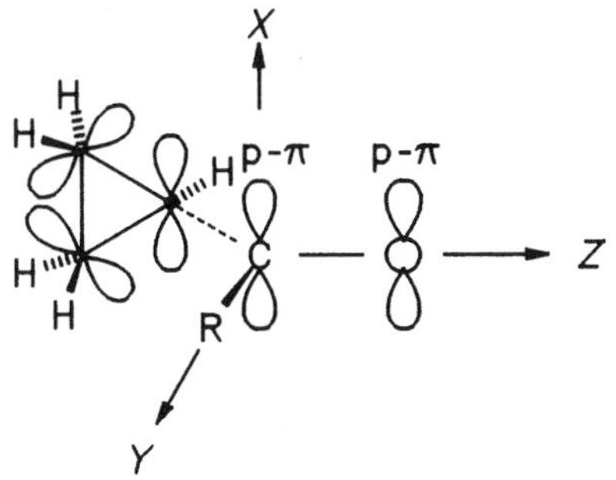

Figure 11-31. π-Orbital overlap in α,β-cyclopropyl ketones. For maximum overlap, the three-membered ring should lie perpendicular to the plane of the R–C(O) and parallel to the π-orbital of the C=O group.

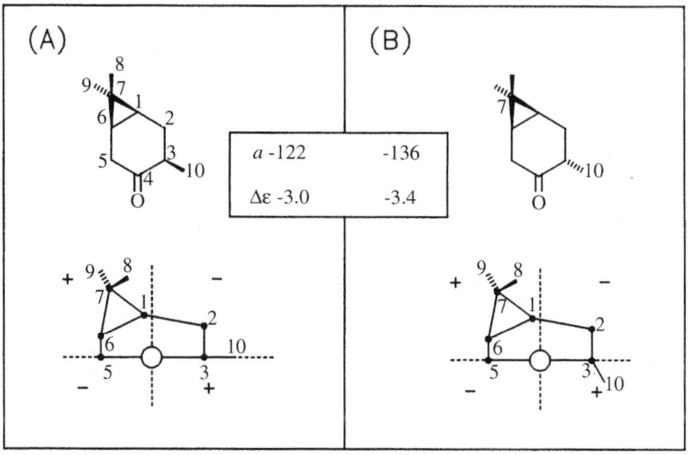

Figure 11-32. (A) Isocaranone and its octant diagram. (B) Caranone and its octant diagram. ORD and CD data are shown in the inset.

effects are predicted (Figure 11-33D) for the corresponding β,γ-cyclopropyl ketones—weakly positive for the *syn*, weakly negative for the *anti*. Yet, a strongly positive Cotton effect is found for the *anti*, and a negative Cotton effect is found for the *syn*. When the cyclopropane group assumes a very different orientation with respect to the ketone carbonyl (Figure 11-34), it appears to exert only a weak homoconjugation effect. And when the cyclopropane is symmetrically disposed about the ketone carbonyl group of β,γ-cyclopropyl ketones, as in bicyclo[3.1.0]hexan-3-ones, static dissymmetric octant perturbers control the sign of the $n \rightarrow \pi^*$ Cotton effect.[86]

Not unexpectedly therefore, the relative orientation of the cyclopropane ring with respect to the carbonyl group plays a crucial role in determining whether a cyclopropyl ketone behaves like an inherently *dissymmetric* chromophore, in which case the octant rule will not dominate, or whether it behaves like an inherently *symmetric* chromophore, to which the octant rule may be applied. Any generalized rule for predicting the sign of the Cotton effect in cyclopropyl ketones must take into account the relative orientation of the two chromophores with respect to each other, and recognize the geometries in which they are weakly or strongly interacting.

The importance of the relative orientation of the cyclopropane and carbonyl chromophores of α,β-cyclopropyl ketones has been discussed briefly[74] in terms of the maximum overlap of the delocalized cyclopropane orbitals[82,83] and the nonbonding (n) orbitals of the carbonyl oxygen. In Figure 11-31, with only slight rotation about the O=C-cyclopropyl single bond, the delocalized orbital of the cyclopropane overlaps the *delocalized* nonbonding $2\overset{\circ}{p}_y$ orbital of the C=O group. Thus, the symmetry-forbidden carbonyl $n \rightarrow \pi^*$ transition takes on aspects of an allowed $\pi \rightarrow \pi^*$ transition by mixing delocalized cyclopropane excited states into the carbonyl $n \rightarrow \pi^*$ state. (See Figure 11-35.) The sign and magnitude of the mixing coefficient are determined by the difference in energy between the cyclopropyl excited states and the carbonyl

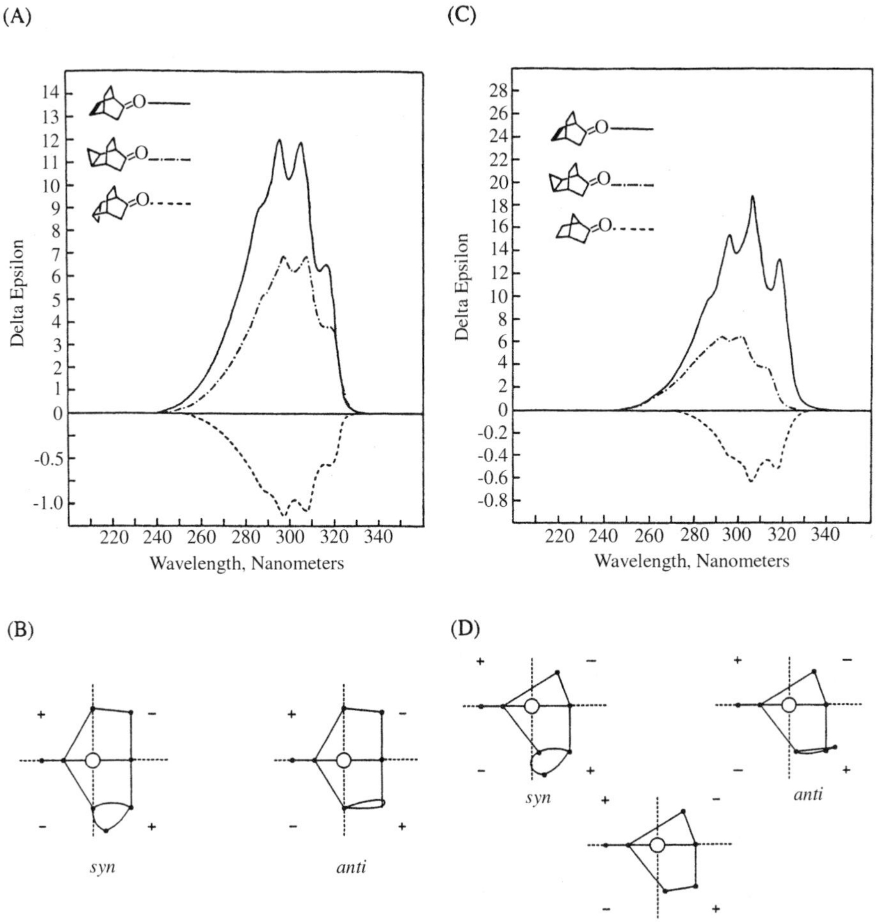

Figure 11-33. (A) Comparison of CD curves of the β,γ-unsaturated ketone bicyclo[2.2.2]oct-5-en-2-one (—) with those of its *syn* (– – –) and *anti* (– · – · – ·) β,γ-cyclopropyl ketone analogs. (B) Octant diagram for the *syn* (left) and *anti* (right) β,γ-cyclopropyl ketones. (C) Comparison of CD curves of the β,γ-unsaturated ketone 2-norbornenone (—) with those of 2-norbornanone (– – –) and the *anti* β,γ-cyclopropyl ketone analog (– · · – · · – ·). (D) Octant diagrams for 2-norbornanone (center) and its *syn* (left) and *anti* (right) β,γ-cyclopropyl ketone analogs. [(A) and (C) are reprinted with permission from ref. 85. Copyright© 1971, American Chemical Society.]

$n \rightarrow \pi^*$ state and the relative orientation of the two chromophores. Using these considerations to understand the magnitudes of the observed Cotton effects, one readily expects that the mixing will be greater and lead to larger rotatory strengths (cf. the *anti*-cyclopropyl ketones of Figure 11-33) when the plane of the carbonyl *n*-system lies parallel with the cyclopropane ring (in which the cyclopropane delocalized orbitals lie). However, when the cyclopropane ring and carbonyl *n* system approach orthogonality, the independent chromophores are expected to exhibit less mixing, mani-

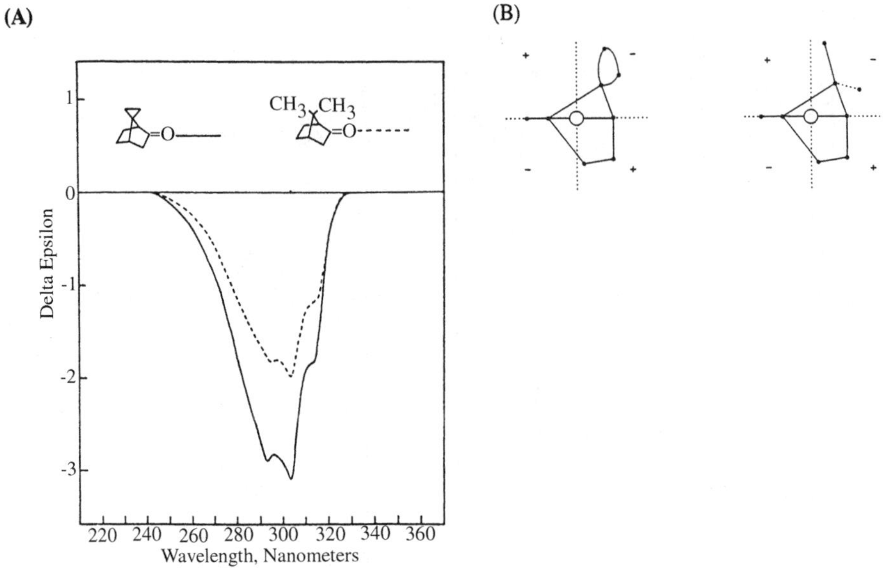

Figure 11-34. (A) CD spectra of spiro-(cyclopropane-1,7′-norbornan-2-one) (—) and α-fenchocamphorone (– – – –). (B) Octant diagrams for spiro(cyclopropane-1,7′-norbornan-2-one) (left) and α-fenchocamphorone (right). [(A) is reprinted with permission from ref. 85. Copyright© 1971 American Chemical Society.]

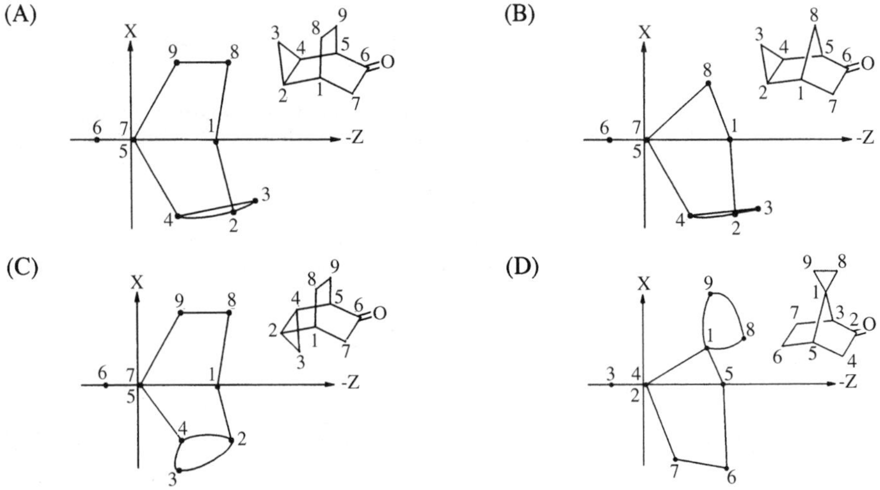

Figure 11-35. Projections in the *XZ* plane for (A) *anti*-tricyclo[3.2.2.0^{2,4}]nonan-6-one, (B) *anti*-tricyclo[3.2.1.0^{2,4}]octan-6-one, (C) *syn*-tricyclo[3.2.2.0^{2,4}]nonan-6-one, and (D) spiro(cyclopropane-1,7′-norbornanone). The coordinate system of Figure 11-31 is used throughout.

fested by much lower rotational strengths, with values perhaps marginal for an inherently dissymmetric chromophore. (cf. *syn*-cyclopropyl ketone of Figure 11-33A and the spirocyclopropyl ketone of Figure 11-34).

Some of the geometric considerations of the foregoing discussion are found in Figure 11-35, which depicts the projections of the cyclopropyl ketones of Figures 11-33 and 34 onto the *XZ* plane. (The carbonyl π orbitals lie in the *XZ* plane; see Figure 11-31.) Thus, for the *anti*-cyclopropyl ketones of Figure 11-33, the plane of the cyclopropane ring (and the delocalized cyclopropane orbitals) is nearly parallel to the plane of the carbonyl *n* system, and the Cotton effects are large. On the other hand, for the *syn*-cyclopropyl ketones of Figures 11-33 and 11-34, the plane of the cyclopropane ring is nearly orthogonal to the carbonyl *n* system, and much smaller rotatory strengths obtain.

In order to generate the *sign* of the Cotton effect from the relative orientation of the pertinent chromophores, an attempt was made[85] to determine the chirality of the extended *n*-system (Figure 11-31) from the skew angle (θ) defined by the intersection of (*i*) a plane orthogonal to the in-plane *p* orbital of the cyclopropyl carbon β to the carbonyl carbon and (*ii*) the plane of the oxygen *n* orbital ($2p_y^{\circ}$ of Figure 11-31). Qualitatively, θ is found to be (arbitrarily) positive for the *anti*-cyclopropyl ketones and negative for the *syn*-cyclopropyl ketone (Figure 11-36) and the spirocyclopropyl ketone. These observations correlate with the experimentally determined Cotton effect signs (Figures 11-33 and 11-34); however, the correlation lacks sufficient theoretical justification and is based merely on an extension of the β,γ-unsaturated ketone approach.[35] Such a simplified treatment neglects the multiplicity of low-lying cyclopropane electronic transitions[87] involving molecular Rydberg states. Still, a correlation *is* found, even though there are as yet too few examples to determine its general applicability. The approach also does not answer the problem posed by α,β-cyclopropyl ketones,[74] whose chromophores are potentially more strongly coupled (except in unfavorable spatial orientations) than those of β,γ-cyclopropyl ketones.

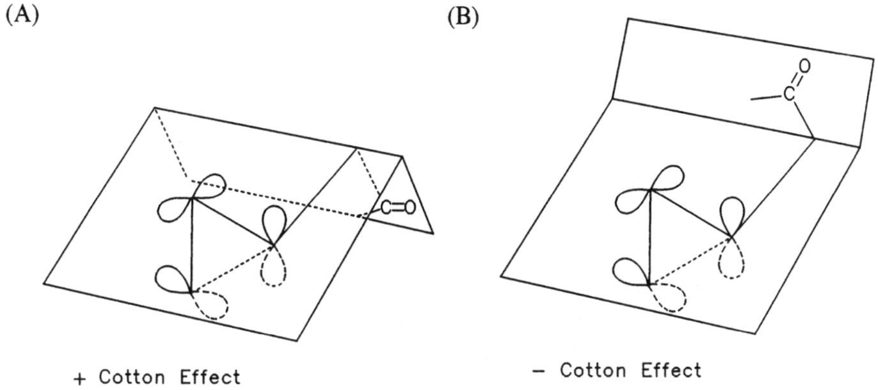

Figure 11-36. Enantiomeric chromophores associated with β,γ-cyclopropyl ketones. From ref. 85.

In the late 1960s, Craig Deutsche at UCLA attempted a more detailed theoretical treatment (unpublished) of the model compounds in which the allowed cyclopropane transitions[88] were mixed with the carbonyl $n \rightarrow \pi^*$ transition to give the excited-state wave function

$$\psi_{ex} = (n \rightarrow \pi^*) + \sum_i \lambda_i(\pi \rightarrow \sigma^*) + \sum_j \lambda_j(\pi \rightarrow \pi^*).$$

In order to obtain qualitative predictions, a perturbation treatment based on point dipole (cyclopropane)–quadrupole (carbonyl) interactions, as well as more extensive dipole–quadrupole interactions, was carried out. However, this approach failed to predict the observed Cotton effect signs, due undoubtedly to an inability at that time to recognize that the cyclopropane transition of lowest energy is a Rydberg transition.[87] The general difficulty in assigning the lowest lying cyclopropane transitions complicate the theoretical treatment, and work was in progress by Lightner and Bouman to extend *ab initio* extended basis set calculations in the random-phase approximation to α,β and β,γ-unsaturated ketones. This as yet unfinished study (due to the premature death of Tom Bouman) satisfactorily explained the electronic spectra of cyclopropane and the chiroptical properties of *trans*-1,2-dimethylcyclopropane,[87] the first step in explaining the UV and CD spectra of cyclopropyl ketones. Other theoretical treatments of the dependence on orientation of the $n \rightarrow \pi^*$ and $\pi \rightarrow \pi^*$ UV transitions of β,γ-cyclopropyl ketones using a modified INDO procedure were carried out by Boerth, Van-Catledge, and Kao.[6e,f] The results showed little dependence of orientation on the $n \rightarrow \pi^*$ transition.

References

1. Stern, E. S., and Timmons, C. J., *Electronic Absorption Spectroscopy in Organic Chemistry*, St. Martin's Press, New York, 1971.
2. (a) Ferguson, L. N., *J. Chem. Ed.* **46** (1969), 404–412.
   (b) Ferguson, L. N. and Nnadi, J. C., *J. Chem. Ed.* **42**, (1965), 529–535.
3. Houk, K. N. *Chem. Rev.* **76**, (1976), 1–74.
4. (a) Moscowitz, A., Hansen, Aa. E., Forster, L. S., and Rosenheck, K., *Biopolymers Sympos. No. 1* (1964), 74–89.
   (b) Hansen, Aa. E., "On the Optical Activity of Compounds Containing Coupled Chromophores," licentiate thesis, H. C. Ørsted Institute, Copenhagen, 1963.
   (c) Moscowitz, A., *Proc. Roy. Soc. London* **A297**, (1967), 40–42.
5. (a) Leonard, N. J., and Owens, F. H., *J. Am. Chem. Soc.* **80**, (1958), 6039–6045.
   (b) Kosower, E. M., Closson, W. D., Goering, H. L., and Gross, J. C., *J. Am. Chem. Soc.* **83**, (1961), 2013–2015.
6. (a) Dauben, W. G., and Berezin, G. H., *J. Am. Chem. Soc.* **89**, (1967), 3449–3452.
   (b) De Meijere, A., *Angew. Chem. Int. Ed. Engl.* **18**, (1979), 809–826.
   (c) Hess, L. D., and Pitts, J. N., Jr., *J. Am. Chem. Soc.* **89**, (1967), 1973–1979.
   (d) Pincock, R. E., and Haywood-Farmer, J., *Tetrahedron Lett.* (1967), 4759–4762.
   (e) Van-Catledge, F. A., Boerth, D. W., and Kao, J., *J. Org. Chem.* **47**, (1982), 4096–4106.
   (f) Boerth, D. W., *J. Org. Chem.* **47**, (1982), 4085–4096.
7. (a) Djerassi, C., *Optical Rotatory Dispersion*, McGraw-Hill, New York, 1960.

(b) Crabbé, P., *Optical Rotatory Dispersion and Circular Dichroism in Organic Chemistry*, Holden-Day, San Francisco, 1965.
(c) For a recent compilation of CD data on ketones, see, "Chiroptical Properties of Compounds Containing C=O Groups," Boiadjiev, S. E., and Lightner, D. A., in *The Chemistry of Double Bond Functional Groups*, (Patai, S., ed.), John Wiley & Sons, Ltd., Chichester, U.K., 1997, Chapter 5, pp. 155–260.
8. (a) For an excellent review, see Gawroński, J. K., "Conformations, Chiroptical and Related Spectral Properties of Enones," in *The Chemistry of Enones. Part 1* (Patai, S., and Rappoport, Z., eds.), John Wiley & Sons, 1989, Chapter 3, pp. 55–103.
(b) Gawroński, J. K., *Tetrahedron* **38**, (1982), 3–26.
9. Kirk, D. N., *Tetrahedron* **42**, (1986), 777–818.
10. (a) Ziffer, H., and Robinson, C. H., *Tetrahedron* **24**, (1968), 5803–5816.
(b) Kuriyama, K., Moriyama, M., and Iwata, T., *Tetrahedron Lett.* 1968, 1661–1664.
11. (a) Snatzke, G., *Tetrahedron* **21**, (1965), 413–419.
(b) Snatzke, G., *Tetrahedron* **21**, (1965), 421–438.
(c) Snatzke, G., *Tetrahedron* **21**, (1965), 439–448.
(d) Snatzke, G., "α,β- and β,γ-Unsaturated Ketones," in *Optical Rotatory Dispersion and Circular Dichroism in Organic Chemistry* (Snatzke, G., ed.), Sadtler Research Labs, Inc., Philadelphia, 1967, Chapter 13, pp. 208–223.
(e) Snatzke, G., and Snatzke, F., "The Carbonyl Chromophore: Unsaturated Ketones and Lactones," in *Fundamental Aspects and Recent Developments in Optical Rotatory Dispersion and Circular Dichroism* (Ciardelli, F., and Salvadori, P., eds.), Heyden & Son, Ltd., New York, 1973, Chapter 3, pp. 109–125.
(f) Snatzke, G., *Angew. Chem. Int. Ed. Engl.* **18**, (1979), 363–377.
12. (a) Djerassi, C., Records, R., Bunnenberg, E., Mislow, K., and Moscowitz, A., *J. Am. Chem. Soc.* **84**, (1962), 870–872.
(b) Whalley, W. B., *Chem. and Ind.* (1962), 1024–1025.
13. (a) Burnett, R. D., and Kirk, D. N., *J. Chem. Soc., Perkin Trans. 1* (1981), 1460–1468.
(b) Beecham, A. F., *Tetrahedron* **27**, (1971), 5207–5216.
14. (a) Kuball, H-G., Neubrech, S., and Schönhofer, A., *Chem. Phys.* **163**, (1992), 115–132.
(b) Frelek, J., Szczepek, W. J., and Weiss, H. P., *Tetrahedron: Asymmetry* **6**, (1995), 1419–1430.
(c) Frelek, J., Szczepek, W. J., and Weiss, H. P., *Tetrahedron: Asymmetry* **4**, (1993), 411–424.
(d) Frelek, J., Szczepek, W. J., Weiss, H. P., Reiss, G. J., Frank, W., Brechtel, J., Schultheis, B., and Kuball, H-G., *J. Am. Chem. Soc.* **120**, (1998), 7010–7019.
15. Hug, W., and Wagnière, G., *Helv. Chim. Acta* **54**, (1971), 633–649.
16. (a) Burgstahler, A. W., and Barkhurst, R. C., *J. Am. Chem. Soc.* **92**, (1970), 7601–7603.
(b) Burgstahler, A. W., Barkhurst, R. C., and Gawroński, J. K., "Cotton Effects and Allylic–Homoallylic Chirality of Steroidal Olefins and Conjugated Dienes and Enones," in *Modern Methods of Steroid Analysis* (Heftmann, E., ed.), Academic Press, Inc., New York, 1973, Chapter 16, pp. 349–379.
17. Determined by molecular mechanics calculations in PCMODEL, version 4.0–7.0, Serena Software, Inc., Bloomington, IN 47402–3076.
18. Beecham, A. F., *Tetrahedron* **27**, (1971), 5207–5216.
19. Lightner, D. A., Bouman, T. D., Gawroński, J. K., Gawrońska, K,, Chappuis, J. L., Crist, B. V., and Hansen, Aa. E., *J. Am. Chem. Soc.* **103**, (1981), 5314–5327.
20. Lightner, D. A., Flores, M. J., and Crist, B. V., *J. Org. Chem.* **45**, (1980), 3518–3522.
21. (a) Burgstahler, A. W., and Naik, N. C., *Helv. Chim. Acta* **54**, (1971), 2920–2924.
(b) Burgstahler, A. W., Boger, D. L., and Naik, N. C., *Tetrahedron* **32**, (1976), 309–315.
22. Barieux, J-J., Gore, J., and Subit, M., *Tetrahedron Lett.* (1975), 1835–1838.
23. Suga, T., and Imamura, K., *Bull. Chem. Soc. Jpn* **45**, (1972), 2060–2064.
24. Suga, T., Imamura, K., and Shishibori, T., *J. Chem. Soc., Chem. Commun.* (1971), 126–127.
25. (a) Sato, T., Tada, M., Takahashi, T., Horibe, I., Ishii, H., Iwata, T., Kuriyama, K., Tamura, Y., and Tori, K., *Chem. Lett.* (1977), 1191–1194.

(b) Tada, M., Sato, T., Takahashi, T., Tori, K., Horibe, I., and Kuriyama, K., *J. Chem. Soc. Perkin Trans 1* (1981), 2695–2701.

26. (a) Asahina, Y., Ishidate, M., and Tukamoto, T. *Chem. Ber.* **69**, (1936), 343–348.
    (b) Asahina, Y., Ishidate, M., and Tukamoto, T., *Chem. Ber.* **69**, (1936), 355–357.
27. (a) Carnelutti, J., and Nasini, R., *Chem. Ber.* **13**, (1880), 2208–2211.
    (b) Nasini, R., *Atti R. Accad. Lincei* 3A, **13**, (1881) 129.
    (c) Nasini, R., *Gazz. Chim. Ital.* **13** (1883), 151.
28. Mitchell, S., and Schwarzwald, K., *J. Chem. Soc.* (1939), 889–893.
29. (a) Woodward, R. B., and Kovach, E. G., *J. Am. Chem. Soc.* **72**, (1950), 1009–1016.
    (b) Woodward, R. B., and Yates, P., *Chem. and Ind.* (1954), 1391–1393.
30. Alpen, E. L., Kumler, W. D., and Strait, L. A., *J. Am. Chem. Soc.* **72**, (1950), 4558–4561.
31. (a) Cookson, R. C., and Wariyar, N. S., *J. Chem. Soc.* (1956), 2302–2311.
    (b) Cookson, R. C., and Lewin, N., *Chem. and Ind.* (1956), 984–985.
32. Labhart, H., and Wagnière, G., *Helv. Chim. Acta* **42**, (1959), 2219–2227.
33. (a) Mislow, K., Glass, M. A. W., O'Brien, R. E., Rutkin, P., Steinberg, D. H., and Djerassi, C., *J. Am. Chem. Soc.* **82**, (1960), 4740–4742.
    (b) Mislow, K., and Djerassi, C., *J. Am. Chem. Soc.* **82**, (1960), 5247.
34. (a) Mislow, K., Glass, M. A. W., O'Brien, R. E., Rutkin, P., Steinberg, D. H., Weiss, J., and Djerassi, C., *J. Am. Chem. Soc.* **84**, (1962), 1455–1478.
    (b) Mislow, K., *Ann. N.Y. Acad. Sci.* **93**, (1962), 459–484.
35. (a) Mislow, K., Glass, M. A. W., Moscowitz, A., and Djerassi, C., *J. Am. Chem. Soc.* **83**, (1961), 2771–2772.
    (b) Moscowitz, A., Mislow, K., Glass, M. A. W., and Djerassi, C., *J. Am. Chem. Soc.* **84**, (1962), 1945–1955.
36. Mislow, K., and Berger, J. G., *J. Am. Chem. Soc.* **84**, (1962), 1956–1961.
37. Bunnenberg, E., Djerassi, C., Mislow, K., and Moscowitz, A., *J. Am. Chem. Soc.* **84**, (1962), 2823–2826; *erratum* **84**, (1962), 5003.
38. Deutsche, C. W., Lightner, D. A., Woody, R. W., and Moscowitz, A., *Ann. Rev. Phys. Chem.* **20**, (1969), 407–448.
39. (a) Cookson, R. C., and McKenzie, S., *Proc. Chem. Soc.* (1961), 423–424.
    (b) Birnbaum, H., Cookson, R. C., and Lewin, N., *J. Chem. Soc.* (1961), 1224–1228.
40. Cookson, R. C., and Hudec, J., *J. Chem. Soc.* (1962), 429–434.
41. Lightner, D. A., Gawroński, J., Hansen, Aa. E., and Bouman, T. D., *J. Am. Chem. Soc.* **103**, (1981), 4291–4296.
42. Mason, S. F., *J. Chem. Soc.* (1962), 3285–3288.
43. (a) Ballard, R. E., Mason, S. F., and Vane, G. W., *Trans. Faraday Soc.* **59**, (1963), 775–782.
    (b) Mason, S. F., *Quart. Rev.* **17**, (1963), 20–66.
    (c) Grinter, R., Mason, S. F., and Vane, G. W., *Trans. Faraday Soc.* **60**, (1964), 285–290.
44. (a) Mason, S. F., *Proc. Chem. Soc.* (1964), 61.
    (b) Moscowitz, A., *Proc. Chem. Soc.* (1964), 60.
45. Longuet-Higgins, H. C., and Murrell, J. N., *Proc. Phys. Soc. London* **A68**, (1955), 601–611.
46. (a) Gorodetsky, M., and Mazur, Y., *Tetrahedron Lett.* (1964), 227–232.
    (b) Gorodetsky, M., Amar, D., and Mazur, Y., *J. Am. Chem. Soc.* **86**, (1964), 5218–5224.
    (c) Witz, P., Herrmann, H., Lehn, J. M., and Ourisson, G., *Bull. Soc. Chim. France* (1963), 1101–1112.
47. (a) Bays, D. E., and Cookson, R. C., *J. Chem. Soc. (B)* (1967), 226–229.
    (b) Bays, D. E., Cookson, R. C., and MacKenzie, S., *J. Chem. Soc. (B)* (1967), 215–226.
    (c) Bays, D. E., Cannon, G. W., and Cookson, R. C., *J. Chem. Soc. (B)* (1966), 885–892.
    (d) Cookson, R. C., *Proc. Royal Soc. London* **A297**, (1967), 29–39.
48. (a) Sandman, D. J., and Mislow, K., *J. Org. Chem.* **33**, (1968), 2924–2926.
    (b) Sandman, D. J., Mislow, K., Giddings, W. P., Dirlam, J., and Hanson, G. C., *J. Am. Chem. Soc.* **90**, (1968), 4877–4884.
49. (a) Hagishita, S., and Kuriyama, K., *J. Chem. Soc., Perkin Trans. 2* (1977), 1937–1941.

(b) Hagishita, S., and Kuriyama, K., *J. Chem. Soc., Perkin Trans. 2* (1978), 58–67.
50. Thomas, H. T., and Mislow, K., *J. Am. Chem. Soc.* **92**, (1970), 6292–6298.
51. Erman, W. F., Treptow, R. S., Bakuzis, P., and Wenkert, E., *J. Am. Chem. Soc.* **93**, (1971), 657–665.
52. Korvola, J., and Mälkönen, P. J., *Suomen Kemist. B.* **45**, (1972), 381–382.
53. Paquette, L. A., Farnham, W. B., and Ley, S. V., *J. Am. Chem. Soc.* **97**, (1975), 7273–7279.
54. (a) Stockis, A., and Weissberger, E., *J. Am. Chem. Soc.* **97**, (1975), 4288–4292.
(b) Weissberger, E., and Laszlo, P., *Acc. Chem. Res.* **9**, (1976), 209–217.
55. (a) Lightner, D. A., Christiansen, G. D., and Melquist, J. L., *Tetrahedron Lett.* (1972), 2045–2048.
(b) Lightner, D. A., Jackman, D. E., Christiansen, G. D., *Tetrahedron Lett.* (1978), 4467–4470.
56. (a) Nakazaki, M., Naemura, K., and Harita, S., *Bull. Chem. Soc. Jpn* **48**, (1975), 1907–1913.
(b) Nakazaki, M., Naemura, K., and Kondo, Y., *J. Org. Chem.* **41**, (1976), 1229–1233.
(c) Nakazaki, M., Naemura, K., Harada, H., and Narutaki, H., *J. Org. Chem.* **47**, (1982), 3470–3474.
(d) Nakazaki, M., Naemura, K., and Arashiba, N., *J. Chem. Soc., Chem. Commun.* (1976), 678–679.
(e) Nakazaki, M., Naemura, K., and Arashiba, N. *J. Org. Chem.* **43**, (1978), 888–891.
57. (a) Kuritani, H., Iwata, F., Sumiyoshi, M., and Shingu, K., *J. Chem. Soc., Chem. Commun.* (1977), 542–543.
(b) Kuritani, H., Shingu, K., and Nakagawa, M., *Tetrahedron Lett.* (1980), 529–530.
58. (a) Sonney, J-M., and Vogel, P., *Helv. Chim. Acta* **63**, (1980), 1034–1044.
(b) Carrupt, P-A., and Vogel, P., *Tetrahedron Lett.* **22**, (1981), 4721–4722.
(c) Lettierri, A., Carrupt, P-A., and Vogel, P., *Chimia* **42**, (1988), 27–28.
59. (a) Schippers, P. H., and Dekkers, H. P. J. M., *J. Am. Chem. Soc.* **105**, (1983), 79–84.
(b) Schippers, P. H., van der Ploeg, J. P. M., and Dekkers, H. P. J. M., *J. Am. Chem. Soc.* **105**, (1983), 84–89.
60. (a) Bouman, T. D., and Hansen, Aa. E., *Croat. Chem. Acta* **62**, (1989), 227–243. (b) For a more current theoretical view of 2-norbornenone, see Itansen, Aa. E., and Bak, K. L. *Enantiomer* (2000) in press. The results of this study show that the μ-m mechanism dominates the rotatory strength, but the μ-μ mechanism contributes about one-third of the total rotatory strength of the $n \rightarrow \pi^*$ transition. In contrast, the μ-μ (polarizability) mechanism almost completely dominates the sign and magnitude of the $\pi \rightarrow \pi^*$ transition rotatory strength.
61. Lightner, D. A., Bouman, T. D., Crist, B. V., Rodgers, S. L., Knobloch, M. A., and Jones, A. M., *J. Am. Chem. Soc.* **109**, (1987), 6248–6259.
62. (a) Bird, C. W., Cookson, R. C., and Hudec, J., *Chem. and Ind.* (1960), 20–21.
(b) Cookson, R. C., Hill, R. R., and Hudec, J., *Chem. and Ind.* (1961), 589–590.
63. Cookson, R. C., Henstock, J., and Hudec, J., *J. Am. Chem. Soc.* **88**, (1966), 1059–1060.
64. (a) Sauers, R. R., and de Paolis, A. M., *J. Org. Chem.* **38**, (1973), 639–641.
(b) Sauers, R. R., and Henderson, T. R. *J. Org. Chem.* **39**, (1974), 1850–1853.
65. Becker, H-D., Skelton, B. W., and White, A. H., *J. Chem. Soc., Perkin Trans. 2* (1981), 442–446.
66. (a) Hudec, J., *J. Chem. Soc., Chem. Commun.* (1967), 539–540.
(b) Powell, G. P., Totty, R. N., and Hudec, J., *J. Chem. Soc., Perkin Trans. 1* (1978), 1015–1019.
67. Pak, C. S., *Circular Dichroism Studies of Bicyclo[3.1.0]hexanones and Bicyclo[2.2.1]heptanones*, University of Nevada, Reno, Ph.D. dissertation, 1975. See also ref. 58b.
68. (a) Toan, V. V., and Lightner, D. A., *Tetrahedron* **43**, (1987), 5769–5774.
(b) Wijekoon, W. M. D., and Lightner, D. A., *J. Org. Chem.* **52**, (1987), 4171–4175.
69. (a) Nicoud, J. F., Eskanazi, C., and Kagan, H. B., *J. Org. Chem.* **42**, (1977), 4270–4272.
(b) Weissberger, E., *J. Am. Chem. Soc.* **96**, (1974), 7219–7221.
70. Robbins, T. A., Toan, V. V., Givens, J. W., III, and Lightner, D. A., *J. Am. Chem. Soc.* **114**, (1992), 10799–10810.
71. Djerassi, C., Riniker, R., Riniker, B., *J. Am. Chem. Soc.* **78**, (1956), 6377–6389.
72. Norin, T., *Acta Chem. Scand.* **17**, (1963), 738–748.
73. Legrand, M., Viennet, R., and Caumartin, J., *Comptes Rendus* **253**, (1961), 2378–2380.
74. Djerassi, C., Klyne, W., Norin, T., Ohloff, G., and Klein, E., *Tetrahedron* **21**, (1965), 163–178.
75. Schaffner, K., and Snatzke, G., *Helv. Chim. Acta* **48**, (1965), 347–361.
76. (a) Gray, R. T., and Smith, H. E., *Tetrahedron* **23**, (1967), 4229–4241.

(b) Hill, R. K., and Morgan, J. W., *J. Org. Chem.* **33**, (1968), 927–928.
(c) Kuriyama, K., Tada, H., Sawa, Y. K., Ito, S., and Itoh, I., *Tetrahedron Lett.* (1968), 2539–2544. These authors also investigated ketone $\pi \rightarrow \pi^*$ Cotton effects and suggested that they follow a normal octant rule.
(d) Butcher, F. K., Coombs, R. A., and Davies, M. T., *Tetrahedron* **24**, (1968), 4041–4050.
77. Leriverend, P., and Conia, J. M., *Bull. Soc. Chim. France* (1966), 121–125.
78. Wynberg, H., and Marsman, B., *J. Org. Chem.* **45**, (1980), 158–161.
79. Lightner, D. A., and Jackman, D. E., *Tetrahedron Lett.* (1975), 3051–3054.
80. Tocanne, J. F., *Tetrahedron* **28**, (1972), 389–416.
81. Mash, E. A., Gregg, T. M., Stahl, M. T., and Walters, W. P., *J. Org. Chem.* **61**, (1996), 2738–2742.
82. Walsh, A. D., *Trans. Faraday Soc.* **45**, (1949), 179–190.
83. Ingraham, L. I., "Steric Effects on Certain Physical Properties," in *Steric Effects in Organic Chemistry* (Newman, M. S., ed.), John Wiley, New York, 1956, pp. 518–521.
84. Brown, H. C., and Suzuki, A., *J. Am. Chem. Soc.* **89**, (1967), 1933–1941.
85. Lightner, D. A., and Beavers, W. A., *J. Am. Chem. Soc.* **93**, (1971), 2677–2681.
86. Lightner, D. A., Pak, C. S., Crist, B. V., Rodgers, S. L., and Givens, J. W., III, *Tetrahedron* **41**, (1985), 4321–4330.
87. Bohan, S., and Bouman, T. D., *J. Am. Chem. Soc.* **108**, (1986), 3261–3266.
88. Raymonds, J. W., and Simpson, W. T., *J. Chem. Phys.* **47**, (1967), 430–448.

CHAPTER

12

Dienes

Based on the success of the work on the ketone chromophore in the late 1950s and early 1960s, olefins and dienes became the subjects of interest of a number of research groups studying stereochemistry via chiroptical measurements such as ORD or CD. Work in this area continues even to the present. An excellent, thorough review of the area up to the work of the early 1970s was published in 1973.[1]

12.1. Olefin Sector Rules

Since most alkene electronic transitions occur below approximately 220 nm, unlike the ketone $n \rightarrow \pi^*$ transition (~290 nm), most of the early olefin ORD and CD measurements were difficult to obtain across the entire absorption band with the use of the then available instrumentation. Either the optics did not permit going to shorter wavelengths or the light absorption was too high (due to large values of ε) when concentrations were increased to obtain an adequate level of optical rotation. However, gas-phase CD spectroscopy at very low concentrations of simple optically active olefins such as *trans*-cyclooctene and α- and β-pinene (Table 12-1) indicated a multiplicity of bands lying between 220 and 140 nm.[2] The gas-phase CD spectrum of *trans*-cyclooctene appears in Figure 12-1.

Of particular interest are three transitions found lying between 180 and 230 nm and assigned as $\pi_x \rightarrow 3s$ Rydberg (~230–210 nm), $\pi_x \rightarrow \pi_x^*$ (~209–190 nm), and $\pi_x \rightarrow \pi_y^*$ (~180 nm). More recently, this observation was confirmed for (3*R*)-methylcyclopentene (Table 12-1) by the vacuum CD studies of Gedanken et al.[3] who found a multiplicity of Rydberg and valence transitions for this planar olefin—at approximately 205 nm (yielding a negative Cotton effect), about 185 nm (positive Cotton effect), approximately 161 nm (negative Cotton effect), and 149 nm (positive Cotton effect). The observations of multiple transitions lying between 220 and 180 nm and their assignments were supported by theoretical calculations (RPA) from Hansen and Bouman,[4,5] who computed multiple transitions for 3-methylcyclopentene ($\pi \rightarrow 3s$, $\pi \rightarrow 3p_x$, $\pi \rightarrow 3p_z$, $\pi \rightarrow \pi^*$) lying in the two longest wavelength bands. When the compound is in solution, the CD spectra can be expected to broaden and shift, but the essential element remains: Inside the spectroscopic window that is normally accessible with modern CD instrumentation, multiple overlapping transitions can be expected to lie within the band envelopes. This presents a potentially far more complicated situation than that found in studies of ketone $n \rightarrow \pi^*$ spectra, for the $n \rightarrow \pi^*$ band of ketones is much more homogeneous, from a spectroscopic standpoint. And it implies considerably more difficulties in developing rational chirality rules for alkenes. Of

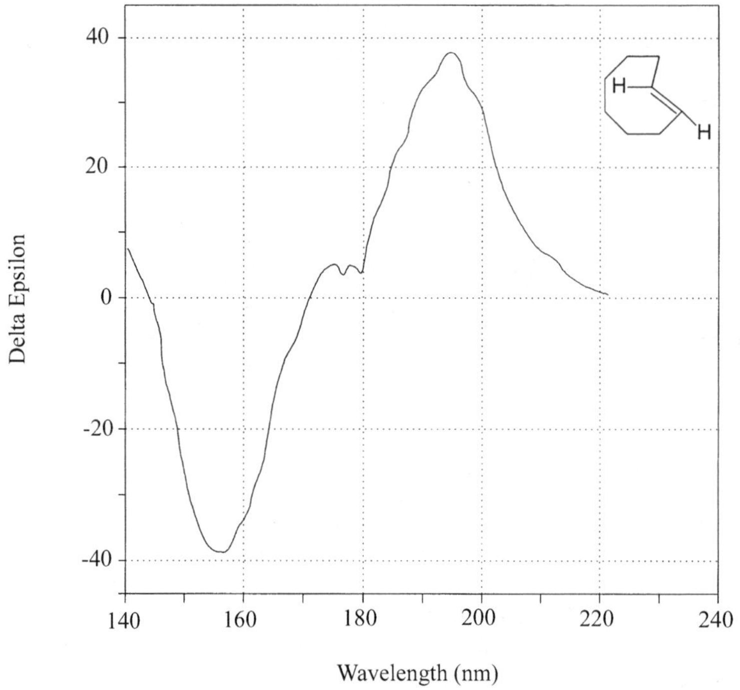

Figure 12-1. The CD spectrum of *trans*-cyclooctene. Replotted using data from ref. 2. Copyright© 1973 American Chemical Society.

course, this fact was not generally appreciated in the early days of olefin ORD and CD studies.

One of the earliest attempts to bring attention to the problems of multiple absorption bands in the chiroptical spectra of alkenes came from Moscowitz and Mislow,[6] who interpreted the chiroptical properties of *trans*-cyclooctene; one conformation, as determined by PCMODEL,[7] is shown in Figure 12-2. Moscowitz and Mislow pointed out that at one extreme the two *p*-orbitals of the sp^2 carbons could be completely co-planar, while at the other extreme they could be perpendicular. In a molecule such as this, it can be expected that neither of these situations would obtain, but that some intermediate condition would be found. Therefore, this chromophore was considered to be inherently dissymmetric, and the resulting optical activity would be derived from the sense of twist, thought to be around 17° from molecular models. (The amount of twist can now be obtained from various computational sources, such as PCMODEL.) While it was recognized that the bridge consisting of the six methylene groups connecting the two vinyl carbons would have a perturbing effect on the chromophore, it was felt that the major contribution to the observed results came from the twisted double bond. At that time, however, there were no experimental data available for

Table 12-1. Examples of Olefins and Dienes

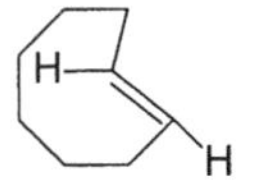

trans-cyclooctene

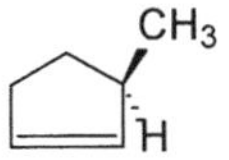

(3*R*)-Methylcyclopentene

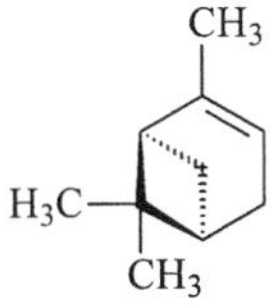

α–Pinene

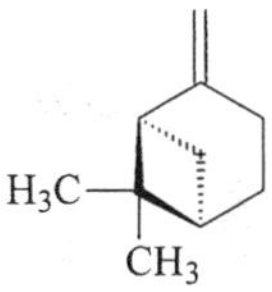

β–Pinene

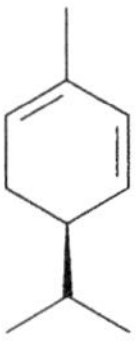

α-Phellandrene

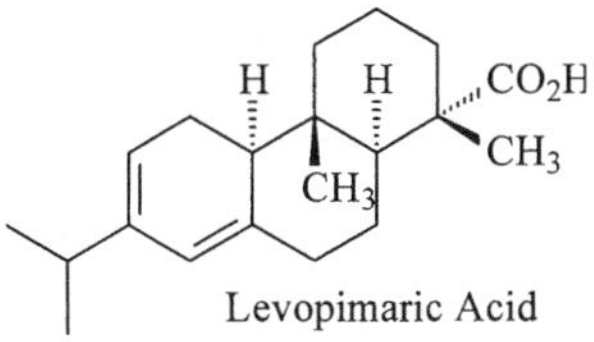

Levopimaric Acid

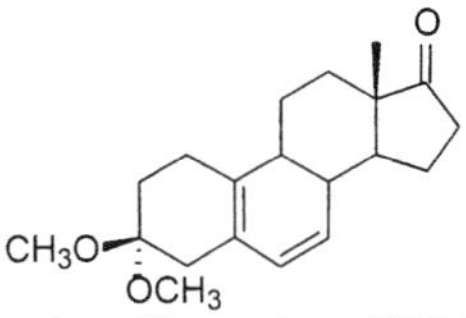

3,3-Dimethoxy-19-norandrosta-5(10),6-dien-17-one

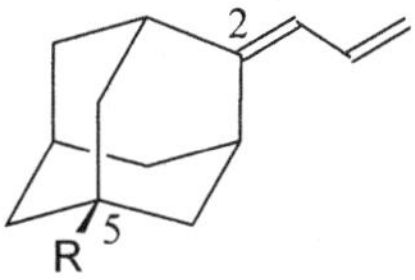

5-substituted 2-adamantylidene

comparison, so the sign of the optical rotation for *trans*-cyclooctene was predicted on the basis of an evaluation of the relevant matrix elements.

By 1967, with considerably more experimental data at hand, Mazur et al.[8] postulated an "allylic quasi-axial hydrogen" rule for the carbon–carbon double-bond chromophore. A reasonably high degree of success in correlating the molecular structure of the compound with chiroptical data by assuming that the contribution of a hydrogen or other substituent located adjacent to a C=C π-system was significant and predictable. After the bond to the substituent was located such that orbital overlap with the π-system was feasible, one could use a model (Figure 12-3) to predict the sign of the Cotton effect curve. In particular, this approach worked best when the carbon–carbon double-bond was not twisted. In that case, the chromophore is inherently symmetric, but asymmetrically perturbed, a situation wherein the *p*-orbitals of the double bond are very close to coplanar and are aligned with the bonding orbital between the allylic carbon and its substituents, particularly the axial or quasi-axial substituent.

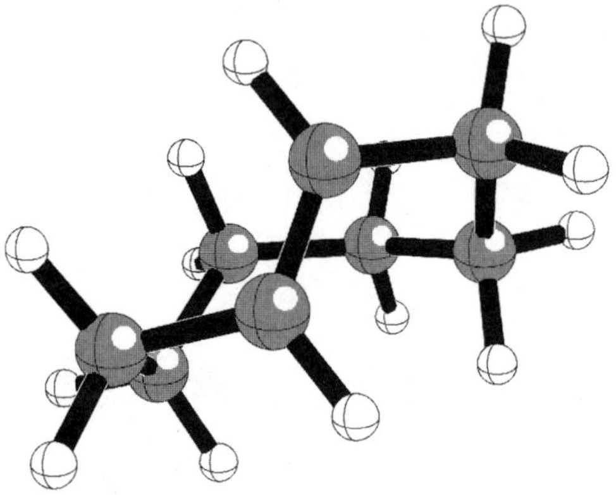

Figure 12-2. Conformation of *trans-cyclo*octene from PCModel.[7] Structure created by M. T. Huggins with Müller and Falk's "Ball and Stick" program (Cherwell Scientific, Oxford, U.K.) for the Macintosh.

In 1970, Scott and Wrixon[9] proposed an "octant rule" for olefins. (See Figure 12-4.) The rule was tested on a wide range of natural products and was found to have a limited usefulness.

By the late 1970s, it was widely accepted that the relative simplicity of the 2π electron system of the C=C bond was not manifested in the ultraviolet and chiroptical spectra of this chromophore. Whereas the electric dipole forbidden $n \rightarrow \pi^*$ transition for the C=O bond of ketones and aldehydes is well separated (~ 280 nm) from other possible transitions, in olefins many overlapping transitions occur in the immediate vicinity of the lower bounds of most commercially available UV, ORD, and CD spectrophotometers.[10–12]

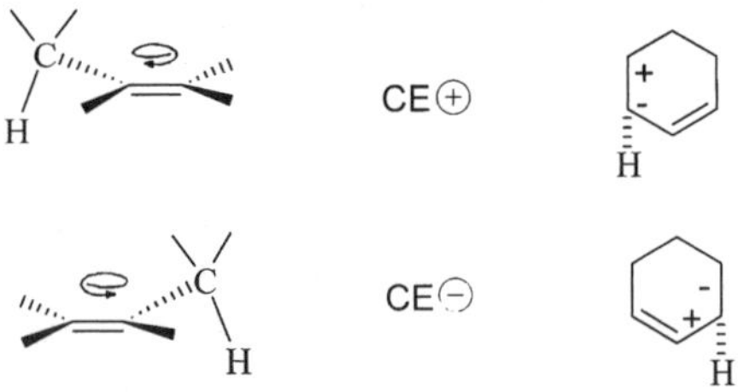

Figure 12-3. Allylic Quasi-axial Rule Adapted with permission from ref. 8. Copyright© 1967 The Royal Society of Chemistry.

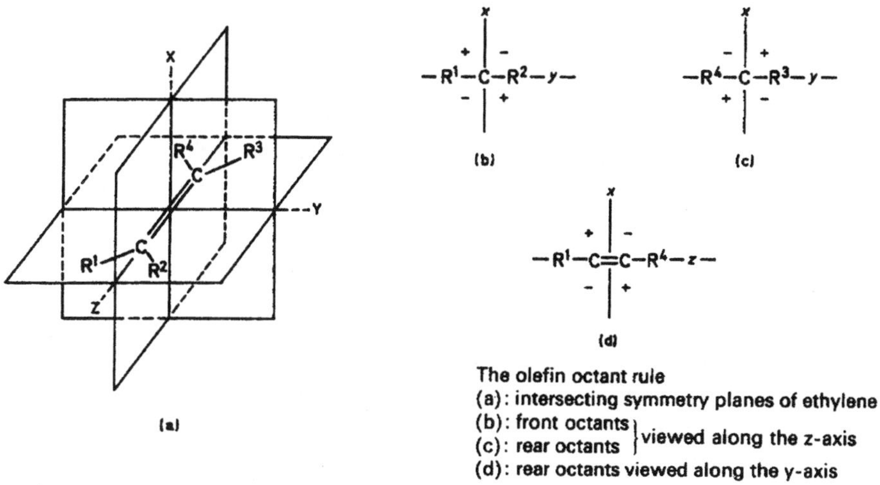

Figure 12-4. An Octant Rule for alkenes. Reproduced with permission from ref. 9. Copyright© 1970 Elsevier Science.

12.2. Alkenes: The Current Situation

The principal chiroptical absorption in alkenes arises from the $\pi_x \to \pi_x^*$ transition, but $\pi_x \to \sigma^*$ and $\pi_x \to \pi_y^*$ transitions occur in this region of the spectrum as well.[2] Often, two CD bands of comparable intensity and opposite sign can be observed with chiral olefins. Furthermore, given the small separations amongst these energy levels, mixing of the levels readily occurs, with small perturbations due to substituent groups in the immediate environment of the chromophore. Such mixing of energy levels means that the observed CD absorption near 200 nm may arise from $\pi_x \to \sigma^*$ or $\pi_x \to \pi_x^*$ transitions.[1]

Furthermore, an additional complexity arises in the evaluation of chiroptical data derived from olefins. If the double bond itself is distorted from planarity, then one should treat the chromophore as inherently dissymmetric. On the other hand, a planar alkene may constitute a symmetric chromophore that is asymmetrically perturbed. This dichotomy in behavior has been referred to in earlier chapters.

There have been three major suggestions for the correlation of chiroptical data with structure: a helicity rule for inherently dissymmetric chromophores, an octant rule for planar alkenes, and the allylic axial rule. The classical example of the first situation is *trans*-cyclooctene. Calculations on this highly strained and rigid molecule have been carried out by several groups.[13–16] As a result of their work, Moscowitz and Mislow[6] proposed a helicity rule for the alkene as an inherently dissymmetric chromophore: A positive sign results from a right-handed helix. However, there are very few compounds known wherein the alkene is skewed to an extent that it cannot be considered to fall into the category of a symmetric chromophore that is asymmetrically perturbed. When

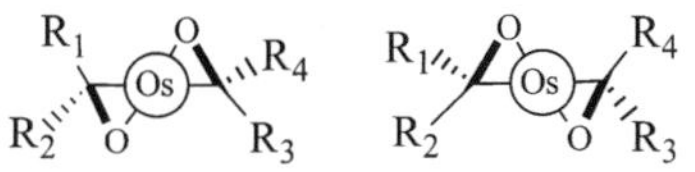

Figure 12-5. Chiral osmate esters from alkenes. Adapted with permission from ref. 20. Copyright© 1969 The Royal Society of Chemistry.

the original results were compared with the experimental evidence,[17] it was seen that a revision was needed.

For the situation of an inherently symmetric, but asymmetrically perturbed, alkene, the octant rule (Figure 12-4) proposed by Scott and Wrixon was employed. With the use of the three symmetry planes of the alkene, and by analyzing compounds of known absolute configuration, signs can be assigned to the spatial array surrounding the chromophore. However, although this rule seems to work well with *endo*-polycyclic alkenes, a couple of problems were recognized: Allylic substitution may play a dominant role, and allylic oxygen (OH, OR, or OCH_3) has an effect opposite to that of carbon or hydrogen.

Indeed, the allylic quasi-axial hydrogen rule for alkenes has been applied quite successfully on a number of occasiions. According to this rule, the optical activity is related to the chirality of the array containing the double bond and any quasi-axial allylic hydrogens: A positive sign correlates with a right-handed helical arrangement. In reality, however, problems arise when one has to sum the effect of several such hydrogens as in Δ^4- and Δ^5-steroids. In such instances, the summation process can be difficult or arbitrary.

Given the foregoing uncertainties, it would appear that the best approach to using chiroptical data with mono-olefins is to compare the unknown with model compounds of known and very similar structure. Alternative suggestions include studying the osmate esters[18–20] formed from chiral alkenes by dihydroxylation and studying platinum(II) complexes[21] of the chiral alkene. In osmate esters, the sense of twist has been associated with the positive or negative Cotton effect curve (Figure 12-5).

Although chiroptical data from alkene chromophores have been used in assigning absolute configurations, such data have not been successfully used for conformational studies.

12.3. Conjugated Dienes

Conjugated dienes, both cisoid and transoid, have been studied throughout the past 40 years. The well-known Woodward's rules[22] for assigning λ^{max} to the parent compounds and to those compounds with substituents directly attached to the chromophore provided a starting point for these studies. Actual measurements can be difficult to make on compounds with large molar extinction coefficients (ε values). When log ε values approach 4 or greater, as they can with conjugated systems, the sample must be prepared at a high level of dilution to allow at least some light to pass through it and reach the detector for a determination of ellipticity (via CD measurements) or rotation (via ORD measurements). Of course, very dilute samples will yield very small responses when the optical rotation (or ellipticity) is measured.

Cisoid Dienes

When a diene—particularly a cisoid diene—is part of a ring system, it is often skewed. In 1965, Butcher, from microwave spectroscopy, determined the degree of non-planarity in 1,3-cyclohexadiene to be 17.5° ± 2°.[23] PCMODEL[7] calculations show that 1,3-cyclohexadiene, which might be considered the parent compound of those discussed in this section, has a skew angle (C1, C2, C3, C4) of approximately 16°. Thus, the conjugated, cisoid diene chromophore of cyclohexadiene is not planar and may be viewed as an inherently dissymmetric chromophore. In the early 1960s, Ziffer, Charney, and Weiss, working at the National Institutes of Health (NIH), often in collaboration with Moscowitz at the University of Minnesota, made significant progress in data collection and interpretation relative to these conjugated dienes. Their collective work[24–28] has served as the basis for later developments.

One of the first problems chemists hoped to solve by chiroptical spectroscopy measurements was that of the configuration and conformation of α-phellandrene and levopimaric acid (Table 12-1).[25,29–32] An important outcome of this work was the demonstration of conformational mobility in molecules such as α-phellandrene. The optical activity of the molecule was clearly shown to be dependent on temperature. It was realized that different conformers might have not only opposite signs, but very different rotational strengths; that is, a small amount of a conformer with a large rotational strength can overwhelm a conformer that is present in greater amounts with the very small rotational strength.[26] The difficulties associated with a complete solution to these problems are nicely summarized by Lane and Allinger.[33]

A helicity rule for *cisoid*-dienes has been stated and thoroughly tested. As shown in Figure 12-6, a skew angle can be defined for *cisoid*-dienes, the helicity of which has been associated with the sign of the Cotton effect: **P** is positive and **M** is negative. It has been pointed out[34] that, due to inaccuracies in the wave functions when the skew angle θ is near 90°, this rule is best applied only when the skew angle is in the range of 0° to 45° or in the range of 135° to 180°. As the rule was being formulated and tested, it appeared that a carbon–oxygen allylic bond was having a profound influence. Beecham and coworkers[35,36] found that the helicity of the O–C–C=C system might have a larger influence that that of the diene itself. Another important early study was that by Gawroński, Liljefors, and Nordén,[37] in which unexpectedly large rotational strengths were exhibited by the planar conformations of the diene chromophore, indicating a very strong influence from substituents.

Lightner and coworkers,[38–40] studied 5-alkyl-1,3-cyclohexadienes from an experimental point of view, including NMR, UV, CD, and, particularly, variable-temperature CD experiments. Particularly interesting was the use of variable-temperature studies analyzing the equilibrium between ψ-axial and ψ-equatorial conformers of 5-methyl

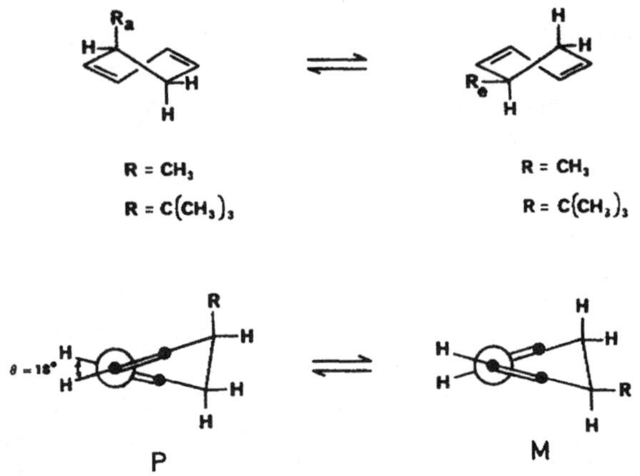

Figure 12-6. Conformational drawings of (+)-(5*R*)-methyl-1,3-cyclohexadiene and (+)-(5*R*)-*tert*-butyl-1,3-cyclohexadiene showing the ψ-axial CH_3 or $C(CH_3)_3$ group associated with **P** diene chirality, and the ψ-equatorial group with **M** diene chirality. The skew angle (ϕ) is indicated to be 18°. Reprinted with permission from ref. 38. Copyright© 1981 American Chemical Society.

and 5-*tert*-butyl-1,3-cyclohexadienes. With the methyl substituent, one sees (Figure 12-7) a virtual invariance of the CD intensity with a significant decrease in temperature. This effect has been interpreted as an indication that there is little difference in energy ($\Delta G^\circ_{ax-eq} < 50$ cal/mol) between the two conformers.[38] On the other hand, the *t*-butyl substituted cyclohexadiene (Figure 12-8) undergoes dramatic changes when the temperature is reduced. In fact, an inversion of sign occurs. From the data, Lightner et al.[38] determined the conformational difference in free energy to be approximately 400 cal/mol. The inversion of sign is associated with the much stronger rotational strength of the ψ-axial substituent ($R_a = +55.6 \times 10^{-40}$ cgs) than the ψ-equatorial group ($R_e = -15.0 \times 10^{-40}$ cgs). Since the ψ-equatorial substituent is only slightly more favored, and the ψ-axial substituent makes a more significant contribution, increasing the temperature to the point where a slight excess of the ψ-axial conformer would give a positive curve.

Lightner et al. also made major contributions to the study of the theoretical aspects of diene optical activity. In their study,[38] they brought together much of the older work along with new data. Perhaps the best way to summarize this important paper is to reproduce its concluding remarks:

> The results of the theoretical analysis presented here illustrate clearly the difficulties involved in defining a model system for the alleged chromophoric parts of a molecule. More specifically, twisted butadiene can definitely not be considered a reliable model system for the long-wavelength circular dichroism of molecules containing a homoannular, cisoid diene system. This is in part due to the observation that the computed rotatory strength associated with the diene group itself appears to be unstable with respect to small variations in structural parameters or computation scheme, and in part because some "non-chromophoric" parts of the molecules play a decisive role in the resulting chiroptical properties. We find that the use of a localized orbital basis provides a convenient and direct procedure for the definition of the proper chromophore, and for a semiquantitative repre-

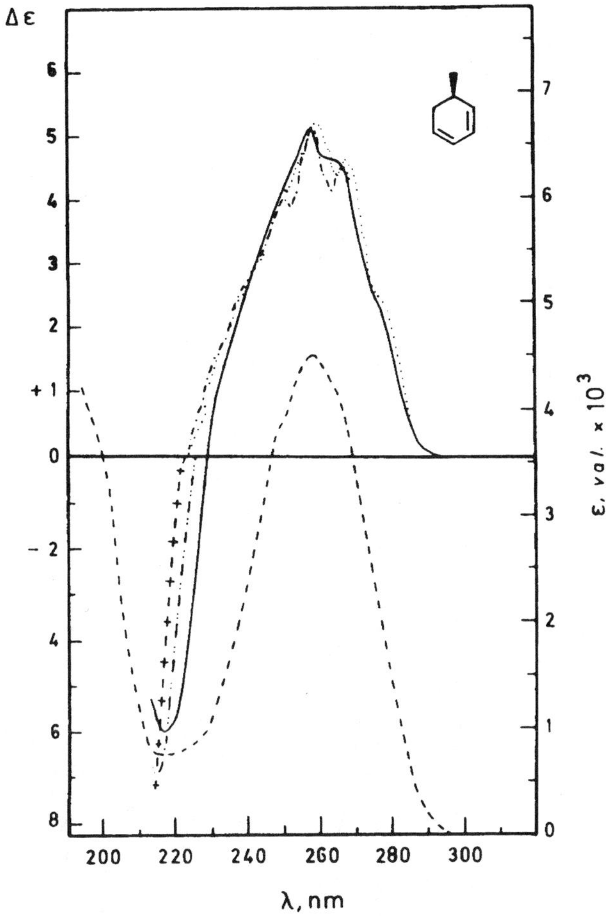

Figure 12-7. Ultraviolet (- - -) spectrum of (+)-(5*R*)-methyl-1,3-cyclohexadiene at 25 °C in methylcyclohexane. CD spectra were measured in 4:1 methylcyclohexane-isopentane at 25.5 °C (—), –44 °C (- ·· -), –70 °C (· · ·), –100 °C (- + - +), and –152 °C (- · · ·) and are uncorrected for solvent contraction. Reprinted with permission from ref. 38. Copyright© 1981 American Chemical Society.

> sentation of the various "helicity rule" and "antihelicity rule" contributions in the form displayed in [Figure 12-9 shown here].

Even so, exceptions have been reported, such as those found with 3,3-dimethoxy-19-norandrostane-5(10),6-dien-17-one (Table 12-1) and several derivatives thereof.[41] These compounds were resynthesized and subjected to X-ray crystallographic analyses to confirm the nature of the structures (at least in the solid state). Lightner et al.[38] suggests that this anomaly can be understood by noting the lack of allylic axial

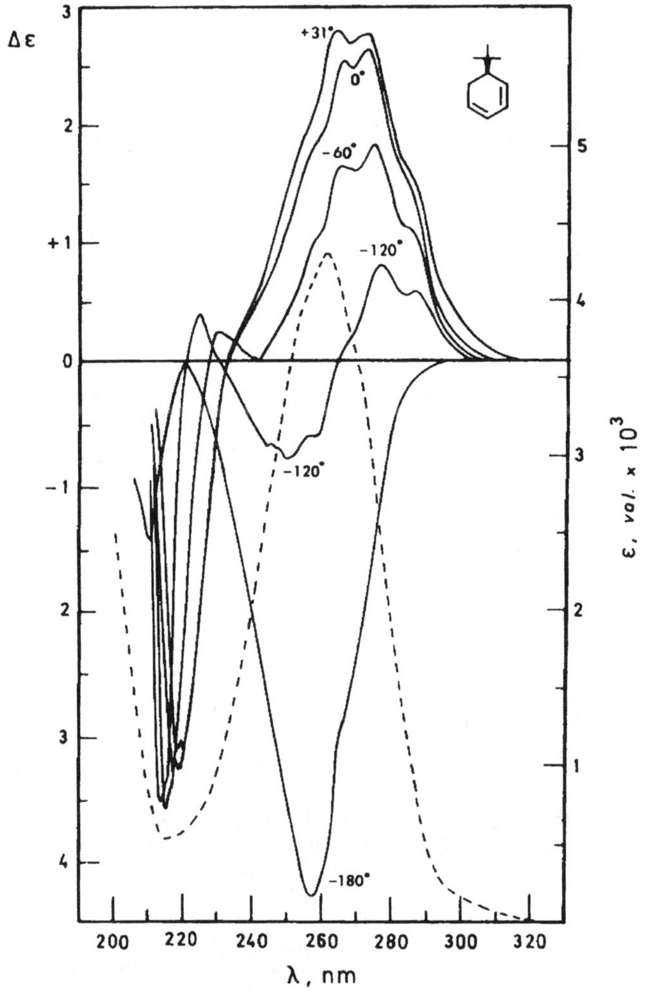

Figure 12-8. Ultraviolet (- - -) spectrum of (+)-(5*R*)-*tert*-butyl-1,3-cyclohexadiene at 28 °C in methylcyclohexane-isopentane (4:1). CD (—) spectra were measured in the same solvent mixture, and temperatures (°C) are indicated on the curves and are uncorrected for solvent contraction. Reprinted with permission from ref. 38. Copyright© 1981 American Chemical Society.

substituents and the combined dissignate effect of three equatorial and ring substituents.

Further contributions commenting on the importance of allylic and homoallylic substitution patterns include the theoretical work of Charney, Lee, and Rosenfield[42] and the work of Burgstahler et al.[43]

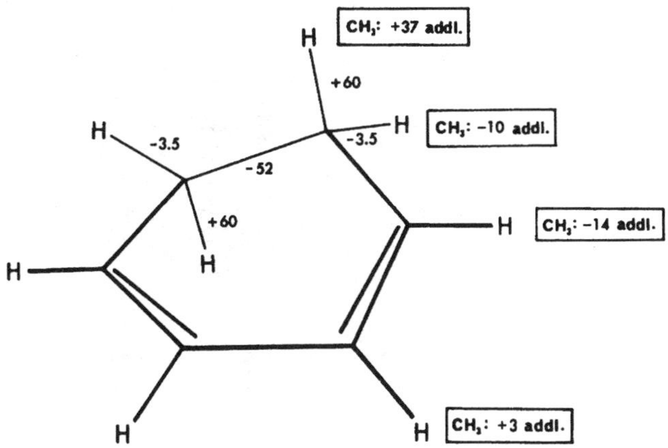

Figure 12-9. Estimated bond contributions to $R^{[r]}$ ($\times 10^{40}$ cgs) for the lowest transition of **P**-helicity 1,3-cyclohexadiene. The total value, R_{CHD}, for 1,3-cyclohexadiene is +48 (**P** helicity), obtained by summing the appropriate contributions shown. Reprinted with permission from ref. 38. Copyright© 1981 American Chemical Society.

Transoid Dienes

Dienes with the *transoid* conformation[24] have been studied by a number of research groups. One of the leading contributors to this field has been Harry Walborsky of Florida State University. In a series of papers in the 1980s, he and his coworkers prepared and obtained spectra from a number of compounds. Among the more interesting examples were planar acyclic 1,3-dienes.[44] From studies of molecules of the alkyl-substituted cyclohexylidene propene family, Walborsky proposed a "planar diene rule."[44] Simply stated, the rule says that "after the 1,3-diene chromophore and all the atoms attached directly to it are placed in a single plane, oriented as shown in Figure 12-10, atoms or groups of atoms falling above the plane will make positive contributions and those falling below will make a negative contribution to the Cotton effect for the long wavelength $\pi \rightarrow \pi^*$ transition."[44]

In 1986, Reddy, Goedken, and Walborsky[45] modified their planar diene rule after studying a number of polyalkyl-substituted cyclohexylidenepropenes. In this version of the rule, it seemed that axial methyl groups near the chromophore were more important than the ring CH_2 groups. In the earlier version of the rule, there was but one nearby axial methyl group and two ring methylenes. In 1987,[46] Walborsky, Gawrońska, and Gawroński prepared and studied several 5-substituted 2-adamantylidene derivatives.[46] From the data, the authors were able to determine that the difference in the contributions of axial and equatorial substituents is positive for transoid chromophores (such as dienes and α,β-unsaturated aldehydes), but negative for cisoid chromophores (such as α,β-unsaturated esters and ketones). In a 1988 publication,[47] a series of chiral 4-axial and 4-equatorial methyl and hydroxyl-substituted *E* and *Z* (1*R*)-2-adamantylidenepropenes (Table 12-1) (and α,β-unsaturated carbonyls) was prepared from a starting material of known absolute configuration.

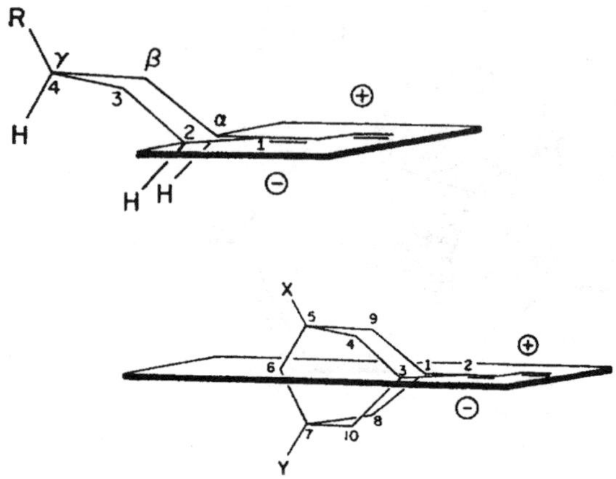

Figure 12-10. A "planar diene rule" as applied to cyclohexylidenepropenes and adamantylidenepropenes. Reprinted with permission from ref. 46. Copyright© 1987 American Chemical Society.

From the results, the authors proposed a "provisional" empirical *bond-centered* "sector rule" for transoid cyclohexylidene and adamantylidine compounds and also presented a "provisional" empirical *atom-centered* "sector rule." (The original manuscript[47] and a review[48] provide a more detailed analysis.) However, suffice it to say that despite the well-designed and excellently performed experimental plan, the answers are not clear cut. A multiplicity of overlapping electronic transitions makes it very difficult to present a correlation rule for these dienes that is as universal as the octant rule for saturated ketones.

As if to underscore the previous statement, a group lead by Piero Salvadori in Pisa has reported on new calculations relevant to Walborsky's earlier work.[49] The newer calculations predicted CD Cotton effect signs opposite to those present in Walborsky's experimental data in every case. Salvadori agrees that, indeed, the dienes are planar, but the two theoretical treatments that he used—the De Voe coupled-oscillators theory and a semiempirical MO-SCF method—gave results inconsistent with these data. Since both treatments have been successful with other chromophores, the origin of the optical activity of planar transoid dienes seems to be an unanswered question.

Contributions to the theoretical understanding of the chiroptical properties of dienes continue to appear. The year 1993 brought a detailed analysis of the interaction of strategically located hydrogens attached to sp^3 carbons and neighboring π-bonds.[50] The authors of the study emphasize the considerable increase expected (on the basis of their calculations) in the intensity of CD bands wherever interaction between C–H groups and the π system can take place. Their results are consistent with the cisoid diene helicity rules currently accepted by most scientists.

While a 1994 paper investigated the conformation of 1,3-cycloheptadiene,[51] and a Ph.D. dissertation also examined the same class of compounds,[52] neither work presented any chiroptical data. However, an interesting paper has appeared describing the

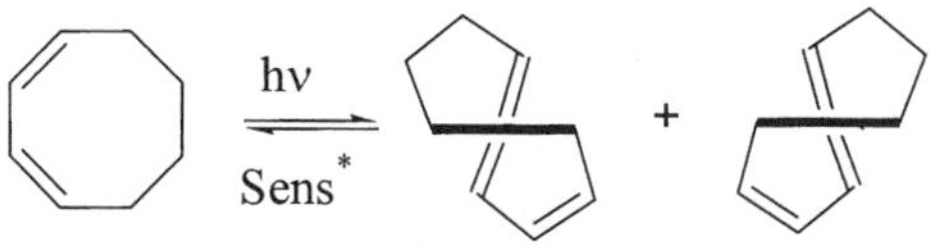

1Z,3Z (R)-(-)-1E,3Z (S)-(+)-1E,3Z

Figure 12-11. Photoisomerization of (Z,Z)-1,3-cyclooctadiene, sensitized by a chiral sensitizer. Adapted from ref. 53. Copyright© 1997 American Chemical Society.

preparation of optically active (*E,Z*)-1,3-cyclooctadiene (using asymmetric photosensitization [see Figure 12-11]) and its chiroptical properties.[53] This work serves to confirm some previous data[54] and presents conclusive evidence for the currently accepted theoretical predictions regarding the optical activity of diene systems. The results are consistent with recent calculations of Bouman and Hansen.[5]

References

1. Burgstahler, A. W., Barkhurst, R. C., and Gawroński, J. K., *Modern Methods of Steroid Analysis* (Heftmann, E., ed.), Academic Press, New York, 1973, pp. 349–369.
2. Mason, M. G., and Schnepp, O., *J. Chem. Phys.* **59** (1973), 1092–1098.
3. Levi, M., Cohen, D., Schurig, V., Basch, H., and Gedanken, V. *J. Amer. Chem. Soc.* **102** (1980), 6972–6975.
4. Hansen, Aa. E., and Bouman, T. D., *J. Am. Chem. Soc.* **107** (1985), 4828–4839.
5. Bouman, T. D., and Hansen, Aa. E., *Croatica Chemica Acta* **62** (1989), 227–243.
6. Moscowitz, A., and Mislow, K., *J. Am. Chem. Soc.* **84** (1962), 4605–4606.
7. PCModel versions 4.0 through 7.0, Serena Software, Inc., Bloomington, IN 47402–3076. PCModel uses the MMX force field, a variation of Allinger's MM2 force field.
8. Yogev, A., Amar, D., and Mazur, Y., *Chem. Commun.*, (1967), 339–341.
9. Scott, A. I., and Wrixon, A. D., *Tetrahedron* **26** (1970), 3695–3715.
10. Mason, S. F., *Molecular Optical Activity and the Chiral Discriminations,* Cambridge University Press, Cambridge, U.K. 1982, pp. 58–63.
11. Legrand, M., and Rougier, M. J., "Applications of Optical Activity to Stereochemical Determinations," in *Stereochemistry*, Vol. 2 (Kagan, H. B., ed.), Thieme, Stuttgart, 1977, pp. 33–182.
12. Snatzke, G., "Chiroptical Properties of Organic Compounds: Some Applications to Unsaturated Systems" in *Optical Activity and Chiral Discrimination,* (Mason, S. F., ed.), D. Reidel, Dordrecht, The Netherlands, 1979, pp. 43–55.
13. Yaris, M., Moscowitz, A., and Berry, J., *J. Chem. Phys.* **49** (1968), 3150–3160.
14. Kovac, S., Solcaniova, E., Beska, E., and Rapos, P., *J. Chem. Soc. Perkin Trans. 2* (1973), 107–109.
15. Robin, M. B., Hart, R. R., and Kuebler, N. A. *J. Chem. Phys.* **44** (1966), 1803–1811.
16. Levin, C. C., and Hoffmann, R., *J. Am. Chem. Soc.* **94** (1972), 3446–3449.
17. Cope, A. C., and Mehta, A. S., *J. Am. Chem. Soc.* **86** (1964), 5626–5630.
18. Djerassi, C., Closson, W., and Lippman, A. E., *J. Am. Chem. Soc.* **78** (1956), 3163–3166.
19. Sakota, N., and Tanaka, S., *Bull. Soc. Chem. Japan* **44** (1971), 485–488.
20. Scott, A. I., and Wrixon, A. D., *Chem. Commun.* (1969), 1184–1186.
21. Scott, A. I., and Wrixon, A. D., *Tetrahedron* **27** (1971), 2339–2369.
22. Woodward, R. B., *J. Am. Chem. Soc.* **63** (1941), 1123–1126; **64** (1942), 72–75, 76–80.
23. Butcher, S. S., *J. Chem. Phys.* **42** (1965), 1830–1832.
24. Weiss, U., Ziffer, H., and Charney, E., *Tetrahedron* **21** (1965), 3105–3120.

25. Charney, E., *Tetrahedron* **21** (1965), 3127–3139.
26. Charney, E., Edwards, J. M., Weiss, U., and Ziffer, H., *Tetrahedron* **28** (1972), 973–979.
27. Moscowitz, A., Charney, E., Weiss, U., and Ziffer, H., *J. Am. Chem. Soc.* **83** (1961), 4661–4663.
28. Ziffer, H., Charney, E., and Weiss, U., *J. Amer. Chem. Soc.* **84** (1962), 2961–2963.
29. Snatzke, G., Kovats, E., and Ohloff, G., *Tetrahedron Letters* (1966), 4551–4553.
30. Burgstahler, A. W., Ziffer, H., Weiss, U., *J. Am. Chem. Soc.* **83**,(1961), 4660–4661.
31. Weiss, U., Ziffer, H., and Charney, E., *Chem. Ind.* (London) (1962), 1286–1287.
32. Burgstahler, A. W., Gawroński, J. K., Niemann, T. F., and Feinberg, B. A., *Chem. Comm.* (1971), 121–122.
33. Lane, G. A., and Allinger, N. L., *J. Amer. Chem. Soc.* **96** (1974), 5825–5830.
34. Hug, W., and Wagniere, G., *Helv. Chim. Acta* **54** (1971), 633–649.
35. Beecham, A. F., Mathieson, A. McL., Johns, S. R., Lamberton, J. A., Sioumis, A. A., Batterham, T. J., and Yound, I. G. *Tetrahedron* **27** (1971), 3725–3738.
36. Beecham, A. J., *Tetrahedron* **27** (1971), 5207–5216.
37. Gawroński, J. K., Liljefors, T., and Nordén, B., *J. Am. Chem. Soc.* **101** (1979), 5515–5522.
38. Lightner, D. A., Bouman, T. D., Gawroński, J. K., Gawrońska, K., Chappuis, J. L., Crist, B. V., and Hansen, Aa. E., *J. Am. Chem. Soc.* **103** (1981), 5314–5327. These results suggest that 1,3-cyclohexadienes substituted in the allylic positions behave like an achiral chromophore that is dissymmetrically perturbed by the substituents. More recent studies by Professor Hansen (personal communication) using DFT-optimized geometries and larger basis sets provide a much more detailed picture of the rotatory strength tensor, with numerical results quite different from the results above. However, the principal conclusions about the role of allylic substituents remains unchanged; a quadrant rule is obeyed and the dissymmetric chromophore concept of a helicity rule is thought to be of minimal importance.
39. Lightner, D. A., and Chappuis, J. L., *J. Chem. Soc. (Chem. Comm.)* (1981), 372–373.
40. Crist, B. V., Rodgers, S. L., Gawroński, J. K., and Lightner, D. A., *Spectros. Int. J.* **4**, (1985), 19–34.
41. Ahmad, R., Carrington, R., Midgley, J. M., Whalley, W. B., Weiss, U., Ferguson, G., and Roberts, P. J., *J. Chem. Soc. Perkin 2* (1978), 263–267.
42. Charney, E., Lee, C.-H., and Rosenfeld, J. S., *J. Am. Chem. Soc.* **101** (1979), 6802–6804.
43. Burgstahler, A. W., Wahl, G., Dang, N., Sanders, M. E., and Nemirovsky, A., *J. Am. Chem. Soc.* **104** (1982), 6873–6874.
44. Duraisamy, M., and Walborsky, H. M., *J. Org. Chem.* **105** (1983), 3264–3269.
45. Reddy, S. M., Goedken, V. L., and Walborsky, H. M., *J. Org. Chem.* **108** (1986), 2691–2699.
46. Walborsky, H. M., Gawrońska, K., and Gawroński, J. K., *J. Am. Chem. Soc.* **109** (1987), 6719–6726.
47. Walborsky, H. M., Reddy, S. M., and Brewster, J. H., *J. Org. Chem.* **53** (1988), 4832–4846.
48. Gawroński, J. K., and Walborsky, H. M., "Diene Chirality" in *Circular Dichroism: Principles and Applications* (Nakanishi, K., Berova, N., and Woody, R. W., eds.), VCH Publishers, Inc., New York, 1994, pp. 301–335.
49. Clericuzio, M., Rosini, C., Persico, M., and Salvadori, P., *J. Org. Chem.* **56** (1991), 4343–4346.
50. Araki, S., Seki, T., Sakakabara, S., Hirota, M., Kodama, Y., and Nishio, M., *Tetrahedron: Asymmetry* **4** (1993), 555–574.
51. Nevins, N., Stewart, E. L., Allinger, N. L., and Bowen, J. P., *J. Phys. Chem.* **98** (1994), 2056–2061.
52. Robbins, T. A., Ph.D. dissertation, University of Nevada, Reno, NV, 1990.
53. Inoue, Y., Tsuneishi, H., Hakushi, T., and Tai, A., *J. Am. Chem. Soc.* **119** (1997), 472–478.
54. Isaksson, R., Roschester, J., Sandström, J., and Wistrand, L.-G., *J. Am. Chem. Soc.* **107** (1985), 4074–4075.

CHAPTER

13

Biaryls and Helicenes

13.1. Benzene Sector Rules

Benzene, the simplest aromatic compound, is an inherently symmetric chromophore (see Section 3.8) that exhibits optical activity when it is perturbed in a chiral sense. Many important compounds, such as phenylalanine (Table 13-1), have chiral centers that are contiguous or homocontiguous (Figure 13-1) with the aromatic ring and, accordingly, have been studied intensely for many years to determine whether correlations can be made between their structures (conformation, configuration, or both) and their chiroptical properties. An excellent and thorough review by Howard Smith of Vanderbilt has recently appeared[1] and accompanies his briefer summary of research in this area.[2] At the National Institute of Health, Elliott Charney gathered much of the relevant theoretical background in his book published more than 20 years ago.[3]

Benzene has three absorption bands in the near ultraviolet, generally ascribed to $\pi \rightarrow \pi^*$ transitions. (See Table 13-2.) In the substituted benzene derivatives that will be considered here (i.e., wherein the substitution is necessary to create the chiral compounds being studied), there is likely to be considerable mixing of these transitions, which will introduce some complexity into the analyses. The first two bands, ^1L_b and ^1L_a, are nominally electric and magnetic dipole-forbidden transitions by spectroscopic symmetry rules. (Recall that the carbonyl $n \rightarrow \pi^*$ transition is electric dipole forbidden, but magnetic dipole allowed.) The third band, ^1B_b, with its large ε value, is clearly electric dipole allowed, but magnetic dipole forbidden. With substitution, the UV transitions tend to shift to longer wavelengths, especially when the substituent atom directly attached to the ring is rich in electrons or is part of a π-system.

Table 13-1. Examples of Chiral Aromatic Compounds

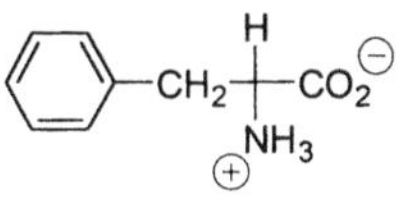	Phenylalanine	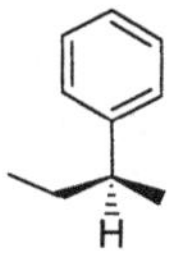	(*S*)-(+)-2-Phenylbutane
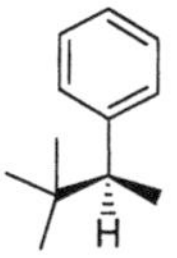	(*S*)-2-Phenyl-3,3-dimethylbutane	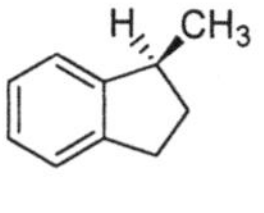	(*S*)-1-methylindane

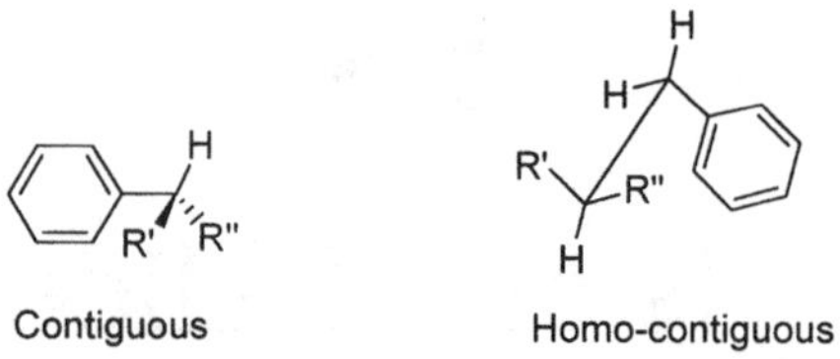

Figure 13-1. Definitions of Contiguous and Homocontiguous Aromatic Substitution.

Furthermore, with ring substitution, degeneracy of the benzene π molecular orbitals is lifted, and transitions are no longer electric dipole forbidden. Magnetic dipole contributions may arise from the mixing of extrachromophoric contributions. Thus, even in the simple chiral derivative (*S*)-(+)-2-phenyl-3,3-dimethylbutane (see Figure 3-11), all three bands exhibit CD intensity, and all have positive Cotton effects. Other, chiral and optically active substituted benzenes have shown similar CD bands for the near UV transitions.

Attempts to correlate aromatic Cotton effect signs with molecular stereochemistry go back to the studies of Verbit[4–8] in the mid-1960s and to Snatzke[9] in the early 1970s. However, only limited success has been achieved by the empirical derivation of sector rules. The choice of sectors depends on the local symmetry that is adopted. Schellman[10] proposed a quadrant rule for monosubstituted benzenes, with one quadrant plane (the plane of the benzene ring) as the nodal plane for all six *p*-orbitals of the ring. The second quadrant plane is orthogonal to the first and passes through ring carbons 1 and 4. DeAngelis and Wildman[11] developed and applied a quadrant rule to a fairly large number of aromatic compounds. These efforts to determine structural properties from optical activity will be examined in this chapter, as will be polyaromatics and biaryl compounds. For derivatives of benzene, it will be useful to discuss the results in four categories: a chiral center contiguous with a benzene ring having no additional ring substituents, a chiral center contiguous with a benzene ring with multiple ring substituents, a chiral center homocontiguous with a benzene ring with no additional substituents, and a chiral center homocontiguous with a polysubstituted benzene.

Table 13-2. Principal Absorption Bands in the UV Spectra of Benzene and Toluene[a]

	Absorption Band Maximum for Benzene (in hexane)		Absorption Band Maximum for Toluene (in hexane)	
Band Designation	λ, nm	ε	λ, nm	ε
^1L_b	254	250	262	260
^1L_a	204	8,800	208	7,900
^1B_b	184	68,000	189	55,000

[a] Data from Lambert, J. B.; Shurvell, H. F.; Lightner, D. A.; Cooks, R. G.,*Organic Structural Spectroscopy,* Prentice-Hall, Inc., Upper Saddle River, NJ, **1998,** 289.

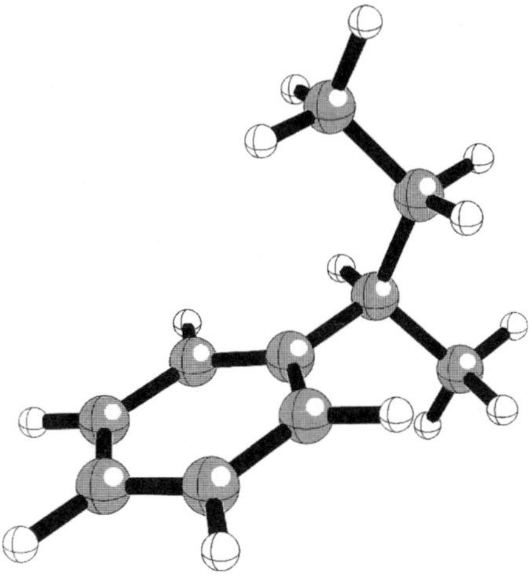

Figure 13-2. Representation of the conformation of 2-phenylbutane as determined by PCModel. Structure created by M. T. Huggins with Müller and Falk's "Ball and Stick" program (Cherwell Scientific, Oxford, U.K.) for the Macintosh.

Molecules with a Chiral Center Contiguous with a Benzene Ring

(*S*)-(+)-2-phenyl-3,3-dimethylbutane, (*S*)-(+)-2-phenylbutane, and (*S*)-(+)-1-methylindane (Table 13-1) are three of the chiral aromatic molecules that have been studied extensively. The first and third of these compounds were used in attempts to find molecules with some degree of conformational rigidity. The simpler molecule, 2-phenylbutane, has a rather different conformation, as determined by molecular modeling.[12] From Figure 13-2, we can see that in the preferred conformation, the benzylic hydrogen lies in the same plane as the aromatic ring itself. For molecules such as this, a fairly simple sector rule—a quadrant rule—has been devised and has met with considerable success. In the benzene quadrant rule, one plane consists of the benzene ring and the attached benzylic carbon and hydrogen. The second intersecting plane (represented by the dashed line in the structural drawing in Table 13-3) passes through the benzylic carbon, the *ipso*- and *para*-carbons of the benzene ring. The signs shown are for the quadrants in front of the plane of the page and reverse for the quadrants behind the plane of the page. In the table, R′ is located in a positive quadrant (behind the plane of the page), and R″ lies in a negative quadrant (in front of the page).

Table 13-3. Benzene Quadrant Rule and Functional Group Priorities

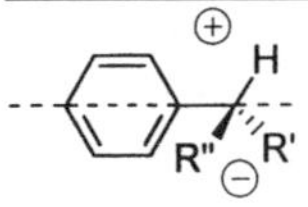	SH, CO_2^-, $C(CH_3)_3$ > CH_3 > NH_2 > $^+NH_3$, $^+N(CH_3)_3$, OH, OCH_3, Cl, I and CH_3 > CO_2H > $^+NH_3$, OH, OCH_3

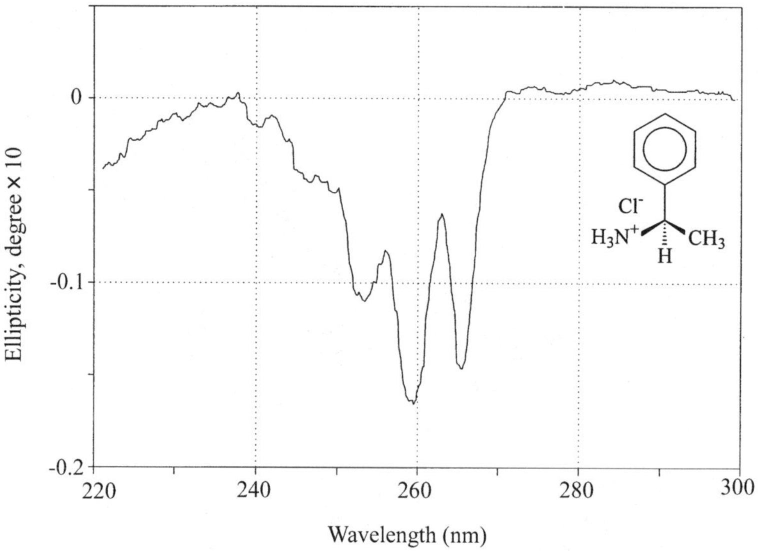

Figure 13-3. Circular Dichroism spectrum of (*R*)-α-phenylethylamine hydrochloride in water (*c* 0.105 g/100 mL, path 1 cm) Replotted using data from Smith, H. E.; Willis, T. C., *J. Amer. Chem. Soc.* **93**, (1971), 2282. Copyright© 1971 American Chemical Society.

From studies with compounds of known absolute configuration, Smith and Fontana[13] prepared sequences that can be used in conjunction with the model shown in Table 13-3 to predict the sign of the ^1L_b Cotton effect:

Thus, if R″ = CH_3 and R′ = OH [(*R*)-α-phenethyl alcohol], the sign of the Cotton effect should be negative, since the methyl group makes a larger contribution than the hydroxyl and the methyl is in a negative quadrant. For similar reasons, (*R*)-α-phenylethylamine hydrochloride (R′ = CH_3, R″ = NH_3Cl) should also have a negative Cotton effect (Figure 13-3). This approach has been shown to have a fairly general usefulness.

Molecules with Substituents on the Aromatic Ring; Contiguous Chiral Center

As can be seen from Figure 13-4, substituents on the aromatic ring can, and do, influence the signs of the Cotton effects that are observed. The nature of the substituent, as well as the relative position i.e., *ortho*, *meta*, or *para*) influence the results. Fortunately, as with similar problems with aromatic multisubstitution, definite patterns have emerged. To use these "rules" for predictive reasons, one must know whether the vibronic contribution of the single chiral substituent is positive or negative and then whether the attached group has a positive or negative spectroscopic moment.[14,15] The reader is directed to Smith's reviews[1,2] for more details.

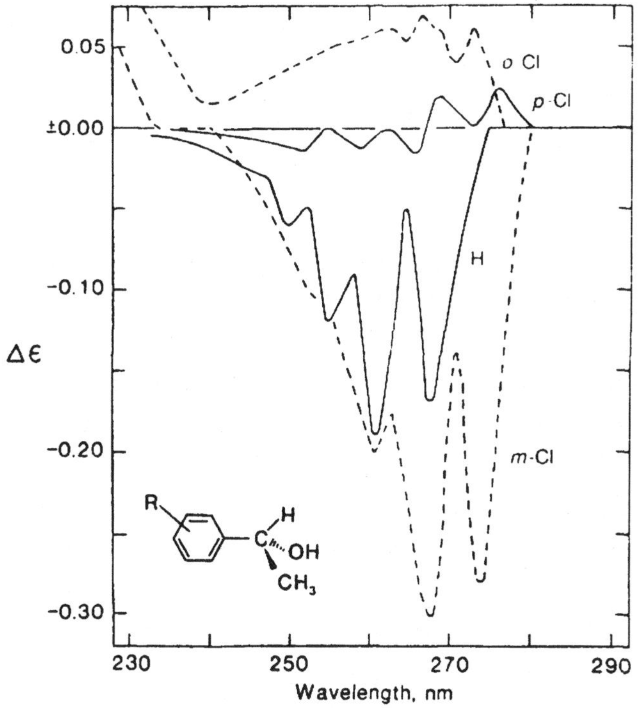

Figure 13-4. Circular Dichroism spectra of (*R*)-α-phenylethyl alcohol, (*R*)-α-(*para*-chlorophenyl)ethyl alcohol, (*R*)-α-(*meta*-chlorophenyl)ethyl alcohol, (*R*)-α-(*ortho*-chlorophenyl)ethyl alcohol in methanol. Reprinted from Pickard, S. T.; Smith, H. E. *J. Amer. Chem. Soc.*, **112**, (1990), 5741–5747. Copyright© 1990 American Chemical Society.

Monosubstituted Aromatics with Homocontiguous Chiral Centers

Unlike the cases discussed earlier, in which only one conformer, that with the benzylic hydrogen eclipsing the aromatic ring (Figure 13-2), had to be considered, situations in which the benzyl group (ϕCH_2–) is substituted on a chiral carbon lead to a di- or tricomponent equilibrium situation as shown in Figure 13-5. Upon a visual analysis, one might assume that the energy of conformer *A* of 1-phenyl-2-hydroxypropane (R′ = OH, R″ = CH_3), wherein the two larger substituents are gauche to the aromatic ring, would be sufficiently higher than that of either conformer *B* or conformer *C* (in which there is one substituent, the OH or the CH_3, and a hydrogen) and, therefore, safely ignored. However, using the rotational energy facility within PCModel,[12] one can predict (Figure 13-6) that while *C* is approximately 2 to 2.5 kcal/mol higher in energy than *B*, *A* is less than about 1 kcal/mol higher than *B*. So there may be significant amounts of all three conformers present in solution at room temperature. Similar results have been obtained by analyzing the energy levels of 1-phenyl-2-aminopropane and its hydrochloride salt. With this ammonium hydrochloride salt, the energy differ-

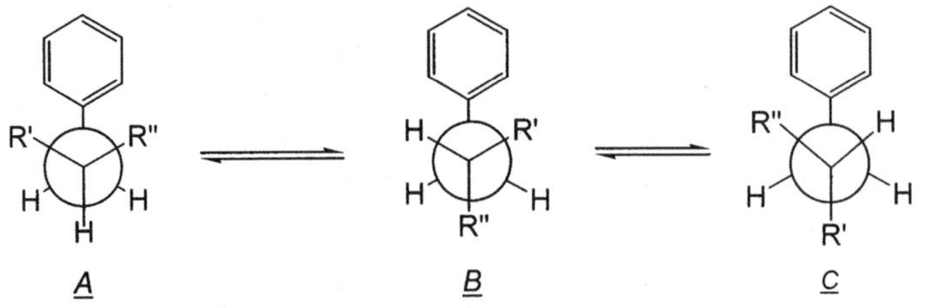

Figure 13-5. Representation of a tricomponent equilibrium mixture involving rotation about the carbon–carbon σ-bond homocontiguous with an aromatic ring.

ence between the two highest energy conformers is estimated to be almost 3 kcal/mol, and one would be justified in treating the chemical as a two-component mixture. The observed Cotton effects will be a function of the concentration and rotatory contribution of each conformer. Thus, to analyze the CD spectrum of a sample of (*R*)-1-phenyl-2-hydroxypropane, it is first necessary to estimate the conformer population of a three-component mixture based on the predicted energy differences. Then, for each

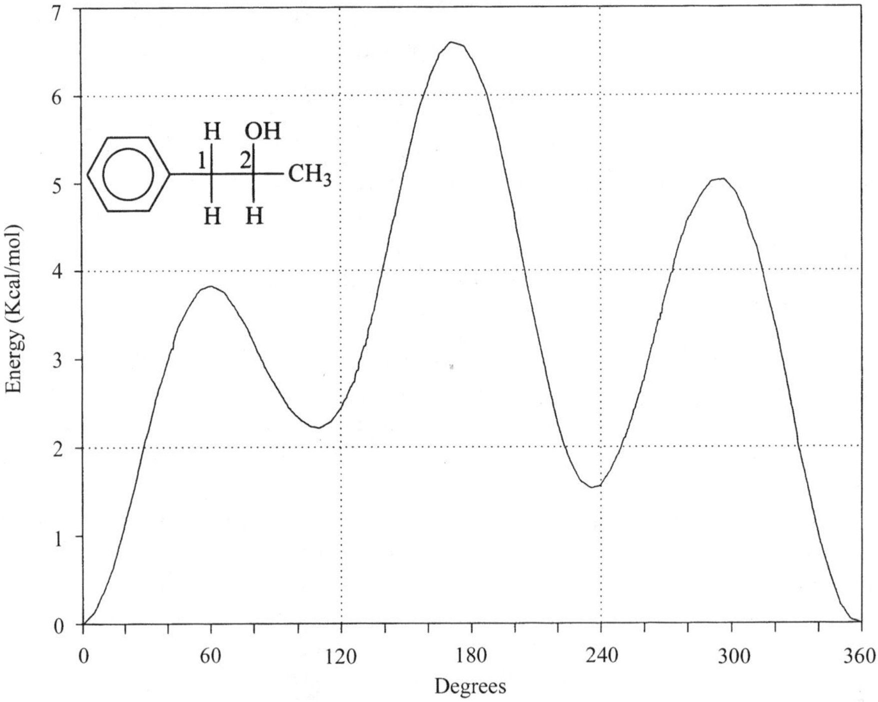

Figure 13-6. Relative energy levels from rotation about the C1-C2 bond of 1-phenyl-2-hydroxypropane as determined from the rotational energy function of PCModel.

conformer, one can use the quadrant rule mentioned earlier (with the caveat that hydrogen atoms make little contribution, as do substituents lying in one of the two symmetry planes) to predict the sign and intensity of that conformer. The summation of these factors should match with the observed data. When one can legitimately ignore the highest energy conformer, the problem becomes much simpler and much more likely to yield reliable data.

CD spectra which are the result of conformational diversity such as that just described would likely be very sensitive to changes in both solvent and temperature. For (*R*)-2-amino-1-phenylpropane, the spectrum is reported to be negative in methanol, ethanol, and 2-propanol, but in cyclohexane it becomes positive, although with a weak negative maximum at 272 nm.[16] No variable-temperature studies on molecules such as these have been reported. As can be seen, extreme care must be taken in making conformational assignments in such cases.

Typically, additional chiral centers located further away (than homocontiguous ones) from the aromatic ring do not affect the ^1L_b band Cotton effects.

Molecules with Homocontiguous Chiral Center; Multisubstitution on the Aromatic Ring

As exemplified in Figure 13-4, the introduction of substituents onto the aromatic ring may cause changes in the observed spectra. This situation also obtains with molecules having a homocontiguous chiral center. Smith[1,2] has gathered considerable information dealing with such molecules, and all those interested should consult his work.

13.2. Biaryls or Dissymmetric Chromophores

Very early on, Van't Hoff[17] recognized that a chiral center—using the language of logic—was neither necessary nor sufficient for optical activity. For sufficiency, we need think only of *meso* compounds—materials with multiple chiral centers, but overall molecular symmetry; for necessity, we look at allenes, twisted biaryls, and the helicenes. In the rest of this section, the focus is on the biaryls. In the next section, we discuss helicenes.

Although it is common to draw biphenyl (or diphenyl) as a planar molecule, it has long been known, and is easily demonstrated by molecular mechanics, that the two rings lie in different planes. The twist angle for biphenyl (Figure 13-7) is predicted by PCModel to be around 40° (dihedral angle 2,1,1′,2′). In 1976, on the basis of X-ray diffraction studies,[18] biphenyl was reported to be planar in the crystal form. For biphenyl in the gas phase, the torsion angle was reported to be 42° from electron diffraction studies.[19] Early molecular mechanics calculations by Kao and Allinger[20] yielded a value of 40.2°, which Kao later revised to 38.4°.[21] When Lii and Allinger introduced MM3 in 1989, the value was again given as 40.2°, and Allinger's calculations suggest that crystal-packing forces favor the planar over the nonplanar structure and force the molecule into a planar conformation in the crystal.[22] While the energy

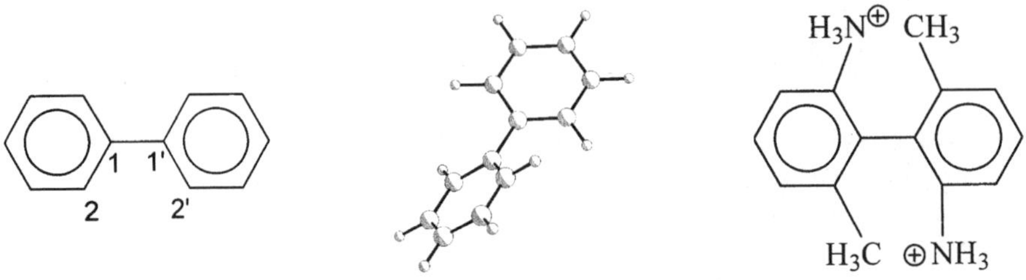

Figure 13-7. Structure and conformation of biphenyl and a tetrasubstituted biphenyl.

of an isolated biphenyl increases about 1.98 kcal/mol going from a twisted to a planar conformation, calculations showed an improvement of about 4.64 kcal/mol from lattice stabilization forces in the crystal which easily outweighs the destabilization of an individual molecule.[22]

In 1944, the CD spectrum of an isomer of 6,6′-dimethyl-2,2′-diaminodiphenyl hydrochloride (Figure 13-7) was measured by Kuhn in Switzerland.[23] Since then, many more spectra have been obtained, and a correlation was found between the helicity of the biaryl system and the sign of the long-wavelength Cotton effect. The theory behind this correlation, which involves exciton chirality, is discussed in more detail in Chapter 14. When the discipline of chiroptical spectroscopy underwent a rejuvenation in the early 1960s with the advent of newer instrumentation, biaryls were among the compounds studied. Mislow, then at New York University (and more recently at Princeton University) had been looking into correlations between compounds of known absolute configuration and visible region rotatory power. With the introduction of the Rudolf spectropolarimeter, he began a collaboration with Djerassi that led to exciting results, using first optical rotatory dispersion and then circular dichroism. In one of the earliest papers from these collaborators,[24] biphenyls, 2,2′-bridged biphenyls, and 1,1′-binaphthyls (Figure 13-8) were studied by ORD. Molecular models (and now molecular mechanics) clearly show the nonplanarity of these materials. From the ORD studies, one could see multiple overlapping Cotton effect curves. From the data, it was concluded that when the long-wavelength Cotton effect for the bridged biphenyls and the binaphthyls was positive, the *R*-configuration was indicated, except in the case of the 2,2′-dinitrobiphenyl, where the sign of the Cotton effect is reversed. The authors comment that the reversal of sign is likely due to the fact this long-wavelength band arises from a different electronic transition, one normally not seen in the UV spectrum, because it is masked by strong end absorption from bands at shorter wavelengths. Thus, another bonus of chiroptical spectroscopy is that some electronic transitions which are difficult to see when they are buried under a large neighboring transition can be seen when they have a large enough rotatory power compared with that of their neighbors.

Later, it became possible to measure CD spectra, which, when examined, validated the earlier ORD results.[25] For inherently dissymmetric chromophores with multiple and intense transitions, as found in biaryls, CD would prove to be the measurement of

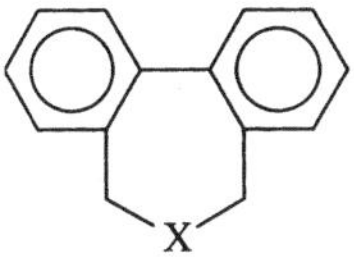

Bridged 2,2'-biphenyl: X= CO; CHOH; O, for example.

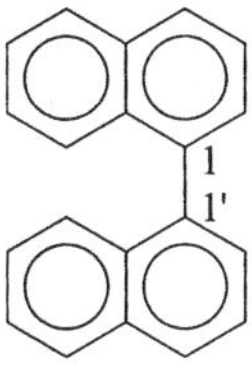

1,1'-binaphthyl

Figure 13-8. Structure of Bridged Biphenyls and Binaphthyl.

choice, yielding a marked reduction of overlapping "tails" compared with ORD curves. Figure 13-9 shows the UV (dashed line) and CD (solid line) spectra of an archetype of these chiral, bridged biaryl compounds. In 1962, Mislow[26] reviewed the various studies correlating biaryl chiroptical spectroscopy with stereochemistry. Also at that time, he and his collaborators had completed measuring the CD spectra of virtually all of the compounds for which ORD spectra had previously been determined.[27] Again,

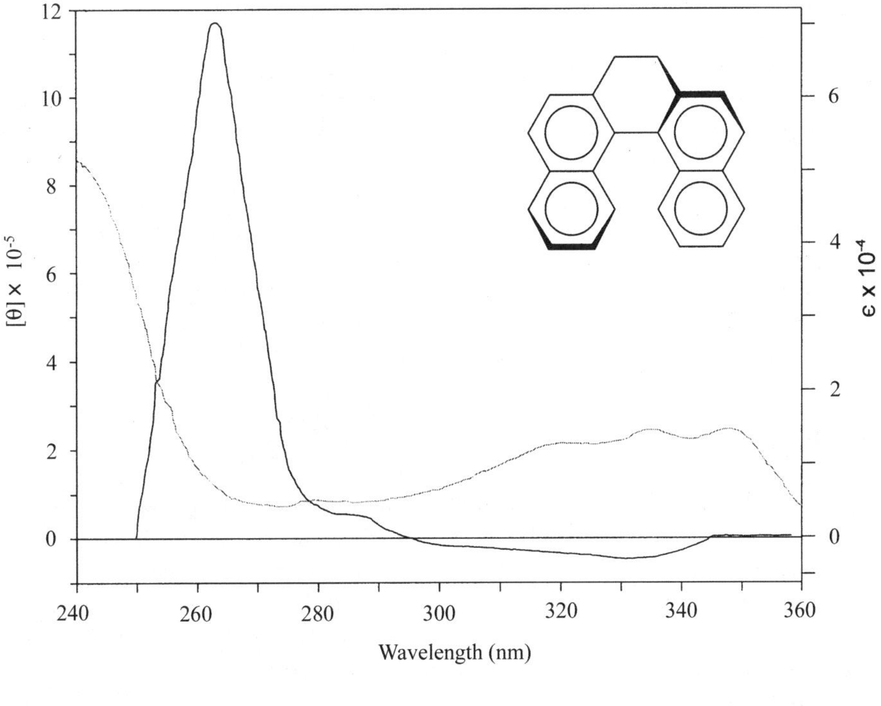

Figure 13-9. The CD (heavy line) and UV (light line) spectra of the 2,2′-bridged 1,1′-binaphthyl. Adapted from ref. 25. Copyright© 1962, American Chemical Society.

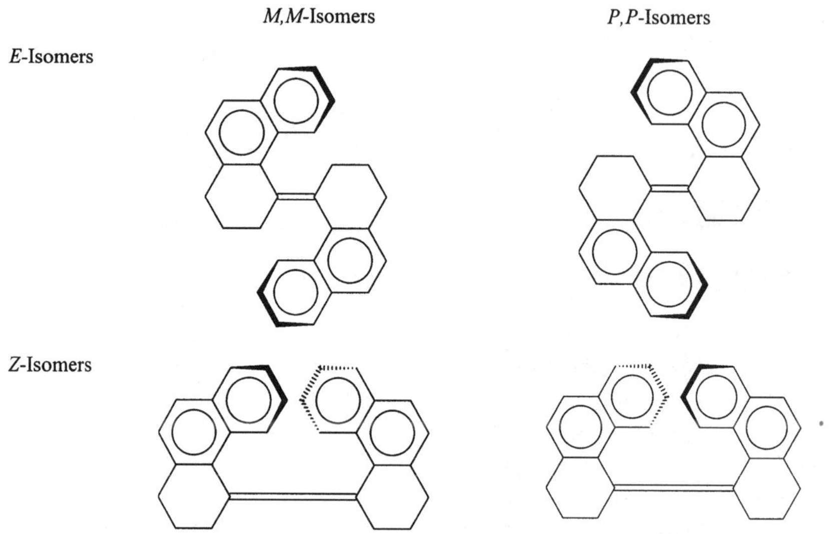

Figure 13-10. Alkenes with Severe Steric Interactions.

all of the earlier data and the interpretations thereof were confirmed and the advantages of CD over ORD pointed out.

Subsequently, in the United Kingdom, Mason played a large role in the application of chiroptical spectroscopy to the prediction of biaryl stereochemistry.[28] His conclusions agreed with the earlier data (i.e., the right-handed helicity led to a positive long-wavelength absorption).

Recently, Harada, Feringa, and coworkers published a series of papers[29] dealing with some unusual olefins (Figure 13-10). In these molecules, severe steric interactions of naphthalene units cause a twisting of a central double bond to the point where this double bond, like that of *trans*-cyclooctene, can be considered to be an inherently dissymmetric chromophore.

13.3. Helicenes

Another molecule of historical interest has been the focus of much research over the years: Hexahelicene is one member of a class of compounds in which steric forces make what might have been considered a planar molecule adopt a helical conformation clearly seen in Figure 13-11.

In this molecule, all the electronic transitions of the chromophore, which is the complete molecule, are electric dipole and magnetic dipole allowed. All the transitions have large rotatory power, and the rotation of the molecule is enormous.[30] In this case, there is no question of conformational analysis, but chiroptical measurements were expected to lead to the assignment of an absolute configuration to the helicenes. The

Hexahelicene as a planar molecule   Hexahelicene as a space-filling molecule

Figure 13-11. Hexahelicene.

family consists of the penta- through nonahelicenes,[31] as well as various derivatives involving ring substitution or the introduction of heterocyclic rings containing nitrogen or sulfur.[32]

One of the earliest publications was in 1961 by Moscowitz,[33] who reported molecular orbital calculations which led to the conclusion that the dextrorotatory hexahelicene would correspond to a left-handed helix. In 1968, Newman at Ohio State University prepared a large amount of (+)-hexahelicene and was able to measure the UV, ORD, and CD spectra of the compound both in chloroform and in methanol.[34] The UV and CD spectra of hexahelicene may be found in Figure 3-11. Besides Moscowitz, others had entered the computational arena. Fitts and Kirkwood[35] approached the problem from polarizability theory, and Tinoco and Woody[36] used a free-electron model. As Hansen[37] points out, both of these approaches predicted that (+)-hexahelicene would correspond to a right-handed helix, in contrast to the original prediction.[33]

The question was answered unequivocally by 1971. The absolute configuration of (–)-2-bromohexahelicene was determined by X-ray crystallography, and the compound was converted to (–)-hexahelicene.[38] Since the levorotatory bromine-substituted hexahelicene has a left-handed helical configuration, the (–)-helicene must also have the same configuration.

Later calculations by Brickell, Brown, Kemp, and Mason[31] and by Wagnière[39] using more extensive semiempirical molecular orbital calculations gave the correct assignment of the absolute configuration and yielded results that were in fair agreement with the observed data. In retrospect,[38,39] it appears that the early calculations by Moscowitz[33] in the late 1950s included approximations and truncations that were necessary to make the problem amenable to solution with the computers available at that time. Hansen and Bak[37] have now performed newer calculations that do not rely on the empirical X-ray data,[38] but that use a geometry derived from theory. Their results are consistent with the accepted idea that the (–)-hexahelicene has the left-handed helical configuration.

It is interesting to note[40] that (–)-pentahelicene, (–)-hexahelicene, (–)-heptahelicene, (–)-octahelicene, and (–)-nonahelicene all have the ***M***-helicity.

References

1. Smith, H. E., *Chem. Rev.* (1998), **98**, 1709–1740.
2. Smith, H. E., "Circular Dichroism of the Benzene Chromophore" in *Circular Dichroism: Principles and Applications* (Nakanishi, K., Berova, N., and Woody, R. W., eds.), VCH Publishers, Inc., New York, 1994, Chapter 15.
3. Charney, E., *The Molecular Basis of Optical Activity: Optical Rotatory Dispersion and Circular Dichroism,* John Wiley & Sons, Inc., New York, 1979, pp. 227–234.
4. Verbit, L., *J. Amer. Chem. Soc.* **87**, (1965), 1617–1619.
5. Verbit, L., *J. Amer. Chem. Soc.* **88**, (1966), 5340.
6. Verbit, L., and Heffron, P. J., *Tetrahedron* **23**, (1967), 3865–3873.
7. Verbit, L., Pfeil, E., and Becker, W., *Tet. Lett.* (1967), 2169–2172.
8. Verbit, L., and Heffron, P. J., *Tetrahedron* **24**, (1968), 1231–1236.
9. Snatzke, G., Kajtar, M., and Werner-Zamojska, F., *Tetrahedron* **28**, (1972), 281–288.
10. Schellman, J. A., *Accts. Chem. Res.* **1**, (1968), 144–151.
11. De Angelis, G. G., and Wildman, W. C., *Tetrahedron* **25**, (1969), 5099–5112.
12. PCModel, Versions 4.0–7.0, Serena Software, Inc., Bloomington, IN 47402–3076.
13. Smith, H. E., Fontana, L. P., *J. Org. Chem.* **56**, (1991), 432-435.
14. Platt, J. R., *J. Chem. Phys.* **19**, (1951), 263–271.
15. Petruska, J., *J. Chem. Phys.* **34**, (1961), 1120–1136.
16. Smith, H. E., Neergard, J. R., de Paulis, T., and Chen, F.-M., *J. Amer. Chem. Soc.* **105**, (1983), 1578–1584.
17. Van't Hoff, J. H., *Chemistry in Space* (Marsh, J. E., trans.), Oxford University Press, Oxford, U.K., 1891.
18. Charbonneau, G. P., and Delugeard, Y., *Acta Crystallogr.* **B32**, (1976), 1420–1423.
19. (a) Almennigen, A., and Bastiansen, O., *Kgl. Norske Videnskab. Selskabs., Skrifter* **4**, (1958), 1–16.
    (b) Bastiansen, O., and Traetteberg, M., *Tetrahedron* **17**, (1962), 147–154.
    (c) Barrett, R. M., and Steele, D., *J. Mol. Struct.* **11**, (1972), 105–125.
20. Kao, J., and Allinger, N. L., *J. Amer. Chem. Soc.* **99**, (1977), 975–986.
21. Kao, J., *J. Am. Chem. Soc.* **109**, (1987), 3817–3829.
22. Lii, J.-H., and Allinger, N. L., *J. Amer. Chem. Soc.* **111**, (1989), 8576–8582.
23. Kuhn, W., and Rometsch, R., *Helv. Chem. Acta* **27**, (1944), 1080–1102, 1346–1371.
24. Mislow, K., Glass, M. A. W., O'Brien, R. E., Rutkin, P., Steinberg, D. H., Weiss, J., and Djerassi, C., *J. Am. Chem. Soc.* **84**, (1962), 1455–1478.
25. Bunnenberg, E., Djerassi, C. Mislow, K., and Moscowitz, A., *J. Am. Chem. Soc.* **84**, (1962), 2823–2826.
26. Mislow, K., *Annals of the NY Acad. Science* **93**, (1962), 457–484.
27. Mislow, K., Bunnenberg, E., Records, R., Wellman, K., and Djerassi, C., *J. Am. Chem. Soc.* **85**, (1963), 1342–1349.
28. Mason, S. F., Seal, R. H., and Roberts, D. R., *Tetrahedron* **30**, (1974), 1671–1682.
29. (a) Harada, N., Saito, A., Koumura, N., Uda, H., de Lange, B., Jager, W. F., Wynberg, H., and Feringa, B. L., *J. Amer. Chem. Soc.* **119**, (1997), 7241–7248.
    (b) Harada, N., Saito, A., Koumura, N., Roe, D. C., Jager, W. F., Zijlstra, R. W. J., de Lange, B., and Feringa, B. L., *J. Am. Chem. Soc.* **119**, (1997), 7249–7255.
    (c) Harada, N., Koumura, N., and Feringa, B. L., *J. Am. Chem. Soc.* **119**, (1997), 7256–7264.
    (d) Zijlstra, R. W. J., Jager, W. F., de Lange, B., van Duijnen, P. Th., Feringa, B. L., Goto, H., Koumura, N., and Harada, N., *J. Org. Chem.* **64**, (1999), 1667–1674.
30. Moscowitz, A., in Djerassi, C., *Optical Rotatory Dispersion: Applications to Organic Chemistry,* McGraw-Hill Book Company, New York, 1960, p. 170.
31. Mason, S. F., Brickell, W. S., Brown, A., and Kemp, C. M., *J. Chem. Soc. A* (1971), 756–760.
32. Wynberg, H., and Groen, M. B., *J. Am. Chem. Soc.* **92**, (1970), 6664–6645.
33. Moscowitz, A., *Tetrahedron* **13**, (1961), 48–56.

34. Newman, M. S., Darlak, R. S., and Tsai, L., *J. Am. Chem. Soc.* **89**, (1967), 6191–6193.
35. Fitts, D. D., and Kirkwood, J. G., *J. Am. Chem. Soc.* **77**, (1955), 4940–4941.
36. Tinoco, I., and Woody, R. W., *J. Chem. Phys.* **40**, (1964), 160–165.
37. Hansen, Aa. E., and Bak, K. L., *Enantiomer*, (2000), in press.
38. (a) Lightner, D. A., Hefelfinger, D. T., Frank, G. W., Powers, T. W., and Trueblood, K. N., *Nature (London)* **233**, (1971), 124.
(b) Lightner, D. A., Hefelfinger, D. T., Frank, G. W., Powers, T. W., and Trueblood, K. N., *J. Am. Chem. Soc.* **94**, (1972), 3492–3497.
39. Wagnière, G., *Jerusalem Symp. Quantum Chem. Biochem.* **3**, (1971), 127–139.
40. Martin, R. H., and Marchant, M. J., *Tetrahedron* **30**, (1974), 343–345.

CHAPTER 14

Exciton Coupling and Exciton Chirality

The process that takes a molecule from its ground state to an electronically excited state or from an excited state to a higher excited state is called an electronic transition and involves the movement of an electron. In near-ultraviolet–visible absorption, this typically means that the electron moves from an occupied low-lying molecular orbital to a higher lying unoccupied molecular orbital, as in an $n \rightarrow \pi^*$ or a $\pi \rightarrow \pi^*$ excitation. In an electronic transition, the movement of an electron creates an instantaneous dipole or polarization of charge called an electric dipole transition moment, or an induced electric dipole, a vector quantity with both a direction (orientation) and a magnitude (intensity) that vary according to the nature of the particular transition and the chromophore involved. When two or more chromophores are brought into proximity, even though they are not conjugated, they may interact with one another when one chromophore is excited. This interaction is called *exciton coupling*,[1,2] and considerable stereochemical information may be extracted from exciton-coupling spectra. When the interacting chromophores are oriented spatially to form a *chiral* array, the spectral band seen in the circular dichroism (CD) spectrum typically takes on a characteristic *bisignate* form. A chirality rule has been postulated for this phenomenon: The *exciton chirality rule* of Harada at Tohoku University and Nakanishi at Columbia University states that the signed order of the bisignate CD can be correlated with the absolute orientation of the interacting chromophores.[2,3]

14.1. Origin of Exciton Coupling

Two or more chromophores may interact with one another, even if their orbital overlap and electron exchange are negligible, when one chromophore is electronically excited. That is, for electronic transitions of similar energy associated with two chromophores, the electric dipole transition on one chromophore can interact with that from the second chromophore, and the excitation energy becomes delocalized over the two (or more) chromophores. Such dipole–dipole coupling of locally excited states produces a delocalized excitation (called an exciton) and results in a splitting (called exciton splitting) of the locally excited states (Figure 14-1). This splitting is manifested in two distinct bands in the UV–visible spectrum, one redshifted from the center of the local excitation and the other blueshifted therefrom. Unless the exciton splitting is large, however, one often observes simply a single broad band when both exciton transitions are allowed (Figure 14-2). When only one exciton transition is allowed, the band may

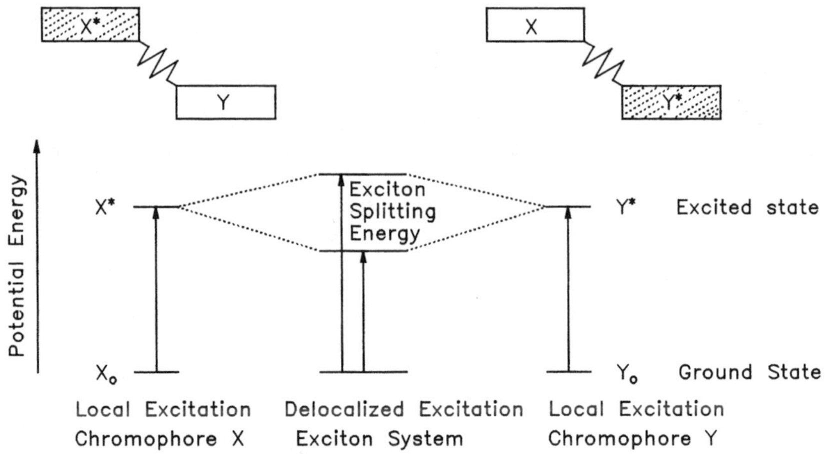

Figure 14-1. Diagrammatic representation of exciton coupling in an array of two chromophores (X and Y) held together by covalent bonding or weak intermolecular forces. Local excitations are shown (left and right) for the chromophores in their locally excited (X^* or Y^*) monomer states. However, when the two chromophores lie sufficiently close to one another, or when the local excitations are sufficiently intense, in the composite system or molecule, excitation is delocalized between the two chromophores, and the excited state (the exciton) is split by resonance interaction of the local excitations. Exciton coupling may take place between identical chromophores (X = Y) or non-identical chromophores (X ≠ Y), but is less effective when the excitation energies are very different (i.e., when the relevant UV–visible bands do not overlap).

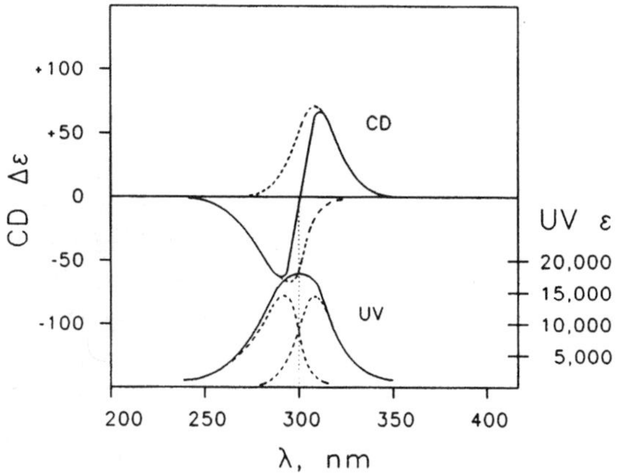

Figure 14-2. Exciton coupling as seen in UV (lower) and CD (upper) spectra. The observed UV curve (——) results from the summation of the UV curves of the two exciton transitions (----). The observed bisignate CD curve (——) results from the summation of two oppositely signed CD curves (----) of the two exciton transitions.

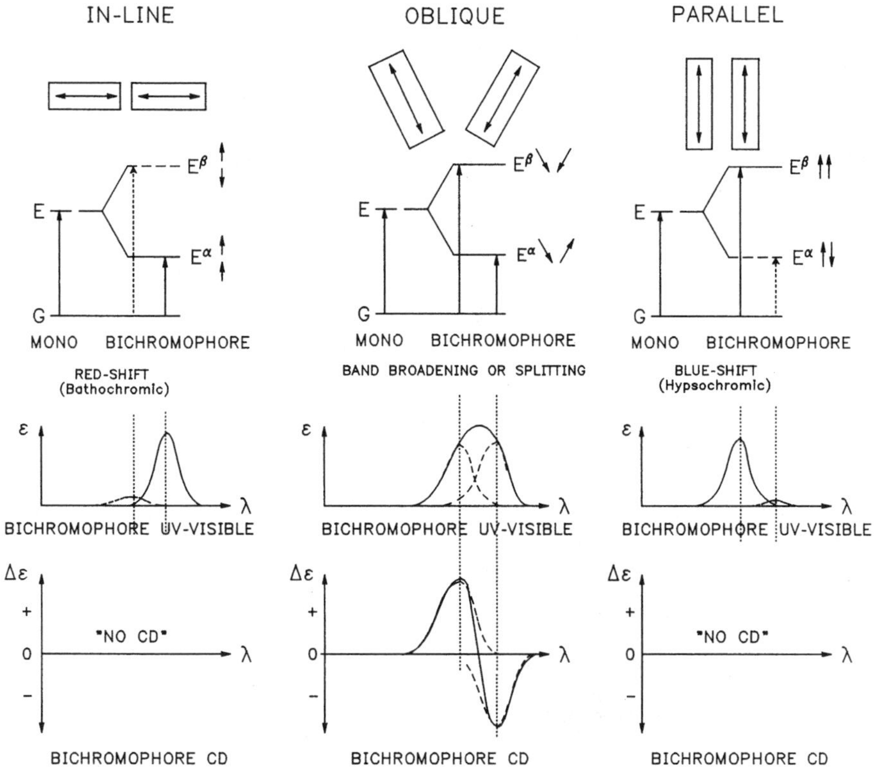

Figure 14-3. Orientation dependence in exciton coupling between two chromophores (rectangular boxes) and their induced electric dipoles (represented by double-headed arrows). The limiting orientations lead to redshifted (in-line), broadened (oblique), or blueshifted (parallel) spectra. The solid arrows connecting ground (G) and excited (E) states represent allowed transitions; the dashed arrows represent forbidden transitions—as noted in the UV–visible spectra. No CD transitions are found when the dipoles have an in-line or a parallel arrangement, neither of which are chiral.

also appear as a single band that is redshifted or blueshifted. Each of the UV–visible exciton bands will have corresponding CD bands if the two chromophores are arranged in a chiral orientation. Since the CD bands from an exciton couplet are always oppositely signed, exciton splitting is usually much easier to detect by CD spectroscopy than by UV–visible spectroscopy.

Exciton UV and CD spectra thus derive from the intrinsic electronic spectral properties of the component chromophores, and they also depend on the distance between the chromophores, as well as on their orientation.[1–4] Since excitons arise from induced electric dipole–dipole coupling during electronic transitions, strongly allowed transitions with large electric dipole transition moments are intrinsically more effective than weakly allowed transitions, especially since dipole–dipole interaction falls off as the inverse cube of the distance between the chromophores. While the interchromophoric distance and the intensity of the electronic transition are important factors in exciton coupling, no less important is the relative orientation of the chromophores—or, more specifically, the relative orientation of the relevant transition dipoles. The orientation is important because it determines whether each of the two exciton transitions is allowed and because it implies stereochemistry and a way to relate absolute stereochemistry to the signed order of the exciton CD couplet. The relative orientation of two dipoles may be broken down into three limiting cases, as shown in Figure 14-3, which correlates the orientation with predicted UV and CD spectra. Thus, stacked chromophores are expected to exhibit a blueshifted UV band, because only the higher energy exciton transition is allowed.[1] Chromophores arranged in-line, on the other hand, exhibit a redshifted UV band, because only the lower energy excitation is allowed. The transition dipoles do not form a chiral array in these two limiting cases; hence, no CD results. In the more general oblique orientation, both exciton transitions are allowed, and the summed UV band may be observed as split or unsplit, depending on whether the exciton splitting energy is large or small. When the dipoles assume a chiral orientation, a bisignate CD curve results from overlapping positive and negative Cotton effects. As has been noted, the signed order of the CD bands correlates with the absolute stereochemistry of the chiral orientation.[2]

14.2. Orientation Dependence in Exciton Coupling

Examples of the orientation dependence in exciton coupling may be seen in the UV–visible spectra of porphyrins[5] (Figure 14-4). When only one chromophore is present, as in Figure 14-4A, the spectra show the typical very intense, sharp Soret band near 400 nm and weak long-wavelength bands near 550 nm. However, when two chromophores are present, as in Figures 14-4B through F, the Soret band is usually split and shifted, which may be taken as evidence for an interchromophoric interaction (by exciton coupling). If the chromophores did not interact, the spectra of Figures 14-4B through F would look exactly like that of Figure 14-4A (at twice the intensity).

When the two porphyrin chromophores lie in a stacked or parallel arrangement (as in Figure 14-4B), a blueshifted single band is observed, exactly as predicted by Figure 14-3 (right). The Soret band is blueshifted and broadened relative to the monochromophore (dashed line, or see Figure 14-4A). At the other extreme, when the porphyrin chromophores are arranged in-line, as in Figures 14-4C and 14-4F, exciton-coupling theory predicts a redshifted band (Figure 14-3, left). In fact two bands are observed, with the more intense band being redshifted. The electric dipole transition moments are probably not arranged perfectly in-line. In other orientations in which the chromophores assume an oblique arrangement, as in Figures 14-4D and E, spectra are predicted to exhibit two bands, which are, in fact, observed. Other examples of exciton coupling in UV–visible spectroscopy may also be found,[1–3] but the splittings are often not as wide as those seen in Figure 14-4. In such cases, when the chromophores

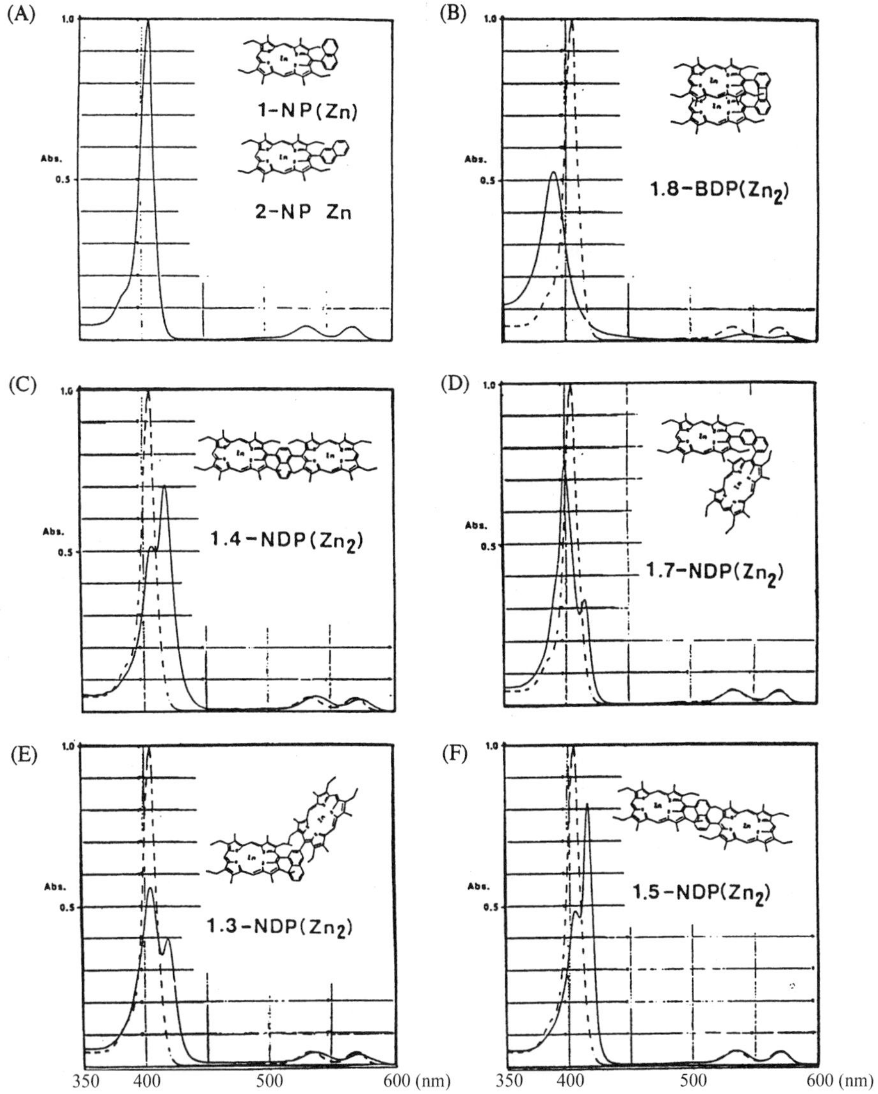

Figure 14-4. UV–visible absorption spectra of zinc etioporphyrins attached covalently to a naphthalene. (A) is the reference spectrum for monomeric 1-naphthyl and 2-naphthyl porphyrins: 1-NP and 2-NP, respectively. (B)–(F) Compare the spectra (dashed line) of the monomer porphyrins of (A) with those of variously oriented dimer porphyrins. [Reprinted with permission from ref. 5. Copyright© 1988 American Chemical Society.]

constitute a chiral array, CD can be an excellent tool for detecting and studying exciton coupling.[2,3,6]

14.3. Exciton Coupling and Circular Dichroism

Anthracene is a particularly good example of a chromophore used in detecting exciton coupling and examining orientation dependence because it has intense UV transitions and thus large electric dipole transition moments, one oriented along its long axis (^1B_b) and another oriented along the short axis (^1L_a), as shown in Figure 14-5. The more intense transition (^1B_b, ε ~200,000) is oriented along the long axis of the anthracene, while the less intense transition (^1L_a, ε ~7,500) is polarized along the short axis. The associated dipole strengths are 88×10^{36} and 2.7×10^{36} cgs, respectively.[2] In the example shown in Figure 14-5, wherein two anthracenes are fused to the bicyclo[2.2.2]octane framework,[2] the intense long-axis polarized induced electric dipoles of the chromophores are oriented neither parallel nor in-line (Figure 14-3), but intersect at an obtuse angle and lie one in each of two intersecting planes (forming dihedral angle of ~120°). The UV spectrum of the compound clearly resembles that of the individual component anthracene chromophores, yet differs in two significant

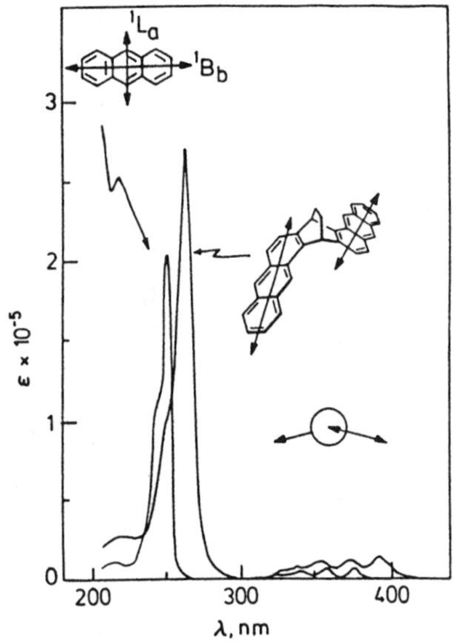

Figure 14-5. UV spectra of anthracene and of bis-anthracene fused to bicyclo[2.2.2]octane. Imposed on the anthracene structure are the in-plane orientations of the electric dipole transition moment vectors associated with the intense short-wavelength UV bands (^1B_b) near 250 nm and the weak long-wavelength bands (^1L_a) near 360 nm. λ^{max} for the exciton bands of bis-anthracene is redshifted. [Reprinted from ref. 6, page 163, Copyright© 1994, with permission from Elsevier Sci.]

ways: It is not simply twice the intensity of one anthracene spectrum, and it is not the sum of two *independent* anthracene transitions, as would be the case if the chromophores did not interact. Rather, the observed spectrum shows a broadened and intensified ^1B_b band redshifted from 252 nm to 267 nm, characteristic of two overlapping *exciton* transitions. The broadening is due to unresolved exciton splitting, and the bathochromic shift is due to an alignment of the ^1B_b transition dipoles at an intersection angle of approximately 150°.

Exciton splitting in the bis-anthracene is far easier to detect by CD than by UV spectroscopy. Thus, in Figure 14-6, two exciton transitions, seen as a single band in the UV spectrum (Figure 14-5), appear as oppositely signed members of a characteristic exciton couplet—just as is predicted by the orientation dependence shown in Figure 14-3. Note, too, that the intensity ($\Delta\varepsilon$) of the CD transitions is very large compared with ordinary CD spectra of noninteracting chromophores (Chapters 5–9). The signed order of the component Cotton effects of the CD exciton couplet may be correlated with the absolute stereochemistry of the molecule, or more exactly, with the helicity of the two relevant electric dipole transition moments.[2,3,6] For the absolute configuration shown, the long-axis transition dipoles make a positive dihedral, or torsion angle, and a positive torsion angle correlates with a long-wavelength positive Cotton effect of the exciton CD couplet, according to the exciton chirality rule.

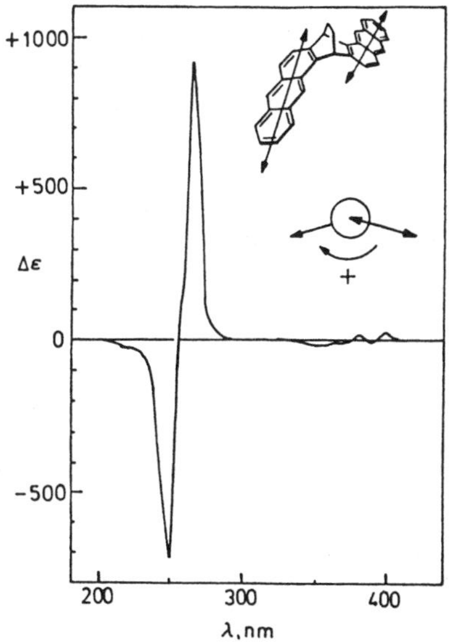

Figure 14-6. CD spectrum of the chiral bis-anthracene of Figure 14-5. The electric transition dipoles (↔) corresponding to the approximately 250-nm UV transition of each chromophore intersect at a positive torsion angle of about 151°. The signed order of the Cotton effects of the 250-nm exciton couplet confirms a (+) helicity of the electric dipole transition moments according to the absolute configuration shown. [Reprinted from ref. 6, page 163, Copyright© 1994, with permission from Elsevier Science.]

14.4. The Exciton Chirality Rule

Exciton coupling leads to a shifted, broadened, and sometimes split band in the UV–visible spectrum when two chromophores are held in close proximity and when there is no orbital overlap. For the composite system or molecule, when the chromophores are held in a chiral orientation, exciton coupling can be detected more easily and more clearly by CD than by UV spectroscopy. In the CD spectrum, two *oppositely signed* Cotton effects are typically observed, corresponding to the relevant UV–visible absorption band(s). The signed order of the CD transitions has been correlated[2] with the relative orientation of the relevant electric dipole transition moments, one from each chromophore.

According to the exciton chirality rule of Harada and Nakanishi,[2,3] when the relevant induced electric dipoles are oriented in a negative (–) torsion angle (negative chirality), the long-wavelength component of the associated exciton couplet can be expected to exhibit a negative Cotton effect (Figure 14-7). When dipoles are oriented to form a positive (+) torsion angle (positive chirality), the long-wavelength Cotton effect is positive, as is found in the bis-anthracene compound of Figure 14-6. For bis-anthracene, the CD spectrum corresponding to the UV spectrum of Figure 14-5 shows an exciton splitting very clearly as two oppositely signed, very intense Cotton effects near 267 nm. In fact, the intensity ($\Delta\varepsilon$) far exceeds any that we have discussed in previous chapters where the Cotton effects ($|\Delta\varepsilon| \sim 1$) are attributable to dissymmetric vicinal action. (One key to recognizing an exciton is a very intense bisignate CD Cotton effect.) The signed order of the Cotton effects can be correlated with the relative orientation of the ^1B_b electric dipole transition moments from the two anthracene

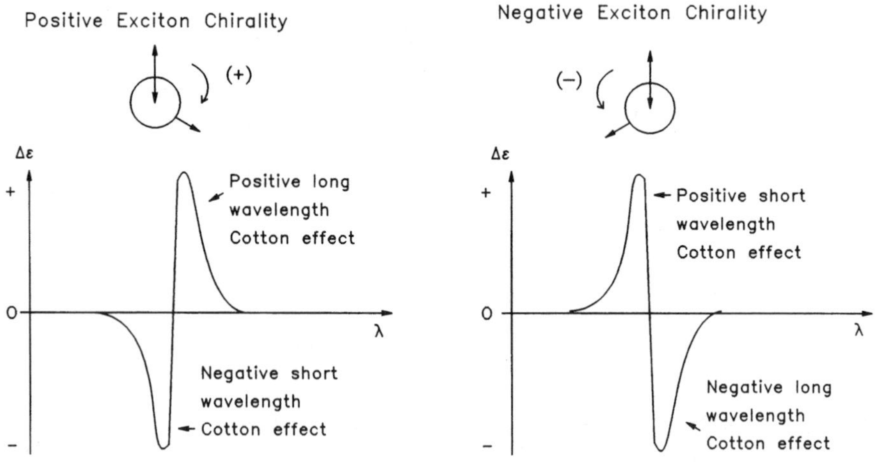

Figure 14-7. Elements of the exciton chirality rule relating the torsion angle or helicity of two interacting electric dipole transition moments (↔) to the signed order of the circular dichroism Cotton effects.

chromophores. For the particular bis-anthracene shown, the extraordinarily intense exciton couplet has a (+) component at 268 nm and a (–) component at 250 nm, corresponding to a (+) helical orientation [positive exciton chirality] of the induced electric dipoles. Thus, from the CD spectrum, one can determine the helical orientation of the transition moments and therefore the absolute configuration of the bichromophore system or molecule.

14.5. Absolute Configuration from Exciton Chirality

There are many examples of applications of the exciton chirality rule to determining absolute configuration.[2,3,7] In most examples, a chiral compound is derivatized to introduce one or more suitable chromophores. Often, hydroxyl groups are derivatized as esters of acids with appropriate chromophores, usually aromatic chromophores such as those of Table 14-1. In some cases the choice of chromophore is such that its intense electric dipole transition moment matches up well with that of a preexisting chromophore, as in the C=C bond of allylic alcohols[2,3] or an α,β-unsaturated ketone.[2,7] In many other cases, such as sugars, diols, or diamines, two or more chromophores are introduced. Here, the ideal chromophore should have a very intense UV–visible electronic transition located in a convenient spectral window, with the orientation of its electric dipole transition moment being well-defined relative to the ester R–O bond. *p*-Dimethylaminobenzoate, which has an intense (ε ~30,000) transition in an easily accessible, generally noninterfering region (near 310 nm) has been one of the chromophore derivatives most often used. In this and in other benzoic acid and cinnamic acid esters (Table 14-1), the relevant induced electric dipole is oriented along the long axis of the molecule, i.e. from nitrogen to carboxyl in *p*-dimethylaminobenzoates. Although the chromophore might in fact adopt a large number of different conformations (relative to its point of attachment on the chiral molecule) by rotating about the ester C–O bonds, one conformation (*s-cis*) predominates. Thus, the relevant induced electric dipoles are aligned parallel to the ester O–R bond (Figure 14-8), which is crucial in applying the exciton chirality rule. With this in mind, one can determine the relative helicity (+ or –) of the electric dipole transition moments by inspection and thus make the assignment of absolute configuration from the CD spectrum. Chromophores with electric dipole transitions not aligned parallel to the ester R–O ester bond, such as 2-naphthoic or 2-anthroic acid, are less satisfactory, but other acids with induced electric dipoles oriented *perpendicular* to the R–O ester bond (e.g., 9-anthroates) have been used satisfactorily.[3] A good derivatizing agent is thus a symmetric acid in which the alignment of a strongly allowed electric dipole transition moment is known with a high degree of certainty. Other carboxylic acid chromophores (Table 14-1), such as *p*-dimethylaminocinnamate ($\varepsilon_{362}^{max} = 44{,}000$) and *p*-methoxycinnamate ($\varepsilon_{306}^{max} = 24{,}000$) have proven quite useful as well.[3]

The application of the exciton chirality rule to determining the absolute configuration of *trans*-1,2-cyclohexanediol is straightforward. Assuming a stable chair conformation with equatorial OH groups, and looking down the C–C bond of the carbons

Table 14-1. Influence of Structure and Substitution on the UV Intensity (ε) and λ_{max} of Aromatic Acids Used as Chromophores in Exciton Chirality[a]

(A)	(B)	(C)
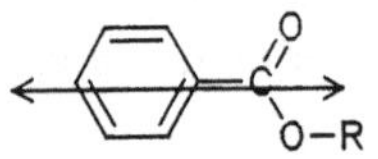	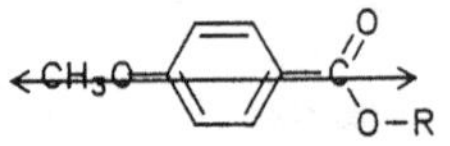	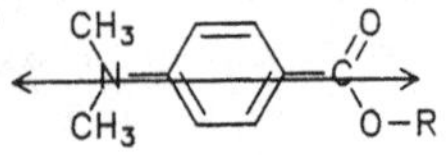
$\varepsilon_{230}^{max} \sim 15,000$	$\varepsilon_{270}^{max} \sim 20,000$	$\varepsilon_{310}^{max} \sim 30,000$
(D)	(E)	(F)
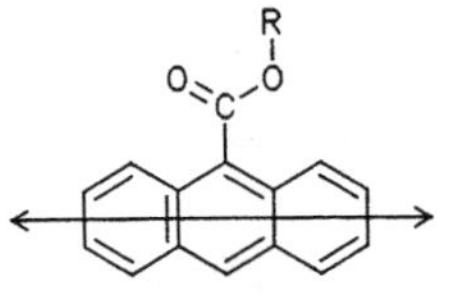	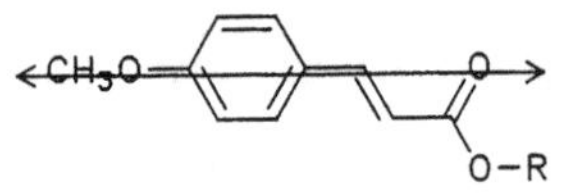	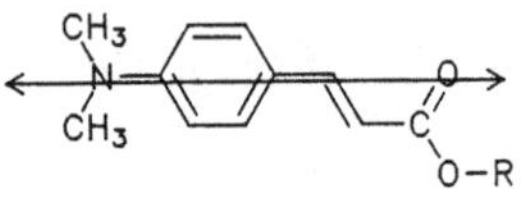
$\varepsilon_{252}^{max} \sim 140,000$	$\varepsilon_{306}^{max} \sim 24,000$	$\varepsilon_{362}^{max} \sim 44,000$

[a]The long axis polarizations of the intense UV transitions are denoted by↔.

bearing the OH groups (Table 14-2), one notes that the O–C–C–O torsion angles have opposite helicities: (+) for the 1(*S*),2(*S*) enantiomer, (–) for the 1(*R*),2(*R*). When a chromophore, such as *p*-dimethylaminobenzoic acid, is introduced in making a diester of the diol, and if the orientation of the relevant electric dipole transition moment of the chromophore is aligned parallel to the C–O bonds of the diol component, then the torsion angle and helicity made by the induced electric dipoles will be (+) for the 1(*S*),2(*S*) and (–) for the 1(*R*),2(*R*). According to the exciton chirality rule, the 1(*S*),2(*S*) diester will exhibit a positive exciton chirality, with a long-wavelength positive and short-wavelength negative couplet for the approximately 310-nm electronic transition(s), whereas the 1(*R*),2(*R*) diester will exhibit a negative exciton chirality with a long-wavelength negative and short-wavelength positive couplet (Table 14-2).

This distinction is seen clearly in Figures 14-9 and 14-10, in which the bis-*p*-dimethylaminobenzoate esters give mirror-image bisignate exciton CD curves,[8] with the signed order of each couplet correlating with the helicity of the diol O–C–C–O torsion

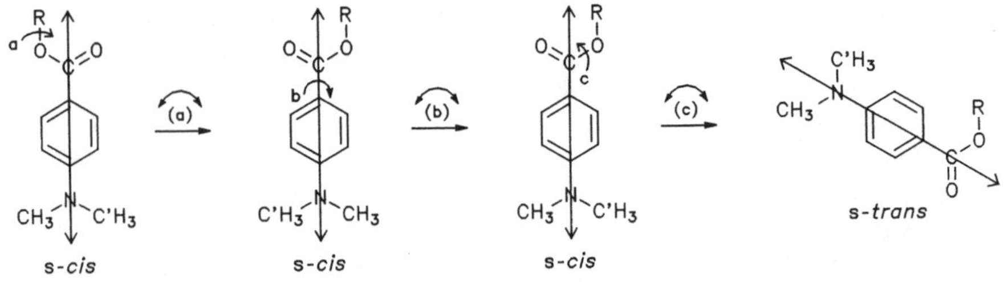

Figure 14-8. Reorientation of the *p*-dimethylaminobenzoate long-wavelength electric diple transition moment (↔) following rotations of the larger group about (a) R–O single bond, (b) the C–C=O single bond, and (c) the O–C=O single bond. Only rotation about (c) reorients the dipole from vertical to inclined.

Table 14-2. Conformational Structure and Newman Projection Diagrams of 1(*S*),2(*S*) and 1(*R*),2(*R*)-cyclohexanediol and Their bis-*p*-Dimethylaminobenzoate Derivatives, and CD Data for the Bisignate Cotton Effects of the Latter

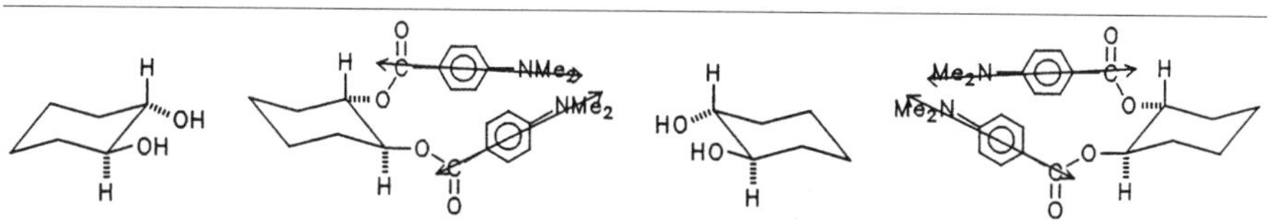

(1*S*,2*S*)	$\Delta\varepsilon_{295}^{max}$ − 44, $\Delta\varepsilon_{320}^{max}$ + 83[a]	(1*R*,2*R*)	$\Delta\varepsilon_{295}^{max}$ + 44, $\Delta\varepsilon_{320}^{max}$ − 83[a]

[a]CD data in CH_3OH solvent from ref. 8.

angle and, more particularly, with its diester. The UV spectrum shows a broadened long-wavelength absorption near 310 nm, lying between the components of the CD couplets, as predicted from Figures 14-2 and 14-3. The CD spectrum of the mono ester (Figure 14-9) is very different and shows only a very weak monosignate curve. The presence of two *p*-dimethylaminobenzoate chromophores makes an enormous difference. The diester CD spectrum is not simply the sum of two monoester CD curves; its striking contrast implies a different origin for the CD transition—one that is diagnostic of, and consistent with, exciton coupling. And the signed order of the bisignate Cotton effects is consistent with the predictions of the exciton chirality rule.[2,3,6]

Thus, orientation, proximity, and the nature of the chromophore are paramount considerations in the exciton chirality rule. Extrachromophoric considerations are relatively unimportant, unless they alter the expected conformation. For example, the CD spectra of bis-*p*-dimethylaminobenzoates of 5α-cholestan-2α,3β-diol and 1(*R*),2(*R*)-cyclohexanediol, both diequatorial diols with the same absolute configuration, are essentially identical (Figure 14-10). Most steroid diols, whether with vicinal hydroxyls or distant hydroxyls, derivatized as *p*-dimethylaminobenzoates give bisignate CD Cotton effects originating from exciton coupling and exhibiting a signed order in the CD couplet that is consistent with the exciton chirality rule (Table 14-3). In most of the examples shown, the relative orientations of the electric dipole transition moments fall into well-defined stereochemical arrangements, and the correlation between a known absolute configuration and the absolute configuration predicted by the exciton chirality rule is excellent. In a few cases, the electric dipole transition moments lie in the same plane, 180° apart (2β,3α) or 0° apart (3β,7β). Although the exciton chirality rule predicts that there is no CD for these orientations, weak bisignate CD spectra are observed. It is unclear whether the very weak Cotton effects are altered

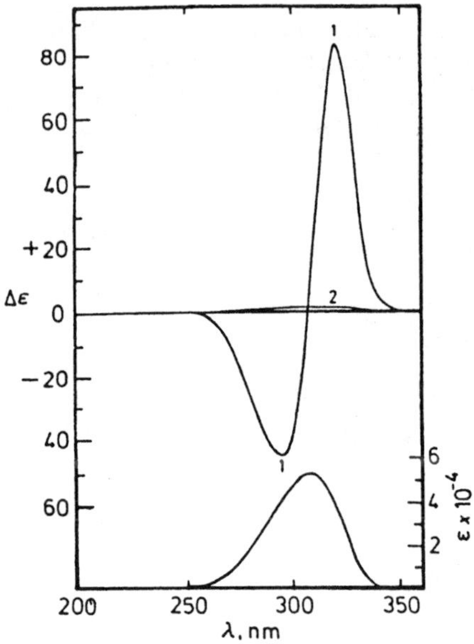

Figure 14-9. CD (upper) and UV (lower) spectra of 1(*S*),2(*S*)-cyclohexanediol bis-*p*-dimethylaminobenzoate (curve 1) and mono-*p*-dimethylaminobenzoate (curve 2) in CH_3OH solvent at 22°C. [Reprinted from ref. 8, page 166, Copyright© 1994, with permission from Elsevier Science.]

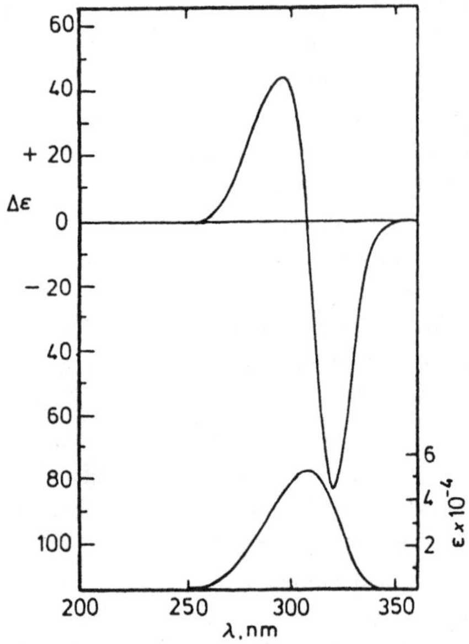

Figure 14-10. CD (upper) and UV (lower) spectra of the bis-*p*-dimethylaminobenzoates of 1(*R*),2(*R*)-cyclohexanediol (c.f. Figure 14-9) and 5α-cholestan-2α, 3β-diol are nearly the same (CH_3OH, 22°C). [Reprinted from ref. 8, page 166, Copyright© 1994, with permission from Elsevier Science.]

(e.g., the sign may be inverted) by changing the solvent, but it is clear that the relevant electric dipole transition vectors cannot strictly be in the same plane. The conformational factors that distort the aforementioned alignments are not understood completely. Relative orientations such as these warrant further study in order to place the exciton chirality rule on a firmer empirical footing.

As expected, the magnitude of the exciton CD Cotton effects tends to decrease with increasing distance between the chromophores.[2,3] On the steroid skeleton, even in the 3,15-diol, in which the chromophores are more than 13Å apart, or the 3,17-diol, in which the chromophores are more than 16Å apart, the $\Delta\varepsilon$ values are still larger than those typically seen in ketones governed by the octant rule.[9] In order to magnify the Cotton effect intensities in long-range interactions, chromophores with more intense electric dipole transition moments have been used. For example, when a porphyrin chromophore is attached to the 5α-cholestan-3β,6α-diol as the diester of 5-(p-carboxyphenyl)-10,15,20-triphenylporphine (structure below), the CD Cotton effects ($\Delta\varepsilon_{423}^{max} = +412$, $\Delta\varepsilon_{414}^{max} = -263$) associated with the porphyrin's intense Soret band ($\varepsilon_{419}^{max} = 350{,}000$) are magnified approximately tenfold above the bis-*p*-dimethylaminobenzoate ester ($\Delta\varepsilon_{319}^{max} = +30$, $\Delta\varepsilon_{294}^{max} = -30$). And in the example of long-range interaction in the 3,17 diol porphyrin diesters, $\Delta\varepsilon_{423}^{max} = +111$ and $\Delta\varepsilon_{414}^{max} = -77$ for 3α,17β; $\Delta\varepsilon_{423}^{max} = -61$ and $\Delta\varepsilon_{414}^{max} = +48$ for 3β,17α; and $\Delta\varepsilon_{423}^{max} = -12$ and $\Delta\varepsilon_{415}^{max} = +15$ for 3β,17β (Figure 14-11).[9]

Even more dramatic long-range exciton interactions have been detected and studied from the porphyrin chromophore attached to a dimeric steroid (Figure 14-12) and to brevitoxin (Table 14-4). In the former, the *Z*-isomer separates the two porphyrin chromophores by about 34Å; yet, a moderately strong bisignate exciton couplet is seen in the CD spectrum. In contrast, when the *E*-isomer separates the two chromophores (by ~29Å), the exciton interaction is weakened because the porphyrin chromophores lie in a nearly in-line arrangement. In the case of brevitoxin, the interchromophoric distance is enormous (40–50Å); yet, a weak exciton couplet is seen in the CD spectrum. These systems constitute the longest range exciton coupling seen to date and amply demonstrate the utility of CD spectroscopy for detecting exciton interactions and the importance of choosing the right chromophore with strongly allowed electronic transitions in creating an observable exciton. Such systems have been used to explore self-assembly and ion transport across lipid bilayers.[9c]

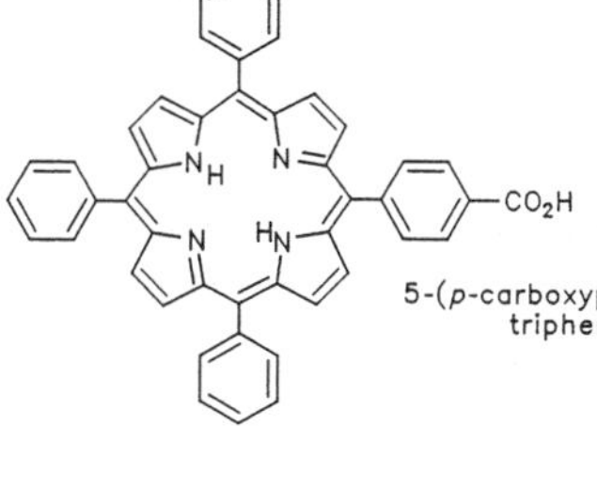

5-(*p*-carboxyphenyl)-10,15,20-triphenylporphine

Table 14-3. Observed and Predicted Exciton Chirality of Steroid Diol bis-*p*-Dimethylaminobenzoates[a]

OH at	Steroid Diol	Bisignate CD Δε(λnm)	Bisignate CD Δε(λnm)	Predicted Exciton Chirality[b]
2α,3β	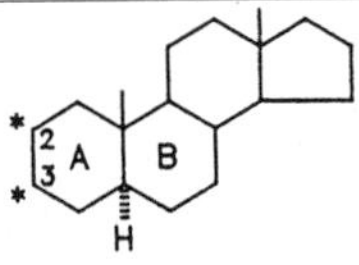	+27(295)	−61(321)	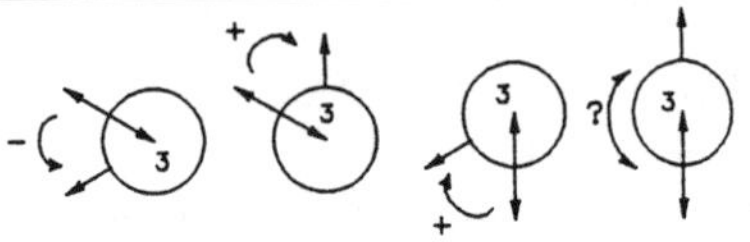 (2α,3β) (2β,3β) (2α,3α) (2β,3α)
2α,3β		−33(295)	+62(320)	
2β,3α		−6(297)	+12(321)	
2α,3α		−9(222)[c]	+20(238)[c]	
3β,4α	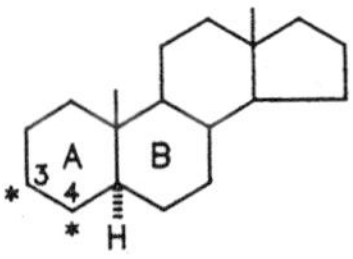	−53(297)	+91(322)	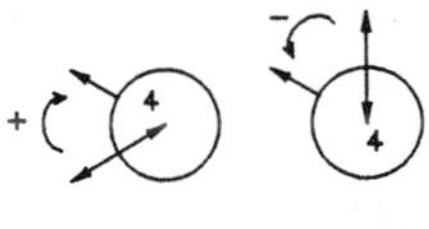 (3β,4α) (3β,4β)
3β,4β		+25(296)	−68(321)	
3β,6β	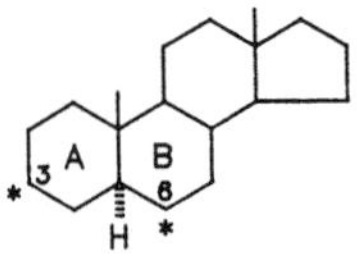	+19(295)	−38(320)	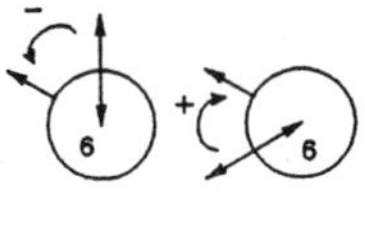 (3β,6β) (3β,6α)
3β,6α		−30(294)	+30(319)	
3β,7α	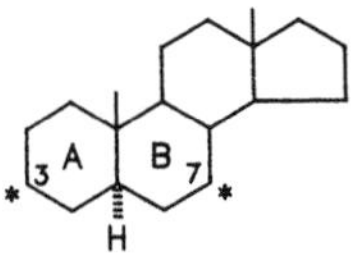	−11(295)	+29(320)	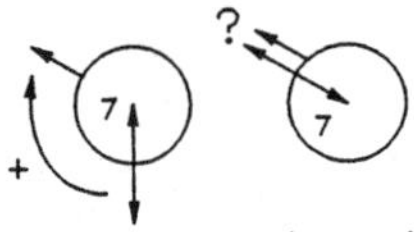 (3β,7α) (3β,7β)
3β,7β		+4(300)	−3(321)	
3β,11β	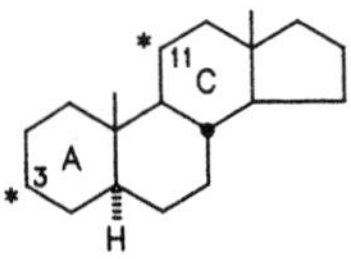	−9(294)	+18(320)	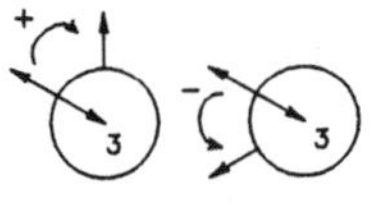 (3β,11β) (3β,11α)
3β,11α		+18(295)	−35(320)	
3β,17β	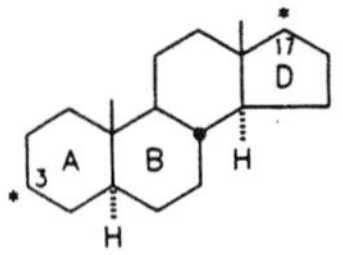	+5(293)	−3(318)	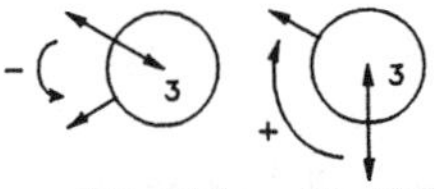 (3β,17β) (3α,17β)
3α,17β		−4(289)	+17(316)	
3β,15β	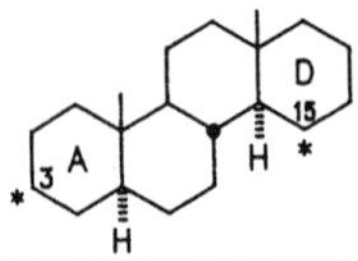	+6(291)	−20(319)	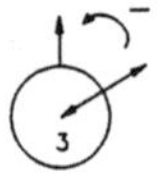 (3β,15β)

[a]The predicted exciton chirality correlates well with the observed CD.
[b]From relative orientations of ^1B_b electric dipole transistion moments of bis-*p*-dimethylaminobenzoates. Epimers not shown have not been studied.
[c]Dibenzoate. Data from refs. 2 and 9.

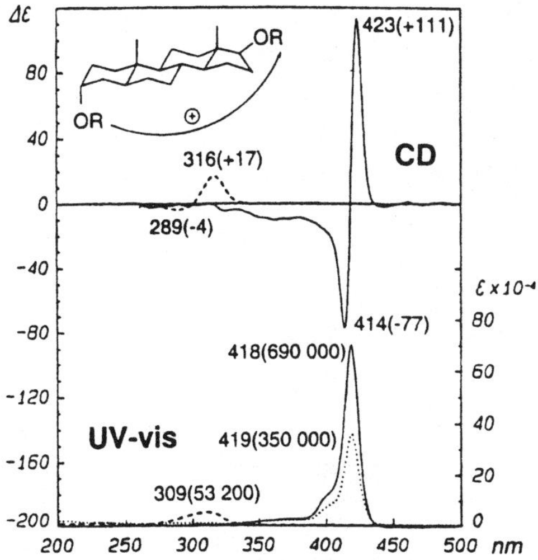

Figure 14-11. Soret UV–visible absorption band of the porphyrin chromophore (·····) CD (upper) and UV–visible (lower) spectra of the bisporphyrin ester (———) and bis-*p*-dimethylamino benzoate ester (- - -) of 5α-androstan-3α,17β-diol in CH_2Cl_2.

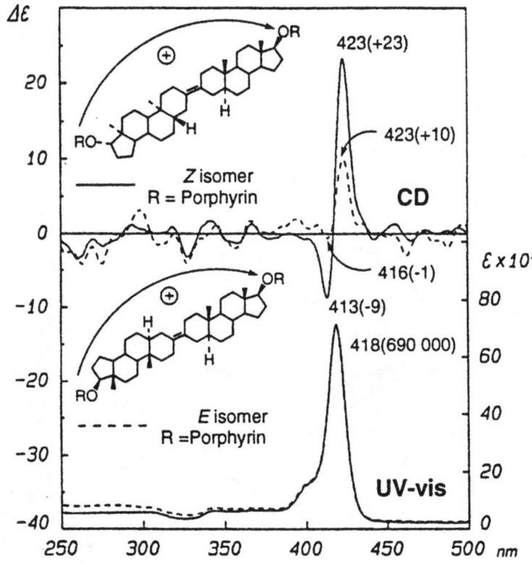

Figure 14-12. CD (upper) and UV–visible (lower) spectra of the bisporphyrin esters of the *Z*-dimeric steroid (———) and the *E*-dimeric steroid (- - -) in CH_2Cl_2. [Figs. 14-11 and 12 are reprinted with permission from refs. 9a and 9b. Copyright© 1995, 1996 American Chemical Soc.]

Table 14-4. Very Long-Range Exciton Coupling between Porphyrin Chromophores,[a] as Detected by CD Spectroscopy

Spacer Unit	Δε (λ,nm)	Δε (λ,nm)
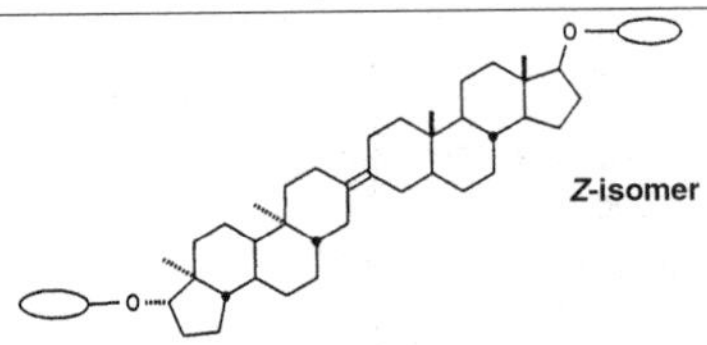	+23 (423)[b]	−9 (413)[b]
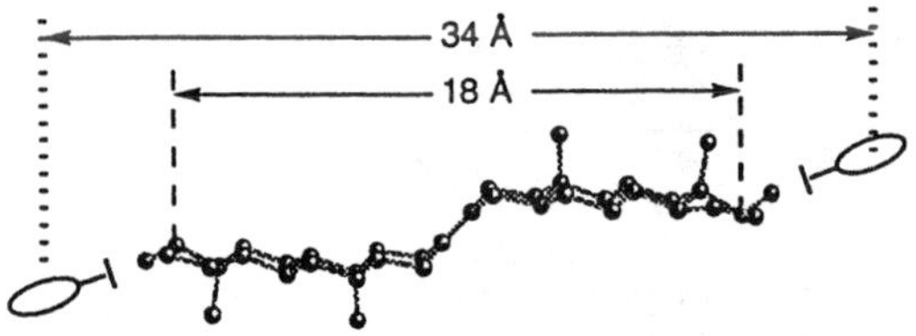	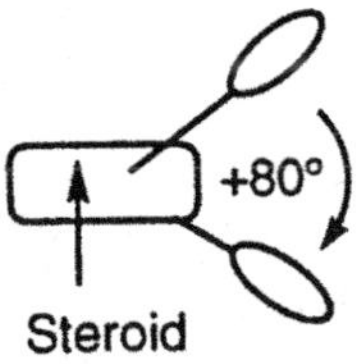	
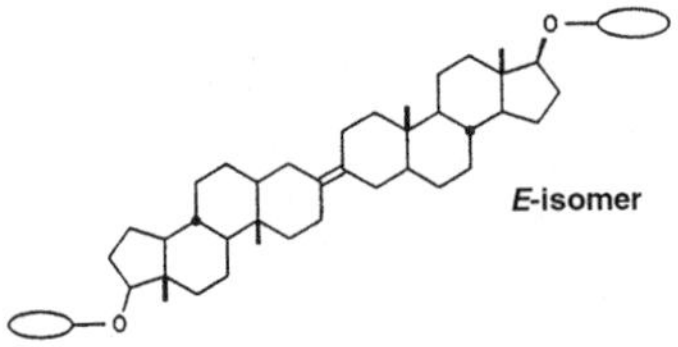	+10 (423)[b]	−1 (416)[b]
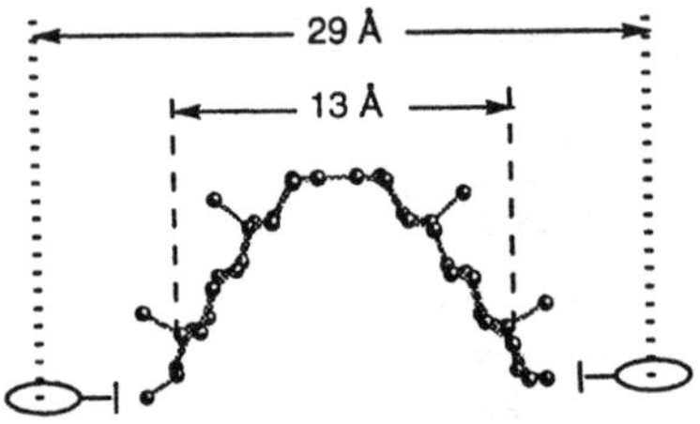		
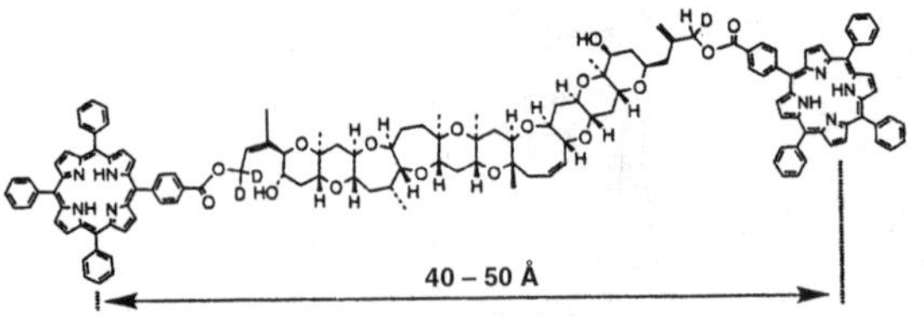	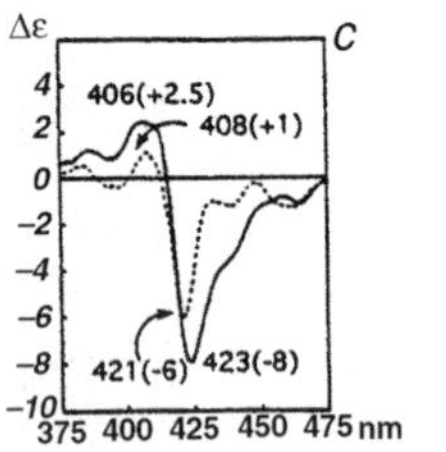	

[a]Represented by ellipses. [b]In CH_2Cl_2. [c]In H_2O-CH_3OH. [Graphics reproduced with permission from ref. 9b. Copyright© 1996 American Chemical Society.]

Table 14-5. Conformational Energy Differences between Diesters of 1(*R*),2(*R*)-Cyclohexanediol and 5-(*p*-Carboxyphenyl)-10,15,20-triphenylporphine[a]

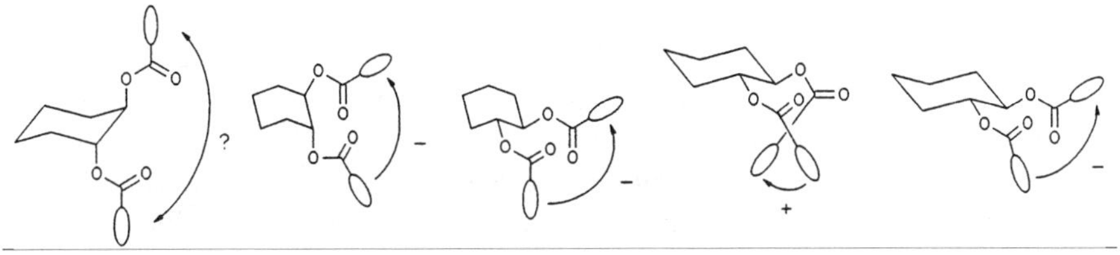

ΔE[b]	0.0	5.7	6.9	7.6	8.1

[a]Porphyrin represented by an ellipse. [b]In kcal/mole (Macromodel ver. 4.5). Data from ref. 9b.

However, even the ideal chromophore may prove less than perfect. The bis-porphyrin ester of 1(*R*),2(*R*)-cyclohexane diol, which brings the porphyrin chromophores into close (vicinal) proximity to each other, was expected to yield extraordinarily strong Cotton effects for the exciton couplet in the Soret band region—especially since the 1,4-diester of 5α-cholestan-3β,6α-diol gave $\Delta\varepsilon_{423}^{max} = -263$. Contrary to expectations, the 1,2-diester of 1(*R*),2(*R*)-cyclohexanediol gave weaker intensities: $\Delta\varepsilon_{421}^{max} = -230$ and $\Delta\varepsilon_{412}^{max} = +170$.[9b] The signed order of the Cotton effects matches the prediction from the exciton chirality rule, but the intensities of the Cotton effects do not. The discrepancy has been explained in terms of conformational changes in the cyclohexane ring.[9b] Molecular mechanics calculations predict that the usually more stable diequatorial chair is some 8 kcal/mole higher in energy than the diaxial chair (Table 14-5). Apparently severe nonbonded steric interactions between the porphyrin units destabilize the diequatorial conformation, with the diaxial chair becoming comparatively much more stable. Even boat conformers are computed to be more stable than the diequatorial chair, although not as stable as the diaxial chair. It is unclear what sort of CD spectrum might be expected from the diaxial chair conformer; the remaining boat and chair conformers predict mainly a negative chirality exciton couplet, which in fact is observed. These results suggest using caution in exciton CD studies of conformationally flexible compounds with vicinal chiral centers.

14.6. Additivity of Pairwise Exciton Couplings in Polychromophore Systems

For compounds with more than two interacting chromophores, it has been shown that the net CD can be approximated by summing the CDs from each pairwise interaction.[3,10] This approximation was demonstrated by Nakanishi et al. for the tetrakis-*p*-bromobenzoates of glucose, galactose, and mannose[11a] and, more recently, with the various bis-*p*-methoxycinnamate esters of α-methyl glucopyranoside (Figure 14-

13A), whose individual CD curves (Figure 14-13B) may be summed to give the observed CD of the tris-*p*-methoxycinnamate ester (Figure 14-13C). Even more impressive is the observation that the observed CD curve (Figure 14-13D) of the tetrakis-*p*-methoxycinnamate ester of α-methyl D-galactopyranoside is reproduced nicely by summing all of the CD curves from pairwise interactions in the bis-esters (Figures 14-13E and F). This observation is impressive because the summation includes many strong Cotton effects, not all of the same sign. (Cf. Figures 14-13E and 14-13F.) The complete spectral library for the three basic pyranoses—glucose, galactose, and mannose—including some 216 characteristic CD curves reflecting a total of 72 permutations of di-, tri- and tetra-acylates, has been catalogued.[11b]

14.7. Redshifted Chromophores

For some compounds with preexisting chromophores such as olefins or α,β-unsaturated ketones, derivatives have been prepared purposefully, with the newly introduced chromophore matching up well in its UV–visible spectral characteristics. Examples include benzoate esters of allylic alcohols[2] and benzoate or *p*-methoxy-, *p*-chloro-, or *p*-bromo-benzoate esters of alcohols containing α,β-unsaturated ketones.[2,7] For many other compounds, however, the preexisting chromophore may complicate analyses of CD curves due to overlap with the introduced chromophores. To avoid such interactions, redshifted chromophores[3] have been proposed whose intense UV–visible absorption is shifted away from that of the preexisting chromophore. The porphyrin chromophore discussed in the previous section is an obvious example. Other redshifted acylating chromophores that have been developed recently include the polyene acids and dipyrrinones of Table 14-6. Still other redshifted chromophores, viz., aldehydes, have been used for derivatives of diamines as their Schiff bases (Table 14-7).[3,12]

14.8. Bisignate CD Sign Inversions

In a very few examples in which the absolute configuration was well established, the exciton CD couplet was unexpectedly reversed in sign.[8b,9b,13] One example, discussed in Section 14.5, is the bis-porphyrin ester of *trans*-cyclohexane-1,2-diol, in which the diaxial chair conformation is thought to predominate. Another example was detected in the bis-Schiff base of 1(*S*),2(*S*)-*trans*-diaminocyclohexane and 7-piperidinohepta-2,4,6-trienal (Figure 14-14), which gave a CD spectrum showing typical exciton-split, bisignate Cotton effects ($\Delta\varepsilon_{405}^{max} = -92$, $\Delta\varepsilon_{358}^{max} = +71$) and a broad UV–visible absorption near 383 nm.[13a] When the trienal above is converted to the bis-cation with trifluoroacetic acid, its UV–visible curve shows two sharp absorptions at 480 and 550 nm, corresponding to a large (2,890 cm^{-1}) exciton splitting (Figure 14-15). Such a large splitting has been observed in very few examples[3,14] and is due in part to the narrow half-widths of the two bands. Again, a negative exciton chirality is seen: $\Delta\varepsilon_{546}^{max} = -232$ and $\Delta\varepsilon_{475}^{max} = +231$. In both the neutral and the bis-cation bis-Schiff base,

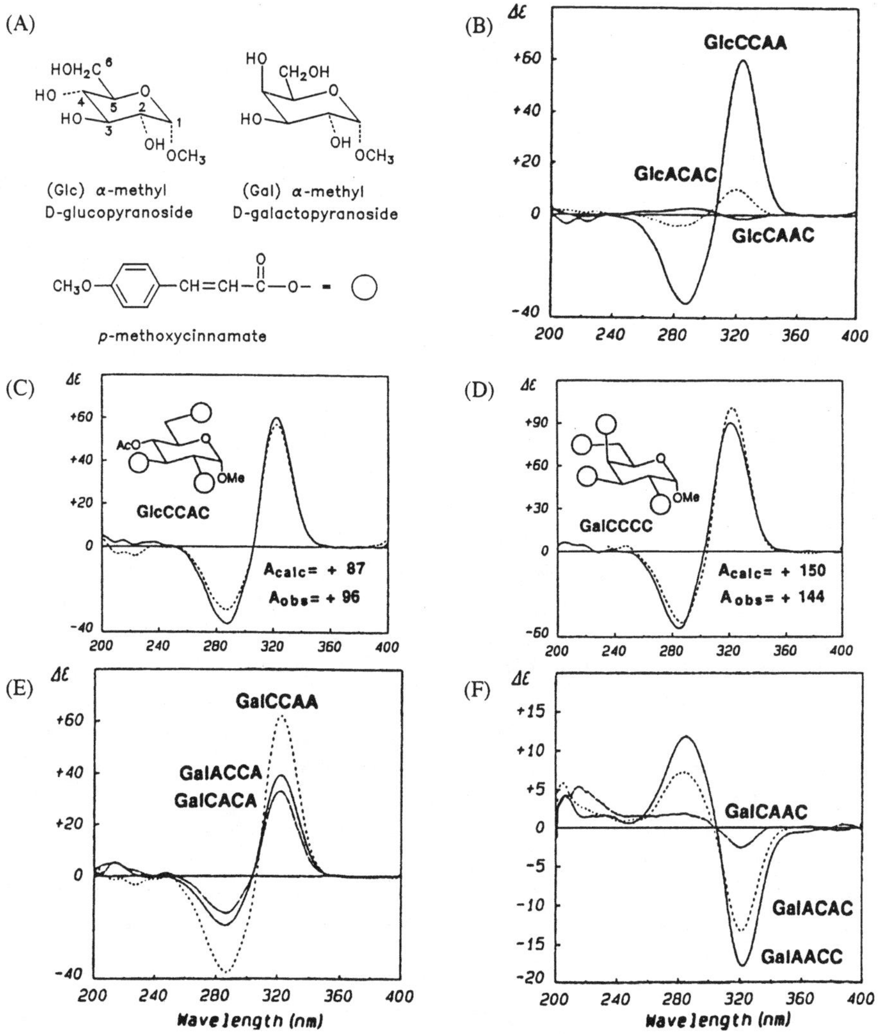

Figure 14-13. (A) Stable chair conformations of α-methyl D-glucopyranoside (Glc), α-methyl D-galactopyranoside (Gal), and *p*-methoxycinnamate (O), the chromophore. (B) CD spectra of three different Glc diacetate-bis-*p*-methoxycinnamate esters. Read the substituent locations as Glc 2,3,4,5 for A = acetate and C = *p*-methoxycinnamate. GlcCCAA has *p*-methoxycinnamate chromophores at C(2) and C(3). (C) Observed (———) and calculated (· · · ·) CD curves of the tris-2,3,5-*p*-methoxycinnamate ester of Glc. A_{obs} and A_{calc} are the sum of the $\Delta\varepsilon^{max}$ values for the couplet, prefixed by the sign of the long-wavelength Cotton effect. (D) Observed (· · · ·) and calculated (———) CD curves for the tetrakis *p*-methoxycinnamate ester of Gal. (E) and (F) CD curves of the various possible bis-*p*-methoxycinnamatediacetate esters of Gal. All CD curves in CH_3CN solvent. [Reprinted from ref. 3a with permission of Wiley–VCH.]

Table 14-6. Red-shifted Acylating Chromophores for Derivatizing Diols and Diamines[a]

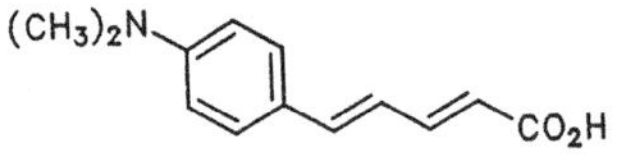	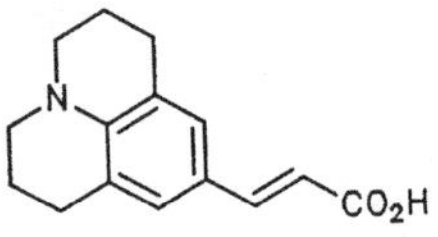	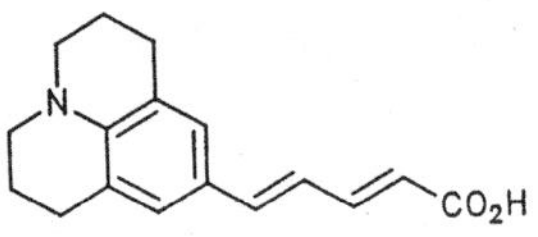
ε_{382}^{max} ~ 34,000	ε_{382}^{max} ~ 27,000	ε_{410}^{max} ~ 37,000
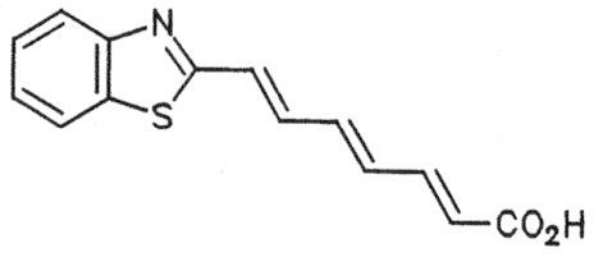	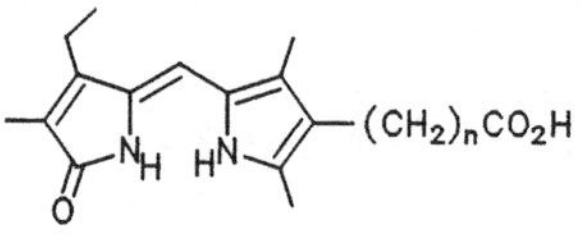	
ε_{358}^{max} ~ 58,000	(n = 1 or 2) ε_{390}^{max} ~ 30,000[b]	

[a]Data from ref. 2b and 10 from Cai, G.; Bozhkova, N.; Odingo, J.; Berova, N.; Nakanishi, K.*J. Am. Chem. Soc.* **1994**, *116*, 3760–3767. [b]Data from ref. 8.

the negative exciton chirality does not correspond to the helicity of the N–C–C–N torsion angle, which is positive. As shown in Figure 14-14, rotations about the cyclohexyl C–N bonds can give two different conformations: one in which the imino-carbon hydrogens are *anti* to the axial hydrogens at C(1) and C(2) of the cyclohexane ring and one in which they are *syn*. C–N bond rotation alters the orientation of the long axis of the chromophore from one that parallels the C–N bond (as in *anti*) to one that does not (as in *syn*). The consequence of such conformational changes is considerable: The *syn* conformer is expected to give a negative exciton chirality CD couplet—opposite to that predicted from the (positive) N–C–C–N torsion angle. The relatively greater stability of the *syn* over the *anti* conformation is supported by molecular mechanics calculations, in addition to the CD spectrum.[3,13] However, no variable-temperature CD measurements have been carried out to explore the conformational equilibrium depicted in Figure 14-14.

Table 14-7. Red shifted Schiff Base Chromophores as Exciton Chirality Derivatives for Amines[a]

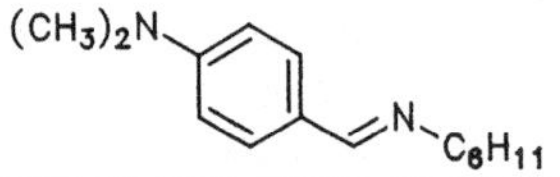	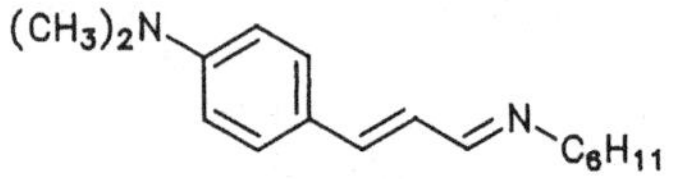	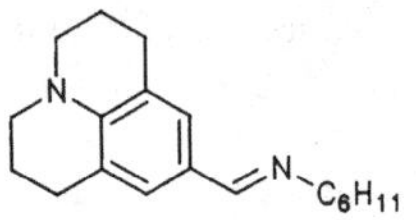
ε_{305}^{max} ~ 24,300 (free base) ε_{395}^{max} ~ 51,700 (+TFA)	ε_{361}^{max} ~ 37,000 (free base) ε_{460}^{max} ~ 64,500 (+TFA)	ε_{331}^{max} ~ 21,400 (free base) ε_{420}^{max} ~ 48,300 (+TFA)

[a]Data from ref. 3 for the free base and the protonated form.

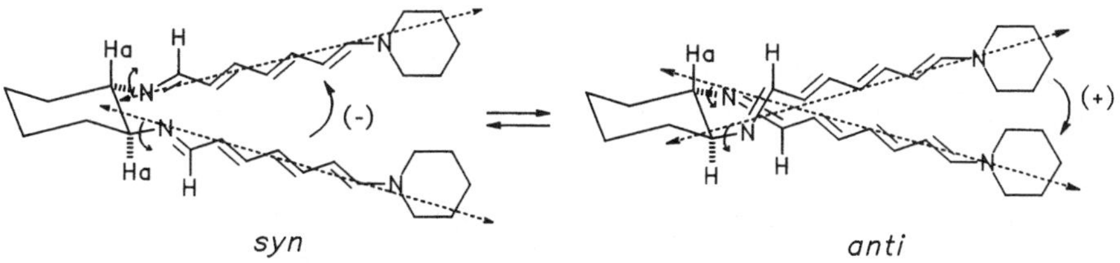

Figure 14-14. Conformational change due to rotations about the C–N bonds of the chair cyclohexane reorients the chromophores of the bis-Schiff base of 1(*S*),2(*S*)-*trans*-diaminocyclohexane and 7-piperidinohepta-2,4,6-trienal so as to invert the relative helicity from negative (left) to positive (right). According to MM2 molecular mechanics calculations and the exciton chirality rule (ref. 13a), the conformation on the left is favored.

Yet another example comes from the diester of 1(*R*),2(*R*)-*trans*-cyclohexanediol and the dipyrrinone acid, 2,3,7,8-tetramethyl-(10*H*)-dipyrrin-1-one-9-carboxylic acid.[8b] The CD spectrum shows the expected bisignate Cotton effects associated with exciton coupling of the dipyrrinone 400-nm transition (Figure 14-16), but the signed order of the Cotton effects changes with the solvent. In CH_2Cl_2 or CH_3CN solvent, the long-wavelength component of the couplet is negative, but the magnitude falls off in CH_3OH and inverts in $(CH_3)_2SO$ solvent. A negative exciton chirality is expected from the sign of the O–C–C–O torsion angle, which corresponds well to the CD data in CH_2Cl_2, CH_3CN, and even CH_3OH—but not in $(CH_3)_2SO$. Apparently, the dipyrrinone chromophores reorient by rotation about the pyrrole–CO_2R bonds, as shown in

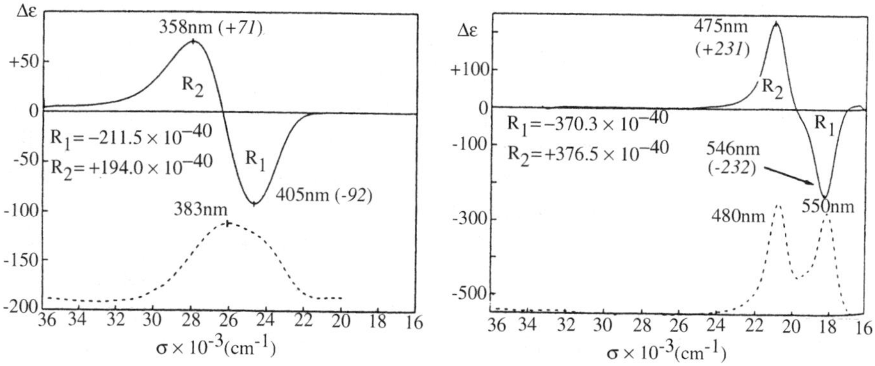

Figure 14-15. CD (———) and UV–visible (- - - -) spectra of the neutral Schiff base (left) and dication (right) of Figure 14-14. The exciton couplets indicates a negative chirality, corresponding to a predominantly *syn* conformer. [Reprinted with permission from ref. 13a. Copyright© 1993 American Chemical Society.]

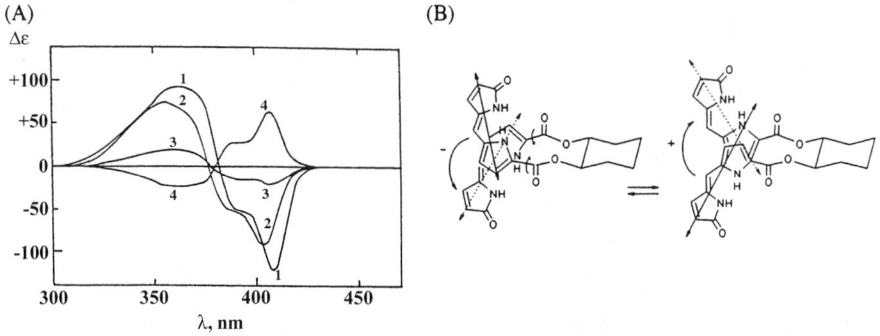

Figure 14-16. (A) CD spectra of 9.65 × 10^{-5}-M solutions of the bis(dipyrrinone ester) shown in (B), in (1) CH_2Cl_2, (2) CH_3CN, (3) CH_3OH, and (4) $(CH_3)_2SO$ at 21°C. (B) Probable conformations of the bis(dipyrrinone ester) of 1(*R*),2(*R*)-*trans*-cyclohexanediol and 2,3,7,8-tetramethyl-(10*H*)-dipyrrin-1-one-9-carboxylic acid in CH_2Cl_2 and CH_3CN (left) and in $(CH_3)_2SO$ (right), showing solvent-induced reorientation of the dipyrrinone chromophores and their long-wavelength electric dipole transition moments from (–) to (+) helicity. The conformational charge is achieved by rotations about the pyrrole – COOR bonds. Methyl groups are removed from the dipyrrinone β carbons for clarity. [(A) is reprinted with permission from ref. 8b. Copyright© 1991 American Chemical Society.]

Figure 14-16, and this reorientation causes the long-axis polarized transition dipoles to invert their helicity.

14.9. Enhanced Sensitivity by Fluorescence-Detected CD

Recently, it was shown that the exciton chirality rule works nicely for bisignate circularly dichroic *fluorescence*.[15] 1(*S*),2(*S*)-Cyclohexanediol was derivatized as its bis(6-methoxy-2-naphthoate) diester (Figure 14-17A), thereby attaching a chromophore with both a strong absorption near 240 nm and a strong fluorescence (Φ_f ~0.64 in CH_3CN, λ_{emiss} ~392 nm, from λ_{exc} ~240 nm). The expected positive-chirality bisignate CD curve that was centered near 240 nm (in the absorption mode) was easily detected at a sample concentration of approximately 10^{-7} M (Figure 14-17B). However, at a concentration of about 10^{-8} M, the detection limit of the instrument (JASCO-J720) had essentially been reached, and only barely detectable CD was evident. In contrast, when the instrument was modified to detect fluorescence, a positive chirality bisignate excitation CD curve was observed at 10^{-8} M, centered near 240 nm (Figure 14-17C). At a concentration of around 10^{-9} M, the fluorescence-detected CD (FDCD) was still strong, and even at approximately 10^{-10} M (200 pg/mL), a bisignate CD could be detected. When no other fluorescent groups are present, FDCD offers a very powerful way to detect nanogram scale quantities of material.

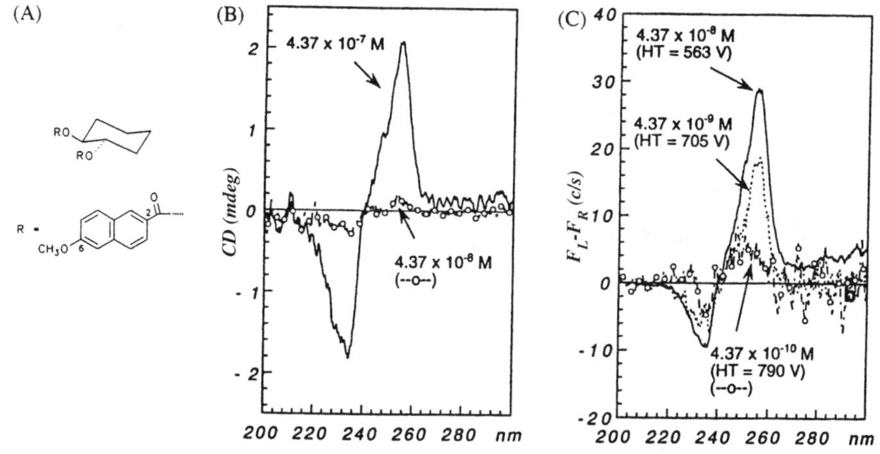

Figure 14-17. (A) 1(*S*),2(*S*)-Cyclohexanediol derivatized as its diester with 6-methoxy-naphthalene-2-carboxylic acid. (B) Ordinary CD spectra of (A) at dilute concentrations in CH_3CN. (C) FDCD spectra of (A) measured at high dilution. [(B) and (C) are reprinted with permission from ref. 15. Copyright© 1997 American Chemical Society.]

14.10. Applications of the Exciton Chirality Rule

The exciton chirality rule of Harada and Nakanishi has been applied often to determine the absolute configuration of a diverse array of compounds.[2,3,7] In closing the chapter, we present but a few examples.

In many instances, the absolute configuration of a molecule would be quite tedious to determine absent the exciton chirality rule. For example, the absolute configuration of (−)-2,2′-bis(bromomethyl)-1,1′-binaphthyl (Figure 14-18A) was determined by the anomalous X-ray diffraction method on a crystal of the resolved (−)-binaphthyl.[16] The approximately 225-nm ^1B_b transition exciton couplet in the CD spectrum of the compound correlates with a (+) chirality, in complete accord with the X-ray results.[2] The exciton chirality rule is thus confirmed by an independent method, but determining absolute configuration by CD is typically faster. The absolute configurations of other optically active binaphthyls, as well as bianthryls, have also been determined from their exciton chirality CD spectra.[3,7] In the 1,1′-bianthryl shown in Figure 14-18B, the ethano bridge fixes the interplanar angle of the two anthracene planes at about 30°. The CD Cotton effects corresponding to the approximately 250-nm ^1B_b transition is extraordinarily strong. Since a (−) exciton chirality is observed, the molecule has the ***M***-helicity or (*S*) absolute configuration shown.

The utility of the exciton chirality rule was demonstrated recently for the bicyclo-[3.3.1]nonanediol with fused naphthalene chromophores (Figure 14-19A).[17] Both the diketone and the diol give intense bisignate π–π^* Cotton effects associated with the naphthalene ^1B_b transition polarized along the long axis of the chromophore. Similarly, the absolute configuration of a biflavone natural product was determined to have an

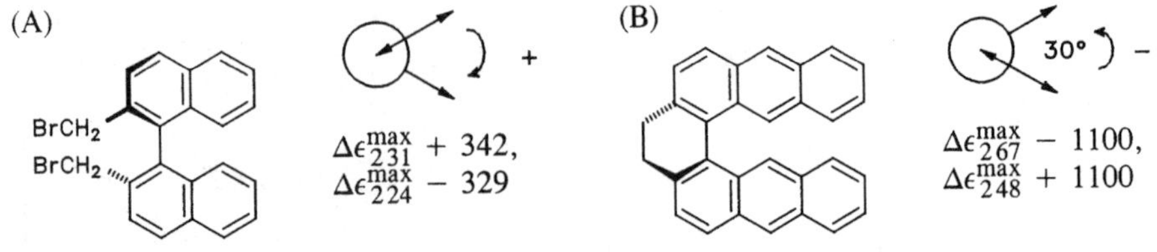

Figure 14-18. (A) Absolute configuration of (–)-2,2′-bis-(bromomethyl)-1,1′-binaphthyl from correlation with positive exciton chirality CD data. (B) Absolute configuration of ethano-bridged 1,1′-bianthryl from CD data corresponding to a negative exciton chirality. Data from ref. 2a.

M-helicity from its exciton CD spectrum (Figure 14-19B) and theoretical calculations that revealed an intense CD couplet (R^{max}_{341} +681, R^{max}_{339} –525) from transitions oriented on the long axis of the *p*-methoxycinnamate, and a dominant CD couplet (R^{max}_{318} –1041, R^{max}_{315} +899) from transitions mainly on the long axis of the *p*-methoxy-benzoyl.[18]

Tryptycenes may be chiral, as in the example of (+)-5,12-dihydro-5,12[1′,2′]benzonaphthacene-1,15-dicarboxylic acid dimethyl ester shown in Figure 14-20A. Determining its absolute configuration would be difficult in the absence of CD. Given the intense (+) chirality exciton couplet ($\Delta\varepsilon^{max}_{243} = +151$, $\Delta\varepsilon^{max}_{220} = -178$) for the 233-nm transition (ε^{max} ~84,000), the exciton chirality rule predicts the absolute configuration shown.[2] Analyzing the three-chromophore system is simple: The pairwise interactions are found, and then the binary couplets are summed. In the case of the benzotryptycene, coupling between the two methyl benzoate chromophores is expected to be small, because the relevant electronic transition moments lie parallel to each other (Figure 14-20B). However, pairwise couplings between each methyl benzoate chromophore and the naphthalene chromophore are expected to be large, given the approximately

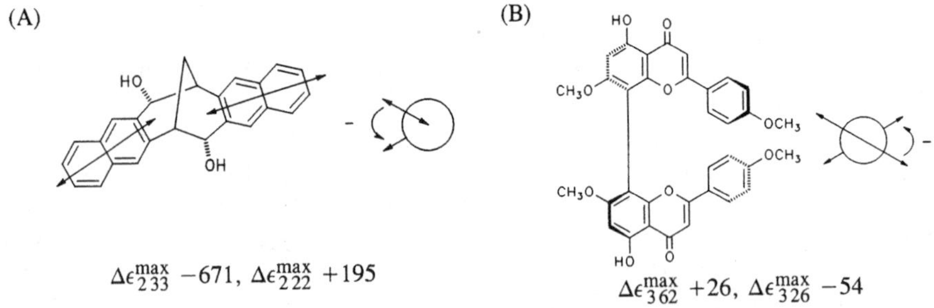

Figure 14-19. Absolute configurations of (A) (–)-7(*S*),8(*R*),15(*S*),16(*R*)-7,8,15,16-tetrahydro-7,15-methanocycloocta[1,2-a:5,6a′]dinaphthalene-8,16-diol, determined from the negative exciton chirality CD couplet, and (B) (*aR*) (–)-4′,4‴,7,7″-tetra-*O*-methylcupressuflavone, determined from its negative exciton chirality *p*-methoxybenzoyl CD couplet. Data from refs. 17 and 18.

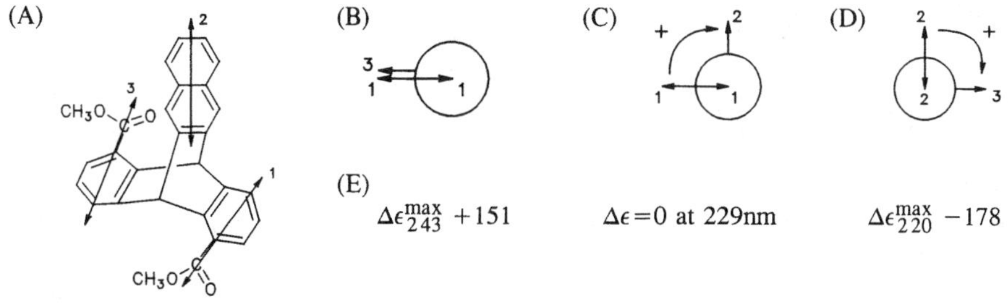

Figure 14-20. (A) Absolute configuration of 5(*S*),12(*S*)-(+)-5,12-Dihydro-5,12[1′,2′]benzonaphthacene-1,15-dicarboxylic acid dimethyl ester. (B), (C), and (D) Orientations of pairs of induced electric dipole moments from the chromophores: 1 and 3 from the methylbenzoate chromophores, 2 from the naphthalene. The 1,3 couplet is predicted to be zero, since the dipoles are parallel. The 1,2 and 2,3 couplets have a (+) chirality, so the net chirality is predicted to be (+). (E) CD Cotton effects for the intense 233-nm transition couplet of the benzotriptycene, showing a (+) exciton chirality. Data from ref. 2a.

90° angle between the transition moment vectors (Figure 14-20C and D). The exciton chirality rule predicts a (+) chirality for each of these two couplings and a net (+) exciton chirality for the enantiomer of Figure 14-20A. Determining the absolute configuration by other methods would doubtless be much more difficult for this compound.

Although the absolute configuration of (–)-spiro[4.4]nonane-1,6-dione had been assigned earlier[19] by applying Horeau's method of optical rotations, the assignment was supported with greater certainty by the exciton chirality rule. The *cis*, *trans*-diol was obtained following reduction of the diketone.[2a] This diol is readily distinguished from the C_2-symmetric *cis,cis* and *trans,trans*-diols by NMR. Since the bis-*p*-dimethylamino benzoate of the *cis-trans*-diol exhibits a negative exciton chirality CD, it follows that the absolute configuration of the dione is that shown in Figure 14-21.

The absolute configuration of *trans*-7,8-dihydroxy-7,8-dihydrobenzo[a]pyrene (Figure 14-22), a carcinogenic metabolite of benzo[a]pyrene, was determined by the exciton chirality method applied to its bis-*p*-dimethylaminobenzoate.[20] Hydroxylation was found to yield the *trans* stereochemistry in the flexible dihydroxylated ring, as determined from the vicinal coupling constant of the hydrogens at C(7) and C(8)

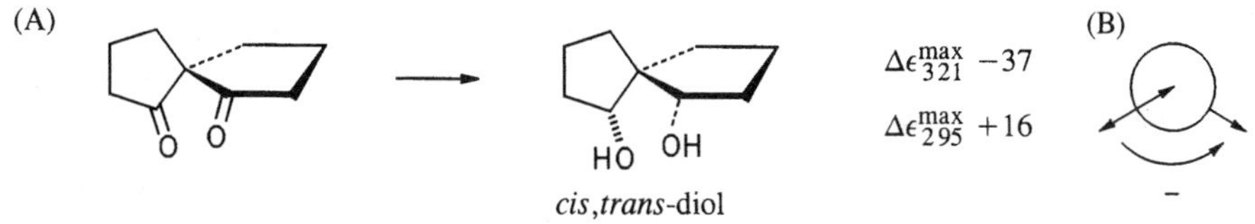

Figure 14-21. (A) Conversion of (–)-spiro[4.4]nonane-1,6-dione to its *cis*, *trans*-diol and the exciton of the *cis*, *trans*-diol bis-*p*-dimethylaminobenzoate. (B) Helicity of the relevant electric dipole transition moments of the *cis*, *trans*-diol bis-*p*-dimethylaminobenzoate. Data from ref. 2a.

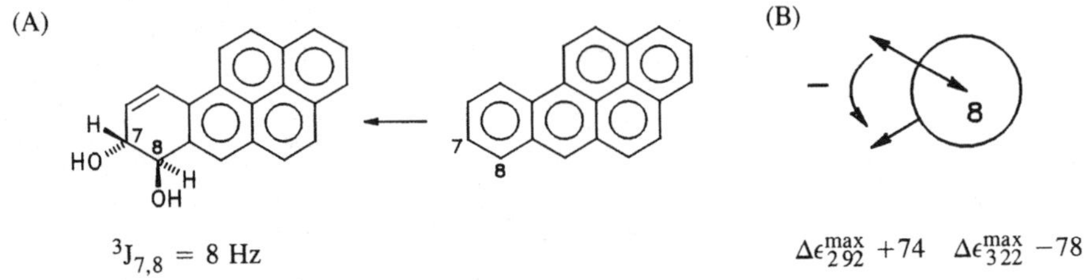

Figure 14-22. (A) *trans*-7,8-Dihydroxy-7,8-dihydrobenzo[a]pyrene formed by enzymic dihydroxylation of benzo[a]pyrene. ${}^3J_{7,8}$ is the vicinal H–H coupling constant. (B) Exciton couplet Cotton effects seen in the CD spectrum of the bis-*p*-dimethylaminobenzoate of the diol in (A), and the helicity of the electric dipole transition moment vectors, showing negative exciton chirality. Data from ref. 19.

($J_{7,8}$ = 8 Hz). With the conformation of the ring known and a *trans*-diol configuration, the CD spectrum of the bis-*p*-dimethylaminobenzoate was used to determine the absolute configuration. The exciton CD couplet corresponded to a negative exciton chirality; thus, the absolute configuration was assigned 7(*R*),8(*R*), as shown.

The absolute configuration of twistane, first determined by applying the octant rule to twistan-2-one (see Chapter 8), was later revised to that shown in Figure 14-23 on the basis of chemical correlations and the CD spectrum of the dibenzoate of the *vic*-diol of the corresponding twistene.[21] Thus, the exciton couplet seen, with the long-wavelength positive Cotton effect, corresponded to a positive exciton chirality and hence the absolute configuration shown.

The enzymic dihydroxylation of toluene by *Pseudomonas putida* yields (+)-*cis*-1,2-dihydroxy-3-methyl-3,5-cyclohexadiene, whose absolute configuration was unknown (Figure 14-24). The relative stereochemistry was found to be *cis*, and the absolute configuration was determined following catalytic reduction of the diene to afford an all-*cis* 3-methyl-1,2-cyclohexanediol.[22] Although there are two chair conformations for the diol, the one with an equatorial methyl (*eae*) is expected to be more stable (by ~2.8 kcal/mole, according to molecular mechanics calculations.)[23] The dibenzoate derivative gave a negative exciton chirality couplet in the CD spectrum. Consequently, the organism produces the diene-diol shown in Figure 14-24D.

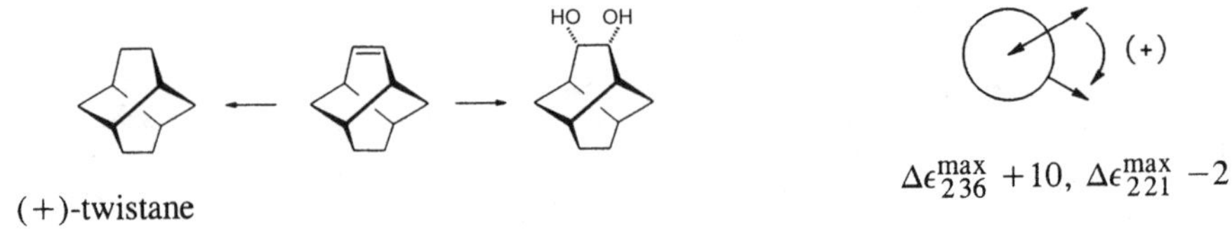

Figure 14-23. Absolute configuration of (+)-twistane, determined by correlation with the *cis*-diol of (+)-twistene. The exciton CD couplet Cotton effects of the diol dibenzoate indicate a (+) helicity of the O–C–C–O torsion angle and, hence, the absolute configuration shown. Data from ref. 21.

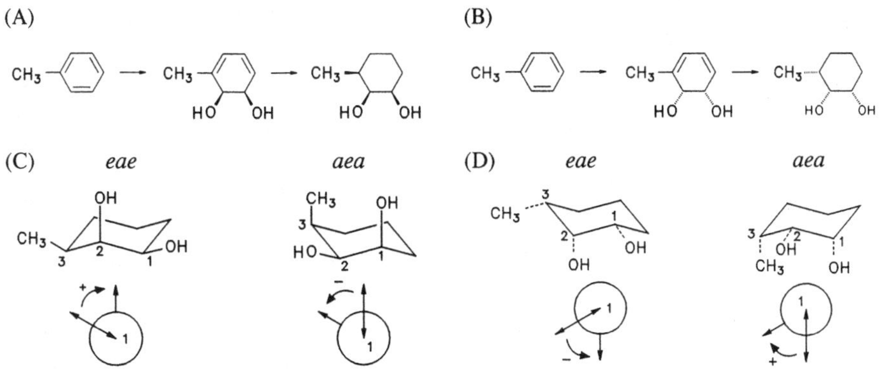

Figure 14-24. (A) and (B) *cis*-Dihydroxylation of toluene gives either of two enantiomeric diene-diols, which, upon catalytic reduction, afford the corresponding enantiomeric cyclohexane-1,2-diols. (C) and (D) Chair methylcyclohexanediol conformers and their predicted exciton chirality. The *eae* conformers with equatorial methyls are expected to be more stable than the *aea* conformers.

Assigning the absolute configuration of acyclic compounds with single stereogenic centers has been a goal in applications of the exciton chirality rule. A new, clever concept for attaining this goal made use of a zinc porphyrin tweezer (Figure 14-25).[24] The two chromophores are covalently linked by a short, flexible alkyl chain. Added L-lysine methyl ester forms an adduct to the tweezer by coordinating its amino groups, one to each zinc (porphyrin). This molecular sandwich, with zinc porphyrin "bread" and diamine filling, has its porphyrin chromophores oriented in a unique arrangement, based on steric differences between large and small groups, so that their relevant electric dipole transition moments give rise to exciton coupling and a bisignate CD that is characteristic of the absolute configuration of the diamine. Covalent binding of

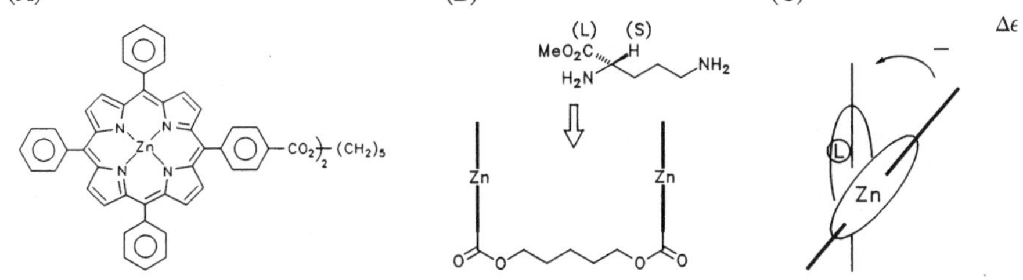

Figure 14-25. (A) Zinc porphyrin tweezer. (B) Porphyrin tweezer, viewed edgewise and ready to complex the diamine L-lysine methyl ester. (C) Top view of the complex showing that the large group (L) imposes a left-handed helicity to the relevant porphyrin electric dipole transition moment for the stacked dimer complex. Data from ref. 24.

porphyrins to amino alcohols with one stereogenic center has also been found to be useful.[3b,10]

The natural pigment of mammalian bile, bilirubin,[25] is usually drawn in an extended shape (Figure 14-26A), which belies its metabolic precursor, heme (Figure 14-26B). Yet, by far the most stable conformation of bilirubin is one in which the pigment folds into a bent, "ridge-tile" shape maintained by a network of intramolecular hydrogen bonds (Figure 14-26C).[26] Either of two mirror-image ridge-tile conformations are possible, and they interconvert rapidly (3–95 s^{-1} at 50–95°C),[27] despite their considerable stability. Thus, isotropic solutions of bilirubin are not optically active. However, when the conformational enantiomerism displayed in Figure 14-26C is displaced toward either enantiomer, solutions become optically active in pigment. This response has been noted for aqueous solutions of bilirubin bound to serum albumins[28] and in organic solvents containing optically active amines[29] and sulfoxides.[30] In such studies, often intense bisignate CD Cotton effects were found for the long-wavelength bilirubin electronic transition near 450 nm. The CD spectrum has been rationalized in terms of a molecular exciton arising from coupling between the two dipyrrinone component chromophores of bilirubin[26,28,31,32] that are held in a chiral orientation by interacting with a chiral complexation agent such as a protein, an optically active amine, or a sulfoxide.[28b–30] Such complexation displaces the equilibrium shown in Figure 14-26C toward either the ***M*** or ***P*** enantiomer. The equilibrium can also be displaced intramo-

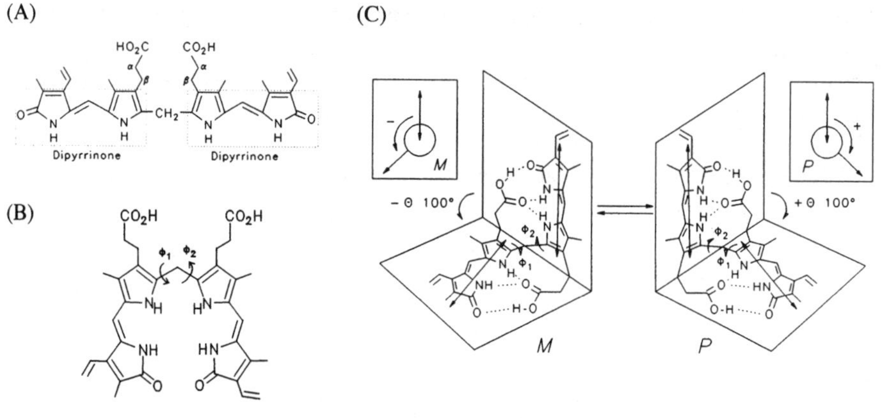

Figure 14-26. (A) Linear representation of bilirubin. (B) Porphyrinlike representation of bilirubin. (C) "Ridge-tile" shaped, folded intramolecularly hydrogen-bonded enantiomeric conformations of bilirubin (***M*** and ***P***). Interconversion (***M*** ⇄ ***P***) is accomplished by rotating about ϕ_1 and ϕ_2. In ***M*** and ***P***, the dipyrrinone chromophores are planar, and the angle of intersection of the two planes (dihedral angle θ) is approximately 100° for $\phi_1 \sim \phi_2 \sim 60°$. The double-headed arrows represent the approximate direction and intensity of the dipyrrinone long-wavelength electric dipole transition moments. The relative orientations or helicities (***M***, minus; ***P***, plus) of the vectors are shown (inset) for each enantiomer. For these conformations, the ***M*** dipole helicity correlates with the ***M*** molecular chirality and the ***P*** helicity with the ***P*** molecular chirality.

lecularly. Thus, the introduction of a methyl group at the α- or the β-carbon of the bilirubin propionic acid groups generates analogs that are potentially optically active: α(*R*),α′(*R*), α(*S*),α′(*S*), β(*R*),β′(*R*), β(*S*),β′(*S*), or the *meso* diastereomers α(*R*),α′(*S*) and β(*R*),β′(*S*)[33] when one uses a synthetic bilirubin analog, mesobilirubin-XIIIα (Figure 14-27). For intramolecularly hydrogen bonded rubins, nonbonded steric interactions between the α or β methyls and the C(10) $-CH_2-$ or the C(7) and C(13) ring methyls, respectively (Figure 14-27C), can be minimized when the bilirubin adopts the ***M***-helicity diastereomer α(*S*) or β(*S*) or when it adopts the ***P***-helicity diastereomer α(*R*) or β(*R*).[33] The net result is a forced resolution of the conformational diastereomers that leads to very intense exciton CD couplets (Figure 14-28).

The diastereoselectivity of hydrogen-bonded ridge-tile conformers is determined by the absolute configuration at the propionic acid α and β carbons, which is known from synthetic precursors.[33] On the basis of the *R* or *S* stereochemistry at these

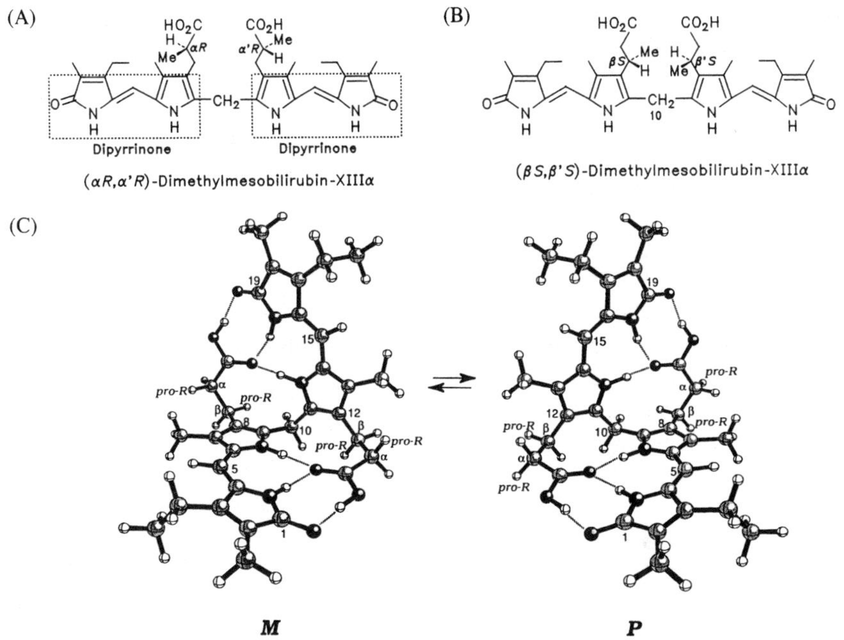

Figure 14-27. Linear representations of α(*R*),α′(*R*)-mesobilirubin-XIIIα (A) and β(*S*),β′(*S*)-dimethylmesobilirubin-XIIIα (B). (C) Ball-and-stick models of folded intramolecularly hydrogen-bonded mesobilirubin-XIIIα. The (*pro-R*) α-hydrogens are sterically compressed into the CH_3 groups at C(7) and C(13), and the (*pro-R*) β-hydrogens are sterically crowded by the $-CH_2-$ at C(10) in the ***M***-helicity conformation. The *pro-S* hydrogens are sterically crowded in the ***P***-helicity conformation. Replacing the *pro-R* hydrogens by CH_3 groups drives the equilibrium toward ***P***, and replacing the *pro-S* by CH_3 groups drives it toward ***M***. However, replacing one *pro-R* and one *pro-S* hydrogen by CH_3 gives a *meso* diastereomer (*R*,*S*) that can adopt either the ***M*** or the ***P*** conformation while leaving one propionic acid ineffectively hydrogen bonded. [Reprinted with permission from ref. 26. Copyright© 1994 American Chemical Society.]

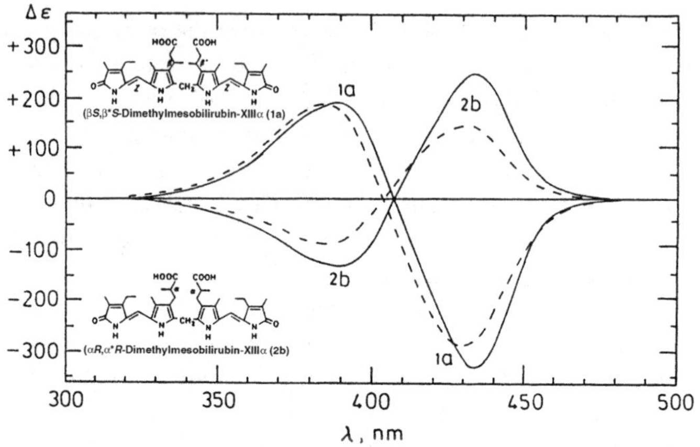

Figure 14-28. Bisignate circular dichroism spectra of 10^{-5} M ($\beta S,\beta' S$)-dimethylmesobilirubin-XIIIα (**1a**) and ($\alpha R,\alpha' R$)-dimethylmesobilirubin-XIIIα (**2b**) in $CHCl_3$ (———) and CH_3OH (- - -) solvents at 22°C. [Reprinted with permission from ref. 26. Copyright© 1994 American Chemical Society.]

stereogenic centers, one can predict which conformer (***M*** or ***P***) should be favored, the prediction may then be confirmed by exciton coupling theory and the exciton chirality rule.[26,29a] Knowledge of pigment stereochemistry is important, because bilirubin is transported as a tightly bound complex with serum albumin and because enzymic glucuronidation is probably enantioselective, with either the ***M*** or the ***P*** conformational enantiomer reacting more rapidly. The avidity for forming hydrogen bonds to dipyrrinones[34] has been shown not only in bilirubin, but also in the parent dipyrrinones (Figures 14-26A and 14-27A).

For C(8)-alkyl-substituted dipyrrinones and for dipyrrinones with esters, dipyrrinone-to-dipyrrinone hydrogen bonds (Figure 14-29A) form strong self-association

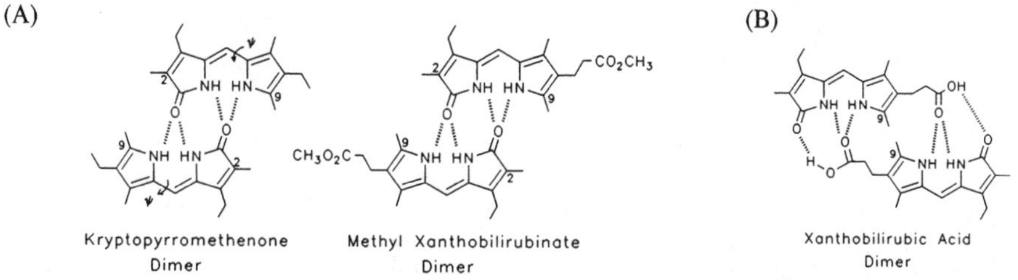

Figure 14-29. (A) Dipyrrinone dimers of kryptopyrromethenone dimer (left) and methyl xanthobilirubinate (right) with four hydrogen bonds. An analog of kryptopyrromethenone with ethyl, rather than methyl, at C(2) is found in the crystal as essentially the same intermolecularly hydrogen-bonded dimer, with $\psi \sim 4°$ (ref. 35). Consistent with this dimeric representation for methyl xanthobilirubinate in $CDCl_3$, ^{1}H-NMR NOEs are found between the methyls at C(2) and C(9). (B) Xanthobilirubic acid dimer in planar representation with six hydrogen bonds.

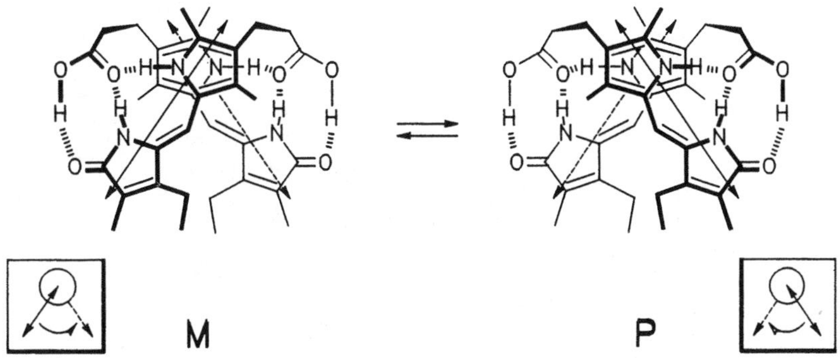

Figure 14-30. Isoenergetic enantiomeric conformations of xanthobilirubic acid interconverting between ***M***- and ***P***-helicities. The inset boxes show the relative orientations of the long-axis polarized $\pi \rightarrow \pi^*$ induced electric dipoles of the dipyrrinones.

dimers (with K_{assoc} ~25,000 M at 25°C).[34] But with dipyrrinone acids, a dimer is formed with six hydrogen bonds (as in bilirubin, Figure 14-26C) between the carboxylic acid and dipyrrinone groups (Figure 14-29B). The planar structure shown would have a severe nonbonded interaction between the methyls at C(9), which is alleviated by allowing the dipyrrinones to stack. Thus, two types of dimers are possible: a planar dimer with four hydrogen bonds and a stacked dimer with six that might be expected to exhibit exciton coupling. The stacked dimer, which is favored in many dipyrrinone acids, adopts either of two mirror-image structures (Figure 14-30) with opposite helicities of the dipyrrinone long-wavelength electric dipole transition moments. Thus, if the equilibrium could be displaced toward ***M*** or ***P***, solutions should exhibit an exciton couplet in the CD spectrum.

Such a displacement is possible with the introduction of a methyl group at the β-carbon of the propionic acid chain.[36] For the β*S*-stereoisomer, the ***M***-helicity dimer

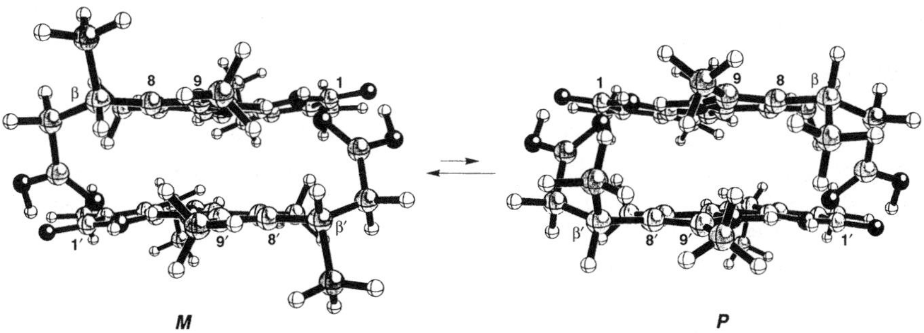

Figure 14-31. Edge view of stacked intermolecularly hydrogen-bonded dimers of the dipyrrinone acid β(*S*)-methylxanthobilirubic acid, held in a left-handed (***M***) chiral orientation (left) and in a right-handed (***P***) chiral orientation (right). [Reproduced with permission from ref. 36b. Copyright© 1995 American Chemical Society.]

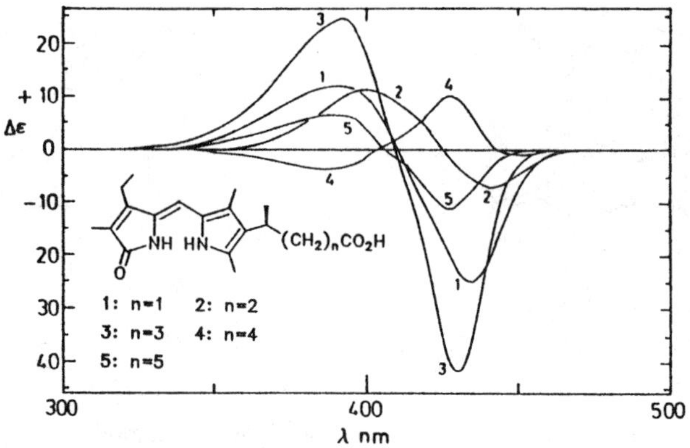

Figure 14-32. Circular dichroism spectra of 5×10^{-5}-M β(*S*)-methylxanthobilirubic acid (**1**) and its homologs in CCl_4 solvent at 23°C. [Reprinted with permission from ref. 36b. Copyright© 1995 American Chemical Society.]

has the β*S* methyl groups located away from the C(9) methyls, but in the ***P***-helicity dimer the β*S* methyls are oriented toward the C(9) methyls, creating a more sterically crowded dimer (Figure 14-31). Thus, one would expect a predominance of the ***M***-helicity dimer and a negative exciton chirality CD spectrum. In fact, moderately strong CD couplets are seen for the chiral β(*S*)-methylxanthobilirubic acid and its homologs (Figure 14-32). All of the couplets have a long-wavelength negative Cotton effect, as predicted by the exciton chirality rule for the ***M***-helicity dimer—except for the hexanoic acid dimer.[36]

Other examples of applications of the exciton chirality rule to the determination of absolute configurations, such as in polyols, are not hard to find,[2,3] and the rule has even been reported to be useful when applied to an on-line high-performance liquid chromatography system.[37]

References

1. (a) Kasha, M., Rawls, H. R., and El-Bayoumi, M. A., *Pure Appl. Chem.* **11** (1965), 371–392. (b) McRae, E. G., and Kasha, M., "The Molecular Exciton Model," in *International Symposium on Physical Processes in Radiation Biology* (Augenstein, L., Mason, R., and Rosenberg, B., eds.), Academic Press, New York 1964, pp. 23–49.
2. (a) Harada, N., and Nakanishi, K., *Circular Dichroic Spectroscopy: Exciton Coupling in Organic Stereochemistry*, University Science Books, Mill Valley, CA, 1983. (b) Berova, N., Harada, N., and Nakanishi, K, "Electronic Spectroscopy: Exciton Coupling, Theory and Applications," in *Encyclopedia of Spectroscopy and Spectrometry* (Lindon, J., Tranter, G., and Holmes, J., eds.), Academic Press, New York, 1999.
3. For leading references, see (a) Nakanishi, K., and Berova, N., in *Circular Dichroism: Principles and Applications* (Nakanishi, K., Berova, N., and Woody, R. W., eds.), VCH Publishers, Inc., Deerfield Beach, FL, 1994, Chap. 13, pp. 361–398. (b) Berova, N., Borhan, B., Dong, J. G., Guo, J., Huang,

X., Karnaukhova, E., Kawamura, A., Lou, J., Matile, S., Nakanishi, K., Rickman, B., Su, J., Tan, Q., and Zanze, I., *Pure & Appl Chem.* **70** (1998), 377–383.
4. Hansen, Aa. E., and Bouman, T. D., *Adv. Chem. Phys.* **44** (1980), 545–644.
5. Osuka, A., and Maruyama, K., *J. Am. Chem. Soc.* **110** (1988), 4454–4456.
6. Lightner, D. A., in *Analytical Applications of Circular Dichroism* (N. Purdie and H. G. Brittain, eds.), Elsevier, Amsterdam, 1994.
7. Boiadjiev, S. E., and Lightner, D. A., in *The Chemistry of Double Bond Functional Groups* (S. Patai and Z. Rappoport, eds.), J. Wiley, Ltd., Chichester, Sussex, U.K., 1997.
8. (a) Byun, Y. S., and Lightner, D. A., *Tetrahedron* **47** (1991), 9759–9772.
(b) Byun, Y. S., and Lightner, D. A., *J. Org. Chem.* **56** (1991), 6027–6033.
9. (a) Matile, S., Berova, N., Nakanishi, K., Novkova, S., Philipova, I., and Blagoev, B., *J. Am. Chem. Soc.* **117**, (1995), 7021–7022.
(b) Matile, S., Berova, N., Nakanishi, K., Fleischhauer, J., and Woody, R. W., *J. Am. Chem. Soc.* **118** (1996), 5198–5206.
(c) Matile, S., Berova, N., and Nakanishi, K., *Chemistry & Biology* **3** (1996), 379–392.
10. Dong, J-G., Akritopoulou-Zanze, I., Guo, J., Berova, N., Nakanishi, K., and Harada, N., *Enantiomer* **2** (1997), 397–409.
11. (a) Liu, H. W., and Nakanishi, K., *J. Am. Chem. Soc.* **103** (1981), 7005–7006.
(b) Wiesler, W. T., Berova, N., Ojika, M., Myers, H. V., Chang, M., Zhou, P., Lo, L-C., Niwa, M., Takeda, R., and Nakanishi, K., *Helv. Chim. Acta* **73** (1990), 509–551.
12. Gargiulo, D., Ikemoto, N., Odingo, J., Bozhkova, N., Iwashita, T., Berova, N., and Nakanishi, K., *J. Am. Chem. Soc.* **116** (1994), 3760–3767.
13. (a) Berova, N., Gargiulo, D., Derguini, F., Nakanishi, K., and Harada, N., *J. Am. Chem. Soc.* **115** (1993), 4769–4775.
(b) Buss, V., Kolster, K., and Görs, B., *Tetrahedron: Asymmetry* **4** (1993), 1–4.
14. (a) Kemp, C. M., and Mason, S. F., *Tetrahedron* **22** (1966), 629–635.
(b) Brown, A., Kemp, C. M., and Mason, S. F., *J. Chem. Soc. A* (1971), 751–755.
15. Dong, J-G., Wada, A., Takakuwa, T., Nakanishi, K., and Berova, N., *J. Am. Chem. Soc.* **119** (1997), 12024–12025.
16. Harada, K., and Tanaka, J., *Bull. Chem. Soc. Japan* **46** (1973), 2747–2751.
17. Harada, N., Hiyoshi, N., and Naemura, K., *Recl. Trav. Chim. Pays-Bas* **114** (1995), 157–162.
18. Harada, N., Ono, H., Uda, H., Parveen, M., Khan, N. U., Achari, B., and Dutta, P. K., *J. Am. Chem. Soc.* **114** (1992), 7687–7692.
19. Gerlach, H., *Helv. Chim. Acta* **51** (1968), 1587–1593.
20. Nakanishi, K., Kasai, H., Cho, H., Harvey, R. G., Jeffrey, A. M., Jennette, K. W., and Weinstein, I. B., *J. Am. Chem. Soc.* **99** (1977), 258–260.
21. Tichý, M., *Coll. Czech. Chem. Comm.* **39** (1974), 2673–2684.
22. Ziffer, H., Jerina, D. M., Gibson, D. T., and Kobal, V. M., *J. Am. Chem. Soc.* **95** (1973), 4048–4049.
23. PCModel version 5.0–7.0. Serena Software, Inc., Bloomington, IN 47402–3076. PCModel uses the MMX force field, a variation of Allinger's MM2 force field.
24. Huang, X., Rickman, B. H., Borhan, B., Berova, N., and Nakanishi, K., *J. Am. Chem. Soc.* **120** (1998), 6185–6186.
25. (a) McDonagh, A. F., in *The Porphyrins,* Vol. VI (Dolphin, D., ed.), Academic Press, New York, pp. 293–491.
(b) For a recent review of the optical activity and stereochemistry of linear oliogopyrroles and bile pigments, see Boiadjiev, S.E., and Lightner, D.A., *Tetrahedron: Asymmetry* **10** (1999), 607–655.
26. For leading references, see Person, R. V., Peterson, B. R., and Lightner, D. A., *J. Am. Chem. Soc.* **116** (1994), 42–59.
27. Kaplan, D., and Navon, G., *Isr. J. Chem.* **23** (1983), 177–186.
28. (a) Blauer, G., *Isr. J. Chem.* **23** (1983), 201–209.
(b) Lightner, D. A., Wijekoon, W. M. D., and Zhang, M. H., *J. Biol. Chem.* **263** (1988), 16669–16676.
29. (a) Lightner, D. A., Gawroński, J. K., and Wijekoon, W. M. D., *J. Am. Chem. Soc.* **109** (1987), 6354–6362.

(b) Pu, Y. M., and Lightner, D. A., *Croatica Chem. Acta* **62** (1989), 301–324.
30. (a) Gawronski, J. K., Połoński, T., and Lightner, D. A., *Tetrahedron* **46** (1990), 8053–8066.
(b) Trull, F. R., Shrout, D. P., and Lightner, D. A., *Tetrahedron* **48** (1992), 8189–8198.
31. Bauman, D., Killet, C., Boiadjiev, S. E., Lightner, D. A., Schönhofer, A., and Kuball, H-G., *J. Phys. Chem.* **100** (1996), 11546–11558.
32. Blauer, G., and Wagnière, G., *J. Am. Chem. Soc.* **97** (1976), 1949–1954.
33. (a) Puzicha, G., Pu, Y-M., and Lightner, D. A., *J. Am. Chem. Soc.* **113** (1991), 3583–3592.
(b) Boiadjiev, S. E., Person, R. V., Puzicha, G., Knobler, C., Maverick, E., Trueblood, K. N., and Lightner, D. A., *J. Am. Chem. Soc.* **114** (1992), 10123–10133.
34. Nogales, D. F., Ma, J-S., and Lightner, D. A., *Tetrahedron* **49** (1993), 2361–2372.
35. Cullen, D. L., Black, P. S., Meyer, E. F., Lightner, D. A., Quistad, G. B., and Pak, C-S., *Tetrahedron* **33** (1977), 477–483.
36. (a) Boiadjiev, S. E., Anstine, D. T., and Lightner, D. A., *J. Am. Chem. Soc.* **117** (1995), 8727–8736.
(b) Boiadjiev, S. E., Anstine, D. T., Maverick, E., and Lightner, D. A., *Tetrahedron: Asymmetry* **6** (1995), 2253–2270.
37. Nishida, Y., Kim, J-H., Ohrui, H., and Meguro, H., *J. Am. Chem. Soc.* **119** (1997), 1484–1485.

Index to Chiroptical Data and Spectra

Subject Index

RETURN TO